Copositive and Completely Positive Matrices

Other World Scientific Titles by the Author

Completely Positive Matrices
ISBN: 978-981-238-368-6

Copositive and Completely Positive Matrices

Naomi Shaked-Monderer
The Max Stern Yezreel Valley College, Israel

Abraham Berman
Technion — Israel Institute of Technology, Israel

World Scientific

NEW JERSEY • LONDON • SINGAPORE • BEIJING • SHANGHAI • HONG KONG • TAIPEI • CHENNAI • TOKYO

Published by

World Scientific Publishing Co. Pte. Ltd.

5 Toh Tuck Link, Singapore 596224

USA office: 27 Warren Street, Suite 401-402, Hackensack, NJ 07601

UK office: 57 Shelton Street, Covent Garden, London WC2H 9HE

Library of Congress Cataloging-in-Publication Data

Names: Shaked-Monderer, Naomi, author. | Berman, Abraham, author.
Title: Copositive and completely positive matrices / Naomi Shaked-Monderer,
The Max Stern Yezreel Valley College, Israel, Abraham Berman,
Technion--Israel Institute of Technology, Israel.
Other titles: Completely positive matrices
Description: Revised edition. | New Jersey : World Scientific, [2021] | Revision of:
Completely positive matrices / Abraham Berman, Naomi Shaked-Monderer. c2003. |
Includes bibliographical references and index.
Identifiers: LCCN 2021003572 | ISBN 9789811204340 (hardcover) |
ISBN 9789811204357 (ebook) | ISBN 9789811204364 (ebook other)
Subjects: LCSH: Matrices.
Classification: LCC QA188 .B465 2021 | DDC 512.9/434--dc23
LC record available at https://lccn.loc.gov/2021003572

British Library Cataloguing-in-Publication Data

A catalogue record for this book is available from the British Library.

For any available supplementary material, please visit
https://www.worldscientific.com/worldscibooks/10.1142/11386#t=suppl

Desk Editor: Liu Yumeng

To our families, who make our lives complete and positive,
and to our many copositive coauthors.

Preface

This book is an updated version of "Completely positive matrices" (World Scientific 2003). But it is more than an updated version, since it contains many topics that were not covered in the 2003 book. In particular, a lot of attention is paid to copositive matrices, hence the name of the new book.

Both copositive matrices and completely positive matrices may be viewed as (nonnegative) variations of real positive semidefinite matrices. Recall that a real positive semidefinite matrix is a real symmetric matrix A all of whose eigenvalues are nonnegative. This is equivalent, among other characterizations, to

$$\mathbf{x}^T A \mathbf{x} \geq 0 \text{ for every vector } \mathbf{x} \text{ of the right order,} \tag{0.1}$$

or to

$$A = BB^T \text{ for some matrix } B. \tag{0.2}$$

If (0.1) holds for nonnegative vectors $\mathbf{x}$, the matrix A is copositive, and if (0.2) holds with a nonnegative B, the matrix A is completely positive. The minimum number of columns in such nonnegative B is called the cp-rank of A. Both the $n \times n$ copositive matrices and the $n \times n$ completely positive matrices form closed convex cones in the space of $n \times n$ symmetric matrices. Both types of matrices have many applications, one of the most prominent of them is to mathematical optimization. The applications are mentioned in the book — see the notes to Sections 2.1 and 3.1, but we stick mainly with the theoretical basis.

Three main problems in the theory of copositive matrices are:

(a) deciding whether a given matrix is copositive,
(b) characterizing extreme rays of the copositive cone,
(c) describing the facial structure of the copositive cone.

For completely positive matrices main problems are:

(d) deciding whether a given matrix is completely positive,
(e) determining the cp-rank of a given completely positive matrix,
(f) describing the facial structure of the completely positive cone.

Though a lot of progress was made regarding these problems, they are still open, see [Berman *et al.* (2015)]. In this book we present a comprehensive up to date theory of copositive and completely positive matrices. At the ends of many sections we highlight the related open questions in italics.

The book consists of five chapters. The second chapter deals with copositivity (problems (a) and (b)); The third with complete positivity (problem (d)); Chapter 4 deals with the cp-rank problem (e); Chapter 5 deals with the structure of both cones (problems (c) and (f)). The first chapter contains the needed background, and can be used by itself as a starting point for a graduate or advanced undergraduate course or seminar in matrix theory. The whole book can be used as a reference for researchers in matrix theory and its applications. Different combinations of parts of the book may be used as a text in a course on topics in matrix theory. This was done at the Technion, and we are grateful to the participants of the course for their valuable remarks and suggestions.

Work on the book was partly motivated by our joint GIF project with Mirjam Dür. In October 2017 we co-organized with Mirjam and Immanuel Bomze an Oberwolfach workshop titled *Copositivity and Complete Positivity*. Many contributors to the study of these topics were present, and we are in debt to all of them, and also to those who could not participate, but whose works are presented in this book. Of course, we are also thankful to our home institutions, the Max Stern Yezreel Valley College and the Technion — Israel Institute of Technology.

Naomi Shaked-Monderer
Abraham Berman
Haifa, September 2020

Contents

Chapter 1

Background

1.1 Matrix theoretic background

In this section we review matrix theoretical notations, terminology and results. Some of these are very basic, and are mentioned mainly to establish the notations of this book. Some are less basic, but still we do not provide any proof, and the readers are invited to either try and prove them (see the exercises at the end of the section), or look them up in matrix analysis books (see the notes at the end of the section). Unless stated otherwise, the matrices in this book are real.

$\mathbb{R}^n$ and Euclidean vector spaces

We denote by $\mathbb{R}^n$ the vector space of real column vectors of length n. The entries of $\mathbf{x} \in \mathbb{R}^n$ are denoted by $x_1, \ldots, x_n$. The *nonnegative orthant* of $\mathbb{R}^n$ is denoted by $\mathbb{R}^n_+$, *i.e.*,

$$\mathbb{R}^n_+ = \{\mathbf{x} \in \mathbb{R}^n \mid x_i \geq 0 \text{ for every } i = 1, \ldots, n\}.$$

If $\mathbf{x} \in \mathbb{R}^n_+$, we say that $\mathbf{x}$ is a *nonnegative vector* and write $\mathbf{x} \geq \mathbf{0}$. If all the entries of $\mathbf{x}$ are positive, we say that the vector $\mathbf{x}$ is *positive* and write $\mathbf{x} > \mathbf{0}$. For a general vector $\mathbf{x} \in \mathbb{R}^n$, the *support* of $\mathbf{x} \in \mathbb{R}^n$ is

$$\operatorname{supp} \mathbf{x} = \{i \mid x_i \neq 0\}.$$

The *absolute value* of $\mathbf{x}$, $|\mathbf{x}|$, is the nonnegative vector whose entries are the absolute values of the entries of $\mathbf{x}$. The *positive part* of $\mathbf{x}$ is the vector in $\mathbb{R}^n$ which consists of all the positive entries of $\mathbf{x}$, and all its other entries are zero. The *negative part* of $\mathbf{x}$ is the vector in $\mathbb{R}^n$ which consists of the absolute values of the negative entries of $\mathbf{x}$, and all its other entries are zero. That is, $\mathbf{x}_+$ and $\mathbf{x}_-$ are two nonnegative vectors with disjoint supports such that

$$\mathbf{x} = \mathbf{x}_+ - \mathbf{x}_- \quad \text{and} \quad |\mathbf{x}| = \mathbf{x}_+ + \mathbf{x}_- \,.$$

If $\alpha \subseteq \{1,\dots,n\}$, we denote by $\mathbf{x}[\alpha]$ the vector obtained from $\mathbf{x}$ by erasing every entry x_i, $i \notin \alpha$. For $S \subseteq \mathbb{R}^n$, $\operatorname{span} S$ denotes the subspace of $\mathbb{R}^n$ spanned by S.

A *Euclidean vector space* is a vector space V over the field of real numbers $\mathbb{R}$, endowed with an inner product $\langle \cdot, \cdot \rangle$. For $V = \mathbb{R}^n$ we denote by $\langle \cdot, \cdot \rangle$ the *standard inner product* on $\mathbb{R}^n$, and by $\| \cdot \|$ the norm it induces, the 2-*norm*. That is, for $\mathbf{x}, \mathbf{y} \in \mathbb{R}^n$,

$$\langle \mathbf{x}, \mathbf{y} \rangle = \mathbf{x}^T\mathbf{y} = \sum_{i=1}^{n} x_i y_i \,,$$

$$\|\mathbf{x}\| = \sqrt{\mathbf{x}^T\mathbf{x}} = \left(\sum_{i=1}^{n} x_i^2 \right)^{1/2} .$$

A *unit vector* is a vector $\mathbf{x}$ such that $\|\mathbf{x}\| = 1$. The *angle* between two nonzero vectors $\mathbf{x}, \mathbf{y}$ is the unique $0 \le \theta \le \pi$ such that

$$\cos\theta = \frac{\langle \mathbf{x}, \mathbf{y} \rangle}{\|\mathbf{x}\| \|\mathbf{y}\|} .$$

The *orthogonal complement* a set $S \subseteq V$ is the subspace

$$S^\perp = \{\mathbf{x} \mid \langle \mathbf{x}, \mathbf{y} \rangle = 0 \text{ for every } \mathbf{y} \in S\} .$$

We denote by $\mathbf{e}_i$, $i = 1, \dots, n$, the vectors of the *standard basis* of $\mathbb{R}^n$. That is, $\mathbf{e}_i$ is the vector all of whose entries are zero, except for 1 in the i-th position. The vector with all entries equal to 1 is usually denoted by $\mathbf{e}$. (When several of these vectors are involved, of different lengths, we may use other ad-hoc notations.) The zero vector in $\mathbb{R}^n$ is denoted by $\mathbf{0}$.

Matrices and determinants

We denote the vector space of $m \times n$ real matrices by $\mathbb{R}^{m\times n}$ ($\mathbb{R}^{n\times 1} = \mathbb{R}^n$). For a matrix A, A^T denotes the transpose of A, $\operatorname{cs} A$ denotes the column space of A, $\operatorname{rs} A$ denotes the row space of A and $\operatorname{ns} A$ denotes the null space of A.

Some special matrices: $0_{m\times n}$ is the zero matrix in $\mathbb{R}^{m\times n}$, $J_{m\times n}$ is the $m \times n$ matrix of all ones; If $m = n$ we also write 0_n and J_n. I_n is the $n \times n$ identity matrix. We omit the subscripts when the orders of the matrices are clear. A square matrix all of whose entries are zero except for 1 in the i, j position will be denoted by E_{ij} (its order will always be clear from the context).

A *permutation matrix* P is a square matrix which has exactly one nonzero entry, equal to 1, in each row and each column. Multiplying a matrix A from the left by P results in a permutation of the rows of A (that is, if $p_{1\,i_1}, p_{2\,i_2} \dots, p_{n\,i_n}$ are the nonzero elements in P, then PA is the matrix whose j-th row is the i_j-th row of A). Multiplying A from the right by P^T has the same effect on the columns of A. Hence for a square matrix A, PAP^T is a matrix obtained from A by permuting both the rows and the columns of A the same way. A *reversal matrix* is a permutation matrix of the form

$$R = \begin{pmatrix} 0 & & 1 \\ & \iddots & \\ 1 & & 0 \end{pmatrix}.$$

Multiplying a vector on the left by this matrix reverses the order of its rows. The $n \times n$ *cyclic permutation matrix* is

$$C = C_n = \begin{pmatrix} \mathbf{e}_2 & \mathbf{e}_3 & \dots & \mathbf{e}_n & \mathbf{e}_1 \end{pmatrix} = \begin{pmatrix} 0 & 0 & \dots & 0 & 1 \\ 1 & 0 & \dots & 0 & 0 \\ 0 & 1 & \ddots & \ddots & \vdots \\ \vdots & \ddots & \ddots & 0 & 0 \\ 0 & \dots & 0 & 1 & 0 \end{pmatrix}.$$

A *circulant matrix* is a square matrix in which each row is obtained from its first row by a circular shift of one step forward. A circulant matrix A is completely determined by its first row (or its first column).

$$A = \begin{pmatrix} a_0 & a_{n-1} & \dots & a_2 & a_1 \\ a_1 & a_0 & a_{n-1} & & a_2 \\ \vdots & a_1 & a_0 & \ddots & \vdots \\ a_{n-2} & & \ddots & \ddots & a_{n-1} \\ a_{n-1} & a_{n-2} & \dots & a_1 & a_0 \end{pmatrix}.$$

A matrix A is circulant if and only if $CAC^T = A$ (or, equivalently, $C^TAC = A$). If A is also symmetric, then $a_{n-i} = a_i$ for $i = 1, \dots, n-1$ (and then also $RAR = A$).

Given a vector $\mathbf{d} \in \mathbb{R}^n$, $\mathbf{d}^T = (d_1, \dots, d_n)$, we denote by Diag($\mathbf{d}$) or Diag$(d_1, \dots, d_n)$ the $n \times n$ diagonal matrix with $d_1, \dots, d_n$ as diagonal elements. Note the capital D. For a square matrix A, diag$A$ (with lower-case d) denotes the vector of diagonal entries of A. A matrix D is a *positive diagonal matrix* if D is diagonal and all its diagonal elements are positive. A square matrix A such that diag$A = \mathbf{0}$ is called a *hollow matrix*. If D is a positive diagonal matrix, we say that DAD is obtained from A by a *positive diagonal congruence*, or by *positive diagonal scaling*.

We denote by $\|\cdot\|$ the operator norm on $\mathbb{R}^{n\times n}$ induced on $\mathbb{R}^{n\times n}$ by the 2-norm on $\mathbb{R}^n$. That is, for $A \in \mathbb{R}^{n\times n}$

$$\|A\| = \max_{\|\mathbf{x}\|=1} \|A\mathbf{x}\|.$$

If A is an $m \times n$ matrix, α and β are sets of indices, $\alpha \subseteq \{1,\ldots,m\}$ and $\beta \subseteq \{1,\ldots,n\}$, then $A[\alpha|\beta]$ is the submatrix of A with rows indexed by the elements of α in increasing order and columns indexed by the elements of β in increasing order. $A[\alpha|\beta)$ is the submatrix of A with rows indexed by α and columns by $\{1,\ldots,n\}\setminus\beta$ (both index sets in increasing order). $A(\alpha|\beta]$ and $A(\alpha|\beta)$ are defined in a similar way. For square matrices, $A[\alpha|\alpha]$ is abbreviated by $A[\alpha]$, and $A(\alpha|\alpha)$ by $A(\alpha)$. We write $A[i_1, i_2, \ldots, i_k]$ instead of $A[\{i_1, i_2, \ldots, i_k\}]$. The matrix $A[\alpha]$ is a *principal submatrix* of A, and if $\alpha = \{1,\ldots,k\}$ for some $1 \le k \le n$, then it is a *leading principal submatrix* of A. The determinants of such submatrices are a *principal minor* and a *leading principal minor*, respectively.

An $n \times n$ matrix A is *reducible* if for some $\alpha \subseteq \{1,\ldots,n\}$ the submatrix $A[\alpha|\alpha)$ is equal to 0. Equivalently, A is reducible if for some permutation matrix P, PAP^T has the block form

$$\begin{pmatrix} A & B \\ 0 & C \end{pmatrix},$$

where A and C are square. A square matrix which is not reducible is *irreducible*.

If A and B are square matrices, their *direct sum* $A \oplus B$ is the matrix in block form

$$\begin{pmatrix} A & 0 \\ 0 & B \end{pmatrix}.$$

If A is any matrix, we denote by $|A|$ the matrix whose elements are the absolute values of the corresponding elements in A, $|A|_{ij} = |a_{ij}|$. The maximum and minimum of two matrices of the same order are also computed entrywise. Hence the matrices $\max(A,B)$ and $\min(A,B)$ have the same order as A and B, and their elements are

$$(\max(A,B))_{ij} = \max(a_{ij}, b_{ij}) \quad , \quad (\min(A,B))_{ij} = \min(a_{ij}, b_{ij}).$$

The *comparison matrix* of a square matrix A is denoted by $M(A)$. It is defined by

$$M(A)_{ij} = \begin{cases} |a_{ij}|, & \text{if } i = j \\ -|a_{ij}|, & \text{if } i \neq j \end{cases}.$$

Another concept that we will need is that of a generalized inverse of a matrix.

Theorem 1.1. *If A is an $m \times n$ matrix, there exists a unique $n \times m$ matrix X such that $AXA = A$, $XAX = X$ and AX and XA are symmetric.*

Definition 1.2. The unique matrix X in Theorem 1.1 is called the *Moore-Penrose generalized inverse* of A, and is denoted by $A^\dagger$.

Of course, if A is a square invertible matrix, $A^\dagger = A^{-1}$. Here is another example.

Example 1.3. $J_n^\dagger = \frac{1}{n^2} J_n$.

The importance of the Moore-Penrose generalized inverse lies in the fact that $A^\dagger \mathbf{b}$ is the best least squares solution of $A\mathbf{x} = \mathbf{b}$. That is, it is the vector of minimal 2-norm for which the 2-norm of $A\mathbf{x} - \mathbf{b}$ is minimal. In other words, the vector $A^\dagger \mathbf{b}$ is the projection of $\mathbf{b}$ on $\operatorname{cs} A$.

We mention now some facts about determinants that will be needed in the book. First, we recall the general form of the Laplace expansion formula, expansion along a the columns indexed by α (it will be needed in Section 2.4). The cardinality of the set α is denoted by $|\alpha|$.

Theorem 1.4. [The Laplace Expansion Formula] *Let A be an $n \times n$ matrix, and $\alpha \subseteq \{1, \ldots, n\}$, $|\alpha| = k$. Then*

$$\det A = \sum_{\substack{|\beta|=k \\ \beta \subseteq \{1,\ldots,n\}}} (-1)^{\Sigma_{i\in\alpha} i + \Sigma_{i\in\beta} i} \det A[\beta|\alpha] \det A(\beta|\alpha).$$

In Section 1.3 we will need the *Cauchy-Binet Formula* for the determinant of the product of two matrices.

Theorem 1.5. [The Cauchy-Binet Formula] *Let A be an $n \times m$ and B an $m \times n$ matrix, $n \leq m$. Then*

$$\det(AB) = \sum_{\substack{|\alpha|=n \\ \alpha \subseteq \{1,\ldots,m\}}} \det A[1, \ldots, n|\alpha] \det B[\alpha|1, \ldots, n]. \tag{1.1}$$

Example 1.6. Let

$$A = \begin{pmatrix} 1 & 2 & 3 \\ 4 & 5 & 6 \end{pmatrix} \text{ and } B = \begin{pmatrix} 7 & 8 \\ 9 & 0 \\ 1 & 2 \end{pmatrix}.$$

Then, by the Cauchy-Binet Formula (1.1),

$$\begin{aligned}\det(AB) &= \det\begin{pmatrix} 1 & 2 \\ 4 & 5 \end{pmatrix}\det\begin{pmatrix} 7 & 8 \\ 9 & 0 \end{pmatrix}+ \\ &\det\begin{pmatrix} 1 & 3 \\ 4 & 6 \end{pmatrix}\det\begin{pmatrix} 7 & 8 \\ 1 & 2 \end{pmatrix}+ \\ &\det\begin{pmatrix} 2 & 3 \\ 5 & 6 \end{pmatrix}\det\begin{pmatrix} 9 & 0 \\ 1 & 2 \end{pmatrix}.\end{aligned}$$

In Section 1.4 we will use the following useful determinantal identity due to Sylvester.

Theorem 1.7. **[Sylvester's Determinantal Identity]** *Let A be an $n\times n$ matrix and let α be a subset of cardinality k of $\{1,\ldots,n\}$. Arrange the $n-k$ numbers in $\{1,\ldots,n\}\setminus\alpha$ in their natural increasing order, $i_1 < \ldots < i_{n-k}$, and let*

$$b_{pq} = \det A[\alpha\cup\{i_p\}|\alpha\cup\{i_q\}],$$

$B = (b_{pq})_{p,q=1}^{n-k}$. *Then*

$$\det B = (\det A[\alpha])^{n-k-1}\det A. \tag{1.2}$$

Example 1.8. Let

$$A = \begin{pmatrix} 1 & 2 & 3 \\ 4 & 5 & 6 \\ 7 & 8 & 9 \end{pmatrix}, \ \alpha = \{2\}.$$

Then

$$b_{11} = \det\begin{pmatrix} 1 & 2 \\ 4 & 5 \end{pmatrix}, \ b_{12} = \det\begin{pmatrix} 2 & 3 \\ 5 & 6 \end{pmatrix}$$
$$b_{21} = \det\begin{pmatrix} 4 & 5 \\ 7 & 8 \end{pmatrix}, \ b_{22} = \det\begin{pmatrix} 5 & 6 \\ 8 & 9 \end{pmatrix}.$$

The minor $\det A[\alpha\cup\{i\} \mid \alpha\cup\{j\}]$ $(i,j\notin\alpha)$ is called a *bordered minor* of $A[\alpha]$ in A.

Another determinantal formula, which will be useful in Section 2.7 is the following special case of the Jacobi's complementary minor formula.

Theorem 1.9. **[Jacobi's Complementary Minor Formula]** *Let A be a nonsingular $n\times n$ matrix and let α be a subset of $\{1,\ldots,n\}$. Then*

$$\det A[\alpha] = \det A \det A^{-1}(\alpha).$$

We conclude this part with the determinant of a Cauchy matrix. An $n \times n$ *Cauchy matrix* is a matrix C such that

$$c_{ij} = \frac{1}{x_i + y_j}, \quad i, j = 1, \dots, n, \tag{1.3}$$

where $x_1, \dots, x_n$ and $y_1, \dots, y_n$ are two sequences of numbers such that $x_i \neq -y_j$ for every $i, j = 1, \dots, n$. If $x_i = y_i$, $i = 1, \dots, n$, C is a *symmetric Cauchy matrix*. A *Hilbert matrix* is a special case of a symmetric Cauchy matrix, where $x_i = y_i = i - 1/2$, $i = 1, \dots, n$. We denote the $n \times n$ Hilbert matrix by $H(n)$. So the i, j element of $H(n)$ is

$$\frac{1}{i + j - 1}.$$

Let C be a Cauchy matrix given by (1.3), then

$$\det C = \frac{\prod_{1 \leq j < i \leq n}(x_i - x_j)\prod_{1 \leq j < i \leq n}(y_i - y_j)}{\prod_{i,j=1}^{n}(x_i + y_j)}. \tag{1.4}$$

Eigenvalues and eigenvectors

We denote the *characteristic polynomial* of an $n \times n$ matrix A by $p_A(x)$, *i.e.*, $p_A(x) = \det(xI - A)$. We will need the following formula for the coefficients of $p_A(x)$.

Theorem 1.10. *For any $n \times n$ matrix A,*

$$p_A(x) = x^n - p_1 x^{n-1} + p_2 x^{n-2} - \dots + (-1)^k p_k x^{n-k} + \dots + (-1)^n p_n,$$

where p_k is the sum of all $k \times k$ principal minors of A.

Recall that if

$$A\mathbf{x} = \lambda \mathbf{x} \tag{1.5}$$

for some vector $\mathbf{x} \neq \mathbf{0}$, then λ is an *eigenvalue* of A, and $\mathbf{x}$ is an *eigenvector* of A associated with λ. The eigenvalues of A are all the roots (real and complex) of the characteristic polynomial $p_A(x)$. The eigenvalue λ is a *simple eigenvalue* of A if the multiplicity of λ as a zero of $p_A(x)$ is 1.

The set of all eigenvalues of A is the *spectrum* of A. The *spectral radius* of A, $\rho(A)$, is the maximum modulus of an eigenvalue of A. Clearly, the closed disc in the complex plane centered at the origin with radius $\rho(A)$ contains all the eigenvalues of A, and it is the smallest such disc. The next result, Geršgorin's Theorem, describes a smaller set that covers the spectrum of $n \times n$ (real or complex) matrix A, consisting of a union of closed discs.

Theorem 1.11. **[Geršgorin's Theorem]** *If A is an $n \times n$ matrix and*

$$R_i = \sum_{\substack{j=1 \\ j \neq i}}^{n} |a_{ij}|, \quad i = 1, \ldots, n,$$

then all the eigenvalues of A are contained in the union of n closed discs D_i, $i = 1, \ldots, n$, where

$$D_i = \{z \mid |z - a_{ii}| \leq R_i\}.$$

Moreover, if A is irreducible and an eigenvalue λ of A lies on the boundary of $\bigcup_{i=1}^{n} D_i$, then λ lies on the boundary of each of the discs D_i, that is, $|\lambda - a_{ii}| = R_i$ for every $i = 1, \ldots, n$.

Since the spectrum of A is the same as that of A^T, a similar result can be stated using the sums of the off-diagonal elements in each column.

Theorem 1.11 can be used to obtain conditions on the elements of a matrix which guarantee invertibility.

Definition 1.12. A matrix A is *diagonally dominant* if

$$|a_{ii}| \geq \sum_{\substack{j=1 \\ j \neq i}}^{n} |a_{ij}| \quad \text{for all} \quad i = 1, \ldots, n.$$

It is *strictly diagonally dominant* if

$$|a_{ii}| > \sum_{\substack{j=1 \\ j \neq i}}^{n} |a_{ij}| \quad \text{for all} \quad i = 1, \ldots, n.$$

It is *irreducibly diagonally dominant* if it is irreducible and diagonally dominant with strict inequality for at least one index i.

Note that there are irreducible diagonally dominant matrices which are not irreducibly diagonally dominant. For example, $(n-2)I_n + J_n$.

Corollary 1.13. *Strictly diagonally dominant matrices and irreducibly diagonally dominant matrices are invertible.*

Since most of the matrices in this book are symmetric, we conclude this section with some special attention to such matrices. First we remind the reader that all the eigenvalues of a real symmetric matrix are real, and the matrix is orthogonally diagonalizable.

Theorem 1.14. *If A is an $n \times n$ real symmetric matrix, then all the eigenvalues of A are real, and there exists an orthogonal $n \times n$ matrix U (that is, $U^T U = I$), such that $U^T A U = D$, where D is a diagonal matrix, with the eigenvalues of A on the diagonal.*

Diagonalizability means that the dimension of the *eigenspace* of a symmetric A corresponding to λ (that is, the subspace consisting of all vectors which satisfy (1.5)) is equal to the multiplicity of λ as a root of $p_A(x)$. So for symmetric matrices, an eigenvalue λ is simple if and only if the dimension of its eigenspace is 1.

As the eigenvalues of a real symmetric matrix A are all real, we may put them in nonincreasing order

$$\lambda_1(A) \geq \lambda_2(A) \geq \ldots \geq \lambda_n(A).$$

We state here the real version of the Rayleigh-Ritz Theorem.

Theorem 1.15. **[Rayleigh-Ritz Theorem]** *Let A be an $n \times n$ real symmetric matrix, with eigenvalues $\lambda_1 \geq \lambda_2 \geq \ldots \geq \lambda_n$. Then*

$$\begin{aligned}\lambda_1 &= \max_{\|\mathbf{x}\|=1} \mathbf{x}^T A \mathbf{x} \\ \lambda_n &= \min_{\|\mathbf{x}\|=1} \mathbf{x}^T A \mathbf{x}.\end{aligned}$$

For a real symmetric matrix we denote by $\operatorname{In} A$ the *inertia* of A. The inertia of A is a triple $(i_+(A), i_-(A), i_0(A))$, where $i_+(A)$ is the number of positive eigenvalues of A, $i_-(A)$ is the number of negative eigenvalues of A and $i_0(A)$ is the multiplicity of 0 as an eigenvalue of A. Sylvester's Law of Inertia states, in the case of real matrices, that the inertia of a symmetric matrix is invariant under congruence.

Theorem 1.16. **[Sylvester's Law of Inertia]** *Let A and B be two real symmetric matrices. Then there is a nonsingular real matrix S such that $B = S^T AS$ if and only if A and B have the same inertia.*

The eigenvalues of a general $n \times n$ matrix depend continuously on the entries of the matrix. In general, making this intuitive statement precise is cumbersome, but in the case of real symmetric $n \times n$ matrices, with the eigenvalues listed in a nonincreasing order, it is quite easy.

Theorem 1.17. *If A_t are all symmetric matrices, and $\lim_{t \to t_0} A_t = A$ (entrywise), then for every $i = 1, \ldots, n$,*

$$\lambda_i(A) = \lim_{t \to t_0} \lambda_i(A_t).$$

We conclude with a theorem describing the interlacing relation between the eigenvalues of a real symmetric matrix A and those of a principal submatrix of A.

Theorem 1.18. [Cauchy's Interlacing Theorem] *Let A be a real symmetric matrix, and B any principal submatrix of A of order m. Then for every $1 \le i \le m$,*

$$\lambda_i(A) \ge \lambda_i(B) \ge \lambda_{n-m+i}(A).$$

In particular, in the case that $m = n - 1$,

$$\lambda_1(A) \ge \lambda_1(B) \ge \lambda_2(A) \ge \lambda_2(B) \ge \ldots \ge \lambda_{n-1}(B) \ge \lambda_n(A).$$

Exercises

1.1. Let A and B be as in Theorem 1.5. Use Theorem 1.10 and the fact that $p_{BA}(x) = x^{n-m}p_{AB}(x)$ to prove the Cauchy-Binet Formula (1.1).

1.2. Prove the formula in Theorem 1.9.
Hint: If $A = \begin{pmatrix} B & C \\ D & E \end{pmatrix}$ and $A^{-1} = \begin{pmatrix} B' & C' \\ D' & E' \end{pmatrix}$, where $B = A[\alpha]$ and $E' = A^{-1}(\alpha)$, then $\begin{pmatrix} B & C \\ D & E \end{pmatrix}\begin{pmatrix} I & C' \\ 0 & E' \end{pmatrix} = \begin{pmatrix} B & 0 \\ D & I \end{pmatrix}$.

1.3. Prove formula (1.4) for the determinant of a Cauchy matrix.
Hint: You can start by subtracting the last column from all other columns and then dividing each row and each column by its common factor.

1.4. Calculate the determinant of $H(n)$, the $n \times n$ Hilbert matrix.

1.5. Use Theorem 1.11 to prove Corollary 1.13.

1.6. Prove the Rayleigh-Ritz Theorem, Theorem 1.15.

1.7. Show that

(a) If $D = \text{Diag}(d_1, \ldots, d_n)$ is a diagonal matrix, then $D^\dagger$ is the diagonal matrix such that

$$(D^\dagger)_{ii} = \begin{cases} 1/d_i, & \text{if } d_i \ne 0, \\ 0, & \text{if } d_i = 0. \end{cases}$$

(b) If A is a real symmetric matrix, $A = UDU^T$ where D is a diagonal matrix and U is an orthogonal matrix, then $A^\dagger = UD^\dagger U^T$.

1.8. Prove the following properties of the Moore-Penrose generalized inverse.

(a) $(A^\dagger)^\dagger = A$

(b) $(A^\dagger)^T = (A^T)^\dagger$
(c) $A^T AA^\dagger = A^T = A^\dagger AA^T$
(d) $(AA^T)^\dagger = (A^T)^\dagger A^\dagger$.
(e) For every $\mathbf{x} \in \operatorname{cs} A$, $AA^\dagger \mathbf{x} = \mathbf{x}$.

Notes. General references on Matrix Theory are [Gantmacher (1959)], [Horn and Johnson (2013)], [Horn and Johnson (1994)] and [Marcus and Minc (1964)]. See also [Hogben (2014)] for a comprehensive coverage of linear algebra topics. A classical reference on generalized inverses is [Ben-Israel and Greville (2003)]. A beautiful expository article on Hilbert matrices is [Choi (1983)]. The Hilbert matrix $H(n)$ has many lovely features. For example, all the entries of its inverse are integers, e.g. [Choi (1983)], [Berman and Gueron (2002)]. It is highly ill conditioned and as such useful in testing numerical algorithms. As we will see later, the Hilbert matrices are totally positive and completely positive.

In the discussion of diagonal dominance we follow [Varga (2000)]. Theorem 1.11 is a combination of the original Geršgorin's Theorem and an improvement due to Taussky.

1.2 Positive semidefinite matrices

In this section we review the definition and properties of positive semidefinite matrices. This is not a comprehensive survey — we restrict our attention to real positive semidefinite matrices, and to properties that will be needed in our discussion of completely positive matrices.

Definition 1.19. A matrix $A \in \mathbb{R}^{n\times n}$ is *positive semidefinite* if it is symmetric and $\mathbf{x}^T A\mathbf{x}$ is nonnegative for every $\mathbf{x} \in \mathbb{R}^n$.

We use $A \succeq 0$ to denote that A is positive semidefinite, and write $A \succeq B$ if $A - B \succeq 0$. Note that an identity matrix is positive semidefinite, and so is a square zero matrix. Another simple example is

$$A = \begin{pmatrix} 1 & -1 \\ -1 & 1 \end{pmatrix},$$

for which $\mathbf{x}^T A\mathbf{x} = (x_1 - x_2)^2$.

Proposition 1.20. *The sum of positive semidefinite matrices of the same order is positive semidefinite. Also, if A is a positive semidefinite matrix and $a > 0$, then aA is also positive semidefinite.*

Proof. The first assertion follows from the equality $\mathbf{x}^T(A+B)\mathbf{x} = \mathbf{x}^T A\mathbf{x} + \mathbf{x}^T B\mathbf{x}$, and the second from the equality $\mathbf{x}^T(aA)\mathbf{x} = a\mathbf{x}^T A\mathbf{x}$. □

Proposition 1.21. *If A and B are $n \times n$ positive semidefinite matrices, then $\operatorname{ns}(A+B) = \operatorname{ns} A \cap \operatorname{ns} B$ and $\operatorname{cs}(A+B) = \operatorname{cs} A + \operatorname{cs} B$.*

The following proposition is also easy to prove.

Proposition 1.22. *If A is an $n \times n$ positive semidefinite matrix, and S is any $n \times m$ matrix, then $S^T AS$ is positive semidefinite.*

Proposition 1.23. *Any principal submatrix of a positive semidefinite matrix is also positive semidefinite.*

Proof. Let A be an $n \times n$ positive semidefinite matrix, and let $A[\alpha]$ be a principal submatrix of A, $\alpha \subseteq \{1, 2, \ldots, n\}$. Consider $\mathbf{y}^T A[\alpha]\mathbf{y}$ for $\mathbf{y} \in \mathbb{R}^{|\alpha|}$. Let $\mathbf{x}$ be a vector of n entries such that $\mathbf{y} = \mathbf{x}[\alpha]$ and $x_i = 0$ for $i \notin \alpha$. Then $\mathbf{y}^T A[\alpha]\mathbf{y} = \mathbf{x}^T A\mathbf{x} \geq 0$. □

Being symmetric, any $n \times n$ real positive semidefinite matrix A has n real eigenvalues (counting multiplicities), and is orthogonally diagonalizable. That is, there exists an orthogonal matrix U such that $U^T AU = D$ is diagonal. In particular,

$$\operatorname{rank} A = \operatorname{rank} D = \text{the number of nonzero eigenvalues of } A.$$

Using these facts (and some others) we derive several characterizations of positive semidefinite matrices. These are stated in Theorem 1.25. For one of them we need the notion of a Gram matrix.

Definition 1.24. Let $\mathbf{v}_1, \ldots, \mathbf{v}_n$ be vectors in an Euclidean space V. The matrix of inner products

$$A = \begin{pmatrix} \langle \mathbf{v}_1, \mathbf{v}_1 \rangle & \langle \mathbf{v}_1, \mathbf{v}_2 \rangle & \ldots & \langle \mathbf{v}_1, \mathbf{v}_n \rangle \\ \langle \mathbf{v}_2, \mathbf{v}_1 \rangle & \langle \mathbf{v}_2, \mathbf{v}_2 \rangle & \ldots & \langle \mathbf{v}_2, \mathbf{v}_n \rangle \\ \vdots & \vdots & & \vdots \\ \langle \mathbf{v}_n, \mathbf{v}_1 \rangle & \langle \mathbf{v}_n, \mathbf{v}_2 \rangle & \ldots & \langle \mathbf{v}_n, \mathbf{v}_n \rangle \end{pmatrix}$$

is called the *Gram matrix* of $\mathbf{v}_1, \ldots, \mathbf{v}_n$, and denoted $\operatorname{Gram}(\mathbf{v}_1, \ldots, \mathbf{v}_n)$.

We can now state the following characterizations.

Theorem 1.25. *Let A be an $n \times n$ real symmetric matrix. Then the following statements are equivalent.*

(a) *A is positive semidefinite.*
(b) *All the eigenvalues of A are nonnegative.*
(c) *All the principal minors of A are nonnegative.*
(d) *There exists an* $n \times k$ *real matrix B such that* $A = BB^T$.
(e) *There exists an* $n \times n$ *lower triangular real matrix L such that* $A = LL^T$.
(f) *There exists an* $n \times n$ *real symmetric matrix C such that* $A = C^2$.
(g) *There exist a k-dimensional Euclidean vector space V and vectors* $\mathbf{v}_1, \dots, \mathbf{v}_n \in V$ *such that* $A = \mathrm{Gram}(\mathbf{v}_1, \dots, \mathbf{v}_n)$.
(h) *There exist k vectors* $\mathbf{b}_1, \dots, \mathbf{b}_k \in \mathbb{R}^n$ *such that* $A = \sum_{i=1}^k \mathbf{b}_i \mathbf{b}_i^T$.

Proof. We prove the following implications: (a)⇒(b), (a)⇒(c)⇒(b) ⇒ (f)⇒(d)⇒(g)⇒(a), (d)⇒(e)⇒(d), and (b)⇔(h).

(a)⇒(b): Let $A\mathbf{x} = \lambda \mathbf{x}$, $\mathbf{x} \neq \mathbf{0}$. Then $\mathbf{x}^T A \mathbf{x} = \lambda \mathbf{x}^T \mathbf{x} \geq 0$ and since $\mathbf{x}^T \mathbf{x} > 0$, $\lambda = \mathbf{x}^T A \mathbf{x} / \mathbf{x}^T \mathbf{x} \geq 0$.

(a)⇒(c): Let $A[\alpha]$ be a principal submatrix of A. By Proposition 1.23, $A[\alpha]$ is also positive semidefinite, and by the implication (a)⇒(b), all its eigenvalues are nonnegative. Since the determinant is the product of the eigenvalues, $\det A[\alpha]$ is nonnegative.

(c)⇒(b): Let

$$p_A(x) = x^n - p_1 x^{n-1} + p_2 x^{n-2} - \dots + (-1)^k p_k x^{n-k} + \dots + (-1)^n p_n$$

be the characteristic polynomial of A. Then p_k is the sum of all $k \times k$ principal minors of A, and since (c) holds, $p_k \geq 0$, $k = 1, 2, \dots, n$. Suppose $x < 0$. If n is even, then x^n is positive and all other terms in $p_A(x)$ are nonnegative. If n is odd, then x^n is negative and all other terms in $p_A(x)$ are nonpositive. This shows that A cannot have negative eigenvalues. Being symmetric, all the eigenvalues of A are real, which means that the eigenvalues are nonnegative.

(b)⇒(f): Since A is symmetric with nonnegative eigenvalues, it is orthogonally similar to a nonnegative diagonal matrix D. That is, $A = UDU^T$ for some orthogonal matrix U and a nonnegative diagonal matrix $D = \mathrm{Diag}(d_1, \dots, d_n)$. But then $A = U\sqrt{D}U^T U \sqrt{D} U^T$, where $\sqrt{D} = \mathrm{Diag}(\sqrt{d_1}, \dots, \sqrt{d_n})$. Hence $A = C^2$ where $C = U\sqrt{D}U^T$.

(d)⇒(e): To prove this implication we use the fact that any real matrix C has a QR factorization, *i.e.*, $C = QR$ where the rows of Q are orthonormal and R is upper triangular. Let $A = BB^T$, and let $B^T = QR$ be a QR factorization of B^T. Then $A = R^T Q^T Q R = LL^T$, where $L = R^T$ is lower triangular. (See also Exercise 1.35 for another proof of this implication.)

(d)⇒(g): Let $A = BB^T$, where B is an $n \times k$ matrix. Let $V = \mathbb{R}^k$, and let $\mathbf{v}_i^T$ be the i-th row of B. Then $A = \mathrm{Gram}(\mathbf{v}_1, \dots, \mathbf{v}_n)$.

(g)⇒(a): Let $A = \text{Gram}(\mathbf{v}_1, \ldots, \mathbf{v}_n)$, and let $\mathbf{x} \in \mathbb{R}^n$. Then

$$\begin{aligned}\mathbf{x}^T A\mathbf{x} = \sum_{i,j=1}^{n} a_{ij}x_i x_j &= \sum_{i,j=1}^{n} \langle \mathbf{v}_i, \mathbf{v}_j \rangle x_i x_j \\ &= \sum_{i,j=1}^{n} \langle x_i\mathbf{v}_i, x_j\mathbf{v}_j \rangle \\ &= \left\langle \sum_{i=1}^{n} x_i\mathbf{v}_i, \sum_{j=1}^{n} x_j\mathbf{v}_j \right\rangle = \left\| \sum_{i=1}^{n} x_i\mathbf{v}_i \right\|^2 \geq 0.\end{aligned}$$

(b)⇔(h): $A = BB^T$ iff $A = \sum_{i=1}^{k} \mathbf{b}_i\mathbf{b}_i^T$, where $\mathbf{b}_i$, $i = 1, \ldots, k$, are the columns of B.

The implications (e)⇒(d) and (f)⇒(d) are clear, so the proof is complete. □

Using the last theorem we may obtain many examples of positive semidefinite matrices. For example, $J = \mathbf{e}\mathbf{e}^T$ is positive semidefinite.

Remark 1.26. By Theorem 1.25(b), the matrix C constructed in our proof of (b)⇒(f) was actually positive semidefinite. It can be shown that for any positive semidefinite matrix A there is a *unique* positive semidefinite matrix C satisfying $A = C^2$. It is called the *square root* of A, and denoted by $\sqrt{A}$. The matrix B in (d), however, is not unique. In particular, if A is a rank r positive semidefinite matrix, there exists an $n \times r$ matrix B such that $A = BB^T$. To see that, observe that there exist a unitary matrix U and a diagonal matrix $D = \text{Diag}(d_1, \ldots, d_n)$ such that $U^TAU = D$. The d_i's are the eigenvalues of A. They are nonnegative and exactly r of them are positive, say $d_1, d_2, \ldots, d_r > 0$. Thus $A = \hat{B}\hat{B}^T$, where $\hat{B} = U\sqrt{D}$. Since $\hat{B}[1, \ldots, n|r+1, \ldots, n] = 0$, $A = BB^T$, where $B = \hat{B}[1, \ldots, n|1, \ldots, r]$.

We leave the proof of the next proposition as an exercise.

Proposition 1.27. *If* $A = BB^T$, *then* $\text{ns}\, A = \text{ns}\, B^T$ *and* $\text{cs}\, A = \text{cs}\, B$.

Corollary 1.28. *If A is positive semidefinite and* $\mathbf{x}^T A\mathbf{x} = 0$, *then* $A\mathbf{x} = \mathbf{0}$.

If a diagonal entry in a positive semidefinite matrix is equal to zero, then all entries in the corresponding row and column are zeros. This observation follows from the fact that for every i and j,

$$\det \begin{pmatrix} a_{ii} & a_{ij} \\ a_{ij} & a_{jj} \end{pmatrix} \geq 0.$$

This is a special case of the following Principal Submatrix Rank Property, whose proof is also left as an exercise.

Proposition 1.29. *Let A be an $n \times n$ positive semidefinite matrix, and let $\alpha \subseteq \{1, \dots, n\}$. Then* $\operatorname{cs} A[\alpha|\alpha) \subseteq \operatorname{cs} A[\alpha]$, *and*

$$\operatorname{rank} A[\alpha|1, \dots, n] = \operatorname{rank} A[1, \dots, n|\alpha] = \operatorname{rank} A[\alpha].$$

Using Theorem 1.25 we deduce some additional results.

Proposition 1.30. *If A is positive semidefinite and k is a positive integer, then A^k is positive semidefinite.*

Proof. By Theorem 1.25(f), $A = C^2$, where C is symmetric. Therefore $A^k = (C^2)^k = (C^k)^2$ is positive semidefinite. □

The next corollary follows from Propositions 1.30 and 1.20.

Corollary 1.31. *If $f(x) = \sum_{i=0}^m a_i x^i$ is a polynomial with nonnegative coefficients, and A is positive semidefinite, then $f(A) = \sum_{i=0}^m a_i A^i$ is also positive semidefinite.*

We also consider two other types of products of matrices.

Definition 1.32. The *Hadamard product* of $A, B \in \mathbb{R}^{m\times n}$ is the matrix $C = A \circ B$, defined by the entry by entry product. That is, $c_{ij} = a_{ij}b_{ij}$.

Definition 1.33. The *Kronecker product* of $A \in \mathbb{R}^{m\times n}$ and $B \in \mathbb{R}^{p\times q}$ is the block matrix $A \otimes B$ whose i, j block is $a_{ij}B$.

Proposition 1.34. *The Hadamard product of positive semidefinite matrices is positive semidefinite.*

Proof. Let $A = \sum_{i=1}^k \mathbf{v}_i \mathbf{v}_i^T$ and $B = \sum_{j=1}^l \mathbf{w}_j \mathbf{w}_j^T$. Then $A \circ B = \sum_{i=1,j=1}^{k,l} (\mathbf{v}_i \circ \mathbf{w}_j)(\mathbf{v}_i \circ \mathbf{w}_j)^T$. □

Definition 1.35. Let k be a positive integer. The *Hadamard k-th power* of a matrix A, $A^{(k)}$, is the Hadamard product of k copies of A,

$$A^{(k)} = A \circ A \circ \ldots \circ A.$$

Corollary 1.36. *If A is positive semidefinite and k is a positive integer, then $A^{(k)}$ is also positive semidefinite.*

To prove similar results for the Kronecker product we need to use two basic properties of this product. These properties are stated in the next lemma, and we leave the proof as an exercise.

Lemma 1.37.

(a) *If AC and BD are defined, then* $(A \otimes B)(C \otimes D) = AC \otimes BD$.
(b) $(A \otimes B)^T = A^T \otimes B^T$.

Corollary 1.38. *The Kronecker product of positive semidefinite matrices is positive semidefinite.*

Proof. If $A = BB^T$ and $C = FF^T$ then, by Lemma 1.37,

$$A \otimes C = BB^T \otimes FF^T = (B \otimes F)(B^T \otimes F^T) = (B \otimes F)(B \otimes F)^T.$$

□

Remark 1.39. Proposition 1.34 follows from the last corollary, since $A \circ C$ is a principal submatrix of $A \otimes C$.

Positive semidefinite matrices which are diagonally dominant will play a role in our discussion of completely positive matrices. Recall that a matrix $A \in \mathbb{R}^{n \times n}$ is diagonally dominant if for every $i = 1, \ldots, n$

$$|a_{ii}| \geq \sum_{\substack{j=1 \\ j \neq i}}^{n} |a_{ij}|.$$

Geršgorin's Theorem (Theorem 1.11) implies that the eigenvalues of a diagonally dominant matrix lie in the right halfplane $\{z \in \mathbb{C} \mid \operatorname{Re} z \geq 0\}$. We state here this result in the case that A is a real symmetric matrix, and provide an alternative proof for the first part in this case.

Proposition 1.40. *A diagonally dominant symmetric matrix with nonnegative diagonal entries is positive semidefinite.*

Moreover, if A is irreducible, and

$$a_{ii} > \sum_{\substack{j=1 \\ j \neq i}}^{n} |a_{ij}|$$

for at least one i, then A is nonsingular.

Proof. We prove the first part. Let A be an $n \times n$ symmetric diagonally dominant matrix with nonnegative diagonal entries. Let A_{ij} be the matrix whose principal submatrix based on rows and columns i, j is

$$\begin{pmatrix} |a_{ij}| & a_{ij} \\ a_{ij} & |a_{ij}| \end{pmatrix},$$

and all its other entries are equal to zero. Then each A_{ij} is a positive semidefinite matrix (of rank at most 1), and $A - \sum_{i<j} A_{ij}$ is a nonnegative diagonal matrix. The matrix A is therefore positive semidefinite, as a sum of positive semidefinite matrices. □

We conclude the section with a discussion of positive definite matrices.

Definition 1.41. A matrix $A \in \mathbb{R}^{n \times n}$ is *positive definite* if it is symmetric and $\mathbf{x}^T A \mathbf{x} > 0$ for every $\mathbf{0} \neq \mathbf{x} \in \mathbb{R}^n$.

Proposition 1.42. *An $n \times n$ matrix A is positive definite if and only if it is a nonsingular positive semidefinite matrix.*

Proof. The "only if " part is obvious. For the "if" part, write the positive semidefinite matrix A as $A = BB^T$. Then $\mathbf{x}^T A \mathbf{x} = \|B^T \mathbf{x}\|^2$. By Proposition 1.27, this implies that $\mathbf{x}^T A \mathbf{x} = 0$ iff $\mathbf{x} \in \operatorname{ns} A$. Hence if A is nonsingular, $\mathbf{x}^T A \mathbf{x} > 0$ for every $\mathbf{0} \neq \mathbf{x} \in \mathbb{R}^n$. □

The following theorem is analogous to Theorem 1.25, but has an additional statement concerning nested minors. We define the notion of nested minors and state the theorem without a proof.

Definition 1.43. Let α_i, $i = 1, \ldots, k$, be subsets of $\{1, \ldots, n\}$ such that for every i, $|\alpha_i| = i$ and $\alpha_i \subseteq \alpha_{i+1}$. Then the minors $\det(A[\alpha_i])$ are *nested principal minors* of A.

The leading minors of A, $\det(A[1, \ldots, i])$, $i = 1, \ldots, n$, are n nested principal minors of A.

Theorem 1.44. *The following statements on an $n \times n$ symmetric matrix A are equivalent.*

(a) *A is positive definite.*
(b) *All the eigenvalues of A are positive.*
(c) *All the principal minors of A are positive.*
(d) *$A = BB^T$ for some nonsingular matrix B.*
(e) *$A = LL^T$, where L is a nonsingular lower triangular matrix.*

(f) $A = C^2$, *where* C *is a nonsingular symmetric matrix.*
(g) A *is the Gram matrix of* n *linearly independent vectors.*
(h) $A = \sum_{i=1}^{n} \mathbf{b}_i \mathbf{b}_i^T$, *where* $\mathbf{b}_1, \ldots, \mathbf{b}_n \in \mathbb{R}^n$ *are linearly independent.*
(i) A *has a set of* n *positive nested principal minors.*

Example 1.45. The $n \times n$ Hilbert matrix $H(n)$, whose i, j entry is $1/(i+j-1)$, is positive definite. To see that, consider $V = C[0,1]$, the space of real continuous functions on $[0,1]$, with the inner product

$$\langle f, g \rangle = \int_0^1 f(x)g(x)\,dx.$$

Then $H(n)$ is the Gram matrix of $f_i(x) = x^{i-1}$, $i = 1 \ldots, n$.

Example 1.46. The matrix

$$\begin{pmatrix} 0 & 0 \\ 0 & -1 \end{pmatrix}$$

shows that an analogous statement to (i)⇒(a) does not hold in the positive semidefinite case. That is, a matrix with a set of nonnegative nested principal minors is not necessarily positive semidefinite.

In the positive semidefinite case we have the following weaker result.

Proposition 1.47. *Let* A *be an* $n \times n$ *symmetric matrix. Let* $A[\alpha_i]$, $i = 1, \ldots, n$, *be nested minors of* A, *such that* $A[\alpha_i]$, $i = 1, 2, \ldots, n-1$, *are positive, and* $\det A \geq 0$. *Then* A *is positive semidefinite.*

Note that if A is a positive semidefinite matrix, then for every $\varepsilon > 0$, $A + \varepsilon I$ is positive definite. Hence every positive semidefinite matrix is a limit of positive definite matrices. The converse is also true: any matrix which is a limit of positive definite (or positive semidefinite) matrices is positive semidefinite.

Exercises

1.9. Prove Proposition 1.21.
Hint: To deduce the second assertion from the first, use the fact that for a symmetric matrix A, $\operatorname{cs} A = (\operatorname{ns} A)^{\perp}$.

1.10. Prove Proposition 1.27.
Hint: Same as in the previous exercise.

1.11. Prove Corollary 1.28.

1.12. Prove Proposition 1.29.

Hint: Suppose that $\alpha = \{1, 2, \ldots, k\}$. Then $A = BB^T$, $B = \begin{pmatrix} X & Y \\ Z & W \end{pmatrix}$, where X is $k \times k$. Use Propositions 1.21 and 1.27.

1.13. Let A be a real symmetric matrix of rank r, and suppose that the leading $r \times r$ principal submatrix A_0 of A is nonsingular. Show that A is positive semidefinite if and only if A_0 is positive definite and there exists a matrix E such that

$$A = \begin{pmatrix} A_0 & A_0 E \\ E^T A_0 & E^T A_0 E \end{pmatrix}.$$

Hint: For the "if" direction see Proposition 1.29.

1.14. Prove Lemma 1.37.

1.15. Prove Proposition 1.47.
Hint: Assume $\alpha_i = \{1, \ldots, i\}$. For every $\varepsilon > 0$, $A_\varepsilon = A + \varepsilon E_{nn}$ is positive definite.

1.16. Show that if A is symmetric, then

$$\exp(A) = \lim_{n \to \infty} \sum_{k=1}^{n} \frac{A^k}{k!}$$

is positive definite.

1.17. Let a_i, $i = 1, \ldots, n$, be positive numbers, $a_i \neq a_j$ for $i \neq j$. Show that the symmetric Cauchy matrix C whose i, j entry is $1/(a_i + a_j)$ is positive definite.

1.18. Prove: A matrix A is positive definite if and only if A^{-1} is positive definite.

1.19. Show that if A is positive definite, then the classical adjoint of A, $\operatorname{adj} A$, is positive definite. Give an example of a symmetric matrix A such that $\operatorname{adj} A$ is positive definite but A is not.

1.20. If A is positive definite, then A can be represented as a polynomial in A^2.

1.21. Show that if A is a positive semidefinite matrix, then

(a) The generalized inverse $A^\dagger$ is also positive semidefinite.

(b) $\sqrt{A^\dagger} = \sqrt{A}^\dagger$.

Hint: See Exercise 1.7.

1.22. Show that if A is a positive semidefinite matrix and $\mathbf{v} \in \operatorname{cs} A$, then there exists $\delta_0 > 0$ such that $A - \delta \mathbf{v}\mathbf{v}^T$ is positive semidefinite for every $0 \leq \delta \leq \delta_0$, and $\operatorname{rank}(A - \delta_0 \mathbf{v}\mathbf{v}^T) = \operatorname{rank} A - 1$.
Hint: $\mathbf{v} = \sqrt{A}\mathbf{u}$, and $A - \delta \mathbf{v}\mathbf{v}^T = \sqrt{A}(I - \delta \mathbf{u}\mathbf{u}^T)\sqrt{A}$. Hence it suffices to consider the case $A = I$.

Notes. As mentioned in the beginning of this section, this is not a comprehensive survey. A good reference on positive semidefinite matrices (real and complex) is [Horn and Johnson (2013)]. Most of Theorem 1.44 follows easily from Theorem 1.25 and Propositions 1.42 and 1.27. The implication (i)⇒(b), however, requires additional theory, such as interlacing (Theorem 1.18). The Hadamard product is also called *Schur product*, since Proposition 1.34 was proved by I. Schur. Apparently the entrywise product of matrices was first used by Moutard in 1894. This fact and other historical remarks on this product can be found in [Horn and Johnson (1994)], which contains an excellent extensive reference on Hadamard products and on Kronecker products.

1.3 Nonnegative matrices and M-matrices

In this section we present some of the main results on nonnegative and related matrices. We include only results which are relevant to the subject of this book, and mostly without proof. The section is divided into three parts: nonnegative matrices, totally nonnegative matrices, and M-matrices.

Nonnegative matrices

Definition 1.48. A matrix A is *nonnegative* if all its entries are nonnegative. A matrix A is *positive* if all its entries are positive.

We write $A \geq 0$ to denote that A is nonnegative, and $A \geq B$ if $A - B$ is nonnegative. $\mathbb{R}_+^{m\times n}$ denotes the set of all nonnegative matrices in $\mathbb{R}^{m\times n}$.

Theorem 1.49. *If A is a nonnegative square matrix, then $\rho(A)$, the spectral radius of A, is an eigenvalue of A, and A has a nonnegative eigenvector corresponding to $\rho(A)$.*

Remark 1.50. In the case that A is an $n \times n$ nonnegative matrix,

$$\min_{1\le i\le n} \sum_{j=1}^{n} a_{ij} \le \rho(A) \le \max_{1\le i\le n} \sum_{j=1}^{n} a_{ij}.$$

And, of course, also

$$\min_{1\le j\le n} \sum_{i=1}^{n} a_{ij} \le \rho(A) \le \max_{1\le j\le n} \sum_{i=1}^{n} a_{ij}.$$

For an irreducible nonnegative A, Theorem 1.49 is just the tip of the iceberg.

Theorem 1.51. **[The Perron-Frobenius Theorem]** *Let A be a square irreducible nonnegative matrix. Then*

(a) *$\rho(A)$ is a positive simple eigenvalue.*
(b) *A has a positive eigenvector corresponding to $\rho(A)$.*
(c) *If $\lambda_0 = \rho(A), \lambda_1 = \rho(A)e^{i\theta_1}, \ldots, \lambda_{k-1} = \rho(A)e^{i\theta_{k-1}}$ are the eigenvalues of A with absolute value $\rho(A)$, then $\lambda_1, \ldots, \lambda_{k-1}$ are also simple, $\theta_i = 2\pi i/k$, $1 = 1, \ldots, k-1$, and A is similar to $e^{2\pi i/k}A$. In particular, the whole spectrum of A is invariant under a rotation of the complex plane by $2\pi/k$.*
(d) *If k in (c) is greater than 1, then there exists a permutation matrix P such that*

$$P^T A P = \begin{pmatrix} 0 & A_{12} & 0 & \ldots & 0 \\ 0 & 0 & A_{23} & \ldots & 0 \\ \vdots & \vdots & \ddots & \ddots & \vdots \\ 0 & & \ldots & 0 & A_{k-1\,k} \\ A_{k1} & 0 & & \ldots & 0 \end{pmatrix} \tag{1.6}$$

where the zero blocks on the diagonal are square.

Definition 1.52. The eigenvalue $\rho(A)$ of an irreducible nonnegative matrix A is called the *Perron root* of A, and a positive eigenvector $\mathbf{x}$ such that $A\mathbf{x} = \rho(A)\mathbf{x}$ is called a *Perron vector* of A. The number of eigenvalues of an irreducible nonnegative matrix A of modulus $\rho(A)$ is called the *index of cyclicity* of the irreducible matrix A.

Example 1.53. The matrix

$$A = \begin{pmatrix} 0 & 1 & 0 & 0 \\ 0 & 0 & 1 & 0 \\ 0 & 0 & 0 & 1 \\ 4 & 0 & 3 & 0 \end{pmatrix}$$

is irreducible. Its spectral radius is 2, the eigenvalues are $2, -2, i, -i$. $(1, 2, 4, 8)^T$ is a Perron vector of A. The index of cyclicity is 2. Switching the second and third rows and columns of A yields the matrix

$$\begin{pmatrix} 0 & 0 & 1 & 0 \\ 0 & 0 & 0 & 1 \\ 0 & 1 & 0 & 0 \\ 4 & 3 & 0 & 0 \end{pmatrix}$$

which has the form (1.6).

In view of Theorem 1.51, the index of cyclicity of any square positive matrix is 1.

Corollary 1.54. *If A is a positive square matrix, then $\rho(A)$ is a simple eigenvalue of A, which is greater than the absolute value of any other eigenvalue, and A has a positive eigenvector corresponding to $\rho(A)$.*

Totally nonnegative matrices

Totally nonnegative matrices are an important class of nonnegative matrices.

Definition 1.55. A square matrix is *totally nonnegative* if all its minors are nonnegative, and *totally positive* if all its minors are positive.

Every zero matrix is totally nonnegative, as is every J_n matrix and every identity matrix. A nontrivial 3×3 example is

$$\begin{pmatrix} 3 & 4 & 1 \\ 2 & 3 & 1 \\ 1 & 2 & 2 \end{pmatrix}.$$

Here is another example.

Example 1.56. A Cauchy matrix

$$C = \left(\frac{1}{x_i + y_j} \right)_{i,j=1}^{n}$$

is totally positive when both the sequences $\{x_i\}_{i=1}^n$ and $\{y_i\}_{i=1}^n$ are positive and strictly increasing. To see that, observe that every submatrix of C is also a Cauchy matrix, and use the formula (1.4) for the determinant of a Cauchy matrix.

Remark 1.57. Note that a symmetric totally nonnegative matrix is both nonnegative and positive semidefinite. A matrix which has both these properties is called *doubly nonnegative.*

While I_n and J_n are totally nonnegative, $J_n + I_n$ is not (for $n \geq 3$). So the sum of totally nonnegative matrices is not necessarily totally nonnegative. However, by the Cauchy-Binet Formula (1.1), the product of totally nonnegative matrices is totally nonnegative.

Proposition 1.58. *If A and B are $n \times n$ totally nonnegative matrices, then AB is also totally nonnegative. If A and B are totally positive, so is AB.*

Any totally nonnegative matrix may be factored into a product of totally nonnegative matrices of a special form.

Definition 1.59. An $n \times n$ upper triangular matrix U is called *triangular totally positive* if for every $k = 1, \ldots, n$ and sets of k indices, $\alpha = \{i_1, i_2, \ldots, i_k\}$ and $\beta = \{j_1, j_2, \ldots, j_k\}$, ordered increasingly and satisfying $1 \leq i_s \leq j_s \leq n$ for $s = 1, \ldots, k$, the minor $\det A[\alpha|\beta]$ is positive.

An $n \times n$ lower triangular matrix L is called *triangular totally positive* if L^T is triangular totally positive.

A *triangular totally nonnegative* matrix is exactly what the name says — a totally nonnegative matrix which is triangular.

The rest of the minors of a triangular totally positive matrix are necessarily zero by the triangular structure of the matrix, hence a triangular totally positive matrix is totally nonnegative, but not totally positive.

Theorem 1.60. *For every totally nonnegative matrix A there exist a lower triangular totally nonnegative matrix L and an upper triangular totally nonnegative matrix U such that $A = LU$.*

If A is totally positive then L and U are both triangular totally positive.

M-matrices

We consider now square matrices whose off-diagonal entries are nonpositive. Such a matrix is called a *Z-matrix.* Of particular interest are the following Z-matrices.

Definition 1.61. A square matrix $A = sI - B$, $B \geq 0$, is an *M-matrix* if $s \geq \rho(B)$.

An M-matrix $A = sI - B$, $B \geq 0$, is nonsingular if and only if $s > \rho(B)$. There are many characterizations of nonsingular M-matrices. Some of them are given in the following theorem.

Theorem 1.62. *Let $A = sI - B$ be an $n \times n$ matrix, $B \geq 0$. Then the following statements are equivalent.*

(a) *A is a nonsingular M-matrix.*
(b) *All the principal minors of A are positive.*
(c) *A is positive stable, i.e., the real parts of the eigenvalues of A are positive.*
(d) *For every permutation matrix P, there exist a lower triangular matrix L with positive diagonal entries and an upper triangular matrix U with positive diagonal entries such that $P^TAP = LU$.*
(e) *There exists a positive diagonal matrix $D = \mathrm{Diag}(d_1, \ldots, d_n)$ such that AD is strictly diagonally dominant, i.e., $d_i a_{ii} > \sum_{j \neq i} -d_j a_{ij}$.*
(f) *For every permutation matrix P the leading principal minors of P^TAP are positive.*
(g) *A^{-1} is nonnegative.*

To property (g) in Theorem 1.62 we add the following.

Theorem 1.63. *If A and B are nonsingular M-matrices, then $A \geq B$ if and only if $B^{-1} \geq A^{-1}$.*

To obtain analogies for some of these characterizations for general, not necessarily nonsingular, M-matrices (see Exercise 1.29), one can use the following observation.

Proposition 1.64. *Let $A = sI - B$, $B \geq 0$. Then A is an M-matrix if and only if $A + \varepsilon I$ is a nonsingular M-matrix for every $\varepsilon > 0$.*

Corollary 1.65. *A matrix $A = sI - B$, where B is a symmetric nonnegative matrix, is an M-matrix if and only if it is positive semidefinite.*

The only theorem that we prove in this section also deals with symmetric M-matrices. The theorem is needed in Section 3.4.

Theorem 1.66. *If A is a symmetric M-matrix, then there exists a positive diagonal matrix D such that DAD is diagonally dominant.*

Proof. If A is not irreducible, then it is a direct sum of irreducible symmetric M-matrices. So it suffices to prove the theorem in the case that A

is irreducible. In that case, let ρ be the Perron root of B, and let $\mathbf{x}$ be a Perron vector of B. Let $D = \text{Diag}(\mathbf{x})$. Then AD is diagonally dominant, since

$$AD\mathbf{e} = s\mathbf{x} - \rho\mathbf{x} \geq \mathbf{0}.$$

Because D is a positive diagonal matrix, this implies that $DAD\mathbf{e} \geq \mathbf{0}$. That is, DAD is a symmetric diagonally dominant matrix. □

Remark 1.67. By the proof of Theorem 1.66, if A is nonsingular, then DAD is strictly diagonally dominant, and if A is singular, then the sum of each row in DAD is zero.

Exercises

1.23. A square nonnegative matrix A is called *primitive* if A^m is positive for some positive integer m. Show that the index of cyclicity of a primitive matrix is 1. (The converse is also true: an irreducible nonnegative matrix with index of cyclicity 1 is primitive.)

1.24. Let A be an $n \times n$ nonsingular nonnegative matrix. Show that if $A^{-1} \geq 0$, then A has a single positive element in every row and every column, hence $A = PD$, where P is a permutation matrix and D is a positive diagonal matrix.
Hint: Let $B = A^{-1} \geq 0$. $AB = I$ implies the existence of k_i such that $a_{i\,k_i} b_{k_i\,i} > 0$, for every $1 \leq i \leq n$. Let $(A')_{ij} = a_{i\,k_i}$ if $j = i$, and $(A')_{ij} = 0$ otherwise. Show that $0 \leq A'B \leq I$ to deduce that $A'B$ is diagonal, and use the equality $A' = (A'B)A$.

1.25. Prove Proposition 1.58.

1.26. The symmetric Pascal matrix $P(n)$ is defined by $p_{1i} = p_{i1} = 1$ for every $1 \leq i \leq n$, and $p_{ij} = p_{i-1\,j} + p_{i\,j-1}$ for $2 \leq i, j \leq n$. Prove that $P(n)$ is totally positive.
Hint: Find a nonsingular triangular totally nonnegative matrix S such that $P(n+1) = S(1 \oplus P(n))S^T$.

1.27. Show that if A is an irreducible symmetric Z-matrix, then its smallest eigenvalue is a simple eigenvalue, with a corresponding positive eigenvector.

1.28. Suppose that $A = sI - B$, $B \geq 0$ is irreducible. Prove that A is a nonsingular M-matrix if and only if A^{-1} is positive.

1.29. Let $A = sI - B$, $B \geq 0$. Prove that the following statements are equivalent.

(a) A is an M-matrix.
(b) All the principal minors of A are nonnegative.
(c) A is nonnegative stable, *i.e.*, the real parts of the eigenvalues of A are nonnegative.
(d) There exist a permutation matrix P, a lower triangular matrix L with nonnegative diagonal entries, and an upper triangular matrix U with nonnegative diagonal entries, such that $P^T AP = LU$.
(e) There exists a positive diagonal matrix $D = \text{Diag}(d_1, \ldots, d_n)$ such that AD is diagonally dominant, *i.e.*, $d_i a_{ii} \geq \sum_{j \neq i} -d_j a_{ij}$.

1.30. Show that if an irreducible $n \times n$ matrix A is a singular M-matrix, then $\operatorname{rank} A = n - 1$.

1.31. Show that if A is an $n \times n$ singular M-matrix, then there exists a positive diagonal matrix D such that $B = AD$ satisfies

$$b_{ii} = \sum_{\substack{j=1 \\ j \neq i}}^{n} -b_{ij}\,, \quad i = 1, \ldots, n.$$

Notes. There are several texts on nonnegative matrices and their applications. For example [Bapat and Raghavan (1997)], [Berman and Plemmons (1994)], [Gantmacher (1959)], [Horn and Johnson (2013)], [Minc (1988)], [Seneta (1981)] and [Varga (2000)]. For the proof of the claim in Exercise 1.24 see [DeMarr (1972)].

Corollary 1.54 was proved by Perron in 1907 and Theorems 1.49 and 1.51 were proved by Frobenius in 1912. Together they form the classical Perron-Frobenius Theorem.

References on totally nonnegative matrices include [Ando (1987)], [Gantmacher and Krein (2002)], [Gasca and Michelli (1996)] and [Karlin (1968)]. Theorem 1.60 is due to [Cryer (1976)]. In [Gantmacher and Krein (2002)] totally nonnegative matrices are called completely positive (see the notes of Section 2.1). More recent books on these matrices are [Pinkus (2010)] and [Fallat and Johnson (2011)].

The M in M-matrices stands for Minkowsky. A symmetric nonsingular M-matrix is called a *Stieltjes matrix*. In other words, a Stieltjes matrix is a positive definite matrix in which the off-diagonal entries are nonpositive. The proof of Theorem 1.66 is taken from [Drew *et al.* (1994)]. More on

M-matrices can be found, e.g. in [Berman and Plemmons (1994)] and in [Horn and Johnson (1994)].

1.4 Schur complements

Definition 1.68. Let $A[\alpha]$ be a nonsingular principal submatrix of A. Then the *Schur complement* of $A[\alpha]$ in A, $A/A[\alpha]$, is the matrix

$$A(\alpha) - A(\alpha|\alpha](A[\alpha])^{-1}A[\alpha|\alpha).$$

For example, if A has block form

$$A = \begin{pmatrix} B & C \\ D & E \end{pmatrix}, \tag{1.7}$$

where E is nonsingular, then $A/E = B - CE^{-1}D$. If $A[\alpha]$ is not the right lower corner of A, we can use simultaneous permutation of the rows and columns of A to transfer $A[\alpha]$ to that corner. We therefore assume from now on that A is an $n \times n$ matrix of the form (1.7), and $A[\alpha] = E$.

Example 1.69. Let

$$A = \begin{pmatrix} 1 & 2 & 3 & 4 \\ 5 & 6 & 7 & 8 \\ 9 & 10 & 11 & 12 \\ 13 & 14 & 15 & 16 \end{pmatrix},$$

$$E_1 = \begin{pmatrix} 11 & 12 \\ 15 & 16 \end{pmatrix},$$

and $E_2 = (16)$. Then

$$\begin{aligned} A/E_1 &= \begin{pmatrix} 1 & 2 \\ 5 & 6 \end{pmatrix} - \begin{pmatrix} 3 & 4 \\ 7 & 8 \end{pmatrix} \begin{pmatrix} 11 & 12 \\ 15 & 16 \end{pmatrix}^{-1} \begin{pmatrix} 9 & 10 \\ 13 & 14 \end{pmatrix} \\ &= \begin{pmatrix} 0 & 0 \\ 0 & 0 \end{pmatrix}, \end{aligned}$$

and

$$\begin{aligned} A/E_2 &= \begin{pmatrix} 1 & 2 & 3 \\ 5 & 6 & 7 \\ 9 & 10 & 11 \end{pmatrix} - \begin{pmatrix} 4 \\ 8 \\ 12 \end{pmatrix} \left(\frac{1}{16}\right) \begin{pmatrix} 13 & 14 & 15 \end{pmatrix} \\ &= -\frac{1}{4} \begin{pmatrix} 9 & 6 & 3 \\ 6 & 4 & 2 \\ 3 & 2 & 1 \end{pmatrix}. \end{aligned}$$

In the case that $a_{ii} \neq 0$ we also denote $A/A[i]$ by $A/[i]$.

Theorem 1.70. *Let A be as in* (1.7). *Then* $\det A = \det(A/E)\det E$ *and* $\operatorname{rank} A = \operatorname{rank}(A/E) + \operatorname{rank} E$.

Proof. Consider the equality

$$\begin{pmatrix} I & -CE^{-1} \\ 0 & I \end{pmatrix}\begin{pmatrix} B & C \\ D & E \end{pmatrix} = \begin{pmatrix} B - CE^{-1}D & 0 \\ D & E \end{pmatrix}. \tag{1.8}$$

The matrix on the right-hand side of (1.8) is

$$\begin{pmatrix} A/E & 0 \\ D & E \end{pmatrix},$$

and by (1.8) it is row equivalent to A. Thus

$$\operatorname{rank}\begin{pmatrix} A/E & 0 \\ D & E \end{pmatrix} = \operatorname{rank}(A/E) + \operatorname{rank} E.$$

Since

$$\det\begin{pmatrix} I & -CE^{-1} \\ 0 & I \end{pmatrix} = 1,$$

(1.8) implies that

$$\det A = \det\begin{pmatrix} A/E & 0 \\ D & E \end{pmatrix} = \det(A/E)\det E. \qquad \square$$

The next proposition provides a formula for the elements of A/E.

Proposition 1.71. *Let A be an $n \times n$ matrix, and let $E = A[\alpha]$ be a nonsingular principal submatrix of A. Then for $i, j \in \{1, \ldots, n\} \setminus \alpha$,*

$$(A/E)_{ij} = \frac{\det A[\{i\} \cup \alpha | \{j\} \cup \alpha]}{\det E}.$$

Proof. Observe that $(A/E)_{ij} = A[\{i\} \cup \alpha | \{j\} \cup \alpha]/E$, and use Theorem 1.70. $\square$

Schur complements have a nice quotient property.

Theorem 1.72. *If*

$$M = \begin{pmatrix} A & B \\ C & D \end{pmatrix} \quad \textit{and } D = \begin{pmatrix} E & F \\ G & H \end{pmatrix},$$

and if D and H are nonsingular, then D/H is nonsingular principal submatrix of M/H, and $M/D = (M/H)/(D/H)$.

Proof. Suppose $D = M[\alpha]$, where $\alpha = \{n-k+1, \dots, n\}$, and $H = D[\beta]$, where $\beta = \{n-l+1, \dots, n\} \subseteq \alpha$. By Proposition 1.71, D/H is a principal submatrix of M/H; $D/H = (M/H)[\alpha \setminus \beta]$. By Theorem 1.70, $\det(D/H) = \det D / \det H$, hence D/H is nonsingular.

Let $U = (\det H)(M/H)$. The elements of U are bordered minors of H in M. Hence by Sylvester's Determinantal Identity (1.2), for every $\gamma, \delta \subseteq \{1, \dots, n-l\}$ such that $|\gamma| = |\delta| = p$,

$$\det U[\gamma|\delta] = (\det H)^{p-1} \det M[\gamma \cup \beta | \delta \cup \beta]. \tag{1.9}$$

Now, for every $i, j \in \{1, \dots, n-k\}$,

$$\begin{aligned} ((M/H)/(D/H))_{ij} &= \frac{\det(M/H)[\{i\} \cup (\alpha \setminus \beta) | \{j\} \cup (\alpha \setminus \beta)]}{\det(D/H)} \\ &= \frac{(\det H) \det(M/H)[\{i\} \cup (\alpha \setminus \beta) | \{j\} \cup (\alpha \setminus \beta)]}{\det D} \\ &= \frac{\det U[\{i\} \cup (\alpha \setminus \beta) | \{j\} \cup (\alpha \setminus \beta)]}{(\det H)^{k-l} \det D}. \end{aligned}$$

But by (1.9),

$$\det U[\{i\} \cup (\alpha \setminus \beta) | \{j\} \cup (\alpha \setminus \beta)] = (\det H)^{k-l} \det M[\{i\} \cup \alpha | \{j\} \cup \alpha],$$

hence

$$((M/H)/(D/H))_{ij} = \frac{\det M[\{i\} \cup \alpha | \{j\} \cup \alpha]}{\det D} = (M/D)_{ij}. \qquad \square$$

We are particularly interested in the case that A is a real symmetric matrix. In this case, we have the following theorem.

Theorem 1.73. *Let A be a real symmetric matrix in block form* (1.10), *with E nonsingular. Then*

$$\operatorname{In} A = \operatorname{In} E + \operatorname{In} A/E.$$

Proof. As in (1.8),

$$\begin{pmatrix} I & -CE^{-1} \\ 0 & I \end{pmatrix} \begin{pmatrix} B & C \\ C^T & E \end{pmatrix} \begin{pmatrix} I & 0 \\ -E^{-1}C^T & I \end{pmatrix} = \begin{pmatrix} B - CE^{-1}C^T & 0 \\ 0 & E \end{pmatrix},$$

and the required equality of inertias follows. $\square$

If A is a real positive semidefinite matrix, we can extend the notion of Schur complement to the case that $A[\alpha]$ is singular, using the Moore-Penrose generalized inverse.

Definition 1.74. Let $A[\alpha]$ be a principal submatrix of a real positive semidefinite matrix A. Then the *generalized Schur complement* of $A[\alpha]$ in A is the matrix

$$A(\alpha) - A(\alpha](A[\alpha])^{\dagger} A[\alpha).$$

It is also denoted by $A/A[\alpha]$.

Observe that no confusion should occur, since $(A[\alpha])^{\dagger} = (A[\alpha])^{-1}$ when $A[\alpha]$ is nonsingular. As before, it would be convenient to assume that $A[\alpha]$ is the lower right corner of A. Hence we assume that A is in block form

$$A = \begin{pmatrix} B & C \\ C^T & E \end{pmatrix}. \tag{1.10}$$

Since A is positive semidefinite, $\operatorname{cs} C^T \subseteq \operatorname{cs} E$ (Proposition 1.29), and by Exercise 1.8(e) this implies that

$$EE^{\dagger}C^T = C^T. \tag{1.11}$$

This equality plays an essential role in proving the properties of the generalized Schur complement.

Theorem 1.75. *If A is a real positive semidefinite matrix in block form* (1.10), *then the matrices $A/E = B - CE^{\dagger}C^T$ and*

$$\begin{pmatrix} B - A/E & C \\ C^T & E \end{pmatrix} = \begin{pmatrix} CE^{\dagger}C^T & C \\ C^T & E \end{pmatrix}$$

are positive semidefinite.

Proof. Using (1.11), it is easy to check that the following equality holds:

$$\begin{pmatrix} I & -CE^{\dagger} \\ 0 & I \end{pmatrix} \begin{pmatrix} B & C \\ C^T & E \end{pmatrix} \begin{pmatrix} I & 0 \\ -E^{\dagger}C^T & I \end{pmatrix} = \begin{pmatrix} B - CE^{\dagger}C^T & 0 \\ 0 & E \end{pmatrix}.$$

Hence, by Propositions 1.22 and 1.23, A/E is positive semidefinite.

Consider now the second matrix. Since E is positive semidefinite, it has a square root and so does $E^{\dagger}$. Use the fact that $\operatorname{cs} C^T \subseteq \operatorname{cs} E = \operatorname{cs} \sqrt{E}$ and therefore $\sqrt{E}\sqrt{E^{\dagger}}C^T = \sqrt{E}\sqrt{E}^{\dagger}C^T = C^T$ to get that

$$\begin{pmatrix} B - A/E & C \\ C^T & E \end{pmatrix} = \begin{pmatrix} CE^{\dagger}C^T & C \\ C^T & E \end{pmatrix} = \begin{pmatrix} C\sqrt{E^{\dagger}} \\ \sqrt{E} \end{pmatrix} \begin{pmatrix} \sqrt{E^{\dagger}}C^T & \sqrt{E} \end{pmatrix}.$$

By Theorem 1.25, this matrix is also positive semidefinite. □

Remark 1.76. For the matrix $\begin{pmatrix} CE^\dagger C^T & C \\ C^T & E \end{pmatrix}$ to be positive semidefinite, it suffices that E is positive semidefinite and $\operatorname{cs} C^T \subseteq \operatorname{cs} E$, since then we have:

$$\begin{pmatrix} CE^\dagger C^T & C \\ C^T & E \end{pmatrix} = \begin{pmatrix} C\sqrt{E^\dagger} \\ \sqrt{E} \end{pmatrix} \begin{pmatrix} \sqrt{E^\dagger} C^T & \sqrt{E} \end{pmatrix}.$$

A result similar to Theorem 1.70 holds for generalized Schur complements of a positive semidefinite matrix.

Theorem 1.77. *Let A be a positive semidefinite matrix in block form* (1.10). *Then*

$$\det A = \det(A/E) \det E, \qquad \operatorname{rank} A = \operatorname{rank}(A/E) + \operatorname{rank} E,$$

and

$$\operatorname{rank} \begin{pmatrix} CE^\dagger C^T & C \\ C^T & E \end{pmatrix} = \operatorname{rank} E.$$

Proof. Prove the first part by imitating the proof of Theorem 1.70 and using (1.11).

For the proof of the second part, use (as in the previous proof)

$$\begin{pmatrix} B - A/E & C \\ C^T & E \end{pmatrix} = \begin{pmatrix} CE^\dagger C^T & C \\ C^T & E \end{pmatrix} = \begin{pmatrix} C\sqrt{E^\dagger} \\ \sqrt{E} \end{pmatrix} \begin{pmatrix} \sqrt{E^\dagger} C^T & \sqrt{E} \end{pmatrix}.$$

Hence

$$\operatorname{rank} \begin{pmatrix} CE^\dagger C^T & C \\ C^T & E \end{pmatrix} = \operatorname{rank} \begin{pmatrix} C\sqrt{E^\dagger} \\ \sqrt{E} \end{pmatrix} = \operatorname{rank} \sqrt{E} = \operatorname{rank} E$$

(see Proposition 1.27). □

Consider the special case of an $n \times n$ positive definite matrix A. In this case, $a_{11} > 0$ and

$$A = \mathbf{a}_1 \mathbf{a}_1^T + A^{(1)},$$

where $\mathbf{a}_1$ is the first column of A multiplied by $\left(\sqrt{a_{11}}\right)^{-1}$, and $A^{(1)} = 0 \oplus (A/[1])$ has rank $n-1$. We may now let $\mathbf{a}_2$ be the second column of $A^{(1)}$ (the first nonzero column in this matrix), multiplied by $\left(\sqrt{a_{22}^{(1)}}\right)^{-1}$, and get that

$$A = \mathbf{a}_1 \mathbf{a}_1^T + \mathbf{a}_2 \mathbf{a}_2^T + A^{(2)},$$

where $A^{(2)} = 0_2 \oplus (A/A[1,2])$ (see Theorem 1.72, noting that $a_{22}^{(1)} = A[1,2]/[1]$). We may repeat the procedure. In the $(k+1)$-th step we start with $A^{(k)} = 0_k \oplus (A/A[1,\ldots,k])$, which has rank $n-k$, and let $\mathbf{a}_{k+1}$ be the $k+1$ column of $A^{(k)}$ multiplied by $\left(\sqrt{a_{k+1\,k+1}^{(k)}}\right)^{-1}$. After n steps we get

$$A = \sum_{i=1}^{n} \mathbf{a}_i \mathbf{a}_i^T = LL^T,$$

where $L = \begin{pmatrix} \mathbf{a}_1 & \mathbf{a}_2 & \ldots & \mathbf{a}_n \end{pmatrix}$ is a lower triangular matrix with a positive diagonal. This is the *Cholesky decomposition* of the positive definite matrix A.

The Cholesky decomposition may be generalized to singular positive semidefinite matrices in a couple of ways. Suppose $\operatorname{rank} A = r$. Suppose the first nonzero column in A is column i and take $\mathbf{a}_1$ to be this column divided by $\left(\sqrt{a_{ii}}\right)^{-1}$. Then

$$A = \mathbf{a}_1 \mathbf{a}_1^T + A^{(1)}.$$

The matrix $A^{(1)}$ is positive semidefinite and $\operatorname{rank} A^{(1)} = \operatorname{rank} A - 1$ (as $A^{(1)}$ is $A/[i]$ with added row and column of zeros). The i-th column in $A^{(1)}$ is zero, and so is any other column that in A was a scalar multiple of the i-th column. Repeat this process by choosing at the $(k+1)$-th step the first nonzero in $A^{(k)} = A - \sum_{i=1}^{k} \mathbf{a}_i \mathbf{a}_i^T$. The matrix $Q = \begin{pmatrix} \mathbf{a}_1 & \mathbf{a}_2 & \ldots & \mathbf{a}_r \end{pmatrix}$ has the property that $\min \operatorname{supp} \mathbf{a}_{i+1} > \min \operatorname{supp} \mathbf{a}_i$ for every i. We can either use the decomposition $A = QQ^T$, or extend Q to a lower triangular matrix L by inserting zero columns between the columns of Q. Then $A = LL^T$ (but the lower triangular L has some zero diagonal entries). Similarly one may obtain a factorization $A = UU^T$, where U is upper triangular — see Exercise 1.36).

Exercises

1.32. Prove that if the matrix A is an M-matrix, and E is a nonsingular principal submatrix of A, then A/E is an M-matrix.

1.33. Prove that if the matrix A is totally nonnegative, and E is a nonsingular principal submatrix of A, then A/E is also totally nonnegative.

1.34. Show that if A is as in (1.7) and E is nonsingular, then

$$\operatorname{rank} \begin{pmatrix} CE^{-1}D & C \\ D & E \end{pmatrix} = \operatorname{rank} E.$$

1.35. Show that if A is a symmetric nonsingular M-matrix, then in the Cholesky decomposition $A = LL^T$ the lower triangular matrix L is an M-matrix (and so is L^T).

1.36. Show that if A is positive semidefinite, then $A = UU^T$ for some upper triangular matrix U.

Notes. The name "Schur complement" was coined by Emilie Haynsworth in [Haynsworth (1968)]. The results on the relations between a matrix and its Schur complement are due to [Schur (1917)]. The quotient property (Theorem 1.72) was proved in [Crabtree and Haynsworth (1969)]. An interesting discussion of properties and applications of the Schur complement is carried out in [Cottle (1974)]. A recent reference to the Schur complement and its applications is [Zhang (2005)].

1.5 Graphs

In this section we establish graph terminology and notations as used in this book, and quote some relevant results. As the list is quite long, we divide this section into subsections, starting, naturally, with the basics.

Graphs and subgraphs

Definition 1.78. A *graph* G is a pair (V, E) in which $V = V(G)$ is a finite set, whose elements are called *vertices*, and $E = E(G)$ is a finite set of *edges*, where each edge is an unordered pair of distinct vertices.

We use drawings to represent graphs. Each vertex is marked by a dot, and each edge by a line connecting exactly two of these dots.

Example 1.79. If

$$V(G) = \{a, b, c, d\} \quad \text{and} \quad E(G) = \{\{a, b\}, \{b, c\}, \{c, d\}, \{b, d\}\},$$

then G can be drawn like that:

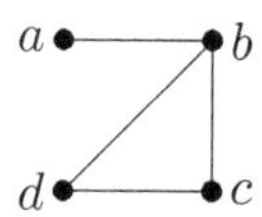

(This is of course just one of many possible drawings — we can place the dots wherever we wish, and draw the lines curved.)

The graphs we are talking about are usually called *simple graphs*, but we omit the "simple", since we will not be using other types of graphs, such as graphs with multiple edges (that is, several identical pairs of vertices, to be drawn like that:) and/or loops (that is, pairs of identical vertices, each to be drawn as). However, in Section 2.14 we will consider also signed simple graphs. A *signed graph* has the form $\mathcal{G} = (V, E, \Sigma)$ where $V = V(\mathcal{G})$ is the vertex set, $E = E(\mathcal{G})$ is the edge set, and $\Sigma : E \to \{-1, 1\}$ signs the edges. When considering signed graphs, we will draw a positive edge as a solid line, and a negative edge as a dashed line.

Example 1.80. If we assign negative signs to the edges $\{b, c\}$ and $\{b, d\}$ of the graph G in example 1.79, we obtain a signed graph $\mathcal{G}$ that can be drawn like that:

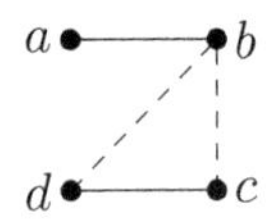

If $e = \{u, v\}$ is an edge of G we say that u and v are the *vertices* or the *ends* of e, that u and v are *adjacent* or *neighbours*, that e is *incident* with u (and with v), and that u is *incident* with e (as is v). Two edges are *adjacent* if they have a common vertex. The *degree* of a vertex $v \in V(G)$ is the number of vertices adjacent to it; it is denoted by $d(v)$. A vertex v of degree 1 is called a *pendant* vertex. A graph in which every vertex is adjacent to every other vertex is a *complete graph*.

Two graphs, $G_1 = (V(G_1), E(G_1))$ and $G_2 = (V(G_2), E(G_2))$, are *isomorphic* if there is a one-to-one function φ from $V(G_1)$ onto $V(G_2)$ such that $\{\varphi(u), \varphi(v)\}$ is an edge of G_2 if and only if $\{u, v\}$ is an edge of G_1. Two isomorphic graphs can be drawn exactly the same way, up to the labelling of the vertices. Since we will be mostly interested in properties of graphs which do not depend on the labels, isomorphic graphs will often be treated as one and the same. For example, we denote any complete graph on n vertices by K_n.

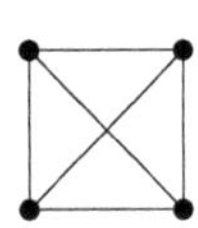

K_4

The *complement* $\overline{G}$ of a graph G has the same vertices as G and two vertices in $V(\overline{G})$ are adjacent in $\overline{G}$ if and only if they are not adjacent in G.

A graph H is a *subgraph* of G if $V(H)$ is a subset of $V(G)$ and $E(H)$ is a subset of $E(G)$. We denote it by $H \subseteq G$, and say also that G *contains* H. If either $V(H)$ is a proper subset of $V(G)$ or $E(H)$ is a proper subset of $E(G)$, we say that H is a *proper subgraph* of G. If $V \subseteq V(G)$, the graph H such that $V(H) = V$ and $E(H)$ consists of all the edges of G that have both ends in V is the subgraph of G *induced* by V. Such a subgraph of G is called an *induced subgraph.* If H is a subgraph of G such that $V(H) = V(G)$, we say that H is a *spanning subgraph* of G.

Example 1.81. In the following drawing, H_1 is an induced subgraph of G, H_2 is a spanning subgraph of G, and H_3 is a subgraph of G which is neither induced nor spanning.

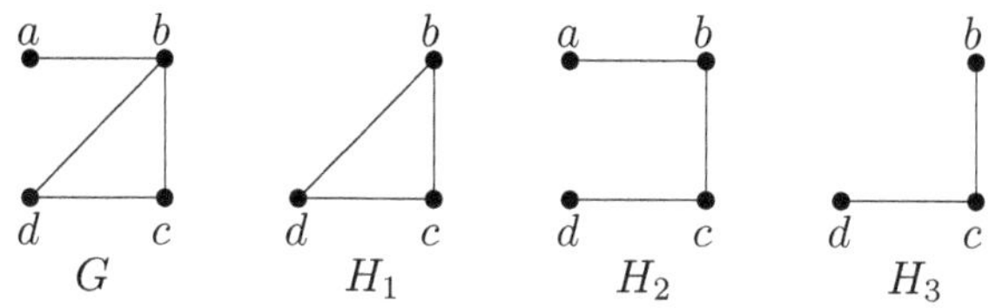

If v is a vertex of G, we denote by $G - v$ the subgraph of G induced by $V(G) \setminus \{v\}$. That is, $G - v$ is the graph obtained by deleting from G the vertex v along with all edges incident with it. If e is an edge of G, we denote by $G - e$ the graph obtained by removing e from G (that is, from $E(G)$). If u and v are vertices of G, and $e = \{u, v\}$ is not an edge of G, we denote by $G + e$ the graph obtained from G by adding the edge e.

If G_1 and G_2 are subgraphs of the same graphs G, the *union* $G_1 \cup G_2$ of the two subgraphs is the subgraph of G with vertex set $V(G_1) \cup V(G_2)$ and edge set $E(G_1) \cup E(G_2)$. G_1 and G_2 are *disjoint* if $V(G_1) \cap V(G_2) = \emptyset$. If G_1 and G_2 have at least one vertex in common, their *intersection* $G_1 \cap G_2$ has vertex set $V(G_1) \cap V(G_2)$ and edge set $E(G_1) \cap E(G_2)$.

A subset of $V(G)$ in which no two vertices are adjacent in G is an *independent set* (or a *stable set*) of vertices. A *clique* is a subset of $V(G)$ which induces a complete subgraph of G. A clique is *maximal* if it is not a proper subset of any other clique. We denote by $\mathrm{cc}\,(G)$ the *edge-clique covering number*, the minimum number of complete subgraphs of G needed to cover all of G's edges.

A sequence of edges $\{v_0, v_1\}, \{v_1, v_2\}, \ldots, \{v_{m-1}, v_m\}$ is called a *walk of length* m from v_0 to v_m. (That is, the length of a walk is the number of edges in it.) The walk is *closed* if $v_0 = v_m$ and *open* if $v_0 \neq v_m$. Such a walk from v_0 to v_m is a *path* if the vertices $v_1, \ldots, v_m$ are distinct. The vertices v_0 and v_m are the *end vertices* of the path, and the vertices $v_1, \ldots, v_{m-1}$ are the *internal vertices* of this path. A finite set of paths is *independent* if no two of them share an internal vertex. A *cycle* is a closed path. We sometimes denote a path or a cycle with edges $\{v_0, v_1\}, \{v_1, v_2\}, \ldots, \{v_{m-1}, v_m\}$ also as $v_0 - v_1 - \cdots - v_{m-1} - v_m$. If H is a subgraph of G, a path P in G, which has at least one edge, is called an *H-path* if $P \cap H$ consists of the two end vertices of P (and no edge). A path whose ends are the vertices x and y is an *xy-path.* A *chord* of a cycle (of length > 3) is an edge which connects two nonconsecutive vertices of the cycle. A graph which is a (chordless) cycle of length n is called an *n-cycle* and denoted by C_n. The graph $C_3 = K_3$ is called a *triangle.* The graph which is a path on n vertices (and length $n-1$) is denoted by P_n.

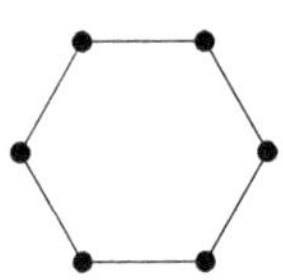

C_6

P_5

If u and v are vertices in G, and there is a path in G connecting u and v, then we say that u and v are *connected* in G, and the *distance between u and v in G*, $d_G(u, v)$, is the length of the shortest path in G between u and v. A graph G is *connected* if any two distinct vertices in G are connected. Otherwise G is *disconnected.* It is easy to see that if G is disconnected, then $G = G_1 \cup \ldots \cup G_k$ for some k, where $G_1, \ldots, G_k$ are pairwise disjoint induced subgraphs of G, and each of them is connected. The subgraphs $G_1, \ldots, G_k$ are the *components* of G.

A vertex v of a graph G is a *cut vertex* of G if $G - v$ has more components than G. If G is connected, then $v \in V(G)$ is a cut vertex of G if and only if $G = G_1 \cup G_2$ where both G_1 and G_2 are connected, and the two subgraphs have only the vertex v in common. A *block* is a connected graph with no cut vertices. A *block of a graph* G is a subgraph of G, which is a block, and is not properly contained in any other block subgraph of G. A graph is *k-connected* if the deletion of any $k-1$ vertices, together with their incident edges, leaves a connected graph with at least one edge. In particular, a 2-connected graph is a block on 3 or more vertices.

Example 1.82. The graph in the next drawing is connected, u and v are cut vertices. The graph has four blocks — a triangle, two single edges, and a 4-cycle.

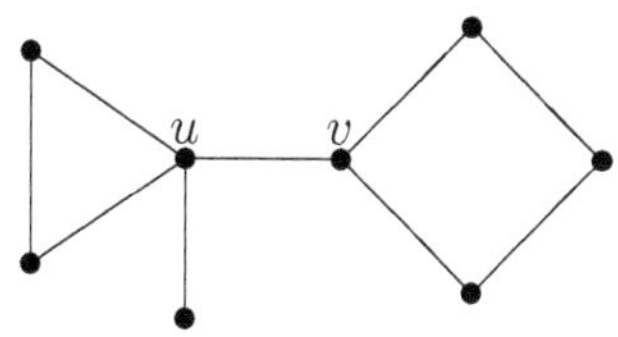

We will need the following two results.

Theorem 1.83. *If G is 2-connected and u and v are vertices of G, then there exist at least two paths connecting u and v, which have no common internal vertex.*

Proposition 1.84. *If H_0 is a 2-connected subgraph of a 2-connected graph G, then G may be constructed by starting from H_0 and successively adding H-paths to subgraphs H already constructed.*

Proof. Let H be a maximal subgraph of G thus constructed from H_0. The subgraph H has to be induced, since by the maximality of H there are no vertices $u, v \in V(H)$ such that $\{u, v\}$ is an H-path, i.e. $\{u, v\} \in E(G) \setminus E(H)$. Suppose on the contrary that $H \neq G$. As G is connected, there exists an edge $\{u, v\}$ of G such that $u \in V(H)$ and $v \in V(G) \setminus V(H)$. Since G is 2-connected, there exists a path P in $G - v$ from some $w \in V(H)$ to u. But then the path consisting of $P \cup \{u, v\}$ is an H-path in G, which may be added to H to construct a larger subgraph of G, in contradiction to the maximality of H. □

Special types of graphs

We describe here some families of graphs. A graph G is *k-regular* if all the vertices in the graph have the same degree k. A graph G is *strongly regular* with parameters (n, k, a, c) if G is a k-regular graph on n vertices such that any two adjacent vertices have a common neighbours and any two non-adjacent vertices have c common neighbours. The strongly regular graphs have a minor role in this book, in Chapter 5.

Trees play a more prominent role in the book. A *tree* is a connected graph that contains no cycles.

Example 1.85. There are three nonisomorphic trees on 5 vertices:

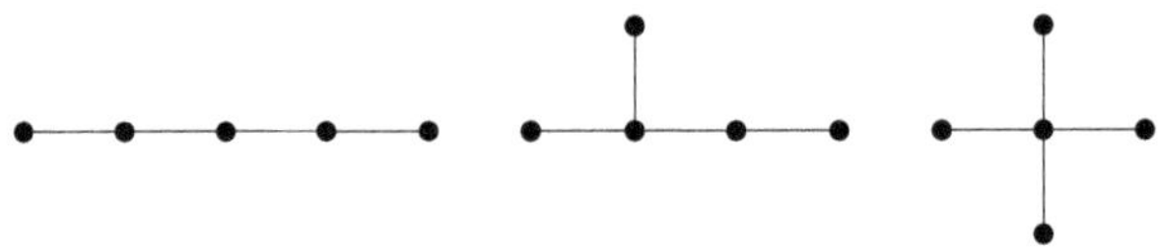

Two basic results concerning trees are stated below.

Theorem 1.86. *A connected graph G is a tree if and only if $|E(G)| = |V(G)| - 1$.*

Theorem 1.87. *If G is a tree and $|V(G)| > 1$, then G has at least two pendant vertices.*

A graph G is *bipartite* if $V(G)$ can be partitioned into two subsets X and Y, such that each edge has one vertex in X and one vertex in Y (X and Y are independent sets). Equivalently, a bipartite graph is a graph that contains no cycle of odd length (an *odd cycle*). In particular, every tree and every even cycle are bipartite, and every odd cycle is not. A *complete bipartite* graph $K_{m,n}$ is a bipartite graph in which X has n vertices, Y has m vertices, and every vertex in X is adjacent to each of the vertices in Y.

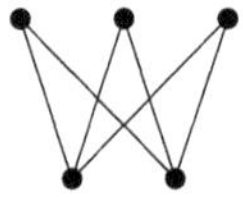

$K_{3,2}$

While trees are bipartite graphs, bipartite graphs are part of a still larger family — the triangle free graphs. A graph G is *triangle free* if, as the name suggests, G contains no triangles. An odd cycle on 5 vertices or more is an example of a triangle free graph, which is not bipartite.

We will need the following result on triangle free graphs.

Theorem 1.88. [**Mantel's Theorem**] *If G is a triangle free graph such that $|V(G)| = n$, then*

$$|E(G)| \leq \lfloor n^2/4 \rfloor.$$

Equality holds if and only if $G = K_{n/2,n/2}$ (if n is even), or $G = K_{(n-1)/2,(n+1)/2}$ (if n is odd).

At the other end of the spectrum are chordal graphs. A graph G is *chordal* if each cycle in G of length greater than 3 has a chord. So G is chordal if it has no induced cycle of length greater than 3. Any complete graph is, of course, chordal, as is every tree. Another example of a chordal graph is the graph we denote by T_n, which is important in the study of completely positive and copositive matrices. The graph T_n consists of $n-2$ triangles with a common base. That is, T_n is a graph on n vertices, where two vertices u and v are adjacent, and all other vertices are adjacent to both u and v and to no other vertex. Another chordal graph is the *fan* graph on n vertices, F_n. It consists of a path on $n-1$ vertices, and an additional vertex adjacent to all the path vertices.

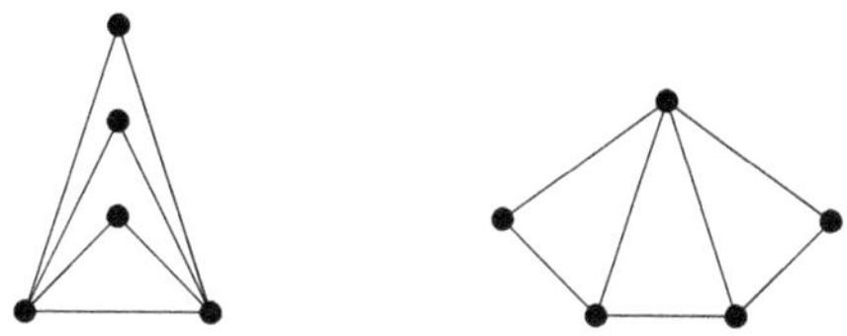

T_5 F_5

A special type of chordal graph is a *block-clique graph*, a connected graph in which each block is a complete graph.

Example 1.89. The drawing shows an example of a block-clique graph.

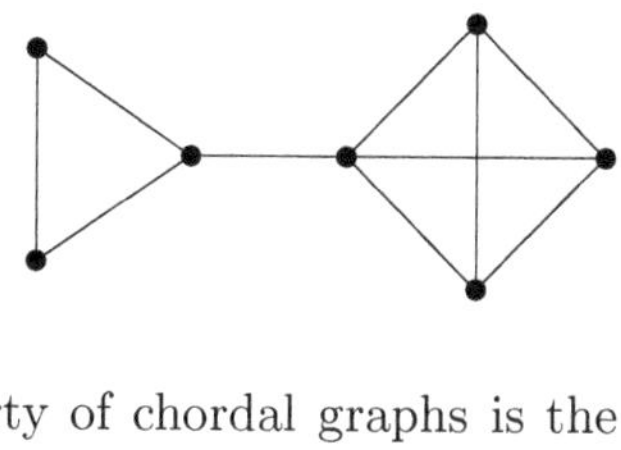

An important property of chordal graphs is the existence of a special ordering of the vertices.

Theorem 1.90. *A graph G on n vertices is chordal if and only if there is an ordering $v_1, \ldots, v_n$ of $V(G)$ such that for each $i = 1, \ldots, n-1$ the set*

$$\{v_j \mid v_j \text{ is adjacent to } v_i \text{ and } j > i\}$$

is a clique.

Such an ordering of the vertices is called a *perfect elimination ordering*, or a *PEO* (for reasons that will be explained later in the section).

Example 1.91. Here is a perfect elimination ordering of the vertices of T_5:

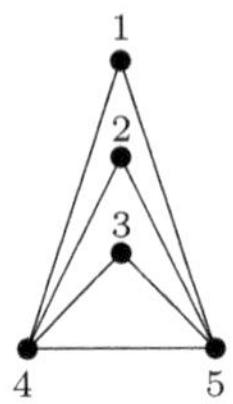

For the next class of graphs we need to introduce first two important numbers associated with a graph, the chromatic number and the clique number. A graph G is k-*colorable* if $V(G)$ can be partitioned into k disjoint subsets, $V_1, \ldots, V_k$, such that adjacent vertices do not belong to the same V_i. We can think of it as coloring the vertices of G in k colors, so that any two adjacent vertices have different colors.

Example 1.92. T_5 is 3-colorable. Suppose the vertices are labeled as in Example 1.91. Clearly, three colors are needed to color the vertices $3, 4, 5$, but 1 and 2 may have the same color as 3.

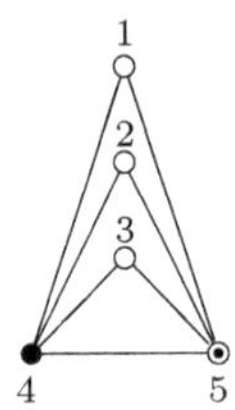

The smallest k such that G is k-colorable is called the *chromatic number* of G and is denoted by $\chi(G)$. For example, $\chi(T_n) = 3$ for any $n \geq 3$. If G contains K_p as a subgraph, its chromatic number is at least p. If G has at least one edge, its chromatic number is 2 if and only if G is bipartite. If G has n vertices, then $1 \leq \chi(G) \leq n$. The upper bound is obtained by K_n and the lower bound by the graph with no edges. A better upper bound for the chromatic number is given in the following theorem.

Theorem 1.93. *If G is connected, and is neither K_n nor an odd cycle C_{2s+1}, then $\chi(G)$ is bounded by the maximal degree of the vertices of G.*

The *clique number* of a graph G is the cardinality of the largest clique; it is denoted by $\omega(G)$. For example $\omega(T_n) = 3$, and $\omega(G) = 2$ for any triangle free graph G with at least one edge. A clique in G of size $\omega(G)$ is called a *maximum clique*. The clique number of $\overline{G}$ is called the *stability number* of G. It is the cardinality of the largest stable set in G, and denoted by $\alpha(G)$.

Clearly, $\chi(G) \geq \omega(G)$. A graph is *perfect* if $\chi(H) = \omega(H)$ for every induced subgraph H of G. Bipartite graphs are obviously perfect. Though it is not as obvious, chordal graphs also turn out to be perfect. An odd cycle on 5 or more vertices is an example of a graph which is not perfect: For $s \geq 2$ we have $\chi(C_{2s+1}) = 3$ while $\omega(C_{2s+1}) = 2$. The next theorem is a characterization of perfect graphs, conjectured by Claude Berge, and proved more than forty years after the conjecture was made.

Theorem 1.94. [The Strong Perfect Graph Theorem] *A graph G is perfect if and only if neither G nor $\overline{G}$ contains an induced odd cycle of length greater than* 3.

The *line graph* of a graph G has the edges of G as its vertices, and two such vertices are adjacent if and only if they are adjacent edges of G. If H is the line graph of G, we say that G is the *root* of H.

Example 1.95. The figure below shows a graph and its line graph.

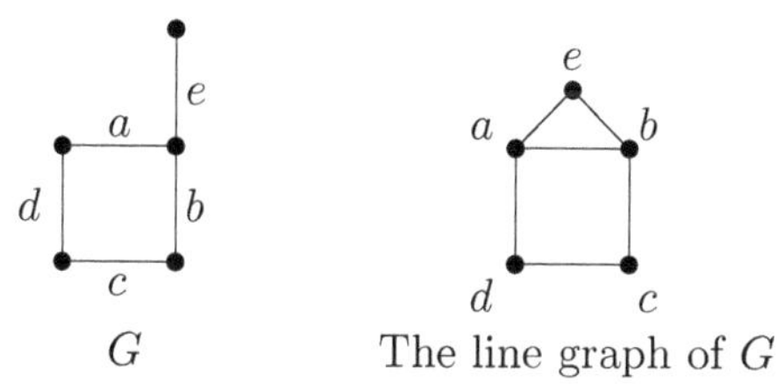

A graph G is said to be *line perfect* if its line graph is perfect. The next theorem gives a characterization of line perfect graphs.

Theorem 1.96. *A graph is line perfect if and only if it contains no odd cycle of length greater than* 3.

It turns out that these graphs have a special role in the theory of completely positive matrices (which seems to have nothing to do with their being line perfect).

If $e = \{u, v\}$ is an edge in a graph G, a *subdivision* of e is an operation where e is replaced by a path whose ends are u and v, and whose internal vertices are not vertices of $G - e$. A graph obtained from G by successively subdividing edges is called a *subdivision* of G. A *contraction* of an edge $e = \{u, v\}$ of a graph G is an operation in which the vertex u and the vertex v are identified, deleting any arising loops and replacing multiple edges by a single edge. A *minor* of graph G is a graph obtained from G by a sequence of the operations of these three types (in any order): isolated vertex deletion, edge deletion and edge contraction. Minors are important in the characterization of graphs with certain properties, due to the following theorem. A family of graphs is *minor-closed* if any minor of a graph in the family is also in the family.

Theorem 1.97. **[The Graph Minor Theorem]** *If a family of graphs is minor-closed, then this family is characterized by a finite set of forbidden minors.*

Graphs and matrices

Matrices are used in graph theory to "store" the specifications of a graph, and graphs are used in matrix theory as a way to describe the zero/nonzero pattern of a matrix.

We associate a matrix with every graph. Given a graph G with n vertices, label the vertices $1, 2, \ldots, n$. The *adjacency matrix*, $A(G) = (a_{ij})$, of G is the symmetric $\{0,1\}$-matrix where $a_{ij} = a_{ji} = 1$ if and only if i and j are adjacent.

Example 1.98. The adjacency matrix of K_n is $J_n - I_n$. The adjacency matrix of C_5, when the vertices are numbered consecutively, is

$$\begin{pmatrix} 0 & 1 & 0 & 0 & 1 \\ 1 & 0 & 1 & 0 & 0 \\ 0 & 1 & 0 & 1 & 0 \\ 0 & 0 & 1 & 0 & 1 \\ 1 & 0 & 0 & 1 & 0 \end{pmatrix}.$$

Another matrix associated with a graph is the Laplacian. The *Laplacian matrix* $L(G)$ of a graph G with vertices $1, 2, \ldots, n$ is $D(G) - A(G)$, where $D(G) = \text{Diag}(d_1, \ldots, d_n)$ is the diagonal matrix with d_i equals the degree of i. Observe that for any graph G, $L(G)$ is a singular M-matrix.

Example 1.99.

$$L(C_5) = \begin{pmatrix} 2 & -1 & 0 & 0 & -1 \\ -1 & 2 & 1 & 0 & 0 \\ 0 & -1 & 2 & -1 & 0 \\ 0 & 0 & -1 & 2 & -1 \\ -1 & 0 & 0 & -1 & 2 \end{pmatrix}.$$

The classical Matrix-Tree Theorem gives the number of spanning trees of a graph. It is a nice demonstration of the relationship between graphs and matrices.

Theorem 1.100. **[The Matrix-Tree Theorem]** $\operatorname{adj} L(G) = kJ$ *where* k *is the number of spanning trees of* G.

The above matrices associated with a given graph were symmetric. Another matrix associated with a graph G is the *incidence matrix.* For a graph G with vertex set $V(G) = \{v_1, \ldots, v_n\}$ and edge set $E(G) = \{e_1, \ldots, e_m\}$ the incidence matrix of G, $B(G)$ is an $n \times m$ matrix with $b_{ij} = 1$ if v_i is one of the vertices of e_j, and $b_{ij} = 0$ otherwise.

Example 1.101. For T_5, with vertices 1 and 2 as the base vertices, and edges

$$e_1 = \{1,2\},\ e_2 = \{1,3\},\ e_3 = \{1,4\},\ e_4 = \{1,5\},$$
$$e_5 = \{2,3\},\ e_6 = \{2,4\},\ e_7 = \{2,5\},$$

we have

$$B(T_5) = \begin{pmatrix} 1 & 1 & 1 & 1 & 0 & 0 & 0 \\ 1 & 0 & 0 & 0 & 1 & 1 & 1 \\ 0 & 1 & 0 & 0 & 1 & 0 & 0 \\ 0 & 0 & 1 & 0 & 0 & 1 & 0 \\ 0 & 0 & 0 & 1 & 0 & 0 & 1 \end{pmatrix}.$$

Theorem 1.102. *If* G *is a graph with* n *vertices and* k *bipartite components, then* $\operatorname{rank} B(G) = n - k$.

Proof. Suppose $\mathbf{x} \in \mathbb{R}^n$ satisfies $\mathbf{x}^T B(G) = \mathbf{0}^T$. Then for any edge $\{v_i, v_j\}$ in G, $x_i + x_j = 0$, so $x_j = -x_i$. By induction, if there is a path of length k between v_i and v_j, then $x_j = (-1)^k x_i$. Hence if i is a vertex on an odd cycle, $x_i = 0$. And for any bipartite component of G, x_i has one value for every v_i in one of the independent bipartition sets, and the opposite value for every i such that v_i is in the other bipartition set. Thus the null space of $B(X)^T$ is spanned by k vectors, each a $\{0, 1, -1\}$-vector supported by the vertex set of one bipartite component. □

In the reverse direction, we associate with each matrix a graph. For general square matrices, it is natural to use directed graphs. A *directed graph (digraph)* D consists of a finite set $V(D)$ of *vertices*, and a set $A(D)$ of ordered pairs of (not necessarily distinct) vertices, called *arcs*. It is drawn with dots for the vertices and arrows for the arcs. In a digraph the analogue of a path is a *directed path* — a sequence of arcs $(v_0, v_1), (v_1, v_2), \ldots, (v_{m-1}, v_m)$ where $v_0, \ldots, v_{m-1}$ are distinct vertices. A digraph D is *strongly connected* if for every two vertices of D there is a directed path from u to v and a directed path from v to u.

Given an $n \times n$ matrix A, the *directed graph of* A, $D(A)$, has n vertices, $1, 2, \ldots, n$, and an arc from i to j if and only if $a_{ij} \neq 0$.

Example 1.103. If

$$A = \begin{pmatrix} 1 & 2 & 0 & 0 \\ 0 & 0 & 3 & 0 \\ 0 & 4 & 0 & 5 \\ 0 & 0 & 0 & 0 \end{pmatrix},$$

then $D(A)$ is

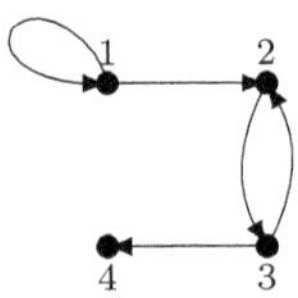

It can be shown that a matrix A is irreducible if and only if $D(A)$ is strongly connected. More can be said if A is nonnegative.

Theorem 1.104. *Let A be an $n \times n$ irreducible nonnegative matrix. Let S_i be the set of all lengths of directed cycles through i in $D(A)$, and let h_i be the greatest common divisor of all the numbers in S_i. Then $h_1 = h_2 = \ldots = h_n = h$, and h is the index of cyclicity of A, i.e., the number of eigenvalues of A of modulus $\rho(A)$.*

The main application of graph theory in this book will be to symmetric matrices. In this case we use graphs instead of digraphs. The graph of an $n \times n$ symmetric matrix A is denoted $G(A)$. The vertices of $G(A)$ are $1, 2, \ldots, n$, and i and j are adjacent in $G(A)$ if and only if $i \neq j$ and $a_{ij} \neq 0$.

Example 1.105. If

$$A = \begin{pmatrix} 1 & 2 & 0 & 0 \\ 2 & 0 & 3 & 4 \\ 0 & 3 & 5 & 6 \\ 0 & 4 & 6 & 0 \end{pmatrix},$$

then $G(A)$ is

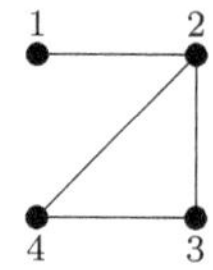

If A is a symmetric matrix with $G(A) = G$, we say that A is a *symmetric matrix realization* of G. If A is also nonnegative then it is a *symmetric nonnegative realization* of G. Note that in the graph of A there is no indication which diagonal entries of A are zero, and which are not. It is easy to see that a symmetric matrix is irreducible if and only if $G(A)$ is connected. If P is a permutation matrix, $G(P^TAP)$ is isomorphic to $G(A)$, since simultaneous permutation of rows and columns changes only the labels of the vertices of the graph.

For a real $n \times n$ symmetric matrix A we denote by $G_-(A)$ the graph on vertices $\{1, \dots, n\}$ in which vertices i and j are adjacent if and only if $a_{ij} < 0$. Similarly, $G_+(A)$ the graph on vertices $\{1, \dots, n\}$ in which vertices i and j are adjacent if and only if $a_{ij} > 0$. And for $c \in \mathbb{R}$ we denote by $G_c(A)$ the graph on vertices $\{1, \dots, n\}$ in which vertices i and j are adjacent if and only if $a_{ij} = c$.

We will also consider the signed graph of a matrix A. The signed graph of an $n \times n$ symmetric matrix A is denoted $\mathcal{G}(A)$. It is the graph $G(A)$ in which each edge $\{i, j\}$ is assigned the sign of a_{ij}.

Example 1.106. If

$$A = \begin{pmatrix} 1 & -2 & 0 & 0 \\ -2 & 1 & -1 & -4 \\ 0 & -1 & 5 & 1 \\ 0 & -4 & 1 & 1 \end{pmatrix},$$

then $\mathcal{G}(A)$, $G_-(A)$ and $G_1(A)$ are shown below. (Recall that in a drawing of a signed graph, a dashed line denotes a negative edge and a solid line denotes a positive one.)

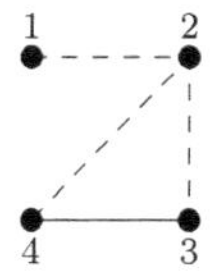

$\mathcal{G}(A)$

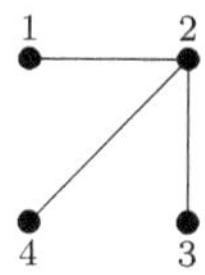

$G_-(A)$

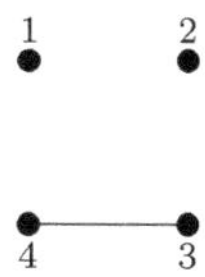

$G_1(A)$

We conclude the section with a few results relating graph properties with the properties of their symmetric matrix realizations. Specifically, we consider chordal graphs, and their positive semidefinite matrix realizations.

The following result explains why perfect elimination orderings of the vertices of a chordal graph are named so.

Proposition 1.107. *Let G be a chordal graph, let $v_1, \ldots, v_n$ be a perfect elimination ordering of its vertices, and let $P = \begin{pmatrix} \mathbf{e}_{v_1} & \ldots & \mathbf{e}_{v_n} \end{pmatrix}$ be the corresponding permutation matrix. If A is any positive semidefinite matrix realization of G, then $P^TAP = LL^T$, where L has r columns, and the support of each column is a clique in G.*

Such a factorization of P^TAP may be obtained by a Cholesky-type decomposition of P^TAP, as described at the end of Section 1.4. In fact, chordal graphs are characterized by this property of their positive semidefinite realizations.

Theorem 1.108. *A graph G is chordal if and only if every real positive semidefinite matrix realization of A is the sum of rank 1 positive semidefinite matrices whose graphs are (complete) subgraphs of G.*

Another spectral graph theory result that will be useful is the following. If $\mathcal{M}$ is a class of real symmetric matrices, we denote by $\mathrm{mr}_{\mathcal{M}}(G)$ the minimum rank of matrices in $\mathcal{M}$ with graph G. In the next theorem $\mathcal{PSD}$ is the family of real positive semidefinite matrices, and $\mathrm{cc}\,(G)$ is the edge-clique covering number defined above.

Theorem 1.109. *If a graph G is chordal, then* $\mathrm{mr}_{\mathcal{PSD}}(G) = \mathrm{cc}\,(G)$.

Exercises

1.37. Let G be a 2-connected graph, V a proper subset of $V(G)$, $|V| \geq 2$, and v a vertex of G which is not in V. Show that there exist distinct vertices u and w in V and a path from u through v to w all of whose internal vertices are not in V.

1.38. Show that if G is a connected graph, which is not a block, then G has a block which contains only one cut vertex (actually, at least two such blocks).

1.39. Prove Theorem 1.88.
Hint: Prove separately for even n and for odd n, use induction on n.

1.40. Show that if G is chordal, and for some $e \in E(G)$ the graph $G - e$ is also chordal, then there exits a unique maximal clique K in G such that both ends of e lie in K.

1.41. Show that if $G \neq K_n$ is a chordal graph on n vertices, then there exists an edge $e \notin E(G)$ such that $G + e$ is also chordal.
Hint: In the case that G is connected, let $v_1, \ldots, v_n$ be a perfect elimination ordering of $V(G)$. Let v_i be the first vertex adjacent to v_1 which has an adjacent vertex v_j that is not adjacent to v_1. Let $e = \{v_1, v_j\}$.

1.42. Use Theorem 1.100 to prove Cayley's Formula: The number of spanning trees of K_n is n^{n-2}.

1.43. Show, without using Theorem 1.94, that if every subgraph of G is perfect, then G contains no odd cycle of length greater than 3.

1.44. For any graph G, show that $B(G)B(G)^T = |L(G)|$. The matrix $Q(G) = |L(G)| = D(G) + A(G)$ is called the *signless Laplacian* of G. Check that $Q(G)$ is diagonally dominant, and $\operatorname{rank} Q(G) = n - c$ where n is the number of vertices of G and c is the number of bipartite components in G.

Notes. Two good basic graph theory books are [Bondy and Murty (2008)] and [Diestel (2018)]. Good sources on strongly regular graphs are [Godsil and Royle (2001)] and [Brouwer and Haemers (2012)]. A reference on perfect graphs, and in particular chordal graphs and their applications, is [Golumbic (2004)]. Theorem 1.83 is a special case of Menger's Theorem, which says that if a graph G with at least $k + 1$ vertices is k-connected, then every two distinct vertices are connected by at least k paths which share no internal vertex. Theorem 1.88 was proved by Mantel in 1907. It is a special case of Turán's Theorem which gives a sharp upper bound for the number of edges in a graph which does not contain a K_p.

Theorem 1.90 was proved by Rose in 1970. Perfect elimination ordering of $G(A)$ is used in reordering the rows and columns of a matrix A in such a way that no nonzero elements are created during Gauss elimination (see [Golumbic (2004)], and Proposition 1.107). The term block-clique graphs is due to [Johnson and Smith (1996)]. Other names for these graphs include Husimi trees (see [Johnson and Smith (1996)]) and 1-chordal graphs (see [Drew *et al.* (2000)]).

Theorem 1.93 was proved by Brooks in 1941. Theorem 1.94 was proved in [Chudnovsky *et al.* (2006)], more than forty years after Claude Berge made the conjecture, which was known as the Strong Perfect Graph Conjecture. The Weak Perfect Graph Conjecture, also made by Berge, was that the complement of a perfect graph is perfect. This was proved by Lovàsz in 1972. Theorem 1.96 is due to Trotter. The graphs that contain no odd cycle of length greater than 3 play a major role in the theory of complete positivity (see Section 3.5). The Strong Perfect Graph Theorem implies that they are all the graphs with the property that all their subgraphs are perfect (one direction does not depend on the theorem, see Exercise 1.43). Theorem 1.97 was proved by Robertson and Seymour in a series of papers that appeared over 20 years, starting in 1983, see [Diestel (2018), Section 12.7].

The Matrix–Tree Theorem (Theorem 1.100) was proved by Kirchhoff in 1847. Theorem 1.104 is due to Romanovsky. For more on spectral graph theory, dealing with matrices associated with graphs, see, e.g., [Brouwer and Haemers (2012)] and [Godsil and Royle (2001)]. Theorem 1.108 was proved in [Agler *et al.* (1988)] and independently in [Paulsen *et al.* (1989)], then again in [Kakimura (2010)]. Theorem 1.109 was proved in [Shaked-Monderer (1993)] for the real case and in [Booth *et al.* (2008)] for both the real and the complex cases.

1.6 Convex cones

This section contains basic results on convex cones in a finite dimensional Euclidean space V, with inner product $\langle\ ,\ \rangle$. The section is divided into several subsections.

Convexity

Definition 1.110. A set K in a Euclidean space V is *convex* if for every $\mathbf{x}, \mathbf{y} \in K$ and $0 \leq a \leq 1$, $a\mathbf{x} + (1-a)\mathbf{y} \in K$.

Obviously, every linear subspace of V is a convex set. Another simple example is the closed unit ball in $\mathbb{R}^n$, $B = \{\mathbf{x} \in \mathbb{R}^n \mid \|\mathbf{x}\| \leq 1\}$. By the triangle inequality, B is convex.

Definition 1.111. A set K in a vector space V is a *cone* if for every $\mathbf{x} \in K$ and $a \geq 0$, $a\mathbf{x} \in K$.

Combining the two definitions we get the following definition.

Definition 1.112. A set K in a Euclidean space V is a *convex cone* if for every $\mathbf{x}, \mathbf{y} \in K$ and nonnegative scalars a, b, $a\mathbf{x} + b\mathbf{y} \in K$.

Example 1.113. Let $K_1 = \mathbb{R}^2_+ = \{(x,y)^T \in \mathbb{R}^2 \mid x \geq 0, y \geq 0\}$ and $K_2 = \{(x,y)^T \in \mathbb{R}^2 \mid 0 \leq y \leq x\}$.

K_1 and K_2 are convex cones in $\mathbb{R}^2$. Both cones are also closed. K_1 is a particular case of a general example. For every positive integer n, the nonnegative orthant of $\mathbb{R}^n$,

$$\mathbb{R}^n_+ = \{\mathbf{x} \in \mathbb{R}^n \mid x_i \geq 0 \text{ for every } i = 1, \ldots, n\},$$

is a closed convex cone.

Example 1.114. Given a unit vector $\mathbf{u} \in \mathbb{R}^n$, and an angle $0 \leq \theta < \pi/2$, let $K_{\mathbf{u},\theta}$ be the set of all vectors $\mathbf{x}$ in $\mathbb{R}^n$ such that the angle between $\mathbf{x}$ and $\mathbf{u}$ is at most θ. That is,

$$K_{\mathbf{u},\theta} = \{\mathbf{x} \in \mathbb{R}^n \mid \langle \mathbf{x}, \mathbf{u} \rangle \geq \|\mathbf{x}\| \cos\theta\}.$$

$K_{\mathbf{u},\theta}$ is a closed convex cone. It is clearly closed, and closed under multiplication by a nonnegative scalar, and if $\mathbf{0} \neq \mathbf{x}, \mathbf{y} \in K_{\mathbf{u},\theta}$ and $\mathbf{z} = \mathbf{x} + \mathbf{y}$, then

$$\begin{aligned} \langle \mathbf{z}, \mathbf{u} \rangle &= \langle \mathbf{x}, \mathbf{u} \rangle + \langle \mathbf{y}, \mathbf{u} \rangle \\ &\geq \|\mathbf{x}\| \cos\theta + \|\mathbf{y}\| \cos\theta \\ &= (\|\mathbf{x}\| + \|\mathbf{y}\|) \cos\theta \\ &\geq \|\mathbf{x} + \mathbf{y}\| \cos\theta \qquad (1.12) \\ &= \|\mathbf{z}\| \cos\theta. \end{aligned}$$

Hence $\mathbf{z} \in K_{\mathbf{u},\theta}$.

We will refer to cones of this type by the name *circular cones.*

A *base* of a convex cone K is a subset $B \subseteq K$ such that every nonzero vector in K is a positive multiple of exactly one vector in B.

Example 1.115. The simplex $\Delta_n = \left\{\mathbf{x} \in \mathbb{R}^n_+ \mid \sum_{i=1}^n x_i = 1\right\}$ is a compact convex set, which is a base of the nonnegative orthant $\mathbb{R}^n_+$. The nonnegative section of the unit sphere, $\Gamma_n = \left\{\mathbf{x} \in \mathbb{R}^n_+ \mid \sum_{i=1}^n x_i^2 = 1\right\}$, is another base of $\mathbb{R}^n_+$, which is compact, but not convex.

For any set $S \subseteq V$, denote by conv$S$ the minimal convex set containing S, called the *convex hull* of S, and by cone$S$ the minimal convex cone containing S. It is easy to see that conv$S$ consists of all the convex combinations of elements in S, that is, vectors which can be represented as $\sum_{i=1}^{m} \lambda_i \mathbf{x}_i$ for some vectors $\mathbf{x}_i \in S$ and nonnegative scalars λ_i satisfying $\sum_{i=1}^{m} \lambda_i = 1$. Similarly, cone$S$ consists of all nonnegative combinations of elements in S, that is, vectors which can be represented as $\sum_{i=1}^{m} \lambda_i \mathbf{x}_i$, where $\mathbf{x}_i \in S$ and λ_i is a nonnegative scalar, $i = 1, \ldots, m$.

The following proposition contains basic observations, which follow easily from Definition 1.112.

Proposition 1.116. *Let K, K_1 and K_2 be convex cones in a Euclidean space V. Then*

(a) cl$K$, *the closure of K, is a convex cone.*
(b) int$K$, *the interior of K, is a convex cone.*
(c) *The intersection $K_1 \cap K_2$ is also a convex cone.*
(d) *The sum $K_1 + K_2 = \{\mathbf{x}_1 + \mathbf{x}_2 \mid \mathbf{x}_1 \in K_1, \mathbf{x}_2 \in K_2\}$ is a convex cone.*

Note, however, that the sum of closed convex cones is not necessarily closed.

Example 1.117. Let K_1 be the cone of vectors in $\mathbb{R}^3$ such that the angle between them and the vector $(1, 0, 1)$ is at most $\pi/4$. That is,

$$K_1 = \{(x_1, x_2, x_3)^T \mid x_1 \geq 0, x_3 \geq 0, 2x_1x_3 \geq x_2^2\}.$$

Let K_2 be the x-axis in $\mathbb{R}^3$. Then

$$(0, 1, 0)^T = \lim_{x \to 0+} [(1/(2x), 1, x)^T + (-1/(2x), 0, 0)^T].$$

Hence $(0, 1, 0)^T \in \text{cl}(K_1 + K_2)$ but $(0, 1, 0)^T \notin K_1 + K_2$.

We leave the proof of the next proposition as exercise.

Proposition 1.118. *If K_1 and K_2 are convex cones, then*

(a) $\text{int}(K_1 \cap K_2) = (\text{int}K_1) \cap (\text{int}K_2)$.
(b) $\text{int}(K_1 + K_2) = (\text{int}K_1) + K_2 = K_1 + (\text{int}K_2)$.

Remark 1.119. If K is a convex cone in V, then $K - K$ is the minimal subspace of V containing K, and $K \cap (-K)$ is the maximal subspace of V contained in K.

Proposition 1.120. *Every convex cone K in V has a nonempty interior in $K - K$.*

Proof. Let $\mathbf{x}_1, \ldots, \mathbf{x}_k, \mathbf{y}_1, \ldots, \mathbf{y}_k$ be vectors in K such that

$$\{\mathbf{x}_1 - \mathbf{y}_1, \ldots, \mathbf{x}_k - \mathbf{y}_k\}$$

is a basis for $K - K$. Then the set $\{\mathbf{x}_1, \ldots, \mathbf{x}_k, \mathbf{y}_1, \ldots, \mathbf{y}_k\}$ spans $K - K$. It therefore contains a basis $\{\mathbf{z}_1, \ldots, \mathbf{z}_k\} \subseteq K$ of $K - K$. Then

$$\left\{\mathbf{z} = \sum_{i=1}^{k} t_i \mathbf{z}_i \mid t_i > 0 \text{ for every } i\right\}$$

is a nonempty open set contained in K. □

Observe that the proof relied on the fact that V, and therefore $K - K$, is finite dimensional.

Definition 1.121. For a convex cone K in a Euclidean space V, the *dimension* of K, $\dim K$, is defined to be $\dim(K - K)$, and denoted by $\dim K$. The interior of a convex cone K in $K - K$ is called the *relative interior* of K, and denoted by $\mathrm{ri}K$. The set $K \setminus \mathrm{ri}K$ is the *relative boundary* of K, and it is denoted by $\mathrm{rb}K$.

By Proposition 1.116(b) $\mathrm{ri}K$ is a convex cone in $K - K$, and therefore in V.

Hyperplanes and separation

Definition 1.122. A *hyperplane* H in a Euclidean space V is a set of the form

$$H = \{\mathbf{x} \in V \mid \langle \mathbf{x}, \mathbf{u} \rangle = a\}, \tag{1.13}$$

where $\mathbf{u}$ is a nonzero vector in V, and a is a real number.

H is a hyperplane through the origin iff $a = 0$. In that case, H is a subspace of V of dimension $\dim V - 1$. If H is as in (1.13) and $a \neq 0$, then H is a translation of such a subspace; that is, $H = H_0 + \mathbf{x}_0$, where $H_0 = \{\mathbf{x} \in V \mid \langle \mathbf{x}, \mathbf{u} \rangle = 0\}$, and $\mathbf{x}_0$ is any vector in H.

Every hyperplane H defines halfspaces of V — the hyperplane in (1.13) borders two *closed halfspaces*.

$$\{\mathbf{x} \in V \mid \langle \mathbf{x}, \mathbf{u} \rangle \geq a\} \text{ and } \{\mathbf{x} \in V \mid \langle \mathbf{x}, \mathbf{u} \rangle \leq a\},$$

and two *open halfspaces*

$$\{\mathbf{x} \in V \mid \langle \mathbf{x}, \mathbf{u} \rangle > a\} \text{ and } \{\mathbf{x} \in V \mid \langle \mathbf{x}, \mathbf{u} \rangle < a\}.$$

In the case that $a = 0$, any of these halfspaces is a convex cone.

A hyperplane H is said to separate two sets S_1 and S_2 if S_1 lies in one of the halfspaces bordered by H, and S_2 lies in the opposite one. There are actually several separation variations — the halfspaces may be closed or open, the sets may be contained in H or not. We need the following version.

Definition 1.123. A hyperplane H *separates properly* sets S_1 and S_2, if S_1 lies in one of the closed halfspaces bordered by H, S_2 lies in the opposite closed halfspace, and at least one of the two sets is not contained in H.

We quote here without proof two classic separation results for convex cones.

Theorem 1.124. *Let K be a closed convex cone in V, and let $\mathbf{b} \in V$, $\mathbf{b} \notin K$. Then there exists a nonzero vector $\mathbf{u} \in V$ such that $\langle \mathbf{x}, \mathbf{u} \rangle \geq 0$ for every $\mathbf{x} \in K$, and $\langle \mathbf{b}, \mathbf{u} \rangle < 0$.*

Theorem 1.125. *Let K_1 and K_2 be convex cones such that $\operatorname{ri} K_1 \cap \operatorname{ri} K_2 = \emptyset$, then there exists a hyperplane H passing through the origin, which separates K_1 and K_2 properly.*

If K is a convex cone, and H is a hyperplane such that K lies in one of the two closed halfspaces defined by H, and $\operatorname{bd} K \cap H \neq \emptyset$, we say that H is a *supporting hyperplane* of K (at $\mathbf{x}_0$, if $\mathbf{x}_0 \in \operatorname{bd} K \cap H$).

The following is a version of the Supporting Hyperplane Theorem stated for convex cones.

Theorem 1.126. [Supporting Hyperplane Theorem] *Let K be a convex cone, and let $\mathbf{x}_0$ be a point on the boundary $\operatorname{bd} K$ of K. Then there exists a supporting hyperplane H of K such that $\mathbf{x}_0 \in H$.*

Note that if K is a closed convex cone, and $H = \{\mathbf{x} \in V \mid \langle \mathbf{x}, \mathbf{u} \rangle = a\}$ is a supporting hyperplane of K at $\mathbf{x}_0 \in \operatorname{bd} K$, then $a = 0$ (see Exercise 1.57).

Extreme rays

If $\mathbf{x} \in V$, then $\operatorname{cone}\{\mathbf{x}\} = \{a\mathbf{x} \mid 0 \leq a \in \mathbb{R}\}$ is the *ray generated by* $\mathbf{x}$. If $\mathbf{x} \in K$, where K is a convex cone, then the ray generated by $\mathbf{x}$ is contained in K.

Definition 1.127. Let K be a convex cone and $\mathbf{0} \neq \mathbf{x} \in K$. If $\mathbf{x} = \mathbf{y} + \mathbf{z}$, where $\mathbf{y}, \mathbf{z} \in K$, implies that $\mathbf{y}$ and $\mathbf{z}$ are both nonnegative scalar multiples of $\mathbf{x}$, then $\mathbf{x}$ is an *extreme vector* of K, and the ray generated by $\mathbf{x}$ is an *extreme ray* of K.

Example 1.128. The extreme rays of the cone K_1 of Example 1.113 are the nonnegative x-axis, generated by $(1,0)^T$, and the nonnegative y-axis, generated by $(0,1)^T$. The extreme rays of the cone K_2 of Example 1.113 are the nonnegative x-axis, and the ray generated by $(1,1)^T$.

Example 1.129. The extreme vectors of a circular cone $K_{\mathbf{u},\theta}$ are the vectors which form an angle θ with $\mathbf{u}$. To see that these are indeed extreme vectors, let $\mathbf{z}$ satisfy $\langle \mathbf{z}, \mathbf{u} \rangle = \|\mathbf{z}\| \cos\theta$ and suppose $\mathbf{z} = \mathbf{x} + \mathbf{y}$, where $\mathbf{x}, \mathbf{y} \in K_{\mathbf{u},\theta}$. Then equality necessarily holds in (1.12), and therefore one of the vectors $\mathbf{x}, \mathbf{y}$ is a nonnegative multiple of the other, and so is $\mathbf{z}$. We leave it for the reader to show that these are all the extreme vectors.

The following proposition follows easily from Definition 1.127.

Proposition 1.130. *If $K = \text{cone}S$, then there exists a subset T of S such that the vectors in T generate all the extreme rays of K.*

The next theorem is a kind of inverse of Proposition 1.130.

Theorem 1.131. [Krein-Milman Theorem] *If T is a set of extreme vectors of a closed convex cone K which generate all the extreme rays of K, then $K = \text{cl}(\text{cone}T)$.*

The next theorem and its proof will be useful in the study of complete positivity, and in particular the cp-rank.

Theorem 1.132. [Carathéodory's Theorem] *If $K = \text{cone}S$ is a convex cone in a Euclidean space of dimension n, then every $\mathbf{x} \in K$ can be represented as a nonnegative combination of at most n elements of S.*

Proof. Let $\mathbf{x} \in K$. Then $\mathbf{x} = \sum_{i=1}^m \lambda_i \mathbf{x}_i$, for some $\mathbf{x}_1, \ldots, \mathbf{x}_m \in S$ and positive $\lambda_1, \ldots, \lambda_m$. We show that if $m > n$, then $\mathbf{x}$ may also be represented as a positive combination of less then m elements. If $m > n$, then $\mathbf{x}_1, \ldots, \mathbf{x}_m$ are linearly dependent, hence there exist real numbers μ_i, $i = 1, \ldots, m$, not all zero, such that $\mathbf{0} = \sum_{i=1}^m \mu_i \mathbf{x}_i$. We may assume that at least one μ_i is positive. Let

$$a = \min\left\{ \frac{\lambda_i}{\mu_i} \;\middle|\; \mu_i > 0 \right\},$$

and $\nu_i = \lambda_i - a\mu_i$, $i = 1, \ldots, m$. Then $\mathbf{x} = \sum_{i=1}^m \nu_i \mathbf{x}_i$, where all the ν_i are nonnegative, and at least one of them is zero. $\square$

Remark 1.133. By the proof of Theorem 1.132, if $\mathbf{x}$ is a positive combination of m linearly dependent vectors $\mathbf{x}_1, \ldots, \mathbf{x}_m$, then there exists a linearly independent subset of $\{\mathbf{x}_1, \ldots, \mathbf{x}_m\}$ such that $\mathbf{x}$ is a positive combination of its elements.

Remark 1.134. By Theorem 1.132, each element of K is a nonnegative combination of at most $d = \dim(K - K)$ elements of S. However, the upper bound of d elements may not be sharp. The minimum number k such that every element in $K = \text{cone}S$ is a nonnegative combination of at most k elements is called the *Carathéodory number* of K.

For example, consider the circular cone $K = K_{\mathbf{e}_3, \pi/4}$ in $\mathbb{R}^3$. Then $\mathbf{x} \in K$ if and only if $x_3^2 \geq x_1^2 + x_2^2$ and $x_3 \geq 0$, and $K = \text{cone}S$ where $S = \left\{\mathbf{x} \in \mathbb{R}^3 \,\middle|\, x_3 = \sqrt{x_1^2 + x_2^2}\right\}$. Every $\mathbf{x} \in K$ is a nonnegative combination of at most two elements in S, i.e., the Carthéodory number of K is 2, although $K - K = \mathbb{R}^3$. We leave the details as an exercise. It is crucial in this example that S is infinite — see next proposition.

Proposition 1.135. *Let $K = \text{cone}S$ be a convex cone in a Euclidean space, where S is finite and $\dim(K - K) = d$. Then there exists an element in K that cannot be represented as a nonnegative combination of less than d elements of S.*

Proof. Suppose $S = \{\mathbf{v}_1, \ldots, \mathbf{v}_k\}$. Let

$$\Omega = \{\alpha \subseteq \{1, \ldots, k\} \mid |\alpha| \leq d - 1\},$$

and for each $\alpha \in \Omega$ let

$$K_\alpha = \text{cone}\{\mathbf{v}_i \mid i \in \alpha\}.$$

Suppose on the contrary that $K = \bigcup_{\alpha\in\Omega} K_\alpha$. Let $K_{\alpha_1}, \ldots, K_{\alpha_m}$, $\alpha_j \in \Omega$, have the property that none of these m cones is contained in the union of the others and $K = \bigcup_{j=1}^m K_{\alpha_j}$. Without loss of generality assume that $\mathbf{v}_1 \notin \text{span}\{\mathbf{v}_i \mid i \in \alpha_1\}$. Let $\mathbf{x} \in K_{\alpha_1} \setminus \bigcup_{j=2}^m K_{\alpha_j}$. Consider vectors of the form $\mathbf{v}_1 + t\mathbf{x}$, $t > 0$. Each such vector is in K, and is not in K_{α_1}, by our assumptions. For $2 \leq j \leq m$, the set of $t > 0$ such that $\mathbf{v}_1 + t\mathbf{x} \in K_{\alpha_j}$ is bounded, since each K_{α_j} is closed, and if there were t_n such that $\lim_{n\to\infty} t_n = \infty$ and $\mathbf{v}_1 + t_n\mathbf{x} \in K_{\alpha_j}$, we would get that

$$\mathbf{x} = \lim_{n\to\infty} \tfrac{1}{t_n}(\mathbf{v}_1 + t_n\mathbf{x}) \in K_{\alpha_j},$$

which contradicts the choice of $\mathbf{x}$. Hence there exists T such that for every $t > T$ the vector $\mathbf{v}_1 + t\mathbf{x} \in K \setminus \bigcup_{j=1}^m K_{\alpha_j}$. A contradiction. □

Faces

An extreme ray is in fact a special case of a face of a cone. This concept is relevant to Chapter 5.

Definition 1.136. A *face* of a closed convex cone K is a convex cone $F \subseteq K$ such that if $\mathbf{x}, \mathbf{y} \in K$ and $\mathbf{x} + \mathbf{y} \in F$, then $\mathbf{x}, \mathbf{y} \in F$.

It is easily seen that $K \cap (-K)$ and K itself are both faces of K. These are the *trivial faces* of K. A nontrivial face is called a *proper face.* An extreme ray of K is a face of dimension 1. On the other end, we say that a proper face of K is a *maximal face* if the only face that strictly contains it is K itself.

Example 1.137. Let $K = \mathbb{R}^n_+$. For $1 \leq k \leq n-1$, $\operatorname{conv}\{\mathbf{e}_{i_1}, \ldots, \mathbf{e}_{i_k}\}$, for fixed $1 \leq i_1 < \cdots < i_k \leq n$, is a k-dimensional face of K. These are all the k-dimensional faces of $\mathbb{R}^n_+$.

Example 1.138. If K is a convex cone in $\mathbb{R}^n$ and $H = \{\mathbf{x} \mid \langle \mathbf{x}, \mathbf{a} \rangle = 0\}$, $\mathbf{0} \neq \mathbf{a} \in K^*$, is a supporting hyperplane of K such that $K \cap H$ is nonempty, then $K \cap H$ is a face of K.

The intersection of faces of K is a face of K. For $\mathbf{x} \in K$ we denote by $\mathcal{F}^{\mathbf{x}}_K$ the intersection of all faces of K containing $\mathbf{x}$. $\mathcal{F}^{\mathbf{x}}_K$ is called the *minimal face* of K containing $\mathbf{x}$. If $\mathbf{x} \in \operatorname{ri} K$, then $\mathcal{F}^{\mathbf{x}}_K = K$ (see Exercise 1.54). Hence if K has a nonempty interior, each non-trivial face of K is contained in the boundary of K.

An *exposed face* of K is a nonempty face of the form $K \cap H$, where H is a hyperplane. An *exposed ray* is an exposed face of dimension 1.

For every $\mathbf{x}$ on the boundary of a convex cone K there exists an exposed face of K containing $\mathbf{x}_0$ by the Supporting Hyperplane Theorem (Theorem 1.126). A nonempty intersection of exposed faces of K is an exposed face of K (see Exercise 1.56). The intersection of all exposed faces of K containing $\mathbf{x} \in K$ is the *minimal exposed face* of K containing $\mathbf{x}$. We denote it by $\widehat{\mathcal{F}}^{\mathbf{x}}_K$. Clearly, for every $\mathbf{x} \in K$ we have $\mathcal{F}^{\mathbf{x}}_K \subseteq \widehat{\mathcal{F}}^{\mathbf{x}}_K$.

Example 1.139. Consider the cone $K = K_{\mathbf{e}_3, \pi/4} \bigcap \mathbb{R}^3_+$ in $\mathbb{R}^3$.

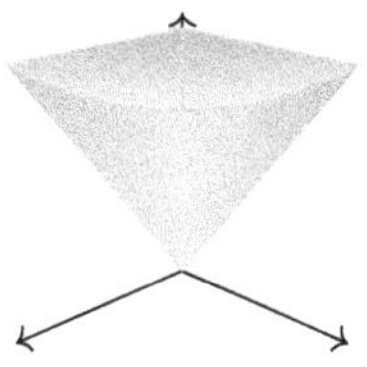

The faces of this cone are

$$F_1 = \text{cone}\{(1,0,1)^T, (0,0,1)^T\}, \quad F_2 = \text{cone}\{(0,1,1)^T, (0,0,1)^T\}$$

(two dimensional), the extreme ray cone$\{\mathbf{e}_3\}$, and the extreme rays generated by the vectors $(x, \sqrt{1-x^2}, 1)^T$, $0 \le x \le 1$ (that lie on the boundary of the cone $K_{\mathbf{e}_3,\pi/4}$ and in the nonnegative orthant). All these faces are exposed.

Example 1.140. Let K be the cone in $\mathbb{R}^3$ whose intersection with the plane $x_3 = 1$, which a base of K, is shown in the figure below.

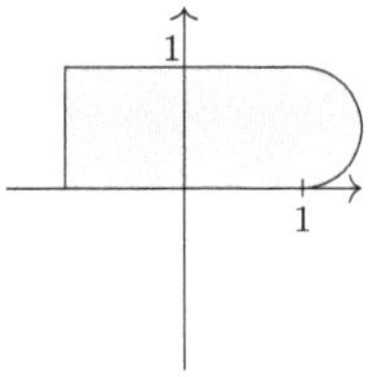

The rays generated by $(1,0,1)^T$ and $(1,1,1)^T$ are not exposed rays of K (they are extreme, though).

We state without proof the following result.

Theorem 1.141. *Let K be a closed convex cone. Every exposed ray of K is an extreme ray of K, and the set of exposed rays of K is dense in the set of extreme rays of K.*

Some more facts on faces may be found in the exercises.

Duality

Definition 1.142. Let S be a set in V. The set

$$S^* = \{\mathbf{y} \in V \mid \langle \mathbf{y}, \mathbf{x} \rangle \ge 0 \text{ for every } \mathbf{x} \in S\}$$

is called the *dual* of S.

Example 1.143. For K_2 of Example 1.113,

$$K_2^* = \{(x,y)^T \mid 0 \le x, -x \le y\}.$$

To see that, note that $(1,1)^T \in K_2$ and $(1,0)^T \in K_2$. Therefore, if $(x,y)^T \in K_2^*$, then $\langle (x,y)^T, (1,1)^T \rangle \ge 0$ and $\langle (x,y)^T, (1,0)^T \rangle \ge 0$, that is, $x + y \ge 0$ and $x \ge 0$. This proves that $K_2^* \subseteq \{(x,y)^T \mid 0 \le x, -x \le y\}$.

To prove the reverse inclusion, suppose $(x_1,y_1)^T \in K_2$ and $(x_2,y_2)^T \in \{(x,y)^T \mid 0 \le x, -x \le y\}$. Then

$$\langle (x_1,y_1)^T, (x_2,y_2)^T \rangle = x_1x_2 + y_1y_2 \ge x_1x_2 - y_1x_2 = (x_1 - y_1)x_2 \ge 0.$$

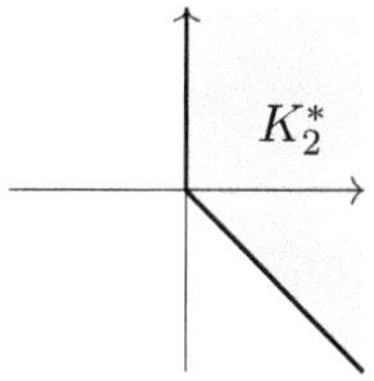

Similarly, it is easy to see that the cone $K_1 = \mathbb{R}^2_+$ of the same example is self dual, *i.e.*, $K_1^* = K_1$.

We leave the next two examples as exercises.

Example 1.144. The dual of the circular cone $K_{\mathbf{u},\theta}$ is the circular cone $K_{\mathbf{u},\pi/2-\theta}$.

Example 1.145. If L is a subspace of V, then $L^* = L^\perp$, the orthogonal complement of L.

The following theorem lists some basic (and easy to prove) duality properties.

Theorem 1.146. *Let S, S_1, S_2 be nonempty subsets of V. Then*

(a) *S^* is a closed convex cone.*
(b) *If $S_1 \subseteq S_2$, then $S_2^* \subseteq S_1^*$.*
(c) *$S \subseteq S^{**}$.*
(d) *$S^* = (\mathrm{cl}\, S)^* = (\mathrm{cone} S)^* = (\mathrm{conv} S)^*$.*
(e) *$S_1^* \cap S_2^* \subseteq (S_1 + S_2)^*$, and if $\mathbf{0} \in S_1 + S_2$, $(S_1 + S_2)^* \subseteq S_1^* \cap S_2^*$.*

Theorem 1.147. *S is a closed convex cone if and only if $S = S^{**}$.*

Proof. "If": Follows from Theorem 1.146(a).

"Only if": By Theorem 1.146(c), it is enough to show that $S^{**} \subseteq S$. Suppose $\mathbf{b} \in S^{**}$, $\mathbf{b} \notin S$. Then by Theorem 1.124 there is a vector $\mathbf{u} \in V$ such that $\langle \mathbf{u}, \mathbf{b} \rangle < 0$, while $\langle \mathbf{u}, \mathbf{x} \rangle \geq 0$ for every $\mathbf{x} \in S$. But then $\mathbf{u} \in S^*$ and thus $\mathbf{b} \notin S^{**}$, a contradiction. □

Corollary 1.148. *If K is a convex cone, then $\mathrm{cl} K = K^{**}$.*

Proof. By Theorem 1.146(d) $K^* = (\mathrm{cl} K)^*$, and therefore $K^{**} = (\mathrm{cl} K)^{**}$. By Theorem 1.147, $(\mathrm{cl} K)^{**} = \mathrm{cl} K$. □

When S_1 and S_2 in Theorem 1.146 are closed convex cones, statement (e) can be strengthened.

Corollary 1.149. *If K_1 and K_2 are closed convex cones then*

$$\mathrm{cl}(K_1^* + K_2^*) = (K_1 \cap K_2)^*.$$

Proof. $(K_1 \cap K_2)^* = (K_1^{**} \cap K_2^{**})^* = (K_1^* + K_2^*)^{**} = \mathrm{cl}(K_1^* + K_2^*)$. The middle equality follows from Theorem 1.146(e) and the right equality follows from Corollary 1.148, since the sum of convex cones is a convex cone. □

Types of convex cones

Definition 1.150. A convex cone K in V is *pointed* if $K \cap (-K) = \{\mathbf{0}\}$, *reproducing* if $K - K = V$, *solid* if $\mathrm{int}\, K \neq \emptyset$, and *full* if it is closed, convex, pointed and solid.

Recall that $K - K$ is the minimal subspace of V containing K, and that $K \cap (-K)$ is the maximal subspace of V contained in K. Hence a pointed cone contains no subspace but $\{\mathbf{0}\}$, and no proper subspace of V contains a reproducing cone K.

Remark 1.151. A pointed closed convex cone in V induces a partial order in V, via $\mathbf{x} \leq \mathbf{y}$ iff $\mathbf{y} - \mathbf{x} \in K$.

Proposition 1.152. *A closed convex cone in V is solid if and only if it is reproducing.*

Proof. "If": See Proposition 1.120.

"Only if": Let $\mathbf{v}$ be an interior point of K. Then for every $\mathbf{x} \in V$ there is a sufficiently small positive d such that $\mathbf{u} = \mathbf{v} + d\mathbf{x} \in \mathrm{int}\, K$. Since $\mathbf{x} = \mathbf{u}/d - \mathbf{v}/d$, $\mathbf{x} \in \mathrm{int}\, K - \mathrm{int}\, K$. Hence $V \subseteq \mathrm{int}\, K - \mathrm{int}\, K$, K is reproducing. □

The next proposition relies on Proposition 1.152, and we leave the proof as an exercise.

Proposition 1.153. *A closed convex cone K is pointed if and only if K^* is solid.*

Recall that a sum of closed convex cones is not necessarily closed (Example 1.117). However, in the following case the sum is closed.

Proposition 1.154. *Let K_1 and K_2 be closed convex cones in V. If the cone $K_1 + K_2$ is pointed, then $K_1 + K_2$ is closed.*

Proof. Let $\mathbf{x}_k \in K_1$ and $\mathbf{y}_k \in K_2$ and $\mathbf{z}_k = \mathbf{x}_k + \mathbf{y}_k$ for every $k \geq 1$. Suppose $\lim_{k\to\infty} \mathbf{z}_k = \mathbf{z}$. If $\mathbf{x}_k$ is bounded, it has a converging subsequence, $\mathbf{x}_{k_j}$. Let $\mathbf{x} = \lim_{j\to\infty} \mathbf{x}_{k_j} \in K_1$. Then

$$\lim_{j\to\infty} \mathbf{y}_{k_j} = \lim_{j\to\infty} \mathbf{z}_{k_j} - \lim_{j\to\infty} \mathbf{x}_{k_j} = \mathbf{z} - \mathbf{x}.$$

Hence $\mathbf{z} - \mathbf{x} \in K_2$, i.e., $\mathbf{z} \in K_1 + K_2$.

If the sequence $\mathbf{x}_k$ is not bounded then, replacing it by a subsequence we may assume that $\lim_{k\to\infty} \|\mathbf{x}_k\| = \infty$. Let

$$\hat{\mathbf{x}}_k = \frac{\mathbf{x}_k}{\|\mathbf{x}_k\|}, \quad \hat{\mathbf{y}}_k = \frac{\mathbf{y}_k}{\|\mathbf{x}_k\|}, \quad \hat{\mathbf{z}}_k = \frac{\mathbf{z}_k}{\|\mathbf{x}_k\|}.$$

As $\mathbf{z}_k$ is bounded, $\lim_{k\to\infty} \hat{\mathbf{z}}_k = \mathbf{0}$. Since $\hat{\mathbf{x}}_k$ is bounded, it has a subsequence converging to $\mathbf{x} \in K_1$, and the corresponding subsequence of $\hat{\mathbf{y}}_k$ converges to $\mathbf{0} - \mathbf{x}$. That is, $-\mathbf{x} \in -K_1$ and also $-\mathbf{x} \in K_2$. This contradicts the assumption that $K_1 + K_2$ is pointed. □

Theorem 1.155. *If K is a closed pointed cone in V, then there is a hyperplane H separating K and $-K$, such that $H \cap K = \{\mathbf{0}\}$.*

Proof. If K is a closed pointed cone, then there exists a pointed cone K' such that $K \setminus \mathbf{0} \subseteq \mathrm{ri} K'$ (see Exercise 1.64). By Theorem 1.125 there exists a hyperplane H through the origin which separates K' and $-K'$ properly. Clearly, $H \cap \mathrm{ri} K' = \emptyset$; hence $H \cap K = \{\mathbf{0}\}$. □

Definition 1.156. A convex cone K is a *polyhedral cone* if it is finitely generated, *i.e.*, $K = \mathrm{cone} S$ for some finite set S.

$\mathbb{R}^n_+$ is an example a polyhedral cone.

Proposition 1.157. *A polyhedral cone is necessarily closed.*

Note that K is a polyhedral cone if and only if K has a finite number of extreme rays (this follows from the last proposition together with Proposition 1.130 and Theorem 1.131).

Remark 1.158. Every closed convex cone in $\mathbb{R}^2$ is polyhedral. But for $n \geq 3$, there are convex cones which are not polyhedral. For example, any circular cone in $\mathbb{R}^n$, $n \geq 3$, has infinitely many extreme rays, and therefore is not polyhedral.

We add the following without proof.

Theorem 1.159. *K is a polyhedral cone if and only if it is the intersection of finitely many closed halfspaces.*

For polyhedral cones we have the following separation theorem.

Theorem 1.160. *Let K be a pointed polyhedral cone in V, and let L be a subspace of V such that $K \cap L = \{\mathbf{0}\}$. Then there exists a supporting hyperplane H of K such that $H \cap K = \{\mathbf{0}\}$ and $H \supseteq L$.*

This theorem is equivalent to a family of results known as *Theorems of the Alternative.* Perhaps the most famous in the family is the following.

Theorem 1.161. **[Farkas' Lemma]** *Let A be an $m \times n$ matrix, and let $\mathbf{b} \in \mathbb{R}^m$. Then exactly one of the following statements is true.*

(1) *There exists $\mathbf{x} \in \mathbb{R}^m$ such that $A\mathbf{x} = \mathbf{b}$ and $\mathbf{x} \geq \mathbf{0}$.*
(2) *There exists $\mathbf{y} \in \mathbb{R}^n$ such that $A^T\mathbf{y} \geq \mathbf{0}$ and $\mathbf{y}^T\mathbf{b} < 0$.*

The next theorem of the alternative will be needed in Section 2.2.

Theorem 1.162. **[Steimke's Theorem]** *Let A be an $m \times n$ matrix, and let $\mathbf{b} \in \mathbb{R}^m$. Then exactly one of the following statements is true.*

(1) *There exists $\mathbf{x} \in \mathbb{R}^m$ such that $A\mathbf{x} \geq \mathbf{0}$, $A\mathbf{x} \neq 0$.*
(2) *There exists $\mathbf{y} \in \mathbb{R}^n$ such that $A^T\mathbf{y} = \mathbf{0}$ and $\mathbf{y} > \mathbf{0}$.*

Proof. First note that (1) and (2) cannot hold together, since then $\mathbf{x}^T A^T \mathbf{y} = (A\mathbf{x})^T \mathbf{y} > 0$, as $A\mathbf{x}$ is a nonzero nonnegative vector, and $\mathbf{y}$ is a positive vector. On the other hand, $\mathbf{x}^T A^T \mathbf{y} = \mathbf{x}^T(A\mathbf{y}) = 0$.

If (1) does not hold, let $K = \mathbb{R}^m_+$ and $L = \operatorname{cs} A$. Then $K \cap L = \{\mathbf{0}\}$, and there exists a supporting hyperplane H of K at $\mathbf{0}$ such that $H \supseteq L = \operatorname{cs} A$. Then $H = \{\mathbf{x} \mid \langle \mathbf{x}, \mathbf{y} \rangle = 0\}$ for some nonzero $\mathbf{y} \in \mathbb{R}^m$, and we may assume that $\langle \mathbf{x}, \mathbf{y} \rangle > 0$ for every $\mathbf{0} \neq \mathbf{x} \in K$. Then $\mathbf{y} > 0$ and $(A\mathbf{x})^T \mathbf{y} = 0$ for every $\mathbf{x} \in \mathbb{R}^n$ implies $A^T\mathbf{y} = \mathbf{0}$. □

Cones of matrices

We will be interested mostly in symmetric matrices, hence we consider $\mathcal{S}_n$ — the space of all symmetric matrices. The vector space $\mathcal{S}_n$ is a subspace of $\mathbb{R}^{n \times n}$, which is isomorphic to $\mathbb{R}^{n(n+1)/2}$. The standard inner product in $\mathbb{R}^{n \times n}$ translates in $\mathcal{S}_n$ into the following: $\langle A, B \rangle = \operatorname{trace}(AB)$. The norm induced by this inner product is the *Frobenius norm*, which on $\mathcal{S}_n$ is

$$\|A\|_F = \sqrt{\operatorname{trace}(A^2)} = \sqrt{\sum_{i=1}^n \sum_{j=1}^n a_{ij}^2} = \sqrt{\sum_{i=1}^n \lambda_i^2},$$

where $\lambda_1, \ldots, \lambda_n$ are all the eigenvalues of A.

Several sets of symmetric matrices form convex cones in $\mathcal{S}_n$ (and therefore in $\mathbb{R}^{n\times n}$).

Let $\mathcal{N}_n$ denote the set of $n \times n$ symmetric nonnegative matrices. The following theorem is easy to prove and is left as an exercise.

Theorem 1.163. *The set $\mathcal{N}_n$ is a full polyhedral cone in $\mathcal{S}_n$. Its interior consists of all symmetric positive matrices. Its extreme rays are all the rays generated by E_{ii}, $i = 1, \ldots, n$, and those generated by the matrices $E_{ij} + E_{ji}$, $1 \le i < j \le n$. As a cone in $\mathcal{S}_n$, $\mathcal{N}_n$ is self dual.*

Remark 1.164. The cone $\mathcal{N}_n$ is a subset of $\mathbb{R}^{n\times n}_+$ — the set of all nonnegative $n \times n$ matrices, which is a convex cone in $\mathbb{R}^{n\times n}$.

Recall that a square matrix with all diagonal entries equal to zero is called a hollow matrix. We denote by $\mathcal{HN}_n$ the set of all hollow symmetric nonnegative $n \times n$ matrices. Clearly $\mathcal{HN}_n \subseteq \mathcal{N}_n$.

Theorem 1.165. *The set $\mathcal{HN}_n$ is a pointed polyhedral cone in $\mathcal{S}_n$, which is not solid. The rays generated by $E_{ij} + E_{ji}$, $1 \le i < j \le n$, are all the extreme rays of this cone. Its dual in $\mathcal{S}_n$ is $\mathcal{HN}_n + \mathcal{D}_n$, where $\mathcal{D}_n$ is the subspace of $\mathcal{S}_n$ consisting of all diagonal matrices.*

Let $\mathcal{PSD}_n$ denote the set of $n \times n$ positive semidefinite matrices.

Theorem 1.166. *The set $\mathcal{PSD}_n$ is a full cone in $\mathcal{S}_n$. The interior of $\mathcal{PSD}_n$ is the open cone $\mathcal{PD}_n$ of positive definite matrices, and its extreme rays are generated by the rank 1 symmetric matrices $\mathbf{x}\mathbf{x}^T$, $\mathbf{0} \ne \mathbf{x} \in \mathbb{R}^n$. $\mathcal{PSD}_n$ is self dual.*

Proof. The set $\mathcal{PSD}_n$ is a convex cone by Proposition 1.20. It is closed, since $A_k \in \mathcal{PSD}_n$ and $\lim_{k\to\infty} A_k = A$ implies that for every $\mathbf{x} \in \mathbb{R}^n$,

$$\mathbf{x}^T A\mathbf{x} = \lim_{k\to\infty} \mathbf{x}^T A_k \mathbf{x} \ge 0.$$

It is easy to see that $\mathcal{PSD}_n$ is pointed. If $\mathbf{x}^T A\mathbf{x} \ge 0$ nd $\mathbf{x}^T(-A)\mathbf{x} \ge 0$ for every $\mathbf{x} \in \mathbb{R}^n$, then $\mathbf{x}^T A\mathbf{x} = 0$ for every $\mathbf{x}$, hence $A = 0$ (consider $\mathbf{x} = \mathbf{e}_i$, $1 \le i \le n$, and then $\mathbf{x} = \mathbf{e}_i + \mathbf{e}_j$, $1 \le i < j \le n$).

We show that $\operatorname{int}\mathcal{PSD}_n = \mathcal{PD}_n$ (in particular, $\mathcal{PSD}_n$ is solid). Suppose A is a positive definite matrix. Then all the leading minors of A are positive, and are continuous functions of the matrix entries. Hence there exists an $\varepsilon > 0$ such that a perturbation of each entry of A by at most ε will still leave all leading principal minors positive. Then $A + \varepsilon B$ is positive definite for every matrix $B \in \mathcal{S}_n$ such that $\|B\|_F \le 1$. On the other hand, if a matrix

A is not positive definite, let $\mathbf{x}$ be an eigenvector of A corresponding to the eigenvalue 0. Then $A + \varepsilon(-\mathbf{x}^T\mathbf{x})$ is not positive semidefinite for every $\varepsilon > 0$: $\mathbf{x}^T(A + \varepsilon(-\mathbf{x}^T\mathbf{x}))\mathbf{x} = -\varepsilon||\mathbf{x}||^2 < 0$.

Next, every rank 1 positive semidefinite matrix is an extreme vector of $\mathcal{PSD}_n$. If $\mathbf{x}\mathbf{x}^T = A + B$, where $A, B \in \mathcal{PSD}_n$, then, by Exercise 1.11, $\operatorname{ns} A \cap \operatorname{ns} B = \operatorname{ns} \mathbf{x}\mathbf{x}^T = \{\mathbf{x}\}^\perp$. Hence either $\operatorname{ns} A = \{\mathbf{x}\}^\perp$, or $\operatorname{ns} A = \mathbb{R}^n$. In any case, A is a scalar multiple of $\mathbf{x}\mathbf{x}^T$. By Theorem 1.25(h), $\mathcal{PSD}_n$ is generated by the rank 1 matrices.

Finally, to see that $\mathcal{PSD}_n^* = \mathcal{PSD}_n$, note that

$$\operatorname{trace}(AB) = \mathbf{e}^T(A \circ B)\mathbf{e}.$$

Since $A \circ B$ is positive semidefinite if A and B are (Proposition 1.34), this equality implies that $\mathcal{PSD}_n \subseteq \mathcal{PSD}_n^*$. For the reverse inclusion, note that if $A \in \mathcal{PSD}_n^*$ then $\operatorname{trace}(\mathbf{x}^T A\mathbf{x}) = \operatorname{trace}(A\mathbf{x}\mathbf{x}^T) \geq 0$ for every $\mathbf{x} \in \mathbb{R}^n$. □

The cone $\mathcal{PSD}_n$ is not polyhedral, since it has infinitely many extreme rays.

Next we consider the sets $\mathcal{DNN}_n$, of $n \times n$ matrices which are both nonnegative and positive semidefinite, and $\mathcal{SPN}_n = \mathcal{PSD}_n + \mathcal{N}_n$.

Theorem 1.167. *The sets $\mathcal{DNN}_n$ and $\mathcal{SPN}_n$ are full cones in $\mathcal{S}_n$ that are each other dual.*

$\operatorname{int}\mathcal{DNN}_n$ consists of the positive definite matrices which are also entrywise positive.

Proof. As $\mathcal{DNN}_n = \mathcal{N}_n \cap \mathcal{PSD}_n$, the last two theorems imply that $\mathcal{DNN}_n$ is a closed pointed convex cone.

$$\operatorname{int}\mathcal{DNN}_n = \operatorname{int}\mathcal{N}_n \cap \operatorname{int}\mathcal{PSD}_n \neq \emptyset,$$

as, for example, $J + I$ is a positive definite positive matrix. By Proposition 1.116, $\mathcal{SPN}_n$ is a convex cone. By Proposition 1.118(b), $\operatorname{int}\mathcal{SPN}_n$ is not empty. Every $X \in \mathcal{SPN}_n$ has nonnegative diagonal entries. From this it is easy to deduce that $\mathcal{SPN}_n$ is pointed. Hence it is also closed, by Proposition 1.154. By Proposition 1.146(e), $\mathcal{SPN}_n^* = \mathcal{DNN}_n$, and, as $\mathcal{SPN}_n$ is closed, Corollary 1.149 guarantees that $\mathcal{DNN}_n^* = \mathcal{SPN}_n$. □

It is easy to deduce the following from the known extreme rays of $\mathcal{PSD}_n$ and $\mathcal{N}_n$.

Proposition 1.168. *The extreme rays of $\mathcal{SPN}_n$ are the matrices $E_{ij}+E_{ji}$, $1 \leq i < j \leq n$, and the rank 1 matrices of the form $\mathbf{x}\mathbf{x}^T$, where $\mathbf{x}$ has both positive and negative entries.*

Remark 1.169. If A is an SPN matrix, $A = P + N$, where $P \in \mathcal{PSD}_n$ and $N \in \mathcal{N}_n$, then we may also write $A = (P + \mathrm{Diag}(N)) + (N - \mathrm{Diag}(N))$. That is, $\mathcal{SPN}_n = \mathcal{PSD}_n + \mathcal{HN}_n$.

There is no general description of the extreme rays of $\mathcal{DNN}_n$. However, there are some partial results. We state them here without proof.

Proposition 1.170. *For $n \leq 4$, the extreme vectors of $\mathcal{DNN}_n$ are the rank 1 matrices in $\mathcal{DNN}_n$.*

For $n = 5$, $A \in \mathcal{DNN}_5$ is an extreme vector of $\mathcal{DNN}_5$ if and only if $\operatorname{rank} A = 1$, *or* $\operatorname{rank} A = 3$ *and* $G(A) = C_5$.

The protagonists of this book are the copositive matrices and the completely positive matrices. Each of these two types of matrices is obtained by adding nonnegativity constraints to a definition of a positive semidefinite matrix.

A matrix $A \in \mathbb{R}^{n \times n}$ is *copositive* if it is symmetric and $\mathbf{x}^T A \mathbf{x}$ is nonnegative for every nonnegative $\mathbf{x} \in \mathbb{R}^n$.

A matrix $A \in \mathbb{R}^{n \times n}$ is *completely positive* if $A = BB^T$, where $B \in \mathbb{R}^{n \times m}$ (for some m) is nonnegative.

We denote by $\mathcal{COP}_n$ the set of all $n \times n$ copositive matrices, and by $\mathcal{CP}_n$ the set of all $n \times n$ completely positive matrices. These sets are full cones in $\mathcal{S}_n$, which are each other's dual. This, and other properties of these cones, as well as their relations to the previously mentioned cones will be discussed in the coming chapters.

Exercises

1.45. Let C be a closed unbounded convex set in $\mathbb{R}^n$, and let $\mathbf{x}_0 \in \operatorname{ri} C$. Prove: there exists a nonzero $\mathbf{x}_1 \in \mathbb{R}^n$ such that the ray $\{\mathbf{x}_0 + t\mathbf{x}_1 \mid t \geq 0\}$ lies entirely in C.
Hint: By considering $C - \mathbf{x}_0 = \{\mathbf{x} - \mathbf{x}_0 \mid \mathbf{x} \in C\}$ in place of C, and $\operatorname{span} C$ instead of $\mathbb{R}^n$, one may assume $\mathbf{x}_0 = 0 \in \operatorname{int} C$.

1.46. Prove Proposition 1.116.

1.47. Prove Proposition 1.118.

1.48. Show that $K - K$ is the minimal subspace of V containing K, and $K \cap (-K)$ is the maximal subspace of V contained in K.

1.49. Show that if $\mathbf{x}$ forms an angle less than θ with $\mathbf{u}$, then $\mathbf{x}$ is not an extreme vector of $K_{\mathbf{u},\theta}$.

1.50. Complete the details of the example in Remark 1.134.
Hint: For $\mathbf{x} \in K$, $\mathbf{x} \notin S$ and $\mathbf{x} \neq \mathbf{e}_3$ show that $\mathbf{x} = t\mathbf{y} + s\mathbf{z}$ for some $s, t > 0$, where $\mathbf{y}, \mathbf{z} \in S$ lie on the plane through $\mathbf{x}$ and $\mathbf{e}_3$.

1.51. Let K be a closed convex cone in a Euclidean space, and let F be a face of K. Show that F is a closed convex cone.

1.52. Let K be a closed convex cone in a Euclidean space such that $\operatorname{ri} K \neq \emptyset$, and let F be a proper face of K. Show that $F \subseteq \operatorname{rb} K$.
Hint: It may be convenient to consider $K - K$ as the Euclidean space, and K as a solid cone.

1.53. Let K be a closed convex cone in a Euclidean space, and let F be a face of K. Let $F' \subseteq F$. Show that the following are equivalent:

(a) F' is a face of F.
(b) F' is a face of K.

If (a)/(b) hold, F' is called a *subface* of F.

1.54. Prove that if K is convex cone, F is a face of K and $K' \subseteq K$ is a convex cone such that $\operatorname{ri} K' \cap F \neq \emptyset$, then $K' \subseteq F$. Deduce that if $\mathbf{x} \in \operatorname{ri} K$, then $\mathcal{F}_K^{\mathbf{x}} = K$.
Hint: Let $\mathbf{y} \in \operatorname{ri} K' \cap F$. For $\mathbf{x} \in K'$ there exists $\mathbf{z} \in K'$ such that $\mathbf{y}$ is on the line segment between $\mathbf{x}$ and $\mathbf{z}$.

1.55. Prove that if K is closed convex cone, then every maximal face of K is exposed.
Hint: Exercises 1.52 and 1.54.

1.56. Let K be closed convex cone. Show that if the intersection of exposed faces of K is not empty, then it is an exposed face of K.
Hint: It suffices to consider faces on the boundary. If $F_i = K \cap H_i$, $i \in I$, where $H_i = \{\mathbf{x} \mid \langle \mathbf{x}, \mathbf{a}_i \rangle = 0\}$, choose linearly independent $\mathbf{a}_{i_1}, \ldots, \mathbf{a}_{i_k}$ spanning $\operatorname{span}\{\mathbf{a}_i \mid i \in I\}$. Let $\mathbf{a} = \sum_{i=j}^k \mathbf{a}_{i_j}$ and consider $H = \{\mathbf{x} \mid \langle \mathbf{x}, \mathbf{a} \rangle = 0\}$.

1.57. Let K be a closed convex cone. Show that if

$$H = \{\mathbf{x} \in V \mid \langle \mathbf{x}, \mathbf{u} \rangle = a\}$$

is a supporting hyperplane of K at $\mathbf{x}_0 \neq 0$, then $a = 0$ and we may take $\mathbf{u} \in K^*$.
Hint: $t\mathbf{x}_0 \in K$ for every $t > 0$.

1.58. Let K be a full cone in a Euclidean space V.

(a) Show that every maximal (proper) face of K has the form

$$F_{\mathbf{u}} = \{\mathbf{x} \in K \mid \langle \mathbf{x}, \mathbf{u} \rangle = 0\},$$

where $\mathbf{u} \in K^*$ generates an extreme ray of K^*.

(b) Show that if $\mathbf{u} \in K^*$ generates an exposed ray of K^*, then

$$F_{\mathbf{u}} = \{\mathbf{x} \in K \mid \langle \mathbf{x}, \mathbf{u} \rangle = 0\}$$

is a maximal face of K.

Hint: For (b): Let $\mathbf{v} \in K = (K^*)^*$ be such that the

$$\{t\mathbf{u} \mid t \geq 0\} = \{\mathbf{x} \in K^* \mid \langle \mathbf{x}, \mathbf{v} \rangle = 0\}.$$

Show that no maximal face of K, other than $F_{\mathbf{u}}$, contains $\mathbf{v}$.

1.59. Let K be a solid closed convex cone in a Euclidean space. Show that $\mathbf{x} \in \mathrm{bd}K$ if and only if there exists $\mathbf{y} \in K^*$ such that $\langle \mathbf{y}, \mathbf{x} \rangle = 0$.

1.60. Prove the statement in Example 1.144.

1.61. Prove the statement in Example 1.145.

1.62. Prove Theorem 1.146.

1.63. Prove Proposition 1.153

1.64. Prove that if K is a pointed cone, then there exists a pointed cone K' such that $K \setminus \{\mathbf{0}\} \subseteq \mathrm{ri}K'$.

1.65. Show that every closed convex cone in $\mathbb{R}^2$ is polyhedral.
Hint: For every three vectors in $\mathbb{R}^2$, one of the vectors is a nonnegative combination of the other two.

1.66. Prove Proposition 1.157.

1.67. Prove Theorem 1.163.

1.68. Prove Theorem 1.165.

1.69. Prove Proposition 1.168.

Notes. A classical reference on convexity is [Rockafellar (1997)]. Cones in matrix theory are discussed in [Berman (1973)].

Example 1.117 is a counter example to Lemma 16.2.2 of [Hall (1998)]. Theorem 1.131 is a finite dimensional version of the Krein-Milman Theorem. Theorem 1.132 is a version of Carathéodory's Theorem for convex cones. Proposition 1.152 is a theorem of Krein and Rutman, [Kreĭn and Rutman (1950)].

If K is a closed convex cone in $\mathbb{R}^n$, $\pi(K)$ is defined to be the set of $n \times n$ matrices A for which $AK \subseteq K$. It is a closed convex cone. Note that $\mathbb{R}_+^{n\times n} = \pi(\mathbb{R}_+^n)$. The Perron-Frobenius theory of nonnegative matrices has generalizations for matrices in $\pi(K)$, see [Berman and Plemmons (1994)].

The characterization of the extreme rays of $\mathcal{DNN}_n$, $n \leq 5$, was done in [Ycart (1982)] (see also [Hamilton-Jester and Li (1996)] for this and some other results on the extreme rays of $\mathcal{DNN}_n$).

1.7 Optimization and the Karush-Kuhn-Tucker conditions

Suppose f and h_i, $i = 1, \dots, k$, are functions defined on the same open set in $\mathbb{R}^n$, which have continuous partial derivatives there. Consider the following constrained optimization problem

$$\begin{aligned} \text{minimize} \quad & f(\mathbf{x}) \\ \text{subject to} \quad & h_i(\mathbf{x}) = 0, \quad i = 1, \dots, k. \end{aligned} \tag{1.14}$$

(To maximize f, minimize $-f$.) This may be thought of as looking for a minimal value of f over a surface S defined by the constraints. To solve the problem, we may use the method of Lagrange multipliers. In this method, we introduce additional k real variables $\lambda_1, \dots, \lambda_k$, called the *Lagrange multipliers*, and solve $n + k$ equations that form necessary conditions for $\mathbf{x}$ to be a local extremum of f under the k constraints, provided that $\mathbf{x} \in S$ is a regular point for the constraints. A point $\mathbf{x}$ is a *regular point* for the constraints $h_i(\mathbf{x}) = 0$, $i = 1, \dots, k$, if $\mathbf{x} \in S$ and $\nabla h_1(\mathbf{x}), \dots, \nabla h_k(\mathbf{x})$ are k linearly independent vectors in $\mathbb{R}^n$, where ∇h denotes the gradient vector of h. (In particular, if there is a single constraint $h(\mathbf{x}) = 0$, then any point $\mathbf{x} \in S$ at which $\nabla h(\mathbf{x}) \neq \mathbf{0}$ is a regular point.)

Theorem 1.171. *Let* $\mathbf{x}$ *be a local minimum for the problem* (1.14)*, where* f *and* $h_1, \dots, h_k$ *have continuous first order partial derivatives in an open set in* $\mathbb{R}^n$*. If* $\mathbf{x}$ *is a regular point for the constraints of the problem, then* *there exist real numbers* $\lambda_1, \ldots, \lambda_k$ *such that*

$$\nabla f(\mathbf{x}) - \sum_{i=1}^{k} \lambda_i \nabla h_i(\mathbf{x}) = \mathbf{0}.$$

The necessary condition given in Theorem 1.171 consists of n equations, which together with the k constraints form a system of $n+k$ equations in the $n+k$ unknowns $x_1, \ldots, x_n$ and $\lambda_1, \ldots, \lambda_k$. Only points that are part of a solution to this system may yield a local minimum point $\mathbf{x}$.

Now consider a problem were we have both equality constraints and inequality constraints.

$$\begin{aligned} \text{minimize} \quad & f(\mathbf{x}) \\ \text{subject to} \quad & h_i(\mathbf{x}) = 0, \quad i = 1, \ldots, k \\ & g_i(\mathbf{x}) \leq 0, \quad i = 1, \ldots, m. \end{aligned} \tag{1.15}$$

The set S of all $\mathbf{x} \in \mathbb{R}^n$ who satisfy the $m+k$ constraints is called the *feasible set* for the problem. Suppose f, $h_1, \ldots, h_k$ and $g_1, \ldots, g_m$ have continuous first order partial derivatives in an open set containing S. In search for a solution to problem (1.15), two sets of lagrange multipliers are introduced: $\lambda_1, \ldots, \lambda_k$ and $\mu_1, \ldots, \mu_m$. The λ_i's are the multipliers associated with the equality constraint functions, and μ_i's are multipliers associated with the inequality constraint functions.

The next theorem states a special version of the Karush-Kuhn-Tucker (KKT) conditions. In general, these are necessary conditions for the existence of a local minimum in problem (1.15). Various versions differ in the "regularity conditions" replacing the regularity condition on $\mathbf{x}$ that appeared in the previous theorem. We will use the KKT theorem in the case that all the constraint functions are affine, i.e., each has the form $\mathbf{x} \mapsto \mathbf{a}^T\mathbf{x} + c$ for some $\mathbf{a} \in \mathbb{R}^n$ and $c \in \mathbb{R}$. Hence we state the theorem only for this case, in which no regularity conditions are needed.

Theorem 1.172. [Karush-Kuhn-Tucker necessary conditions] *Let* $\mathbf{x}$ *be a local minimum for the problem* (1.15), *where* $h_1, \ldots, h_k$ *and* $g_1, \ldots, g_m$ *are affine functions on* $\mathbb{R}^n$, *and* f *is a function with continuous partial derivatives in an open set containing the feasible region. Then there exist real numbers* $\mu_1, \ldots, \mu_k$ *and nonnegative numbers* $\lambda_1, \ldots, \lambda_m$ *such that*

$$\nabla f(\mathbf{x}) + \sum_{i=1}^{k} \lambda_i \nabla h_i(\mathbf{x}) + \sum_{i=1}^{m} \mu_i \nabla g_i(\mathbf{x}) = \mathbf{0},$$

and

$$\mu_i g_i(\mathbf{x}) = 0, \quad i = 1, \ldots, m.$$

The functions we will need to optimize will be quadratic functions. It is useful to note that if $f(\mathbf{x}) = \mathbf{x}^T A\mathbf{x}$, where A is an $n \times n$ symmetric matrix, then $\nabla f(\mathbf{x}) = 2A\mathbf{x}$. (We may consider instead $f(\mathbf{x}) = \frac{1}{2}\mathbf{x}^T A\mathbf{x}$, without changing the local minimum points.)

Example 1.173. Let $\mathbf{a} \in \mathbb{R}^n_+$, $c > 0$, and let A be an $n \times n$ symmetric matrix. Consider the problem

$$\begin{aligned} \text{minimize} \quad & \frac{1}{2}\mathbf{x}^T A\mathbf{x} \\ \text{subject to} \quad & \mathbf{a}^T\mathbf{x} = c \\ & \mathbf{x} \geq \mathbf{0}. \end{aligned}$$

A minimum definitely exists, as the feasible set defined by the constraints is a nonempty compact set, and $f(\mathbf{x}) = \frac{1}{2}\mathbf{x}^T A\mathbf{x}$ is continuous. In this case, $g_i(\mathbf{x}) = -\mathbf{e}_i^T\mathbf{x}$, $i = 1, \ldots, n$, and $h(\mathbf{x}) = \mathbf{a}^T\mathbf{x}$. The minimum is therefore attained at an $\mathbf{x}$ which satisfies the Karush-Kuhn-Tucker necessary conditions

$$A\mathbf{x} + \lambda\mathbf{a} - \sum_{i=1}^{n} \mu_i \mathbf{e}_i = \mathbf{0}$$

and $\mu_i x_i = 0$ for every $i = 1, \ldots, n$, for some nonnegative $\mu_1, \ldots, \mu_n$ and a real λ. Denoting

$$\boldsymbol{\mu} = \begin{pmatrix} \mu_1 \\ \vdots \\ \mu_n \end{pmatrix},$$

these necessary conditions may be written in vector notation: If $\mathbf{x}$ is a point in the feasible set where the minimum is attained,

$$A\mathbf{x} + \lambda\mathbf{a} - \boldsymbol{\mu} = \mathbf{0} \quad \text{and} \quad \boldsymbol{\mu}^T\mathbf{x} = 0. \tag{1.16}$$

Exercises

1.70. Use Lagrange multipliers to prove the Rayleigh-Ritz Theorem (Theorem 1.15).
Hint: Show that the minimum and maximum of $\mathbf{x}^T A\mathbf{x}$ subject to $\sum_{i=1}^n x_i^2 - 1 = 0$ are attained at eigenvectors of A.

1.71. Let A be an $n \times n$ symmetric matrix, and suppose the minimization problem

$$\begin{aligned} \text{minimize} \quad & \frac{1}{2}\mathbf{x}^T A\mathbf{x} \\ \text{subject to} \quad & \mathbf{a}^T\mathbf{x} = c \\ & \mathbf{x} \geq \mathbf{0} \end{aligned}$$

has a solution $\mathbf{u}^T > 0$. Show that

(a) There exists $\lambda > 0$ such that $A\mathbf{u} = \lambda\mathbf{e}$.
(b) If for $\mathbf{v}^T \in \mathbb{R}^n$ satisfies $A\mathbf{v} = \mu\mathbf{e}$ and $\mathbf{e}^T\mathbf{v} = 1$, then $\mu = \lambda$.
(c) $\mathbf{u}^T A\mathbf{u} = \lambda$.

Hint: $\mathbf{u}$ is a local minimum of $\frac{1}{2}\mathbf{x}^T A\mathbf{x}$ on $\mathbf{x}^T\mathbf{e} = 1$, use Lagrange multipliers.

Notes. Kuhn and Tucker published their result in 1951, and the theorem was known as the Kuhn-Tucker Theorem. The name of Karush was added when it was discovered that the latter's master's thesis from 1939 contained such a result. Regularity conditions are also called *constraints qualifications*. The observation that these are not needed in the case that the constraint functions are affine may be seen by deriving the KKT conditions from the Fritz John necessary conditions. For details on this, and for more on optimization, see, e.g., [Bertsekas (2016)].

1.8 The PSD completion problem

Definition 1.174. A *partial matrix* is a "matrix" where some of the entries are not specified. We denote the unspecified entries by a question mark. A *completion* of a partial matrix is a real matrix obtained by replacing the unspecified entries by real numbers.

Example 1.175.

$$\begin{pmatrix} e & -1 & ? \\ \pi & ? & 1 \\ 0 & ? & ? \end{pmatrix}$$

is a partial matrix.

Definition 1.176. A *partial symmetric matrix* A is a partial matrix in which the diagonal entries are specified, and if a_{ij} is specified, then a_{ji} is also specified and equal to a_{ji}.

Example 1.177.

$$\begin{pmatrix} e & \pi & 0 \\ \pi & -1 & ? \\ 0 & ? & 1 \end{pmatrix}$$

is a partial symmetric matrix.

The reason for the requirement in Definition 1.176 that the diagonal entries be specified will be explained in the notes at the end of the section (with the aid of the exercises).

Definition 1.178. The *specification graph* of an $n \times n$ partial symmetric matrix A is a graph with vertices $\{1, 2, \ldots, n\}$, in which i and j are adjacent if and only if $i \neq j$ and a_{ij} is specified. If G is the specification graph of a partial symmetric matrix A we say that A is a *G-partial* matrix.

Example 1.179. The specification graph of the matrix in Example 1.177 is

Definition 1.180. A *partial PSD matrix* is a partial symmetric matrix in which all the fully specified principal submatrices are positive semidefinite. If A is a partial PSD matrix with specification graph G, we say that A is a *G-partial PSD* matrix. A *PSD completion* of A is a completion of A which is positive semidefinite.

Note that by the definition, the diagonal entries of a partial PSD matrix are nonnegative.

Example 1.181. Let A be the $n \times n$ partial matrix with all entries specified and equal to 1, except for a couple of unspecified entries in positions $1, n$ and $n, 1$. The matrix A has a completion which is positive semidefinite, the matrix J_n. In fact, J_n is the only possible PSD completion of A. If in such a completion B the $1n$ and $n1$ entries are equal to a, then

$$B[1, 2, n] = \begin{pmatrix} 1 & 1 & a \\ 1 & 1 & 1 \\ a & 1 & 1 \end{pmatrix}.$$

Hence $\det(B[1, 2, n]) = -(1 - a)^2$, and necessarily $a = 1$.

In fact, Example 1.181 is a special case of the following proposition, whose proof is left as an exercise.

Proposition 1.182. *If A is a partial PSD matrix with only one pair of unspecified entries, then A has a PSD completion.*

Consider the following question: Given a partial PSD matrix A, how can one tell if it can be completed to a positive semidefinite matrix? Clearly, if A is not partial PSD, it cannot be completed to a positive semidefinite matrix. But even if A is partial PSD, it does not guarantee PSD completability.

Example 1.183. Let A be a $k \times k$ partial symmetric matrix, $k > 3$, in which the diagonal, first superdiagonal and first subdiagonal entries are equal to 1, the $1,k$ and the $k,1$ entries are equal to 0, and all other entries are unspecified,

$$A = \begin{pmatrix} 1 & 1 & ? & & \dots & ? & 0 \\ 1 & 1 & 1 & ? & & ? & ? \\ ? & 1 & 1 & \ddots & \ddots & & \vdots \\ & ? & \ddots & \ddots & \ddots & ? & \\ \vdots & & \ddots & \ddots & 1 & 1 & ? \\ ? & ? & & ? & 1 & 1 & 1 \\ 0 & ? & \dots & \dots & ? & 1 & 1 \end{pmatrix}.$$

A is partial PSD, since the fully specified principal submatrices are the diagonal entries, which are positive, and the 2×2 principal submatrices, which are either J_2 or I_2. However, A cannot be completed to a positive semidefinite matrix. To see that, observe that by a repeated use of Example 1.181, the only value that can be placed instead of any of the "?" signs is 1. But then

$$\operatorname{rank} B[1,2] = 1 \neq 2 = \operatorname{rank} B[1,2|1,\dots,n],$$

a contradiction the Principal Submatrix Rank Property (Proposition 1.29).

It is natural to define:

Definition 1.184. A graph G is *PSD completable* if any G-partial PSD matrix A can be completed to a positive semidefinite matrix.

Since positive semidefiniteness is preserved by a simultaneous permutation of rows and columns, the labelling of the vertices of a graph does not affect its PSD completability. The specification graph of the matrix in Example 1.183 is C_k, so C_k, $k > 3$, is not PSD completable. By Proposition 1.182 the graph on n vertices which is one edge short of being the complete graph is an example of a PSD completable graph.

The following theorem is a full characterization of PSD completable graphs.

Theorem 1.185. *A graph G is PSD completable if and only if it is chordal.*

Proof. We prove first that if G is not chordal, then it is not PSD completable. If G is not chordal, then it contains an induced cycle of length $k > 3$. Construct a G-partial PSD matrix A. Let the principal submatrix corresponding to the cycle be the matrix in Example 1.182, set all other diagonal entries equal to 1, and all other specified entries equal to 0. Then A is partial PSD matrix that cannot be completed to a positive semidefinite matrix.

Now suppose G is a chordal graph on n vertices. Then there is a sequence of chordal graphs $G_0 = G, G_1, \ldots, G_m = K_n$ such that each G_i is chordal, and is obtained from its predecessor by addition of one edge (see Exercise 1.41). It is easy to see that every G_i-partial PSD matrix A_i can be "completed" to a G_{i+1}-partial PSD matrix: Suppose $e = \{p, q\}$ is the edge in G_{i+1} which is not an edge of G_i. Let K be the maximal clique in G_{i+1} containing the edge e (see Exercise 1.40). By Proposition 1.182, the principal (partial) submatrix of A_i corresponding to K has a positive semidefinite completion. Use it to specify a value in the p, q and q, p positions. By induction this shows that every G-partial PSD matrix has a PSD completion. □

Exercises

1.72. Show that if the diagonal entries of a partial matrix A are unspecified, and $a_{ij} = a_{ji}$ whenever both are specified, then A can be completed to a positive semidefinite matrix.

1.73. Consider the partial (but not partial symmetric) matrices

$$A = \begin{pmatrix} 2 & 2 & 2 \\ 2 & 2 & 1 \\ 2 & 1 & ? \end{pmatrix} \quad \text{and } B = \begin{pmatrix} 1 & 1 & 2 \\ 1 & 2 & 2 \\ 2 & 2 & ? \end{pmatrix}.$$

Show that in both of these partial matrices, each fully specified principal submatrix is positive semidefinite, but one of them has a completion which is positive semidefinite, and the other does not.

1.74. Give an example of a C_4-partial PSD matrix which has a PSD completion.

1.75. Prove Proposition 1.182 (directly, without using Theorem 1.185). *Hint:* Assume

$$A = \begin{pmatrix} d_1 & ? & \mathbf{u}^T \\ ? & d_2 & \mathbf{v}^T \\ \mathbf{u} & \mathbf{v} & E \end{pmatrix}.$$

Show that

$$\begin{pmatrix} d_1 & \mathbf{u}^T E^\dagger \mathbf{v} & \mathbf{u}^T \\ \mathbf{v}^T E^\dagger \mathbf{u} & d_2 & \mathbf{v}^T \\ \mathbf{u} & \mathbf{v} & E \end{pmatrix}$$

is a positive semidefinite completion of A.

1.76. A *G-partial PD* matrix is defined in a similar way to a G-partial PSD matrix. It is a partial symmetric matrix with specification graph G, such that each fully specified principal submatrix is positive definite. A graph G is *PD completable* if any G-partial PD matrix can be completed to a positive definite matrix. Prove that a graph G is PD completable if and only if it is PSD completable.

Notes. The section is based on [Grone *et al.* (1984)]. The results in this classical paper are proved for Hermitian matrices, but we present them here in the real case. An interesting and important result in that paper is that if a partial symmetric matrix A has a positive definite completion, then there is a unique matrix in the class of all positive definite completions of A whose determinant is maximal. This is the matrix whose inverse has zeros in the positions corresponding to the unspecified entries of A.

A partial PSD matrix A with a non-chordal specification graph may have a PSD completion (see Exercise 1.74). This depends on the numerical values of the specified entries of A. An extensive discussion of this problem is carried out in [Barrett *et al.* (1996)] and [Barrett *et al.* (1998)].

In the definition of a partial symmetric matrix it was required that the diagonal entries are specified. Exercises 1.72 and 1.73 give a taste of what may happen if this requirement is dropped.

Results on DNN completions and on SPN completions appear in Sections 3.10 and 2.13, respectively.

Chapter 2

Copositivity

2.1 Definition and basic properties

Consider the following generalization of real positive semidefinite matrices: real symmetric matrices whose quadratic forms are nonnegative whenever the input vector is nonnegative.

Definition 2.1. A symmetric $n \times n$ matrix A is *copositive* if $\mathbf{x}^T A\mathbf{x} \geq 0$ for every $\mathbf{x} \in \mathbb{R}^n_+$. The matrix A is *strictly copositive* if $\mathbf{x}^T A\mathbf{x} > 0$ for every nonzero $\mathbf{x} \in \mathbb{R}^n_+$.

Obviously, every positive semidefinite matrix is an example of a copositive matrix, and so is every symmetric nonnegative matrix (while positive definite matrices and symmetric nonnegative matrices with a positive diagonal are examples of strictly copositive matrices, see Exercises 2.5 and 2.6). For matrices of order 2, this covers all copositive matrices (and strictly copositive ones).

Example 2.2. Let

$$A = \begin{pmatrix} a & b \\ b & c \end{pmatrix}$$

be a copositive matrix. Then $a = \mathbf{e}_1^T A\mathbf{e}_1 \geq 0$ and $c = \mathbf{e}_2^T A\mathbf{e}_2 \geq 0$. For $\mathbf{x} = \mathbf{e}_1 + t\mathbf{e}_2$, $t > 0$,

$$\mathbf{x}^T A\mathbf{x} = a + 2bt + ct^2 \geq 0.$$

If $a = 0$, $\mathbf{x}^T A\mathbf{x} = t(2b + tc^2) \geq 0$ implies $2b + tc^2 \geq 0$ for every $t > 0$ and thus $b \geq 0$. Hence in this case A is nonnegative. If A is not nonnegative, then $a, c > 0$ and $b < 0$. For $t = -b/c > 0$ the inequality $\mathbf{x}^T A\mathbf{x} \geq 0$ implies $\det A \geq 0$. This, together with the nonnegativity of the diagonal entries, implies that A is positive semidefinite.

For orders larger than 2 there exist copositive matrices that are neither positive semidefinite nor nonnegative, as we shall soon see.

Copositive matrices appear in various applications, of which the most prominent is probably mathematical optimization — see details in the notes at the end of this section. However, determining whether a given matrix is copositive is a difficult problem (in complexity terms, it is co-NP-hard). Characterizing copositive matrices is a major problem in the study of copositivity. The copositive matrices of order n form a cone, and the geometric structure of this cone is another topic of interest. Even the description of the extreme rays of this cone is a difficult open problem, and an important one. The cone of copositive matrices contains the cone of SPN matrices, which is easier to handle. It is therefore interesting to understand the gap between these two cones.

In this chapter we present known properties of copositive matrices, some partial results on these questions and tools that help in dealing with them. These provide some insight into the problems at hand, but also highlight their difficulty. We begin in this section, with some initial observations and simple properties of copositive matrices.

Remark 2.3. To check copositivity of a matrix A, it suffices to check that $\mathbf{x}^T A\mathbf{x} \geq 0$ for every $\mathbf{x}$ in a set, which is a base of the cone $\mathbb{R}^n_+$. Recall that the compact convex simplex $\Delta_n = \{\mathbf{x} \in \mathbb{R}^n_+ \mid \sum_{i=1}^n x_i = 1\}$ is such a base (Example 1.115). For every $n\times n$ matrix A the quadratic form $\mathbf{x}^T A\mathbf{x}$ attains a minimum on Δ_n, and A is copositive if and only if $\min_{\mathbf{x}\in\Delta_n} \mathbf{x}^T A\mathbf{x} \geq 0$. The matrix is strictly copositive if this minimum is positive. Alternatively, $\min_{\mathbf{x}\in\Gamma_n} \mathbf{x}^T A\mathbf{x}$ may be considered, where $\Gamma_n = \{\mathbf{x} \in \mathbb{R}^n_+ \mid \sum_{i=1}^n x_i^2 = 1\}$ is the nonnegative section of the unit sphere. We use Δ and Γ where the dimension n is clear from the context.

For ease of reading, we consider copositive matrices. The corresponding results for strictly copositive matrices are mentioned later in the section.

Proposition 2.4. *Every principal submatrix of a copositive matrix is copositive.*

Proof. Let A be an $n \times n$ copositive matrix and $\alpha \subseteq \{1, \ldots, n\}$. For every $\mathbf{x} \in \mathbb{R}^{|\alpha|}_+$ let $\mathbf{y} \in \mathbb{R}^n_+$ have $\mathbf{y}[\alpha] = \mathbf{x}$ and all other entries equal to zero. Then $\mathbf{x}^T A[\alpha]\mathbf{x} = \mathbf{y}^T A\mathbf{y} \geq 0$. □

Combining Proposition 2.4 and the discussion in Example 2.2 we get the following result. Part (b) follows from (c), but deserves a separate mention.

Corollary 2.5. *Let A be a copositive matrix of order n. Then*

(a) *for every $1 \le i \le n$, $a_{ii} \ge 0$,*
(b) *if $a_{ii} = 0$, then $a_{ij} = a_{ji} \ge 0$ for every $j \ne i$,*
(c) *for every $1 \le i \ne j \le n$, $a_{ij} \ge -\sqrt{a_{ii}a_{jj}}$.*

Corollary 2.6. *The sum of all the eigenvalues of a copositive matrix A is nonnegative.*

The conditions in Corollary 2.5 are necessary conditions for copositivity, but are far from sufficient.

Example 2.7. The matrix

$$A = \begin{pmatrix} 1 & -1 & -1 \\ -1 & 1 & -1 \\ -1 & -1 & 1 \end{pmatrix}$$

is not copositive, since the sum of its entries is negative, i.e., $\mathbf{e}^T A\mathbf{e} < 0$. Note that A does satisfy the conditions in Corollary 2.5.

By the next result, the set of all $n \times n$ copositive matrices, $\mathcal{COP}_n$, is a convex cone. We discuss this cone in Section 2.3, and then again in Chapter 5. The simple proof of Proposition 2.8 is left as an exercise.

Proposition 2.8. *Let A and B be copositive matrices of the same order, and let $t > 0$. Then $A + B$ and tA are copositive matrices.*

Recall that a matrix that is a sum of a positive semidefinite matrix and a symmetric nonnegative one is called an SPN matrix. By Proposition 2.8 any SPN matrix is an example of a copositive matrix.

Example 2.9. The SPN matrix

$$\begin{pmatrix} 1 & -1 & 1 \\ -1 & 1 & 0 \\ 1 & 0 & 1 \end{pmatrix} = \begin{pmatrix} 1 & -1 & 1 \\ -1 & 1 & -1 \\ 1 & -1 & 1 \end{pmatrix} + \begin{pmatrix} 0 & 0 & 0 \\ 0 & 0 & 1 \\ 0 & 1 & 0 \end{pmatrix}$$

is copositive. It is neither positive semidefinite nor nonnegative.

Not all copositive matrices are SPN. A copositive matrix which is not SPN is called *exceptional*. The first example of an exceptional copositive matrix was given by Alfred Horn.

Example 2.10. The matrix

$$H = \begin{pmatrix} 1 & -1 & 1 & 1 & -1 \\ -1 & 1 & -1 & 1 & 1 \\ 1 & -1 & 1 & -1 & 1 \\ 1 & 1 & -1 & 1 & -1 \\ -1 & 1 & 1 & -1 & 1 \end{pmatrix},$$

is called the *Horn matrix*. To see that H is copositive, observe that

$$\begin{aligned} \mathbf{x}^T H \mathbf{x} &= (x_1 - x_2 + x_3 + x_4 - x_5)^2 + 4x_2x_4 + 4x_3(x_5 - x_4) \quad (2.1) \\ &= (x_1 - x_2 + x_3 - x_4 + x_5)^2 + 4x_2x_5 + 4x_1(x_4 - x_5). \quad (2.2) \end{aligned}$$

(2.1) implies that $\mathbf{x}^T H \mathbf{x} \geq 0$ for any nonnegative vector $\mathbf{x}$ with $x_5 \geq x_4$, and (2.2) implies that $\mathbf{x}^T H \mathbf{x} \geq 0$ for any nonnegative vector $\mathbf{x}$ with $x_5 < x_4$. The reader is invited to try showing that it is not SPN. A proof is discussed in Section 2.3 (and once more in Section 2.8).

We continue with a corollary to Proposition 2.8, that may also be proved directly from the definition of copositivity.

Corollary 2.11. *If A is copositive and $B \geq A$ then B is copositive.*

Proposition 2.12. *Let A be an $n \times n$ copositive matrix, and let S be an $n \times m$ nonnegative matrix. Then S^TAS is also copositive.*

Proof. For $\mathbf{x} \in \mathbb{R}^m_+$, $S\mathbf{x} \geq \mathbf{0}$ and thus $\mathbf{x}^T(S^TAS)\mathbf{x} = (S\mathbf{x})^TA(S\mathbf{x}) \geq 0$. □

Corollary 2.13. *Let A be a symmetric $n \times n$ matrix. Then*

(a) *If P is an $n \times n$ permutation matrix, then A is copositive if and only if P^TAP is copositive.*
(b) *If D is an $n \times n$ positive diagonal matrix, then A is copositive if and only if DAD is copositive.*

Corollary 2.13 is very helpful in the study of copositive matrices. We often permute simultaneously rows and columns of a matrix for convenience, and scale any matrix with a positive diagonal to have a diagonal of ones.

Corollary 2.14. *Let A is an $n \times n$ copositive matrix, and let E an $n \times m$ nonnegative matrix. Then*

$$B = \begin{pmatrix} A & AE \\ E^TA & E^TAE \end{pmatrix}$$

is copositive.

Proof. This follows from Proposition 2.12, as

$$B = \begin{pmatrix} I_n \\ E^T \end{pmatrix} A \begin{pmatrix} I_n & E \end{pmatrix}. \qquad \square$$

We leave the next observation as an exercise.

Proposition 2.15. *If $A = A_1 \oplus A_2 \oplus \cdots \oplus A_m$, then A is copositive if and only if each A_i is.*

The next result is easier to state and prove for matrices whose diagonal entries are all equal to 1. Note that if all the diagonal entries of a symmetric matrix A are positive, we can apply the proposition to an appropriate diagonal scaling of A.

Proposition 2.16. *Let A be a symmetric matrix with diagonal equal to* **e**. *Then A is copositive if and only if every principal submatrix of A all of whose off-diagonal entries are strictly less than 1 is copositive.*

Proof. It suffices to show that if $a_{ij} \geq 1$, then A is copositive if and only if $A(i)$ and $A(j)$ are copositive. By Proposition 2.4, we only need to prove the "if" part, and by Corollary 2.13(a) we may consider the case that $i = 1$ and $j = 2$. So suppose $a_{12} \geq 1$ and both $A(1)$ and $A(2)$ are copositive. Given $\mathbf{x} \in \mathbb{R}^n_+$, let $c = x_1 + x_2$, and $\mathbf{y}(t) = (t, c - t, x_3, \ldots, x_n)^T$. For every $0 \leq t \leq c$, $\mathbf{y}(t) \in \mathbb{R}^n_+$. For a fixed $\mathbf{x}$, the function $\mathbf{y}(t)^T A\mathbf{y}(t)$ is a quadratic function of t, in which the coefficient of t^2 is $2 - 2a_{12} \leq 0$, that is, a concave function (linear if $a_{12} = 1$). This function therefore attains its minimum on the closed interval $[0, c]$ either at $t = 0$ or at $t = c$. The minimum is nonnegative, since $\mathbf{y}(0)^T A\mathbf{y}(0) \geq 0$ by the copositivity of $A(1)$ and $\mathbf{y}(c)^T A\mathbf{y}(c) \geq 0$ by the copositivity of $A(2)$. As $\mathbf{x} = \mathbf{y}(x_1)$, we get that $\mathbf{x}^T A\mathbf{x} \geq 0$. $\square$

The proof of the next corollary is left as an exercise (recall that the minimum of two matrices is taken entrywise, $\min(A, B)_{ij} = \min(a_{ij}, b_{ij})$).

Corollary 2.17. *Let A be a symmetric matrix with diagonal equal to* **e**. *Then A is copositive if and only if $\bar{A} = \min(A, J)$ is copositive.*

Remark 2.18. In Corollary 2.5, if A is strictly copositive then (a) and (c) hold with strict inequality (and (b) is irrelevant). Most of the other results in this section (that is, Propositions 2.4, 2.15, 2.16, and Corollaries 2.11, 2.13 and 2.14) remain valid if "copositive" is replaced by "strictly copositive". In Proposition 2.8, it suffices that one of the copositive matrices in a sum is strictly copositive for the sum to be strictly copositive, and Proposition 2.12 remains valid if "copositive" is replaced by "strictly copositive", and the nonnegative S has a full column rank. All proofs are left as exercises.

The reader is also invited to check that all the results so far, except for Proposition 2.16, trivially hold if we replace "copositive" by "SPN". Regarding the latter, see Exercise 2.12.

Remark 2.19. There is another notion of copositivity. An $n \times n$ matrix A is *copositive-plus* if A is copositive and $\mathbf{x}^T A\mathbf{x} = 0$ for $\mathbf{x} \in \mathbb{R}^n_+$ implies that $A\mathbf{x} = 0$. Every strictly copositive matrix is copositive-plus, but the converse does not hold. For example, every positive semidefinite matrix is copositive-plus by Corollary 1.28, but is not necessarily strictly copositive (see Exercise 2.6).

Exercises

2.1. Show that the following matrices are copositive.

(a) $$\begin{pmatrix} 1 & 2 & 0 & 0 \\ 2 & 1 & 0 & 0 \\ 0 & 0 & 1 & -1 \\ 0 & 0 & -1 & 1 \end{pmatrix}$$

(b) $$\begin{pmatrix} 1 & -1 & 2 & 3 \\ -1 & 1 & 0 & 4 \\ 2 & 0 & 1 & -1 \\ 3 & 4 & -1 & 1 \end{pmatrix}$$

(c) $$\begin{pmatrix} 1 & -1 & 1 & -1 \\ -1 & 1 & 0 & 1 \\ 1 & 0 & 1 & 0 \\ -1 & 1 & 0 & 1 \end{pmatrix}$$

2.2. Prove Proposition 2.4 for the strictly copositive case.

2.3. Prove Corollary 2.5. Show also that if A is strictly copositive then (a) and (c) hold with strict inequality.

2.4. Prove Corollary 2.6.

2.5. Show that a symmetric nonnegative matrix is strictly copositive if and only if it has a positive diagonal.

2.6. Show that a positive semidefinite matrix A is strictly copositive if and only if $A\mathbf{x} \neq \mathbf{0}$ for every $\mathbf{0} \neq \mathbf{x} \in \mathbb{R}^n_+$ (i.e., there is no nonzero nonnegative vector in $\operatorname{ns} A$). In particular, every positive definite matrix is strictly copositive.

2.7. Prove Proposition 2.8. Show also that if A is strictly positive, B is copositive and $t > 0$, then $A + B$ and tA are strictly copositive.

2.8. Prove Corollary 2.11. Show also that if A is strictly copositive, and $B \geq A$, then B is strictly copositive.

2.9. Prove that in Corollary 2.12, if A is strictly copositive and the nonnegative S has full column rank, then $S^T AS$ is strictly copositive.

2.10. Prove the following strictly copositive version of Corollary 2.14. Show that for every strictly copositive A, and every nonnegative E of full column rank, the matrix

$$B = \begin{pmatrix} A & AE \\ E^T A & E^T AE \end{pmatrix}$$

is strictly copositive.

2.11. Prove Proposition 2.15. Show that the proposition remains valid if “copositive” is replaced by “strictly copositive”.

2.12. Use Proposition 2.16 to provide another proof that the Horn matrix is copositive. Check that, as H is not SPN, this shows that Proposition 2.16 does not hold if we replace “copositive” by “SPN”.

2.13. Prove Proposition 2.16 for strictly copositive matrices: Show that a matrix A with diagonal equal to $\mathbf{e}$ is strictly copositive if and only if every principal submatrix of A with off-diagonal entries less than 1 is strictly copositive.

2.14. Prove Corollary 2.17.

2.15. Prove that a symmetric matrix with diagonal $\mathbf{e}$ is SPN if and only if $\bar{A} = \min(A, J)$ is SPN. That is, Corollary 2.17 holds if “copositive” is replaced by “SPN”.

2.16. Let

$$A = \begin{pmatrix} 1 & -1 & t & 0 & 0 & 0 \\ -1 & 1 & -t & 0 & 1 & 0 \\ t & -t & 1 & -t & 0 & 0 \\ 0 & 0 & -t & 1 & -t & t \\ 0 & 1 & 0 & -t & 1 & -1 \\ 0 & 0 & 0 & t & -1 & 1 \end{pmatrix},$$

where $t = \sqrt{0.5}$. Show that A is copositive.

2.17. Generalize Proposition 2.16 to a symmetric A with a positive diagonal.
Hint: Consider DAD, where $D = \text{Diag}\left(1/\sqrt{a_{11}}, \ldots, 1/\sqrt{a_{nn}}\right)$.

2.18. Let A be an $n \times n$ symmetric matrix with nonnegative diagonal elements. Let $B = \mathbf{a}\mathbf{a}^T$, where $a_i = \sqrt{a_{ii}}$. Show that A is copositive if and only if the matrix $\min(A, B)$ is copositive.

2.19. Give an example of copositive matrices A and B of the same order, such that their Hadamard product $A \circ B$ is not copositive.

2.20. Prove that a symmetric nonnegative matrix is copositive-plus if and only if it is a direct sum of a strictly copositive matrix and, possibly, a zero matrix.

Notes. The first result on copositivity, due to Motzkin, appeared in [Motzkin (1952)]. It was a reported determinantal necessary and sufficient condition for a quadratic form to be nonnegative on nonnegative variables. (The result itself was not totally accurate. The reported condition is indeed sufficient, but not necessary, as pointed out in [Cottle *et al.* (1970b)], where this observation and a counter example to necessity are attributed to Baumert. Correct determinantal criteria of this flavor may be found in Section 2.7.) The next to address such quadratic forms was [Gaddum (1958)], where a recursive copositivity/strict copositivity characterization is discussed. Then came [Diananda (1961)], where it was shown that small copositive matrices (of order up to 4) are SPN (see Section 2.9). The first example of a non-SPN copositive matrix, Example 2.10, is due to Alfred Horn, and was communicated by Marshal Hall Jr. [Diananda (1961); Hall and Newman (1963)]. Most of the basic results in this section appeared already in these two papers, either explicitly or implicitly. The very useful Proposition 2.16 is due to [Hoffman and Pereira (1973)].

The first appearance of the term "copositive" (and the term "completely positive") is in [Hall and Newman (1963)], still in the context of quadratic forms. A quadratic form

$$Q(\mathbf{x}) = \mathbf{x}^T A \mathbf{x} = \sum_{i,j=1}^{n} a_{ij} x_i x_j$$

is copositive if $Q(\mathbf{x}) \geq 0$ whenever $\mathbf{x} = (x_1, \ldots, x_n)^T \geq 0$. That is, Q is copositive if and only if the matrix $A = (a_{ij})_{i,j=1}^{n}$ is copositive. The origin of the term "copositive" is not clear, it may stand for "conditional positivity". It seems that the shift from copositive quadratic forms to copositive matrices was made in [Cottle *et al.* (1970b)]. There one can find also the first appearance of the term "strictly copositive". The term "copositive-plus" was suggested by Cottle and first used in [Cottle and Dantzig (1968)]. Initially, copositive matrices were not necessarily symmetric, but as the quadratic form associated with M is the same as the one associated with $\frac{1}{2}\left(M + M^T\right)$, this is not a crucial difference. A more general concept is that of copositivity with respect to a cone. An $n \times n$ symmetric matrix A is *copositive with respect to a cone* $K \subseteq \mathbb{R}^n$ if $\mathbf{x}^T A \mathbf{x} \geq 0$ for every $\mathbf{x} \in K$. We mostly do not discuss this more general concept in this book. Another extension we do not go into, is to copositive forms of degree d, or *copositive tensors* of order d. These were introduced in [Qi (2013)].

Originally, the interest in copositive matrices/forms arose in the theory of inequalities. Other uses were discovered since then. We mention some of these, elaborating only where going into details is not too lengthy.

Mathematical Programming. One of the first applications to copositivity (plus) was in the Linear Complementarity Problem. In this problem a square matrix M and a vector $\mathbf{q}$ are given, and one seeks nonnegative vectors $\mathbf{z}$ and $\mathbf{w}$ such that

$$M\mathbf{z} - \mathbf{w} = \mathbf{q}$$

and $\mathbf{z}^T \mathbf{w} = 0$. An algorithm for solving this problem was suggested in [Lemke (1964/1965)]. It was shown there that if $M + M^T$ is (what was subsequently called) copositive-plus, then in the conclusion of the algorithm either a solution is obtained, or it is determined that such a solution does not exist. The algorithm was proposed in the context of finding a Nash equilibrium of a bimatrix game. However, solution of Linear Complementarity Problems (and generalizations thereof) is used in various applications. and the algorithm was shown to be applicable, for example, to a problem in visual labeling and artificial neural networks in [Miller and Zucker (1991)], and a to dynamic multi-rigid-body contact problems, e.g., in [Anitescu and Potra (1997)].

Bézier patches. In computer aided design, a designed surface object (such as a panel of an airplane or a car) may be approximated by a net of small triangular patches. These are bivariate quadratic functions defined on triangles. Among various considerations, such as smoothness of the combined patches, there may be a need for approximating nonnegative data by a nonnegative function, when the interpolating surface should be on the same side of a plane as the data points (e.g., in designing a car panel that should not interfere with other parts of the car). A triangular Bézier patch is a quadratic function $Q(\mathbf{x})$ on a base triangle, expressed in the Bernstein-Bézier form as a quadratic function in the barycentric coordinates of $\mathbf{x}$ in the base triangle (connecting $\mathbf{x}$ to the vertices of the triangle divides the triangle into three smaller ones, and each barycentric coordinate is the ratio between the area of one of these smaller triangles and the area of the whole triangle). Expressed this way, the nonnegativity issue translates into the question of nonnegativity of a quadratic function on the simplex Δ_3, i.e., to the copositivity of a 3×3 matrix. This is pointed out (for multivariate quadratics) in [Micchelli and Pinkus (1989)], where an algorithm for checking copositivity is suggested, and in [Nadler (1992)], where a characterization of 3×3 copositive matrices by their entries is (re-)established.

Simpson's paradox. Consider a population C, partitioned into n mutually exclusive subpopulations, $C_1, \ldots, C_n$. Each member of the population C may have or not have attribute A, and have or not have attribute B. For each attribute X, X denotes the set of members with this attribute, and $\overline{X}$ the set of those without it. Simpson's Paradox occurs when the following conditional probabilities inequalities hold:

$$p(A|B) \leq p(\overline{A}|B)$$

("the attributes are negatively associated or independent in the population") and

$$p(A|B \cap C_i) > p(\overline{A}|B \cap C_i) \quad \text{for every } i = 1, \ldots, n$$

("the attributes are positively associated in each subpopulation"), or vice versa.

[Hadjicostas (1997)] considers the possible occurrence of the Simpson's Paradox in the following setting. Suppose the conditional probabilities

$$a_i = p(A \cap B|C_i), \quad b_i = p(A \cap \overline{B}|C_i), \quad c_i = p(\overline{A} \cap B|C_i), \quad d_i = p(\overline{A} \cap \overline{B}|C_i)$$

are all known and positive, $i = 1, \dots, n$, but the probabilities $p_i = p(C_i)$, $i = 1, \dots, n$, are not known. What should an analyst check in order to determine whether Simpson's Paradox may occur in this setting?

Consider the symmetric matrix K with entries

$$k_{ij} = \frac{1}{2} \cdot (a_i d_j + a_j d_i - b_i c_j - b_i c_j), \quad 1 \leq i, j \leq n.$$

It is shown that if the attributes are positively associated in each C_i, then the diagonal entries of K are positive. If K is strictly copositive, then the attributes cannot be negatively associated or independent in the whole population. If K is copositive, Simpson's Paradox holds (with the attributes independent in the population) if and only if the positive vector $\mathbf{p} = (p_1, \dots, p_n)^T$ is a zero of K (i.e., $\mathbf{p}^T K \mathbf{p} = 0$). (In particular, if K is copositive but not positive semidefinite, the paradox will not occur — see Proposition 2.47.) If K is not copositive, the paradox may occur. It is also shown that for such a matrix K of order $n \geq 3$, checking (strict) copositivity amounts to checking (strict) copositivity of the 2×2 principal submatrices. (The analysis uses recursive copositivity characterization results that can be found in Section 2.7.)

The 123 Theorem and similar Probability Theory results. In [Alon and Yuster (1995)] the following result was proved: Let $b > a > 0$ be real numbers. Then for every i.i.d. real random variables X and Y,

$$\text{Prob}(|X - Y| \leq b) < \left(2\left\lceil \frac{b}{a} \right\rceil - 1\right) \text{Prob}(|X - Y| \leq a). \tag{2.3}$$

If $a = 1$ and $b = 2$ the inequality becomes "the 123 Theorem":

$$\text{Prob}(|X - Y| \leq 2) < 3\,\text{Prob}(|X - Y| \leq 1).$$

Other results of this type include the case that X and Y are vector-valued i.i.d. variables in [Alon and Yuster (1995)], and a result relating $\text{Prob}(|X + Y| \leq b)$ and $\text{Prob}(|X - Y| \leq a)$ for $a \leq b$, in [Siegmund-Schultze and von Weizsäcker (2007)] (the case $a = b = 1$) and in [Dong *et al.* (2015)] (general case).

In [Kovačec *et al.* (2013)] it is observed that each of these results is equivalent to the strict copositivity of a certain matrix. For example, in the first result consider the case that the i.i.d. random variables X and Y assume only a finite number of real values, $a_1, \dots, a_m$ with probabilities $p_1, \dots, p_m$, respectively. For a property P, let $\chi(P) = 1$ if the property holds, and $\chi(P) = 0$ otherwise. Then

$$\begin{aligned}\operatorname{Prob}(|X-Y| \leq r) &\leq \sum_{i,j=1}^{m} \operatorname{Prob}(|a_i - a_j| \leq r, X = a_i, Y = a_j) \\ &= \sum_{i,j=1}^{m} \chi(|a_i - a_j| \leq r)\operatorname{Prob}(X = a_i, Y = a_j) \\ &= \sum_{i,j=1}^{m} \chi(|a_i - a_j| \leq r) p_i p_j \\ &= \mathbf{p}^T A(r)\mathbf{p},\end{aligned}$$

where $\mathbf{p} = (p_1, \ldots, p_m)^T$ is the vector of probabilities, and $(A(r))_{ij} = \chi(|a_i - a_j| \leq r)$. For $b > a > 1$ let $C = \left(2\left\lceil \frac{b}{a}\right\rceil - 1\right)A(a) - A(b)$. Then (2.3) holds for every two i.i.d. variables assuming only the values $a_1, \ldots, a_m$ if and only if the matrix C is strictly copositive.

It is proved in [Kovačec *et al.* (2013)] that the matrix C is indeed strictly copositive under the conditions of the theorem. (Here too the proof of strict copositivity uses the recursive copositivity characterization results that can be found in Section 2.7.) Similar matrices are constructed in the other cases.

Neural Computation. The field of neural computation attempts to figure out principles of information processing in biological nervous systems by mathematical modeling. A certain conductance-based model for recurrent neuron networks may be reduced to a dynamic of the form

$$\dot{\mathbf{x}} + \mathbf{x} = (\mathbf{b} + W\mathbf{x})_+ \,,$$

where $\mathbf{b}$ is the input vector, W is a symmetric matrix (the *synaptic weight matrix*), and the output is a steady state vector $\mathbf{x}$ (i.e., $\dot{\mathbf{x}} = \mathbf{0}$). The properties of the synaptic weight matrix W determine the existence of outputs and their relationships to the input. In [Hahnloser *et al.* (2003)] it is shown that a symmetric network of this form converges to an equilibrium point for every $\mathbf{b}$ and every initial conditions if and only if $I - W$ is strictly copositive.

Stability of Switched Positive Linear Systems. Positive dynamical systems are dynamical systems whose states are confined to the positive orthant. Such systems arise in important applications in various fields, such as biology, probability, economics, and communication. The stability of these systems is a topic of much interest. In [Mason and Shorten (2007)] stability of certain such systems is related to copositivity of the Lyapunov function.

Mathematical Optimization, again. In the last two decades there is a renewed interest in the mathematical optimization community in copositive matrices (and completely positive ones). This is due to the fact that many nonconvex and combinatorial optimization problems may be reformulated as linear problems in matrix variables, over the copositive cone (or the completely positive one). The difficulty of such NP-hard problem is thus shifted into the cone (indeed, checking that a matrix is copositive is known to be co-NP-complete [Murty and Kabadi (1987)]). A *copositive program* is a linear optimization problem with a matrix variable, of the form

$$\begin{aligned}&\min\langle C, X\rangle\\ &\text{s.t. } \langle A_i, X\rangle = b_i, \;\; i = 1, \ldots, m\;,\\ &X \in \mathcal{COP}_n\end{aligned}$$

where C and A_i, $i = 1, \ldots, n$, are given symmetric matrices, $b_i \in \mathbb{R}$, $\mathcal{COP}_n$ is the convex cone of $n \times n$ copositive matrices (see Section 2.3), and the inner product is that of $\mathcal{S}_n$.

In [Motzkin and Straus (1965)] it was shown that if G is a simple graph with adjacency matrix A, then $\max_{\mathbf{x}\in\Delta} \mathbf{x}^T A\mathbf{x}$ is equal to $\frac{\omega(G)-1}{\omega(G)}$, where $\omega(G)$ is the clique number of G. This may be formulated as a copositive program,

$$\frac{1}{\omega(G)} = \min\{\lambda \mid \lambda J - (J - A) \in \mathcal{COP}_n\},$$

where A is the adjacency matrix of a graph G on n vertices, and J is the all-ones $n \times n$ matrix. In [de Klerk and Pasechnik (2002)] a similar representation of the stability number of the graph $\alpha(G)$ was presented:

$$\alpha(G) = \min\{\lambda \mid \lambda(I + A) - J \in \mathcal{COP}_n\}.$$

The term "copositive programming" was coined in the paper [Bomze *et al.* (2000)], where the standard quadratic problem was formulated as a copositive program. It was shown in [Burer (2009)] that every quadratic problem with linear and binary constraints may be formulated as a copositive program, under a mild condition. This shows the wide applicability of copositive programming.

For more on copositive programming and its history see [Dür (2010)] and [Bomze (2012)]. Two previous surveys on copositivity are [Ikramov and Savel′eva (2000)], which considers also copositivity with respect to a subspace, and [Hiriart-Urruty and Seeger (2010)], which takes a variational approach. [Bomze *et al.* (2012)] considers basic properties of the copositive and completely positive cones and related cones, as well as relations between matrices in these cones. It also contains a comprehensive bibliography on copositivity, complete positivity and copositive optimization, and on more applications of copositivity.

2.2 Spectral properties of copositive matrices

There is one spectral property that is shared by both positive semidefinite matrices and nonnegative matrices. It turns out that this property is shared by all copositive matrices.

Theorem 2.20. *The spectral radius $\rho(A)$ of a copositive matrix A is an eigenvalue of A.*

Proof. Let λ be the largest eigenvalue of A. Assume on the contrary, that $\lambda \neq \rho = \rho(A)$. Then $-\rho$ is an eigenvalue of A with a corresponding eigenvector $\mathbf{x}$ of norm 1, and $\mathbf{x}^T A\mathbf{x} = -\rho$. Write $\mathbf{x} = \mathbf{x}_+ - \mathbf{x}_-$, and let $\mathbf{y} = |\mathbf{x}| = \mathbf{x}_+ + \mathbf{x}_-$. Then $\|\mathbf{y}\| = \|\mathbf{x}\| = 1$. By the copositivity of A and the nonnegativity of $\mathbf{x}_+$ and $\mathbf{x}_-$, we have

$$\mathbf{x}^T A\mathbf{x} + \mathbf{y}^T A\mathbf{y} = 2(\mathbf{x}_+^T A\mathbf{x}_+ + \mathbf{x}_-^T A\mathbf{x}_-) \geq 0,$$

that is, $\mathbf{y}^T A\mathbf{y} \geq -\mathbf{x}^T A\mathbf{x} = \rho$. On the other hand, by the Rayleigh-Ritz Theorem (Theorem 1.15), $\mathbf{y}^T A\mathbf{y} \leq \lambda < \rho$, a contradiction. □

Note that while the eigenvalue $\rho(A)$ of an irreducible nonnegative matrix has a corresponding positive eigenvector, this property obviously does not extend to copositive matrices, as it does not hold for positive semidefinite matrices (consider, for example, the matrix $\begin{pmatrix} 1 & -1 \\ -1 & 1 \end{pmatrix}$). However, looking more carefully at the proof of Theorem 2.20 we may deduce a condition that guarantees the existence of a nonnegative eigenvector corresponding to $\rho(A)$.

Corollary 2.21. *Let A be a nonzero copositive matrix with spectral radius $\rho = \rho(A)$. If $-\rho$ is an eigenvalue of A, then A has a nonnegative eigenvector corresponding to the eigenvalue ρ.*

Moreover, for every eigenvector $\mathbf{x}$ in the eigenspace of $-\rho$, $|\mathbf{x}|$ is an eigenvector corresponding to ρ, and $\|\mathbf{x}_-\| = \|\mathbf{x}_+\|$. The multiplicity of ρ as an eigenvalue of A is greater or equal to the multiplicity of $-\rho$.

Proof. By the argument in the proof of Theorem 2.20, if $\mathbf{x}$ is any unit eigenvector corresponding to the eigenvalue $-\rho$, and $\mathbf{y} = |\mathbf{x}| = \mathbf{x}_+ + \mathbf{x}_-$, then $\mathbf{y}^T A\mathbf{y} \geq \rho$. As $\mathbf{y}$ is a unit vector, this implies that $\mathbf{y}^T A\mathbf{y} = \rho$ and $\mathbf{y}$ is an eigenvector corresponding to ρ.

Since the eigenvector $\mathbf{x}$ corresponding to $-\rho$ and the eigenvector $\mathbf{y}$ corresponding to ρ are orthogonal,

$$0 = \mathbf{x}^T\mathbf{y} = \|\mathbf{x}_+\|^2 - \|\mathbf{x}_-\|^2.$$

This holds also for scalar multiples of the unit vector $\mathbf{x}$.

If the multiplicity of $-\rho$ is k, let $\mathbf{x}_1, \mathbf{x}_2, \ldots, \mathbf{x}_k$ form a basis of the eigenspace of $-\rho$ such that the matrix I_k is a submatrix of $(\mathbf{x}_1\ \mathbf{x}_2\ \ldots\ \mathbf{x}_k)$. Then the vectors $\mathbf{y}_i = |\mathbf{x}_i|$, $i = 1, \ldots, k$, are in the eigenspace of ρ. As I_k is a submatrix of $(\mathbf{y}_1\ \mathbf{y}_2\ \ldots\ \mathbf{y}_k)$, these k vectors are linearly independent. □

Another case where there is a nonnegative eigenvector corresponding to $\rho(A)$ is given in the next proposition. (See, however, Theorem 2.76.)

Proposition 2.22. *Let A be an $n \times n$ copositive matrix, $n \geq 2$, whose single positive eigenvalue is $\rho(A)$. If A is not positive semidefinite, then A has a nonnegative eigenvector corresponding to $\rho(A)$.*

Proof. The eigenvalues of A are

$$\rho = \lambda_1 > 0 \geq \lambda_2 \geq \cdots \geq \lambda_n,$$

with corresponding orthonormal eigenvectors $\mathbf{v}_1, \mathbf{v}_2, \ldots, \mathbf{v}_n$, and $\lambda_n < 0$. Suppose on the contrary that the eigenvector $\mathbf{v}_1$ corresponding to $\rho = \rho(A)$ has both positive and negative entries. Choose any vector $\mathbf{x} > \mathbf{0}$, which is orthogonal to $\mathbf{v}_1$. Either $\mathbf{x}$ is not orthogonal to $\mathbf{v}_n$, or for some small $\varepsilon > 0$ the vector $\mathbf{x} + \varepsilon\mathbf{v}_n$, which is orthogonal to $\mathbf{v}_1$ and not orthogonal to $\mathbf{v}_n$, is positive. Thus there exists $\mathbf{y} > \mathbf{0}$ orthogonal to $\mathbf{v}_1$ and not orthogonal to $\mathbf{v}_n$. That is, $\mathbf{y} = \sum_{i=2}^{n} a_i\mathbf{v}_i$, $a_n \neq 0$. But then

$$\mathbf{y}^T A\mathbf{y} = \sum_{i=2}^{n} a_i^2\lambda_i \leq a_n^2\lambda_n < 0,$$

contrary to the assumption that A is copositive. □

In the general case, however, only the following weaker result holds.

Proposition 2.23. *Let A be an $n \times n$ copositive matrix. Then there exists a positive vector in the subspace spanned by the eigenvectors corresponding to the nonnegative eigenvalues.*

Proof. The claim obviously holds for a positive semidefinite A. So suppose that the eigenvalues of A are $\lambda_1 \geq \ldots \lambda_k \geq 0 > \lambda_{k+1} \geq \cdots \geq \lambda_n$, $k < n$. Let $\mathbf{v}_1, \ldots, \mathbf{v}_k, \mathbf{v}_{k+1}, \ldots, \mathbf{v}_n$ be corresponding orthonormal eigenvectors. We first show that each nonzero vector in the subspace spanned by $\{\mathbf{v}_{k+1}, \ldots, \mathbf{v}_n\}$ has both positive and negative entries. Suppose on the contrary that $\mathbf{x} = \sum_{i=k+1}^{n} a_i \mathbf{v}_i$ is nonnegative and nonzero. Then $a_i \neq 0$ for at least one i, and

$$0 \leq \mathbf{x}^T A \mathbf{x} = \sum_{i=k+1}^{n} a_i^2 \lambda_i < 0,$$

a contradiction.

Now let

$$B = \begin{pmatrix} \mathbf{v}_{k+1}^T \\ \vdots \\ \mathbf{v}_n^T \end{pmatrix}.$$

By the above, there is no nonzero vector $\mathbf{y} \in \mathbb{R}^{n-k}$ such that $B^T \mathbf{y} \geq 0$. Thus by Stiemke's Theorem of the Alternative (Theorem 1.162), there is a positive $\mathbf{x} \in \mathbb{R}^n$ such that $B\mathbf{x} = \mathbf{0}$. Obviously, $\mathbf{x} \in \operatorname{span}\{\mathbf{v}_1, \ldots, \mathbf{v}_k\}$. □

Exercises

2.21. Let A be a copositive matrix and let $\rho = \rho(A)$ be its spectral radius. Show that if an eigenvector $\mathbf{x}$ of A corresponding to ρ satisfies $\mathbf{x}_-^T A \mathbf{x}_+ \geq 0$, then there exists a nonnegative eigenvector of A corresponding to ρ.
Hint: Let $\mathbf{y} = \mathbf{x}_+ + \mathbf{x}_-$ and consider $\mathbf{y}^T A \mathbf{y} - \mathbf{x}^T A \mathbf{x}$.

2.22. Let

$$A = \begin{pmatrix} 33 & -15 & -15 \\ -15 & 8 & 10 \\ -15 & 10 & 8 \end{pmatrix}.$$

Show that A is copositive. Check that 3 is an eigenvalue of A with a corresponding positive eigenvector. Show that $\rho(A) > 3$, and there is no nonnegative eigenvector corresponding to $\rho(A)$.
Hint: Proposition 2.16, and check $A\mathbf{e}$.

2.23. Let A be a strictly copositive matrix. Show that A has a nonzero nonnegative vector in the subspace spanned by the eigenvectors corresponding to the positive eigenvalues.

2.24. Let A be an $n \times n$ copositive matrix, with eigenvalues

$$\lambda_1 \geq \cdots \geq \lambda_k \geq 0 > \lambda_{k+1} \geq \cdots \geq \lambda_n,$$

$k < n$. Let $\mathbf{v}_1, \ldots, \mathbf{v}_k, \mathbf{v}_{k+1}, \ldots, \mathbf{v}_n$ be corresponding orthonormal eigenvectors. Show that if V is a subspace of $\mathbb{R}^n$ such that $\mathbf{x}^T A\mathbf{x} \geq 0$ for every $\mathbf{x} \in V$, then $\dim V \leq k$.
Hint: Similarly to the proof of Proposition 2.23, show that $V \cap E_- = \{\mathbf{0}\}$, where $E_- = \operatorname{span}\{\mathbf{v}_{k+1}, \ldots, \mathbf{v}_n\}$.

2.25. Let A be an $n \times n$ copositive matrix, with eigenvalues

$$\lambda_1 \geq \cdots \geq \lambda_k \geq 0 > \lambda_{k+1} \geq \cdots \geq \lambda_n,$$

$k < n$. Let $\mathbf{v}_1, \ldots, \mathbf{v}_k, \mathbf{v}_{k+1}, \ldots, \mathbf{v}_n$ be corresponding orthonormal eigenvectors. Let $E_+ = \operatorname{span}\{\mathbf{v}_1, \ldots, \mathbf{v}_k\}$ and $E_- = \operatorname{span}\{\mathbf{v}_{k+1}, \ldots, \mathbf{v}_n\}$, and denote by P_+ and P_- the matrices of the orthogonal projection on E_+ and E_-, respectively. Show that for every $\mathbf{x} \in \mathbb{R}^n_+$,

$$\|P_-\mathbf{x}\|^2 \leq -\frac{\lambda_1}{\lambda_{k+1}}\|P_+\mathbf{x}\|^2.$$

Notes. Historically, Theorem 2.20, proved in [Haynsworth and Hoffman (1969)], is the first result on the spectral properties of copositive matrices. In fact, a stronger result was proved there: The spectral radius $\rho(A)$ of a symmetric matrix A is an eigenvalue of A if and only if A is copositive with respect to some self-dual cone. Corollary 2.21 was proved in [Bomze (2008)] (with the second part observed in [Shaked-Monderer (2018)]). Propositions 2.22 and 2.23 are due to [Johnson and Reams (2005)], and so is Exercise 2.23. The result in Exercise 2.21 is due to [Bomze (2008)]. Exercise 2.24 is a result of [Baccari and Naffouti (2016)], and Exercise 2.25 is one of [Naffouti and Baccari (2016)].

2.3 Cones of copositive matrices

The set of $n \times n$ copositive matrices is denoted by $\mathcal{COP}_n$. In this section we show that $\mathcal{COP}_n$ is a closed pointed convex cone, describe its interior and dual, as well as some of its extreme rays. Characterizing the extreme rays of $\mathcal{COP}_n$ is a major open problem, and more results on this problem will appear in subsequent sections of this chapter. Other aspects of the geometry of $\mathcal{COP}_n$ are presented in Section 5.1.

Theorem 2.24. *$\mathcal{COP}_n$ is a full cone.*

Proof. The proof that $\mathcal{COP}_n$ is a closed convex cone is similar to the proof that $\mathcal{PSD}_n$ is a closed convex cone (Theorem 1.166), and is left as an exercise. Since $\mathcal{COP}_n$ contains the solid cone $\mathcal{PSD}_n$, it is also solid.

Finally, if $A \in \mathcal{COP}_n \cap (-\mathcal{COP}_n)$, applying Proposition 2.5 to A and $-A$ we get that for every $i = 1, \ldots, n$ both $a_{ii} \geq 0$ and $-a_{ii} \geq 0$, hence $a_{ii} = 0$. This implies in turn that for every $i \neq j$ both $a_{ij} \geq 0$ and $-a_{ij} \geq 0$, hence $a_{ij} = 0$. □

Theorem 2.25. *A matrix A is in the interior of $\mathcal{COP}_n$ if and only if A is strictly copositive.*

Proof. For $A \in \mathcal{COP}_n$ let $\mu = \min_{\mathbf{x}\in\Gamma} \mathbf{x}^T A\mathbf{x}$, and let $\mathbf{u}$ be a vector in Γ at which this minimum is attained. For any nonzero $\mathbf{x} \geq 0$

$$\mathbf{x}^T A\mathbf{x} \geq \mu\|\mathbf{x}\|^2,$$

and for $B \in \mathcal{S}_n$ with $\|B\| \leq 1$ we have

$$\mathbf{x}^T(A + \varepsilon B)\mathbf{x} \geq \mu\|\mathbf{x}\|^2 - \varepsilon\|\mathbf{x}\|^2 = (\mu - \varepsilon)\|\mathbf{x}\|^2.$$

If A is strictly copositive, then $\mu > 0$ and the above inequality implies that $A + \varepsilon B$ is copositive for every $\|B\| \leq 1$ and sufficiently small $\varepsilon > 0$. If A is not strictly copositive, then $\mu = 0$, attained at $\mathbf{u} \in \Gamma$. For $B = -\mathbf{u}^T\mathbf{u}$ we get that $\mathbf{u}^T(A + \varepsilon B)\mathbf{u} = -\varepsilon < 0$ for every $\varepsilon > 0$. Thus in this case A is not in $\operatorname{int}\mathcal{COP}_n$. □

Theorem 2.26. *The dual of $\mathcal{COP}_n$ is the cone*

$$\mathcal{CP}_n = \operatorname{cone}\{\mathbf{x}\mathbf{x}^T \mid \mathbf{x} \in \mathbb{R}^n_+\}.$$

Proof. In Section 3.2 we will show that $\mathcal{CP}_n$ is a closed convex cone, and its dual is $\mathcal{COP}_n$. This will imply that $\mathcal{COP}_n^* = \mathcal{CP}_n^{**} = \mathcal{CP}_n$. □

Remark 2.27. By Corollary 2.13, the cone $\mathcal{COP}_n$ is invariant under the transformations $A \mapsto P^T AP$, where P is a permutation matrix, and $A \mapsto DAD$, where D is a positive diagonal matrix. If $B = DP^T APD$, where P is a permutation matrix, D is a positive diagonal matrix, we say that B is *in the orbit of A*.

In fact it can, and will, be shown that these are the only cone automorphisms of $\mathcal{COP}_n$. That is, if $\Phi : \mathcal{S}_n \to \mathcal{S}_n$ is an invertible linear transformation mapping $\mathcal{COP}_n$ onto itself, then Φ has the form

$$\Phi(A) = M^T AM,$$

where $M = PD$, where P is a permutation matrix, and D is a positive diagonal matrix — see Section 5.1.

The extreme rays of $\mathcal{COP}_n$ are not fully known, but we list some of them in this section. We refer to a matrix that generates an extreme ray of $\mathcal{COP}_n$ as an *extremal copositive matrix*. But first we mention in the next proposition the basic fact that A is an extremal copositive matrix if and only if any matrix in its orbit is extremal.

Proposition 2.28. *Let A be a symmetric $n \times n$ matrix. Then*

(a) *For an $n \times n$ permutation matrix P, A is an extremal copositive matrix if and only if $P^T AP$ is.*
(b) *For an $n \times n$ positive diagonal matrix D, A is an extremal copositive matrix if and only if DAD is.*

The proof is left as an exercise.

Theorem 2.29. *The following matrices generate extreme rays of $\mathcal{COP}_n$.*

(a) E_{ii}, $i = 1, \ldots, n$.
(b) $E_{ij} + E_{ji}$, $1 \le i < j \le n$.
(c) $\mathbf{x}\mathbf{x}^T$, *where* $\mathbf{x} \in \mathbb{R}^n$ *has some positive entries and some negative entries.*

Proof. If $E_{ii} = A + B$, where A and B are copositive, then for every $j \neq i$,

$$a_{jj} + b_{jj} = 0, \quad a_{jj} \ge 0, \quad b_{jj} \ge 0.$$

Hence $a_{jj} = b_{jj} = 0$. Then, by Corollary 2.5(b), for every $j \neq i$ and any k,

$$a_{jk} + b_{jk} = 0, \quad a_{jk} \ge 0, \quad b_{jk} \ge 0,$$

which implies that $a_{jk} = b_{jk} = 0$. Hence

$$A = a_{ii} E_{ii}, \quad B = b_{ii} E_{ii}.$$

This proves that E_{ii} generates an extreme ray. The proof that $E_{ij} + E_{ji}$, $1 \le i < j \le n$, generates an extreme ray is similar, and is left as an exercise.

We now turn to examine matrices of the form (c). By Proposition 2.28 it suffices to show that $\mathbf{x}\mathbf{x}^T$ generates an extreme ray of $\mathcal{COP}_n$ for every vector $\mathbf{x}$ such that

$$x_1 = \ldots = x_k = 1, \quad x_{k+1} = \ldots = x_m = -1, \quad x_{m+1} = \ldots = x_n = 0. \tag{2.4}$$

Consider first the special case that $\mathbf{x}^T = (1, -1)$. We show that in this case $\mathbf{x}\mathbf{x}^T$ is an extremal matrix in $\mathcal{COP}_2$. Suppose

$$\mathbf{x}\mathbf{x}^T = \begin{pmatrix} 1 & -1 \\ -1 & 1 \end{pmatrix} = A + B,$$

where A and B are copositive. Then without loss of generality we may assume that

$$A = \begin{pmatrix} a & b \\ b & c \end{pmatrix},$$

where $b < 0$ and $a, c \geq 0$. Since $\mathbf{e}^T\mathbf{x}\mathbf{x}^T\mathbf{e} = 0$ and $\mathbf{e}^T A\mathbf{e}$ and $\mathbf{e}^T B\mathbf{e}$ are both nonnegative, we necessarily have $\mathbf{e}^T A\mathbf{e} = 0$. That is,

$$a + 2b + c = 0. \tag{2.5}$$

Hence for every $\mathbf{y} \in \mathbb{R}^2$

$$\mathbf{y}^T A\mathbf{y} = ay_1^2 + 2by_1y_2 - (a+2b)y_2^2 = (y_1 - y_2)(a(y_1 + y_2) + 2by_2). \tag{2.6}$$

Now, if $a + b > 0$, let $\mathbf{y}^T = (1 - \varepsilon, 1)$. By (2.6), $\mathbf{y}^T A\mathbf{y} = -\varepsilon(2a - \varepsilon a + 2b)$ is negative when $\varepsilon > 0$ is sufficiently small, which contradicts the copositivity of A. On the other hand, if $a + b < 0$, then $\mathbf{y}^T = (1 + \varepsilon, 1)$ yields a negative $\mathbf{y}^T A\mathbf{y}$ when $\varepsilon > 0$ is sufficiently small, again a contradiction. Therefore $a + b = 0$ and, due to (2.5), $b + c = 0$. This implies that A (and hence also B) is a scalar multiple of $\mathbf{x}\mathbf{x}^T$.

In the general case $\mathbf{x}$ is as in (2.4) and

$$\mathbf{x}\mathbf{x}^T = \begin{pmatrix} X_{11} & X_{12} & 0 \\ X_{12}^T & X_{22} & 0 \\ 0 & 0 & 0 \end{pmatrix},$$

where $X_{11} = J_k$, $X_{22} = J_{m-k}$, X_{12} is a $k \times (m-k)$ matrix with all entries equal to -1, and each 0 stands for a zero matrix of appropriate order. If $\mathbf{x}\mathbf{x}^T = A + B$ where A and B are copositive, then using the same arguments as in the beginning of the proof, $a_{ij} = b_{ij} = 0$ for every i and every $j > m$. For every $i \leq k$ and $k < j \leq m$,

$$\mathbf{x}\mathbf{x}^T[i,j] = \begin{pmatrix} 1 & -1 \\ -1 & 1 \end{pmatrix}$$

is an extreme matrix in $\mathcal{COP}_2$, and hence

$$A[i,j] = a_{ii} \begin{pmatrix} 1 & -1 \\ -1 & 1 \end{pmatrix}.$$

That is, $a_{ii} = a_{jj} = -a_{ij}$ for every $i < k$ and $k < j \leq m$. Combining all these equalities we get that $A = a_{11}\mathbf{x}\mathbf{x}^T$, and thus $B = (1 - a_{11})\mathbf{x}\mathbf{x}^T$. □

Not every rank 1 positive semidefinite matrix is extremal.

Example 2.30. The matrix $J_n = \mathbf{e}^T\mathbf{e}$ does not generate an extreme ray of $\mathcal{COP}_n$. Indeed,

$$J_n = \mathbf{x}\mathbf{x}^T + 2\sum_{j=2}^{n}(E_{1j} + E_{j1}),$$

where $\mathbf{x}$ is the vector in $\mathbb{R}^n$ with first entry is -1 and all other $n-1$ entries equal to 1.

By Proposition 2.28 all the extreme rays of $\mathcal{N}_n$ and some of the extreme rays of $\mathcal{PSD}_n$ are also extreme rays of $\mathcal{COP}_n$. However, there may be extremal matrices of $\mathcal{COP}_n$ that are not extremal matrices of one of the cones $\mathcal{N}_n$ and $\mathcal{PSD}_n$.

Example 2.31. The Horn matrix (see Example 2.10) generates an extreme ray of $\mathcal{COP}_5$. The proof uses arguments similar to those in the proof of Theorem 2.29, and is left as an exercise.

Remark 2.32. As mentioned in the previous section, every SPN matrix is copositive, thus

$$\mathcal{COP}_n \supseteq \mathcal{SPN}_n = \mathcal{PSD}_n + \mathcal{N}_n.$$

The matrices listed in Theorem 2.29 generate *all* the extreme rays of the cone $\mathcal{SPN}_n$ (see Proposition 1.168).

For $n = 5$, the Horn matrix H is neither nonnegative nor positive semidefinite, and it is extremal, so it cannot be represented as a sum of a positive semidefinite matrix and a symmetric nonnegative matrix. Thus $\mathcal{COP}_5 \neq \mathcal{SPN}_5$. We shall see in Section 2.9 that the equality $\mathcal{COP}_n = \mathcal{SPN}_n$ does hold for $n \leq 4$.

We continue this initial look at extremal copositive matrices with a few observations on such matrices.

Proposition 2.33. *Let $A \in \mathcal{COP}_n$ be an extremal copositive matrix, which has at least one positive diagonal entry. Then*

(a) *If $a_{ii} = 0$ then $a_{ji} = a_{ij} = 0$ for every $j \neq i$.*
(b) *For every $i \neq j$, $a_{ij} = a_{ji} \leq \sqrt{a_{ii}a_{jj}}$.*

Proof. By the assumption, A is not in the orbit of any $E_{ij} + E_{ji}$, $i \neq j$. Let $\alpha = \{1 \leq i \leq n \mid a_{ii} > 0\}$. Without loss of generality we may assume that $\alpha = \{1, \ldots, k\}$, where $k \geq 1$ and $A[\alpha]$ has a positive diagonal.

(a) By Proposition 2.5(a), $A \geq A[\alpha] \oplus 0_{n-k}$. That is, $A = (A[\alpha] \oplus 0_{n-k}) + N$, $N \geq 0$. This contradicts the extremality of A, unless $N = 0$.

(b) By part (a), $A = A[\alpha] \oplus 0$, where $A[\alpha]$ has a positive diagonal. If either $a_{ii} = 0$ or $a_{jj} = 0$, then $a_{ij} = \sqrt{a_{ii}a_{jj}}$ by part (a). Thus it suffices to consider the case that A has a positive diagonal. Suppose $a_{ij} > \sqrt{a_{ii}a_{jj}}$ for some $i \neq j$. Let $D = \operatorname{Diag}\left(1/\sqrt{a_{11}}, \ldots, 1/\sqrt{a_{nn}}\right)$. Then the copositive matrix DAD has diagonal $\mathbf{e}$, and its i, j element is greater than 1. Let $\delta = (DAD)_{ij} - 1 > 0$. By Proposition 2.16 the matrix $DAD - \delta(E_{ij} + E_{ji})$ is also copositive. Thus $A - \delta D^{-1}(E_{ij} + E_{ji})D^{-1}$ is copositive. But since A is not in the orbit of $E_{ij} + E_{ji}$, this contradicts the extremality of A. □

Any extremal matrix in $\mathcal{COP}_n$ can be extended to an extremal matrix in $\mathcal{COP}_m$, $m > n$, in a trivial way. The proof of the next proposition is left as an exercise.

Proposition 2.34. *Let $A \in \mathcal{COP}_n$ be extremal, then $A \oplus 0_m$ is extremal in $\mathcal{COP}_{n+m}$.*

For another way to extend an extremal matrix in $\mathcal{COP}_n$ to an extremal matrix in $\mathcal{COP}_{n+m}$ see Proposition 2.73.

The extremal matrices listed in Proposition 2.28 are either positive semidefinite matrices of rank 1, or nonnegative matrices of rank 2. By the next observation all other extremal matrices have higher ranks (indeed, the Horn matrix has rank 5).

Proposition 2.35. *Let A be an extremal matrix in $\mathcal{COP}_n$. Then*

(a) *If* $\operatorname{rank} A = 1$, *then A is positive semidefinite.*

(b) *If* $\operatorname{rank} A = 2$, *then A is nonnegative.*

Proof. (a) If $\operatorname{rank} A = 1$, the symmetric A is either $\mathbf{x}\mathbf{x}^T$ or $-\mathbf{x}\mathbf{x}^T$ for some $\mathbf{x}$. As A is copositive, its diagonal elements are nonnegative, hence $A = \mathbf{x}\mathbf{x}^T$.

(b) Suppose $\operatorname{rank} A = 2$. By its rank, the extremal A cannot be in $\mathcal{PSD}_n$. Hence it has one negative eigenvalue and one positive eigenvalue, and $A = \mathbf{u}\mathbf{u}^T - \mathbf{v}\mathbf{v}^T$ for some orthogonal $\mathbf{u}, \mathbf{v} \in \mathbb{R}^n$. Thus there are $\mathbf{x}, \mathbf{y} \in \mathbb{R}^n$ such that $A = \mathbf{x}\mathbf{y}^T + \mathbf{y}\mathbf{x}^T$ (e.g., $\mathbf{x} = \frac{1}{2}(\mathbf{u} + \mathbf{v})$ and $\mathbf{y} = \mathbf{u} - \mathbf{v}$). In particular, $a_{ii} \geq 0$ implies that $x_i y_i \geq 0$ for every i. Without loss of generality we may assume that there exists some $k \leq n$ such that we have $x_i y_i > 0$ for $i \leq k$, and $x_i y_i = 0$ for $i > k$. Let $\alpha = \{1, \ldots, k\}$.

By Proposition 2.33, $A = A[\alpha] \oplus 0_{n-k}$. Considering $A[\alpha]$ instead of A we may assume that the diagonal of A is positive, that is, $x_i y_i > 0$ for every i. Using a diagonal scaling of A, we may assume that $x_i y_i = 1$ for every i. Then $\operatorname{supp} \mathbf{x}_+ = \operatorname{supp} \mathbf{y}_+$, $\operatorname{supp} \mathbf{x}_- = \operatorname{supp} \mathbf{y}_-$, $y_i = 1/x_i$ for every i, and we may assume that $\operatorname{supp} \mathbf{x}_+ \neq \emptyset$. If also $\operatorname{supp} \mathbf{x}_- \neq \emptyset$, given any $i \in \operatorname{supp} \mathbf{x}_+$ and $j \in \operatorname{supp} \mathbf{x}_-$, let $t = x_i y_j (< 0)$. Then $y_i x_j = \frac{1}{t}$ and

$$A[i,j] = \begin{pmatrix} 2 & t + \frac{1}{t} \\ t + \frac{1}{t} & 2 \end{pmatrix}.$$

This is a 2×2 copositive matrix, and thus $t + \frac{1}{t} \geq -2$ (Proposition 2.5(c)). Multiplying by the negative t we get that $(t+1)^2 \leq 0$, and thus $t = -1$, and

$$A[i,j] = \begin{pmatrix} 2 & -2 \\ -2 & 2 \end{pmatrix}.$$

Since this holds for every such i and j we get that $\operatorname{rank} A = 1 < 2$, a contradiction. Thus $\mathbf{x}, \mathbf{y} \in \mathbb{R}^n_+$ and $A \geq 0$. $\square$

As the problem of characterizing extremal copositive matrices is hard, it is useful to consider the following concept, which is weaker than extremality.

Definition 2.36. Let $\mathcal{M}$ be a set of $n \times n$ copositive matrices. A matrix $A \in \mathcal{COP}_n$ is *irreducible with respect to* $\mathcal{M}$, or $\mathcal{M}$*-irreducible*, if for every $\varepsilon > 0$ and $M \in \mathcal{M}$ the matrix $A - \varepsilon M \notin \mathcal{COP}_n$.

If $\mathcal{M}$ consists of a single matrix M we also say that A is *irreducible with respect to* M, or M*-irreducible*. When $M = E_{ij} + E_{ji}$ we also say that A is $\{i, j\}$*-irreducible*. When $\mathcal{M}$ is one of the cones $\mathcal{N}_n$, $\mathcal{HN}_n$ or $\mathcal{PSD}_n$, we omit the subscript n. All these notions should not be confused with irreducibility of the matrix in the usual sense. Obviously, $A \in \mathcal{COP}_n$ is extremal if and only if it is M-irreducible for every $M \in \mathcal{COP}_n$ which is not a scalar multiple of A.

Example 2.37. Consider a 2×2 copositive matrix

$$A = \begin{pmatrix} a & b \\ b & c \end{pmatrix}.$$

The matrix A is $\{1, 1\}$-irreducible if and only if either $a = 0$ and $b, c \geq 0$, or $a, c > 0 > b$ and $ac = b^2$; that is, if and only if A is either nonnegative with $a = 0$, or positive semidefinite of rank 1 with $b < 0$. To see that, suppose that A is $\{1,1\}$-irreducible. Recall that A is either nonnegative or positive semidefinite. In the first case it is obvious that $\{1,1\}$-irreducibility is equivalent to $a=0$. If A is not nonnegative, then A is positive semidefinite with $b<0$ and $a,c>0$. For every $\varepsilon>0$ the matrix $A-\varepsilon E_{11}$ is not copositive, and in particular not positive semidefinite, thus $(a-\varepsilon)c-b^2<0$, that is $ac-b^2<\varepsilon c$ for every $\varepsilon>0$, so $\det A=0$ and A has rank 1. The reverse implication is simple, as a rank 1 positive semidefinite matrix A, which is not nonnegative, is extremal and thus $\{1,1\}$-irreducible.

Similarly, it can be seen that A is $\{1,2\}$-irreducible if and only if $b=-\sqrt{ac}$, where $a,c\geq 0$; that is, either A has a zero row (column), or it is a rank 1 positive semidefinite matrix with $b<0$.

We leave the proof of the next two observations as exercises.

Proposition 2.38. *Let $A\in\mathcal{COP}_n$, and let $\mathcal{M}$ a set of copositive matrices. If $\mathcal{M}'\subseteq\mathcal{M}$ and A is $\mathcal{M}$-irreducible, then A is also $\mathcal{M}'$-irreducible.*

Proposition 2.39. *Let $A\in\mathcal{COP}_n$, and let $\mathcal{K}\subseteq\mathcal{COP}_n$ be a convex cone. Then A is $\mathcal{K}$-irreducible if and only if A is X-irreducible for every matrix X, which generates an extreme ray of $\mathcal{K}$.*

Remark 2.40. If a matrix $A\in\mathcal{COP}_n$ is extremal, or M-irreducible for some nonzero symmetric $M\in\mathcal{COP}_n$, then A is not in the interior of $\mathcal{COP}_n$. By Proposition 2.25, such A is not strictly copositive. Hence there exists a nonzero vector $\mathbf{u}\in\mathbb{R}^n_+$ such that $\mathbf{u}^TA\mathbf{u}=0$.

It turns out that nonnegative vectors at which the quadratic form vanishes are very useful in the study of copositive matrices and extremal copositive matrices. We discuss these vectors and their relation to extremality and $\mathcal{M}$-irreducibility in detail in the next sections.

Exercises

2.26. Show that for $1\leq i<j\leq n$ the matrix $E_{ij}+E_{ji}$ generates an extreme ray of $\mathcal{COP}_n$.

2.27. Using the fact that the matrix

$$\begin{pmatrix} 1 & -1 & 1 \\ -1 & 1 & -1 \\ 1 & -1 & 1 \end{pmatrix}$$

is extremal in $\mathcal{COP}_3$, show that the Horn matrix H of Example 2.10 generates an extreme ray of $\mathcal{COP}_5$.

2.28. Prove Proposition 2.34.

2.29. Prove Proposition 2.38.

2.30. Prove Proposition 2.39.

2.31. Show that a nonnegative matrix A is $\{i,i\}$-irreducible if and only if $a_{ii} = 0$.

2.32. Complete the missing details in Example 2.37. In particular, prove the characterization of $\{1,2\}$-irreducible 2×2 copositive matrices.

2.33. Show that if A is a copositive matrix, $\{i,j\} \subseteq \alpha$, and $A[\alpha]$ is $\{i,j\}$-irreducible, then A is $\{i,j\}$-irreducible.

2.34. Show that the matrix

$$A = \begin{pmatrix} 1 & 0 & 0 \\ 0 & 1 & -1 \\ 0 & -1 & 1 \end{pmatrix}$$

is $\{1,2\}$-irreducible.
Hint: For every $\varepsilon > 0$ let $A_\varepsilon = A - \varepsilon(E_{12} + E_{21})$ and consider $\mathbf{u}_\varepsilon = (\varepsilon, 1, 1)^T$.

2.35. Show that if an SPN matrix is $\mathcal{HN}$-irreducible, then it is positive semidefinite. Give an example of a positive semidefinite matrix that is not $\mathcal{HN}$-irreducible.

2.36. Let A be an $n \times n$ copositive matrix, $n \geq 3$. Show that if $A(i)$, $A(j)$ and $A(k)$ are $\mathcal{HN}$-irreducible for three different indices i, j, k, then A is $\mathcal{HN}$-irreducible.
Hint: Exercise 2.33.

Notes. The discussion of the basic facts regarding $\mathcal{COP}_n$, including its duality to $\mathcal{CP}_n$ and the description of some of its extremal rays (the positive semidefinite ones, the nonnegative ones, and the one generated by the Horn matrix in $\mathcal{COP}_5$), are based on [Hall and Newman (1963)] and [Hall (1998)]. Proposition 2.33 appeared first in [Diananda (1962)]. The terminology of an exceptional matrix is due to [Johnson and Reams (2008)]. The terminology of a copositive matrix being "in the orbit" of another copositive matrix, first used in [Hildebrand (2012)], stems from viewing the transformations $A \mapsto P^TAP$, where P is a permutation matrix, and $A \mapsto DAD$, where D is a positive diagonal matrix, as actions of the automorphism group of $\mathcal{COP}_n$ on the matrices in this cone. This automorphism group was first described in [Gowda *et al.* (2013)] (in the general setting of a cone of matrices copositive with respect to a given cone). See also [Shitov (2019)]. Proposition 2.35 was proved in [Shaked-Monderer *et al.* (2015)].

The terminology of $\mathcal{M}$-irreducibility was introduced in [Dickinson *et al.* (2013a)], but the idea of considering $\mathcal{HN}$-irreducibility and $\mathcal{N}$-irreducibility instead of the harder notion of extremality dates back to [Diananda (1962)], [Hall and Newman (1963)] and [Baumert (1965, 1966, 1967)].

2.4 Zeros of copositive matrices

Who do you call a zero? In our terminology, a zero of a matrix A is a nonzero nonnegative vector $\mathbf{x}$ at which the quadratic form $\mathbf{x}^T A\mathbf{x}$ vanishes.

Definition 2.41. Let A be an $n \times n$ symmetric matrix. A *zero of* A is a nonzero vector $\mathbf{u} \in \mathbb{R}^n_+$ such that $\mathbf{u}^T A\mathbf{u} = 0$. The set of all zeros of A is denoted by $\mathcal{V}^A$.

We mentioned in the previous section that every matrix A on the boundary of the cone $\mathcal{COP}_n$ has a zero, and that zeros are instrumental in the attempt to characterize extremal copositive matrices. But note that while our main interest is in copositive matrices and their zeros, we use the term and the notation for general symmetric matrices.

Example 2.42. The zeros of the Horn matrix

$$H = \begin{pmatrix} 1 & -1 & 1 & 1 & -1 \\ -1 & 1 & -1 & 1 & 1 \\ 1 & -1 & 1 & -1 & 1 \\ 1 & 1 & -1 & 1 & -1 \\ -1 & 1 & 1 & -1 & 1 \end{pmatrix}$$

include the vectors $\mathbf{u}_i = \mathbf{e}_i + \mathbf{e}_{i \dotplus 1}$, $i = 1, \dots, 5$, where $\dotplus$ denotes addition modulo 5 on $\{1, \dots, 5\}$. Of course, every positive scalar multiple of $\mathbf{u}_i$ is also a zero of H, but in addition, for every $t, s > 0$ the vector $s\mathbf{u}_i + t\mathbf{u}_{i \dotplus 1}$ is also a zero of H, $i = 1, \dots, 5$.

Remark 2.43. For any copositive matrix A and $t > 0$, $\mathbf{u}$ is a zero of A if and only if $t\mathbf{u}$ is a zero of A. Thus $\mathcal{V}^A$ can be split into equivalence classes with respect to multiplication by a positive scalar. The set $\mathcal{V}^A \cap \Delta$ is a complete set of representatives for $\mathcal{V}^A$, but other sets of representatives may be considered. Note that some copositive matrices may have infinitely many non-equivalent zeros. For example, the Horn matrix H has infinitely many zeros in $\mathcal{V}^H \cap \Delta$.

Remark 2.44. If A is an $n \times n$ symmetric matrix, and $\mathbf{u} \in \mathbb{R}^n_+$ has $\operatorname{supp} \mathbf{u} \subseteq \alpha$, then $\mathbf{u}^T A \mathbf{u} = \mathbf{u}[\alpha]^T A[\alpha] \mathbf{u}[\alpha]$. Hence $\mathbf{u}$ is a zero of A if and only if $\mathbf{u}[\alpha]$ is a zero of $A[\alpha]$.

We continue with some basic facts about zeros of copositive matrices.

Proposition 2.45. *Let A be an $n \times n$ copositive matrix, and let $\mathbf{u} \in \mathbb{R}^n_+$ be a zero of A. Then $A\mathbf{u} \geq \mathbf{0}$. Moreover, $(A\mathbf{u})_i = 0$ for every $i \in \operatorname{supp} \mathbf{u}$.*

Proof. For any fixed $1 \leq i \leq n$, $(\mathbf{u}^T + t\mathbf{e}_i^T)A(\mathbf{u} + t\mathbf{e}_i) \geq 0$ for every $t > 0$, i.e., $2t\mathbf{e}_i^T A\mathbf{u} + t^2 a_{ii} \geq 0$. Thus $\mathbf{e}_i^T A\mathbf{u} \geq -\frac{t}{2} a_{ii}$ for every $t > 0$, implying that $(A\mathbf{u})_i \geq 0$.

The second assertion follows trivially from the fact that $\mathbf{u}^T A\mathbf{u} = 0$ is the inner product of the two nonnegative vectors $\mathbf{u}$ and $A\mathbf{u}$. □

The next proposition is a strengthening of the last part of the Proposition 2.45. The proof is left as an exercise.

Proposition 2.46. *Let $\mathbf{u}$ be a zero of a copositive matrix A. If $\operatorname{supp} \mathbf{u} \subseteq \alpha$ and the submatrix $A[\alpha]$ is positive semidefinite, then $(A\mathbf{u})_i = 0$ for every $i \in \alpha$.*

To see why this is indeed a strengthening, see the following proposition and its corollary.

Proposition 2.47. *If a copositive matrix A has a positive zero, then A is positive semidefinite and singular.*

Proof. Let $\mathbf{u}$ be a positive zero of A. Then $A\mathbf{u} = \mathbf{0}$ by Proposition 2.45, which proves the singularity of A. For every $\mathbf{x} \in \mathbb{R}^n$, $t\mathbf{u} + \mathbf{x} > 0$ if $t > 0$ is sufficiently large, and thus

$$0 \leq (t\mathbf{u}^T + \mathbf{x}^T)A(t\mathbf{u} + \mathbf{x}) = t^2 \mathbf{u}^T A\mathbf{u} + 2t\mathbf{x}^T A\mathbf{u} + \mathbf{x}^T A\mathbf{x} = \mathbf{x}^T A\mathbf{x}. \qquad \square$$

From Proposition 2.47 and Remark 2.44 we get the following.

Corollary 2.48. *Let A be an $n \times n$ copositive matrix, and let $\mathbf{u} \in \mathbb{R}^n_+$ be a zero of A. Then $A[\operatorname{supp} \mathbf{u}]$ is positive semidefinite and singular.*

Remark 2.49. There is another approach that can be used to prove Proposition 2.45. Denote $\eta = \min_{\mathbf{x} \in \Delta} \frac{1}{2}\mathbf{x}^T A\mathbf{x}$. This minimum is attained at some

$\mathbf{u} \in \Delta$. By the Karush-Kuhn-Tucker Theorem (see Example 1.173 in Section 1.7), there exist $\boldsymbol{\mu} \in \mathbb{R}^n_+$ and $\lambda \in \mathbb{R}$ such that $\boldsymbol{\mu}^T\mathbf{u} = 0$ and

$$A\mathbf{u} + \lambda\mathbf{e} - \boldsymbol{\mu} = \mathbf{0}.$$

Multiplying by $\mathbf{u}^T$ we get that

$$\mathbf{u}^T A\mathbf{u} + \lambda = 0,$$

so $\lambda = -2\eta \leq 0$, as A is copositive, and

$$A\mathbf{u} = \boldsymbol{\mu} - \lambda\mathbf{e} \geq 2\eta\mathbf{e} \geq \mathbf{0}.$$

Moreover, for every $i \in \operatorname{supp}\mathbf{u}$ we have $\mu_i = 0$, and therefore $(A\mathbf{u})_i = 2\eta$. Under the conditions of Proposition 2.45, $\mathbf{u}$ is a zero of the copositive matrix A, so $\eta = 0$.

This also proves Proposition 2.47. However, for the latter we can also argue as follows. A positive zero $\mathbf{u}$ is in the interior of $\mathbb{R}^n$. Hence if $\mathbf{u}$ is a zero of the copositive matrix A, it is a local minimum of $\frac{1}{2}\mathbf{x}^T A\mathbf{x}$. The gradient $\nabla\left(\frac{1}{2}\mathbf{x}^T A\mathbf{x}\right) = A\mathbf{x}$ therefore vanishes at $\mathbf{u}$, i.e. $A\mathbf{u} = \mathbf{0}$.

By Corollary 2.48, each zero of a copositive A is associated with a positive semidefinite principal submatrix of A. Now let A be a copositive matrix, and suppose $A[\alpha]$ is positive semidefinite. Let

$$K_\alpha = \{\mathbf{x} \in \mathcal{V}^A \mid \operatorname{supp}\mathbf{x} \subseteq \alpha\} \cup \{\mathbf{0}\},$$

the set of all zeros of A whose support is contained in α, and the zero vector.

Proposition 2.50. *Let A be an $n \times n$ symmetric matrix, and suppose $A[\alpha]$ is positive semidefinite. Then every nonempty K_α is a closed convex cone.*

Proof. Assume without loss of generality that $\alpha = \{1, \dots, m\}$. Note that $\mathbf{x} \in K_\alpha$ if and only if $\mathbf{x}[\alpha] \in \mathbb{R}^m_+ \cap \operatorname{ns} A[\alpha]$, and all the other entries of $\mathbf{x}$ are zeros. As $\mathbb{R}^m_+ \cap \operatorname{ns} A[\alpha]$ is a closed convex cone , so is K_α. (It is a polyhedral cone, as $\mathbb{R}^m_+ \cap \operatorname{ns} A[\alpha]$ is the intersection of the polyhedral cone $\mathbb{R}^m_+$ with a linear subspace.) □

Remark 2.51. By Corollary 2.48, each zero of a copositive A belongs to K_α for some set α such that $A[\alpha]$ is positive semidefinite. As A has a finite number of positive semidefinite principal submatrices, this means that $\mathcal{V}^A \cup \{\mathbf{0}\}$ is the union of a finite number of closed convex cones. Note that for a copositive A the set $\mathcal{V}^A \cup \{\mathbf{0}\}$ itself is not necessarily convex (for example, $\mathbf{u}_1$ and $\mathbf{u}_3$ of Example 2.42 are zeros of the Horn matrix, but $\mathbf{u}_1 + \mathbf{u}_3$ is not). Compare this to the case that A is an $n \times n$ positive semidefinite matrix, in which $\mathcal{V}^A \cup \{\mathbf{0}\}$ is a closed convex cone (in fact, a polyhedral cone), as the intersection of the closed convex cone $\mathbb{R}^n_+$ with the null space of A, $\operatorname{ns} A$.

Minimal zeros

Zeros of copositive matrices whose support is minimal, in the sense described below, are of particular interest.

Definition 2.52. Let A be an $n \times n$ copositive matrix. A zero $\mathbf{u}$ of A is called a *minimal zero of A* if there exists no zero of A whose support is a proper subset of $\operatorname{supp} \mathbf{u}$. The set of all minimal zeros of A is denoted by $\mathcal{V}^A_{\min}$.

A set $\sigma \subseteq \{1, \dots, n\}$ that is the support set of a minimal zero is called a *minimal support* of A.

Example 2.53. Each of the vectors $\mathbf{u}_i$, $i = 1, \dots, 5$, in Example 2.42 is a zero of the Horn matrix H with support of size 2. Since the diagonal entries of H are all positive, H has no zero with support of size 1. Thus each of the $\mathbf{u}_i$'s is a minimal zero of H. The index sets $\{i, i \oplus 1\}$, $i = 1, \dots, 5$, are minimal supports of H.

In the following proposition $K_\alpha = \{\mathbf{x} \in \mathcal{V}^A \mid \operatorname{supp} \mathbf{x} \subseteq \alpha\} \cup \{\mathbf{0}\}$, as defined above.

Proposition 2.54. *Let A be an $n \times n$ copositive matrix, and $A[\alpha]$ a positive semidefinite principal submatrix of A. Let $\mathbf{u}$ be a zero of A with* $\operatorname{supp} \mathbf{u} = \sigma \subseteq \alpha$. *Then the following are equivalent.*

(a) $\mathbf{u}$ *is a minimal zero.*
(b) *If $\mathbf{v}$ is a zero such that* $\operatorname{supp} \mathbf{v} \subseteq \operatorname{supp} \mathbf{u}$, *then $\mathbf{v} = t\mathbf{u}$ for some $t > 0$.*
(c) *$A[\sigma]$ is positive semidefinite of nullity 1, and its null space is spanned by a single positive vector.*
(d) $\mathbf{u}$ *generates an extreme ray of the cone K_α.*

Proof. (a) $\Rightarrow$ (b): Let $\mathbf{v}$ be any zero whose support is a subset of σ. By the minimality of $\mathbf{u}$, $\operatorname{supp} \mathbf{v} = \sigma$. Let $t = \min_{i \in \sigma} \frac{u_i}{v_i} > 0$. The vector $\mathbf{w} = \mathbf{u} - t\mathbf{v}$ is nonnegative and $w_k = 0$ for some $k \in \sigma$. That is, $\operatorname{supp} \mathbf{w}$ is a proper subset of σ. As $(A\mathbf{v})_i = 0$ for every $i \in \sigma$ by Proposition 2.46, $\mathbf{v}^T A\mathbf{u} = 0$, and therefore

$$\mathbf{w}^T A\mathbf{w} = \mathbf{u}^T A\mathbf{u} - 2t\mathbf{v}^T A\mathbf{u} + t^2\mathbf{v}^T A\mathbf{v} = 0.$$

Since the support of any zero of A cannot be a proper subset of σ, $\operatorname{supp}\mathbf{w}$ has to be empty and $\mathbf{w} = \mathbf{0}$, implying that $\mathbf{v}$ is a scalar multiple of $\mathbf{u}$.

(b) $\Rightarrow$ (c): By Proposition 2.48 the principal submatrix $A[\sigma]$ is positive semidefinite and singular, and $\mathbf{u}[\sigma]$ is a positive eigenvector corresponding to the zero eigenvalue. To see that 0 is a simple eigenvalue of $A[\sigma]$, suppose $\mathbf{x} \in \operatorname{ns} A[\sigma]$. Then there exists a scalar t such that the vector $\mathbf{u}[\sigma] - t\mathbf{x} \in \operatorname{ns} A[\sigma]$ is nonnegative, and has at least one zero entry. Let $\mathbf{v}$ be the vector in $\mathbb{R}^n$ obtained by appending zero entries to $\mathbf{u}[\sigma] - t\mathbf{x}$. Then $\mathbf{v}^T A\mathbf{v} = 0$ and $\operatorname{supp}\mathbf{v}$ is a proper subset of σ. By the minimality of $\mathbf{u}$, $\mathbf{v} = \mathbf{0}$ and $\mathbf{x}$ is a multiple of $\mathbf{u}[\sigma]$.

(c) $\Rightarrow$ (d): Suppose $\mathbf{u} = \mathbf{v} + \mathbf{w}$, where $\mathbf{v}, \mathbf{w} \in K_\alpha$. Then $\operatorname{supp}\mathbf{v}$ and $\operatorname{supp}\mathbf{w}$ are subsets of σ. Since $A[\sigma]$ is positive semidefinite, Proposition 2.46 implies that $A[\sigma]\mathbf{v}[\sigma] = A[\sigma]\mathbf{w}[\sigma] = A[\sigma]\mathbf{u}[\sigma] = \mathbf{0}$. As $\mathbf{u}[\sigma]$ spans $\operatorname{ns} A[\sigma]$, $\mathbf{v}[\sigma]$ and $\mathbf{w}[\sigma]$ are scalar multiples of $\mathbf{u}[\sigma]$. That is, both $\mathbf{v}$ and $\mathbf{w}$ are scalar multiples of $\mathbf{u}$.

(d) $\Rightarrow$ (a): Suppose $\mathbf{v}$ is a zero of A with $\operatorname{supp}\mathbf{v}$ is a proper subset of σ, then $(A\mathbf{v})_i = 0$ for every $i \in \sigma$ by Proposition 2.46. Hence $\mathbf{u}^T A\mathbf{v} = 0$ and $(\mathbf{u} + t\mathbf{v})^T A(\mathbf{u} + t\mathbf{v}) = 0$ for every real t. For small enough $\varepsilon > 0$ each of $\mathbf{u} \pm \varepsilon\mathbf{v}$ is a nonnegative vector, whose support is a subset of α. For such ε $\mathbf{u} \pm \varepsilon\mathbf{v} \in K_\alpha$ and their sum is $\mathbf{u}$. By the extremality of $\mathbf{u}$, this implies that both $\mathbf{u} + \varepsilon\mathbf{v}$ and $\mathbf{u} - \varepsilon\mathbf{v}$, and therefore also $\mathbf{v}$, are scalar multiples of $\mathbf{u}$. This contradicts the assumption that $\operatorname{supp}\mathbf{v}$ is a proper subset of σ. □

We stress some of the implications of the last proposition for a copositive A. By the equivalence of (a) and (c), the minimal supports completely determine (up to scalar multiplication) the minimal zeros, and they correspond to singular positive semidefinite principal submatrices of nullity 1, with a positive eigenvector corresponding to the zero eigenvalue. The following two corollaries therefore follow easily from Proposition 2.54. We leave both proofs as an exercise.

Corollary 2.55. *A copositive matrix A has a finite number of minimal zeros, up to multiplication by scalar.*

For a zero $\mathbf{u}$ of a copositive matrix A, we may apply Proposition 2.54 to $\alpha = \operatorname{supp}\mathbf{u}$, and use Carathéodory's Theorem (Theorem 1.132) to deduce the following.

Corollary 2.56. *If $\mathbf{u}$ is a zero of a copositive matrix A, then $\mathbf{u}$ is a finite sum of minimal zeros of A.*

The two corollaries above provide a clue to the importance of minimal zeros. The next proposition generalizes, in a sense, part (b) in Proposition 2.54.

Proposition 2.57. *Let A be an $n \times n$ copositive matrix, and let $A[\alpha]$ be positive definite for some $\alpha \subseteq \{1,\ldots,n\}$. Let $\mathbf{u}$ be a zero of A such that $|\operatorname{supp}\mathbf{u} \setminus \alpha| = 1$. Then $\mathbf{u}$ is a minimal zero, and if $\mathbf{v}$ is a zero of A such that $\operatorname{supp}\mathbf{v} \setminus \alpha = \operatorname{supp}\mathbf{u} \setminus \alpha$, then $\mathbf{v}$ is a scalar multiple of $\mathbf{u}$.*

Proof. We may assume without loss of generality that $\alpha = \{1,\ldots,k\}$ and $\operatorname{supp}\mathbf{u} \setminus \alpha = \{k+1\}$, and that $u_{k+1} = 1$.

$$A[\alpha \cup \{k+1\}] = \begin{pmatrix} B & \mathbf{c} \\ \mathbf{c}^T & d \end{pmatrix},$$

where $B = A[\alpha]$. Denote $\mathbf{x} = \mathbf{u}[\alpha]$. Since $\mathbf{u}$ is a zero of A and $u_{k+1} = 1$, we have $(A\mathbf{u})_{k+1} = \mathbf{c}^T\mathbf{x} + d = 0$ and therefore

$$0 = \mathbf{u}^T A\mathbf{u} = \mathbf{x}^T B\mathbf{x} + 2\mathbf{c}^T\mathbf{x} + d = \mathbf{x}^T B\mathbf{x} - d,$$

that is, $d = \mathbf{x}^T B\mathbf{x}$.

Let $\sigma = \operatorname{supp}\mathbf{u}$. As $A[\sigma \cap \alpha]$ is positive definite, $\mathbf{u}$ is a minimal zero by the equivalence of (a) and (c) in Proposition 2.54.

Now suppose a zero $\mathbf{v}$ satisfies $\operatorname{supp}\mathbf{v} \setminus \alpha = \{k+1\}$. It suffices to show that if $v_{k+1} = 1$, then $\mathbf{v} = \mathbf{u}$. So assume $v_{k+1} = 1$ and let $\mathbf{y} = \mathbf{v}[\alpha]$. As above, $\mathbf{y}^T B\mathbf{y} = d$. Also $(A\mathbf{v})[\alpha] = B\mathbf{y} + \mathbf{c} \geq \mathbf{0}$, and as $\mathbf{x}$ is nonnegative, this implies that $\mathbf{c}^T\mathbf{x} \geq -\mathbf{y}^T B\mathbf{x}$. Therefore

$$\begin{aligned} 0 = \mathbf{u}^T A\mathbf{u} &= \mathbf{x}^T B\mathbf{x} + 2\mathbf{c}^T\mathbf{x} + d \\ &\geq \mathbf{x}^T B\mathbf{x} - 2\mathbf{y}^T B\mathbf{x} + \mathbf{y}^T B\mathbf{y} = (\mathbf{x}-\mathbf{y})^T B(\mathbf{x}-\mathbf{y}) \geq 0. \end{aligned}$$

Since B is positive definite, this implies that $\mathbf{x} = \mathbf{y}$. Let $\mathbf{w} = \mathbf{u} - \mathbf{v}$. Then $\operatorname{supp}\mathbf{w} \subseteq \alpha$ and $\mathbf{w}[\alpha] = \mathbf{x} - \mathbf{y} = \mathbf{0}$, so $\mathbf{w} = \mathbf{0}$. □

The next proposition extends the previous result.

Proposition 2.58. *Let A be an $n \times n$ copositive matrix, and let $A[\alpha]$ be positive definite for some $\alpha \subseteq \{1,\ldots,n\}$. Suppose that $\mathbf{u}_1,\ldots,\mathbf{u}_m$ are mutually different zeros of A such that $\operatorname{supp}\mathbf{u}_i \setminus \alpha = p_i$, $i = 1,\ldots,m$, and $\operatorname{supp}\mathbf{u}_i \cap \alpha \subseteq \operatorname{supp}\mathbf{u}_{i+1} \cap \alpha$. Then $\mathbf{u}_1,\ldots,\mathbf{u}_m$ are minimal zeros, $p_1,\ldots,p_m$ are mutually different indices, and any zero $\mathbf{v}$ of A such that $\operatorname{supp}\mathbf{v} \subseteq \alpha \cup \{p_1,\ldots,p_m\}$ is a nonnegative linear combination of $\mathbf{u}_1,\ldots,\mathbf{u}_m$.*

In particular, $\mathbf{u}_1,\ldots,\mathbf{u}_m$ are the only minimal zeros whose support is contained in $\alpha \cup \{p_1,\ldots,p_m\}$.

Proof. The vectors $\mathbf{u}_1, \ldots, \mathbf{u}_m$ are minimal zeros and $p_1, \ldots, p_m$ are mutually different by the previous proposition. We may assume without loss of generality that $(\mathbf{u}_i)_{p_i} = 1$, $i = 1, \ldots, m$. Since we are considering zeros with support contained in $\alpha \cup \{p_1, \ldots, p_m\}$, we may also assume that A is of order $|\alpha| + m$, and reorder its rows and columns so that $\alpha = \{1, \ldots, k\}$ and $p_i = k + i$ for every i. Then

$$A = \begin{pmatrix} B & C \\ C^T & E \end{pmatrix},$$

where $B = A[\alpha]$ is $k \times k$, E is $m \times m$ and C is $k \times m$, and we may write

$$\mathbf{u}_i = \begin{pmatrix} \mathbf{x}_i \\ \mathbf{e}_i \end{pmatrix}, \quad i = 1, \ldots, m,$$

where $\mathbf{x}_i \in \mathbb{R}^k_+$ and $\mathbf{e}_i$ are the standard basis vectors in $\mathbb{R}^m$. We also denote $X = (\mathbf{x}_1 \ldots \mathbf{x}_m)$.

Since the columns of $\begin{pmatrix} X \\ I_m \end{pmatrix}$ are zeros of A, we have $A \begin{pmatrix} X \\ I_m \end{pmatrix} \geq 0$. Thus $BX + C \geq 0$, and for every $\mathbf{x} \in \mathbb{R}^k_+$ and $\mathbf{y} \in \mathbb{R}^m_+$ this implies that

$$\mathbf{x}^T C \mathbf{y} \geq -\mathbf{x}^T B X \mathbf{y}. \tag{2.7}$$

For every $l \in \operatorname{supp} \mathbf{x}_j$,

$$(A\mathbf{u}_j)_l = (B\mathbf{x}_j + C\mathbf{e}_j)_l = 0.$$

Hence for $1 \leq i \leq j \leq m$ the inclusion $\operatorname{supp} \mathbf{x}_i \subseteq \operatorname{supp} \mathbf{x}_j$ implies

$$\mathbf{x}_i^T (B\mathbf{x}_j + C\mathbf{e}_j) = 0,$$

that is,

$$-\mathbf{x}_i^T C \mathbf{e}_j = \mathbf{x}_i^T B \mathbf{x}_j. \tag{2.8}$$

On the other hand, $(A\mathbf{u}_i)_{k+j} \geq 0$ means that

$$(C^T \mathbf{x}_i + E\mathbf{e}_i)_j \geq 0,$$

that is,

$$e_{ij} \geq -\mathbf{x}_i^T C \mathbf{e}_j. \tag{2.9}$$

Combining (2.8) and (2.9) we get that $e_{ij} \geq \mathbf{x}_i^T B \mathbf{x}_j$ for $1 \leq i \leq j \leq m$, and since E and $X^T B X$ are both symmetric we conclude that

$$E \geq X^T B X. \tag{2.10}$$

Let $\mathbf{v}$ be a zero of A, and write it as

$$\mathbf{v} = \begin{pmatrix} \mathbf{x} \\ \mathbf{y} \end{pmatrix},$$

where $\mathbf{x} \in \mathbb{R}^k_+$ and $\mathbf{y} \in \mathbb{R}^m_+$. Then

$$\begin{aligned} 0 = \mathbf{v}^T A \mathbf{v} &= \mathbf{x}^T B \mathbf{x} + 2\mathbf{x}^T C \mathbf{y} + \mathbf{y}^T E \mathbf{y} \\ &\geq \mathbf{x}^T B \mathbf{x} - 2\mathbf{x}^T B X \mathbf{y} + \mathbf{y}^T X^T B X \mathbf{y} = (\mathbf{x} - X\mathbf{y})^T B (\mathbf{x} - X\mathbf{y}), \end{aligned}$$

where the inequality follows from (2.7) and (2.10). Since B is positive definite, this last inequality implies that $\mathbf{x} = X\mathbf{y}$. Hence

$$\mathbf{v} = \begin{pmatrix} \mathbf{x} \\ \mathbf{y} \end{pmatrix} = \begin{pmatrix} X\mathbf{y} \\ \mathbf{y} \end{pmatrix} = \begin{pmatrix} X \\ I \end{pmatrix} \mathbf{y} = \sum_{i=1}^{m} y_i \mathbf{u}_i. \qquad \square$$

The proof of the following corollary is left as an exercise.

Corollary 2.59. *Let A be a copositive matrix and $\mathbf{u}_1, \mathbf{u}_2$ minimal zeros of A with supports σ_1 and σ_2, respectively. If $\sigma_2 \setminus \sigma_1$ consists of a single index, then every zero $\mathbf{u}$ of A with support contained in $\sigma_1 \cup \sigma_2$ is a nonnegative combination of $\mathbf{u}_1$ and $\mathbf{u}_2$. In particular, up to multiplication by a positive scalar, $\mathbf{u}_1$ and $\mathbf{u}_2$ are the only minimal zeros with support contained in $\sigma_1 \cup \sigma_2$.*

Consider an $n \times n$ copositive matrix A with a positive diagonal. By diagonal scaling we may assume that A has diagonal $\mathbf{e}$. Each minimal support of A is of size at least 2. There exists a zero with support $\{i, j\}$ if and only if $a_{ij} = -1$, and then we may take the zero to be $\mathbf{e}_i + \mathbf{e}_j$. If $\mathbf{u}_1, \ldots, \mathbf{u}_r$ are all such minimal zeros of A with support of size 2, then the matrix $U = (\mathbf{u}_1 \ \ldots \ \mathbf{u}_r)$ is the incidence matrix of the graph $G_{-1}(A)$ whose vertices are $\{1, \ldots, n\}$ and its edges correspond to the -1 entries of A. Thus $\operatorname{rank} U = n - k$, where k is the number of bipartite components of $G_{-1}(A)$ (see Theorem 1.102). The next proposition adds some more information in the case that the minimal zeros of A (with support of any size) span $\mathbb{R}^n$.

Proposition 2.60. *Let A be an $n \times n$ copositive matrix with diagonal $\mathbf{e}$, and suppose the minimal zeros of A span $\mathbb{R}^n$. Let $\tau_1, \ldots, \tau_k$ be the vertex sets of the components of $G_{-1}(A)$ that are bipartite. Then there exist $p_j \in \tau_j$, $j = 1, \ldots, k$, and k different minimal supports $\sigma_1, \ldots, \sigma_k$ of A, each of size strictly greater than 2, such that $p_j \in \sigma_j$, $j = 1, \ldots, k$.*

Proof. Let $\mathbf{u}_1, \ldots, \mathbf{u}_s$ be s linearly independent minimal zeros of A with support of size 2, where $s = n - k$, and the nonzero entries of each $\mathbf{u}_i$ are equal to 1. Complete these s vectors to a basis of $\mathbb{R}^n$ by minimal zeros with supports of size greater than 2, $\mathbf{u}_{s+1}, \ldots, \mathbf{u}_n$. Then $U = (\mathbf{u}_1 \ \ldots \ \mathbf{u}_n)$ is a nonsingular matrix.

Let $\alpha = \{1, \dots, s\}$ and expand $\det U$ along the columns α. Then

$$\det U = (-1)^{\frac{s(s+1)}{2}} \sum_{\substack{\beta \subseteq \{1,\dots,n\} \\ |\beta|=s}} (-1)^{s(\beta)} \det U[\beta|\alpha] \det U(\beta|\alpha),$$

where $s(\beta)$ is the sum of the elements of β. As $\det U \neq 0$, there exists $\beta \subseteq \{1, \dots, n\}$, $|\beta| = s$, such that $\det U[\beta|\alpha] \det U(\beta|\alpha) \neq 0$. By permutation of the rows of U we may assume that $\beta = \alpha$, so $U[\alpha]$ and $U(\alpha)$ are nonsingular. The first s columns of U form a submatrix of the incidence matrix of $G_{-1}(A)$, and the vertex set α of linearly independent rows cannot contain any vertex set of a connected bipartite component of $G_{-1}(A)$. By permuting the last $k = n - s$ rows we may assume that for $j = 1, \dots, k$, $p_j = s+j \in \tau_j$. As $\det U(\alpha) \neq 0$, we may now permute the last $k = n - s$ columns of U so that $u_{ii} \neq 0$ for $i = s+1, \dots, n$. Let $\sigma_1, \dots, \sigma_k$ be the supports of last $n - s$ columns of U. Then for each $j = 1, \dots, k$ $u_{p_j p_j} = \left(\mathbf{u}_{p_j}\right)_{p_j} \neq 0$ means that $p_j \in \operatorname{supp} \sigma_j$. □

Exercises

2.37. Show that the matrix

$$\begin{pmatrix} 1 & -1 & -1 \\ -1 & 1 & t \\ -1 & t & 1 \end{pmatrix}$$

is copositive if and only if $t \geq 1$.
Hint: Consider the zero $\mathbf{u} = \mathbf{e}_1 + \mathbf{e}_2$.

2.38. Let A be an $n \times n$ symmetric matrix. Let D be a positive diagonal matrix and P a permutation matrix, of the same order as A. Check that $\mathbf{u}$ is a zero of A if and only if $\mathbf{v} = P^T D^{-1} \mathbf{u}$ is a zero of the matrix $P^T DADP$, and $\operatorname{supp} \mathbf{v} = \operatorname{supp} \mathbf{u}$.

2.39. Let A and B be $n \times n$ copositive matrices, $A \geq B$. Show that $\mathcal{V}^A \subseteq \mathcal{V}^B$.

2.40. Let A and B be copositive matrices, or orders n and m, respectively. Show that $\mathbf{v} \in \mathbb{R}^{n+m}$ is a zero of $A \oplus B$ if and only if $\mathbf{v} = \begin{pmatrix} \mathbf{u} \\ \mathbf{w} \end{pmatrix}$, where $\mathbf{u} \in \mathbb{R}^n$ is a zero of A or $\mathbf{u} = \mathbf{0}$, $\mathbf{w} \in \mathbb{R}^m$ is a zero of B or $\mathbf{w} = \mathbf{0}$, and $\mathbf{u}$ and $\mathbf{w}$ are not both zero vectors.

2.41. Let A be an $n \times n$ symmetric matrix with diagonal equal to $\mathbf{e}$. Show that if $\mathbf{u}$ is a zero of A with support of size 2, $\operatorname{supp}\mathbf{u} = \{i, j\}$, then $u_i = u_j$ and $\mathbf{u}$ is a minimal zero.

2.42. Prove Proposition 2.46.
Hint: In this case, $\mathbf{u}^T A\mathbf{u} = \mathbf{u}[\alpha]^T A[\alpha]\mathbf{u}[\alpha]$, and $(A\mathbf{u})[\alpha] = A[\alpha]\mathbf{u}[\alpha]$.

2.43. Let A be an $n \times n$ copositive matrix, and let $\mathbf{u}$ be a zero of A, with $|\operatorname{supp}\mathbf{u}| = m$. Let

$$U = \{\mathbf{x} \in \mathbb{R}^n \mid \mathbf{e}^T\mathbf{x} = 0 \text{ and } \operatorname{supp}\mathbf{x} \subseteq \operatorname{supp}\mathbf{u}\},$$

and $V = U + \operatorname{span}\{\mathbf{u}\}$. Prove:

(a) $\dim V = m$.
(b) $\mathbf{x}^T A\mathbf{u} = 0$ for every $\mathbf{x} \in U$.
(c) $\mathbf{x}^T A\mathbf{x} \geq 0$ for every $\mathbf{x} \in V$.
(d) If A has k nonnegative eigenvalues, then $m \leq k$.

Hint: For (b), use Remark 2.49. In (c) use (b) to deduce that $\mathbf{x}^T A\mathbf{x} = (\mathbf{x} + t\mathbf{u})^T A(\mathbf{x} + t\mathbf{u})$ for $\mathbf{x} \in U$, $t \in \mathbb{R}$. To prove (d) see Exercise 2.24 (or Corollary 2.48 and interlacing).

2.44. Prove: If σ is a minimal support of a copositive matrix A, then $A[\alpha]$ is positive definite for every proper subset α of σ.

2.45. Prove Corollary 2.55.

2.46. Prove Corollary 2.56.

2.47. Let A be a copositive matrix. Show that if $\sigma_1, \ldots, \sigma_m$ are minimal supports of A, and for each $i \neq j$, $\sigma_i \cup \sigma_j$ is the support set of a zero of A, then $\sigma_1 \cup \cdots \cup \sigma_m$ is a support of a zero of A.
Hint: Suppose $\sigma_i = \operatorname{supp}\mathbf{u}_i$, $i = 1, \ldots, m$, then $A[\sigma_i \cup \sigma_j]$ is positive semidefinite, and $\mathbf{u}_i^T A\mathbf{u}_j = 0$ for every i and j.

2.48. Prove Corollary 2.59.

2.49. Show that if σ is a zero support, and $\sigma' \subseteq \sigma$ is a minimal support such that $|\sigma \setminus \sigma'| = 1$, then there exists exactly one more minimal support $\sigma'' \subseteq \sigma$ other than σ', and $\sigma = \sigma' \cup \sigma''$.

2.50. Let A be a copositive matrix, and let $\mathbf{u}$ and $\mathbf{v}$ be two minimal zeros of A. Show that the following are equivalent.

(a) $(A\mathbf{u})_i = 0$ for every $i \in \operatorname{supp} \mathbf{v}$.
(b) $(A\mathbf{v})_i = 0$ for every $i \in \operatorname{supp} \mathbf{u}$.
(c) $\mathbf{u}^T A\mathbf{v} = 0$.
(d) $\mathbf{u} + \mathbf{v}$ is a zero of A.

2.51. Let A be a copositive matrix, and let $\mathbf{u}$ be a zero of A. Let

$$\mathcal{V}^1 = \{\mathbf{v} \in \mathcal{V}^A \mid \operatorname{supp} \mathbf{v} \supseteq \operatorname{supp} \mathbf{u}\} \text{ and } \mathcal{V}^2 = \{\mathbf{v} \in \mathcal{V}^A \mid \mathbf{v}^T A\mathbf{u} = 0\}.$$

Prove that

$$\bigcup_{\mathbf{v} \in \mathcal{V}^1} \operatorname{supp} \mathbf{v} = \bigcup_{\mathbf{v} \in \mathcal{V}^2} \operatorname{supp} \mathbf{v}. \tag{2.11}$$

Hint: Show that if $\mathbf{v} \in \mathcal{V}^2$, then $\mathbf{u} + \mathbf{v} \in \mathcal{V}^1$.

2.52. Let A be a copositive matrix and let $\mathbf{u}$ be a zero of A. Let

$$\mathcal{V}^2_{\min} = \{\mathbf{v} \in \mathcal{V}^A_{\min} \mid \mathbf{v}^T A\mathbf{u} = 0\}.$$

Show that the equality (2.11) holds also when $\mathcal{V}^2$ is replaced by $\mathcal{V}^2_{\min}$.

Notes. In [Dickinson (2010)] the term zero was coined, in the more accurate (but cumbersome) form "zeros in the nonnegative orthant", and the notation $\mathcal{V}^A$ was introduced to include all nonnegative zeros of the form $\mathbf{x}^T A\mathbf{x}$ (including $\mathbf{0}$). The definition here follows the later terminology and notation of [Dickinson *et al.* (2013a)], which is by now quite established. However, zeros were used since the beginning of copositivity study. Proposition 2.45 is mentioned already in [Diananda (1961)] and then in Baumert's work (in terms of quadratic forms and their gradient). The proof here follows [Hoffman and Pereira (1973)]. See [Johnson and Reams (2008)] for a different proof. The extension in Proposition 2.46 was observed in [Dickinson *et al.* (2013a)]. Proposition 2.47 first appeared in [Diananda (1962)].

The concepts of minimal zeros and minimal supports are due to [Hildebrand (2014)]. The results on their properties, including the discussion of K_α, the Propositions and Corollaries numbered 2.47–2.60, were first proved in that paper. Exercise 2.43 is based on [Baccari and Naffouti (2016)] and Exercise 2.47 on [Dickinson (2017b)]. Exercises 2.51–2.52 are based on [Dickinson and Hildebrand (2016)].

2.5 $\mathcal{M}$-irreducibility, exceptionality and extremality

In Section 2.3 we mentioned that extremality and $\mathcal{M}$-irreducibility of a copositive matrix A are associated with the existence of zeros of A. We now explore these connections in more detail. We will see, in this section and the ones following it, that zeros are useful also for determining copositivity, and in determining whether a given matrix is SPN or exceptional. We start with a proposition that characterizes, for a given $A \in \mathcal{COP}_n$, the symmetric matrices M such that small perturbations of A in the direction of M remain in the cone $\mathcal{COP}_n$.

Proposition 2.61. *Let A be an $n \times n$ copositive matrix, and let M be an $n \times n$ symmetric matrix. Then there exists $\varepsilon > 0$ such that $A + \varepsilon M$ is copositive if and only if the following two conditions hold.*

(a) $\mathbf{u}^T M \mathbf{u} \geq 0$ *for every zero* $\mathbf{u}$ *of* A.
(b) *If* $\mathbf{u}$ *is a zero of both* A *and* M, *then* $(M\mathbf{u})_i \geq 0$ *for every* i *such that* $(A\mathbf{u})_i = 0$.

Proof. Suppose $A+\varepsilon M$ is copositive for $\varepsilon > 0$, and $\mathbf{u}$ is a zero of A. Then

$$0 \leq \mathbf{u}^T(A + \varepsilon M)\mathbf{u} = \varepsilon \mathbf{u}^T M \mathbf{u},$$

and thus (a) holds. If $\mathbf{u}$ is a zero of both A and M, then $\mathbf{u}^T(A+\varepsilon M)\mathbf{u} = 0$, and therefore

$$(A + \varepsilon M)\mathbf{u} \geq 0.$$

Hence for every i such that $(A\mathbf{u})_i = 0$, $((A + \varepsilon M)\mathbf{u})_i = \varepsilon(M\mathbf{u})_i \geq 0$.

Conversely, suppose (a) and (b) hold. Assume on the contrary that for every $\varepsilon > 0$ the matrix $A+\varepsilon M$ is not copositive. Choose a positive sequence ε_ℓ converging to 0, and let $A_\ell = A + \varepsilon_\ell M$. For every ℓ let $\mathbf{x}_\ell \in \Delta$ be a vector at which $\min_{\mathbf{x}\in\Delta} \frac{1}{2}\mathbf{x}^T A_\ell \mathbf{x}$ is attained. Since A_ℓ is not copositive, this minimum is negative. Replacing $\mathbf{x}_\ell$ by a converging subsequence, if necessary, we may assume that $\lim_{\ell\to\infty} \mathbf{x}_\ell = \mathbf{u} \in \Delta$. Then

$$0 \leq \mathbf{u}^T A \mathbf{u} = \lim_{\ell\to\infty} \mathbf{x}_\ell^T A_\ell \mathbf{x}_\ell \leq 0,$$

i.e., $\mathbf{u}$ is a zero of A. By (a), $\mathbf{u}^T M \mathbf{u} \geq 0$. On the other hand,

$$\mathbf{x}_\ell^T M \mathbf{x}_\ell \leq -\frac{1}{\varepsilon_\ell}\mathbf{x}_\ell^T A \mathbf{x}_\ell \leq 0$$

for every ℓ, and therefore

$$\mathbf{u}^T M \mathbf{u} = \lim_{\ell\to\infty} \mathbf{x}_\ell^T M \mathbf{x}_\ell \leq 0.$$

Hence $\mathbf{u}^T M \mathbf{u} = 0$, and $\mathbf{u}$ is also a zero of M.

By the Karush-Kuhn-Tucker optimality conditions (see Example 1.173), for any fixed ℓ there exist Lagrange multipliers $\boldsymbol{\mu}_\ell \in \mathbb{R}^n_+$ and $\lambda_\ell \in \mathbb{R}$ such that $\mathbf{x}_\ell^T \boldsymbol{\mu}_\ell = 0$ and

$$A_\ell \mathbf{x}_\ell + \lambda_\ell \mathbf{e} - \boldsymbol{\mu}_\ell = \mathbf{0}. \tag{2.12}$$

Multiplying (2.12) by $\mathbf{x}_\ell^T$ we get that

$$\mathbf{x}_\ell^T A_\ell \mathbf{x}_\ell + \lambda_\ell = 0,$$

so $\lambda_\ell = -\mathbf{x}_\ell^T A_\ell \mathbf{x}_\ell > 0$, and $A_\ell \mathbf{x}_\ell = \boldsymbol{\mu}_\ell - \lambda_\ell \mathbf{e} < \boldsymbol{\mu}_\ell$. Thus

$$(A_\ell \mathbf{x}_\ell)_i > 0 \quad \Rightarrow \quad (\boldsymbol{\mu}_\ell)_i > 0 \quad \Rightarrow \quad (\mathbf{x}_\ell)_i = 0. \tag{2.13}$$

As $\mathbf{u} = \lim_{\ell \to \infty} \mathbf{x}_\ell$, we have that for sufficiently large ℓ

$$u_i > 0 \quad \Rightarrow \quad (\mathbf{x}_\ell)_i > 0 \quad \Rightarrow \quad (\boldsymbol{\mu}_\ell)_i = 0, \tag{2.14}$$

and thus $\mathbf{u}^T \boldsymbol{\mu}_\ell = 0$ for such ℓ. Also, $A\mathbf{u} = \lim_{\ell \to \infty} A_\ell \mathbf{x}_\ell$ implies that for large enough ℓ

$$(A\mathbf{u})_i > 0 \quad \Rightarrow \quad (A_\ell \mathbf{x}_\ell)_i > 0, \tag{2.15}$$

which together with (2.13) implies that $(A\mathbf{u})^T \mathbf{x}_\ell = 0$. Therefore for large enough ℓ both $\mathbf{u}^T \boldsymbol{\mu}_\ell = 0$ and $\mathbf{u}^T A \mathbf{x}_\ell = 0$. For such ℓ, multiply (2.12) by $\mathbf{u}^T$. This yields

$$0 = \mathbf{u}^T A_\ell \mathbf{x}_\ell + \lambda_\ell = \varepsilon_\ell \mathbf{u}^T M \mathbf{x}_\ell + \lambda_\ell,$$

and thus $\mathbf{u}^T M \mathbf{x}_\ell < 0$. Hence there exists i for which $(\mathbf{x}_\ell)_i > 0$, while $(M\mathbf{u})_i < 0$. However, as $\mathbf{x}_\ell^T A\mathbf{u} = 0$, $(A\mathbf{u})_i = 0$ for such i. Since $\mathbf{u}$ is a zero of both A and M, this contradicts the assumption that M satisfies (b). □

Note that if A is copositive, and $A + \varepsilon M$ is copositive for some $\varepsilon > 0$, then so is $A + \delta M$ for every $0 < \delta < \varepsilon$ (by the convexity of $\mathcal{COP}_n$).

Proposition 2.61 yields the following characterizations of M-irreducibility with respect to a copositive M.

Theorem 2.62. *Let $A, M \in \mathcal{COP}_n$. Then the following are equivalent.*

(a) *A is M-irreducible.*
(b) *There exist a zero $\mathbf{u}$ of A and a vector $\mathbf{x} \in \mathbb{R}^n_+$, such that $\mathbf{x}^T A\mathbf{u} = 0$ and $\mathbf{x}^T M\mathbf{u} \neq 0$.*
(c) *There exist a zero $\mathbf{u}$ of A and $1 \le i \le n$ such that $(A\mathbf{u})_i = 0$ and $(M\mathbf{u})_i \neq 0$.*
(d) *There exist a minimal zero $\mathbf{u}$ of A and $1 \le i \le n$ such that $(A\mathbf{u})_i = 0$ and $(M\mathbf{u})_i \neq 0$.*

Proof. (a) $\Rightarrow$ (b): Let A be M-irreducible. If there exists a zero $\mathbf{u}$ of A, which is not a zero of M, then $\mathbf{u}$ and $\mathbf{x} = \mathbf{u}$ satisfy (b). Otherwise, $\mathcal{V}^A \subseteq \mathcal{V}^M$, and by Proposition 2.61 applied to $-M$ there exists a zero $\mathbf{u}$ of A and an index i such that $(A\mathbf{u})_i = 0$ but $(-M\mathbf{u})_i < 0$. In this case $\mathbf{x} = \mathbf{e}_i$ satisfies (b) together with $\mathbf{u}$.

(b) $\Rightarrow$ (c): If $\mathbf{x}$ and $\mathbf{u}$ are as in (b), there exists i such that $x_i > 0$ and $(M\mathbf{u})_i \neq 0$. As $\mathbf{x} \geq \mathbf{0}$, $A\mathbf{u} \geq \mathbf{0}$ and their inner product is 0, $(A\mathbf{u})_i = 0$.

(c) $\Rightarrow$ (d): The zero $\mathbf{u}$ is a sum of minimal zeros $\mathbf{u}_1, \ldots, \mathbf{u}_m$, hence $0 = (A\mathbf{u})_i = (A\sum_{j=1}^m \mathbf{u}_j)_i = \sum_{j=1}^m (A\mathbf{u}_j)_i$. Since $A\mathbf{u}_j \geq \mathbf{0}$ for every j, $(A\mathbf{u}_j)_i = 0$ for every j. And $(M\sum_{j=1}^m \mathbf{u}_j)_i = (M\mathbf{u})_i \neq 0$ implies that $(M\mathbf{u}_j)_i \neq 0$ at least for one j.

(d) $\Rightarrow$ (a): Let $\mathbf{u}$ be a minimal zero of A such that $(A\mathbf{u})_i = 0$ and $(M\mathbf{u})_i \neq 0$ for a certain index i. Suppose that for some $\varepsilon > 0$ the matrix $A - \varepsilon M$ is copositive. Then

$$0 \leq \mathbf{u}^T(A - \varepsilon M)\mathbf{u} = -\varepsilon \mathbf{u}^T M \mathbf{u} \leq 0,$$

so $\mathbf{u}$ is also a zero of M. But then, by Proposition 2.61 applied to $-M$, $(A\mathbf{u})_i = 0$ implies $(-M\mathbf{u})_i = 0$, contrary to the assumption on $\mathbf{u}$. □

We can now characterize $\{i, j\}$-irreducibility, both for the case that $i = j$ and the case $i \neq j$.

Proposition 2.63. *Let $A \in \mathcal{COP}_n$ and $1 \leq i, j \leq n$. Then A is $\{i, j\}$-irreducible if and only if there exists a zero $\mathbf{u}$ of A such that $u_i + u_j > 0$ and $(A\mathbf{u})_i = (A\mathbf{u})_j = 0$.*

Proof. Observe that for $i = j$ the proposition means that A is $\{i, i\}$-irreducible if and only if there exists a zero $\mathbf{u}$ of A such that $u_i > 0$. Since the cases $i = j$ and $i \neq j$ have similar proofs, we consider both cases together. Let $M = E_{ii}$ if $i = j$, $M = E_{ij} + E_{ji}$ if $i \neq j$. Note that

$$(M\mathbf{u})_k = \begin{cases} u_j & \text{if } k = i \\ u_i & \text{if } k = j \\ 0 & \text{otherwise} \end{cases},$$

and by Theorem 2.62 A is M-irreducible if and only if there exists a zero $\mathbf{u}$ of A and $1 \leq k \leq n$ such that $(A\mathbf{u})_k = 0$ and $(M\mathbf{u})_k \neq 0$.

Suppose there exists a zero $\mathbf{u}$ of A such that $u_i + u_j > 0$ and $(A\mathbf{u})_i = (A\mathbf{u})_j = 0$. Without loss of generality we assume $u_j > 0$, then for $k = i$ $(A\mathbf{u})_k = 0$ and $(M\mathbf{u})_k = u_j > 0$.

Conversely, if there exists a zero $\mathbf{u}$ of A and $1 \leq k \leq n$ such that $(A\mathbf{u})_k = 0$ and $(M\mathbf{u})_k \neq 0$, then without loss of generality $k = i$. Then

$(A\mathbf{u})_i = 0$ and $u_i + u_j \geq u_j = (M\mathbf{u})_k > 0$. As $j \in \operatorname{supp} \mathbf{u}$, we also have $(A\mathbf{u})_j = 0$. □

Note that in Proposition 2.63 the zero $\mathbf{u}$ may have both u_i and u_j positive, or only one of them (see Exercise 2.56).

A similar result characterizes M-irreducibility for a rank 1 positive semidefinite matrix M.

Proposition 2.64. *Let $A \in \mathcal{COP}_n$ and $\mathbf{w} \in \mathbb{R}^n$. Then A is $\mathbf{w}\mathbf{w}^T$-irreducible if and only if there exists a zero $\mathbf{u}$ of A such that $\mathbf{u}^T\mathbf{w} \neq 0$.*

Proof. The "if" part is easy and we leave it as an exercise. For the "only if" part apply Theorem 2.62 to A and $M = \mathbf{w}\mathbf{w}^T$. By (b) we have that $\mathbf{u}^T M\mathbf{x} > 0$ for some $\mathbf{x} \in \mathbb{R}^n_+$. That is, $\mathbf{u}^T\mathbf{w} \cdot \mathbf{w}^T\mathbf{x} > 0$, and therefore $\mathbf{u}^T\mathbf{w} \neq 0$. □

In Propositions 2.63 and 2.64 we can replace "zero" by "minimal zero".

Proposition 2.65. *Let $A \in \mathcal{COP}_n$ and $1 \leq i, j \leq n$. Then A is $\{i, j\}$-irreducible if and only if there exists a minimal zero $\mathbf{u}$ of A such that $u_i + u_j > 0$ and $(A\mathbf{u})_i = (A\mathbf{u})_j = 0$.*

Proof. Only the "only if" needs to be proved. If A is $\{i, j\}$-irreducible, by Proposition 2.63 there exists a zero $\mathbf{u}$ of A such that $u_i + u_j > 0$ and $(A\mathbf{u})_i = (A\mathbf{u})_j = 0$. Then $\mathbf{u}$ is a sum of minimal zeros of A. For each minimal zero $\mathbf{v}$ in this sum we necessarily have $(A\mathbf{v})_i = (A\mathbf{v})_j = 0$, and for at least one such minimal zero $\mathbf{v}$ we have $v_i + v_j > 0$. □

The next proposition is proved similarly.

Proposition 2.66. *Let $A \in \mathcal{COP}_n$ and $\mathbf{w} \in \mathbb{R}^n$. Then A is $\mathbf{w}\mathbf{w}^T$-irreducible if and only if there exists a minimal zero $\mathbf{u}$ of A such that $\mathbf{u}^T\mathbf{w} \neq 0$.*

This leads to the following result.

Theorem 2.67. *Let $A \in \mathcal{COP}_n$. Then A is $\mathcal{PSD}$-irreducible if and only if its minimal zeros span $\mathbb{R}^n$.*

Proof. The matrix A is $\mathcal{PSD}$-irreducible if and only it is irreducible with respect to every matrix that generates an extreme ray of the cone $\mathcal{PSD}_n$ (see Proposition 2.39). That is, A is $\mathcal{PSD}$-irreducible if and only if it is $\mathbf{w}\mathbf{w}^T$-irreducible for every nonzero $\mathbf{w} \in \mathbb{R}^n$. By Proposition 2.66, this holds if and only if for every nonzero $\mathbf{w} \in \mathbb{R}^n$ there exists a minimal zero $\mathbf{u}$ such that $\mathbf{u}^T\mathbf{w} \neq 0$. This, in turn, is equivalent to $\text{span}\{\mathbf{u} \mid \mathbf{u} \in \mathcal{V}^A_{\min}\} = \mathbb{R}^n$. $\square$

Corollary 2.68. *If $A \in \mathcal{COP}_n$ is $\mathcal{PSD}$-irreducible, then A has at least n linearly independent minimal zeros.*

SPN matrices and exceptional matrices

The above results about irreducibility with respect to nonnegative matrices, or positive semidefinite matrices, may help determine whether a given matrix is SPN or exceptional. The next proposition and its corollary demonstrate that.

Proposition 2.69. *Let A be an $n \times n$ copositive matrix that has a positive semidefinite principal submatrix of order $n-1$. Then A is SPN.*

Proof. Without loss of generality

$$A = \begin{pmatrix} B & \mathbf{c} \\ \mathbf{c}^T & a \end{pmatrix},$$

where B is positive semidefinite. Let $a' \leq a$ be such that

$$A' = \begin{pmatrix} B & \mathbf{c} \\ \mathbf{c}^T & a' \end{pmatrix}$$

is $\{n, n\}$-irreducible. Then there exists a zero $\mathbf{u}$ of A' such that $u_n > 0$. Without loss of generality $\mathbf{u} = \begin{pmatrix} \mathbf{x}^T & 1 \end{pmatrix}^T$. Then

$$A' = \begin{pmatrix} B & -B\mathbf{x} \\ -\mathbf{x}^T B & \mathbf{x}^T B\mathbf{x} \end{pmatrix} + \begin{pmatrix} 0 & B\mathbf{x} + \mathbf{c} \\ \mathbf{x}^T B + \mathbf{c}^T & a' - \mathbf{x}^T B\mathbf{x} \end{pmatrix}.$$

As $A'\mathbf{u} \geq \mathbf{0}$ and $(A'\mathbf{u})_i = 0$ whenever $u_i > 0$,

$$B\mathbf{x} + \mathbf{c} \geq \mathbf{0} \quad \text{and} \quad \mathbf{c}^T\mathbf{x} + a' = 0.$$

As $0 = \mathbf{u}^T A'\mathbf{u} = \mathbf{x}^T B\mathbf{x} + 2\mathbf{c}^T\mathbf{x} + a'$ and $-\mathbf{c}^T\mathbf{x} = a'$, $a' - \mathbf{x}^T B\mathbf{x} = 0$. That is,

$$A' = \begin{pmatrix} B & -B\mathbf{x} \\ -\mathbf{x}^T B & \mathbf{x}^T B\mathbf{x} \end{pmatrix} + \begin{pmatrix} 0 & B\mathbf{x} + \mathbf{c} \\ \mathbf{x}^T B + \mathbf{c}^T & 0 \end{pmatrix}$$

is SPN, and $A \geq A'$ is also SPN. $\square$

Corollary 2.70. *Let A be an $n \times n$ exceptional copositive matrix. Then the support of each zero of A is of size at most $n-2$.*

Proof. If $\mathbf{u}$ is a zero of A with support σ, then $A[\sigma]$ is positive semidefinite. If $|\sigma|$ were greater than $n-2$, A would SPN by Proposition 2.69. $\square$

More results in which the existence of certain zeros implies that a matrix is SPN may be found in the next section — see Proposition 2.84, Corollary 2.85 and Remark 2.86; Proposition 2.99 and Corollaries 2.100 and 2.101.

Extremal matrices

The results on zeros, minimal zeros and M-irreducibility are important for the study of the difficult problem of characterizing and constructing exceptional extremal matrices. One such result of particular importance is a characterization of extremal copositive matrices in terms of their zeros, which follows from Proposition 2.61.

Theorem 2.71. *Let $A \in \mathcal{COP}_n$. Then the following are equivalent.*

(a) *A is extremal.*
(b) *Every symmetric matrix X that satisfies $(X\mathbf{u})_i = 0$ whenever $\mathbf{u} \in \mathcal{V}^A$ and $(A\mathbf{u})_i = 0$ is a scalar multiple of A.*
(c) *Every symmetric matrix X that satisfies $(X\mathbf{u})_i = 0$ whenever $\mathbf{u} \in \mathcal{V}^A_{\min}$ and $(A\mathbf{u})_i = 0$ is a scalar multiple of A.*

Proof. Obviously (b) implies (c).

(a) $\Rightarrow$ (b): Suppose A is extremal, and let a symmetric matrix X satisfy that $(X\mathbf{u})_i = 0$ whenever $\mathbf{u}$ is a zero of A and $(A\mathbf{u})_i = 0$. If $\mathbf{u}$ is a zero of A, then $(A\mathbf{u})_i = 0$ for every $i \in \operatorname{supp} \mathbf{u}$, and then also $(X\mathbf{u})_i = 0$ for such i, so $\mathbf{u}^T X\mathbf{u} = 0$. So every zero of A is a zero of X, and X obviously satisfies conditions (a) and (b) of Proposition 2.61. Similarly, $-X$ also satisfies these two conditions. Thus for some $\varepsilon > 0$, $A \pm \varepsilon X$ are both copositive, and $A = \frac{1}{2}(A + \varepsilon X) + \frac{1}{2}(A - \varepsilon X)$. By the extremality of A, both $A \pm \varepsilon X$ are scalar multiples of A, and so is X.

(c) $\Rightarrow$ (a): Suppose (c) holds, and $A = B + C$, where B and C are copositive. If $\mathbf{u}$ is a minimal zero of A, then $\mathbf{u}^T B\mathbf{u} \geq 0$ and $\mathbf{u}^T C\mathbf{u} \geq 0$. As

$$0 = \mathbf{u}^T A\mathbf{u} = \mathbf{u}^T B\mathbf{u} + \mathbf{u}^T C\mathbf{u},$$

we get that $\mathbf{u}$ is also a zero of B and C. In particular, for every such $\mathbf{u}$, $B\mathbf{u} \geq \mathbf{0}$ and $C\mathbf{u} \geq \mathbf{0}$, and therefore

$$\mathbf{0} \leq B\mathbf{u} = A\mathbf{u} - C\mathbf{u} \leq A\mathbf{u}.$$

Thus $(B\mathbf{u})_i = 0$ whenever $(A\mathbf{u})_i = 0$ and $\mathbf{u} \in \mathcal{V}^A_{\min}$. By (c), B is a (nonnegative) scalar multiple of A (and so is C). $\square$

Remark 2.72. A copositive matrix A has a finite number of minimal zeros, up to scalar multiplication. Hence the extermality criterion in (c) means that if the minimal zeros of A are known, and $\{\mathbf{u}_1, \ldots, \mathbf{u}_m\}$ is a representing set of the minimal zeros, checking whether A is extremal amounts to solving the homogeneous linear system (in the symmetric matrix X),

$$(X\mathbf{u}_j)_i = 0 \text{ for every } i \text{ such that } (A\mathbf{u}_j)_i = 0, \quad j = 1, \ldots, m,$$

and determining whether the solution space is one dimensional.

We use this characterization to show another way to extend an extremal matrix in $\mathcal{COP}_n$ to an extremal matrix in $\mathcal{COP}_k$ for $k > n$ (other than adding zero rows and columns).

Proposition 2.73. *Let $A \in \mathcal{COP}_n$ be an extremal matrix, which is not nonnegative. Let $E \geq 0$ be an $n \times m$ matrix with exactly one positive entry in each column, then*

$$\begin{pmatrix} A & AE \\ E^T A & E^T AE \end{pmatrix}$$

is an extremal matrix in $\mathcal{COP}_{n+m}$. If A is exceptional, so is the extended matrix.

Proof. It suffices to consider the case that $m = 1$ (the general case follows by induction on m, see Exercise 2.60). By scaling and permuting rows and columns we may assume that $E = \mathbf{e}_n$, and thus $E^T AE = a_{nn}$. Let

$$A' = \begin{pmatrix} A & A\mathbf{e}_n \\ \mathbf{e}_n^T A & a_{nn} \end{pmatrix}.$$

This is a copositive matrix by Proposition 2.14. Note that $A(n) = A(n+1) = A$.

Every zero of A has the form

$$\begin{pmatrix} \mathbf{v} \\ x \end{pmatrix},$$

where $\mathbf{v} \in \mathbb{R}_+^{n-1}$ and $x \geq 0$, and then the zeros of A' include the vectors

$$\begin{pmatrix} \mathbf{v} \\ x \\ 0 \end{pmatrix} \text{ and } \begin{pmatrix} \mathbf{v} \\ 0 \\ x \end{pmatrix}.$$

If $X \in \mathcal{COP}_{n+1}$ satisfies $(X\mathbf{u})_i = 0$ whenever $\mathbf{u}$ is a zero of A and $(A\mathbf{u})_i = 0$, then the extremality of A implies that $X(n+1) = tA(n+1) = tA$, and $X(n) = sA(n) = sA$ for some $t, s \geq 0$. Since A is an extremal matrix which is not nonnegative, there exists $1 \leq i \leq n-1$ such that

$a_{ii} > 0$. As $x_{ii} = ta_{ii} = sa_{ii}$, we get that $t = s$. Hence $x_{ij} = ta_{ij}$ for every $1 \le i \le j \le n+1$, $(i,j) \neq (n, n+1)$.

The extremal A is $\{n,n\}$-irreducible, and therefore has a zero $\mathbf{u}$ with $u_{nn} > 0$. We may assume that $\mathbf{u} = (\mathbf{x}^T \;\; 1)^T$. Let

$$\mathbf{w} = \begin{pmatrix} \mathbf{x} \\ \frac{1}{2} \\ \frac{1}{2} \end{pmatrix} \in \mathbb{R}^{n+1}_{+}.$$

It is easy to see that $\mathbf{w}$ is a zero of A', with both n and $n+1$ in its support. Hence $(A'\mathbf{w})_n = (A'\mathbf{w})_{n+1} = 0$, and therefore also $(X\mathbf{w})_n = (X\mathbf{w})_{n+1} = 0$. Suppose the n-th row of A is $\begin{pmatrix} \mathbf{a}^T & p \end{pmatrix}$. Then $(A\mathbf{u})_n = 0$ implies $p = \mathbf{a}^T\mathbf{x}$, and the n-th row of X is

$$\begin{pmatrix} t\mathbf{a}^T & tp & q \end{pmatrix},$$

where $q = x_{n\,n+1}$. From $(X\mathbf{w})_n = 0$ we deduce that

$$t\mathbf{a}^T\mathbf{x} + \frac{1}{2}tp + \frac{1}{2}q = 0$$

and therefore $q = tp$. This completes the proof that $X = tA'$. By Remark 2.72, this shows that A' is extremal. □

The requirement that exactly one element in each column of E is positive is essential, see Exercise 2.62.

Exercises

2.53. Let $A \in \mathcal{COP}_n$, and let

$$\mathcal{K}^A = \{X \in \mathcal{S}_n \mid \exists \varepsilon > 0 \text{ s.t. } A + \varepsilon X \in \mathcal{COP}_n\}.$$

Show that $\mathcal{K}^A$ is a convex cone, which is not pointed, and $\mathcal{COP}_n \subseteq \mathcal{K}^A$. ($\mathcal{K}^A$ is the *cone of feasible directions.*)

2.54. Consider the $n \times n$ copositive matrix $A = \sum_{i=2}^{n} \mathbf{e}_i^T\mathbf{e}_i$. Compute $\mathcal{K}^A$ (defined in Exercise 2.53), and show that it is not closed.

2.55. Let $A \in \mathcal{COP}_n$. Show that

$$\mathrm{cl}\mathcal{K}^A = \{X \in \mathcal{S}_n \mid \mathbf{u}^T X\mathbf{u} \ge 0 \text{ for every } \mathbf{u} \in \mathcal{V}^A\}.$$

Hint: Use Proposition 2.61 to show that if $\mathbf{u}^T X\mathbf{u} \ge 0$ for every zero $\mathbf{u}$ of A, then $X + \varepsilon I \in \mathcal{K}^A$ for every $\varepsilon > 0$.

2.56. Consider the copositive matrix

$$A = \begin{pmatrix} 1 & 0 & 0 \\ 0 & 1 & -1 \\ 0 & -1 & 1 \end{pmatrix},$$

which is $\{1,2\}$-irreducible according to Exercise 2.34. Show that the only zero of A, up to scalar multiplication, is $\mathbf{u} = \mathbf{e}_2 + \mathbf{e}_3$. (This shows that A may be $\{i,j\}$-irreducible, and not have a zero $\mathbf{u}$ such that $u_i u_j > 0$.)

2.57. Let A be an $n \times n$ copositive matrix, and let $\mathbf{w} \in \mathbb{R}^n$. Show that if there exists a zero $\mathbf{u}$ of A such that $\mathbf{u}^T\mathbf{w} \neq 0$, then A is $\mathbf{w}\mathbf{w}^T$-irreducible. (This is the "if" part of Proposition 2.64.)

2.58. Prove Proposition 2.66.

2.59. Let A be an $n \times n$ copositive matrix, and let $A[\alpha]$ be a positive semidefinite principal submatrix of A. Show that if $\mathbf{u}$ is a zero of A and $|\operatorname{supp}\mathbf{u} \setminus \alpha| = 1$, then $A[\alpha \cup \operatorname{supp}\mathbf{u}]$ is SPN.

2.60. Let A be an $n \times n$ copositive matrix such that $A(1)$ and $A(n)$ are positive semidefinite, A has a zero $\mathbf{v}$ with $\operatorname{supp}\mathbf{v} = \{2,\dots,n-1\}$, and A is $\{1,n\}$-irreducible. Show that A is positive semidefinite.
Hint: $A = P + \delta(E_{1n} + E_{n1})$, where P is a positive semidefinite completion of the partial matrix obtained by making the $1,n$ and $n,1$ entries of A unspecified. Show that $A\mathbf{v} = P\mathbf{v} = \mathbf{0}$. Conclude that if $\mathbf{u}$ is a zero of A with $u_1 + u_n > 0$ such that $(A\mathbf{u})_1 = (A\mathbf{u})_n = 0$, then $\mathbf{u} + \mathbf{v}$ is a zero of A. Consider the cases $u_1 u_n = 0$ and $u_1 u_n > 0$.

2.61. Prove Proposition 2.73 for any $m \geq 1$.
Hint: If $E = (F\ \mathbf{e}_i)$ and $A' = \begin{pmatrix} A & AF \\ F^TA & F^TAF \end{pmatrix}$, then $\begin{pmatrix} A & AE \\ E^TA & E^TAE \end{pmatrix} = \begin{pmatrix} A' & A'\mathbf{e}_i \\ \mathbf{e}_i^TA' & \mathbf{e}_i^TA'\mathbf{e}_i \end{pmatrix}$.

2.62. Show that if $\mathbf{v} = \mathbf{e}_4 + \mathbf{e}_5$ and $H' = \begin{pmatrix} H & H\mathbf{v} \\ \mathbf{v}^TH & \mathbf{v}^TH\mathbf{v} \end{pmatrix}$, where H is the Horn matrix, then H' is not extremal in $\mathcal{COP}_6$.

Notes. In the discussion of zeros and $\mathcal{M}$-irreducibility we follow [Dickinson *et al.* (2013a)], [Hildebrand (2014)] and [Dickinson and Hildebrand

(2016)]. In particular, Theorem 2.71 is due to the latter, and Exercises 2.53–2.55 are also based on results in that paper. The characterization of $\{i,i\}$-irreducibility was first proved in [Baumert (1965, 1966)], but the characterization of $\{i,j\}$-irreducibility, $i \neq j$, in [Baumert (1965)] was incorrect, and the mistake was pointed out and corrected in [Dickinson *et al.* (2013a)] — see Exercise 2.56 and Proposition 2.63. Propositions 2.64–2.66 and Corollary 2.68 are due to [Hildebrand (2014)]. Proposition 2.69 was proved in [Shaked-Monderer (2016)], following a related result in [Dickinson *et al.* (2013a)]. Proposition 2.73 is due to [Baumert (1966)] (the case $m = 1$, another proof). The result in Exercise 2.60 is due to [Johnson and Reams (2008)].

2.6 Methods for determining copositivity

At this point, we know more about copositive matrices, but do not know many such matrices (other than the SPN ones). In fact, the only specific example of an exceptional copositive matrix presented so far is the Horn matrix. So how can we tell whether a given matrix is copositive? As mentioned before, checking copositivity is difficult, and in complexity terms it is a co-NP-complete problem. In this section we present some methods for copositivity detection, using spectral properties, zeros, Schur complements, or the matrix sign pattern. In the next section recursive characterizations of copositivity are considered. None of these methods is as effective as the methods for detecting positive semidefiniteness, some of them only help reduce the problem to one of a smaller order.

Copositivity by spectral properties

Consider a symmetric matrix A. Calculating $\min_\Delta \mathbf{x}^T A\mathbf{x}$ or $\min_\Gamma \mathbf{x}^T A\mathbf{x}$ ($\Delta = \Delta_n = \{\mathbf{x} \in \mathbb{R}^n_+ \mid \mathbf{e}^T\mathbf{x} = 1\}$, $\Gamma = \Gamma_n = \{\mathbf{x} \in \mathbb{R}^n_+ \mid \|\mathbf{x}\| = 1\}$) is not a simple task. Note that if $\min_{\mathbf{x}\in\Gamma} \mathbf{x}^T A\mathbf{x}$ is attained at a positive $\mathbf{v} \in \Gamma$, using Lagrange multipliers we get that $\nabla \mathbf{x}^T A\mathbf{x} = \lambda\nabla(\sum_{i=1}^n x_i^2 - 1)$ at $\mathbf{x} = \mathbf{v}$, that is, $A\mathbf{v} = \lambda\mathbf{v}$. Hence A has an eigenvalue λ with a positive corresponding eigenvector $\mathbf{v}$. In this case, as $\mathbf{v}^T A\mathbf{v} = \lambda\mathbf{v}^T\mathbf{v} = \lambda$, $\min_{\mathbf{x}\in\Gamma} \mathbf{x}^T A\mathbf{x}$ is equal to the eigenvalue λ. This observation is used in the proof of the following theorem.

Theorem 2.74. *A symmetric matrix A is copositive if and only if no principal submatrix has a positive eigenvector corresponding to a negative eigenvalue.*

Proof. Suppose that A is $n \times n$. If for some $\alpha \subseteq \{1, \ldots, n\}$ the submatrix $A[\alpha]$ has a positive eigenvector $\mathbf{u}$ corresponding to a negative eigenvalue λ, let $\mathbf{v} \in \mathbb{R}^n_+$ be the vector obtained by prepending and appending zero entries to $\mathbf{u}$ so that $\mathbf{v}[\alpha] = \mathbf{u}$. Then

$$\mathbf{v}^T A\mathbf{v} = \mathbf{v}[\alpha]A[\alpha]\mathbf{v}[\alpha] = \mathbf{u}^T A[\alpha]\mathbf{u} = \lambda\|\mathbf{u}\|^2 < 0,$$

and A is not copositive.

Conversely, if A is not copositive, then $\min_{i\in\Gamma_n} \mathbf{x}^T A\mathbf{x} = \lambda < 0$. Let $\mathbf{v} \in \Gamma_n$ be a vector at which this minimum is attained. Denote $\sigma = \operatorname{supp} \mathbf{v}$ and $m = |\sigma|$. Then

$$\lambda = \mathbf{v}^T A\mathbf{v} = \mathbf{v}[\sigma]^T A[\sigma]\mathbf{v}[\sigma] \geq \min_{\mathbf{x}\in\Gamma_m} \mathbf{x}^T A[\sigma]\mathbf{x} \geq \min_{\mathbf{x}\in\Gamma_n} \mathbf{x}^T A\mathbf{x} = \lambda,$$

i.e., $\lambda = \min_{\mathbf{x}\in\Gamma_m} \mathbf{x}^T A[\sigma]\mathbf{x}$ is attained at $\mathbf{v}[\sigma]$. By the argument preceding this theorem, $\mathbf{v}[\sigma]$ is a positive eigenvector of $A[\sigma]$ corresponding to the negative eigenvalue λ. □

Remark 2.75. Similarly, a symmetric matrix A is strictly copositive if and only if no principal submatrix has a positive eigenvector corresponding to a nonpositive eigenvalue.

Note that checking copositivity using Theorem 2.74 requires consideration of all principal submatrices, their minimal eigenvalues and corresponding eigenvectors. Compare that to checking positive semidefiniteness of a matrix by considering all its eigenvalues.

There is a special case in which spectral considerations lead to a simple copositivity characterization.

Theorem 2.76. *Let A be a symmetric matrix, which is not positive semidefinite and has exactly one positive eigenvalue. Then A is copositive if and only if A is nonnegative.*

Proof. Let A be copositive with exactly one positive eigenvalue. We need to show that $a_{pq} \geq 0$ for every $1 \leq p \leq q \leq n$. Suppose the eigenvalues of A are $\lambda_1 > 0 \geq \lambda_2 \geq \cdots \geq \lambda_n$. Let $\Lambda = \operatorname{Diag}(\lambda_1, \ldots, \lambda_n)$, and let U be an orthogonal matrix such that $A = U^T \Lambda U$. Since the columns of U^T are eigenvectors of A, and by Proposition 2.22 there is a nonnegative eigenvector corresponding to λ_1, we may assume that the first row of U is nonnegative. That is, the columns $\mathbf{u}_1, \mathbf{u}_2, \ldots, \mathbf{u}_n$ of U satisfy $(\mathbf{u}_i)_1 \geq 0$ for every i.

For $1 \leq p \leq q \leq n$,

$$a_{pq} = \mathbf{u}_p^T \Lambda \mathbf{u}_q = \sum_{i=1}^n \lambda_i (\mathbf{u}_p)_i (\mathbf{u}_q)_i.$$

As A is copositive, for every p

$$0 \le a_{pp} = \sum_{i=1}^{n} \lambda_i (\mathbf{u}_p)_i^2,$$

that is,

$$\lambda_1 (\mathbf{u}_p)_1^2 \ge -\sum_{i=2}^{n} \lambda_i (\mathbf{u}_p)_i^2 = \sum_{i=2}^{n} |\lambda_i| (\mathbf{u}_p)_i^2. \tag{2.16}$$

For $p < q$, the Cauchy-Schwarz inequality yields

$$\sum_{i=2}^{n} |\lambda_i| (\mathbf{u}_p)_i (\mathbf{u}_q)_i \le \left(\sum_{i=2}^{n} |\lambda_i| (\mathbf{u}_p)_i^2 \right)^{\frac{1}{2}} \left(\sum_{i=2}^{n} |\lambda_i| (\mathbf{u}_q)_i^2 \right)^{\frac{1}{2}}$$

and by (2.16),

$$\left(\sum_{i=2}^{n} |\lambda_i| (\mathbf{u}_p)_i^2 \right)^{\frac{1}{2}} \left(\sum_{i=2}^{n} |\lambda_i| (\mathbf{u}_q)_i^2 \right)^{\frac{1}{2}} \le \left(\lambda_1 (\mathbf{u}_p)_1^2 \right)^{\frac{1}{2}} \left(\lambda_1 (\mathbf{u}_q)_1^2 \right)^{\frac{1}{2}} = \lambda_1 (\mathbf{u}_p)_1 (\mathbf{u}_q)_1.$$

Hence

$$\sum_{i=2}^{n} |\lambda_i| (\mathbf{u}_p)_i (\mathbf{u}_q)_i \le \lambda_1 (\mathbf{u}_p)_1 (\mathbf{u}_q)_1,$$

and

$$a_{pq} = \sum_{i=1}^{n} \lambda_i (\mathbf{u}_p)_i (\mathbf{u}_q)_i = \lambda_1 (\mathbf{u}_p)_1 (\mathbf{u}_q)_1 - \sum_{i=2}^{n} |\lambda_i| (\mathbf{u}_p)_i (\mathbf{u}_q)_i \ge 0. \qquad \square$$

Another special case in which copositivity of a matrix may be determined by spectral considerations is when $-\rho$ is an eigenvalue of a symmetric matrix A with spectral radius $\rho = \rho(A)$. For the relevant theorem, we need to first prove a preliminary lemma. Recall that by Corollary 2.21, if $-\rho$ is an eigenvalue of a copositive A and U is the eigenspace of $-\rho$, then each $\mathbf{x} \in U$ satisfies $\|\mathbf{x}_+\| = \|\mathbf{x}_-\|$, and the eigenspace of A corresponding to ρ includes all the vectors $|\mathbf{x}| = \mathbf{x}_+ + \mathbf{x}_-$, $\mathbf{x} \in U$. We call a vector $\mathbf{x}$ *balanced* if $\|\mathbf{x}_+\| = \|\mathbf{x}_-\|$. Note that $\mathbf{x}$ is balanced if and only if $\mathbf{x}$ is orthogonal to $|\mathbf{x}|$. We say that a subspace U of $\mathbb{R}^n$ is *balanced* if every vector in U is balanced.

Lemma 2.77. *Let U be a subspace of $\mathbb{R}^n$. Then U is balanced if and only if for every $\mathbf{x}, \mathbf{y} \in U$, $\mathbf{x}$ is orthogonal to $|\mathbf{y}|$.*

Proof. If $\mathbf{x}$ is orthogonal to $|\mathbf{y}|$ for every $\mathbf{x}, \mathbf{y} \in U$, then in particular every $\mathbf{x} \in U$ is orthogonal to $|\mathbf{x}|$, and is therefore balanced.

For the converse, suppose every vector in U is balanced. Let $\mathbf{x}, \mathbf{y} \in U$. To show that $\mathbf{x}$ is orthogonal to $|\mathbf{y}|$, we may assume without loss of generality that $\operatorname{supp}\mathbf{y} = \{1, 2, \dots, n\}$, the first k entries of $\mathbf{y}$ are positive, and the next $m - k$ entries are negative. Let $\gamma = \{m+1, \dots, n\}$ ($m = n$ and $\gamma = \emptyset$ possible). For every t the vector $\mathbf{x} + t\mathbf{y} \in U$ is balanced, and for $t > 0$ sufficiently large the signs of its first m entries are the same as those of $\mathbf{y}$. Hence for such t,

$$\sum_{i=1}^{k}(x_i + ty_i)^2 - \sum_{i=k+1}^{m}(x_i + ty_i)^2 + \|\mathbf{x}_+\|_\gamma^2 - \|\mathbf{x}_-\|_\gamma^2 = 0.$$

Differentiating we get that for large t

$$\sum_{i=1}^{k}(x_i + ty_i)2y_i - \sum_{i=k+1}^{m}(x_i + ty_i)2y_i = 0. \tag{2.17}$$

Since $\mathbf{y}$ is balanced,

$$\sum_{i=1}^{k} y_i^2 - \sum_{i=k+1}^{m} y_i^2 = 0. \tag{2.18}$$

Substituting (2.18) in (2.17) we get that

$$\sum_{i=1}^{k} x_i y_i - \sum_{i=k+1}^{m} x_i y_i = 0,$$

that is, $\mathbf{x}$ is orthogonal to $|\mathbf{y}|$. □

Theorem 2.78. *Let $A \neq 0$ be a symmetric matrix whose spectral radius is $\rho = \rho(A)$, and all its negative eigenvalues are equal to $-\rho$. Then A is copositive if and only if for every eigenvector $\mathbf{x}$ corresponding to $-\rho$, $|\mathbf{x}| = \mathbf{x}_+ + \mathbf{x}_-$ is an eigenvector corresponding to the eigenvalue ρ.*

Proof. The "only if" part was proved in Corollary 2.21.

For the "if" part, denote by U the eigenspace of $-\rho$, and let

$$W = \operatorname{span}\{|\mathbf{x}| \mid \mathbf{x} \in U\}.$$

For every $\mathbf{x} \in U$ the vector $|\mathbf{x}|$ is an eigenvector of ρ, hence it is orthogonal to $\mathbf{x}$. That is, U is balanced. By Lemma 2.77, W is orthogonal to U. Let $V = (U \oplus W)^\perp$. Denote by P be the matrix of the orthogonal projection on U, and by Q the matrix of the orthogonal projection on W. We will show that $B = Q - P$ is copositive. Since the eigenspace of A corresponding to ρ contains W and all eigenvalues of A other than $-\rho$ are nonnegative, $A - \rho B$ is positive semidefinite. Thus the copositivity of B implies that A is copositive.

To show that B is copositive, consider $\mathbf{v} \in \mathbb{R}^n_+$. Write $\mathbf{v} = \mathbf{x} + \mathbf{y} + \mathbf{z}$, where $\mathbf{x} \in W$, $\mathbf{y} \in V$ and $\mathbf{z} \in U$. Without loss of generality

$$\mathbf{z} = \begin{pmatrix} \mathbf{z}_1 \\ -\mathbf{z}_2 \end{pmatrix}, \ \mathbf{z}_1 \in \mathbb{R}^k_+, \ \mathbf{z}_2 \in \mathbb{R}^m_+, \ k + m = n, \quad \text{and} \quad \|\mathbf{z}_1\| = \|\mathbf{z}_2\|.$$

Divide $\mathbf{x}$ and $\mathbf{y}$ into two segments each, conformally with $\mathbf{z}$:

$$\mathbf{x} = \begin{pmatrix} \mathbf{x}_1 \\ \mathbf{x}_2 \end{pmatrix} \quad \text{and} \quad \mathbf{y} = \begin{pmatrix} \mathbf{y}_1 \\ \mathbf{y}_2 \end{pmatrix}, \quad \mathbf{x}_1, \mathbf{y}_1 \in \mathbb{R}^k, \ \mathbf{x}_2, \mathbf{y}_2 \in \mathbb{R}^m.$$

As $\mathbf{y} \in (U \oplus W)^\perp$, $\mathbf{y}^T\mathbf{z} = \mathbf{y}^T|\mathbf{z}| = 0$. That is,

$$\mathbf{y}_1^T\mathbf{z}_1 - \mathbf{y}_2^T\mathbf{z}_2 = 0 \quad \text{and} \quad \mathbf{y}_1^T\mathbf{z}_1 + \mathbf{y}_2^T\mathbf{z}_2 = 0.$$

Hence

$$\mathbf{y}_1^T\mathbf{z}_1 = \mathbf{y}_2^T\mathbf{z}_2 = 0. \tag{2.19}$$

By Lemma 2.77, $\mathbf{x}^T\mathbf{z} = 0$, i.e., $\mathbf{x}_1^T\mathbf{z}_1 - \mathbf{x}_2^T\mathbf{z}_2 = 0$,

$$\mathbf{x}_1^T\mathbf{z}_1 = \mathbf{x}_2^T\mathbf{z}_2. \tag{2.20}$$

Since $\mathbf{v} \geq \mathbf{0}$, $\mathbf{x}_2 + \mathbf{y}_2 - \mathbf{z}_2 \geq \mathbf{0}$, and thus

$$\mathbf{x}_2 + \mathbf{y}_2 \geq \mathbf{z}_2. \tag{2.21}$$

Transpose (2.21), multiply by $\mathbf{z}_2 \geq \mathbf{0}$, and use (2.19) and Cauchy-Schwarz to obtain

$$\|\mathbf{x}_2\| \cdot \|\mathbf{z}_2\| \geq \mathbf{x}_2^T\mathbf{z}_2 = (\mathbf{x}_2 + \mathbf{y}_2)^T\mathbf{z}_2 \geq \mathbf{z}_2^T\mathbf{z}_2 = \|\mathbf{z}_2\|^2. \tag{2.22}$$

That is,

$$\|\mathbf{x}_2\| \geq \|\mathbf{z}_2\|. \tag{2.23}$$

As $\mathbf{x}_1^T\mathbf{z}_1 = \mathbf{x}_2^T\mathbf{z}_2$ and $\|\mathbf{z}_1\| = \|\mathbf{z}_2\|$, (2.22) and another use of Cauchy-Schwarz yield also

$$\|\mathbf{x}_1\| \cdot \|\mathbf{z}_1\| \geq \mathbf{x}_1^T\mathbf{z}_1 \geq \|\mathbf{z}_1\|^2,$$

so that

$$\|\mathbf{x}_1\| \geq \|\mathbf{z}_1\|. \tag{2.24}$$

Together (2.23) and (2.24) imply that

$$\|\mathbf{x}\|^2 = \|\mathbf{x}_1\|^2 + \|\mathbf{x}_2\|^2 \geq \|\mathbf{z}_1\|^2 + \|\mathbf{z}_2\|^2 = \|\mathbf{z}\|^2. \tag{2.25}$$

For every $\mathbf{v} \geq \mathbf{0}$, $B\mathbf{v} = Q\mathbf{v} - P\mathbf{v} = \mathbf{x} - \mathbf{z}$ and by (2.25)

$$\mathbf{v}^TB\mathbf{v} = (\mathbf{x} + \mathbf{y} + \mathbf{z})^T(\mathbf{x} - \mathbf{z}) = \|\mathbf{x}\|^2 - \|\mathbf{z}\|^2 \geq 0.$$

That is, B is copositive (and therefore so is A). □

Remark 2.79. If A is as in Theorem 2.78, and the multiplicity of $-\rho$ as an eigenvalue of A is 1, then A is SPN. For if $\mathbf{v}$ is a unit eigenvector corresponding to $-\rho$, and if P and Q are as in the proof, then $Q - P = |\mathbf{v}||\mathbf{v}|^T - \mathbf{v}\mathbf{v}^T \geq 0$. Hence A, which is the sum of $\rho(Q - P)$ and a positive semidefinite matrix, is SPN.

Example 2.80. For every balanced subspace U of $\mathbb{R}^n$, if P is the orthogonal projection matrix on U, then $A = I - 2P$ is an orthogonal copositive matrix. Since every orthogonal copositive matrix other than I has eigenvalues -1 and 1, Corollary 2.21 implies that these are all the orthogonal copositive matrices, other than I.

Remark 2.81. For a balanced subspace U of $\mathbb{R}^n$, let W be defined as in the proof of Theorem 2.78. Then it can be shown that $\dim W \geq \dim U$, and thus $\dim U \leq \lfloor \frac{n}{2} \rfloor$. It can also be shown that for every $k \leq \lfloor \frac{n}{2} \rfloor$ there exist balanced subspaces of $\mathbb{R}^n$ of dimension k. The proofs are left as exercises.

Copositivity by zeros

A copositive matrix A satisfies $A\mathbf{u} \geq \mathbf{0}$ for every zero $\mathbf{u}$ of A. This property pretty much characterizes copositive matrices.

Theorem 2.82. *Let A be a symmetric matrix with nonnegative diagonal elements. Then A is copositive if and only if $A\mathbf{u} \geq \mathbf{0}$ for every zero $\mathbf{u}$ of A.*

Proof. The "only if" was proved in Proposition 2.45. For the "if" part, consider first the case that all the diagonal entries of A are positive, and $A\mathbf{u} \geq \mathbf{0}$ for every zero $\mathbf{u}$ of A. Suppose on the contrary that A is not copositive. Let $\mathbf{v}$ be a nonnegative vector such that $\mathbf{v}^T A\mathbf{v} < 0$, and suppose $|\operatorname{supp} \mathbf{v}|$ is minimal among such vectors. By the assumption on the diagonal elements, $|\operatorname{supp} \mathbf{v}| > 1$. Let $i \in \operatorname{supp} \mathbf{v}$ such that $(A\mathbf{v})_i < 0$, and set $\mathbf{u}(t) = \mathbf{v} - t\mathbf{e}_i$. Then $\mathbf{u}(t) \geq \mathbf{0}$, and $\mathbf{u}(t) \neq \mathbf{0}$ for every $0 \leq t \leq v_i$. Denote $\phi(t) = \mathbf{u}(t)^T A\mathbf{u}(t)$. By the choice of $\mathbf{v}$, $\phi(0) < 0$ and $\phi(v_i) \geq 0$. As $\phi(t)$ is continuous on $[0, v_i]$, there exists $0 < t_0 \leq v_i$ such that $\phi(t_0) = 0$. Then $\mathbf{u}(t_0)$ is a zero of A with

$$(A\mathbf{u}(t_0))_i = (A\mathbf{v})_i - t_0 a_{ii} < 0,$$

contrary to the assumption.

Now suppose some of the diagonal elements of A are zeros. Let

$$\alpha = \{1 \leq i \leq n \mid a_{ii} > 0\}.$$

For every zero $\mathbf{u}$ of $A[\alpha]$, prepending and appending zeros to $\mathbf{u}$ yields a zero $\mathbf{v}$ of A (see Remark 2.44), and hence $A[\alpha]\mathbf{u} = (A\mathbf{v})[\alpha] \geq \mathbf{0}$. By the first part of the proof, $A[\alpha]$ is copositive. For every $i \notin \alpha$ the vector $\mathbf{e}_i$ is a zero of A, and therefore $A\mathbf{e}_i \geq \mathbf{0}$. Thus $A \geq A[\alpha] \oplus 0$ is copositive. □

Remark 2.83. In particular, a symmetric matrix with all diagonal elements positive and no zeros is strictly copositive.

The next result is a special case in which copositivity of a matrix is determined by the existence of certain zeros.

Proposition 2.84. *Let A be an $n \times n$ symmetric matrix, $n \geq 2$. Let $\mathbf{u}$ and $\mathbf{v}$ be zeros of A such that* $\operatorname{supp} \mathbf{u} = \{1, \ldots, n\} \setminus \{i\}$, $\operatorname{supp} \mathbf{v} = \{1, \ldots, n\} \setminus \{j\}$, *$j \neq i$, and $A(i)$, $A(j)$ are both positive semidefinite. Then the following are equivalent.*

(a) *A is copositive.*
(b) *A is SPN.*
(c) $\mathbf{u}^T A \mathbf{v} \geq 0$.

Moreover, A is positive semidefinite if and only if $\mathbf{u}^T A \mathbf{v} = 0$.

Proof. As (b) obviously implies (a), we only need to prove two of the implications. Without loss of generality, we may assume $i = n$ and $j = 1$.

(a) $\Rightarrow$ (c): As $\mathbf{u}$ and $\mathbf{v}$ are zeros of the copositive A, $\mathbf{u} \geq \mathbf{0}$ and $A\mathbf{v} \geq \mathbf{0}$, hence $\mathbf{u}^T A \mathbf{v} \geq 0$.

(c) $\Rightarrow$ (b): If $n = 2$, $\operatorname{supp} \mathbf{u} = \{1\}$, $\operatorname{supp} \mathbf{v} = \{2\}$. Hence we may assume that $\mathbf{u} = \mathbf{e}_1$, $\mathbf{v} = \mathbf{e}_2$, and the diagonal of A is zero. Then (c) means that $a_{12} = \mathbf{e}_1^T A \mathbf{e}_2 \geq 0$, i.e., A is nonnegative.

For $n \geq 3$, the graph $K_n - e$, where e is the edge $\{1, n\}$, is a connected chordal graph. Since $A(n)$ and $A(1)$ are both positive semidefinite there exists a positive semidefinite P such that $P(1) = A(1)$ and $P(n) = A(n)$. The matrices A and P may differ only in the $1, n$ and $n, 1$ entries, hence $A = P + \delta\,(E_{1n} + E_{n1})$. To prove that A is SPN it suffices to show that $\delta \geq 0$. By the assumption on $\mathbf{u}$ and $\mathbf{v}$, $\mathbf{u}^T P \mathbf{u} = \mathbf{u}^T A \mathbf{u} = 0$, and similarly $\mathbf{v}^T P \mathbf{v} = 0$. Since P is positive semidefinite, both $\mathbf{u}$ and $\mathbf{v}$ are in the null space of P, hence $\mathbf{u}^T P \mathbf{v} = 0$. Then $\mathbf{u}^T A \mathbf{v} = \mathbf{u}^T P \mathbf{v} + \delta \mathbf{u}^T E_{1n} \mathbf{v} = \delta u_1 v_n$. As $u_1, v_n > 0$ and $\mathbf{u}^T A \mathbf{v} \geq 0$, $\delta \geq 0$.

Note that $\mathbf{u} + \mathbf{v}$ is a zero of A if and only if $\mathbf{u}^T A \mathbf{v} = 0$. Thus if A is copositive and $\mathbf{u}^T A \mathbf{v} = 0$, then $\mathbf{u} + \mathbf{v}$ is a positive zero of A and A is positive semidefinite. Conversely, if A is positive semidefinite, then $\mathbf{u}$ and $\mathbf{v}$ are in $\operatorname{ns} A$, hence $\mathbf{u}^T A \mathbf{v} = 0$. □

In the next corollary we consider another special case. Suppose a symmetric matrix A has n zeros, $\mathbf{u}_1, \ldots, \mathbf{u}_n$, each with supports of size $n-2$, such that supports change cyclically: up to permutation similarity, the support of $\mathbf{u}_{i+1}$ is $\sigma_i = \{1, \ldots, n\} \setminus \{i, i \mathbin{\text{\textcircled{+}}} 1\}$,where $\mathbin{\text{\textcircled{+}}}$ denotes addition modulo n on $\{1, \ldots, n\}$ (i.e., $n \mathbin{\text{\textcircled{+}}} 1 = 1$).

Corollary 2.85. *Let A be an $n \times n$ symmetric matrix, $n \geq 3$. Let $\mathbf{u}_1, \ldots, \mathbf{u}_n$ be zeros of A, whose supports are*

$$\sigma_i = \operatorname{supp} \mathbf{u}_i = \{1, \ldots, n\} \setminus \{i, i \mathbin{\text{\textcircled{+}}} 1\}, \quad i = 1, \ldots, n,$$

where $\mathbin{\text{\textcircled{+}}}$ denotes addition modulo n on $\{1, \ldots, n\}$. Then the following conditions are equivalent.

(a) *A is copositive.*
(b) *$A(i)$ is copositive for every $1 \leq i \leq n$.*
(c) *$A(i)$ is SPN for every $1 \leq i \leq n$.*
(d) *$\mathbf{u}_i^T A \mathbf{u}_{i \mathbin{\text{\textcircled{+}}} 1} \geq 0$ for every $1 \leq i \leq n$.*

Proof. (a) $\Rightarrow$ (d): If A is copositive, (d) follows from $\mathbf{u}_i \geq \mathbf{0}$, $A\mathbf{u}_{i \mathbin{\text{\textcircled{+}}} 1} \geq \mathbf{0}$.

(d) $\Rightarrow$ (c): For each $1 \leq i \leq n$, apply Proposition 2.84 to $A(i \mathbin{\text{\textcircled{+}}} 1)$ in place of A, with its principal submatrices $A[\sigma_i]$ and $A[\sigma_{i \mathbin{\text{\textcircled{+}}} 1}]$, and the zeros $\mathbf{u}_i$, $\mathbf{u}_{i \mathbin{\text{\textcircled{+}}} 1}$ with the $(i \mathbin{\text{\textcircled{+}}} 1)$-th (zero) entry omitted. This implies that $A(i)$ is an SPN matrix for every i.

(c) $\Rightarrow$ (b) is obvious.

(b) $\Rightarrow$ (a): Let $\overline{\mathbf{u}}_1 \in \mathbb{R}_+^{n-1}$ be obtained from $\mathbf{u}_1$ by omitting the first entry. That is, $\overline{\mathbf{u}}_1$ consists of the positive entries of $\mathbf{u}_1$, preceded by one zero entry. Then $\overline{\mathbf{u}}_1$ is a zero of the copositive $A(1)$, so $A(1)\overline{\mathbf{u}}_1 \geq \mathbf{0}$, and therefore $(A\mathbf{u}_1)_i \geq 0$ for every $i \neq 1$. As $\overline{\mathbf{u}}_1$ is also a zero of the copositive $A(2)$, $A(2)\overline{\mathbf{u}}_1 \geq \mathbf{0}$, and $(A\mathbf{u}_1)_i \geq 0$ for every $i \neq 2$. Therefore $A\mathbf{u}_1 \geq 0$.

For every vector $\mathbf{u} \in \mathbb{R}_+^n$ with at least one zero entry, $\mathbf{u}^T A \mathbf{u} \geq 0$. If $\mathbf{v} \in \mathbb{R}_+^n$ is a positive vector, there exists $t > 0$ such that $\mathbf{w} = \mathbf{v} - t\mathbf{u}_1 \geq \mathbf{0}$, and $\mathbf{w}$ has at least one zero entry. Then

$$0 \leq \mathbf{w}^T A \mathbf{w} = \mathbf{v}^T A \mathbf{v} - 2t\mathbf{v}^T A \mathbf{u}_1,$$

i.e., $\mathbf{v}^T A \mathbf{v} \geq 2t\mathbf{v}^T A \mathbf{u}_1$. As $A\mathbf{u}_1 \geq \mathbf{0}$, $\mathbf{v} > \mathbf{0}$ and $t > 0$, this implies that $\mathbf{v}^T A \mathbf{v} \geq 0$. $\square$

Remark 2.86. Let $n \geq 5$, and assume the rest of the conditions and notations of Corollary 2.85. Suppose (a)–(d) hold (A is copositive). Then the following are equivalent.

(i) A is SPN.
(ii) A is positive semidefinite.
(iii) $\mathbf{u}_i^T A\mathbf{u}_{i+1} = 0$ for every $1 \le i \le n$.
(iv) $\mathbf{u}_i^T A\mathbf{u}_{i+1} = 0$ for at least one $1 \le i \le n$.

Since $n \ge 5$, for each $1 \le i \le j \le n$ there exists $\mathbf{u}_k$ such that $\{i, j\} \subseteq \operatorname{supp}\mathbf{u}_k$, hence A is $\mathcal{N}$-irreducible. If A is SPN, it is therefore positive semidefinite. That is, (i) implies (ii). Also, (ii) implies (iii), since for a positive semidefinite A each of the zeros $\mathbf{u}_i$ is in the null space of A. Clearly, (iii) implies (iv). Finally, if (iv) holds for a certain i, then $\mathbf{u}_i + \mathbf{u}_{i+1}$ is also a zero of A, with support of size $n-1$. Hence A is SPN by Proposition 2.69.

Copositivity and the Schur complement

Computationally, the most efficient way to find out whether a given matrix is positive definite is using the Cholesky decomposition to obtain a LL^T-decomposition of the matrix, with L lower triangular. This decomposition uses the fact that for $a_{ii} > 0$, A is positive semidefinite if and only if $A/[i]$ is positive semidefinite. Here is where a key difference between positive semidefinite matrices and copositive matrices lies. The Schur complement $A/[i]$ of a copositive A with $a_{ii} > 0$ is not necessarily copositive. For example,

$$A = \begin{pmatrix} 1 & 1 \\ 1 & 0 \end{pmatrix}$$

is copositive, but $A/[1] = (-1)$ is not copositive. This is yet another evidence of the fact that characterization of copositivity is much more difficult than the characterization of positive semidefiniteness. In some cases, however, it is possible to use Schur complement to reduce the problem of checking a given symmetric matrix for copositivity, to checking a matrix of a smaller order. For convenience we phrase the results in this subsection for the Schur complement with respect to a principal submatrix, which forms the lower right block in the matrix.

The following is a rather general Schur complement result on copositivity.

Theorem 2.87. *Let*

$$A = \begin{pmatrix} B & C \\ C^T & E \end{pmatrix}$$

be an $n \times n$ symmetric matrix, where B is $k \times k$, E is a positive semidefinite matrix, $\operatorname{cs} C^T \subseteq \operatorname{cs} E$, and $E^\dagger C^T \le 0$. Then A is copositive if and only if A/E is copositive.

Proof. If A/E is copositive, then

$$A = \begin{pmatrix} CE^\dagger C^T & C \\ C^T & E \end{pmatrix} + \begin{pmatrix} A/E & 0 \\ 0 & 0 \end{pmatrix}$$

is a sum of the positive semidefinite matrix

$$\begin{pmatrix} CE^\dagger C^T & C \\ C^T & E \end{pmatrix} = \begin{pmatrix} C\sqrt{E^\dagger} \\ \sqrt{E} \end{pmatrix} \begin{pmatrix} \sqrt{E^\dagger} C^T & \sqrt{E} \end{pmatrix}$$

and the copositive matrix $(A/E) \oplus 0$, and is therefore copositive.

Now suppose that A is copositive. To show that A/E is copositive, for any $\mathbf{x} \in \mathbb{R}^k_+$, let $\mathbf{y} = -E^\dagger C^T \mathbf{x} \geq \mathbf{0}$. Then

$$\begin{aligned} 0 \leq \begin{pmatrix} \mathbf{x}^T & \mathbf{y}^T \end{pmatrix} A \begin{pmatrix} \mathbf{x} \\ \mathbf{y} \end{pmatrix} &= \mathbf{x}^T B\mathbf{x} + 2\mathbf{x}^T C\mathbf{y} + \mathbf{y}^T E\mathbf{y} \\ &= \mathbf{x}^T B\mathbf{x} - 2\mathbf{x}^T CE^\dagger C^T \mathbf{x} + \mathbf{x}^T CE^\dagger EE^\dagger C^T \mathbf{x} \\ &= \mathbf{x}^T B\mathbf{x} - 2\mathbf{x}^T CE^\dagger C^T \mathbf{x} + \mathbf{x}^T CE^\dagger C^T \mathbf{x} \\ &= \mathbf{x}^T (B - CE^\dagger C^T)\mathbf{x} \end{aligned}$$

that is, $A/E = B - CE^\dagger C^T$ is copositive. □

(Note that in the proof of Theorem 2.87 the nonpositivity of $E^\dagger C$ was needed only in one direction, to deduce that A/E is copositive when A is copositive. See Exercise 2.71.)

Particularly useful are some special cases of Theorem 2.87.

Corollary 2.88. *Let*

$$A = \begin{pmatrix} B & C \\ C^T & E \end{pmatrix}$$

be an $n \times n$ symmetric matrix, where B is $k \times k$, E is positive definite, and $E^{-1} C^T \leq 0$. Then A is copositive if and only if A/E is copositive.

Remark 2.89. Corollary 2.88 holds if "copositive" is replaced by "strictly copositive". The proof of the strictly copositive version is left as an exercise.

A particular instance of Corollary 2.88 is the following.

Corollary 2.90. *Let*

$$A = \begin{pmatrix} B & C \\ C^T & E \end{pmatrix}$$

be an $n \times n$ symmetric matrix, such that B is $k \times k$, E is positive definite with $E^{-1} \geq 0$, and $C \leq 0$. Then A is copositive if and only if A/E is copositive.

Recall that any nonsingular M-matrix is an example of an E satisfying the condition of Corollary 2.90. For the special case is when E is a single positive diagonal entry, this yields the following.

Corollary 2.91. *Let A be an $n \times n$ symmetric matrix with $a_{ii} > 0$, such that all the off-diagonal elements in row i are nonpositive. Then A is copositive if and only if $A/[i]$ is copositive.*

To complete the picture we consider the case that a diagonal entry in A is zero.

Theorem 2.92. *Let an $n \times n$ symmetric matrix A have $a_{ii} = 0$. Then A is copositive if and only if its i-th row and column are nonnegative and $A(i)$ is copositive.*

Proof. We may assume that $i = n$, and

$$A = \begin{pmatrix} B & \mathbf{c} \\ \mathbf{c}^T & 0 \end{pmatrix}.$$

If A is copositive, then $\mathbf{c} \geq \mathbf{0}$ (by Proposition 2.5(b)), and the principal submatrix $A(n) = B$ is also copositive. Conversely, if B is copositive and $\mathbf{c} \geq \mathbf{0}$, then $A \geq B \oplus 0$ is also copositive (see Propositions 2.15 and 2.11). □

Remark 2.93. In Theorem 2.87 and its corollaries, if the Schur complement is SPN, then A is SPN (see Exercise 2.71). For a result about a case in which the Schur complement of an SPN matrix is SPN, see Section 2.14.

For completeness we mention a variation on Corollary 2.91, which is more general, but seems to be of less practical value. It states that if in an $n \times n$ symmetric matrix A the entry a_{ii} is positive, then A is copositive if and only if $A(i)$ is copositive and $A/[i]$ is copositive with respect to a certain polyhedral subcone of $\mathbb{R}^{n-1}_+$. It is convenient to state this result for the case $i = n$.

Proposition 2.94. *Let*

$$A = \begin{pmatrix} B & \mathbf{c} \\ \mathbf{c}^T & a \end{pmatrix}$$

be an $n \times n$ symmetric matrix, and $a > 0$. Then A is copositive if and only if B is copositive and $\mathbf{x}^T(A/[n])\mathbf{x} \geq 0$ for every $\mathbf{x} \in \mathbb{R}^{n-1}_+$ such that $\mathbf{c}^T\mathbf{x} \leq 0$.

Proof. The matrix A is copositive if and only if

$$\begin{pmatrix} t\mathbf{x}^T & s \end{pmatrix} A \begin{pmatrix} t\mathbf{x} \\ s \end{pmatrix} \geq 0$$

for every $t, s \geq 0$ and $\mathbf{x} \in \mathbb{R}_+^{n-1}$. That is, if and only if

$$\begin{pmatrix} t & s \end{pmatrix} \begin{pmatrix} \mathbf{x}^T B\mathbf{x} & \mathbf{x}^T\mathbf{c} \\ \mathbf{c}^T\mathbf{x} & a \end{pmatrix} \begin{pmatrix} t \\ s \end{pmatrix} \geq 0$$

for every $t, s \geq 0$. So A is copositive if and only if the 2×2 matrix $\begin{pmatrix} \mathbf{x}^T B\mathbf{x} & \mathbf{x}^T\mathbf{c} \\ \mathbf{c}^T\mathbf{x} & a \end{pmatrix}$ is copositive for every $\mathbf{x} \in \mathbb{R}_+^{n-1}$. Recall that a 2×2 symmetric matrix is copositive if and only if it is either nonnegative or positive semidefinite. Thus A is copositive if and only if $\mathbf{x}^T B\mathbf{x} \geq 0$ for every $\mathbf{x} \in \mathbb{R}_+^{n-1}$, and the 2×2 matrix is positive semidefinite for every $\mathbf{x} \in \mathbb{R}_+^{n-1}$ such that $\mathbf{c}^T\mathbf{x} \leq 0$. This last requirement is equivalent (by the nonnegativity of a and $\mathbf{x}^T B\mathbf{x}$) to

$$\det \begin{pmatrix} \mathbf{x}^T B\mathbf{x} & \mathbf{x}^T\mathbf{c} \\ \mathbf{c}^T\mathbf{x} & a \end{pmatrix} = \mathbf{x}^T B\mathbf{x} \cdot a - \mathbf{x}^T\mathbf{c}\mathbf{c}^T\mathbf{x} = a\,\mathbf{x}^T(A/[n])\mathbf{x} \geq 0. \qquad \square$$

Remark 2.95. More generally, let

$$A = \begin{pmatrix} B & \mathbf{c} \\ \mathbf{c}^T & a \end{pmatrix}$$

be an $n \times n$ symmetric matrix. Then A is copositive if and only if $a \geq 0$, B is copositive, and $\mathbf{x}^T \left(aB - \mathbf{c}\mathbf{c}^T\right) \mathbf{x} \geq 0$ for every $\mathbf{x} \in \mathbb{R}_+^{n-1}$ such that $\mathbf{c}^T\mathbf{x} \leq 0$.

The next few results combine zeros of a matrix and the Schur complement in some special cases. First, a symmetric $n \times n$ matrix that has zeros with support of size $n-2$. Under some conditions on these zeros we discuss copositivity of such a matrix, and the possibility of it being exceptional and extremal. It is convenient to introduce the conditions of the next proposition using a graph. Let $G_{\mathcal{V}}^{n-2}$ be a graph on vertices $\{1, \ldots, n\}$, where $\{i, j\}$ is an edge if and only if there exists a zero of A whose support is $\{1, \ldots, n\} \setminus \{i, j\}$.

Proposition 2.96. *Let A be an $n \times n$ symmetric matrix. Let the graph $G_{\mathcal{V}}^{n-2}$ contain a cycle on k vertices, and let α consist of the remaining $n-k$ vertices. Then A is copositive if and only if $A/A[\alpha]$ is copositive.*

Proof. By the assumptions, $n \geq k \geq 3$. We may assume without loss of generality that the cycle is on vertices $\{1, \ldots, k\}$. That is, there exists zeros $\mathbf{u}_1, \ldots, \mathbf{u}_k$ such that

$$\sigma_i = \operatorname{supp} \mathbf{u}_i = \{1, \ldots, n\} \setminus \{i, i \mathbin{\hat{+}} 1\}, \quad i = 1, \ldots, k,$$

where $\hat{+}$ denotes here addition modulo k on $\{1, \ldots, k\}$. Let

$$\alpha = \{k+1, \ldots, n\} = \bigcap_{i=1}^{k} \sigma_i.$$

Then

$$A = \begin{pmatrix} B & C \\ C^T & E \end{pmatrix},$$

where $E = A[\alpha]$ is positive semidefinite, as α is a subset of a zero support of A.

We first show that $\operatorname{ns} E \subseteq \operatorname{ns} C$ (which is equivalent to $\operatorname{cs} C^T \subseteq \operatorname{cs} E$). Suppose $\mathbf{x} \in \operatorname{ns} E$. Let $\mathbf{y} \in \mathbb{R}^n$ be obtained by prepending k zero entries to $\mathbf{x}$. For every i let $\mathbf{y}_i = \mathbf{y}[\sigma_i]$. Then $\mathbf{y}_i^T A[\sigma_i]\mathbf{y}_i = \mathbf{x}^T A[\alpha]\mathbf{x} = 0$. As $A[\sigma_i]$ is positive semidefinite, $A[\sigma_i]\mathbf{y}_i = \mathbf{0}$, hence $(A\mathbf{y})[\sigma_i] = \mathbf{0}$. Since $\bigcup_{i=1}^{k} \sigma_i = \{1, \ldots, n\}$ this implies that $A\mathbf{y} = 0$, and therefore $C\mathbf{x} = \mathbf{0}$. Hence $A = P + Q$, where

$$P = \begin{pmatrix} CE^\dagger C^T & C \\ C^T & E \end{pmatrix}, \qquad Q = \begin{pmatrix} A/E & 0 \\ 0 & 0 \end{pmatrix},$$

P is positive semidefinite and A/E is a $k \times k$ matrix. From this decomposition of A it is clear that if $A/A[\alpha]$ is copositive, then so is A.

Now suppose A is copositive. As $\alpha \subseteq \sigma_i = \operatorname{supp} \mathbf{u}_i$, $(A\mathbf{u}_i)[\alpha] = \mathbf{0}$. That is,

$$\begin{pmatrix} C^T & E \end{pmatrix} \mathbf{u}_i = \mathbf{0}.$$

Hence also (see (1.11) on page 30)

$$\begin{pmatrix} CE^\dagger C^T & C \end{pmatrix} \mathbf{u}_i = CE^\dagger \begin{pmatrix} C^T & E \end{pmatrix} \mathbf{u}_i = \mathbf{0},$$

and

$$P\mathbf{u}_i = \begin{pmatrix} CE^\dagger C^T & C \\ C^T & E \end{pmatrix} \mathbf{u}_i = \mathbf{0}.$$

Therefore each $\mathbf{u}_i$ is a zero of both A and P, and thus also a zero of Q. Moreover, as $\mathbf{u}_i$ is a zero of the copositive A, $A\mathbf{u}_i \geq \mathbf{0}$, which together with $P\mathbf{u}_i = \mathbf{0}$ implies that $Q\mathbf{u}_i \geq \mathbf{0}$. Denote $\beta = \{1, \ldots, k\}$ and $\mathbf{v}_i = \mathbf{u}_i[\beta]$. Each $\mathbf{v}_i$ is a zero of $A/A[\alpha]$, and $\operatorname{supp} \mathbf{v}_i = \sigma_i \setminus \alpha$ is of size $k-2$. For every $i = 1, \ldots, k$,

$$\mathbf{v}_i^T (A/A[\alpha])\, \mathbf{v}_{i \hat{+} 1} \geq 0,$$

since $\mathbf{v}_i \geq \mathbf{0}$ and $(A/A[\alpha])\, \mathbf{v}_{i \hat{+} 1} = (Q\mathbf{u}_{i \hat{+} 1})[\beta] \geq 0$. By Corollary 2.85, applied to $A/A[\alpha]$ and the zeros $\mathbf{v}_1, \ldots, \mathbf{v}_k$, $A/A[\alpha]$ is copositive. $\square$

Proposition 2.96 not only reduces the problem of checking copositivity of such a matrix, it also gives a necessary condition for it to be an exceptional extremal copositive matrix.

Corollary 2.97. *If A is an exceptional extremal copositive matrix with a positive diagonal, then $G_{\mathcal{V}}^{n-2}$ does not contain a k-cycle with $k < n$.*

Proof. Suppose on the contrary that there exists a cycle of length $k < n$ in $G_{\mathcal{V}}^{n-2}$. In the notations of Proposition 2.96, $A = P + Q$, where P is positive semidefinite and $Q = (A/A[\alpha]) \oplus 0$. Since A is extremal, both P and Q are multiples of A. As A is exceptional, $P = 0$. Thus $Q = A$. This is a contradiction, since Q has $n - k$ zero rows and A has a positive diagonal. $\square$

Corollary 2.98. *If A is an $n \times n$ exceptional copositive matrix, and all its zeros with support of size $n - 2$ are minimal, then the degree of each vertex in $G_{\mathcal{V}}^{n-2}$ is at most 2.*

If A is also extremal, and all the minimal zeros have support of size $n - 2$, then there are n minimal zeros and $G_{\mathcal{V}}^{n-2}$ is a cycle on n vertices.

Proof. If $\mathbf{u}$ and $\mathbf{v}$ are minimal zeros of A such that

$$\sigma_1 = \operatorname{supp} \mathbf{u} = \{1, \ldots, n\} \setminus \{i, j\} \quad \text{and} \quad \sigma_2 = \operatorname{supp} \mathbf{v} = \{1, \ldots, n\} \setminus \{j, k\},$$

then $\sigma_2 \setminus \sigma_1 = \{i\}$. By Corollary 2.59, $\mathbf{u}$ and $\mathbf{v}$ are the only minimal zeros with support contained in $\sigma_1 \cup \sigma_2$, and every zero whose support is contained in $\sigma_1 \cup \sigma_2$ is a nonnegative combination of $\mathbf{u}$ and $\mathbf{v}$. Thus there is no zero of A with support $\{1, \ldots, n\} \setminus \{j, l\}$, $l \neq i, k$. That is, the degree of every vertex in $G_{\mathcal{V}}^{n-2}$ is at most 2.

Note that since A is an exceptional matrix, the support of each of its zeros has size at most $n - 2$ (Corollary 2.70). Hence if all the minimal zeros have support of size $n - 2$, each zero is a minimal zero. If A is also extremal, it is $\mathcal{PSD}$-irreducible and therefore has at least n linearly independent minimal zeros (Corollary 2.68). Thus in this case $G_{\mathcal{V}}^{n-2}$ is a graph on n vertices, each with degree at most 2, and it has at least n edges. It therefore has exactly n edges, and must contain a cycle. By the previous corollary this cycle cannot be of length smaller than n, hence $G_{\mathcal{V}}^{n-2}$ has to be the n-cycle. $\square$

Here is one more case in which the Schur complement of a copositive matrix is copositive.

Proposition 2.99. *Let A be a copositive matrix. If A has a zero $\mathbf{u}$ such that $A\mathbf{u} = \mathbf{0}$, and $\sigma = \operatorname{supp} \mathbf{u}$, then $A/A[\sigma]$ is copositive.*

Proof. Suppose A is an $n \times n$ matrix and $|\sigma| = m$. By using a permutation similarity we may assume that $\sigma = \{n-m+1, \dots, n\}$, and $A[\sigma]$ is positive semidefinite of rank $r < m$. By further permuting the last m rows and columns of A we may also assume that for $\alpha = \{n-r+1, \dots, n\}$ the matrix $A[\alpha]$ is positive definite ($|\alpha| = r$). We first show that $A/A[\alpha]$ is copositive. Let

$$A = \begin{pmatrix} B & C \\ C^T & E \end{pmatrix},$$

where $E = A[\alpha]$. Then $A = P + Q$, where

$$P = \begin{pmatrix} CE^{-1}C^T & C \\ C^T & E \end{pmatrix}$$

is positive semidefinite and

$$Q = A/E \oplus 0_r.$$

For every $\mathbf{x} \in \mathbb{R}_+^{n-r}$, let $\mathbf{y} = -E^{-1}C^T\mathbf{x} \in \mathbb{R}^r$ and

$$\mathbf{z} = \begin{pmatrix} \mathbf{x} \\ \mathbf{y} \end{pmatrix} \in \mathbb{R}^n.$$

Since $\mathbf{u}[\alpha] > \mathbf{0}$, we may choose $t > 0$ such that $\mathbf{z} + t\mathbf{u} \geq \mathbf{0}$. By the copositivity of A, $(\mathbf{z} + t\mathbf{u})^T A(\mathbf{z} + t\mathbf{u}) \geq 0$, and since $A\mathbf{u} = \mathbf{0}$,

$$0 \leq (\mathbf{z} + t\mathbf{u})^T A(\mathbf{z} + t\mathbf{u}) = \mathbf{z}^T A\mathbf{z} + 2t\mathbf{z}^T A\mathbf{u} + t^2\mathbf{u}^T A\mathbf{u} = \mathbf{z}^T A\mathbf{z}.$$

As $\mathbf{z}$ was chosen so that $P\mathbf{z} = \mathbf{0}$,

$$0 \leq \mathbf{z}^T A\mathbf{z} = \mathbf{z}^T A\mathbf{z} - \mathbf{z}^T P\mathbf{z} = \mathbf{z}^T Q\mathbf{z}.$$

Since the last r rows and columns in Q are zero,

$$0 \leq \mathbf{z}^T Q\mathbf{z} = \begin{pmatrix} \mathbf{x}^T & \mathbf{y}^T \end{pmatrix} Q \begin{pmatrix} \mathbf{x} \\ \mathbf{y} \end{pmatrix} = \mathbf{x}^T(A/E)\mathbf{x}.$$

This shows that A/E is copositive (and so is Q).

The zero $\mathbf{u}$ of $A = P + Q$ is also a zero of P and Q. Since P is positive semidefinite, $P\mathbf{u} = \mathbf{0}$. Together with $A\mathbf{u} = \mathbf{0}$ this implies $Q\mathbf{u} = \mathbf{0}$. The positive semidefinite matrices $A[\sigma]$ and $P[\sigma]$ both share the positive definite $r \times r$ principal submatrix E, and both have rank r. Hence $P[\sigma] = A[\sigma]$, and therefore $Q[\sigma] = 0$. This in turn implies that the corresponding m rows and columns in the copositive Q are nonnegative. As $\mathbf{u}[\sigma] > \mathbf{0}$ and $Q\mathbf{u} = \mathbf{0}$, we get that the last m entries in each row of Q are zero. So $Q = Q(\sigma) \oplus 0_m$, and we get that $A/A[\sigma] = P/P[\sigma] + Q(\sigma)$. Since P is positive semidefinite and $\operatorname{rank} P = \operatorname{rank} P[\sigma]$, we have that $P/P[\sigma] = 0$, and $A/A[\sigma] = Q(\sigma)$. In particular, $A/A[\sigma]$ is copositive. □

Proposition 2.99 is not an if and only if one. Some additional requirements are needed to guarantee that if $A/A[\sigma]$ is copositive, then so is A. See Exercises 2.76 for one such case.

We state several implications of Proposition 2.99.

Corollary 2.100. *If an $n \times n$ copositive matrix A has a zero $\mathbf{u}$ such that $|\operatorname{supp} \mathbf{u}| = n - 1$ and $A\mathbf{u} = \mathbf{0}$, then A is positive semidefinite.*

Proof. By the proof of Proposition 2.99, denoting $\sigma = \operatorname{supp} \mathbf{u}$, we have $A = P + Q$, where P is positive semidefinite and $Q = (A/A[\sigma]) \oplus 0_{n-1}$ is copositive. As Q is zero except for, possibly, a 1×1 principal submatrix, it is also positive semidefinite. Hence A is positive semidefinite. □

Corollary 2.101. *If an $n \times n$ copositive matrix has a zero $\mathbf{u}$ such that $|\operatorname{supp} \mathbf{u}| = n - 2$, and $(A\mathbf{u})_i = 0$ for some $i \notin \operatorname{supp} \mathbf{u}$, then A is SPN.*

Proof. Applying the previous corollary to $A[\alpha]$, where $\alpha = \operatorname{supp} \mathbf{u} \cup \{i\}$, we get that $A[\alpha]$ is a positive semidefinite principal submatrix of A of order $n - 1$. By Proposition 2.69, A is SPN. □

Copositivity by sign pattern and graph

One special case in which copositivity can be easily determined is when the given symmetric matrix is a Z-matrix.

Theorem 2.102. *A symmetric matrix A all of whose off-diagonal entries are nonpositive is copositive if and only if it is positive semidefinite.*

Proof. Clearly only the "only if" needs to be proved. Suppose A is an $n \times n$ copositive matrix. For any $\mathbf{x} \in \mathbb{R}^n$, write $\mathbf{x} = \mathbf{x}_+ - \mathbf{x}_-$. Let $\alpha = \operatorname{supp} \mathbf{x}_+$ and $\beta = \{1, \ldots, n\} \setminus \alpha \supseteq \operatorname{supp} \mathbf{x}_-$. Then $A[\alpha]$ and $A[\beta]$ are copositive, $\mathbf{x}[\alpha] > \mathbf{0}$ and $\mathbf{x}[\beta] \leq 0$ and therefore, $\mathbf{x}[\alpha]^T A[\alpha]\mathbf{x}[\alpha] \geq 0$ and $\mathbf{x}[\beta]^T A[\beta]\mathbf{x}[\beta] \geq 0$. As $A[\alpha|\beta] \leq 0$, $\mathbf{x}[\alpha]^T A[\alpha|\beta]\mathbf{x}[\beta] \geq 0$. Thus

$$\mathbf{x}^T A\mathbf{x} \geq \mathbf{x}[\alpha]^T A[\alpha]\mathbf{x}[\alpha] + \mathbf{x}[\beta]^T A[\beta]\mathbf{x}[\beta] \geq 0. \qquad \square$$

Remark 2.103. Similarly, suppose A is a symmetric matrix all whose off-diagonal elements are nonpositive. Then A is strictly copositive if and only if it is positive definite. We leave the proof as an exercise.

Sign pattern and graph may have a role in determining copositivity. For a start, we phrase Proposition 2.15 in terms of graphs, and leave its proof as an exercise.

Proposition 2.104. *An $n \times n$ symmetric matrix A is copositive if and only if $A[\alpha]$ is copositive for every set α, which is the vertex set of a component of $G(A)$.*

Recall that for a symmetric $n \times n$ matrix A, $G_-(A)$ denotes the graph on vertices $\{1, \ldots, n\}$ in which $\{i, j\}$ is an edge if and only if $a_{ij} < 0$.

Proposition 2.105. *An $n \times n$ symmetric matrix A is copositive if and only if $A[\alpha]$ is copositive for every set α, which is the vertex set of a component of $G_-(A)$.*

Proof. If A is copositive, then every principal submatrix of A is copositive. For the other implication, let $\alpha_1, \ldots, \alpha_k$ be the vertex sets of the components of $G_-(A)$. Let $\beta = \{1, \ldots, n\} \setminus (\alpha_1 \cup \cdots \cup \alpha_k)$. Then $A[\beta] \geq 0$ is copositive (if nonempty), and $A \geq A[\alpha_1] \oplus \cdots \oplus A[\alpha_k] \oplus A[\beta]$. The result follows by Propositions 2.11 and 2.15. □

Finally, we consider the case that the graph $G(A)$ has a cut vertex.

Theorem 2.106. *Let A be an $n \times n$ symmetric matrix. Let $G(A) = G_1 \cup G_2$, where $G_1 \cap G_2$ consists of a single vertex. Let $V(G_1) = \{1, \ldots, k\}$ and $V(G_2) = \{k, \ldots, n\}$. Then A is copositive if and only if there exist copositive matrices A_1 and A_2 such that $G(A_1) = G_1$, $G(A_2) = G_2$ and*

$$A = (A_1 \oplus 0_{n-k}) + (0_{k-1} \oplus A_2).$$

Proof. It is obvious that if such A_1 and A_2 exist, then A is copositive.

For the main implication, it suffices to show that if A is copositive, and is $\{k, k\}$-irreducible, then such A_1 and A_2 exist. So suppose

$$A = \begin{pmatrix} B_1 & \mathbf{b} & 0 \\ \mathbf{b}^T & a & \mathbf{c}^T \\ 0 & \mathbf{c} & B_2 \end{pmatrix},$$

where $B_1 \in \mathcal{S}_{k-1}$ and $B_2 \in \mathcal{S}_{n-k}$ are both copositive, $a > 0$, and A is $\{k, k\}$-irreducible. By Proposition 2.63 there exists $\mathbf{u} \in \mathbb{R}^n_+$ such that $\mathbf{u}^T A \mathbf{u} = 0$ and $u_k = 1$. Let $\mathbf{u}^T = (\mathbf{v}^T \, 1 \, \mathbf{w}^T)$ for some $\mathbf{v} \in \mathbb{R}^{k-1}_+$ and $\mathbf{w} \in \mathbb{R}^{n-k}_+$.

Let

$$A_1 = \begin{pmatrix} B_1 & \mathbf{b} \\ \mathbf{b}^T & p \end{pmatrix},$$

where p is such that $(\mathbf{v}^T \, 1) A_1 \begin{pmatrix} \mathbf{v} \\ 1 \end{pmatrix} = 0$. By the copositivity of $A[1, \ldots, k]$ such p exists, and $p \leq a$. Let $q = a - p$ and

$$A_2 = \begin{pmatrix} q & \mathbf{c}^T \\ \mathbf{c} & B_2 \end{pmatrix}.$$

Since $A = (A_1 \oplus 0_{n-k}) + (0_{k-1} \oplus A_2)$,

$$0 = \mathbf{u}^T A\mathbf{u} = (\mathbf{v}^T\ 1)A_1 \begin{pmatrix} \mathbf{v} \\ 1 \end{pmatrix} + (1\ \mathbf{w}^T)A_2 \begin{pmatrix} 1 \\ \mathbf{w} \end{pmatrix},$$

and thus also $(1\ \mathbf{w}^T)A_2 \begin{pmatrix} 1 \\ \mathbf{w} \end{pmatrix} = 0$.

We now show that the matrix A_1 is copositive. If $\mathbf{x} \in \mathbb{R}^k_+$ has the form $\mathbf{x}^T = (\mathbf{y}^T\ 0)$, then $\mathbf{x}^T A_1 \mathbf{x} = \mathbf{y}^T B_1 \mathbf{y} \geq 0$. Otherwise, $\mathbf{x}^T = (\mathbf{y}^T\ \gamma)$, $\gamma > 0$, and we may assume $\gamma = 1$. Then for $\mathbf{z} = (\mathbf{y}^T\ 1\ \mathbf{w}^T)^T$ we have

$$0 \leq \mathbf{z}^T A\mathbf{z} = (\mathbf{y}^T\ 1)A_1 \begin{pmatrix} \mathbf{y} \\ 1 \end{pmatrix} + (1\ \mathbf{w}^T)A_2 \begin{pmatrix} 1 \\ \mathbf{w} \end{pmatrix} = \mathbf{x}^T A_1 \mathbf{x}.$$

As A_1 is copositive, $p \geq 0$, and thus $0 \leq q \leq a$. By a similar argument A_2 is also copositive. □

Remark 2.107. No such result exists for the case that $G(A) = G_1 \cup G_2$, where $G_1 \cap G_2$ is more than a single vertex. See Exercise 2.84.

A special class of matrices for which copositivity may be completely characterized combinatorially is the class of $\{0, 1, -1\}$-matrices. The positive semidefinite and the SPN matrices in this class are also fully characterized, as well as the extremal copositive matrices. We therefore dedicate a short section to this class later in this chapter, see Section 2.8.

Exercises

2.63. Prove the statement in Remark 2.75.

2.64. Let U be a balanced subspace of $\mathbb{R}^n$. Let $W = \text{span}\{|\mathbf{x}| \mid \mathbf{x} \in U\}$. Show that $\dim W \geq \dim U$. Deduce that $\dim U \leq \lfloor \frac{n}{2} \rfloor$.
Hint: Consider a basis $\mathbf{u}_1, \mathbf{u}_2, \ldots, \mathbf{u}_k$ of U such that I_k is a submatrix if $(\mathbf{u}_1\ \mathbf{u}_2\ \ldots\ \mathbf{u}_k)$.

2.65. Show that for every $1 \leq k \leq \lfloor \frac{n}{2} \rfloor$ there exists a k-dimensional balanced subspace of $\mathbb{R}^n$.
Hint: Let S be an $\lfloor \frac{n}{2} \rfloor \times k$ matrix of full column rank, and consider the column space of $\begin{pmatrix} S \\ -S \end{pmatrix}$.

2.66. Let A be a nonsingular symmetric matrix all of whose positive eigenvalues are equal to $\rho = \rho(A)$, and all of its negative eigenvalues are equal to λ. Let Q and P be the matrices of the orthogonal projections on the eigenspace of ρ and that of λ, respectively. Show that A is copositive if and only if for every $\mathbf{x} \in \mathbb{R}^n_+$,

$$\|P\mathbf{x}\|^2 \le -\frac{\rho}{\lambda}\|Q\mathbf{x}\|^2.$$

Hint: $A = \rho Q + \lambda P$. See Exercise 2.25 for the "only if" part.

2.67. Use Theorem 2.82 to show that the matrix

$$\begin{pmatrix} 1 & -1 & \frac{1}{2} & -\frac{1}{4} \\ -1 & 1 & -\frac{1}{4} & \frac{1}{2} \\ \frac{1}{2} & -\frac{1}{4} & 1 & -1 \\ -\frac{1}{4} & \frac{1}{2} & -1 & 1 \end{pmatrix}$$

is copositive.
Hint: Show that A and each of its 3×3 principal submatrices have nonzero determinants. Deduce that $\mathbf{e}_1 + \mathbf{e}_2$ ad $\mathbf{e}_3 + \mathbf{e}_4$ are the only zeros of A (up to scalar multiplication).

2.68. Prove the following variation on Theorem 2.82: Let A be a symmetric $n \times n$ matrix such that $\mathbf{v}^T A \mathbf{v} > 0$ for some $\mathbf{v} \in \mathbb{R}^n_+$. Show that A is copositive if and only if $A\mathbf{u} \ge \mathbf{0}$ for every zero $\mathbf{u}$ of A.

2.69. Prove the following characterization of strict-copositivity: Let A be symmetric $n \times n$ matrix. Show that A is strictly copositive if and only if for every $\mathbf{x} \in \mathbb{R}^n_+$, $(A\mathbf{x})_i > 0$ for at least one $i \in \operatorname{supp} \mathbf{x}$.
Hint: For the "if" implication, suppose $\min_{\mathbf{x} \in \Delta} \mathbf{x}^T A \mathbf{x}$ is attained at $\mathbf{v}$, and use Remark 2.49.

2.70. Check the validity of Corollary 2.88 and Remark 2.91.

2.71. Let

$$A = \begin{pmatrix} B & C \\ C^T & E \end{pmatrix}$$

be a symmetric matrix, where the submatrix E is positive semidefinite and $\operatorname{cs} C^T \subseteq \operatorname{cs} E$. Show that if A/E is copositive, then A is copositive. Show that if A/E is SPN, then A is SPN.

2.72. Let

$$A = \begin{pmatrix} B & C \\ C^T & E \end{pmatrix}$$

be a symmetric matrix, where the submatrix E is positive definite with $E^{-1} \ge 0$, and $C \le 0$. Show that A is strictly copositive if and only if A/E is strictly copositive.

2.73. Prove the claim in Remark 2.95.

2.74. Let

$$A = \begin{pmatrix} B & \mathbf{c} \\ \mathbf{c}^T & a \end{pmatrix}$$

be a symmetric matrix with $\mathbf{c} \leq \mathbf{0}$. Show that A is copositive if and only if $a \geq 0$ and B and $aB - \mathbf{c}\mathbf{c}^T$ are both copositive.

2.75. Let A be a symmetric $n \times n$ matrix,

$$A = \begin{pmatrix} B & C \\ C^T & E \end{pmatrix},$$

where E is 2×2. Let

$$\mathbf{v}(t) = \begin{pmatrix} t \\ 1-t \end{pmatrix} \quad \text{and} \quad M(t) = \begin{pmatrix} B & C\mathbf{v}(t) \\ \mathbf{v}(t)^T C^T & \mathbf{v}(t)^T E \mathbf{v}(t) \end{pmatrix}.$$

Show that A is copositive if and only if B is copositive and $M(t)$ is copositive for every $0 \leq t \leq 1$.

Hint: $\begin{pmatrix} \mathbf{x}^T & \mathbf{v}(t)^T \end{pmatrix} A \begin{pmatrix} \mathbf{x}^T \\ \mathbf{v}(t) \end{pmatrix} = \begin{pmatrix} \mathbf{x}^T & 1 \end{pmatrix} M(t) \begin{pmatrix} \mathbf{x}^T \\ 1 \end{pmatrix}$.

2.76. Let A be a symmetric matrix, and suppose $\mathbf{u}$ is a minimal zero of A such that $A\mathbf{u} = \mathbf{0}$. Let $\sigma = \operatorname{supp} \mathbf{u}$. Show that if $A/A[\sigma]$ is copositive, then A is copositive, and if $A/A[\sigma]$ is SPN, then A is SPN.

Hint: Proof of Proposition 2.99, and Exercise 2.71.

2.77. Show that an extremal copositive-plus matrix A is positive semidefinite.

Hint: Consider first the case that A has a positive diagonal.

2.78. Give another proof of Theorem 2.102 using Corollary 2.91 and induction.

2.79. Give yet another proof of Theorem 2.102, using the fact that the smallest eigenvalue of an irreducible symmetric Z-matrix has a corresponding positive eigenvector.

2.80. Prove the claim in Remark 2.103.

2.81. Prove Proposition 2.104.

2.82. Let A be an $n \times n$ copositive matrix, and let α be a set of indices of rows of A such that for every $i \in \alpha$ all the off-diagonal elements in row i are nonpositive. Show that if $A[\alpha]$ is singular, then A is reducible (i.e., $G(A)$ is not connected).
Hint: Without loss of generality, $A[\alpha] = M_1 \oplus M_2$, where M_1 is a singular M-matrix and M_2 is a nonsingular M-matrix. Let $\mathbf{v} > \mathbf{0}$ be a zero of M_1, and consider $A\mathbf{u}$, where $\mathbf{u} \in \mathbb{R}^n$ is obtained by appending zeros to $\mathbf{v}$.

2.83. Let $\mathbf{v} \in \mathbb{R}^n$. Show that the following are equivalent.

(a) $I - \mathbf{v}\mathbf{v}^T$ is copositive.
(b) $I - \mathbf{v}\mathbf{v}^T$ is SPN.
(c) $\|\mathbf{v}_-\| \le 1$ and $\|\mathbf{v}_+\| \le 1$.

2.84. Let

$$A = \begin{pmatrix} 1 & -1 & 1 & 0 \\ -1 & 1 & -1 & 1 \\ 1 & -1 & 1 & -1 \\ 0 & 1 & -1 & 1 \end{pmatrix}.$$

Use Proposition 2.16 to show that A is copositive. Note that $G(A) = G_1 \cup G_2$, where G_1 and G_2 are triangles on vertices $\{1, 2, 3\}$ and $\{2, 3, 4\}$, respectively. Show that there do not exist copositive A_1 and A_2 such that $A = (A_1 \oplus 0) + (0 \oplus A_2)$ and $G(A_i) = G_i$, $i = 1, 2$.
Hint: If $A = B_1 + B_2$, B_1 and B_2 copositive, then every zero $\mathbf{u}$ of A is also a zero of B_1 and B_2.

Notes. Theorem 2.74 and its strictly copositive version in Remark 2.75 are due to [Kaplan (2000)]. These are in fact succinct formulations of previous results — see next section. Theorem 2.76 was proved in [Jargalsaikhan (2013)]; Another proof was presented in [Baccari and Naffouti (2016)]. Theorem 2.78, the terminology of a balanced subspace, and Exercises 2.64 and 2.65, are due to [Shaked-Monderer (2018)]. Exercise 2.66 is based on [Naffouti and Baccari (2016)].

Theorem 2.82 was mentioned without proof (for the case of positive diagonal elements) in [Hoffman and Pereira (1973)]. The variation in Exercise 2.68 appears in [Väliaho (1986)] (previously proved in the Ph.D. dissertation [Pereira (1972)], which is not readily available; the obvious copositive-plus version is proved in the general context of copositivity with respect to a cone in [Cottle *et al.* (1970b)]). Proposition 2.84, Corollary 2.85 and Remark 2.86 are due to [Hildebrand (2017)].

The use of the Schur complement in detecting copositivity dates back to [Bomze (1987)], see also [Bomze (1989, 1996, 2008)]. Proposition 2.94 is based on the latter two. Other versions appear in [Andersson *et al.* (1995)] and [Li and Feng (1993)] (Exercise 2.74, hence Corollary 2.91 and Theorem 2.92), reproved in [Johnson and Reams (2005)] (Corollary 2.91), and [Shaked-Monderer (2016)] (the special case of Corollary 2.90 where E is an M-matrix). The version in Theorem 2.87 appears in [Bisht and Sivakumar (2017)]. The proof of Proposition 2.94 is based on the one in [Li and Feng (1993)], and Exercise 2.75 is also based on that paper. Propositions 2.96 and Corollaries 2.97 and 2.98 were proved in [Hildebrand (2018b)], and Proposition 2.99 in [Baumert (1966)] (in terms of quadratic forms).

Theorem 2.102 was suggested as an exercise in [Murty (1988)], and [Hiriart-Urruty and Seeger (2010)] suggest the proof outlined in Exercise 2.78. Proposition 2.105 states in terms of graphs an observation of [Li and Feng (1993)]. Theorem 2.106 was proved in [Shaked-Monderer (2016)], and the statements in Exercises 2.82 and 2.83 as well. Exercise 2.69 is based on [Siegmund-Schultze and von Weizsäcker (2007), Proof of Proposition 2].

2.7 Almost copositive matrices

If all the proper principal submatrices of a symmetric matrix are positive semidefinite, then the matrix itself is positive semidefinite if and only if its determinant is nonnegative. In this section we consider this kind of recursive characterizations for copositivity. Some of the results are easier to present in a negative form — characterizing the non-copositive matrices among the symmetric matrices all of whose proper principal submatrices are copositive.

Definition 2.108. Let A be a symmetric $n \times n$ matrix. If all the principal submatrices of A of order r (or less) are copositive, we say that A is *copositive of order* r. If A is copositive of order r but not copositive of order $r+1$, we say that A is *copositive of exact order* r. An $n \times n$ matrix A, which is copositive of exact order $n-1$ is also called *almost copositive.*

The notions of a matrix A that is copositive semidefinite/positive definite/strictly copositive/copositive-plus of order r/exact order r are defined similarly.

The following characterization of $n \times n$ matrices that are almost copositive may be viewed as a recursive characterization of copositive matrices, in a negative form.

Theorem 2.109. *Let A be an $n \times n$ matrix that is copositive of order $n-1$. Then the following are equivalent.*

(a) *A is not copositive.*
(b) *There exists a negative eigenvalue of A with a corresponding positive eigenvector.*
(c) *For every vector $\mathbf{a} > \mathbf{0}$ in $\mathbb{R}^n$ there exists a vector $\mathbf{b} < \mathbf{0}$ in $\mathbb{R}^n$, such that $A\mathbf{b} = \mathbf{a}$.*
(d) *A is nonsingular and $A^{-1} \leq 0$.*
(e) $\det A < 0$ *and* $\operatorname{adj} A \geq 0$.

Proof. (a) $\Leftrightarrow$ (b): By Theorem 2.74, A is not copositive if and only if a principal submatrix of A has a negative eigenvalue with a corresponding positive eigenvector. As A is copositive of order $n-1$, such a principal submatrix has to be A itself.

(a) $\Leftrightarrow$ (c): If (c) holds, then for $\mathbf{a} > \mathbf{0}$ and $\mathbf{b} < \mathbf{0}$ such that $A\mathbf{b} = \mathbf{a}$, $(-\mathbf{b})^T A(-\mathbf{b}) = \mathbf{b}^T\mathbf{a} < 0$, and A is not copositive. Conversely, suppose A is not copositive, and that $\mathbf{a} > \mathbf{0}$. The set $\Delta_{\mathbf{a}} = \{\mathbf{x} \in \mathbb{R}^n_+ \mid \mathbf{a}^T\mathbf{x} = 1\}$ is a compact convex set, and $\min_{\mathbf{x}\in\Delta_{\mathbf{a}}} \frac{1}{2}\mathbf{x}^T A\mathbf{x} < 0$. Since A is copositive of order $n-1$, this minimum is necessarily attained at a positive $\mathbf{u} \in \Delta_{\mathbf{a}}$, i.e., in the interior of $\Delta_{\mathbf{a}}$. Using Lagrange multipliers we get that at $\mathbf{u}$, $\nabla\left(\frac{1}{2}\mathbf{x}^T A\mathbf{x}\right) = \lambda\nabla(\mathbf{a}^T\mathbf{x})$ for some λ, i.e. $A\mathbf{u} = \lambda\mathbf{a}$. Hence $0 > \mathbf{u}^T A\mathbf{u} = \lambda\mathbf{u}^T\mathbf{a} = \lambda$. The vector $\mathbf{b} = \frac{1}{\lambda}\mathbf{u} < \mathbf{0}$ satisfies $A\mathbf{b} = \mathbf{a}$.

(b) $\Rightarrow$ (d): Let $\mathbf{v}$ be a unit positive eigenvector of A corresponding to a negative eigenvalue λ. If $\mathbf{x}$ is a unit eigenvector orthogonal to $\mathbf{v}$ and corresponding to an eigenvalue μ, then $\mathbf{x}$ has both positive and negative entries. Thus for some $s > 0$, the vector $\mathbf{v} + s\mathbf{x}$ is nonnegative and has a zero entry. As A is copositive of order $n-1$,

$$\begin{aligned} 0 \leq (\mathbf{v}^T + s\mathbf{x}^T)A(\mathbf{v} + s\mathbf{x}) &= \mathbf{v}^T A\mathbf{v} + 2s\mathbf{x}^T A\mathbf{v} + s^2\mathbf{x}^T A\mathbf{x} \\ &= \lambda + 2s\lambda\mathbf{x}^T\mathbf{v} + s^2\mu = \lambda + s^2\mu. \end{aligned}$$

That is $\mu \geq \frac{-\lambda}{s^2} > 0$. This shows that λ is a simple eigenvalue, and all other eigenvalues are positive. Therefore $\det A < 0$, and A is nonsingular. By the equivalence of (b) and (c), $A^{-1}\mathbf{a} < \mathbf{0}$ for every positive vector $\mathbf{a}$. In particular, for every $s > 0$

$$A^{-1}(s\mathbf{v} + \mathbf{e}_i) < 0.$$

Thus $A^{-1}\mathbf{e}_i < -\frac{s}{\lambda}\mathbf{v}$ for every $s > 0$, and the i-th column of A^{-1} is nonpositive.

(d) $\Rightarrow$ (c): Suppose $A^{-1} \leq 0$. As there is no zero row in A^{-1}, each row has a negative entry. Hence for every $\mathbf{a} > \mathbf{0}$ the vector $A^{-1}\mathbf{a} < 0$.

So (a), (b), (c) and (d) are equivalent. (e) trivially implies (d). Finally, if (a)–(d) hold, then, as shown above, A is nonsingular with a single negative eigenvalue and $A^{-1} \leq 0$. Hence $\det A < 0$ and $\operatorname{adj} A \geq 0$. □

The following theorem states some necessary conditions for a matrix A to be almost copositive.

Theorem 2.110. *Let A be an $n \times n$ matrix, $n \geq 2$, which is copositive of exact order $n-1$. Then*

(a) $\operatorname{In} A = (n-1, 1, 0)$.
(b) *A is positive semidefinite of exact order $n-1$.*
(c) *A is positive definite of order $n-2$.*
(d) *All the principal minors of A^{-1} of order at least 2 are negative.*
(e) *$A^{-1} \leq 0$, and its off-diagonal entries are all negative.*

Proof. Part (a) was shown in the previous proof.

To prove (b) we only need to show that every principal submatrix of A of order $n-1$ is positive semidefinite. As A is copositive of order $n-1$, $\mathbf{x}^T A\mathbf{x} \geq 0$ for every $\mathbf{x} \in \mathbb{R}^n$, which is either nonnegative or nonpositive, and has at least one zero entry. To complete the proof it suffices to show that if $\mathbf{x} \in \mathbb{R}^n$ is any vector with both positive and negative entries, then $\mathbf{x}^T A\mathbf{x} \geq 0$. Let $\mathbf{v}$ be a unit positive eigenvector corresponding to a negative eigenvalue λ of A. Given any vector $\mathbf{x}$ with both positive and negative entries, let

$$\phi(t) = (\mathbf{v}^T + t\mathbf{x})A(\mathbf{v} + t\mathbf{x}) = \mathbf{v}^T A\mathbf{v} + 2t\mathbf{x}^T A\mathbf{v} + t^2\mathbf{x}^T A\mathbf{x}.$$

There exist $t_1 < 0$ and $t_2 > 0$ such that each of $\mathbf{v} + t_i\mathbf{x}$, $i = 1, 2$, is a nonnegative vector with a zero entry and thus $\phi(t_1), \phi(t_2) \geq 0$. Together with $\phi(0) < 0$ this implies that the coefficient of t^2 in the quadratic function $\phi(t)$ is positive, i.e., $\mathbf{x}^T A\mathbf{x} > 0$.

Now for (c). Note that if A is a 2×2 almost copositive matrix, we may consider (c) to hold, considering the determinant of a 0×0 matrix to be 1. Suppose $n \geq 3$. The $n \times n$ almost copositive matrix A has positive diagonal entries (since $a_{ii} = 0$ would imply, by the copositivity of 2×2 principal submatrices, that the i-th row and column of A are nonnegative, which together with the copositivity of $A(i)$ would mean that A is copositive).

If A is not positive definite of order $n-2$, suppose A is positive definite of order k, $1 \leq k \leq n-3$. Let $A[\alpha]$ be a singular positive semidefinite principal submatrix of order $k+1$, whose principal submatrices of order k are positive definite. Without loss of generality we may assume that

$$A = \begin{pmatrix} B & C \\ C^T & E \end{pmatrix},$$

where $E = A[n-k+1,\dots,n]$ is a $k \times k$ positive definite matrix, and $\alpha = \{n-k,\dots,n\}$. For $i = n-k$ we have $(A/E)_{ii} = 0$. As

$$0 > \det A = \det E \; \det(A/E),$$

$\det(A/E) < 0$, and therefore the i-th row of A/E is not all zero. Suppose $(A/E)_{ij} \neq 0$, $j < i$. Let $\beta = \{j\} \cup \alpha$. Then $|\beta| = k+2 < n$ and by part (b) it is a positive semidefinite matrix. Hence $A[\beta]/E$ is a positive semidefinite matrix. However, $(A[\beta]/E)\,[j,i] = (A/E)[j,i]$ is a 2×2 matrix with a zero diagonal element and nonzero off-diagonal elements, and therefore $\det(A[\beta]/E) < 0$, a contradiction.

To show that (d) holds, recall that the $n \times n$ nonsingular A satisfies for every $\alpha \subseteq \{1,\dots,n\}$,

$$\det A \; \det A^{-1}(\alpha) = \det A[\alpha] \tag{2.26}$$

(see Theorem 1.9). By Theorem 2.109 we have $\det A < 0$, and by (d) $\det A[\alpha] > 0$ for $|\alpha| \le n-2$. Thus by the above equation the determinant of every principal submatrix of A^{-1} of order at least 2 is negative.

Finally we prove (e): By Theorem 2.109, $B = A^{-1} \le 0$. Suppose on the contrary that $b_{ij} = 0$ for some $i \neq j$. Then $\det B[i,j] = b_{ii}b_{jj} \ge 0$, contrary to (d). $\square$

Remark 2.111. None of conditions (a)–(d) in Theorem 2.110 is sufficient for almost copositivity, and neither are all four conditions together. For example, the 3×3 matrix

$$A = \begin{pmatrix} 1 & 1 & 0 \\ 1 & 1 & 1 \\ 0 & 1 & 1 \end{pmatrix}$$

is positive semidefinite of order 2 and positive definite of order 1. Its eigenvalues 1 and $1 \pm \sqrt{2}$ and therefore $\operatorname{In} A = (2,1,0)$. The matrix A^{-1} and its principal submatrices of order 2 have negative determinants. The matrix A is, of course, copositive. Condition (e), however, guarantees the non-copositivity of A, as the next theorem shows.

Theorem 2.112. *Let A be a nonsingular copositive matrix. Then each column of A^{-1} has at least one positive entry.*

Proof. Suppose on the contrary that the i-th column in A^{-1} is nonpositive, and let $\mathbf{b} = -A^{-1}\mathbf{e}_i \geq 0$. Then

$$0 \leq \mathbf{b}^T A\mathbf{b} = \mathbf{e}_i^T A^{-1}\mathbf{e}_i \leq 0,$$

implying that $\mathbf{b}$ is a zero of A. But $A\mathbf{b} = -AA^{-1}\mathbf{e}_i = -\mathbf{e}_i$ so $A\mathbf{b} \geq \mathbf{0}$ does not hold, in contradiction to Proposition 2.45. □

Corollary 2.113. *A symmetric matrix A is not copositive if and only if there exists a nonsingular principal submatrix $A[\alpha]$ such that $A[\alpha]^{-1}$ has a nonpositive column.*

Proof. By Theorem 2.112, if $A[\alpha]^{-1}$ has a nonpositive column, then $A[\alpha]$ is not copositive, and therefore neither is A.

If A is not copositive, it is copositive of exact order k for some $0 \leq k < n$. Let $A[\alpha]$ be a non-copositive matrix with $|\alpha| = k+1$. Then, by Theorem 2.109, $A[\alpha]$ is nonsingular and $A[\alpha]^{-1} \leq 0$. (Note that if $k = 0$, then $a_{ii} < 0$ for some i, and $\alpha = \{i\}$.) □

A variation on this is the following result, whose proof is left as an exercise.

Corollary 2.114. *A symmetric matrix A is not copositive if and only if there exists a nonsingular principal submatrix $A[\alpha]$ such that* $\det A[\alpha] < 0$ *and all the cofactors of the last column in $A[\alpha]$ are nonnegative.*

The next theorem contains a characterization of almost copositivity that does not require checking *all* the proper principal submatrices for copositivity.

Theorem 2.115. *Let A be an $n \times n$ symmetric matrix. Then the following are equivalent*

(a) *A is copositive of exact order $n-1$.*
(b) In $A = (n-1, 1, 0)$ *and* $A^{-1} \leq 0$.
(c) $\det A < 0$, $A^{-1} \leq 0$, *and A has a set of $n-2$ positive nested principal minors.*

Proof. By Theorems 2.109 and 2.110, (a) implies (b) and (c).

(c) $\Rightarrow$ (b): If (c) holds, the positive nested minors condition simply means that A has a positive definite principal submatrix of order $n-2$. So we may assume that

$$A = \begin{pmatrix} B & C \\ C^T & E \end{pmatrix},$$

where E is an $(n-2)\times(n-2)$ positive definite submatrix. Then

$$0 > \det A = \det E \det(A/E)$$

implies that the 2×2 matrix A/E has a negative determinant, and therefore $\operatorname{In} A/E = (1,1,0)$. As $\operatorname{In} E = (n-2,0,0)$, we get

$$\operatorname{In} A = \operatorname{In} E + \operatorname{In} A/E = (n-1,1,0).$$

Finally, (b) $\Rightarrow$ (a): If (b) holds, then $\det A < 0$. Together with $A^{-1} \geq 0$ this implies, by (2.26),

$$\det A(i) = \left(A^{-1}\right)_{ii} \det A \geq 0$$

for every $1 \leq i \leq n$. But by the Interlacing Theorem (Theorem 1.18), each principal submatrix of A of order $n-1$ has at most one negative eigenvalue. Therefore all the eigenvalues of $A(i)$ are nonnegative. That is, every principal submatrix of order at most $n-1$ is positive semidefinite. Hence A is copositive of order $n-1$, and nonsingular with $A^{-1} \leq 0$. By Theorem 2.109 it is not copositive. □

A recursive characterization of copositivity in a positive form is the following.

Theorem 2.116. *Let A be an $n\times n$ symmetric matrix, which is copositive of order $n-1$. Then the following are equivalent.*

(a) *A is copositive.*
(b) *There exists an $n\times n$ strictly copositive matrix B such that $A\mathbf{x} = \lambda B\mathbf{x}$ for a positive vector $\mathbf{x}$ implies that $\lambda \geq 0$.*
(c) *For every $n\times n$ strictly copositive matrix B, if $A\mathbf{x} = \lambda B\mathbf{x}$ for a positive vector $\mathbf{x}$, then $\lambda \geq 0$.*

Proof. Clearly (c) $\Rightarrow$ (b).

We show first the implication (a) $\Rightarrow$ (c). If B is any strictly copositive matrix, and $A\mathbf{x} = \lambda B\mathbf{x}$ for some positive vector $\mathbf{x}$, then $0 \leq \mathbf{x}^T A\mathbf{x} = \lambda \mathbf{x}^T B\mathbf{x}$, and as $\mathbf{x}^T B\mathbf{x} > 0$, this implies that $\lambda \geq 0$.

Finally, (b) $\Rightarrow$ (a): Consider the set $D = \{\mathbf{x} \in \mathbb{R}^n_+ \mid \mathbf{x}^T B\mathbf{x} = 1\}$. By the strict copositivity of B, each nonzero vector in $\mathbb{R}^n_+$ is a positive scalar multiple of a vector in D, and thus A is copositive if and only if $\min_{\mathbf{x}\in D} \mathbf{x}^T A\mathbf{x}$ is nonnegative (D is compact). Suppose on the contrary that this minimum is negative, and let $\mathbf{v} \in D$ be a vector at which this minimum is attained. As A is copositive of order $n-1$, $\mathbf{v}$ is a positive vector, and therefore using the Lagrange multipliers method, we get that $\nabla(\mathbf{x}^T A\mathbf{x}) = \lambda \nabla(\mathbf{x}^T B\mathbf{x})$ at $\mathbf{v}$. That is, $A\mathbf{v} = \lambda B\mathbf{v}$. As (b) holds, $\lambda \geq 0$. But then $\mathbf{v}^T A\mathbf{v} = \lambda \mathbf{v}^T B\mathbf{v} = \lambda \geq 0$, a contradiction. □

Remark 2.117. Let A and B be matrices of the same order, A symmetric and B strictly copositive. If A and B satisfy

$$A\mathbf{x} = \lambda B\mathbf{x}\ ,\quad \mathbf{x} > \mathbf{0} \qquad \Rightarrow \qquad \lambda \geq 0,$$

the pair A, B is called a *codefinite pair.* In using Theorem 2.116 to characterize copositivity, it is useful to recall that the strictly copositive B may be taken to be any positive definite matrix, or any nonnegative matrix with a positive diagonal. Theorem 2.116 could then be used to give alternative proofs to parts of Theorem 2.109.

Another recursive characterization of copositivity is given in the following theorem.

Theorem 2.118. *Let A be an $n \times n$ symmetric matrix, which is copositive of order $n-1$. Then the following are equivalent.*

(a) *A is copositive.*
(b) *There exists $\mathbf{x} \geq \mathbf{0}$ such that $A\mathbf{x} \geq \mathbf{0}$.*
(c) $\min_{\mathbf{y}\in\Delta} \max_{\mathbf{x}\in\Delta} \mathbf{x}^T A\mathbf{y} = \max_{\mathbf{x}\in\Delta} \min_{\mathbf{y}\in\Delta} \mathbf{x}^T A\mathbf{y} \geq 0$.

Note that the equality $\min_{\mathbf{y}\in\Delta} \max_{\mathbf{x}\in\Delta} \mathbf{x}^T A\mathbf{y} = \max_{\mathbf{x}\in\Delta} \min_{\mathbf{y}\in\Delta} \mathbf{x}^T A\mathbf{y}$ holds for every matrix A (see the notes at the end of this section).

Proof. (a) implies (b): Consider $\mathbf{u} \in \Delta$ where $\min_{\mathbf{x}\in\Delta} \mathbf{x}^T A\mathbf{x} \geq 0$ is attained. By Remark 2.49, $A\mathbf{u} \geq \mathbf{0}$.

(b) implies (c): Suppose for $\mathbf{u} \in \Delta$ we have $A\mathbf{u} \geq \mathbf{0}$. Then for this $\mathbf{u}$, $\mathbf{u}^T A\mathbf{y} \geq 0$ for every $\mathbf{y} \in \Delta$, and thus

$$\max_{\mathbf{x}\in\Delta} \min_{\mathbf{y}\in\Delta} \mathbf{x}^T A\mathbf{y} \geq \min_{\mathbf{y}\in\Delta} \mathbf{u}^T A\mathbf{y} \geq 0.$$

(c) implies (a): By Theorem 2.109, if A is not copositive, then some $\mathbf{v} > \mathbf{0}$ in Δ is an eigenvector of A corresponding to a negative eigenvalue λ. Then $\mathbf{x}^T A\mathbf{v} < 0$ for every $\mathbf{x} \in \Delta$, and therefore

$$\max_{\mathbf{x}\in\Delta} \min_{\mathbf{y}\in\Delta} \mathbf{x}^T A\mathbf{y} \leq \max_{\mathbf{x}\in\Delta} \mathbf{x}^T A\mathbf{v} < 0,$$

a contradiction. □

The results on almost copositive matrices lead to a characterization of copositivity (or actually, non-copositivity) by Schur complements. Here too we consider the empty 0×0 matrix as having determinant 1 (and therefore being positive definite). The Schur complement of A with respect to such an empty submatrix is A itself.

Theorem 2.119. *Let A be a symmetric $n \times n$ matrix, $n \geq 2$. Then A is not copositive if and only if there exists a principal submatrix $A[\alpha]$ such that, up to simultaneous permutation of rows and columns,*

$$A[\alpha] = \begin{pmatrix} B & C \\ C^T & E \end{pmatrix},$$

where E (of order $\geq |\alpha| - 2$) is positive definite, $E^{-1}C \leq 0$ and $A[\alpha]/E$ has either a negative diagonal element, or a zero diagonal element with a negative off-diagonal element in the same row.

Proof. If such a principal submatrix exists, then $A[\alpha]/E$ is not copositive, by Corollary 2.5. Then $A[\alpha]$ is not copositive by Corollary 2.88, and therefore A is not copositive.

Conversely, if A is not copositive, suppose that it is copositive of exact order k, $0 \leq k \leq n-1$. (If $k = 0$ this means that $a_{ii} < 0$ for some i, and then $\alpha = \{i\}$ and E the empty principal submatrix satisfy the requirements.) Let $A[\alpha]$ be a $(k+1) \times (k+1)$ principal submatrix of A, which is not copositive. Then $\det A[\alpha] < 0$ and $A[\alpha]^{-1}$ is nonpositive with negative off-diagonal elements, all the principal submatrices of A of order k are positive semidefinite, and all the principal submatrices of order $k-1$ are positive definite. If a $k \times k$ principal submatrix is positive definite, then up to permutation similarity

$$A[\alpha] = \begin{pmatrix} p & \mathbf{b}^T \\ \mathbf{b} & E \end{pmatrix},$$

where E is positive definite. Let

$$A[\alpha]^{-1} = \begin{pmatrix} q & \mathbf{d}^T \\ \mathbf{d} & F \end{pmatrix}.$$

Then $\mathbf{d} < \mathbf{0}$, and by (2.26) $\det A[\alpha] < 0$ and $\det E > 0$ imply that $q < 0$. From $A[\alpha]A[\alpha]^{-1} = I$ we get in particular

$$q\mathbf{b} + E\mathbf{d} = \mathbf{0}$$

and therefore

$$E^{-1}\mathbf{b} = -\frac{1}{q}\mathbf{d} < \mathbf{0}.$$

As $A[\alpha]$ is not positive semidefinite, $p < \mathbf{b}^T E\mathbf{b}$. That is,

$$A[\alpha]/E = p - \mathbf{b}^T E\mathbf{b} < 0.$$

If in $A[\alpha]$ all the $k\times k$ principal submatrices are singular, then in $A[\alpha]^{-1}$ all the diagonal elements are zero (by (2.26)). Write

$$A[\alpha]=\begin{pmatrix} B & C \\ C^T & E \end{pmatrix},$$

where E is a positive definite matrix of order $k-1$ and B is 2×2 (if $k=1$, E is the empty matrix). Split $A[\alpha]^{-1}$ into block conformally,

$$A[\alpha]^{-1}=\begin{pmatrix} L & M \\ M^T & N \end{pmatrix},$$

where $M<0$ and L is a 2×2 matrix with zero diagonal and negative off-diagonal elements, hence invertible with $L^{-1}\leq 0$. From $A[\alpha]A[\alpha]^{-1}=I$ we get in particular

$$C^TL+EM^T=0,$$

and therefore

$$E^{-1}C^T=-M^TL^{-1}\leq 0.$$

By Corollary 2.88, the matrix $A[\alpha]/E$ is not copositive. Since the $k\times k$ principal submatrices of $A[\alpha]$ are positive semidefinite of rank $k-1$, the diagonal elements of $A[\alpha]/E$ are zeros. So $A[\alpha]/E$ is 2×2 matrix with zero diagonal, which is not copositive, and therefore its off-diagonal element is negative. □

Corollary 2.120. *Let A be a symmetric $n\times n$ matrix, $n\geq 2$, with $p<n$ positive eigenvalues. Then A is copositive if and only if A is copositive of order $p+1$.*

Proof. The "only if" is obvious. So suppose A is not copositive. There exists a principal submatrix

$$A[\alpha]=\begin{pmatrix} B & C \\ C^T & E \end{pmatrix},$$

where E (of order $\geq|\alpha|-2$) is positive definite, $E^{-1}C\leq 0$ and $A[\alpha]/E$ has either a negative diagonal element, or a zero diagonal element with a negative off-diagonal element in the same row (Theorem 2.119). Suppose $|\alpha|=m$. Then $\operatorname{In}A\geq\operatorname{In}A[\alpha]=\operatorname{In}E+\operatorname{In}(A[\alpha]/E)$. If E is $(m-1)\times(m-1)$, then $A[\alpha]/E$ is a negative element, and we get $\operatorname{In}A\geq(m-1,0,0)+(0,1,0)$, implying that $p\geq m-1$. In the second case, E is of order $m-2$, and the 2×2 Schur complement A/E has zero diagonal elements and negative off-diagonal elements, so that $\operatorname{In}A\geq(m-2,0,0)+(1,1,0)$, and again we get that $p\geq m-1$. In both cases, this means that A has a principal submatrix of order $m\leq p+1$ which is not copositive. □

The theorems in this subsection have "strictly copositive versions". We state them here, and leave the proofs as exercises. Keep in mind that an almost strictly copositive may be either copositive or not copositive. Versions of these theorems for copositive-plus matrices can be found in the exercises.

Theorem 2.121. *Let an $n \times n$ matrix A be strictly copositive of order $n-1$. Then the following are equivalent.*

(a) *A is not strictly copositive.*
(b) *There exists a nonpositive eigenvalue of A with a corresponding positive eigenvector.*
(c) *For every vector $\mathbf{a} > \mathbf{0}$ in $\mathbb{R}^n$ there exist a vector $\mathbf{b} > \mathbf{0}$ in $\mathbb{R}^n$ such that $A\mathbf{b} = \lambda\mathbf{a}$, where $\lambda \leq 0$.*
(d) $\det A \leq 0$ *and* $\operatorname{adj} A > 0$.

Theorem 2.122. *Let an $n \times n$ matrix A be strictly copositive of order $n-1$. Then the following are equivalent*

(a) *A is copositive.*
(b) $\det A = 0$ *and A is positive definite of exact order $n-1$.*
(c) $\operatorname{In} A = (n-1, 0, 1)$ *and A has a positive eigenvector corresponding to the zero eigenvalue.*

Theorem 2.123. *Let an $n \times n$ matrix A be strictly copositive of order $n-1$. Then the following are equivalent*

(a) *A is not copositive.*
(b) $\det A < 0$ *and A is positive definite of exact order $n-1$.*
(c) $\operatorname{In} A = (n-1, 1, 0)$ *and* $A^{-1} < 0$.

Corollary 2.124. *A symmetric matrix A is not strictly copositive if and only if at least one of the following conditions is satisfied.*

(a) *There exists a principal submatrix $A[\alpha]$, which is nonsingular and $A[\alpha]^{-1}$ has a nonpositive column.*
(b) *There exists a principal submatrix $A[\alpha]$, which is singular and has a nonnegative eigenvector corresponding to the zero eigenvalue.*

Corollary 2.125. *A symmetric matrix A is not strictly copositive if and only if it has a principal submatrix $A[\alpha]$ with* $\det A[\alpha] \leq 0$ *such that the cofactors of its last column are all positive.*

The next corollary is a variation on the previous, and is the earliest reported result on copositivity.

Corollary 2.126. *A symmetric matrix A is not strictly copositive if and only if it has a principal submatrix $A[\alpha]$ such that its proper leading principal minors are all positive,* $\det A[\alpha] \le 0$, *and the cofactors of its last column are all positive.*

Theorem 2.127. *Let A be an $n \times n$ symmetric matrix, which is strictly copositive of order $n-1$. Then the following are equivalent.*

(a) *A is strictly copositive.*
(b) *There exists $\mathbf{x} \ge \mathbf{0}$ such that $A\mathbf{x} > \mathbf{0}$.*
(c) $\min_{\mathbf{y}\in\Delta} \max_{\mathbf{x}\in\Delta} \mathbf{x}^T A\mathbf{y} = \max_{\mathbf{x}\in\Delta} \min_{\mathbf{y}\in\Delta} \mathbf{x}^T A\mathbf{y} > 0.$

Theorem 2.128. *Let A be an $n \times n$ symmetric matrix, $n \ge 2$. Then A is not strictly copositive if and only if there exists a principal submatrix $A[\alpha]$ such that, up to simultaneous permutation of rows and columns,*

$$A[\alpha] = \begin{pmatrix} p & \mathbf{b}^T \\ \mathbf{b} & E \end{pmatrix},$$

where E (of order $|\alpha|-1$, possibly empty) is positive definite, $E^{-1}\mathbf{b} \le 0$ and $A[\alpha]/E \le 0$.

Corollary 2.129. *Let A be a symmetric $n \times n$ matrix, $n \ge 2$, with $p < n$ positive eigenvalues. Then A is strictly copositive if and only if A is strictly copositive of order $p+1$.*

Exercises

2.85. Let a symmetric $n \times n$ matrix A be copositive of order $n-1$. Prove that the following are equivalent.

(a) A is not copositive.
(b) For every nonnegative vector $\mathbf{a}$ in $\mathbb{R}^n$ there exist a nonnegative vector $\mathbf{x}$ in $\mathbb{R}^n$, such that $A\mathbf{x} = -\mathbf{a}$.
(c) There exists a nonnegative vector $\mathbf{x}$ in $\mathbb{R}^n$ such that $A\mathbf{x} = -\mathbf{e}$.
(d) There exists a nonnegative vector $\mathbf{x}$ in $\mathbb{R}^n$ such that $A\mathbf{x} < \mathbf{0}$.

Hint: Theorem 2.109.

2.86. Let A be a symmetric $n \times n$ matrix. Prove that the following are equivalent.

(a) A is not copositive.
(b) There exists $\alpha \subseteq \{1, \dots, n\}$ such that the system $A[\alpha]\mathbf{x} = -\mathbf{e}$ has a nonnegative solution.
(c) There exists $\alpha \subseteq \{1, \dots, n\}$ and a nonnegative vector $\mathbf{x}$ in $\mathbb{R}^{|\alpha|}$ such that $A\mathbf{x} < \mathbf{0}$.

(Compare this result to Theorem 2.74. The required computation for each $\alpha \subseteq \{1, \dots, n\}$ is simpler here.)
Hint: Exercise 2.85.

2.87. Let a symmetric $n \times n$ matrix A be copositive of order $n-1$. Prove that the following are equivalent.

(a) A is copositive.
(b) There exists $\mathbf{y} \in \mathbb{R}^n$ with at least one positive entry such that $A\mathbf{y} \geq \mathbf{0}$.

Hint: Use Farkas' Lemma (Theorem 1.161) on Exercise 2.85(b).

2.88. Let A be a symmetric $n \times n$ matrix. Prove that A is copositive if and only if there exists a set $\mathcal{U}$ of vectors in $\mathbb{R}^n$, each with at least one positive entry, such that for every $\alpha \subseteq \{1, \dots, n\}$ there exists $\mathbf{y} \in \mathcal{U}$ with $\operatorname{supp} \mathbf{y} \subseteq \alpha$ and $(A\mathbf{y})[\alpha] \geq \mathbf{0}$.
Hint: Exercise 2.87.

2.89. Show that if A is an $n \times n$ copositive matrix, and $\mathcal{U}$ is a set of vectors as in Exercise 2.88, then for any minimal support α of A, there exists in $\mathcal{U}$ a minimal zero $\mathbf{u}$ of A such that $\operatorname{supp} \mathbf{u} = \alpha$.
Hint: For a minimal zero $\mathbf{v}$ with $\operatorname{supp} \mathbf{v} = \alpha$ consider $\mathbf{y} \in \mathcal{U}$ such that $\operatorname{supp} \mathbf{y} \subseteq \alpha$ and $(A\mathbf{y})[\alpha] \geq \mathbf{0}$.

2.90. Let A be an $n \times n$ symmetric matrix. For every $\sigma \subseteq \{1, \dots, n\}$ consider the system of equations in $(\mathbf{x}, t)$, $\mathbf{x} \in \mathbb{R}^n$, $t \in \mathbb{R}$,

$$A[\sigma]\mathbf{x} = t\mathbf{e} \quad \text{and} \quad \mathbf{e}^T\mathbf{x} = 1 \tag{2.27}$$

(effectively, a linear system of $n+1$ equations in $n+1$ real variables). Let

$$\Omega = \{\sigma \subseteq \{1, \dots, n\} \mid (2.27) \text{ has a solution with } \mathbf{x} > \mathbf{0}\}.$$

For every $\sigma \in \Omega$ denote by $(\mathbf{x}_\sigma, \lambda_\sigma)$ one solution of (2.27) with $\mathbf{x}_\sigma > \mathbf{0}$. Prove:

(a) $\min_{\mathbf{x} \in \Delta_n} \mathbf{x}^T A\mathbf{x} \geq \lambda_\sigma$ for every $\sigma \in \Omega$.
(b) A is copositive if and only if $\lambda_\sigma \geq 0$ for every $\sigma \in \Omega$.

(c) If A is copositive, and $\mathbf{u} \in \Delta_n$ is any zero of A, then $\operatorname{supp} \mathbf{u} \in \Omega$, and $(\mathbf{u}[\sigma], \lambda_\sigma)$ is a solution to (2.27).

Hint: See Exercise 1.71. Note that if $\min_{\mathbf{x} \in \Delta_n} \frac{1}{2}\mathbf{x}^T A\mathbf{x}$ is attained at $\mathbf{v}$, then $\operatorname{supp} \mathbf{v} \in \Omega$.

2.91. Let

$$A = \begin{pmatrix} 162 & -153 & 127 & -27 & -27 & 127 & -153 \\ -153 & 162 & -153 & 127 & -27 & -27 & 127 \\ 127 & -153 & 162 & -153 & 127 & -27 & -27 \\ -27 & 127 & -153 & 162 & -153 & 127 & -27 \\ -27 & -27 & 127 & -153 & 162 & -153 & 127 \\ 127 & -27 & -27 & 127 & -153 & 162 & -153 \\ -153 & 127 & -27 & -27 & 127 & -153 & 162 \end{pmatrix}.$$

Use Exercise 2.91 to check by exact arithmetic computations that A is copositive, and to find its zeros.
Hint: A has exactly 3 positive eigenvalues. Use Corollary 2.120 and the circulant nature of A to save computations. By interlacing, what is the maximum size of a zero support?

2.92. Let A be a nonsingular copositive matrix. Show that there exists a row in A^{-1} the sum of whose elements is positive.
Hint: Otherwise, $A^{-1}\mathbf{e} \leq \mathbf{0}$.

2.93. Let A be a nonsingular strictly copositive matrix. Show that there is no nonnegative column in A^{-1} with a zero diagonal entry.

2.94. Prove that if A is a copositive matrix and $AA^\dagger \geq 0$, then each column of A has a positive entry.

2.95. Prove Theorem 2.121.

2.96. Prove Theorem 2.122.

2.97. Prove Theorem 2.123.

2.98. Prove Corollary 2.124.

2.99. Prove Corollary 2.125.

2.100. Prove Corollary 2.126.

2.101. Prove Theorem 2.127.

2.102. Prove Theorem 2.128.

2.103. Prove Corollary 2.129.

2.104. Let a symmetric matrix A be copositive-plus of order $n-1$, and not strictly copositive. Show that A is copositive-plus if and only if it is singular.

2.105. Let A be a symmetric $n \times n$ matrix. Then A is not copositive-plus if and only if there exists a principal submatrix $A[\alpha]$ such that, up to simultaneous permutation of rows and columns,

$$A[\alpha] = \begin{pmatrix} B & C \\ C^T & E \end{pmatrix},$$

where E is positive definite (of order $\geq |\alpha|-2$, possibly empty), $E^{-1}C^T \leq 0$ and if $A[\alpha]/E$ has a zero diagonal entry, then it has a nonzero off-diagonal entry in the same row.

2.106. Let an $n \times n$ symmetric matrix A, $n \geq 2$, have diagonal equal to $\mathbf{e}$ and exactly one positive eigenvalue. Show that A is copositive if and only if $a_{ij} \geq -1$ for every $i \neq j$.

2.107. Let A be an $n \times n$ symmetric matrix, $n \geq 2$, with $p < n$ positive eigenvalues. Prove that A is copositive-plus if and only if A is copositive-plus of order $p+1$.

2.108. Let A be a symmetric matrix A. Then A is not copositive-plus if and only if at least one of the following conditions is satisfied.

(a) There exists a principal submatrix $A[\alpha]$, which is nonsingular and $A[\alpha]^{-1}$ has a nonpositive column.
(b) There exists a principal submatrix $A[\alpha]$, which is nonsingular and $A[\alpha]^{-1}$ has a nonnegative column with a zero diagonal element.

2.109. Let A be an $n \times n$ symmetric matrix, $n \geq 2$, with $\operatorname{rank} A = r$, $r < n$. Show that A is copositive if and only if A is copositive of order $r+1$.

Notes. The term "copositive (strictly copositive/copositive-plus/positive semidefinite...) of order r" first appeared in [Cottle *et al.* (1970a)]. The terms "exact order r" and "almost copositive matrix" were coined in [Väliaho (1989a)]. The results in this chapter are mostly from these two papers, as well as [Hadeler (1983)] and [Väliaho (1986)]. Specifically, Theorems 2.109 and 2.110 combine results from [Cottle *et al.* (1970a)], [Hadeler (1983)], and [Väliaho (1986, 1989a)]. Theorem 2.112 and Corollaries 2.113, 2.114 were proved in [Väliaho (1986)] (the latter was cited without proof in [Cottle *et al.* (1970a)] and attributed there to E. Keller). The strictly copositive version, Corollary 2.125, was proved in [Motzkin (1967)] and then in [Cottle *et al.* (1970a)]. Theorem 2.115 appeared in [Väliaho (1989a)], and Theorem 2.116 is due to [Hadeler (1983)]. Theorem 2.118 (and its strictly positive version, Theorem 2.127) is due to [Gaddum (1958)], where a direct proof is given (while our proof relies on later results). The equality $\min_{\mathbf{y}\in\Delta}\max_{\mathbf{x}\in\Delta}\mathbf{x}^TA\mathbf{y} = \max_{\mathbf{x}\in\Delta}\min_{\mathbf{y}\in\Delta}\mathbf{x}^TA\mathbf{y}$ is the famous von Neumann's Minimax Theorem, which is an important early result in game theory. Theorem 2.119, and similar results for the strictly copositive and copositive-plus cases, were proved in [Väliaho (1986)], in terms of the principal pivot transform. The *principal pivot transform* with pivot E of $A = \begin{pmatrix} B & C \\ C^T & E \end{pmatrix}$, E nonsingular, is $\begin{pmatrix} A/E & CE^{-1} \\ E^{-1}C^T & E^{-1} \end{pmatrix}$. We bypass here the use of the principal pivot transform, and use only the Schur complement. In [Väliaho (1989b)] these results are used to suggest an algorithm for copositivity detection, for matrices with a relatively large positive definite submatrix. The papers [Väliaho (1986, 1989a)] contain some more results on (almost) copositive-plus matrices that are not included here.

The remaining results in the section are based on [Väliaho (1986)], as are Exercises 2.93, 2.104–2.105 and 2.107–2.109. According to this paper, Corollaries 2.120 and 2.129 and the result in Exercise 2.109 appeared previously in the Ph.D. thesis [Pereira (1972)]. Corollary 2.126 is a corrected version of the result reported in [Motzkin (1952)]. Exercises 2.85–2.89 are based on [Dickinson (2019)], where these results were used to suggest algorithms for finding certificates of copositivity or non-copositivity of a given matrix. Exercises 2.90–2.91 are based on [Bomze *et al.* (2014)]. Exercise 2.94 is based on [Bisht and Sivakumar (2017)].

2.8 Copositive $\{0, 1, -1\}$-matrices and related matrices

Consider matrices in which every entry is from the set $\{0, 1, -1\}$, and all the diagonal elements are equal to 1. While this class of matrices seems very restricted, it is of interest since there is a simple combinatorial characterization of the copositive matrices in this class, as well as the positive semidefinite, SPN and extremal copositive matrices among these matrices.

We present most of these results also in a slightly generalized form — considering matrices that have a diagonal $\mathbf{e}$ and "enough" -1 entries (the meaning of "enough" varies from one result to the other). For the characterizations, recall that for an $n \times n$ symmetric matrix A and $c \in \mathbb{R}$, $G_c(A)$ is the graph on vertices $\{1, \ldots, n\}$ whose edges correspond to the off-diagonal entries equal to c.

Theorem 2.130. *Let A be an $n \times n$ symmetric matrix with diagonal equal to $\mathbf{e}$, such that each of its entries is from the set $\{0, 1, -1\}$. Then A is copositive if and only if $G_{-1}(A)$ is triangle free, and $a_{ij} = 1$ for every $i \neq j$ such that $d_{G_{-1}(A)}(i, j) = 2$.*

Proof. Assume first that A is copositive. The matrix

$$M = \begin{pmatrix} 1 & -1 & -1 \\ -1 & 1 & -1 \\ -1 & -1 & 1 \end{pmatrix}$$

is not copositive, as $\mathbf{e}^T M \mathbf{e} = -3 < 0$. Thus M cannot be a principal submatrix of A, and the graph $G_{-1}(A)$ is triangle free. If $d_{G_{-1}(A)}(i, j) = 2$ for some $i < j$, then there exists k such that $a_{ik} = a_{kj} = -1$. Without loss of generality, $i < k < j$. Then

$$A[i, k, j] = \begin{pmatrix} 1 & -1 & a \\ -1 & 1 & -1 \\ a & -1 & 1 \end{pmatrix}.$$

As $\mathbf{u} = \mathbf{e}_i + \mathbf{e}_k$ is a zero of A, $A\mathbf{u} \geq \mathbf{0}$, and in particular $a_{ij} = a \geq 1$. Since A is a $\{0, 1, -1\}$-matrix, $a = 1$.

Conversely, suppose $G_{-1}(A)$ is triangle free and $a_{ij} = 1$ for all $i \neq j$ such that $d_{G_{-1}(A)}(i, j) = 2$. By Proposition 2.16 it suffices to check copositivity of every principal submatrix of A, in which all the off-diagonal elements are less than 1. Let $A[\alpha]$ be such a principal submatrix. No row in $A[\alpha]$ has two -1 off-diagonal entries, otherwise $a_{ij} = a_{ik} = -1$ for three indices $i, j, k \in \alpha$, and then $d_{G_{-1}(A)}(j, k) = 2$, hence $a_{jk} = 1$, contrary to the assumption on $A[\alpha]$. As $A[\alpha]$ has at most one -1 off-diagonal entry in each row, it is a direct sum of 1×1 matrices equal to (1), and 2×2 matrices equal to

$$\begin{pmatrix} 1 & -1 \\ -1 & 1 \end{pmatrix},$$

and is therefore copositive (even positive semidefinite). □

Example 2.131. The Horn matrix H is a $\{\pm 1\}$-matrix. The graph $G_{-1}(H)$ is the 5-cycle C_5, and whenever $d_{G_{-1}(H)}(i,j) = 2$ we have $h_{ij} = 1$. By Theorem 2.131 H is copositive.

In fact, from Theorem 2.130 we see that the following holds for $\{\pm 1\}$-matrices.

Corollary 2.132. *Let A be an $n \times n$ symmetric matrix with diagonal equal to $\mathbf{e}$, such that each of its entries is from the set $\{1, -1\}$. Then A is copositive if and only if $G_{-1}(A)$ is triangle free.*

Remark 2.133. From the proof of Theorem 2.130 it can be seen that if A is any $n \times n$ copositive matrix with diagonal equal to $\mathbf{e}$, then $G_{-1}(A)$ is triangle free, and $a_{ij} \geq 1$ for every $i \neq j$ such that $d_{G_{-1}(A)}(i,j) = 2$. The converse, however, requires additional conditions.

Here is one slight generalization of Theorem 2.130. The proof is left as an exercise.

Corollary 2.134. *Let A be an $n \times n$ symmetric matrix with diagonal equal to $\mathbf{e}$, such that all its negative entries are equal to -1. Then A is copositive if and only if $G_{-1}(A)$ is triangle free, and $a_{ij} \geq 1$ for every $i \neq j$ such that $d_{G_{-1}(A)}(i,j) = 2$.*

The next corollary is a strictly copositive version of Theorem 2.130, and is also left as an exercise.

Corollary 2.135. *Let A be an $n \times n$ symmetric matrix with diagonal $\mathbf{e}$, such that each of its entries is from the set $\{0, 1, -1\}$. Then A is strictly copositive if and only if it is nonnegative.*

Now consider positive semidefinite matrices. For convenience we assume first that the graph of -1 entries is connected.

Theorem 2.136. *Let A be an $n \times n$ symmetric matrix with diagonal equal to $\mathbf{e}$, such that $G_{-1}(A)$ is connected. Then the following conditions are equivalent.*

(a) *A is positive semidefinite.*
(b) *A is a $\{\pm 1\}$-matrix and* $\operatorname{rank} A = 1$.
(c) *$G_{-1}(A)$ is a complete bipartite graph, and $a_{ij} = 1$ for every $i \neq j$ such that $d_{G_{-1}(A)}(i,j) = 2$.*

Proof. (b) $\Rightarrow$ (c): Let

$$\alpha_1 = \{1 \le i \le n \mid a_{1i} = 1\} \text{ and } \alpha_2 = \{1 \le i \le n \mid a_{1i} = -1\}.$$

Since each row of A is a multiple of the first row and $a_{ii} = 1$ for every i, $a_{ij} = 1$ if $i, j \in \alpha_1$ or $i, j \in \alpha_2$. If $i \in \alpha_1$ and $j \in \alpha_2$ then $a_{1i} = 1$ and $a_{ii} = 1$ imply that the i-th row is equal to the first, and $a_{1j} = -1$, $a_{jj} = 1$ imply that $a_{ij} = -1$. Thus $G_{-1}(A)$ is a complete bipartite graph, with independent sets α_1 and α_2. If $d_{G_{-1}(A)}(i,j) = 2$, then $i, j \in \alpha_1$ or $i, j \in \alpha_2$, hence $a_{ij} = 1$.

(c) $\Rightarrow$ (a): Let $G_{-1}(A) = K_{m,k}$, with α_1 and α_2 the bipartition sets of this graph, $|\alpha_1| = m$ and $|\alpha_2| = k$. By Theorem 2.130 the matrix A is copositive. Let $\mathbf{u} \in \mathbb{R}^n$ be defined by

$$u_i = \begin{cases} k, & \text{if } i \in \alpha_1 \\ m, & \text{if } i \in \alpha_2 \end{cases}.$$

Then $\mathbf{u}$ is a positive zero of the copositive A (in fact, $A\mathbf{u} = \mathbf{0}$, since each row of A has m entries equal to 1 and k entries equal to -1, or vice versa). Therefore A is positive semidefinite.

(a) $\Rightarrow$ (b): We prove this implication by induction on n. For $n = 1$ it is obviously true. Let $n \ge 2$, and suppose the claim holds for $k < n$. Let m be the maximal order of a rank 1 principal submatrix of A ($m \ge 1$, since the diagonal of A is $\mathbf{e}$). If $m < n$, without loss of generality $A[1, \dots, m] = \mathbf{y}\mathbf{y}^T$, where $\mathbf{y}$ is a $\{\pm 1\}$-vector, and we may assume $y_1 = 1$. Since $G_{-1}(A)$ is connected, there exists $1 \le i \le m$ and $j \ge m+1$ such that $a_{ij} = -1$. By permuting rows and columns we may assume that $i = 1$ and $j = m+1$. Then $\mathbf{u} = \mathbf{e}_1 + \mathbf{e}_{m+1}$ is a zero of the positive semidefinite A, and thus $A\mathbf{u} = \mathbf{0}$. In particular, $a_{i\,m+1} = -a_{i1} = -y_i$ for every $1 \le i \le m$. Hence

$$A[1, \dots, m+1] = \begin{pmatrix} \mathbf{y}\mathbf{y}^T & \mathbf{z} \\ \mathbf{z}^T & 1 \end{pmatrix},$$

where $\mathbf{z} = -\mathbf{y}$. But then $A[1, \dots, m+1]$ is a rank 1 matrix, contrary to the maximality of m. Therefore $m = n$, and A is a rank 1 matrix. Since the diagonal is $\mathbf{e}$, it is a $\{\pm 1\}$-matrix. $\square$

In the case that $G_{-1}(A)$ is not necessarily connected, and A is a $\{0, 1, -1\}$-matrix, the following result holds.

Corollary 2.137. *Let A be an $n \times n$ symmetric matrix with diagonal equal to $\mathbf{e}$, such that all its entries are in $\{0, 1, -1\}$. Then A is positive semidefinite if and only if each component of $G(A)$ is a complete graph, and for every vertex set α of such a component, the subgraph of $G_{-1}(A)$ induced on α is either edgeless, or a complete bipartite graph.*

Proof. Let A be positive semidefinite. By Theorem 2.130, $G_{-1}(A)$ is triangle free. Let α a vertex set of a component of $G(A)$ with $|\alpha| \geq 2$. If $|\alpha| = 2$, $A[\alpha]$ cannot have a zero off-diagonal entry since $G(A[\alpha])$ is connected. Hence $A[\alpha]$ is

$$\begin{pmatrix} 1 & 1 \\ 1 & 1 \end{pmatrix} \quad \text{or} \quad \begin{pmatrix} 1 & -1 \\ -1 & 1 \end{pmatrix},$$

and the subgraph of $G_{-1}(A)$ induced on α is either edgeless, or a K_2.

If $|\alpha| \geq 3$, then $A[\alpha]$ cannot have an off-diagonal zero, since otherwise one of the following would be a principal submatrix of $A[\alpha]$ (up to row and column permutation)

$$\begin{pmatrix} 1 & 1 & 0 \\ 1 & 1 & 1 \\ 0 & 1 & 1 \end{pmatrix}, \quad \begin{pmatrix} 1 & 1 & 0 \\ 1 & 1 & -1 \\ 0 & -1 & 1 \end{pmatrix}, \quad \begin{pmatrix} 1 & -1 & 0 \\ -1 & 1 & -1 \\ 0 & -1 & 1 \end{pmatrix},$$

but none of these 3×3 matrices is positive semidefinite. Hence the component of $G(A)$ on vertices α is a complete graph. And since the matrix

$$\begin{pmatrix} 1 & 1 & -1 \\ 1 & 1 & 1 \\ -1 & 1 & 1 \end{pmatrix}$$

is not positive semidefinite, it is not a principal submatrix of $A[\alpha]$. Therefore the subgraph of $G_{-1}(A)$ induced on α is either edgeless or connected. In the latter case, by Theorem 2.136, this induced subgraph has to be complete bipartite.

The converse follows from Theorem 2.136 and is left as an exercise. □

Remark 2.138. The convex hull of $\{\pm 1\}$ positive semidefinite matrices in $\mathbb{R}^n$ is known as the *MAXCUT polytope*. We denote it by $\mathcal{MC}_n$.

From the above we may deduce a characterization of $\{0, 1, -1\}$-matrices that are copositive-plus. The proof is left as an exercise.

Corollary 2.139. *Let A be an $n \times n$ matrix with diagonal equal to* $\mathbf{e}$, *and all entries in* $\{0, 1, -1\}$. *Then the following are equivalent.*

(a) *A is copositive-plus.*
(b) *A is a direct sum of a nonnegative matrix and a positive semidefinite matrix.*
(c) *For every vertex set α of a component of $G(A)$, the subgraph of $G_{-1}(A)$ induced on α is either edgeless, or a complete bipartite graph.*

We can now characterize the SPN matrices with diagonal $\mathbf{e}$ and "enough" -1 entries. Here "enough" means $G_{-1}(A)$ is connected.

Theorem 2.140. *Let A be an $n \times n$ matrix with diagonal equal to $\mathbf{e}$, such that $a_{ij} \geq -1$ for every i, j and $G_{-1}(A)$ is connected. Then A is SPN if and only if $G_{-1}(A)$ is a bipartite graph, and $a_{ij} \geq 1$ whenever $d_{G_{-1}(A)}(i,j)$ is even.*

Proof. If A is SPN, then $A = P + N$, where P is positive semidefinite, N is a hollow nonnegative matrix, and the diagonal of P is $\mathbf{e}$. Then $p_{ij} \geq -1$ for every i, and $G_{-1}(A) \subseteq G_{-1}(P)$. Since $G_{-1}(A)$ is connected, $G_{-1}(P)$ is also connected. As the diagonal of P is $\mathbf{e}$, by Theorem 2.136 the graph $G_{-1}(P)$ is a complete bipartite graph, and P is a $\{\pm 1\}$-matrix of rank 1. In particular, $p_{ij} = 1$ whenever i and j are in the same independent set of $G_{-1}(P)$. Hence each component of $G_{-1}(A)$ is bipartite, and if $d_{G_{-1}(A)}(i,j)$ is even, then $a_{ij} \geq p_{ij} = 1$

Conversely, suppose $G_{-1}(A)$ is a connected bipartite graph and $a_{ij} \geq 1$ whenever $d_{G_{-1}(A)}(i,j)$ is even. Let α_1 and α_2 be the independent sets of $G_{-1}(A)$. Define a matrix P by

$$p_{ij} = \begin{cases} 1, & \text{if } i,j \in \alpha_1, \text{ or } i,j \in \alpha_2 \\ -1, & \text{if } i \in \alpha_1, j \in \alpha_2, \text{ or vice versa} \end{cases}.$$

By the assumptions on A, $A \geq P$, and by Theorem 2.136, P is positive semidefinite. That is, A is SPN. □

Example 2.141. Let H be the Horn matrix. As $G_{-1}(H)$ is the odd cycle C_5, Theorem 2.140 implies that H is not SPN.

Example 2.142. Let

$$A = \begin{pmatrix} 1 & -1 & 1 & 0 & 0 \\ -1 & 1 & -1 & 1 & 0 \\ 1 & -1 & 1 & -1 & 1 \\ 0 & 1 & -1 & 1 & -1 \\ 0 & 0 & 1 & -1 & 1 \end{pmatrix}.$$

The graph $G_{-1}(A)$ is a path on 5 vertices, which is bipartite (see $G(A)$ and $G_{-1}(A)$ below). As $d_{G_{-1}(A)}(1,5) = 4$ is even and $a_{15} = 0$, A is not SPN. By Corollary 2.134 the matrix A is copositive (for $i < j$ we have $d_{G_{-1}(A)}(i,j) = 2$ if and only if $j = i+2$, and $a_{i\,i+2} = 1$).

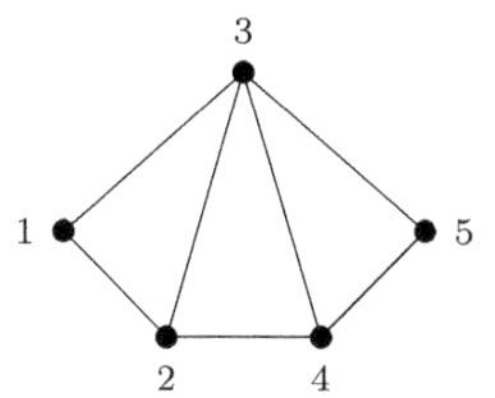

$G(A)$

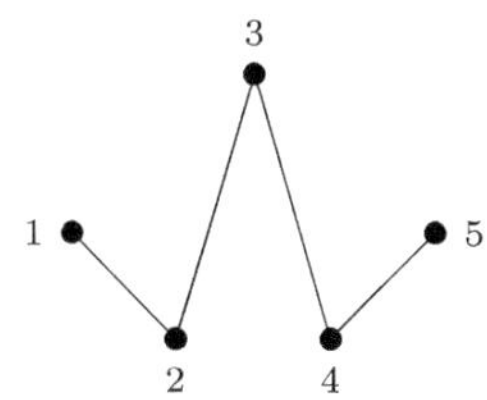

$G_{-1}(A)$

We continue with a characterization of the extremal copositive matrices with diagonal $\mathbf{e}$ and elements from $\{0, 1, -1\}$. We will use Theorem 2.71 in the proof, and therefore need to know all the minimal zeros of such a matrix A. If $a_{ij} = -1$, then $\mathbf{e}_i + \mathbf{e}_j$ is a zero with support $\{i, j\}$, and is a minimal zero since $\text{diag}A = \mathbf{e}$. By the positivity of the diagonal elements, if $A[\alpha] \geq 0$ then α is not the support of a zero of A. Hence if α is a zero support, $A[\alpha]$ must have a -1 element, and either $|\alpha| = 2$, or the zero is not minimal.

We also need to define an additional graph associated with the copositive matrix A, $L^0(A)$, which is a subgraph of the line graph of $G_0(A)$. Each edge of $G_0(A)$ is a vertex of $L^0(A)$, i.e., $\{i, j\}$ is a vertex if and only if $a_{ij} = 0$. Two vertices of $L^0(A)$ are adjacent in $L^0(A)$ if the corresponding edges in $G_0(A)$ share one vertex, and the two other ends are adjacent in $G_{-1}(A)$. That is, if $a_{ij} = a_{jk} = 0$ and $a_{ik} = -1$, then $\{i, j\}$ and $\{j, k\}$ are adjacent in $L^0(A)$.

Example 2.143. If A is the matrix in Example 2.142, $L^0(A)$ is a path on three vertices, $\{1, 4\} - \{1, 5\} - \{2, 5\}$. See figure below.

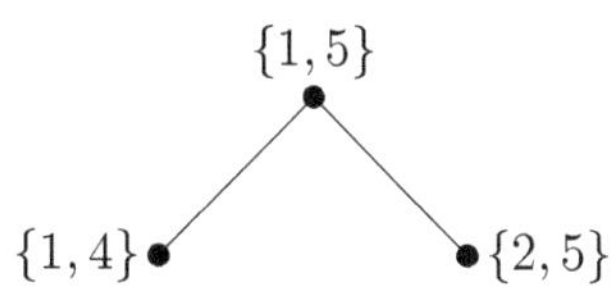

$L^0(A)$

Theorem 2.144. *Let A be a symmetric matrix with diagonal $\mathbf{e}$ such that all its entries are from $\{0, 1, -1\}$. Then A is an extremal copositive matrix if and only if the following conditions hold.*

(a) *$G_{-1}(A)$ is connected and triangle free.*
(b) *$a_{ij} = 1$ if and only if $d_{G_{-1}(A)}(i, j) = 2$.*

(c) *$L^0(A)$ has no isolated vertex, and no component that is a bipartite graph.*

Proof. We first show that the three conditions are necessary. Let A be an extremal copositive matrix. The copositivity implies that $G_{-1}(A)$ is triangle free, and $a_{ij} = 1$ whenever $d_{G_{-1}(A)}(i,j) = 2$. If $G_{-1}(A)$ were not connected, then we would have

$$A = \begin{pmatrix} B & C \\ C^T & E \end{pmatrix},$$

where B and E are copositive, and $C \geq 0$, so

$$A = \begin{pmatrix} B & C \\ C^T & 0 \end{pmatrix} + \begin{pmatrix} 0 & 0 \\ 0 & E \end{pmatrix},$$

so A would not be extremal.

If $a_{ij} = 1$ for some $i \neq j$ such that $d_{G_{-1}(A)}(i,j) \neq 2$, replacing the a_{ij} and a_{ji} entries by 0 would yield a copositive matrix, and thus A would not be extremal.

To prove the necessity of condition (c), we show that if $L^0(A)$ has an isolated vertex, or a bipartite component, then A is not extremal.

First suppose $\{p,q\}$ is an isolated vertex of $L^0(A)$. Let $X = E_{pq} + E_{qp}$. Let $\mathbf{u}$ be a minimal zero of A with $\operatorname{supp}\mathbf{u} = \{j,k\}$. If $\{j,k\} \cap \{p,q\} = \emptyset$, then $X\mathbf{u} = \mathbf{0}$. Otherwise, without loss of generality, $j = p$. Then $(X\mathbf{u})_i = 0$ for every $i \neq q$ (this is obvious for $i \neq p, q$, and $(X\mathbf{u})_p = u_q = 0$ since $k \neq q$). As $\{p,q\}$ is an isolated vertex of $L^0(A)$, $(A\mathbf{u})_q = a_{qp}u_p + a_{qk}u_k = a_{qk}u_k \neq 0$, this means that $(X\mathbf{u})_i = 0$ whenever $(A\mathbf{u})_i = 0$. Since X is clearly not a scalar multiple of A, by Theorem 2.71 the matrix A is not extremal.

Now suppose a component of $L^0(A)$ is bipartite, with independent sets S_1 and S_2. Let X be the symmetric matrix such that

$$x_{ij} = \begin{cases} -1, & \text{if } \{i,j\} \in S_1 \\ 1, & \text{if } \{i,j\} \in S_2 \\ 0, & \text{otherwise} \end{cases}.$$

Let $\mathbf{u}$ be a minimal zero of A, with $\operatorname{supp}\mathbf{u} = \{j,k\}$. Then $a_{jk} = -1$ and $u_j = u_k$. As $(A\mathbf{u})_i = a_{ij}u_j + a_{ik}u_k$, $(A\mathbf{u})_i = 0$ implies that either $a_{ik} = -a_{ij} \neq 0$, or $a_{ik} = a_{ij} = 0$. If the former occurs, $\{i,j\}$ and $\{i,k\}$ are not in S. If the latter occurs, then, since $\{i,k\}$ and $\{i,j\}$ are connected by an edge in $L^0(A)$, either both are in S, one in S_1 and the other in S_2, or both are not in S. Either way, $x_{ij} + x_{ik} = 0$. Hence $(X\mathbf{u})_i = 0$ whenever $(A\mathbf{u})_i = 0$. As X is not a scalar multiple of A (diag$X = \mathbf{0}$ and $X \neq 0$), by Theorem 2.71, A is not extremal.

For the sufficiency of the conditions, suppose A satisfies conditions (a)–(c). By parts of (a) and (b), A is copositive. Let X be a symmetric matrix such that for every minimal zero $\mathbf{u}$ of A, $(X\mathbf{u})_i = 0$ whenever $(A\mathbf{u})_i = 0$. We need to show that X is a scalar multiple of A. Let $x = x_{11}$. As $G_{-1}(A)$ is connected, there exists j such that $a_{1j} = -1$. Then $\mathbf{u} = \mathbf{e}_1 + \mathbf{e}_j$ is a minimal zero of A, and $(A\mathbf{u})_1 = (A\mathbf{u})_j = 0$ implies $(X\mathbf{u})_1 = (X\mathbf{u})_j = 0$. That is, $x_{1j} = -x$ and then $x_{jj} = x$. Continuing in this way we get by the connectivity of $G_{-1}(A)$ that $x_{ii} = x$ for every i, and $x_{ij} = -x$ whenever $a_{ij} = -1$. Moreover, if $a_{ij} = 1$ for $i \neq j$, then there exists k such that $a_{ik} = a_{jk} = -1$. As $\mathbf{u} = \mathbf{e}_j + \mathbf{e}_k$ is a minimal zero of A and $(A\mathbf{u})_i = a_{ij} + a_{ik} = 0$, also $(X\mathbf{u})_i = x_{ij} + x_{ik} = x_{ij} - x = 0$, and therefore $x_{ij} = x$. So far we have shown that for every $a_{ij} \neq 0$, $x_{ij} = xa_{ij}$.

We need to show that $x_{ij} = 0$ whenever $a_{ij} = 0$. Suppose this does not hold. Let S be the vertex set of a connected component of $L^0(A)$, containing at least one $\{i,j\}$ such that $a_{ij} = 0$ and $x_{ij} = s \neq 0$. As $\{i,j\}$ is not an isolated vertex, there exists k such that $a_{ik} = 0$ and $a_{jk} = -1$. Without loss of generality, $i < j < k$. Then

$$A[i,j,k] = \begin{pmatrix} 1 & 0 & 0 \\ 0 & 1 & -1 \\ 0 & -1 & 1 \end{pmatrix} \quad \text{and} \quad X[i,j,k] = \begin{pmatrix} x & s & t \\ s & x & -x \\ t & -x & x \end{pmatrix}.$$

Since $\mathbf{u} = \mathbf{e}_j + \mathbf{e}_k$ is a minimal zero of A and $(A\mathbf{u})_i = 0$, $(X\mathbf{u})_i = s + t = 0$, implying that $t = -s$. As S is the vertex set of a connected component, for every $\{i,j\} \in S$ we have $x_{ij} \neq 0$. We split S into

$$S_1 = \{\{i,j\} \in S \mid a_{ij} = 0 \text{ and } x_{ij} > 0\}$$
$$S_2 = \{\{i,j\} \in S \mid a_{ij} = 0 \text{ and } x_{ij} < 0\}.$$

By the above argument, these are independent sets in $L^0(A)$. This contradicts the assumption that A satisfies (c). Hence $X = xA$, and the proof is complete. □

Remark 2.145. In particular, a copositive matrix A with diagonal $\mathbf{e}$ and all entries in $\{1,-1\}$ is extremal if and only if $G_{-1}(A)$ is connected and $a_{ij} = 1$ if and only if $d_{G_{-1}(A)}(i,j) = 2$.

Example 2.146. The matrix A in Example 2.142 is not extremal, since $L^0(A)$ is a bipartite graph (see Example 2.143).

Example 2.147. Let H_n be a $\{0,1,-1\}$-matrix of order n, n odd, such that $h_{ij} = -1$ if and only if $j = i \dotplus 1$, where $\dotplus$ denotes addition modulo n

on $\{1,\ldots,n\}$, and $h_{ij}=1$ if and only if $d_{G_{-1}(H)}(i,j)=2$. The matrix H_5 is the Horn matrix, and for $n=7$ the matrix is

$$H_7=\begin{pmatrix} 1 & -1 & 1 & 0 & 0 & 1 & -1\\ -1 & 1 & -1 & 1 & 0 & 0 & 1\\ 1 & -1 & 1 & -1 & 1 & 0 & 0\\ 0 & 1 & -1 & 1 & -1 & 1 & 0\\ 0 & 0 & 1 & -1 & 1 & -1 & 1\\ 1 & 0 & 0 & 1 & -1 & 1 & -1\\ -1 & 1 & 0 & 0 & 1 & -1 & 1 \end{pmatrix},$$

occasionally referred to as a *Hoffman-Pereira matrix*. The matrix H_7 is an extremal copositive matrix, as $L^0(A)$ is a 7-cycle. We leave it as an exercise to check that for every odd n, the matrix H_n is an exceptional extremal copositive matrix.

The requirement for an odd n in the last example is not arbitrary.

Corollary 2.148. *If A is an exceptional extremal copositive $\{0,1,-1\}$-matrix, then $G_{-1}(A)$ is not bipartite.*

Proof. Suppose on the contrary that there exists an exceptional extremal $\{0,1,-1\}$-matrix A with $G_{-1}(A)$ bipartite. Since A is extremal, $G_{-1}(A)$ is connected and $a_{ij}=1$ if and only if $d_{G_{-1}(A)}(i,j)=2$. Since A is exceptional, by Theorem 2.136, $G_{-1}(A)$ is not a complete bipartite graph. Thus there exist $p\neq q$ such that $d_{G_{-1}(A)}(p,q)$ is even but greater than 2. For such p and q we have $a_{pq}=0$. By the extremality of A, $\{p,q\}$ cannot be an isolated vertex of $L^0(A)$, hence it has a neighbour, say $\{p,l\}$. Note that if $\{i,j\}$ and $\{i,k\}$ are neighbours in $L^0(A)$, then $d_{G_{-1}(A)}(j,k)=1$, and therefore $d_{G_{-1}(A)}(i,k)$ is odd if and only if $d_{G_{-1}(A)}(i,j)$ is even. Consider the component of $L^0(A)$ containing the vertices $\{p,q\}$ and $\{p,l\}$, and denote its vertex set by S. Let

$$S_1=\{\{i,j\}\in S\mid d_{G_{-1}(A)}(i,j)\text{ is even }\},$$
$$S_2=\{\{i,j\}\in S\mid d_{G_{-1}(A)}(i,j)\text{ is odd }\}.$$

Both are nonempty and, by the above, S_1 and S_2 are independent and this component of $L^0(A)$ is bipartite. This contradicts the extremality of A. □

By the next result, certain extremal copositive matrices have to be $\{0,1,-1\}$-matrices.

Theorem 2.149. *Let A be an $n\times n$ extremal copositive matrix with diagonal $\mathbf{e}$, such that all its minimal supports have size 2. Then A is a $\{0,1,-1\}$-matrix.*

Proof. Since the diagonal of A is $\mathbf{e}$, $\{p,q\}$ is a support of a minimal zero if and only if $a_{pq}=-1$, and a corresponding minimal zero is $\mathbf{u}=\mathbf{e}_p+\mathbf{e}_q$. For every i, $(A\mathbf{u})_i=a_{ip}+a_{iq}$.

Define a graph G whose vertices are the $\frac{n(n+1)}{2}$ unordered pairs $\{i,j\}$, $i,j\in\{1,\ldots,n\}$ (including the pairs $\{i,i\}$). Two such pairs are adjacent in the G if and only if one pair is $\{i,p\}$ and the other is $\{i,q\}$, where $a_{pq}=-1$ and $a_{ip}+a_{iq}=0$. In particular, the entries of A corresponding to adjacent vertices in G are each other's opposite.

Let S be the vertex set of a component of G. If $\{i,j\}$ is a vertex in this component, and $a_{ij}=a$, then for every vertex $\{k,l\}$ in the same component $a_{kl}=-a$ (if there is a path of odd length from $\{i,j\}$ to $\{k,l\}$) or $a_{kl}=a$ (if there is a path of even length from $\{i,j\}$ to $\{k,l\}$). Hence if the component contains an odd cycle, $a_{ij}=0$ for every $\{i,j\}\in S$. If a is nonzero, then the component is bipartite, and each entry of A corresponding to a vertex of S is either a or $-a$. Let $\{p_0,q_0\}$ be the support of a minimal zero. Then $a_{p_0q_0}=-1$ and $a_{p_0p_0}=1$, so $\{p_0,p_0\}$ and $\{p_0,q_0\}$ are adjacent in G. Let S be the vertex set of the component containing this edge. This component is necessarily bipartite, and its independent sets are

$$S_1=\{\{i,j\}\in S\mid a_{ij}=1\},\quad S_2=\{\{i,j\}\in S\mid a_{ij}=-1\}.$$

Define a matrix X by

$$x_{ij}=\begin{cases}1, & \{i,j\}\in S_1\\ -1, & \{i,j\}\in S_2\\ 0, & \text{otherwise}\end{cases}.$$

Let $\mathbf{u}$ be a minimal zero of A with support $\{p,q\}$, and suppose

$$(A\mathbf{u})_i=a_{ip}+a_{iq}=0.$$

If $\{i,p\}\in S_1$, then $\{i,q\}\in S_2$, and vice versa. So also if $\{i,p\}\notin S$, then $\{i,q\}\notin S$. In any case, $(X\mathbf{u})_i=0$. That is, $(X\mathbf{u})_i=0$ whenever $\mathbf{u}$ is a minimal zero of A and $(A\mathbf{u})_i=0$. As A is extremal, X has to be a multiple of A. Since $x_{ij}=a_{ij}$ for $\{i,j\}\in S$, $X=A$, and A is a $\{0,1,-1\}$-matrix. □

The proof of the following corollary is left as an exercise.

Corollary 2.150. *Let A be an $n\times n$ extremal copositive matrix such that all its minimal supports are of size 2. Then A is congruent by a positive diagonal matrix to an extremal $\{0,1,-1\}$-matrix.*

Exercises

2.110. Prove Corollary 2.134.
Hint: To prove the "if" part, consider B defined by

$$b_{ij} = \begin{cases} 1, & \text{if } a_{ij} \geq 1 \\ 0, & \text{if } 0 \leq a_{ij} < 1 \\ -1, & \text{if } a_{ij} = -1 \end{cases}.$$

2.111. Let A be a symmetric matrix with diagonal $\mathbf{e}$, such that $A \geq -J$ and $G_-(A)$ is triangle free. Show that if $a_{ij} \geq 1$ for every $i \neq j$ such that $d_{G_-(A)}(i,j) = 2$, then A is copositive.

2.112. Prove Corollary 2.135.

2.113. Complete the proof of Corollary 2.137.

2.114. Show that if $A \in \mathcal{MC}_n$ then $\sum_{1 \leq i < j \leq n} a_{ij} \geq -\left\lfloor \frac{n}{2} \right\rfloor$, and the inequality is strict if $A \in \operatorname{ri} \mathcal{MC}_n$.
Hint: For $A = \mathbf{v}\mathbf{v}^T$, $\mathbf{v} \in \{\pm 1\}^n$, use Theorem 1.88. Each $A \in \mathcal{MC}_n$ is a convex combination of such rank 1 matrices.

2.115. Prove Corollary 2.139.
Hint: To prove (a) $\Rightarrow$ (c), show that if α is the vertex set of a component of $G(A)$, and $A[\alpha]$ has some negative entries, then $G_{-1}(A[\alpha])$ is connected.

2.116. Let A be an $n \times n$ matrix with diagonal equal to $\mathbf{e}$, such that all its negative off-diagonal entries are equal to -1. Then each minimal zero of A has support of size 2.

2.117. Show that the matrix H_n defined in Example 2.147 is an extremal copositive matrix for every odd n.
Hint: Let $n = 2k - 1$ and $\dotplus$ denote addition modulo n on $\{1, \ldots, n\}$. To show that $L^0(H_n)$ is not bipartite, show that $\{v_1, v_2\}, \{v_2, v_3\}, \ldots \{v_n, v_1\}$ is an odd cycle in $L^0(H_n)$, where $v_{2j-1} = \{j, j \dotplus (k-1)\}$ and $v_{2j} = \{j, j \dotplus k\}$.

2.118. Show that the matrix H_n defined in Example 2.147 is exceptional for every odd n.

2.119. Prove Corollary 2.150.

2.120. Let A be a copositive matrix with diagonal $\mathbf{e}$ such that all its entries are from $\{0, 1, -1\}$. Prove that A is $\mathcal{HN}$-irreducible if and only if the following conditions hold.

(a) $a_{ij} = 1$ if and only if $d_{G_{-1}(A)}(i,j) = 2$.
(b) $L^0(A)$ has no isolated vertices.

Hint: Use Proposition 2.65 to show that if $a_{ij} = 1$, then A is $\{i,j\}$-irreducible if and only if $d_{G_{-1}(A)}(i,j) = 2$, and if $a_{ij} = 0$, then A is $\{i,j\}$-irreducible if and only if $\{i,j\}$ is not an isolated vertex of $L^0(A)$.

2.121. Let A be an $n \times n$ matrix, $n \geq 6$, such that all its entries are from the set $\{1, -1\}$. Show that A is copositive if and only if it has diagonal $\mathbf{e}$ and $G_{-1}(A)$ is triangle free. Show that such a copositive A is extremal if and only if the diameter of $G_{-1}(A)$ is 2.

2.122. Let A be an $n \times n$ symmetric matrix with diag$A = k\mathbf{e}$, k a positive integer, and all entries in $\{k, -1\}$. For $\alpha \subseteq \{1, \ldots, n\}$ denote $\mathbf{e}_\alpha = \sum_{i \in \alpha} \mathbf{e}_i$. Let $G = G_{-1}(A)$.

(a) A is copositive if and only if $k + 1 \geq \omega(G)$.
(b) Let $k = \omega(G) - 1$. Then the set $\{\mathbf{e}_\alpha \mid \alpha \text{ is a maximum clique in } G\}$ is a representing set of all its minimal zeros.
(c) Let $k = \omega(G) - 1$. Show that if A is $\{i,j\}$-irreducible for some $i \neq j$ such that $a_{ij} = k$, then adding $\{i,j\}$ to $G = G_{-1}(A)$ increases the clique number by 1 (i.e., it yields a graph G' with $\omega(G') = \omega(G) + 1$).
(d) Let $k = \omega(G) - 1$. Show that if A is $\mathcal{PSD}$-irreducible, then $\overline{G}$ is connected.

Hint: For (a): Consider $B = (1/k)A$ and use Proposition 2.16. For a clique α in G, the submatrix $B[\alpha]$ is copositive if and only if it is positive semidefinite. For (c) and (d), try proof by contradiction.

2.123. Let G be a connected graph with vertex set $\{1, \ldots, n\}$, and let $k = \omega(G) - 1$. Let A_G be the matrix such that $G_{-1}(A_G) = G$, and $a_{ij} = k$ for every $i \neq j$ such that $\{i,j\}$ is an edge of $\overline{G}$. Show that A_G is copositive and it is $\mathcal{SPN}$-irreducible if and only if the following three conditions hold.

(a) $\overline{G}$ is connected.
(b) There exist $i \neq j$ such that the addition of the edge $\{i,j\}$ to G will increase the clique number.
(c) Every edge of G is contained in a maximum clique of G (a clique of size $\omega(G)$).

Hint: For necessity see the previous exercise. For sufficiency, show that (a) and (b) guarantee $\mathcal{PSD}$-irreducibility, while (b) and (c) guarantee $\mathcal{N}$-irreducibility.

Notes. The characterizations of copositive, copositive-plus and positive semidefinite $\{0,1,-1\}$-matrices, and extremal such matrices are based on [Hoffman and Pereira (1973)]. This work was preceded by characterization of such $\{\pm 1\}$-matrices in [Haynsworth and Hoffman (1969)]. Theorem 2.136 is based partly on that paper, and partly on [Shaked-Monderer *et al.* (2016)]. The latter is also the source of Theorem 2.140. Corollary 2.134 appeared in [Dickinson (2019)]. The result in Exercise 2.121 is due to [Haynsworth and Hoffman (1969)]. Prior to that paper extremal $\{\pm 1\}$-copositive matrices (forms) were considered in [Baumert (1967)] and [Baston (1968/1969)]. Theorem 2.149 and Corollary 2.150 were proved in [Hildebrand (2018a)]. Exercises 2.122–2.123 are based on [Dickinson and de Zeeuw (2020)]. Exercises 2.122(a) is a statement of the fact, proved in [de Klerk and Pasechnik (2002)], that $\alpha(\overline{G}) = \min\{\lambda \mid \lambda(I+A) - J \in \mathcal{COP}_n\}$, where A is the adjacency matrix of $\overline{G}$. In [Dickinson and de Zeeuw (2020)] the results are proved in terms of the stability number of $\overline{G}$, and all graphs with up to 13 vertices which give rise to an $\mathcal{SPN}$-irreducible A_G, as defined in Exercise 2.123, were found. It turns out that there are tens of thousands of such matrices. These were then checked for extremality, using Theorem 2.71. In all but (essentially) three cases these matrices were found to be extremal.

2.9 Small copositive matrices

By small copositive matrices we mean those of order $n \leq 4$. As hinted in Section 2.3, the Horn matrix is not only the earliest example of a copositive matrix that is not SPN, it is also the smallest such example. Up to $n = 4$, $\mathcal{COP}_n = \mathcal{SPN}_n$, and we begin by proving that.

Theorem 2.151. *If A is a copositive matrix of order n, $n \leq 4$, then A is SPN.*

For the proof recall that for an $n \times n$ matrix $G_-(A)$ denotes the graph on all n vertices whose edges represent all the negative entries in A.

Proof. For $n = 1$ this is obvious. The case $n = 2$ follows from Example 2.2, where it was shown that every 2×2 copositive matrix is either nonnegative or positive semidefinite.

Let $n = 3$. If $G_-(A)$ is not connected, then it has an isolated vertex. This means that A has a nonnegative row (and column). Without loss of generality, suppose this is the third row. Then $A \geq A(3) \oplus 0$, and since the 2×2 copositive matrix $A(3)$ is SPN, so is A. So suppose $G_-(A)$ is connected. In this case, $G_-(A)$ has a vertex with degree 2, and thus A has a row with two negative off-diagonal entries. Again suppose this is the third row. By Corollary 2.91 the matrix $A/[3]$ is copositive, and by the case $n = 2$ it is SPN. Thus A is SPN (see Remark 2.93 and Exercise 2.71).

Finally, consider $n = 4$. Since it suffices to show that an $\mathcal{HN}$-irreducible matrix $B \leq A$ is SPN, we assume A itself is $\mathcal{HN}$-irreducible. We may also assume that $G_-(A)$ is connected (otherwise, $A \geq B$ where B is a direct sum of smaller copositive matrices, which are SPN by the previous cases). If in row i of A all the off-diagonal entries are nonpositive, then $A/[i]$ is copositive and therefore SPN, and thus A is a sum of a positive semidefinite matrix and an SPN matrix, and is SPN. If there is no such row, the connected $G_-(A)$ is necessarily triangle free. That is, $G_-(A)$ is either a path on 4 vertices, or a 4-cycle. Without loss of generality, A has the following sign pattern

$$A = \begin{pmatrix} + & - & + & * \\ - & + & - & + \\ + & - & + & - \\ * & + & - & + \end{pmatrix},$$

where the $*$ stands for an entry which may be negative, positive, or zero. As A is $\{1, 3\}$-irreducible, there exists a zero $\mathbf{u}$ of A such that $u_1 + u_3 > 0$, $(A\mathbf{u})_1 = (A\mathbf{u})_3 = 0$. From $(A\mathbf{u})_3 = 0$ and $u_1 + u_3 > 0$ we deduce that $u_2 + u_4 > 0$. Thus $|\operatorname{supp} \mathbf{u}| \geq 2$ and $\alpha = \{1 \leq i \leq 4 \mid (A\mathbf{u})_i = 0\}$, which satisfies $\operatorname{supp} \mathbf{u} \subseteq \alpha$, has $|\alpha| \geq 3$. By Corollaries 2.100 and 2.101, A is SPN. □

Can one tell whether a small matrix is copositive (SPN) just by considering the entries? For 2×2 matrices and 3×3 matrices the answer is positive.

Theorem 2.152. *A 2×2 symmetric matrix*

$$A = \begin{pmatrix} a & b \\ b & c \end{pmatrix}$$

is copositive if and only if $a \geq 0$, $c \geq 0$, and $b + \sqrt{ac} \geq 0$.

Proof. The necessity of these conditions was already shown (see Example 2.2, Corollary 2.5(c)). If $a, c \geq 0$ and $b + \sqrt{ac} \geq 0$, then either $b \geq \sqrt{ac}$, in which case A is nonnegative, or $|b| \leq \sqrt{ac}$, in which case all the principal minors of A are nonnegative and A is positive semidefinite. □

Theorem 2.153. *A* 3×3 *symmetric matrix*

$$A = \begin{pmatrix} a_{11} & a_{12} & a_{13} \\ a_{12} & a_{22} & a_{23} \\ a_{13} & a_{23} & a_{33} \end{pmatrix}$$

is copositive if and only if all the following conditions hold.

$$a_{11} \geq 0, \quad a_{22} \geq 0, \quad a_{33} \geq 0, \tag{2.28}$$

$$\alpha_{12} = a_{12} + \sqrt{a_{11}a_{22}} \geq 0, \tag{2.29}$$

$$\alpha_{13} = a_{13} + \sqrt{a_{11}a_{33}} \geq 0, \tag{2.30}$$

$$\alpha_{23} = a_{23} + \sqrt{a_{22}a_{33}} \geq 0, \tag{2.31}$$

$$\sqrt{a_{11}a_{22}a_{33}} + a_{12}\sqrt{a_{33}} + a_{13}\sqrt{a_{22}} + a_{23}\sqrt{a_{11}} + \sqrt{2\alpha_{12}\alpha_{13}\alpha_{23}} \geq 0. \tag{2.32}$$

The inequalities (2.28)–(2.31) simply mean that all proper principal submatrices of A are copositive, i.e., A is copositive of order 2. Such a matrix is not copositive if and only if $\det A < 0$ and $\operatorname{adj} A \geq 0$ (Theorem 2.109). Thus the theorem will be proved if we show that when (2.28)–(2.31) hold, (2.32) is equivalent to "if $\operatorname{adj} A \geq 0$, then $\det A \geq 0$".

We will need the following lemma in the proof.

Lemma 2.154. *Let*

$$A = \begin{pmatrix} 1 & r & s \\ r & 1 & t \\ s & t & 1 \end{pmatrix}.$$

Denote $\rho = r+1$, $\sigma = s+1$, $\tau = t+1$, *and* $q = \rho + \sigma + \tau - 2$. *If* $\rho, \sigma, \tau \geq 0$ *and* $\operatorname{adj} A \geq 0$, *then*

$$\sqrt{2\rho\sigma\tau} + q \geq 0.$$

Moreover, if in addition all the off-diagonal elements of $\operatorname{adj} A$ *are positive, then* $\sqrt{2\rho\sigma\tau} - q > 0$.

Proof. In terms of ρ, σ, τ and q,

$$\operatorname{adj} A = \begin{pmatrix} \tau(2-\tau) & \sigma\tau - q & \rho\sigma - q \\ \sigma\tau - q & \sigma(2-\sigma) & \rho\tau - q \\ \rho\sigma - q & \rho\tau - q & \rho(2-\rho) \end{pmatrix}.$$

If $\rho, \sigma, \tau \geq 0$ and all the elements of $\operatorname{adj} A$ are nonnegative, then $\rho, \sigma, \tau \leq 2$ and

$$\sigma\tau \geq q, \quad \rho\sigma \geq q, \quad \rho\tau \geq q.$$

Hence $\rho\sigma\tau \leq 8$ and

$$(2\rho\sigma\tau)^{\frac{1}{2}} \geq (\rho\sigma\tau)^{\frac{2}{3}} \geq q.$$

If the off-diagonal elements of $\operatorname{adj} A$ are positive, then $\rho\sigma\tau > \sqrt{q^3}$, and the last inequality is strict. □

Proof of Theorem 2.153. It is easy to see that if a diagonal element of A is zero, then A is copositive if and only if (2.28)–(2.32) hold (see Exercise 2.124).

So suppose all diagonal elements are positive. By considering DAD instead of A, where $D = \operatorname{Diag}\left(1/\sqrt{a_{11}}, 1/\sqrt{a_{22}}, 1/\sqrt{a_{33}}\right)$, we may assume that

$$A = \begin{pmatrix} 1 & r & s \\ r & 1 & t \\ s & t & 1 \end{pmatrix},$$

and prove that A is copositive if and only if the conditions (2.33)–(2.34) below hold.

$$\rho = r + 1 \geq 0, \qquad \sigma = s + 1 \geq 0, \qquad \tau = t + 1 \geq 0 \tag{2.33}$$

$$1 + r + s + t + \sqrt{2\rho\sigma\tau} \geq 0. \tag{2.34}$$

The inequality (2.34) is equivalent to

$$\rho + \sigma + \tau + \sqrt{2\rho\sigma\tau} - 2 \geq 0,$$

or, denoting

$$q = \rho + \sigma + \tau - 2, \tag{2.35}$$

to

$$\sqrt{2\rho\sigma\tau} + q \geq 0. \tag{2.36}$$

By Theorem 2.109 the scaled A is copositive if and only if the inequalities (2.33) hold and in addition

$$\operatorname{adj} A \geq 0 \quad \Rightarrow \quad \det A \geq 0. \tag{2.37}$$

We therefore need to show that when the inequalities (2.33) hold, (2.36) is equivalent to (2.37). Express $\det A$ in terms of ρ, σ, τ and q.

$$\det A = 2\rho\sigma\tau - q^2 = \left(\sqrt{2\rho\sigma\tau} + q\right)\left(\sqrt{2\rho\sigma\tau} - q\right), \tag{2.38}$$

and suppose (2.33) holds.

We first assume that (2.36) is satisfied, and show that then (2.37) holds. By Lemma 2.154, if $\operatorname{adj} A \geq 0$, then $\sqrt{2\rho\sigma\tau} - q \geq 0$, which together with (2.36) implies by (2.38) that $\det A \geq 0$.

Conversely, assume (2.37) holds, and suppose on the contrary that (2.36) does not hold, i.e., $\sqrt{2\rho\sigma\tau} + q < 0$. Since $\rho, \sigma, \tau \geq 0$, it follows that $q < 0$. Together these inequalities imply all the off-diagonal elements in $\operatorname{adj} A$ are positive. By (2.35), $\rho, \sigma, \tau < 2$, hence the diagonal elements of $\operatorname{adj} A$ are nonnegative, and by Lemma 2.154, $\sqrt{2\rho\sigma\tau} - q > 0$. Then (2.38) implies that $\det A < 0$ although $\operatorname{adj} A \geq 0$, in contradiction to the assumption that (2.37) holds. □

The following variation on Theorem 2.153 is left as an exercise.

Theorem 2.155. *A* 3×3 *symmetric matrix*

$$A = \begin{pmatrix} a_{11} & a_{12} & a_{13} \\ a_{12} & a_{22} & a_{23} \\ a_{13} & a_{23} & a_{33} \end{pmatrix}$$

is copositive if and only if conditions (2.28)–(2.31) *and*

$$\det A \geq 0 \quad \text{or} \quad \sqrt{a_{11}a_{22}a_{33}} + a_{12}\sqrt{a_{33}} + a_{13}\sqrt{a_{22}} + a_{23}\sqrt{a_{11}} \geq 0 \qquad (2.39)$$

all hold.

Remark 2.156. The strictly copositive versions of Theorems 2.153 and 2.155 are obtained by replacing all inequalities in the conditions by strict inequalities.

For 4×4 matrices things turn messier. Copositivity may be checked by checking copositivity of smaller matrices with the aid of the previous results. These smaller matrices are either submatrices of A, or the Schur complement of a submatrix, or some other matrices generated from the entries of A. We list in the next theorem the simpler cases, which follow from previous results, and leave the proof as an exercise. We assume that A satisfies the obvious necessary condition of the diagonal being nonnegative.

Theorem 2.157. *Let* A *be an* 4×4 *symmetric matrix*

$$A = \begin{pmatrix} a_{11} & a_{12} & a_{13} & a_{14} \\ a_{12} & a_{22} & a_{23} & a_{24} \\ a_{13} & a_{23} & a_{33} & a_{34} \\ a_{14} & a_{24} & a_{34} & a_{44} \end{pmatrix},$$

and suppose $a_{ii} \geq 0$ *for every* $1 \leq i \leq 4$.

(a) *If* $a_{ii} = 0$, A *is copositive if and only if row* i *is nonnegative and* $A(i)$ *is copositive.*
(b) *If row* i *is nonnegative,* A *is copositive if and only if* $A(i)$ *is copositive.*
(c) *If* $a_{ii} > 0$ *and all the off-diagonal entries in row* i *are nonpositive,* A *is copositive if and only if* $A/[i]$ *is copositive.*
(d) *If the diagonal of* A *is positive and* $a_{ij} \geq \sqrt{a_{ii}a_{jj}}$, A *is copositive if and only if* $A(i)$ *and* $A(j)$ *are copositive.*
(e) *If* $G_-(A)$ *consists of two disjoint edges,* $\{i, j\}$ *and* $\{p, q\}$, A *is copositive if and only if* $A[i, j]$ *and* $A[p, q]$ *are positive semidefinite.*

Theorem 2.157 leaves two cases to consider. One is the case that $G_-(A)$ is a path on 4 vertices and A has no zero elements. The other includes two subcases that may be treated as one: $G_-(A)$ is either a 4-cycle or a path on 4 vertices whose ends are not connected by an edge in $G(A)$, and all other entries in A are positive. In each of these cases it is possible to reduce checking copositivity of A to checking copositivity of a certain 3×3 matrix. The interested reader is referred to the two original papers where such reductions were suggested (see the notes at the end of the section for references).

As $\mathcal{COP}_n = \mathcal{SPN}_n$ for $n \leq 4$, the extreme rays of these cones of small copositive matrices are fully known. These are generated by the rank 1 positive semidefinite matrices of the form $\mathbf{x}\mathbf{x}^T$, where $\mathbf{x}$ has both positive and negative entries, and by the matrices $E_{ij} + E_{ji}$, $1 \leq i, j \leq n$ (see Theorem 2.29). If $A \in \mathcal{COP}_n$, $n \leq 4$, then $A = P + N$, where P is positive semidefinite and N is a hollow nonnegative matrix. Hence if A is $\mathcal{HN}$-irreducible, it is positive semidefinite. (The converse is clearly not true, as the all-ones matrix shows.)

A trigonometric representation

We now add an alternative description of such small copositive/SPN matrices, and the $\mathcal{HN}$-irreducible among them. This description turns out to be useful in studying larger copositive matrices, by parameterizing principal small submatrices. We will consider matrices whose diagonal is equal to $\mathbf{e}$. (The assumption on the diagonal is not very restrictive, since every matrix with a positive diagonal may be scaled by diagonal congruence to a matrix with diagonal $\mathbf{e}$, and treatment of matrices with a zero diagonal element may be reduced to treating a principal submatrix with a positive diagonal.) For A with diagonal $\mathbf{e}$, let $\bar{A} = \min(A, J)$. That is, in $\bar{A}$ each off-diagonal entry of A that is greater than 1 is replaced by 1. The matrix A is copositive if and only if $\bar{A}$ is copositive, and it is SPN if and only if $\bar{A}$ is SPN (see Corollary 2.17 and Exercise 2.15). If A is copositive, then in $\bar{A}$ each off-diagonal entry is in $[-1, 1]$ (see Theorem 2.152), and can therefore be represented as $-\cos\theta$ for some $\theta \in [0, \pi]$.

Theorem 2.158. *A 2×2 symmetric matrix A with diagonal $\mathbf{e}$ is copositive if and only if*

$$\bar{A} = \begin{pmatrix} 1 & -\cos\theta \\ -\cos\theta & 1 \end{pmatrix}$$

for some $\theta \in [0, \pi]$.

If A is copositive, it is $\mathcal{HN}$-irreducible if and only if $A = \bar{A}$ and $\theta = 0$, in which case A is positive semidefinite of rank 1 and $\mathbf{e}$ is a positive zero of A.

Proof. The matrix A is copositive if and only if $\bar{A}$ is copositive, i.e., if and only if $\bar{a}_{12} \geq -1$. Since $\bar{a}_{12} \leq 1$, we may write $\bar{a}_{12} = -\cos\theta$, $\theta \in [0,\pi]$. Then

$$\bar{A} \geq \begin{pmatrix} 1 & -1 \\ -1 & 1 \end{pmatrix} = \begin{pmatrix} 1 & -\cos 0 \\ -\cos 0 & 1 \end{pmatrix}.$$

Hence $\bar{A}$ is an $\mathcal{HN}$-irreducible copositive matrix if and only if $\theta = 0$. Clearly $\mathbf{e} \in \mathbb{R}^2$ is a positive zero of such $\mathcal{HN}$-irreducible A. □

Theorem 2.159. *A 3×3 symmetric matrix A with diagonal $\mathbf{e}$ is copositive if and only if*

$$\bar{A} = \begin{pmatrix} 1 & -\cos\theta & -\cos\varphi \\ -\cos\theta & 1 & -\cos\xi \\ -\cos\varphi & -\cos\xi & 1 \end{pmatrix} \tag{2.40}$$

for some $\theta, \varphi, \xi \in [0,\pi]$ such that $\theta + \varphi + \xi \geq \pi$.

The matrix A is $\mathcal{HN}$-irreducible if and only if $A = \bar{A}$ and $\theta+\varphi+\xi = \pi$, in which case A is positive semidefinite.

Moreover, exactly one of the following holds for an $\mathcal{HN}$-irreducible A:

- $0 < \theta, \varphi, \xi < \pi$, $(\sin\xi, \sin\varphi, \sin\theta)^T = (\sin\xi, \sin(\theta+\xi), \sin\theta)^T$ *is a positive zero of A, and* $\operatorname{rank} A = 2$.
- *Two of θ, φ, ξ are zero, there are two minimal zeros with support of size 2 (up to multiplication by a positive scalar), and* $\operatorname{rank} A = 1$.
- *Exactly one of θ, φ, ξ is zero, there exists a unique (up to multiplication by a positive scalar) minimal zero with support of size 2, and* $\operatorname{rank} A = 2$.

Note that any minimal zero with support $\{i, j\}$ of a matrix with diagonal $\mathbf{e}$ is a multiple of $\mathbf{e}_i + \mathbf{e}_j$.

Proof. The matrix A is copositive if and only if $\bar{A}$ is copositive. As in the previous theorem we may assume that

$$\bar{a}_{12} = -\cos\theta, \quad \bar{a}_{13} = -\cos\varphi, \quad \bar{a}_{23} = -\cos\xi,$$

where $\theta, \varphi, \xi \in [0,\pi]$, and characterize the copositive matrices among matrices of the form (2.40).

We begin with characterizing the $\mathcal{HN}$-irreducible such matrices. Suppose first that A is copositive and $\mathcal{HN}$-irreducible. In particular, $A = \bar{A}$ has the form (2.40), it is positive semidefinite and has a zero support of size at least 2 and is therefore singular. We have:

$$\det\begin{pmatrix} 1 & -\cos\theta & t \\ -\cos\theta & 1 & -\cos\xi \\ t & -\cos\xi & 1 \end{pmatrix} = -(t-\cos\theta\cos\xi)^2 + \sin^2\theta\sin^2\xi = 0$$

if and only if $t = \cos\theta\cos\xi \pm \sin\theta\sin\xi$. Hence, as $\cos\theta\cos\xi + \sin\theta\sin\xi \geq \cos\theta\cos\xi - \sin\theta\sin\xi$ and A is $\mathcal{HN}$-irreducible, $a_{13} = \cos\theta\cos\xi - \sin\theta\sin\xi$. That is, $-\cos\varphi = \cos(\theta+\xi)$. Thus if we show that $\theta+\xi \leq \pi$ we will get the desired equality $\varphi = \pi - (\theta+\xi)$.

Suppose on the contrary that $\theta + \xi > \pi$. Then both θ and ξ have to be positive. Since $\theta = \xi = \pi$ would imply that $A = J$, and J is obviously not $\mathcal{HN}$-irreducible, we must have $\theta + \xi < 2\pi$, so either $\theta < \pi$ or $\xi < \pi$. As $-\cos\varphi = \cos(\theta+\xi)$ we get that

$$\begin{pmatrix} 1 & -\cos\theta & -\cos\varphi \\ -\cos\theta & 1 & -\cos\xi \\ -\cos\varphi & -\cos\xi & 1 \end{pmatrix}$$
$$= \begin{pmatrix} -\cos\theta \\ 1 \\ -\cos\xi \end{pmatrix}\begin{pmatrix} -\cos\theta \\ 1 \\ -\cos\xi \end{pmatrix}^T + \begin{pmatrix} \sin\theta \\ 0 \\ -\sin\xi \end{pmatrix}\begin{pmatrix} \sin\theta \\ 0 \\ -\sin\xi \end{pmatrix}^T. \quad (2.41)$$

Since $0 < \theta < \pi$ or $0 < \xi < \pi$, this means that A is positive semidefinite of rank 2 and

$$\mathbf{u} = \begin{pmatrix} \sin\xi \\ \sin(\theta+\xi) \\ \sin\theta \end{pmatrix} \quad (2.42)$$

spans the null space of A. However, by the assumption that $\theta+\xi > \pi$, $\mathbf{u}$ has both positive and negative entries, and that means that A has no zero, contrary to the assumption that it is $\mathcal{HN}$-irreducible. Hence $\theta + \xi \leq \pi$, and $\varphi = \pi - \theta - \xi$.

Conversely, suppose $\theta + \varphi + \xi = \pi$. Then $-\cos\varphi = \cos(\theta+\xi)$ and $\sin\varphi = \sin(\theta+\xi)$, and therefore (2.41) holds and $\bar{A}$ is positive semidefinite. The vector $\mathbf{u}$ in (2.42) is nonnegative, and is nonzero if and only if at least one of θ, ξ, φ is positive. In this latter case $\mathbf{u}$ is a zero of $\bar{A}$, and, as $\bar{A}$ is positive semidefinite, $\bar{A}\mathbf{u} = \mathbf{0}$.

If two of θ, φ, ξ are zero, by (2.41) the matrix $\bar{A}$ is $\mathbf{x}\mathbf{x}^T$ where $|\mathbf{x}| = \mathbf{e}$ and $\mathbf{x}$ has both 1 and -1 entries. In this case $\bar{A}$ is clearly $\mathcal{HN}$-irreducible (even extremal). If exactly one of θ, φ, ξ is zero, then the zero $\mathbf{u}$ has support of size 2. As $\mathbf{u}$ is also in the null space of $\bar{A}$, Proposition 2.65 implies that $\bar{A}$ is $\mathcal{HN}$-irreducible (and it is easy to see that it is a positive semidefinite matrix of rank 2). In the case that θ, φ, ξ are all positive, $\mathbf{u}$ is a positive zero, $\bar{A}$ is $\mathcal{HN}$-irreducible and $\operatorname{rank} \bar{A} = 2$. Hence if $\theta + \varphi + \xi = \pi$, A is $\mathcal{HN}$-irreducible if and only if $A = \bar{A}$. This completes the characterization of $\mathcal{HN}$-irreducible 3×3 matrices with diagonal $\mathbf{e}$.

For the characterization of the copositive matrices with diagonal $\mathbf{e}$, suppose $\bar{A}$ is as in (2.40) with $\theta, \varphi, \xi \in [0, \pi]$. The matrix $\bar{A}$ is copositive if and only if there exist $\theta', \varphi', \xi' \in [0, \pi]$ such that $\theta' + \varphi' + \xi' = \pi$ and

$$\bar{A} = \begin{pmatrix} 1 & -\cos\theta & -\cos\varphi \\ -\cos\theta & 1 & -\cos\xi \\ -\cos\varphi & -\cos\xi & 1 \end{pmatrix} \geq \begin{pmatrix} 1 & -\cos\theta' & -\cos\varphi' \\ -\cos\theta' & 1 & -\cos\xi' \\ -\cos\varphi' & -\cos\xi' & 1 \end{pmatrix}.$$

As $-\cos t : [0, \pi] \to [-1, 1]$ is strictly increasing, the matrix inequality holds if and only if $\theta' \leq \theta$, $\varphi' \leq \varphi$ and $\xi' \leq \xi$. Hence such θ', φ', ξ' exist if and only if $\theta + \varphi + \xi \geq \pi$. □

Remark 2.160. By the above, $A \in \mathcal{COP}_3$ with a positive diagonal is $\mathcal{HN}$-irreducible if and only if

$$DAD = \begin{pmatrix} 1 & -\cos\theta & \cos(\theta+\xi) \\ -\cos\theta & 1 & -\cos\xi \\ \cos(\theta+\xi) & -\cos\xi & 1 \end{pmatrix},$$

where $\theta, \xi \geq 0$ and $\theta + \xi \leq \pi$, and $D = \operatorname{Diag}\left(1/\sqrt{a_{11}}, 1/\sqrt{a_{22}}, 1/\sqrt{a_{33}}\right)$.

It seems that a similar copositivity characterization of 4×4 copositive matrices with diagonal $\mathbf{e}$ is more cumbersome, as is a description of the $\mathcal{HN}$-irreducible among these. One reason is that there are several possible combinations of minimal supports: an $\mathcal{HN}$-irreducible A may have minimal supports of sizes 2, 3 or 4. Some such cases are considered in the exercises.

Exercises

2.124. Prove Theorem 2.153 in the case that A has a zero diagonal element. *Hint:* In this case (2.28)–(2.31) imply (2.32).

2.125. Prove the strictly copositive version of Theorem 2.153 (see Remark 2.156).

2.126. Prove the strictly copositive version of Theorem 2.155 (see Remark 2.156).

2.127. An $n \times n$ symmetric matrix, $n \le 3$, is copositive-plus if and only if it is either positive semidefinite, or a direct sum of a strictly copositive matrix and a zero matrix.
Hint: For $n = 3$, if A has all diagonal entries positive, scale it to have a diagonal $\mathbf{e}$. Consider the cases (1) $a_{ij} \ge 0$ for every $i \ne j$, (2) $|a_{ij}| \le 1$ for every $i \ne j$. In the latter, consider the case that $|a_{ij}| < 1$ for some $i \ne j$, and the case that $a_{ij} = -1$ for some $i \ne j$.

2.128. Prove Theorem 2.157

2.129. Let

$$A = \begin{pmatrix} 1 & -\cos\theta & -\cos\varphi \\ -\cos\theta & 1 & -\cos\xi \\ -\cos\varphi & -\cos\xi & 1 \end{pmatrix}$$

where $\theta, \varphi, \xi \in [0, \pi]$. Show that if A is positive definite, then $\theta, \varphi, \xi \in (0, \pi)$ and $|\pi - (\theta + \xi)| < \varphi$.

Hint: By the proof of Theorem 2.159, $\det \begin{pmatrix} 1 & -\cos\theta & t \\ -\cos\theta & 1 & -\cos\xi \\ t & -\cos\xi & 1 \end{pmatrix} > 0$ iff $\cos(\theta + \xi) < t < \cos(\theta - \xi)$. Consider separately the case $\theta + \xi \le \pi$ and the case $\pi < \theta + \xi < 2\pi$.

2.130. Let A be a 4×4 copositive matrix with diagonal $\mathbf{e}$, which is $\mathcal{HN}$-irreducible. Suppose all the minimal zeros of A are of size 2. Show that A has one of the following forms, up to permutation similarity:

$$\begin{pmatrix} 1 & -1 & 1 & -t \\ -1 & 1 & -1 & t \\ 1 & -1 & 1 & -t \\ -t & t & -t & 1 \end{pmatrix} \quad \text{or} \quad \begin{pmatrix} 1 & -1 & t & -t \\ -1 & 1 & -t & t \\ t & -t & 1 & -1 \\ -t & t & -1 & 1 \end{pmatrix}, \quad |t| \le 1 .$$

2.131. Let A be a 4×4 copositive matrix with diagonal $\mathbf{e}$, which is $\mathcal{HN}$-irreducible. Suppose that A has a minimal support of size at least 3. Show that the possible sets of minimal supports of A are, up to permutation:

(1) $\{1, 2, 3, 4\}$,
(2) $\{1, 2, 3\}$,
(3) $\{1, 2, 3\}, \{2, 3, 4\}$,
(4) $\{1, 2, 3\}, \{3, 4\}$.

Find the (trigonometric) form of A in cases (3) and (4).
Hint: Corollary 2.59.

2.132. Let

$$A = \begin{pmatrix} 1 & -\cos\varphi_{12} & -\cos\varphi_{13} & -\cos\varphi_{14} \\ -\cos\varphi_{12} & 1 & -\cos\varphi_{23} & -\cos\varphi_{24} \\ -\cos\varphi_{13} & -\cos\varphi_{23} & 1 & -\cos\varphi_{34} \\ -\cos\varphi_{14} & -\cos\varphi_{24} & -\cos\varphi_{34} & 1 \end{pmatrix},$$

$\varphi_{ij} \in [0, \pi]$ for every $1 \le i < j \le 4$. Suppose that A is positive semidefinite with $\operatorname{rank} A = 3$. Show that if $\varphi_{12} + \varphi_{23} < \pi$ and $\operatorname{ns} A$ is spanned by a positive vector $\mathbf{u}$, then

$$\sin\varphi_{23}\cos\varphi_{14} + \sin(\varphi_{12} + \varphi_{23})\cos_{24} + \sin\varphi_{12}\cos\varphi_{34} > 0.$$

Hint: Use Exercise 2.129 to show that $a_{13} - \cos(\varphi_{12} + \varphi_{23}) > 0$. Let $\mathbf{v} = (\sin\varphi_{23}, \sin(\varphi_{12} + \varphi_{23}), \sin\varphi_{12}, 0)^T$. Find the first three elements in $A\mathbf{v}$, and use the fact that $\mathbf{u}^T A\mathbf{v} = 0$ to show that $(A\mathbf{v})_4 < 0$.

2.133. Let A be an $n \times n$ copositive matrix. Show that if A has a zero $\mathbf{u}$ with $|\operatorname{supp} \mathbf{u}| \ge n - 4$ such that $A\mathbf{u} = \mathbf{0}$, then A is SPN.
Hint: Exercise 2.76.

Notes. The fact that $\mathcal{COP}_n = \mathcal{SPN}_n$ for $n \le 4$ was first proved in [Diananda (1962)], in terms of quadratic forms. The proof here is based on [Shaked-Monderer (2016)]. In Section 3.3 we will give a direct proof that $\mathcal{CP}_n = \mathcal{DNN}_n$ for $n \le 4$. The duality between $\mathcal{COP}_n$ and $\mathcal{CP}_n$ and between $\mathcal{SPN}_n$ and $\mathcal{DNN}_n$, implies that either of these results may be deduced from the other. Theorem 2.153 is due to [Chang and Sederberg (1994)]. Chronologically, it was preceded by Theorem 2.155, proved in [Hadeler (1983)]. The strictly copositive version of Theorem 2.153 was proved in [Simpson and Spector (1983)]. In [Li and Feng (1993)] criteria for determining copositivity or strictly copositive of matrices of order 3 and 4 was presented, by considering several cases, and reducing the problem to checking copositivity of smaller matrices. The cases left out in Theorem 2.157 may be found there. These cases are also treated in [Andersson *et al.* (1995)], by a different approach.

The trigonometric description of small copositive matrices and the $\mathcal{HN}$-irreducible ones in the remainder of the section (Theorem 2.158 and onwards) is based mainly on [Dickinson *et al.* (2013a)]. Exercise 2.132 is based on [Afonin *et al.* (2020)].

2.10 $\mathcal{COP}_5$

The cone $\mathcal{COP}_5$ is where everything begins "in earnest": it is the first $\mathcal{COP}_n$ cone that is not equal to $\mathcal{SPN}_n$. The two major problems, of determining copositivity and characterizing extremal copositive matrices, become increasingly difficult as the order of the matrices considered grows. Still, as the smallest of the copositive cones that contain exceptional matrices, $\mathcal{COP}_5$ may give us some insight into these cones. In this section we describe all the extremal matrices in $\mathcal{COP}_5$, and the difference between $\mathcal{COP}_5$ and $\mathcal{SPN}_5$. We conclude with a characterization of matrices in $\mathcal{COP}_5$ as matrices that when scaled to have a $\{0,1\}$-diagonal satisfy that a certain polynomial is a sum of squares. For more on the structure of this cone see Chapter 5.

Since the extremal SPN matrices are already known, we concentrate on describing the exceptional extremal matrices, and for the description of $\mathcal{COP}_5 \setminus \mathcal{SPN}_5$ we consider exceptional $\mathcal{HN}$-irreducible matrices.

The following matrices play a role in both descriptions. For every $\boldsymbol{\theta} = (\theta_1, \theta_2, \theta_3, \theta_4, \theta_5)^T$ with $\theta_i \in [0, \pi)$ let

$$S(\boldsymbol{\theta}) =$$
$$\begin{pmatrix} 1 & -\cos\theta_1 & \cos(\theta_1+\theta_2) & \cos(\theta_4+\theta_5) & -\cos\theta_5 \\ -\cos\theta_1 & 1 & -\cos\theta_2 & \cos(\theta_2+\theta_3) & \cos(\theta_5+\theta_1) \\ \cos(\theta_1+\theta_2) & -\cos\theta_2 & 1 & -\cos\theta_3 & \cos(\theta_3+\theta_4) \\ \cos(\theta_4+\theta_5) & \cos(\theta_2+\theta_3) & -\cos\theta_3 & 1 & -\cos\theta_4 \\ -\cos\theta_5 & \cos(\theta_5+\theta_1) & \cos(\theta_3+\theta_4) & -\cos\theta_4 & 1 \end{pmatrix}. \tag{2.43}$$

Note that $S(\mathbf{0})$ is the Horn matrix. A matrix $S(\boldsymbol{\theta})$ with every $\theta_i \in (0, \pi)$ and $\sum_{i=1}^5 \theta_i < \pi$ is called a *Hildebrand matrix*.

We state first the characterization of exceptional extremal matrices in $\mathcal{COP}_5$, and only then go into the proof.

Theorem 2.161. *A matrix $A \in \mathcal{COP}_5$ is exceptional and extremal if and only if A is in the orbit of $S(\boldsymbol{\theta})$ for some $\boldsymbol{\theta} \in \mathbb{R}^5_+$, where either $\boldsymbol{\theta} = \mathbf{0}$, or $\boldsymbol{\theta} = (\theta_1, \theta_2, \theta_3, \theta_4, \theta_5)^T$ with $\theta_i \in (0, \pi)$ for every i and $\sum_{i=1}^5 \theta_i < \pi$. That is, the exceptional extremal matrices in $\mathcal{COP}_5$ are exactly the matrices in the orbit of the Horn matrix, and those in the orbit of a Hildebrand matrix.*

For the proof of Theorem 2.161, recall that the Horn matrix $S(\mathbf{0})$ is already known to be copositive, exceptional and extremal. We will show that so is any Hildebrand matrix. But first we consider the copositivity of all $S(\boldsymbol{\theta})$ matrices. Throughout this section $\dotplus$ denotes addition modulo 5 on $\{1, \ldots, 5\}$.

Proposition 2.162. *Let $S(\boldsymbol{\theta})$ be as in (2.43). Then*

(a) *If $0 \leq \theta_i < \pi$ for every i, and $\sum_{i=1}^{5} \theta_i = \pi$, then $S(\boldsymbol{\theta})$ is positive semidefinite.*
(b) *If $0 \leq \theta_i < \pi$ for every i, and $\sum_{i=1}^{5} \theta_i < \pi$, then $S(\boldsymbol{\theta})$ is copositive, $\mathcal{N}$-irreducible, and exceptional.*
(c) *If $0 < \theta_i < \pi$, $\theta_i + \theta_{i \dotplus 1} < \pi$ for every i, and $\sum_{i=1}^{5} \theta_i > \pi$, then $S(\boldsymbol{\theta})$ is not copositive.*

Proof. To prove (a), note that for any $\boldsymbol{\theta} = (\theta_1, \theta_2, \theta_3, \theta_4, \theta_5)^T \in \mathbb{R}^5$ with $\sum_{i=1}^{5} \theta_i = \pi$,

$$S(\boldsymbol{\theta}) =$$

$$\begin{pmatrix} \cos(\theta_1+\theta_2) \\ -\cos\theta_2 \\ 1 \\ -\cos\theta_3 \\ \cos(\theta_3+\theta_4) \end{pmatrix} \begin{pmatrix} \cos(\theta_1+\theta_2) \\ -\cos\theta_2 \\ 1 \\ -\cos\theta_3 \\ \cos(\theta_3+\theta_4) \end{pmatrix}^T + \begin{pmatrix} \sin(\theta_1+\theta_2) \\ -\sin\theta_2 \\ 0 \\ \sin\theta_3 \\ -\sin(\theta_3+\theta_4) \end{pmatrix} \begin{pmatrix} \sin(\theta_1+\theta_2) \\ -\sin\theta_2 \\ 0 \\ \sin\theta_3 \\ -\sin(\theta_3+\theta_4) \end{pmatrix}^T.$$

Hence under the conditions of (a) the matrix $S(\boldsymbol{\theta})$ is positive semidefinite, and has rank 2.

The proof of (b) is split into two cases.

Case 1. Suppose $\theta_i \in (0, \pi)$ for every i and $\sum_{i=1}^{5} \theta_i < \pi$, and let $S = S(\boldsymbol{\theta})$. (This is the case that $S(\boldsymbol{\theta})$ is a Hildebrand matrix.) Denote

$$\sigma_i = \{i, i \dotplus 1, i \dotplus 2\}, \quad i = 1, \ldots, 5, \tag{2.44}$$

and let $\mathbf{w}_i \in \mathbb{R}^5$ be the vector with support σ_i such that

$$\mathbf{w}_i[\sigma_i] = \begin{pmatrix} \sin\theta_{i \dotplus 1} \\ \sin(\theta_i + \theta_{i \dotplus 1}) \\ \sin\theta_i \end{pmatrix}. \tag{2.45}$$

Each $\mathbf{w}_i \geq \mathbf{0}$ is a zero of S with support of size $3 = 5 - 2$ (see Theorem 2.159). By Corollary 2.85, S is copositive if and only if $\mathbf{w}_i^T S \mathbf{w}_{i \dotplus 1} \geq 0$ for every i. In our case,

$$\mathbf{w}_i^T S \mathbf{w}_{i \dotplus 1} = \frac{1}{2}(1 - \cos(2\theta_{i \dotplus 1}))(\cos(\theta_i + \theta_{i \dotplus 1} + \theta_{i \dotplus 2}) + \cos(\theta_{i \dotplus 3} + \theta_{i \dotplus 4})).$$

As $0 < 2\theta_{i \dotplus 1} < 2\pi$, S is copositive if and only if

$$\cos(\theta_i + \theta_{i \dotplus 1} + \theta_{i \dotplus 2}) + \cos(\theta_{i \dotplus 3} + \theta_{i \dotplus 4}) \geq 0. \tag{2.46}$$

Indeed, $\sum_{i=1}^{5} \theta_i < \pi$ implies that $\alpha_i = \theta_i + \theta_{i \dotplus 1} + \theta_{i \dotplus 2}$ and $\beta_i = \theta_{i \dotplus 3} + \theta_{i \dotplus 4}$ satisfy

$$0 \leq \frac{|\alpha_i - \beta_i|}{2} < \frac{\alpha_i + \beta_i}{2} = \frac{\sum_{i=1}^{\pi} \theta_i}{2} < \frac{\pi}{2}.$$

Thus $\cos\alpha_i + \cos\beta_i = 2\cos\frac{|\alpha_i-\beta_i|}{2}\cos\frac{\alpha_i+\beta_i}{2} > 0$ and $\mathbf{w}_i^T S\mathbf{w}_{i+1} > 0$ for every i. We get that S is copositive and, by Remark 2.86, exceptional. Note that for every $1 \le i,j \le 5$ there exists a zero of S whose support contains both i and j, therefore a Hildebrand matrix is $\mathcal{N}$-irreducible.

Case 2. At least one of the θ_i's is zero, and $\sum_{i=1}^5 \theta_i < \pi$. Without loss of generality, suppose $\theta_1 = 0$. Let $S = S(\boldsymbol{\theta})$. If $\boldsymbol{\theta} = \mathbf{0}$, S is the Horn matrix and it is already known that it is exceptional and extremal. So suppose $\boldsymbol{\theta} \ne \mathbf{0}$. Then

$$S = \mathbf{a}\mathbf{a}^T + DHD, \tag{2.47}$$

where

$$\mathbf{a} = \begin{pmatrix} -\sin\left(\frac{1}{2}(\theta_2+\theta_3+\theta_4+\theta_5)\right) \\ \sin\left(\frac{1}{2}(\theta_2+\theta_3+\theta_4+\theta_5)\right) \\ -\sin\left(\frac{1}{2}(-\theta_2+\theta_3+\theta_4+\theta_5)\right) \\ \sin\left(\frac{1}{2}(-\theta_2-\theta_3+\theta_4+\theta_5)\right) \\ -\sin\left(\frac{1}{2}(-\theta_2-\theta_3-\theta_4+\theta_5)\right) \end{pmatrix},$$

$H = S(\mathbf{0})$ is the Horn matrix, and $D = \mathrm{Diag}(\mathbf{d})$,

$$\mathbf{d} = \begin{pmatrix} \cos\left(\frac{1}{2}(\theta_2+\theta_3+\theta_4+\theta_5)\right) \\ \cos\left(\frac{1}{2}(\theta_2+\theta_3+\theta_4+\theta_5)\right) \\ \cos\left(\frac{1}{2}(-\theta_2+\theta_3+\theta_4+\theta_5)\right) \\ \cos\left(\frac{1}{2}(-\theta_2-\theta_3+\theta_4+\theta_5)\right) \\ \cos\left(\frac{1}{2}(-\theta_2-\theta_3-\theta_4+\theta_5)\right) \end{pmatrix}.$$

As $\sum_{i=1}^5 \theta_i < \pi$, $\mathbf{d} > \mathbf{0}$ and $\mathbf{a} \ne \mathbf{0}$. That is, $S = S(\boldsymbol{\theta})$ is a sum of a rank 1 positive semidefinite matrix and a matrix in the orbit of the Horn matrix, and is therefore copositive (and not extremal!). The vector $\mathbf{v} = \mathbf{e}_1 + \mathbf{e}_2$ is a zero of S, and $(S\mathbf{v})_4 = \cos(\theta_4+\theta_5) + \cos(\theta_2+\theta_3) \ne 0$, since $0 \le \theta_4+\theta_5 < \pi - (\theta_2+\theta_3) < \pi$. As the zero $\mathbf{v}$ is not in the null space of S, the matrix S is not positive semidefinite. A proof that S is $\mathcal{N}$-irreducible will therefore also show that S is exceptional.

To prove that S is $\mathcal{N}$-irreducible, we need to show that for every i,j in $\{1,\ldots,5\}$ there exists a zero $\mathbf{u}$ of S such that $u_i + u_j > 0$ and $(S\mathbf{u})_i = (S\mathbf{u})_j = 0$. A vector $\mathbf{u}$ is a zero of S if and only if $\mathbf{u}$ is both a zero of $\mathbf{a}\mathbf{a}^T$ and a zero of DHD. That is, if and only if $\mathbf{u}^T\mathbf{a} = 0$ and $D\mathbf{u}$ is a zero of H. In particular, if $\mathbf{u}$ is a zero of S, then

$$S\mathbf{u} = DHD\mathbf{u}. \tag{2.48}$$

Recall that the zeros of H are $s\mathbf{u}_i + t\mathbf{u}_{i\dotplus 1}$, for $s, t \geq 0$ and $i = 1, \ldots, 5$, where $\mathbf{u}_i = \mathbf{e}_i + \mathbf{e}_{i\dotplus 1}$ (see Example 2.42). Hence each zero $\mathbf{u}$ of S has support of size 2 or 3, consisting of consecutive indices. Combining these observations with the formula (2.43) for S we can see that any zero $\mathbf{u}$ of S satisfies:

$$\operatorname{supp}\mathbf{u} = \{i, i \dotplus 1\} \quad \Rightarrow \quad (S\mathbf{u})_j = 0 \text{ if and only if } j \neq i \dotplus 3, \tag{2.49}$$

$$\operatorname{supp}\mathbf{u} = \{i, i \dotplus 1, i \dotplus 2\} \quad \Rightarrow \quad (S\mathbf{u})_j = 0 \text{ if and only if } j \in \operatorname{supp}\mathbf{u}. \tag{2.50}$$

Let $\alpha = \{1 \leq i \leq 5 \mid \theta_i = 0\}$. Then $1 \leq |\alpha| \leq 4$. There are six cases to check, up to cyclic permutation:
$|\alpha| = 1$;
$|\alpha| = 2$ and $\theta_1 = \theta_2 = 0$;
$|\alpha| = 2$ and $\theta_1 = \theta_3 = 0$;
$|\alpha| = 3$ and $\theta_1 = \theta_2 = \theta_3 = 0$;
$|\alpha| = 3$ and $\theta_1 = \theta_2 = \theta_4 = 0$;
$|\alpha| = 4$.

If, for example, $|\alpha| = 1$, then $\theta_i > 0$ for $i \geq 2$. By (2.49), S is $\{1, j\}$-irreducible for every $j \neq 4$, and by (2.50) the matrix S is $\{i, j\}$-irreducible for every $2 \leq i \leq 5$ and $j \in \{i, i \dotplus 1, i \dotplus 2\}$, hence in this case S is $\mathcal{N}$-irreducible. Check similarly that S is $\mathcal{N}$-irreducible in the other five cases.

Finally, we prove (c). Note that under the conditions of (c), $\mathbf{w}_i \geq \mathbf{0}$. Suppose on the contrary that the matrix $S = S(\boldsymbol{\theta})$ is copositive. Then (2.46) holds, which means that

$$\cos(\theta_i + \theta_{i\dotplus 1} + \theta_{i\dotplus 2}) = -\cos|\theta_i + \theta_{i\dotplus 1} + \theta_{i\dotplus 2} - \pi| \geq -\cos(\theta_{i\dotplus 3} + \theta_{i\dotplus 4}).$$

By the assumptions on $\boldsymbol{\theta}$, both $\theta_{i\dotplus 3} + \theta_{i\dotplus 4}$ and $|\theta_i + \theta_{i\dotplus 1} + \theta_{i\dotplus 2} - \pi|$ are in $(0, \pi]$. The monotonicity of $-\cos t$ on this interval then implies that for every i

$$|\theta_i + \theta_{i\dotplus 1} + \theta_{i\dotplus 2} - \pi| \geq \theta_{i\dotplus 3} + \theta_{i\dotplus 4},$$

i.e., either $\pi \geq \sum_{i=1}^{5} \theta_i$, or $\theta_i + \theta_{i\dotplus 1} + \theta_{i\dotplus 2} - \pi \geq \theta_{i\dotplus 3} + \theta_{i\dotplus 4}$. Since $\sum_{i=1}^{5} \theta_i > \pi$, we get that for every $1 \leq i \leq 5$

$$\theta_i + \theta_{i\dotplus 1} + \theta_{i\dotplus 2} - \theta_{i\dotplus 3} - \theta_{i\dotplus 4} \geq \pi.$$

However, summing these five inequalities we get that

$$\sum_{i=1}^{5} \theta_i \geq 5\pi,$$

in contradiction to the assumption that $0 < \theta_i < \pi$ for every i. $\square$

We can now prove the extremality of Hildebrand matrices.

Proposition 2.163. *A Hildebrand matrix is an exceptional extremal matrix in $\mathcal{COP}_5$.*

Proof. By Proposition 2.162(b) above, a Hildebrand matrix H is an exceptional copositive matrix. To show extremality, suppose $H = S(\boldsymbol{\theta})$, where $\boldsymbol{\theta} = (\theta_1, \theta_2, \theta_3, \theta_4, \theta_5)^T \in \mathbb{R}^5_+$ satisfy $\theta_i > 0$ for every i and $\sum_{i=1}^5 \theta_i < \pi$. Let σ_i and $\mathbf{w}_i$ be as in (2.44) and (2.45). The vectors $\mathbf{w}_i$ are zeros of H, and they are minimal, as H has a positive diagonal and no minimal zeros with support of size 2 ($\theta_i > 0$ and $\theta_i + \theta_{i+1} < \pi$ for every i). These are the only minimal zeros of H, up to scalar multiplication (by Corollary 2.59). For every $1 \le i \le 5$ and $(H\mathbf{w}_j)_i = 0$ if and only if $j \in \sigma_i$ (this can be verified directly, but follows also from Corollary 2.101 and the exceptionality of H). By Remark 2.72, to show that H is extremal, it suffices to show that the any symmetric 5×5 matrix X that satisfies

$$(X\mathbf{w}_j)_i = 0, \quad \text{for every} \;\; i \in \sigma_j, \; j = 1, \ldots, 5$$

is a scalar multiple of H. Using a computer algebra system it is easily verified that the solution space of this system has dimension one. □

To complete the proof of Theorem 2.161 it remains to show that every extremal exceptional matrix in $\mathcal{COP}_5$ is either in the orbit of the Horn matrix or in the orbit of a Hildebrand matrix. So suppose $A \in \mathcal{COP}_5$ is exceptional. By diagonal scaling we may assume that $\text{diag}A = \mathbf{e}$. If A is extremal, or even just $\mathcal{HN}$-irreducible, then $a_{ij} \in [-1, 1]$ for every i, j. The minimal supports of A are all of size 2 or 3 (by Corollary 2.70). By the assumption on the diagonal, $\{i, j\}$ is a minimal support of A if and only if $a_{ij} = -1$. Hence the edges of $G_{-1}(A)$ represent the minimal supports of size 2. We define a graph $G^3_{\min}(A)$ to represent the minimal supports of size 3. The vertex set of $G^3_{\min}(A)$ is $\{1, \ldots, 5\}$, and $\{i, j\}$ is an edge if and only if $\{1, \ldots, 5\} \setminus \{i, j\}$ is the support of a minimal zero of A. We may describe properties of the minimal zeros of an exceptional $A \in \mathcal{COP}_5$ in terms of these two graphs.

Lemma 2.164. *Let $A \in \mathcal{COP}_5$ be an exceptional matrix with diagonal* $\mathbf{e}$ *and entries in $[-1, 1]$. Then*

(1) *The graph $G_{-1}(A)$ is a subgraph of a 5-cycle.*
(2) *The graph $G^3_{\min}(A)$ is a subgraph of a 5-cycle.*

(3) *If $\{i, j\}$ is an edge of $G_{-1}(A)$, each edge of $G^3_{\min}(A)$ has at least one of i or j as an end vertex. Moreover, for every $k \neq i, j$ at most one of $\{i, k\}$ and $\{j, k\}$ is an edge of $G^3_{\min}(A)$.*

(4) *If $K_{2,1}$ is a subgraph of $G_{-1}(A)$ with vertex set α, then $G^3_{\min}(A)$ has at most two edges, each with both ends in α.*

(5) *If $K_2 \cup K_2$ is a subgraph of $G_{-1}(A)$ with vertex set α, then $G^3_{\min}(A)$ has at most two edges that do not share a vertex, each with both ends in α.*

Proof. (1) From the copositivity of A it follows that $G_{-1}(A)$ is triangle free — see Remark 2.133. The graph $G_{-1}(A)$ cannot contain a complete bipartite graph on 4 vertices, since this would mean, by Theorem 2.136, that A has a positive semidefinite principal submatrix of order 4. As A is exceptional, it cannot have such a submatrix by Proposition 2.69. So the triangle free $G_{-1}(A)$ does not contain a 4-cycle or a vertex of degree 3, and it is either a 5-cycle or a disjoint union of paths.

(2) By Corollary 2.98, the degree of each vertex in $G^3_{\min}(A)$ is at most 2. Also, as A is exceptional, $G^3_{\min}(A)$ contains no cycle on 3 or 4 vertices by Corollary 2.97. The two assertions combined prove the claim.

(3) The first assertion is simply the statement that a minimal support of size 3 cannot contain a minimal support of size 2. For the second assertion, suppose without loss of generality that $i = 1$, $j = 2$, $k = 3$. If both $\{1, 3\}$ and $\{2, 3\}$ are edges of $G^3_{\min}(A)$, then $\{2, 4, 5\}$ and $\{1, 4, 5\}$ are minimal supports of A, but then by Corollary 2.59 there are no other minimal supports contained in $\{1, 2, 4, 5\}$, contrary to the assumption that $\{1, 2\}$ is a minimal support.

(4) In this case, $A[\alpha]$ is a $\{\pm 1\}$-submatrix of A of order 3. If σ is a minimal support of size 3, $A[\sigma]$ is a rank 2 positive semidefinite matrix with a positive zero. By Theorem 2.159, $|a_{ij}| < 1$ for every $i, j \in \sigma$, $i \neq j$. Hence $|\sigma \cap \alpha| \leq 1$, and since both σ and α are of size 3, $|\sigma \cap \alpha| = 1$. So the edge in $G^3_{\min}(A)$ representing σ has both ends in α. As $G^3_{\min}(A)$ is triangle free by (2), it has at most two such edges.

(5) This follows from the two assertions in (3). $\square$

For an extremal exceptional A we have two additional observations.

Lemma 2.165. *Let $A \in \mathcal{COP}_5$ be an exceptional extremal matrix with diagonal* $\mathbf{e}$*. Then*

(6) *A has at least 5 linearly independent minimal zeros.*

(7) *For every bipartite component of $G_{-1}(A)$ there exists a vertex i in the component and a minimal support σ of size 3 such that $i \in \sigma$ (different σ's for different bipartite components).*

Proof. As A is obviously $\mathcal{PSD}$-irreducible, (6) follows from Corollary 2.68. Then (7) follows from (6) and Proposition 2.60. □

By the above observations, we can prove the following.

Proposition 2.166. *Let A be a 5×5 exceptional extremal copositive matrix with diagonal* $\mathbf{e}$. *Then either $G_{-1}(A)$ is the 5-cycle or $G^3_{\min}(A)$ is the 5-cycle.*

Proof. We first show that for such A the graph $G_{-1}(A)$ cannot proper subgraph of the 5-cycle, unless it has no edges. Up to isomorphism, the graphs shown in the figure below are all the proper subgraphs of the 5-cycle that have edges. Suppose on the contrary that $G_{-1}(A)$ is one of these graphs. We show that then the total number of minimal zeros is less than 5, contradicting Lemma 2.165(6).

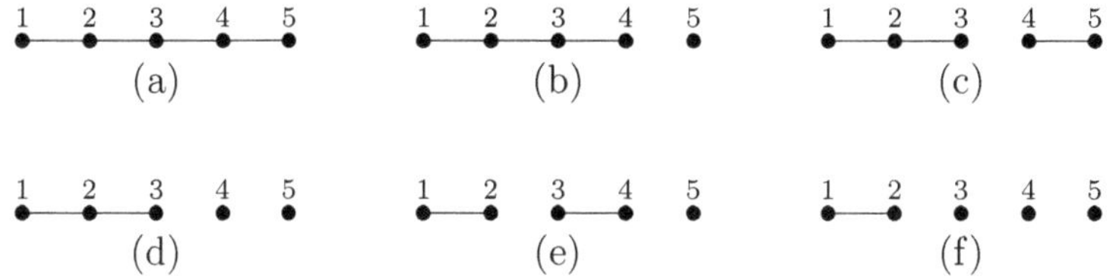

Indeed, if the graph $G_{-1}(A)$ is the graph (a), Lemma 2.164(4) implies that any edge of $G^3_{\min}(A)$ has both ends in $\{1,2,3\}$ and also both ends in $\{3,4,5\}$. Hence in this case $G^3_{\min}(A)$ has no edges, and A has 4 minimal zeros. If $G_{-1}(A)$ is the graph (b), Lemma 2.164(4) implies that the only possible edge in $G^3_{\min}(A)$ is $\{2,3\}$, for a total of at most 4 minimal zeros. If $G_{-1}(A)$ is the graph (c), then Lemma 2.164(3) requires each edge of $G^3_{\min}(A)$ to have an end in $\{4,5\}$, and Lemma 2.164(4) requires each such edge to have both ends in $\{1,2,3\}$, leading to $G^3_{\min}(A)$ being empty, and a total of 3 minimal zeros. If $G_{-1}(A)$ is the graph (d), there are at most two edges in $G^3_{\min}(A)$ by Lemma 2.164(4), and therefore a total of at most 4 minimal zeros. If $G_{-1}(A)$ is the graph (e), then by Lemma 2.164(5) the graph $G^3_{\min}(A)$ has at most 2 edges, for a total of at most 4 minimal zeros. Finally, if $G_{-1}(A)$ is the graph (f), each edge of $G^3_{\min}(A)$ has at least one end in $\{1,2\}$. Then by parts (2) and (3) of Lemma 2.164 there are at most

3 edges in $G^3_{\min}(A)$ (whether $\{1,2\}$ is one of them or not). Again A has at most 4 minimal zeros.

Hence the graph $G_{-1}(A)$ is either the 5-cycle, or the edgeless graph. In the latter case, all the minimal zeros are of size 3, and by Corollary 2.98 the graph $G^3_{\mathcal{V}} = G^3_{\min}(A)$ is the 5-cycle. □

We now describe the matrices A such that $G_{-1}(A)$ is the 5-cycle, and those for which $G^3_{\min}(A)$ is the 5-cycle.

Proposition 2.167. *Let A be a 5×5 copositive matrix with diagonal* $\mathbf{e}$. *If $G_{-1}(A)$ is the 5-cycle, then A is the Horn matrix, up to simultaneous permutation of rows and columns.*

Proof. By permuting rows and columns we may assume that the edges of the 5-cycle are $\{i, i \dotplus 1\}$, $1 \leq i \leq 5$. Then $a_{i\, i\dotplus 1} = -1$ for every i, and $a_{i,i\dotplus 2} = 1$ for every i by Corollary 2.131. □

Proposition 2.168. *Let A be a 5×5 exceptional copositive matrix with diagonal* $\mathbf{e}$. *If $G^3_{\min}(A)$ is the 5-cycle, then A is a Hildebrand matrix, up to simultaneous permutation of rows and columns.*

Proof. Permute rows and columns so that $G^3_{\min}(A)$ is a 5-cycle with edges $\{i, i\dotplus 1\}$, $i = 1, \ldots, 5$. Then $\sigma_i = \{i, i\dotplus 1, i\dotplus 2\}$, $i = 1, \ldots, 5$, are the minimal supports of A. The principal submatrix $A[\sigma_i]$ is a positive semidefinite matrix of rank 2 that has a positive zero. By Theorem 2.159,

$$A[\sigma_i] = \begin{pmatrix} 1 & -\cos\theta_i & -\cos\varphi_i \\ -\cos\theta_i & 1 & -\cos\theta_{i\dotplus 1} \\ -\cos\varphi_i & -\cos\theta_{i\dotplus 1} & 1 \end{pmatrix},$$

where $\theta_i, \theta_{i\dotplus 1}, \varphi_i \in (0, \pi)$, and $\theta_i + \theta_{i\dotplus 1} + \varphi_i = \pi$. Hence $-\cos\varphi_i = \cos(\theta_i + \theta_{i\dotplus 1})$ and $A = S(\boldsymbol{\theta})$, where $\boldsymbol{\theta} = (\theta_1, \theta_2, \theta_3, \theta_4, \theta_5)$, $\theta_i \in (0, \pi)$ for every i, and $\theta_i + \theta_{i\dotplus 1} < \pi$. By Proposition 2.162, as A is copositive and exceptional we necessarily have $\sum_{i=1}^5 \theta_i < \pi$. That is, A is a Hildebrand matrix. □

We can now prove Theorem 2.161.

Proof of Theorem 2.161. We already know that the Horn matrix is an exceptional extremal copositive matrix. By Propositions 2.162 and 2.163 a Hildebrand matrix is an exceptional extremal copositive matrix. Thus matrices in the orbits of these matrices are exceptional and extremal.

Conversely, as any exceptional extremal matrix can be scaled by diagonal congruence to have diagonal $\mathbf{e}$, Propositions 2.166–2.168 imply that such a matrix is in the orbit of the Horn matrix or in the orbit of a Hildebrand matrix. □

Remark 2.169. The complete list of extremal matrices in $\mathcal{COP}_5$ consists of the following matrices.

- Matrices in the orbit of $E_{12}+E_{21} = \mathbf{e}_1\mathbf{e}_2^T+\mathbf{e}_2\mathbf{e}_1^T$, and those in the orbit of $E_{11} = \mathbf{e}_1\mathbf{e}_1^T$. These are the positive scalar multiples of $E_{ij} + E_{ji}$, $1 \le i \le j \le 5$, forming one dimensional algebraic sets parameterized by the scalar.
- Matrices in the orbit of $\mathbf{x}\mathbf{x}^T$, where $\mathbf{x} \in \{0, 1, -1\}^5$ has both 1 and -1 entries. For each $\mathbf{x}$ this is a five dimensional algebraic set, parameterized by the diagonal entries of the scaling matrix.
- Matrices in the orbit of the Horn matrix. These also form a five dimensional algebraic set, parameterized by the diagonal entries of the scaling matrix.
- Matrices in the orbit of a Hildebrand matrix. These form a ten dimensional algebraic set, parameterized by the five diagonal entries of the scaling matrix, and the five entries of the vector $\boldsymbol{\theta}$.

In each of these cases, the permutation matrix determines the component.

The difference between $\mathcal{COP}_5$ and $\mathcal{SPN}_5$

For a description of exceptional matrices in $\mathcal{COP}_5$, we consider first the $\mathcal{HN}$-irreducible such matrices. If $A \in \mathcal{COP}_5$ is $\mathcal{HN}$-irreducible, then for every $i \ne j$ there exists a minimal zero $\mathbf{u}$ such that $u_i + u_j > 0$ and $(A\mathbf{u})_i = (A\mathbf{u})_j = 0$. Either both u_i and u_j are positive, or only one is. Accordingly, we consider two cases of exceptional $\mathcal{HN}$-irreducible matrices: either for every $i \ne j$ there exists a minimal zero $\mathbf{u}$ such that $u_iu_j > 0$, or for some $i \ne j$ there exists no such minimal zero.

Proposition 2.170. *Let $A \in \mathcal{COP}_5$ be an exceptional $\mathcal{HN}$-irreducible matrix with diagonal $\mathbf{e}$. If for every $i, j \in \{1, \ldots, 5\}$, $i \ne j$, there exists a minimal zero $\mathbf{u}$ such that $u_iu_j > 0$ and $(A\mathbf{u})_i = (A\mathbf{u})_j = 0$, then A is permutationally similar to a Hildebrand matrix (in particular, it is extremal).*

Proof. We will show that in that case A has no minimal supports of size 2. First, the minimal zeros of A cannot all be of size 2, since this would imply that for every $i \ne j$ the set $\{i, j\}$ is a minimal support, which is impossible, see, e.g., Lemma 2.164(1). Thus there exists a minimal support of size 3. Without loss of generality suppose $\{1, 2, 3\}$ is a minimal support.

If $\{4, 5\}$ is a minimal support, then $\{1, 4\}$ and $\{1, 5\}$ cannot both be minimal supports, since $G_{-1}(A)$ is triangle free. Without loss of generality, suppose $\{1, 4\}$ is not a minimal support. By the assumption in the proposition, $\{1, 4\}$-irreducibility implies that $\{1, i, 4\}$ is a minimal support, where $i = 2$ or $i = 3$. Without loss of generality, $\{1, 2, 4\}$ is a minimal support. As $|\{1, 2, 4\} \setminus \{1, 2, 3\}| = 1$, there is no minimal support contained in $\{1, 2, 3, 4\}$ other than $\{1, 2, 3\}$ and $\{1, 2, 4\}$ (Corollary 2.59). But then there is no minimal zero $\mathbf{u}$ such that $u_3u_4 > 0$, a contradiction. Hence $\{4, 5\}$ is not a minimal support.

If for $i \in \{1, 2, 3\}$ and $j \in \{4, 5\}$ the set $\{i, j\}$ is a minimal support, then without loss of generality, $i = 3$ and $j = 4$. As $\{3, 4\} \setminus \{1, 2, 3\}$ consists of a single element, $\{1, 2, 3\}$ and $\{3, 4\}$ are the only minimal supports contained in $\{1, 2, 3, 4\}$ (Corollary 2.59 again). Hence a minimal zero $\mathbf{u}$ such that $u_1u_4 > 0$ has to have $\operatorname{supp} \mathbf{u} = \{1, 4, 5\}$. Similarly, $\{2, 4\}$-irreducibility implies that $\{2, 4, 5\}$ is a minimal support of A. Then, as $\{3, 4\} \setminus \{1, 4, 5\}$ consists of a single index, there is no minimal support contained in $\{1, 3, 4, 5\}$ that contains both 3 and 5. The minimal support containing 3 and 5 has to be $\{2, 3, 5\}$. This is a contradiction, as $\{3, 4\} \setminus \{2, 4, 5\}$ consists of a single index, and thus $\{2, 4, 5\}$ and $\{3, 4\}$ are the only minimal supports contained in $\{2, 3, 4, 5\}$. This concludes the proof that A has no minimal supports of size 2.

The matrix A is therefore an exceptional copositive matrix whose only minimal supports are of size 3. By Corollary 2.166, the graph $G^3_{\min}(A)$ is a subgraph of the 5-cycle. As each pair $\{i, j\}$, where $i \neq j$ are in $\{1, \ldots, 5\}$, is contained in one of the minimal supports, the graph $G^3_{\min}(A)$ has to be the whole 5-cycle. By Proposition 2.168 A is permutationally similar to a Hildebrand matrix. □

The next proposition considers the case that A is $\mathcal{HN}$-irreducible, but does not satisfy the assumption of the previous proposition.

Proposition 2.171. *Let $A \in \mathcal{COP}_5$ be an exceptional $\mathcal{HN}$-irreducible matrix with diagonal $\mathbf{e}$. If there exist $p, q \in \{1, \ldots, 5\}$, $p \neq q$, for which there does not exist a minimal zero $\mathbf{u}$ such that $u_pu_q > 0$, then A is permutationally similar to $S(\boldsymbol{\theta})$ for some $\boldsymbol{\theta} = (\theta_1, \theta_2, \theta_3, \theta_4, \theta_5)^T$ such that $\theta_i \in [0, \pi)$ for every i, $\theta_i = 0$ for at least one i, and $\sum_{i=1}^5 \theta_i < \pi$.*

Proof. First note that, as A is exceptional, if $\mathbf{u}$ is a zero with support σ of size 3, then $(A\mathbf{u})_i = 0$ if and only if $i \in \sigma$ (see Corollary 2.101). The assumptions of the proposition therefore imply that A has a minimal support of size 2. Also, no minimal zero $\mathbf{u}$ with support of size 2 can satisfy $A\mathbf{u} = \mathbf{0}$ (see Exercise 2.133).

We consider the possible minimal supports of the matrix A by drawing the possible $G_{-1}(A)$ and $G^3_{\text{min}}(A)$ together, the edges of $G_{-1}(A)$ by dashed lines, and the edges of $G^3_{\text{min}}(A)$ by solid lines. Recall that each of these graphs is a subgraph of a 5-cycle. The graph $G^3_{\text{min}}(A)$ cannot be a full cycle, since in this case for every $i, j \in \{1, \ldots, 5\}$, $i \neq j$, there exists a minimal zero $\mathbf{u}$ of support 3 for which $u_i u_j > 0$. Several configurations cannot appear in the joint drawing, by Lemma 2.164:  and 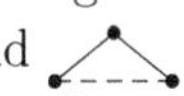 by part (3), and 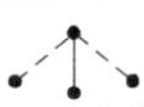 by part (4). Note that $\mathcal{HN}$-irreducibility implies that for every $i \neq j$, either there exists a dashed edge with at least one end in $\{i, j\}$, or there exists a solid edge with both ends not in $\{i, j\}$ (corresponding to whether the zero $\mathbf{u}$ such that $u_i + u_j > 0$ and $(A\mathbf{u})_i = (A\mathbf{v})_j = 0$ has support 2 or 3). In particular, if A has only minimal supports of size 2, at most one vertex is not an end of any dashed edge. Also, the assumption in the proposition is that there exist $p \neq q$ that do not belong together to the same minimal support. This means that these p and q are not joined by a dashed edge, and every solid edge (if exists) has at least one end in $\{p, q\}$.

This leaves us with the following possible joint drawings to consider (up to isomorphism).

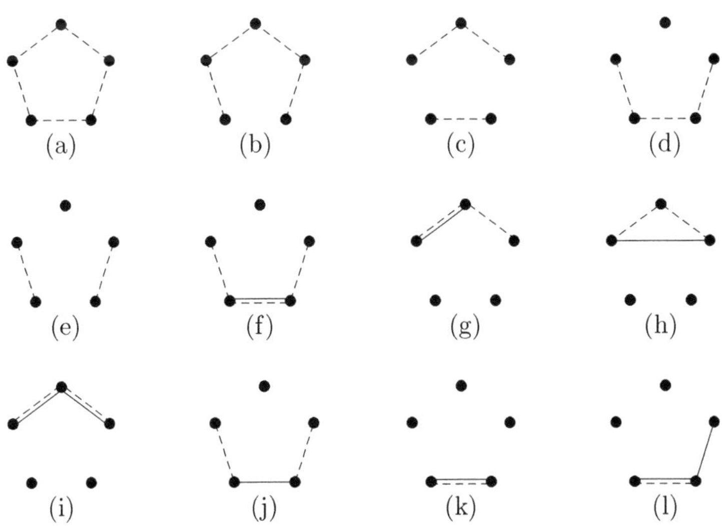

As we go over these configurations, it turns out that (d), (e), (g), (h), (k) and (l) are impossible under our assumptions, and each of the other corresponds to some $S(\boldsymbol{\theta})$ with $\sum_{i=1}^{5} \theta_i < \pi$.

(a) In this case, A is, up to permutation, the Horn matrix (i.e., $S(\mathbf{0})$) by Proposition 2.167.

(b) Without loss of generality, suppose the dashed path is $1-2-3-4-5$. That is,

$$A = \begin{pmatrix} 1 & -1 & 1 & a_{14} & a_{15} \\ -1 & 1 & -1 & 1 & a_{25} \\ 1 & -1 & 1 & -1 & 1 \\ a_{14} & 1 & -1 & 1 & -1 \\ a_{15} & a_{25} & 1 & -1 & 1 \end{pmatrix},$$

and $a_{15} = -\cos\theta_5$, where $\theta_5 \in (0, \pi]$ (since $a_{15} \neq -1$). There exists a minimal zero $\mathbf{u}$ such that $u_1 + u_5 > 0$, and $(A\mathbf{u})_1 = (A\mathbf{u})_5 = 0$. Either $\mathbf{u} = \mathbf{e}_4 + \mathbf{e}_5$ or $\mathbf{u} = \mathbf{e}_1 + \mathbf{e}_2$. Up to a permutation reversing the order of the path we may assume the latter, so $(A\mathbf{u})_5 = a_{15} + a_{25} = 0$. That is, $a_{25} = \cos\theta_5$ (in particular, as $a_{25} \neq -1$, $\theta_5 < \pi$). For the zero $\mathbf{v} = \mathbf{e}_4 + \mathbf{e}_5$ we get that $(A\mathbf{v})_1 \geq 0$, i.e. $a_{14} \geq \cos\theta_5$. Thus $A \geq S(\boldsymbol{\theta})$, where $\boldsymbol{\theta} = (0,0,0,0,\theta_5)^T$, and since A is $\mathcal{HN}$-irreducible we must have $A = S(\boldsymbol{\theta})$ and $\sum_{i=1}^{5} \theta_i = \theta_5 < \pi$.

(c) Suppose the minimal supports are $\{1,2\}$, $\{2,3\}$ and $\{4,5\}$. Then

$$A = \begin{pmatrix} 1 & -1 & 1 & a_{14} & a_{15} \\ -1 & 1 & -1 & a_{24} & a_{25} \\ 1 & -1 & 1 & a_{34} & a_{35} \\ a_{14} & a_{24} & a_{34} & 1 & -1 \\ a_{15} & a_{25} & a_{35} & -1 & 1 \end{pmatrix}.$$

Since $\mathbf{u} = \mathbf{e}_1 + \mathbf{e}_2$ satisfies $(A\mathbf{u})_i = 0$ for $i = 1,2,3$, $(A\mathbf{u})_i \neq 0$ for at least one $i \in \{4,5\}$. Without loss of generality, $(A\mathbf{u})_4 = a_{14} + a_{24} > 0$. Then necessarily $\mathbf{v} = \mathbf{e}_4 + \mathbf{e}_5$ is the minimal zero satisfying $v_1 + v_4 > 0$ and $(A\mathbf{v})_1 = (A\mathbf{v})_4 = 0$. That is, $a_{14} + a_{15} = 0$. Similarly, for $\mathbf{w} = \mathbf{e}_2 + \mathbf{e}_3$ there exists $i \in \{4,5\}$ such that $(A\mathbf{w})_i > 0$. As A is $\{3,i\}$-irreducible, the associated minimal zero has to be $\mathbf{v} = \mathbf{e}_4 + \mathbf{e}_5$, and $(A\mathbf{v})_3 = a_{34} + a_{35} = 0$. Since $A\mathbf{v} \neq \mathbf{0}$, we necessarily have $(A\mathbf{v})_2 > 0$. The minimal zero associated with the $\{2,4\}$-irreducibility of A is therefore not $\mathbf{v}$, and not $\mathbf{u} = \mathbf{e}_1 + \mathbf{e}_2$. That is, $(A\mathbf{w})_4 = a_{24} + a_{34} = 0$. Hence $(A\mathbf{w})_5 > 0$, and $\{2,5\}$-irreducibility is associated with the zero $\mathbf{u} = \mathbf{e}_1 + \mathbf{e}_2$, i.e.,

$(A\mathbf{u})_5 = a_{15} + a_{25} = 0$. We get that

$$A = \begin{pmatrix} 1 & -1 & 1 & -a_{15} & a_{15} \\ -1 & 1 & -1 & -a_{34} & -a_{15} \\ 1 & -1 & 1 & a_{34} & -a_{34} \\ -a_{15} & -a_{34} & a_{34} & 1 & -1 \\ a_{15} & -a_{15} & -a_{34} & -1 & 1 \end{pmatrix}.$$

Letting $a_{15} = -\cos\theta_5$ and $a_{34} = -\cos\theta_3$ we get that $A = S(\boldsymbol{\theta})$, where $\boldsymbol{\theta} = (0, 0, \theta_3, 0, \theta_5)^T$, $\theta_3, \theta_5 \in (0, \pi)$. As

$$0 < a_{14} + a_{24} = \cos\theta_5 + \cos\theta_3 = 2\cos\frac{1}{2}(\theta_3 - \theta_5)\cos\frac{1}{2}(\theta_3 + \theta_5),$$

and $0 \leq \frac{1}{2}|\theta_3 - \theta_5| < \frac{\pi}{2}$, this implies that $\theta_3 + \theta_5 < \pi$, and thus $\sum_{i=1}^{5}\theta_i < \pi$.

(d) Suppose the dashed path is $2 - 3 - 4 - 5$. Then

$$A = \begin{pmatrix} 1 & a_{12} & a_{13} & a_{14} & a_{15} \\ a_{12} & 1 & -1 & 1 & a_{25} \\ a_{13} & -1 & 1 & -1 & 1 \\ a_{14} & 1 & -1 & 1 & -1 \\ a_{15} & a_{25} & 1 & -1 & 1 \end{pmatrix}.$$

There exists a minimal zero $\mathbf{u}$ such that $u_2 + u_5 > 0$, and $(A\mathbf{u})_2 = (A\mathbf{u})_5 = 0$. Either $\mathbf{u} = \mathbf{e}_4 + \mathbf{e}_5$ or $\mathbf{u} = \mathbf{e}_2 + \mathbf{e}_3$. Without loss of generality, assume the latter. Then $(A\mathbf{u})_5 = 0$ implies that $a_{25} + 1 = 0$, $a_{25} = -1$. This contradicts the assumption that $\{2, 5\}$ is not a zero support. Hence (d) is impossible

(e) Suppose the two minimal supports are $\{2, 3\}$ and $\{4, 5\}$, and the corresponding minimal zeros are $\mathbf{u} = \mathbf{e}_2 + \mathbf{e}_3$ and $\mathbf{v} = \mathbf{e}_4 + \mathbf{e}_5$. In particular, $(A\mathbf{u})_2 = (A\mathbf{u})_3 = 0$ and $(A\mathbf{v})_4 = (A\mathbf{v})_5 = 0$. Also, $\{1, 2\}$-irreducibility necessarily means that $(A\mathbf{u})_1 = 0$, and $\{1, 4\}$-irreducibility means that $(A\mathbf{v})_1 = 0$. Without loss of generality, $\{3, 4\}$-irreducibility is associated with the zero $\mathbf{u}$, and $(A\mathbf{u})_4 = 0$. Then, as A is exceptional, $(A\mathbf{u})_5 > 0$ (see Exercise 2.133). Therefore $\{2, 5\}$-irreducibility means that $(A\mathbf{v})_2 = 0$. Now the exceptionality of A implies that $(A\mathbf{v})_3 > 0$ (again by Exercise 2.133). But then neither $(A\mathbf{u})_3 = (A\mathbf{u})_5 = 0$, nor $(A\mathbf{v})_3 = (A\mathbf{v})_5 = 0$. Contradicting the $\{3, 5\}$-irreducibility. So (e) is also impossible.

(f) Let the minimal supports be $\{1,2,3\}$, $\{3,4\}$, $\{4,5\}$ and $\{1,5\}$. Then

$$A = \begin{pmatrix} 1 & -\cos\theta_1 & \cos(\theta_1+\theta_2) & 1 & -1 \\ -\cos\theta_1 & 1 & -\cos\theta_2 & a_{24} & a_{25} \\ \cos(\theta_1+\theta_2) & -\cos\theta_2 & 1 & -1 & 1 \\ 1 & a_{24} & -1 & 1 & -1 \\ -1 & a_{25} & 1 & -1 & 1 \end{pmatrix},$$

where $\theta_1, \theta_2, \theta_1+\theta_2 \in (0,\pi)$. For the zero $\mathbf{u} = \mathbf{e}_3 + \mathbf{e}_4$ we get that $(A\mathbf{u})_2 \geq 0$, and for $\mathbf{v} = \mathbf{e}_1 + \mathbf{e}_5$ we get that $(A\mathbf{v})_2 \geq 0$. That is, $a_{24} \geq \cos\theta_2$ and $a_{25} \geq \cos\theta_1$. Thus $A \geq S(\boldsymbol{\theta})$, where $\boldsymbol{\theta} = (\theta_1, \theta_2, 0, 0, 0)^T$, and by the $\mathcal{HN}$-irreducibility of A, $A = S(\boldsymbol{\theta})$. Note that also in this case $\sum_{i=1}^5 \theta_i = \theta_1 + \theta_2 < \pi$.

(g) Let the minimal supports be $\{1,2\}$, $\{2,3\}$, $\{3,4,5\}$. Then the minimal zero $\mathbf{u} = \mathbf{e}_1 + \mathbf{e}_2$ satisfies $(A\mathbf{u})_i = 0$ for $i = 1,2,3$. Both $\{1,4\}$-irreducibility and $\{1,5\}$-irreducibility have to be associated with the zero $\mathbf{u}$. That is, $(A\mathbf{u})_4 = (A\mathbf{u})_5 = 0$. Therefore $A\mathbf{u} = \mathbf{0}$, contrary to the assumption that A is exceptional. Thus (g) is impossible under the assumptions in the proposition.

The remaining cases are checked similarly, and we leave them as an exercise. □

We can now state and prove the characterization of the $\mathcal{HN}$-irreducible exceptional matrices in $\mathcal{COP}_5$.

Theorem 2.172. *A matrix $A \in \mathcal{COP}_5$ is exceptional and $\mathcal{HN}$-irreducible if and only if A is in the orbit of $S(\boldsymbol{\theta})$ for some $\boldsymbol{\theta} = (\theta_1, \theta_2, \theta_3, \theta_4, \theta_5)^T$ such that $\theta_i \in [0,\pi)$ for every i and $\sum_{i=1}^5 \theta_i < \pi$.*

Proof. For every $\boldsymbol{\theta} = (\theta_1, \theta_2, \theta_3, \theta_4, \theta_5)^T$ such that $\theta_i \in [0,\pi)$ for every i and $\sum_{i=1}^5 \theta_i < \pi$ the matrix $S(\boldsymbol{\theta})$ is an exceptional $\mathcal{HN}$-irreducible copositive matrix by Proposition 2.162. Thus every matrix in the orbit of such $S(\boldsymbol{\theta})$ is also an exceptional $\mathcal{HN}$-irreducible copositive matrix.

Conversely, every exceptional $\mathcal{HN}$-irreducible matrix in $\mathcal{COP}_5$ has a positive diagonal, and is therefore congruent by a positive diagonal matrix to a matrix with diagonal $\mathbf{e}$. The latter is exceptional and $\mathcal{HN}$-irreducible, hence by Propositions 2.170 and 2.171 it has to be, up to permutation of rows and columns, $S(\boldsymbol{\theta})$ for some $\boldsymbol{\theta} = (\theta_1, \theta_2, \theta_3, \theta_4, \theta_5)^T$ such that $\theta_i \in [0,\pi)$ for every i and $\sum_{i=1}^5 \theta_i < \pi$. □

Remark 2.173. Note that the Horn matrix is $\mathcal{HN}$-irreducible, and for every $i \neq j$ there exists a (possibly not minimal) zero $\mathbf{u}$ such that $u_i u_j > 0$. The same holds for any Hildebrand matrix, by Proposition 2.170. Thus it is true for all matrices in the orbit of these matrices, the exceptional extremal matrices. It is no longer true for any other exceptional $\mathcal{HN}$-irreducible matrix — see Exercise 2.136.

Theorem 2.172 yields the following description of the matrices in $\mathcal{COP}_5 \setminus \mathcal{SPN}_5$.

Corollary 2.174. *Every exceptional matrix $A \in \mathcal{COP}_5$ can be decomposed as $A = S + N$, where S is in the orbit of $S(\boldsymbol{\theta})$ for some $\boldsymbol{\theta} \geq \mathbf{0}$ such that $\sum_{i=1}^{5} \theta_i < \pi$, and N a hollow nonnegative matrix.*

We conclude this section with two observations regarding matrices in $\mathcal{COP}_5$. First, combining Theorem 2.172 and Proposition 2.162 we see that any exceptional $\mathcal{HN}$-irreducible matrix in $\mathcal{COP}_5$ is actually $\mathcal{N}$-irreducible. This yields the following corollary.

Corollary 2.175. *If $A \in \mathcal{COP}_5$ is $\mathcal{HN}$-irreducible, but not $\mathcal{N}$-irreducible, then A is positive semidefinite.*

And finally, a decomposition result for all matrices in $\mathcal{COP}_5$.

Corollary 2.176. *Every matrix $A \in \mathcal{COP}_5$ has the form $A = S + N$, where N is a hollow nonnegative matrix, and S is either positive semidefinite, or in the orbit of $S(\boldsymbol{\theta})$ for some $\boldsymbol{\theta} \geq \mathbf{0}$ such that $\sum_{i=1}^{5} \theta_i < \pi$.*

This Corollary has implications for checking copositivity of 5×5 matrices. We describe the idea briefly now, without going into all the details of the background theory, or the relevant optimization methods.

Copositivity of 5×5 matrices

With all this information on $\mathcal{COP}_5$, determining whether a given symmetric $n \times n$ matrix is copositive is still hard, even for $n = 5$. In copositive optimization, there are several approaches to treat this difficulty. In one approach, a hierarchy of cones approximating $\mathcal{COP}_n$ from within was introduced:

$$\mathcal{K}_n^0 \subseteq \mathcal{K}_n^1 \subseteq \mathcal{K}_n^2 \subseteq \cdots \subseteq \mathcal{COP}_n, \quad \text{with} \quad \mathrm{cl}\left(\bigcup_{r \geq 0} \mathcal{K}_n^r\right) = \mathcal{COP}_n,$$

where

$$\mathcal{K}_n^r = \left\{ A \in \mathcal{S}_n \,\middle|\, \left(\sum_{i,j=1}^n a_{ij} x_i^2 x_j^2 \right) \left(\sum_{i=1}^n x_i^2 \right)^r \text{ is SOS} \right\}.$$

Here a polynomial $p(\mathbf{x})$ is *SOS* (Sum Of Squares) if there exists a decomposition of $p(\mathbf{x})$ into a sum of squares of polynomials in $\mathbf{x}$. Checking that a polynomial is SOS may be tackled by optimization methods, and of course implies nonnegativity of the polynomial on every $\mathbf{x} \in \mathbb{R}^n$. In our case, we are interested in nonnegativity on nonnegative vectors, hence our polynomial is expressed in terms of the nonnegative entries of $\mathbf{x} \circ \mathbf{x}$. When $\mathbf{x}$ varies over all of $\mathbb{R}^n$, $\mathbf{x} \circ \mathbf{x}$ varies over all of $\mathbb{R}^n_+$. Note that if $A \in \mathcal{K}_n^r$, then in particular $\sum_{i,j=1}^n a_{ij} x_i^2 x_j^2 \geq 0$ for every $\mathbf{x} \in \mathbb{R}$, hence A is copositive. This explains the inclusion $\mathcal{K}_n^r \subseteq \mathcal{COP}_n$. It was shown that $\mathcal{K}_n^0 = \mathcal{SPN}_n$ for every n. For $n > 4$ the inclusion $\mathcal{K}_n^0 \subseteq \mathcal{COP}_n$ is therefore strict. In fact, it turns out that for $n \geq 5$ all the hierarchy cones are strictly contained in $\mathcal{COP}_n$. This follows from the fact that for $n \geq 5$ and $r > 0$, not all diagonal scalings $X \mapsto DXD$, where D is a positive diagonal matrix, leave the cone $\mathcal{K}_n^r$ invariant. We leave this assertion as an exercise.

While $\mathcal{COP}_5 \neq \mathcal{K}_5^r$ for every r, it can be shown that any matrix $S(\boldsymbol{\theta})$, where $\boldsymbol{\theta} \geq \mathbf{0}$ and $\sum_{i=1}^5 \theta_i < \pi$, is in $\mathcal{K}_5^1$. Using Corollary 2.176 and the inclusion $\mathcal{SPN}_5 = \mathcal{K}_5^0 \subseteq \mathcal{K}_5^1$ this implies the following.

Proposition 2.177. *A matrix $A \in \mathcal{COP}_5$ whose diagonal elements are all 1 or 0 is in $\mathcal{K}_5^1$.*

Since every symmetric matrix with nonnegative diagonal elements may be scaled to have a diagonal of ones and zeros, this yields the following result.

Theorem 2.178. *Let A be a 5×5 symmetric matrix with nonnegative diagonal elements. Then A is copositive if and only if $DAD \in \mathcal{K}_5^1$, where $D = \operatorname{Diag}(d_1, \ldots, d_5)$,*

$$d_i = \begin{cases} \frac{1}{\sqrt{a_{ii}}}, & \text{if } a_{ii} > 0 \\ 1, & \text{if } a_{ii} = 0 \end{cases}.$$

Hence we get the following procedure for determining copositivity of a 5×5 symmetric matrix A. First check the simple necessary condition of having nonnegative diagonal elements. Then scale the matrix as in the theorem and obtain a matrix A' with zero and one diagonal entries, and check whether A' is in $\mathcal{K}_5^1$, i.e., whether $\left(\sum_{i,j=1}^n a'_{ij} x_i^2 x_j^2\right)\left(\sum_{i=1}^n x_i^2\right)$ is SOS. Checking that a polynomial is SOS may be translated to checking the existence of a solution to a system of LMIs (Linear Matrix Inequalities). Specifically, checking that $X \in \mathcal{K}_n^1$ amounts to checking that the following system of LMIs has a feasible solution $M_1, \ldots, M_n$.

$$\begin{array}{ll} X - M_i \in \mathcal{PSD}_n & i = 1, \ldots, n \\ (M_i)_{ii} = 0 & i = 1, \ldots, n \\ (M_i)_{jj} + 2(M_j)_{ij} = 0 & i, j = 1, \ldots, n \text{ and } i \neq j \\ (M_i)_{jk} + (M_j)_{ik} + (M_k)_{ij} \geq 0 & i, j, k = 1, \ldots, n \text{ and } i < j < k \end{array}.$$

To check whether a system of LMIs has a feasible solution, interior point algorithms may be used. Membership of the $S(\boldsymbol{\theta})$ matrices in $\mathcal{K}_n^1$ was verified by finding an explicit solution $M_i(\boldsymbol{\theta})$, $1 \leq i \leq 5$.

Exercises

2.134. Let $\boldsymbol{\theta} = (\pi, 0, 0, 0, 0)^T$. Show that $S(\boldsymbol{\theta})$ is an extremal positive semidefinite matrix.

2.135. Use a computer algebra system to check that the five minimal zeros of any Hildebrand matrix are indeed linearly independent (as guaranteed by Corollary 2.68).

2.136. Complete the proof of Proposition 2.162(b) by checking the remaining five cases. Show that if at least one of the θ_i's is zero, and $\sum_{i=1}^5 \theta_i < \pi$, there always exist $p \neq q$ for which there is no zero $\mathbf{u}$ of S satisfying $u_p u_q > 0$.

2.137. Complete the proof of Proposition 2.171 by checking the remaining cases. Show that (h),(k),(l) are impossible due to the exceptionality of the matrix, and show that $A \geq S(\boldsymbol{\theta})$ for some $\boldsymbol{\theta} \geq \mathbf{0}$ with $\sum_{i=1}^5 \theta_i < \pi$ in each of the cases (i) and (j).

2.138. Show that for every $i \neq j$ in $\{1, 2, 3, 4, 5\}$ there exists a zero $\mathbf{u}$ of the Horn matrix such that $u_i u_j > 0$.

2.139. Suppose that for $A \in \mathcal{COP}_5$ the graph $G^3_{\min}(A)$ contains a 4-cycle. Show that the four minimal zeros corresponding to the 4-cycle are linearly dependent, and each of these four zeros is a linear combination of the other three.

Hint: Suppose $\sigma_1 = \{1, 2, 3\}, \sigma_2 = \{1, 2, 4\}, \sigma_3 = \{1, 4, 5\}, \sigma_4 = \{1, 3, 5\}$ are the four minimal supports. The positive zero of the submatrix $A[\sigma_i]$ is given by Theorem 2.159.

2.140. Let

$$A=\begin{pmatrix} 1 & -\cos\varphi_{12} & -\cos\varphi_{13} & -\cos\varphi_{14} & -\cos\varphi_{15} \\ -\cos\varphi_{12} & 1 & -\cos\varphi_{23} & -\cos\varphi_{24} & -\cos\varphi_{25} \\ -\cos\varphi_{13} & -\cos\varphi_{23} & 1 & -\cos\varphi_{34} & -\cos\varphi_{35} \\ -\cos\varphi_{14} & -\cos\varphi_{24} & -\cos\varphi_{34} & 1 & -\cos\varphi_{45} \\ -\cos\varphi_{15} & -\cos\varphi_{25} & -\cos\varphi_{35} & -\cos\varphi_{45} & 1 \end{pmatrix},$$

$\varphi_{ij} \in [0,\pi]$ for every $1 \le i < j \le 5$. Show that if A is an $\mathcal{HN}$-irreducible exceptional copositive matrix, then $\sum_{1\le i<j\le 5}\varphi_{ij} > 4\pi$.

2.141. Show that the matrix

$$A=\begin{pmatrix} \frac{7}{18} & -\frac{5}{6} & \frac{7}{18} & \frac{7}{18} & -\frac{5}{6} \\ -\frac{5}{6} & \frac{7}{18} & -\frac{5}{6} & \frac{7}{18} & \frac{7}{18} \\ \frac{7}{18} & -\frac{5}{6} & \frac{7}{18} & -\frac{5}{6} & \frac{7}{18} \\ \frac{7}{18} & \frac{7}{18} & -\frac{5}{6} & \frac{7}{18} & -\frac{5}{6} \\ -\frac{5}{6} & \frac{7}{18} & \frac{7}{18} & -\frac{5}{6} & \frac{7}{18} \end{pmatrix}$$

is a copositive matrix, but its Hadamard third power $A^{(3)} = A \circ A \circ A$ is not a copositive matrix.
Hint: A is a Hildebrand matrix. Compute $\mathbf{e}^T A^{(3)}\mathbf{e}$.

2.142. Show that if $\theta_i > 0$ for every $1 \le i \le 4$ and $\sum_{i=1}^4 \theta_i < \frac{\pi}{2}$, the matrix

$$A=\begin{pmatrix} 1 & -\cos\theta_1 & \cos(\theta_1+\theta_2) & 0 & 0 \\ -\cos\theta_1 & 1 & -\cos\theta_2 & \cos(\theta_2+\theta_3) & \cos(\pi/2+\theta_1) \\ \cos(\theta_1+\theta_2) & -\cos\theta_2 & 1 & -\cos\theta_3 & \cos(\theta_3+\theta_4) \\ 0 & \cos(\theta_2+\theta_3) & -\cos\theta_3 & 1 & -\cos\theta_4 \\ 0 & \cos(\pi/2+\theta_1) & \cos(\theta_3+\theta_4) & -\cos\theta_4 & 1 \end{pmatrix}$$

is copositive, and exceptional.
Hint: $A \ge S(\boldsymbol{\theta})$, where $\boldsymbol{\theta}^T = (\theta_1, \theta_2, \theta_3, \theta_4, \frac{\pi}{2})$. Show that $A(1)$ is $\mathcal{HN}$-irreducible, but is not positive semidefinite (see Proposition 2.84).

2.143. Let $A \in \mathcal{COP}_5$ have diagonal $\mathbf{e}$ and the following signed graph. Show that A is SPN. (Recall that in a signed graph of a matrix, dashed edges correspond to negative entries, and solid edges to positive entries.)

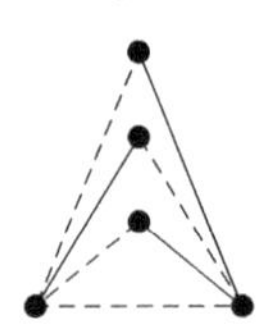

Hint: Show that if for some permutation matrix P and $P^TAP \geq S(\boldsymbol{\theta})$, where $\boldsymbol{\theta} \geq \mathbf{0}$ and $\mathbf{e}^T\boldsymbol{\theta} < \pi$, then there exists $\boldsymbol{\theta}' \geq \mathbf{0}$ with $\mathbf{e}^T\boldsymbol{\theta}' = \pi$ such that $P^TAP \geq S(\boldsymbol{\theta}')$.

2.144. Let $n \geq 1$ and $r \geq 1$. Follow the following steps to prove that if $A \in \mathcal{COP}_n$ satisfies

$$DAD \in \mathcal{K}_n^r \text{ for every positive diagonal } D, \tag{2.51}$$

then $A \in \mathcal{K}_n^0 = \mathcal{SPN}_n$.

(a) If $DAD \in \mathcal{K}_n^r$, where $D = \mathrm{Diag}(d_1, \ldots, d_n)$, and d_i are all positive, then $\left(\sum_{i,j=1}^n a_{ij}y_i^2y_j^2\right)\left(\sum_{i=1}^n d_i^{-1}y_i^2\right)^r$ is SOS.
(b) If $\left(\sum_{i,j=1}^n a_{ij}y_i^2y_j^2\right)\left(\sum_{i=1}^n d_i^{-1}y_i^2\right)^r$ is SOS whenever $d_1, \ldots, d_n$ are all positive, then $\left(\sum_{i,j=1}^n a_{ij}y_i^2y_j^2\right)y_1^{2r}$ is SOS.
(c) If $\left(\sum_{i,j=1}^n a_{ij}y_i^2y_j^2\right)y_1^{2r}$ is SOS, then $\left(\sum_{i,j=1}^n a_{ij}y_i^2y_j^2\right)y_1^{2(r-1)}$ is SOS. Repeat to show that $\sum_{i,j=1}^n a_{ij}y_i^2y_j^2$ is SOS.
(d) Combine the previous steps to deduce that if $DAD \in \mathcal{K}_n^r$ for every positive diagonal D, then $A \in \mathcal{K}_n^0$.

Hint: For (b), let $d_1 = 1$ and $d_i \to \infty$. The cone of SOS polynomials in $\mathbf{x} \in \mathbb{R}^n$ is closed.

Notes. In [Hall and Newman (1963)] the positive semidefinite and nonnegative extremal copositive matrices were identified, and the first example of an exceptional extremal matrix, due to Horn, was presented. The study of extremal copositive matrices continued in the Ph.D. Thesis [Baumert (1965)], resulting in the papers [Baumert (1966)], where some general results regarding extremal copositive forms and their zeros were proved, and [Baumert (1967)], where some extremals in $\mathcal{COP}_5$ were presented. All this was done using the terminology of copositive forms. The circular pattern of the zero supports of the non-Horn exceptional extremal matrices in $\mathcal{COP}_5$ was identified in [Baumert (1965)], although the arguments there contained an error, which was corrected in [Dickinson *et al.* (2013a)]. The characterization of exceptional extremal matrices in $\mathcal{COP}_5$ is due to [Hildebrand (2012)]. The $\mathcal{HN}$-irreducible matrices in $\mathcal{COP}_5$ were characterized in [Dickinson *et al.* (2013a)]. The hierarchy of approximating cones $\mathcal{K}_n^r$ was introduced in [Parrilo (2000)], and the fact that a matrix is in $\mathcal{COP}_5$ if and only if its scaling to a matrix with $\{0,1\}$-diagonal is in $\mathcal{K}_5^1$ was proved in

[Dickinson *et al.* (2013b)]. The details of the proofs, and references to the theory of checking membership in $\mathcal{K}_n^1$ can be found there. This paper also contains several interesting conjectures and open problems, whose essence is: *Can the result in Theorem 2.178 be generalized to* $\mathcal{COP}_n$*,* $n > 5$*, and if so, how?*

This section is based on the three papers [Hildebrand (2012); Dickinson *et al.* (2013a,b)], but the proofs use also the concepts of a minimal zero and a minimal support, and results regarding these concepts developed in [Hildebrand (2014)] and in [Hildebrand (2018a)], and the extremality characterization of [Dickinson and Hildebrand (2016)].

Exercise 2.139 is based on [Afonin *et al.* (2020)], and the example in Exercise 2.141 is due to [Hildebrand (2015)]. Exercise 2.143 is based on [Shaked-Monderer (2018)].

2.11 Exceptional extremal copositive matrices

In $\mathcal{COP}_5$ we encountered the first exceptional extremal copositive matrices. How does the $\mathcal{COP}_5$ example generalize to higher orders? We consider here two generalizations. In one, we look at exceptional matrices in $\mathcal{COP}_n$ whose zero supports have the circular nature that we encountered in the zero supports of the Horn and the Hildebrand matrices. In the other, we describe how one can look for exceptional extremal matrices by identifying first their possible minimal supports. In particular, we discuss the characterization of all extremal matrices in $\mathcal{COP}_6$. In both cases we omit most of the details. In the study of matrices with circular zero support structure, the proofs involve Floquet theory for periodic dynamical systems, which is beyond the scope of this book, and we therefore only quote some of the end results. In the case of $\mathcal{COP}_6$, the methods continue those already mentioned in previous sections, with some additional ideas. As the proofs involve many technical details, we only outline the steps of the proof, and do not state the complete end result — the list all the extremal matrices in $\mathcal{COP}_6$. See the notes at the end of the section for references to the full results.

Extremal matrices with circulant zero supports of size $n-2$.

Exceptional extermal copositive matrices with all minimal supports of size 2 were completely characterized in Corollary 2.150, in combination with Theorem 2.144 and Theorem 2.140. On the other end of the spectrum are $n \times n$ matrices all of whose minimal zeros have support of size $n-2$. Recall that by Corollary 2.98, the graph $G^3_{\min}(A)$ representing these minimal zeros of a copositive A has to be an n-cycle if A is an exceptional extremal matrix.

In $\mathcal{COP}_5$ all exceptional extremal matrices were in the orbit of one of two types of matrices: the Horn matrix or a Hildebrand matrix. These two types of matrices have one thing in common: Each such matrix has a "circulant" set of zero supports of size $3 = 5 - 2$ each: five zero supports, each obtained by removing one element from its predecessor, and adding another instead, cyclically. In the case of a Hildebrand matrix the five zero supports are minimal, and in the case the Horn matrix, these are non-minimal zero supports.

With these two examples in mind, consider exceptional copositive matrices in $\mathcal{COP}_n$, $n \geq 5$, whose set of zeros include n nonnegative vectors $\mathbf{u}_1, \ldots, \mathbf{u}_n$, such that $\operatorname{supp} \mathbf{u}_i = \{1, \ldots n\} \setminus \{i, i \mathbin{+\!\!\!\!\!\circ} 1\}$, where $\mathbin{+\!\!\!\!\!\circ}$ denotes addition modulo n on $\{1, \ldots, n\}$. Denote $\sigma_i = \operatorname{supp} \mathbf{u}_i$, $1 \leq i \leq n$. We refer to the set of supports $\{\sigma_1, \ldots, \sigma_n\}$ as a *circulant zero support set*. If all the vectors $\mathbf{u}_1, \ldots, \mathbf{u}_n$ are minimal zeros of the matrix A, we say it is a *minimal circulant zero support set*, otherwise, it is a *non-minimal circulant zero support set*.

The following results hold for such matrices, and their circulant zero supports set.

Theorem 2.179. *Let A be an $n \times n$ exceptional copositive matrix, with zeros $\mathbf{u}_1, \ldots, \mathbf{u}_n$, such that* $\operatorname{supp} \mathbf{u}_i = \sigma_i$, $i = 1, \ldots, n$. *Then exactly one of the following holds:*

(a) *$\{\sigma_1, \ldots, \sigma_n\}$ is the complete set of minimal supports of A, and*

   (i) *if n is even, then A is a sum of an exceptional copositive matrix with a non-minimal circulant zero supports set, and a rank* 1 *positive semidefinite matrix. In particular, A is not extremal.*

   (ii) *if n is odd, A may be extremal.*

(b) *None of the sets $\sigma_1, \ldots, \sigma_n$ is a minimal support of A. Each minimal support is a proper subset of one of these n sets, and each σ_i contains exactly two minimal supports. Every non-minimal zero of A has support equal to one of $\sigma_1, \ldots, \sigma_n$, and is a sum of two minimal zeros of A. The matrix A is extremal.*

A Hildebrand matrix is a prototype of an extremal copositive matrix satisfying (a), and the Horn matrix is a prototype of an extremal copositive matrix satisfying (b).

Not every set of n vectors $\mathbf{u}_1, \ldots, \mathbf{u}_n$ with supports $\sigma_1, \ldots, \sigma_n$ is a set of zeros of some exceptional A. However, explicit examples of exceptional extremal matrices for which $\sigma_1, \ldots, \sigma_n$ is a non-minimal circulant zero supports set have been constructed for every n. For every odd n, explicit examples of exceptional extremal matrices for which $\sigma_1, \ldots, \sigma_n$ is a minimal circulant zero supports set have also been constructed. Note that Theorem 2.179 does not guarantee that every exceptional A for which $\sigma_1, \ldots, \sigma_n$ are all minimal supports is extremal in the case that n is odd, yet it is not known whether such non-extremal matrices exist.

Exceptional extremal matrices with positive diagonal.

Some numerical details may give a hint on how the problem of characterizing extremal matrices becomes more complicated as the order grows, counting the number of possible minimal support sets of such matrices as an indication of the growing complexity. In the previous section we saw that each exceptional extremal matrix in $\mathcal{COP}_5$ has, up to permutation, one of two sets of minimal supports: either $\{\{1,2\},\{2,3\},\{3,4\},\{4,5\},\{5,1\}\}$ or $\{\{1,2,3\},\{2,3,4\},\{3,4,5\},\{4,5,1\},\{5,1,2\}\}$. A search for all possible sets of minimal supports of exceptional matrices with positive diagonal in $\mathcal{COP}_6$, which are both $\mathcal{PSD}$- and $\mathcal{N}$-irreducible, yielded 44 such sets, up to permutation. A methodical case by case check revealed that exactly 19 of these are sets of minimal supports of extremal matrices. In $\mathcal{COP}_7$ a similar initial search yielded 12,378 possible sets of minimal supports of exceptional $\mathcal{PSD}$- and $\mathcal{N}$-irreducible matrices. It is not known how many of these actually correspond to extremal matrices, but the rate of growth of these numbers seems indicative.

A copositive matrix with a positive diagonal is in the orbit of a matrix with diagonal $\mathbf{e}$. To search for exceptional extremal such matrices, we may start by compiling a list of sets of subsets of $\{1, \ldots, n\}$, which includes all sets of minimal supports of exceptional matrices in $\mathcal{COP}_n$ with diagonal $\mathbf{e}$ that are $\mathcal{PSD}$- and $\mathcal{N}$-irreducible, up to permutation. These are potentially minimal support sets of exceptional extremal matrices. In addition to previous results regarding such matrices, the following lemma and its corollary help to shorten the list. We state the lemma without proof, and leave the corollary as an exercise. For the definition of the MAXCUT polytope $\mathcal{MC}_n$ see Remark 2.138.

Lemma 2.180. *Let A be a positive semidefinite matrix with diagonal $\mathbf{e}$, and let B be the matrix with entries $b_{ij} = \frac{2}{\pi} \arcsin a_{ij}$, $1 \le i, j \le n$. Then $B \in \mathcal{MC}_n$. If A is positive definite, then $B \in \operatorname{int} \mathcal{MC}_n$.*

The polytope $\mathcal{MC}_n$ may be described by linear inequalities, where matrices in its interior satisfy all these inequalities strictly. In the above lemma, if we let $a_{ij} = -\cos\varphi_{ij}$, $\varphi_{ij} \in [0, \pi]$, for every $1 \le i, j \le n$, then $B \in \mathcal{MC}_n$ translates to φ_{ij} satisfying certain linear inequalities, and $B \in \operatorname{int}\mathcal{MC}_n$ translates to φ_{ij} satisfying these inequalities strictly. We will refer to these set of inequalities on φ_{ij} as the $\mathcal{MC}_n$ *inequalities.* The number of inequalities describing $\mathcal{MC}_n$ grows rapidly with n, but for small values of n a complete set of describing inequalities is known.

Corollary 2.181. *Let an $\mathcal{N}$-irreducible $A \in \mathcal{COP}_5$ have diagonal* $\mathbf{e}$ *and $a_{ij} = -\cos\varphi_{ij}$ for every $1 \le i, j \le 5$. Then $\sum_{1\le i<j\le 5} \varphi_{ij} \ge 4\pi$.*

To state the proposition that is the basis for the search for exceptional extremal copositive matrices by their minimal supports, we define an ad-hoc graph $\mathfrak{G}(A)$ for a $\mathcal{PSD}$- and $\mathcal{N}$-irreducible copositive A with diagonal $\mathbf{e}$. Suppose Ω is the set of minimal supports of A of size greater than 2, and that r of the components of the graph $G_{-1}(A)$ (whose edges represent the minimal supports of size 2) are bipartite. Then the vertex set of $\mathfrak{G}(A)$ consists of the pairs (i, γ) where $i \in \{1, \dots, r\}$ and $\gamma \in \Omega$, and an edge connects (i, γ) and (j, γ') if there exists a vertex in the i-th bipartite component of $G_{-1}(A)$ that is an element of γ', or there exists a vertex in the j-th bipartite component of $G_{-1}(A)$ that is an element of γ.

Proposition 2.182. *Let $A \in \mathcal{COP}_n$ be a $\mathcal{PSD}$- and $\mathcal{N}$-irreducible matrix with diagonal* $\mathbf{e}$. *Let $a_{ij} = -\cos\varphi_{ij}$, $\varphi_{ij} \in [0, \pi]$. Let $\Sigma = \{\sigma_1, \dots, \sigma_m\}$ be the set of all minimal supports of A, ordered so that $|\sigma_i| \le |\sigma_{i+1}|$ for every $1 \le i \le m-1$. Then the following hold.*

(a) *For every $1 \le i \le m$, $2 \le |\sigma_i| \le n-2$.*
(b) *For every $1 \le i < j \le m$, σ_i is not a subset of σ_j.*
(c) *Let an index set α be a proper subset of σ_i for some $1 \le i \le m$. If there exist $i_1 < \cdots < i_k$ such that $|\sigma_{i_\ell} \setminus \alpha| = 1$ for every $1 \le \ell \le k$, $\sigma_{i_l} \cap \alpha \subseteq \sigma_{i_{l+1}} \cap \alpha$ for every $1 \le l \le k-1$, and $\sigma_j \subseteq \bigcup_{l=1}^{k} \sigma_{i_l}$, then $j \in \{i_1, \dots, i_k\}$.*
(d) *The graph $\mathfrak{G}(A)$ has a matching of size r (that is, r edges with no common vertices).*
(e) *The following eight conditions hold.*

 (1) *If $\{i, j\} \in \Sigma$, then $\varphi_{ij} = 0$.*
 (2) *If $\{i, j\} \in \Sigma$, then $\varphi_{ik} + \varphi_{jk} \ge \pi$ for every $1 \le k \le n$.*
 (3) *If $\{i, j\} \notin \Sigma$, then $\varphi_{ij} > 0$.*

(4) *If* $\{i, j, k\} \in \Sigma$, *then* $\varphi_{ij} + \varphi_{ik} + \varphi_{jk} = \pi$.
(5) *If no* $\sigma \in \Sigma$ *is a subset of* $\{i, j, k\}$, *then* $\varphi_{ij} + \varphi_{ik} + \varphi_{jk} > \pi$.
(6) *If* $\sigma \in \Sigma$, $|\sigma| = p$, *then* φ_{ij}, $i, j \in \sigma$, *satisfy the* $\mathcal{MC}_p$ *inequalities.*
(7) *If* $\alpha \subsetneq \sigma$ *for* $\sigma \in \Sigma$, $|\alpha| = p$, *then* φ_{ij}, $i, j \in \alpha$, *satisfy the* $\mathcal{MC}_p$ *inequalities strictly.*
(8) *If* i_1, i_2, i_3, i_4, i_5 *are five distinct indices in* $\{1, \ldots, n\}$, *then* $\sum_{1 \le j < k \le 5} \varphi_{i_j i_k} \ge 4\pi$.

Proof. In (a) the left inequality follows from the positivity of the diagonal, and the right one from exceptionality, by Corollary 2.70. Part (b) simply states the minimality of the zero supports. Part (c) follows from Proposition 2.58, and part (d) from Theorem 2.67 and Proposition 2.60.

For conditions (1) and (3) in (e), see Theorem 2.158. For condition (3) note that in this case $\mathbf{u} = \mathbf{e}_i + \mathbf{e}_j$ is a minimal zero of A, hence for every k

$$0 \le (A\mathbf{u})_k = a_{ki} + a_{kj} = -\cos\varphi_{ik} - \cos\varphi_{jk} = \cos(\pi - \varphi_{ik}) - \cos\varphi_{jk}.$$

As $\pi - \varphi_{ik}, \varphi_{jk} \in [0, \pi]$, and $\cos t$ is decreasing on that interval, this implies

$$\pi - \varphi_{ik} \le \varphi_{jk}.$$

For conditions (4) and (5) in (e), see Theorem 2.159.

Conditions (6) and (7) in (e) follow by Lemma 2.180 from the fact that for $\sigma \in \Sigma$ the submatrix $A[\sigma]$ is positive semidefinite, and all its proper principal submatrices are positive definite. Finally, condition (8) in (e) follows from Corollary 2.181. □

The search for all possible sets of minimal supports of $\mathcal{PSD}$- and $\mathcal{HN}$-irreducible matrices in $\mathcal{COP}_n$ begins by searching for sets of subsets of $\{1, \ldots, n\}$ for which (a)–(d) hold, and the set of equalities and inequalities in the variables φ_{ij} (some strict and some non-strict) in conditions (1)–(8) of (e) has a solution. Checking that (a)–(d) can be done by combinatorial algorithms. To check the eight conditions in (e), introduce the variables $\psi_{ij} = \frac{\varphi_{ij}}{\pi}$, $1 \le i \le j \le n$. The eight conditions translate into equalities and inequalities in ψ_{ij}. Introduce a single additional slack variable ε on the smaller side in all strict inequalities, converting them to non-strict inequalities. Solve a linear programming problem of maximizing ε subject to the equality and inequality constraints thus derived from the conditions in (e). This a linear problem in the $\frac{n(n-1)}{2} + 1$ variables $\varepsilon, \psi_{12}, \ldots, \psi_{n-1\,n}$. If the problem is infeasible, or the optimal value of ε is nonpositive, then (e) does not hold. If ε is unbounded, or its optimal value is positive, (e) holds. (With the change of variables from φ_{ij} to ψ_{ij}, all the coefficients of the equations and inequalities in the problem become integers. Thus the solution is rational, and an exact solution may be found, using the simplex method, and it is possible to determine whether it is indeed positive.)

This way we obtain a collection of sets of subsets of $\{1, \ldots, n\}$, for which a copositive A satisfying the conditions in Proposition 2.182 may exist. Any permutation of $\{1, \ldots, n\}$ applied to one of these sets of index sets yields another set that satisfies the conditions, hence the collection is split into equivalence classes. The final list contains one representative from each equivalence class. This is how the lists of 44 potential sets of minimal supports of exceptional extremal matrices with a positive diagonal in $\mathcal{COP}_6$ and 12, 378 such sets for $\mathcal{COP}_7$ were compiled.

Exceptional extremal matrices in $\mathcal{COP}_6$.

Checking which of the 44 sets of indices is indeed a set of minimal supports of an exceptional extremal matrix in $\mathcal{COP}_6$, and describing all such extremal matrices, involves a lot of technical details. We skip most of these, and only outline the process. First, thirteen of the 44 sets of indices may be omitted from the list, by showing that a set of corresponding minimal zeros cannot span $\mathbb{R}^6$. This means that a copositive matrix with these zeros is not $\mathcal{PSD}$-irreducible, and thus is not exceptional and extremal. Exercise 2.139 is the tool in most of these cases. One case is of a "circulant" set of six sets of indices, each with four elements. That such set cannot be the set of minimal supports of an exceptional extremal matrix follows from Theorem 2.179(a).

All the remaining sets, except one, include only subsets of size 2 and 3 of $\{1, \ldots, 6\}$. Hence it is possible to parameterize the corresponding matrices based on their minimal supports of size 2 and 3, and their minimal zeros (see Remark 2.160). If such A is copositive and $\mathbf{u}$ is any of its (minimal) zeros, then $A\mathbf{u} \geq \mathbf{0}$. In two cases, these requirements on the various zeros are incompatible, and two more sets may be omitted from the list.

For each of the 29 remaining sets, we need to check the corresponding matrix for copositivity, or guarantee its copositivity by imposing some inequality constraints on the parameters of the matrix. The idea in Exercise 2.88 is used for that. In ten of the cases, it turns out that the resulting copositive matrix has additional minimal zeros, which means that the considered set of subsets of $\{1, \ldots, 6\}$ is not a complete set of minimal supports, and should be omitted from the list.

In each of the remaining 19 cases, one needs to check extremality of the matrix. This can be done by Theorem 2.71. That is, given the parameterized matrix A, and a representing set $\mathbf{u}_1, \ldots, \mathbf{u}_m$ of all its minimal zeros, one has to check that the solution space to the system

$$(X\mathbf{u}_j)_i = 0 \text{ for every } i \text{ such that } (A\mathbf{u}_j)_i = 0, \quad j = 1, \dots, m,$$

is one dimensional. This is system in 21 variables (the entries of X on and above the diagonal) with coefficients depending on several parameters, and it turns out to be too large for a direct computation of the rank of the coefficient matrix. A method was developed for overcoming this difficulty.

For most of these cases, the end description of the exceptional extremal matrix with the given set of minimal supports is a mostly trigonometric representation, with constraints on the arguments. For example, every exceptional extremal matrix in $\mathcal{COP}_6$ with minimal supports $\{1,5\}$, $\{2,6\}$, $\{1,2,3\}$, $\{2,3,4\}$, $\{3,4,5\}$, $\{4,5,6\}$ has the form

$$\begin{pmatrix} 1 & -\cos\varphi_1 & \cos(\varphi_1+\varphi_2) & \cos\varphi_4 & -1 & \cos\varphi_1 \\ -\cos\varphi_1 & 1 & -\cos\varphi_2 & \cos(\varphi_2+\varphi_3) & \cos\varphi_1 & -1 \\ \cos(\varphi_1+\varphi_2) & -\cos\varphi_2 & 1 & -\cos\varphi_3 & \cos(\varphi_3+\varphi_4) & \cos\varphi_2 \\ \cos\varphi_4 & \cos(\varphi_2+\varphi_3) & -\cos\varphi_3 & 1 & -\cos\varphi_4 & \cos(\varphi_4+\varphi_5) \\ -1 & \cos\varphi_1 & \cos(\varphi_3+\varphi_4) & -\cos\varphi_4 & 1 & -\cos\varphi_5 \\ \cos\varphi_1 & -1 & \cos\varphi_2 & \cos(\varphi_4+\varphi_5) & -\cos\varphi_5 & 1 \end{pmatrix},$$

$$\varphi_i > 0,\ \varphi_1 \le \varphi_5,\ \varphi_1+\varphi_2+\varphi_3+\varphi_4 < \pi,\ \varphi_2+\varphi_3+\varphi_4+\varphi_5 < \pi,\ \varphi_1+\varphi_5 \ne \pi.$$

In some cases, there are pairs $1 \le i < j \le 6$ such that $\{i,j\}$ is not a subset of any of the minimal supports. When this happens, the i,j entry of A may be determined by the requirement that $(A\mathbf{u})_i = (A\mathbf{u})_j = 0$ for a minimal zero $\mathbf{u}$ such that $u_i + u_j > 0$. Such off-diagonal entries may also be expressed as a trigonometric function of the parameterizing angles (as is the case with entries $1,4$, $2,5$, $3,6$ and $1,6$ in the above example), but in some cases they are expressed as a more complicated expression involving ratios of such trigonometric functions.

In a couple of cases, a set of minimal supports yields a certain parametrization of the exceptional extremal matrix, but with two options for the set of constraints on the angles, depending on the identity of the minimal zero $\mathbf{u}$ for which $(A\mathbf{u})_i = (A\mathbf{u})_j = 0$. That is, in these cases there are two non-isomorphic submanifolds of exceptional extremal matrices with the same minimal supports.

One of the 19 cases is a set of six subsets of $\{1,\dots,6\}$, of which one has size 4. This case is particularly challenging, since we have no easy parametrization of the minimal zero with support of size 4, and the corresponding 4×4 principal submatrix.

What can we learn from the characterization of exceptional extremal matrices in $\mathcal{COP}_6$ (except for the fact that things become more complicated very quickly as the order grows)? One important observation is that it is not just the minimal supports of A that are important, but also, for each minimal zero $\mathbf{u}$, the set of indices i such that $(A\mathbf{u})_i = 0$. This turns out to be true for the whole of $\mathcal{COP}_n$, and not just for the extremal matrices in it — see Chapter 5.

Exercises

2.145. Show that for every positive semidefinite A with diagonal $\mathbf{e}$, the matrix B defined in Lemma 2.180, is positive semidefinite. And if A is positive definite, so is B.
Hint: Consider the Taylor series of $\arcsin x$ about 0, and Proposition 1.34.

2.146. Let A be an $n \times n$ positive semidefinite matrix with diagonal $\mathbf{e}$ and $a_{ij} = -\cos\varphi_{ij}$, $\varphi_{ij} \in [0, \pi]$. Show that $\sum_{1\le i<j\le n} \varphi_{ij} \ge \pi \left\lfloor \frac{(n-1)^2}{4} \right\rfloor$, and the inequality is strict if A is positive definite.
Hint: Exercise 2.114 and Lemma 2.180.

2.147. Prove Corollary 2.181.
Hint: Exercises 2.140 and 2.146.

Notes. The study of exceptional extremal matrices is very much lead by Hildebrand, starting with the characterization of 5×5 such matrices in [Hildebrand (2012)]. Copositive matrices of order n with n circulant zero supports of size $n-2$ were investigated in [Hildebrand (2017)], and explicit examples of exceptional extremal matrices with these zero supports were constructed there. Theorem 2.179 is based on that paper. Corollary 2.181 and Proposition 2.182 are based on [Hildebrand (2014)], Lemma 2.180 can be also found there, attributed to [Hirschfeld (2004)]. The full characterization of extremal copositive 6×6 matrices can be found in [Afonin *et al.* (2020)]. See there also details on the structure of the manifold of extremal matrices in $\mathcal{COP}_6$, its various submanifolds and their dimensions. For yet more details on this manifold see [Hildebrand and Afonin (2020)].

2.12 The inverse of a copositive matrix

The inverse of a positive definite matrix is also positive definite. What can be said about the inverse of a nonsingular copositive matrix? We have already mentioned one property of such inverse: each of its columns has at least one positive entry, and a row with a positive sum (Theorem 2.112,

Exercise 2.92). We collect in this section some results about the inverses of copositive matrices, including some partial answers to the question: when is the inverse of a copositive matrix also copositive?

Not every nonsingular copositive matrix has a copositive inverse. For example,

$$\begin{pmatrix} 1 & 2 \\ 2 & 3 \end{pmatrix}^{-1} = \begin{pmatrix} -3 & 2 \\ 2 & -1 \end{pmatrix}.$$

The first theorem in this section tells us that there are no strictly copositive matrices with a copositive inverse, other than the positive definite ones.

Theorem 2.183. *Let A be a nonsingular symmetric $n \times n$ matrix. Then the following are equivalent.*

(a) *A is strictly copositive and A^{-1} is copositive.*
(b) *A is copositive and A^{-1} is strictly copositive.*
(c) *A is positive definite.*

Proof. Obviously (c) implies both (a) and (b). We prove (a) $\Rightarrow$ (c) (this will also prove (b) $\Rightarrow$ (c), by considering A^{-1} in place of A). We show first that for a strictly copositive A, every vector $\mathbf{x} \in \mathbb{R}^n$ may be represented as

$$\mathbf{x} = \mathbf{y} + \mathbf{z}, \quad \text{where } \mathbf{y} \geq \mathbf{0}, \quad A\mathbf{z} \leq 0 \text{ and } \mathbf{y}^T A\mathbf{z} = 0. \tag{2.52}$$

To see that, suppose $\mathbf{x}_0 \in \mathbb{R}^n$ is fixed. Let $f(\mathbf{x}) = \frac{1}{2}(\mathbf{x} - \mathbf{x}_0)^T A(\mathbf{x} - \mathbf{x}_0)$, and consider the minimization problem $\min_{\mathbf{x} \in \mathbb{R}^n_+} f(\mathbf{x})$.

As A is strictly copositive, $\min_{\mathbf{x} \in \Omega} \frac{1}{2}\mathbf{x}^T A\mathbf{x} = \lambda > 0$. Let

$$\Omega = \left\{ \mathbf{x} \in \mathbb{R}^n_+ \;\middle|\; \|\mathbf{x}\| \leq \frac{1}{\lambda}\|A\mathbf{x}_0\| \right\}.$$

For every $\mathbf{x} \in \mathbb{R}^n_+ \setminus \Omega$,

$$\begin{aligned} f(\mathbf{x}) &= \frac{1}{2}(\mathbf{x}^T A\mathbf{x} - 2\mathbf{x}^T A\mathbf{x}_0 + \mathbf{x}_0^T A\mathbf{x}_0) \\ &\geq \lambda\|\mathbf{x}\|^2 - \|\mathbf{x}\|\|A\mathbf{x}_0\| + f(\mathbf{0}) \\ &= \|\mathbf{x}\|(\lambda\|\mathbf{x}\| - \|A\mathbf{x}_0\|) + f(\mathbf{0}) > f(\mathbf{0}). \end{aligned}$$

As $\mathbf{0} \in \Omega$, the above inequality shows that the minimum f attains on the compact Ω is also the minimum of f in $\mathbb{R}^n_+$. Let the minimum be attained at $\mathbf{y} \geq \mathbf{0}$. Then the Karush-Kuhn-Tucker optimality condition for the problem $\min_{\mathbf{x} \in \mathbb{R}^n_+} f(\mathbf{x})$ is satisfied at $\mathbf{y}$. There exists $\boldsymbol{\mu} \in \mathbb{R}^n_+$ such that $\boldsymbol{\mu}^T\mathbf{y} = 0$ and $\nabla f(\mathbf{y}) - \boldsymbol{\mu} = 0$. That is,

$$A(\mathbf{y} - \mathbf{x}_0) = \boldsymbol{\mu} \geq 0 \text{ and } \mathbf{y}^T A(\mathbf{y} - \mathbf{x}_0) = 0.$$

Let $\mathbf{z} = \mathbf{x}_0 - \mathbf{y}$. Then $\mathbf{x} = \mathbf{x}_0$, $\mathbf{y}$ and $\mathbf{z}$ satisfy (2.52).

Now for every $\mathbf{x} \in \mathbb{R}^n$, let $\mathbf{x} = \mathbf{y} + \mathbf{z}$ as in (2.52). Then

$$\begin{aligned}\mathbf{x}^T A\mathbf{x} &= \mathbf{y}^T A\mathbf{y} + 2\mathbf{y}^T A\mathbf{z} + \mathbf{z}^T A\mathbf{z} \\ &= \mathbf{y}^T A\mathbf{y} + \mathbf{z}^T A A^{-1} A\mathbf{z} \\ &= \mathbf{y}^T A\mathbf{y} + (-A\mathbf{z})^T A^{-1}(-A\mathbf{z}) \geq 0,\end{aligned}$$

since A and A^{-1} are copositive, $\mathbf{y} \geq \mathbf{0}$ and $-A\mathbf{z} \geq 0$. Hence the nonsingular A is positive semidefinite, i.e., positive definite. □

The strict copositivity assumption is essential here, as the following example shows.

Example 2.184. Our go-to example of an exceptional copositive matrix, the Horn matrix, has a copositive inverse. Indeed, its inverse is a hollow nonnegative matrix:

$$\begin{pmatrix} 1 & -1 & 1 & 1 & -1 \\ -1 & 1 & -1 & 1 & 1 \\ 1 & -1 & 1 & -1 & 1 \\ 1 & 1 & -1 & 1 & -1 \\ -1 & 1 & 1 & -1 & 1 \end{pmatrix}^{-1} = \frac{1}{2}\begin{pmatrix} 0 & 0 & 1 & 1 & 0 \\ 0 & 0 & 0 & 1 & 1 \\ 1 & 0 & 0 & 0 & 1 \\ 1 & 1 & 0 & 0 & 0 \\ 0 & 1 & 1 & 0 & 0 \end{pmatrix}.$$

The next theorem gives a sufficient condition for a copositive matrix to have a copositive inverse. In fact, for a singular copositive matrix this condition guarantees copositivity of the Moore-Penrose generalized inverse.

Theorem 2.185. *Let A be a copositive matrix. If there exists $t \geq 0$ such that $A - tA^3$ is positive semidefinite, then $A^\dagger$ is also copositive.*

Proof. As A is symmetric, $AA^\dagger = A^\dagger A$ and therefore $A^\dagger(A - tA^3)A^\dagger = A^\dagger - tA$. Hence if the condition holds, $A^\dagger - tA$ is positive semidefinite for some $t \geq 0$, and $A^\dagger$ is copositive as the sum of the copositive tA and the positive semidefinite $A^\dagger - tA$. □

The sufficient condition in Theorem 2.185 is satisfied with $t = 0$ if and only if A is positive semidefinite. In the case that A is not positive semidefinite, $t > 0$ and this condition may be replaced by an equivalent one.

Corollary 2.186. *Let A be a copositive matrix which is not positive semidefinite, and let $\rho = \rho(A)$ be its spectral radius. Then the following two conditions are equivalent, and imply that $A^\dagger$ is also copositive.*

(a) *There exists $t > 0$ such that $A - tA^3$ is positive semidefinite.*
(b) *All the negative eigenvalues of A are equal to $-\rho$.*

Proof. By Theorem 2.185 it suffices to show that (a) and (b) are equivalent. There exists $t > 0$ such that $A - tA^3$ is positive semidefinite if and only if for every eigenvalue λ_i of A, $i = 1, \ldots, n$,

$$\lambda_i - t\lambda_i^3 \geq 0.$$

That is, if and only if every positive eigenvalue satisfies $\lambda_i^2 \leq \frac{1}{t}$ and every negative eigenvalue satisfies $\lambda_i^2 \geq \frac{1}{t}$. This is equivalent to the inequality

$$\min_{\lambda_i<0} |\lambda_i| \geq \max_{\lambda_i>0} \lambda_i.$$

As $\lambda_1 = \rho$ is the spectral radius of A, this may only occur if $|\lambda_i| = \rho$ for every negative eigenvalue λ_i, i.e. all the negative eigenvalues are equal to $-\rho$. □

Example 2.80 of an orthogonal copositive matrices is an example of a matrix with a copositive inverse (but since the inverse is equal to the matrix itself it is not surprising that it is copositive...). If A satisfies the conditions of Corollary 2.186 and has a positive eigenvalue other than $\rho(A)$, then A^{-1} is a copositive matrix that has a copositive inverse, although it does not satisfy the same condition (its only negative eigenvalue is not $-\rho(A^{-1})$).

Example 2.187. Let $A = 20|\mathbf{v}||\mathbf{v}|^T - 20\mathbf{v}\mathbf{v}^T + 10\mathbf{v}_1\mathbf{v}_1^T + 5\mathbf{v}_2\mathbf{v}_2^T$, where

$$\mathbf{v} = \frac{1}{10}\begin{pmatrix} 1 \\ 2 \\ -1 \\ -2 \end{pmatrix}, \quad \mathbf{v}_1 = \frac{1}{5}\begin{pmatrix} -2 \\ 1 \\ 0 \\ 0 \end{pmatrix}, \quad \mathbf{v}_2 = \frac{1}{5}\begin{pmatrix} 0 \\ 0 \\ -2 \\ 1 \end{pmatrix}.$$

Then

$$A^{-1} = \begin{pmatrix} 8 & -4 & 4 & 8 \\ -4 & 2 & 8 & 16 \\ 4 & 8 & 4 & -2 \\ 8 & 16 & -2 & 1 \end{pmatrix}^{-1} = \frac{1}{100}\begin{pmatrix} 8 & -4 & 1 & 2 \\ -4 & 2 & 2 & 4 \\ 1 & 2 & 16 & -8 \\ 2 & 4 & -8 & 4 \end{pmatrix}$$

are both copositive (in this case, SPN).

We next make a couple of observations regarding copositive matrices with a nonnegative hollow inverse (recall that the Horn matrix is one such copositive matrix). The first observation is that an SPN matrix may have a hollow nonnegative inverse only if it is itself nonnegative and hollow.

Theorem 2.188. *Let A be an $n \times n$ SPN matrix, whose inverse is nonnegative and hollow. Then A is a hollow nonnegative matrix. In particular, A has exactly one positive entry in each row and in each column.*

Proof. Let $A = P + N$, where P is a positive semidefinite matrix and N is a hollow nonnegative matrix. Denote the i-th columns of $B = A^{-1}$ by $\mathbf{b}_i$. For every $i = 1, \ldots, n$,

$$0 = \mathbf{e}_i^T A^{-1} \mathbf{e}_i = \mathbf{e}_i^T A^{-1} A A^{-1} \mathbf{e}_i = \mathbf{b}_i^T A \mathbf{b}_i.$$

Hence each $\mathbf{b}_i \geq \mathbf{0}$ is a zero of $A = P + N$, and is therefore also a zero of P. Since P is positive semidefinite, this implies that $P\mathbf{b}_i = \mathbf{0}$ for $i = 1, \ldots, n$. As the columns of A^{-1} are a basis for $\mathbb{R}^n$, $P = 0$ and $A = N$.

It is well known that the inverse of a nonnegative matrix is nonnegative if and only if A has exactly one positive entry in each row and in each column, see Exercise 1.24. □

Remark 2.189. Note that by the last part of Theorem 2.188 an $n \times n$ hollow symmetric nonnegative matrix N may have a hollow nonnegative inverse only if n is even and N is congruent by a positive diagonal matrix to a symmetric permutation matrix, which is a direct sum of 2×2 matrices of the form $\begin{pmatrix} 0 & 1 \\ 1 & 0 \end{pmatrix}$. Such a matrix N of order $n > 3$ is necessarily reducible.

Now consider an irreducible copositive A (i.e., A has a connected graph) of order $n \geq 3$, which has a hollow nonnegative inverse. By Theorem 2.188 and Remark 2.189, A is exceptional. It turns out that it is also $\mathcal{HN}$-irreducible (i.e., any reduction of off-diagonal elements results in a matrix which is not copositive).

Theorem 2.190. *Let $n \geq 3$, and let A be an $n \times n$ irreducible copositive matrix. If A is nonsingular and has a hollow nonnegative inverse, then A is $\mathcal{HN}$-irreducible.*

Proof. We need to show that for every $1 \leq i < j \leq n$ the matrix A is $\{i, j\}$-irreducible. By permuting rows and columns it suffices to show that A is $\{1, 2\}$-irreducible. As A is irreducible, $B = A^{-1}$ is also irreducible. Thus $b_{ij} \neq 0$ for some $1 \leq i \leq 2$ and $3 \leq j \leq n$. By permuting the last $n - 2$ rows and columns of A and $B = A^{-1}$ we may assume that either $b_{1n} \neq 0$ or $b_{2n} \neq 0$, or both. We split A and A^{-1} conformally into blocks.

$$A = \begin{pmatrix} A' & \mathbf{a} \\ \mathbf{a}^T & c \end{pmatrix} \quad \text{and} \quad A^{-1} = \begin{pmatrix} B' & \mathbf{b} \\ \mathbf{b}^T & 0 \end{pmatrix},$$

where A' and B' are $(n-1)\times(n-1)$ (hence at least 2×2), $\mathbf{a},\mathbf{b}\in\mathbb{R}^{n-1}$, and at least one of the two first entries of $\mathbf{b}$ is nonzero. Since $AA^{-1}=I$, $A'\mathbf{b}=\mathbf{0}$. Therefore the nonnegative vector $\mathbf{b}$ is a zero of A' satisfying that $b_1+b_2>0$ and $(A\mathbf{b})_1=(A\mathbf{b})_2=0$. By Proposition 2.63 the matrix A' is $\{1,2\}$-irreducible, and therefore A is $\{1,2\}$-irreducible. □

Note that the Horn matrix whose inverse is nonnegative and hollow is indeed $\mathcal{HN}$-irreducible (it is even extremal).

Exercises

2.148. Let A be an $n\times n$ nonsingular copositive matrix. Show that if in A^{-1} there is a row with a zero diagonal entry such that the other entries in this row are positive, then A is SPN.
Hint: If the last column of A^{-1} is $\begin{pmatrix}\mathbf{b}\\0\end{pmatrix}$, $\mathbf{b}>\mathbf{0}$, then $\mathbf{b}$ is a positive zero of $A(n)$. Use Proposition 2.69.

2.149. Let A be an $n\times n$ nonsingular copositive matrix, $n\geq 3$. Show that there cannot be three different rows in A^{-1}, each with a zero diagonal entry and all other entries positive.
Hint: See Exercises 2.148 and 2.36.

2.150. Let A be an $n\times n$ nonsingular SPN matrix. If there exists a nonsingular matrix S such that $A^{-1}S\geq 0$ and $S^TA^{-1}S$ is hollow, then A is nonnegative.
Hint: Show that $A^{-1}S\mathbf{e}_i$ is a zero of A for every i.

2.151. Let A be an $n\times n$ nonsingular copositive matrix. If there exists a nonsingular matrix S such that $A^{-1}S\geq 0$ and $S^TA^{-1}S$ is hollow, then S is nonnegative.
Hint: If $A^{-1}S\mathbf{e}_i$ is a zero of A, then $A(A^{-1}S\mathbf{e}_i)\geq\mathbf{0}$.

2.152. Let

$$A=\begin{pmatrix}1&-1&1&0&0&1&-1\\-1&1&-1&1&0&0&1\\1&-1&1&-1&1&0&0\\0&1&-1&1&-1&1&0\\0&0&1&-1&1&-1&1\\1&0&0&1&-1&1&-1\\-1&1&0&0&1&-1&1\end{pmatrix}\quad\text{and}\quad S=\begin{pmatrix}1&0&0&0&0&1&0\\0&1&0&0&0&0&1\\1&0&1&0&0&0&0\\0&1&0&1&0&0&0\\0&0&1&0&1&0&0\\0&0&0&1&0&1&0\\0&0&0&0&1&0&1\end{pmatrix}.$$

The matrix A is copositive by Theorem 2.130. Use Exercise 2.150 with the above S to show that A is exceptional.

2.153. Let A be an $n \times n$ strictly copositive matrix, $n \leq 3$. Show that A^{-1} cannot have all diagonal entries equal to zero.
Hint: For $n = 3$, if $\operatorname{diag} A = \mathbf{e}$, then $(\operatorname{adj} A)_{ii} = 0$ for every i implies $|a_{ij}| = 1$ for every $i < j$.

Notes. Theorem 2.183 is due to [Han and Mangasarian (1984)], where it is proved in a more general context of copositivity of a matrix A, which is not necessarily symmetric, with respect to a convex cone

Theorem 2.185 was proved for the inverse of a copositive matrix in [Jacobson (1976)], as a corollary of a more general result. In [Bisht and Sivakumar (2017)] it was proved that if A is copositive and for some $r \geq 0$ the matrix $A - rA^3$ is positive semidefinite on $\operatorname{cs} A$, then $A^\dagger$ is copositive. The statement and proof of Theorem 2.185 here follow [Shaked-Monderer (2019)].

Theorem 2.188 was proved in [Johnson and Reams (2016)]. Exercises 2.148–2.152 present results of that paper. (However, Exercise 2.149 shows that there exists no matrix satisfying the conditions of [Johnson and Reams (2016), Theorem 5].) The result in Exercise 2.153 is due to [Väliaho (1986)].

2.13 The COP and SPN completion problems

The PSD completion problem was described in Section 1.8. Here we consider the corresponding completion problems for copositive matrices and for SPN matrices. We will use the terminology and results of Section 1.8 (a partial symmetric matrix, partial PSD matrix, specification graph), as well as the characterization of PSD completable graphs. Recall that a partial symmetric matrix has all diagonal entries specified.

Consider first the copositive completion problem.

Definition 2.191. A *partial COP matrix* is a partial symmetric matrix A such that all its fully specified principal submatrices are copositive. A *COP completion* of A is a completion of A that is copositive. If A is a partial COP matrix with specification graph G, we say that A is a *G-partial COP matrix.*

The question is: which partial COP matrices have a COP completion? And for graphs: which graphs have the property that every COP partial matrix with this specification graph has a copositive completion? It turns out that every partial COP matrix has a COP completion, no matter what the specification graph is.

Theorem 2.192. *Every partial COP matrix has a COP completion.*

Corollary 2.193. *Every graph G is COP completable, i.e., has the property that every G-partial COP matrix can be completed to a copositive matrix.*

Proof of Theorem 2.192. First consider the case that all diagonal elements a_{ii} of A are nonzero. As A is partial COP, this means that $a_{ii} > 0$ for every i. We will show that if exactly one pair of off-diagonal entries is unspecified, then A has a COP completion. From this the result follows by induction. Suppose the only unspecified entries are in position i, j and j, i. Let $D = \mathrm{Diag}(1/\sqrt{a_{11}}, \ldots, 1/\sqrt{a_{nn}})$. Consider $B = DAD$. Then B has diagonal $\mathbf{e}$, $B(i)$ and $B(j)$ are fully specified and copositive. By Proposition 2.16, setting $b_{ij} = b_{ji} = 1$ yields a copositive matrix. Thus $A = D^{-1}BD^{-1}$ is a copositive completion of the partial A. Note that $a_{ij} = \sqrt{a_{ii}a_{jj}}$.

Now suppose A has some zero diagonal entries. Say $a_{ii} > 0$ for $i \in \alpha$ and $a_{ii} = 0$ for $i \notin \alpha$ (α possibly empty). Then $A[\alpha]$ has a COP completion by the first part of the proof, and for every $i \notin \alpha$ each specified off-diagonal entry in the i-th row and column is nonnegative. Assigning arbitrary nonnegative value to any unspecified entry in every row and column $i \notin \alpha$ yields a matrix which is nonnegative outside the copositive $A[\alpha]$, and is therefore copositive. □

Remark 2.194. Consider the partial matrix

$$A = \begin{pmatrix} 1 & -1 & ? \\ -1 & 1 & -1 \\ ? & -1 & 1 \end{pmatrix}.$$

Any copositive completion of A has a zero $\mathbf{u} = \mathbf{e}_1 + \mathbf{e}_2$, hence $A\mathbf{u} \geq \mathbf{0}$ implies that $a_{13} \geq 1$. This shows that, in general, $a_{ij} = \sqrt{a_{ii}a_{jj}}$ is the minimal value a_{ij} that guarantees copositivity of the completion. (Although in some instances a partial COP matrix may be completed to a copositive matrix by $a_{ij} < \sqrt{a_{ii}a_{jj}}$ — see Exercise 2.154.)

Remark 2.195. Define a *partial SCOP matrix* to be a partial symmetric matrix such that each of its fully specified principal submatrices is strictly copositive. It is also true that every partial SCOP matrix has a strictly copositive completion. The proof is left as an exercise.

Now consider the same questions for SPN matrices.

Definition 2.196. A *partial SPN matrix* is a partial symmetric matrix A such that all its fully specified principal submatrices are SPN. An *SPN completion* of A is a completion of A that is SPN. If A is a partial SPN matrix with specification graph G, we say that A is a *G-partial SPN matrix.* If every G-partial SPN matrix has an SPN completion, we say that G is an *SPN completable graph.*

Note that a complete graph of any order is obviously SPN completable. Below is an example of graphs that are not SPN completable graphs.

Example 2.197. Let G be a graph on $n \geq 5$ vertices, n odd, that is not the complete graph, and has a spanning cycle. Then G is not SPN completable.

Indeed, choose a spanning cycle C of G, and consider the G-partial matrix A with diagonal equal to $\mathbf{e}$, $a_{ij} = -1$ whenever $\{i, j\}$ is an edge of C, and $a_{ij} = 1$ for every other edge $\{i, j\}$ of G. This is a partial SPN matrix: If B is a fully specified submatrix of A, then it is of order at most $n - 1$. Hence $G_{-1}(B)$ consists of some disjoint paths and isolated vertices. If B is the all ones matrix, it is clearly SPN. If $G_{-1}(B)$ has some edges, and $i \neq j$ are both in the same component of $G_{-1}(B)$ with $d_{G_{-1}(B)}(i, j)$ even, then $b_{ij} = 1$. Hence the $\{\pm 1\}$-matrix B is SPN by Theorem 2.140.

The G-partial SPN matrix A does not have an SPN completion since $G_{-1}(A)$ is connected and not bipartite (again by Theorem 2.140). Therefore the graph G is not SPN completable.

To characterize all the SPN completable graphs a couple of observations are in order.

Remark 2.198. If A is a partial SPN matrix and $a_{ii} = 0$ for some i, then A has an SPN completion if and only if $A(i)$ has an SPN completion. One implication is trivial, and for the other note that any SPN completion of $A(i)$ may be extended to an SPN completion of A by assigning 0 (or any nonnegative value) to any unspecified entry in the i-th row and column.

This implies that if G is an SPN completable graph, then every induced subgraph H of G is SPN completable. Any H-partial SPN matrix A may be extended to a G-partial SPN matrix by setting $a_{ii} = 0$ for every $i \in V(G) \setminus V(H)$, and $a_{ij} = 0$ whenever $\{i, j\} \in E(G) \setminus E(H)$.

By this comment, combined with the fact that a direct sum of SPN matrices is SPN, G is SPN completable if and only if every component of G is SPN completable.

Proposition 2.199. *A graph G is SPN completable if and only if every G-partial SPN matrix A with diagonal* $\mathbf{e}$ *and all entries in the interval $[-1,1]$ has an SPN completion.*

Proof. The "only if" direction is obvious.

For the converse, note that, by Remark 2.198, a G-partial SPN matrix A has an SPN completion if and only if $A[\alpha]$ has an SPN completion, where $\alpha = \{1 \le i \le n \mid a_{ii} > 0\}$. Thus it suffices to consider the case that all the diagonal elements of the G-partial SPN matrix A are positive. In this case, let $D = \mathrm{Diag}\left(1/\sqrt{a_{11}}, \dots, 1/\sqrt{a_{nn}}\right)$. The matrix DAD is a G-partial SPN matrix with diagonal $\mathbf{e}$, and if we replace by 1 any specified entry of DAD that is greater than 1, we obtain a G-partial SPN matrix B (see Exercise 2.15). By the assumption, B has an SPN completion, and scaling this completion back by pre- and post-multiplying it by D^{-1}, we get an SPN matrix A'. Replace every unspecified entry of A by the corresponding entry of A' to obtain an SPN completion of A. □

Example 2.200. The graph $G = T_n$ is SPN completable for every $n \ge 4$. To see that, consider T_n-partial SPN matrices with diagonal $\mathbf{e}$ and entries in $[-1,1]$. By Theorem 2.159 such a matrix has the form

$$A = \begin{pmatrix} 1 & -\cos\theta & -\cos\varphi_3 & \cdots & \cdots & -\cos\varphi_n \\ -\cos\theta & 1 & -\cos\xi_3 & \cdots & \cdots & -\cos\xi_n \\ -\cos\varphi_3 & -\cos\xi_3 & 1 & ? & \cdots & ? \\ \vdots & \vdots & ? & \ddots & \ddots & \vdots \\ \vdots & \vdots & \vdots & \ddots & \ddots & ? \\ -\cos\varphi_n & -\cos\xi_n & ? & \cdots & ? & 1 \end{pmatrix},$$

where

$$\theta, \varphi_i, \xi_i \in [0,\pi] \quad \text{and} \quad \theta + \varphi_i + \xi_i \ge \pi \quad \text{for every} \quad 3 \le i \le n.$$

Then there exist $\hat{\varphi}_i, \hat{\xi}_i$, $3 \le i \le n$, such that

$$0 \le \hat{\varphi}_i \le \varphi_i, \quad 0 \le \hat{\xi}_i \le \xi_i \quad \text{and} \quad \theta + \hat{\varphi}_i + \hat{\xi}_i = \pi \quad \text{for} \quad i = 3, \dots, n.$$

By Theorem 2.159 the matrix

$$P = \begin{pmatrix} 1 & -\cos\theta & -\cos\hat{\varphi}_3 & \cdots & \cdots & -\cos\hat{\varphi}_n \\ -\cos\theta & 1 & -\cos\hat{\xi}_3 & \cdots & \cdots & -\cos\hat{\xi}_n \\ -\cos\hat{\varphi}_3 & -\cos\hat{\xi}_3 & 1 & ? & \cdots & ? \\ \vdots & \vdots & ? & \ddots & \ddots & \vdots \\ \vdots & \vdots & \vdots & \ddots & \ddots & ? \\ -\cos\hat{\varphi}_n & -\cos\hat{\xi}_n & ? & \cdots & ? & 1 \end{pmatrix}$$

is a T_n-partial positive semidefinite matrix. As T_n is a chordal graph, there exists a positive semidefinite completion of P. Replace unspecified entries in A by the corresponding entries in the positive semidefinite completion of P. By the monotonicity of the function $-\cos t$ on $[0, \pi]$, the resulting matrix is greater than or equal to the PSD completion of P, and is therefore an SPN completion of the T_n-partial matrix A.

Proposition 2.201. *Let $G = G_1 \cup G_2$, where $G_1 \cap G_2$ is a cut vertex of G. Then G is SPN completable if and only if G_1 and G_2 are both SPN completable.*

Proof. The "only if" follows from one of the observations in Remark 2.198. For the "if" part, suppose A is a G-partial SPN matrix. Assume without loss of generality that the set of vertices of G_1 is $\alpha = \{1, \dots, m\}$, and the set of vertices of G_2 is $\beta = \{m, \dots, n\}$. Then $A[\alpha]$ is a G_1-partial SPN matrix, and $A[\beta]$ is a G_2-partial SPN matrix. By the assumption, $A[\alpha]$ and $A[\beta]$ have SPN completions $P_1 + N_1$ and $P_2 + N_2$, respectively, where P_1 and P_2 are positive semidefinite, and N_1 and N_2 are hollow nonnegative matrices. Consider the partial matrix P with $P[\alpha] = P_1$ and $P[\beta] = P_2$, and all other entries unspecified. Note that P is well defined, since there is no ambiguity regarding its m, m entry. The matrix P is a partial PSD matrix whose graph is chordal, and therefore has a PSD completion. Add to this completion the nonnegative matrix N with $N[\alpha] = N_1$ and $N[\beta] = N_2$, and all other entries equal to zero. The resulting matrix is an SPN completion of A. □

Proposition 2.202. *Let G be a connected triangle free graph. Then G is SPN completable if and only if G is bipartite.*

Proof. Since an odd cycle of length $n \geq 5$ is not SPN completable by Example 2.197, Remark 2.198 implies that a triangle free SPN completable graph cannot contain an odd cycle.

Conversely, suppose G is bipartite. Since G is connected, there is a unique partition of its vertex set into two independent sets α and β. By Theorem 2.136 there is a unique positive semidefinite matrix P such that $p_{ij} = -1$ for every edge $\{i, j\}$ of G (the graph $G_{-1}(P)$ is the complete bipartite graph with independent sets α and β). Let A be any G-partial SPN matrix with diagonal $\mathbf{e}$ whose entries lie in the interval $[-1, 1]$. Then the matrix $P + N$, where N is the nonnegative matrix

$$n_{ij} = \begin{cases} a_{ij} + 1, & \text{if } \{i, j\} \text{ is an edge of } G \\ 0, & \text{otherwise} \end{cases}$$

is an SPN completion of A. □

We can now state and prove the characterization of SPN completable graphs.

Theorem 2.203. *Let G be a graph on n vertices. Then the following are equivalent.*

(a) *G is SPN completable.*
(b) *Every odd cycle in G induces a complete subgraph of G.*
(c) *Every block of G is either complete, or bipartite, or a T_k.*

Proof. (a) $\Rightarrow$ (b): Suppose G is SPN completable. Let C be an odd cycle in G. If C is a 3-cycle, it is complete. If C has $n \geq 5$ vertices, then the subgraph of G induced by C is SPN completable by Remark 2.198, and is therefore complete by Example 2.197.

(b) $\Rightarrow$ (c): Suppose every odd cycle in G induces a complete subgraph of G. Let G' be a block of G which is not bipartite or a T_k. Then either G' is a K_4, or it contains an odd cycle on $n \geq 5$ vertices. In the first case we are done, so assume the latter. Since we assume (b), G' has a complete subgraph on 5 vertices or more. Let K be a maximal complete subgraph of G'. Suppose to the contrary that $K \neq G'$. Then, since G' is connected, there exists an edge $\{p, q\}$, where $p \in V(K)$ and $q \in V(G') \setminus V(K)$. Since G' is a block, $G' - p$ is connected, so there exists a path P from q to a vertex $p' \in V(K) \setminus \{p\}$ all of whose internal vertices are in $V(G') \setminus V(K)$. The length of P is $\ell \geq 1$. The complete graph K has at least 5 vertices, hence for every $k \in V(K) \setminus \{p, p'\}$ there exist paths Q^m in K from p' to p through k of length m, for $m = 2, 3$. The edge $\{p, q\}$ followed by the path P and then by the path Q^m form a cycle in G' of length $1 + \ell + m$ through p, q, k and p'. Choose either $m = 2$ or $m = 3$ so that this cycle is an odd cycle of length 5 or more. By the assumption, the subgraph induced on the vertices of this cycle is complete. In particular, q is adjacent also to p' and to k.

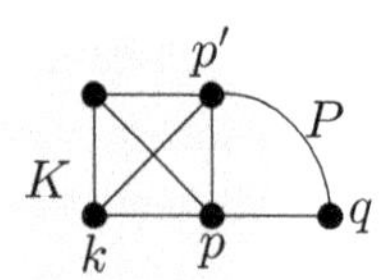

As $k \in V(K) \setminus \{p, p'\}$ was arbitrary, we get that q is adjacent in G' to all the vertices of K. That is, the subgraph of G' induced on $V(K) \cup \{q\}$ is a complete graph, contradicting the maximality of K. Therefore, $K = G'$, and G' is complete.

(c) $\Rightarrow$ (a): Follows from Proposition 2.201 and the fact that complete graphs, bipartite graphs and any T_k are all SPN completable. $\square$

Exercises

2.154. Consider an $n \times n$ partial COP matrix

$$A = \begin{pmatrix} a & \mathbf{b}^T & ? \\ \mathbf{b} & A' & \mathbf{c} \\ ? & \mathbf{c}^T & a \end{pmatrix}.$$

Show that if $\mathbf{b} \geq \mathbf{0}$ and $\mathbf{c} \geq \mathbf{0}$, then

$$A = \begin{pmatrix} a & \mathbf{b}^T & -\sqrt{ad} \\ \mathbf{b} & A' & \mathbf{c} \\ -\sqrt{ad} & \mathbf{c}^T & d \end{pmatrix}$$

is a COP completion. (This is a case in which a COP completion may be achieved with an entry that is smaller than the "generic" completion entry $\sqrt{ad}$.)

2.155. Show that

$$\begin{pmatrix} ? & -1 \\ -1 & 0 \end{pmatrix}$$

has no copositive completion. Hence the assumption in Theorem 2.192, that all the diagonal entries of the partial matrix are specified, is essential.

2.156. Prove the claim in Remark 2.195.

2.157. Show that if every fully specified principal submatrix of

$$A = \begin{pmatrix} ? & \mathbf{b}^T \\ \mathbf{b} & A' \end{pmatrix}$$

is strictly copositive, then A has a strictly copositive completion.
Hint: Let $p = \min_{\mathbf{y}\in\Gamma_{n-1}} \mathbf{b}^T\mathbf{y}$ and $q = \min_{\mathbf{y}\in\Gamma_{n-1}} \mathbf{y}^T A'\mathbf{y} > 0$. Complete by $a_{11} > \frac{p^2}{q}$.

2.158. Show that if A' is copositive but not strictly copositive, then there exists $\mathbf{b} \in \mathbb{R}^{n-1}$ such that

$$A = \begin{pmatrix} ? & \mathbf{b}^T \\ \mathbf{b} & A' \end{pmatrix}$$

has no copositive completion.

2.159. Let A be a partial matrix with all diagonal elements unspecified and $a_{ij} = a_{ji}$ whenever both are specified. Show that A has a COP completion. *Hint:* Complete A to be diagonally dominant.

Notes. The copositive completion result, Theorem 2.192, was stated and proved in [Hogben *et al.* (2005)], but the proof here uses Proposition 2.16, which is due to [Hoffman and Pereira (1973)]. The SPN completion result, Theorem 2.203, and the results leading up to it, were stated and proved in [Shaked-Monderer *et al.* (2016)]. Exercises 2.157–2.159 are based on [Hogben (2007)]. There is a copositive-plus completion result that was not included here, and can be found in [Liu and Smith (2015)]. It states that a graph is copositive-plus completable if and only if it is a disjoint union of pairwise disjoint complete graphs.

2.14 SPN graphs

In this section we consider the question: when is the sufficient condition for copositivity of being SPN also necessary? More precisely, we consider a qualitative version of this question.

Definition 2.204. A graph G is *SPN* if every copositive matrix whose graph is G is SPN. Similarly, a signed graph $\mathcal{G}$ is *SPN* if every copositive matrix whose signed graph is $\mathcal{G}$ is SPN.

A graph that is not an SPN graph is a *non-SPN graph.* Similarly, a signed graph is *non-SPN* if it is not an SPN signed graph.

We consider here the characterization of SPN graphs, which is not yet fully complete. In the process we will touch on signed graphs. A copositive matrix A such that $G(A) = G$ ($\mathcal{G}(A) = \mathcal{G}$) will be called a *COP matrix realization* of the graph G (signed graph $\mathcal{G}$). So a graph, or a signed graph, is SPN if every COP matrix realization of it is SPN. An SPN matrix A such that $G(A) = G$ ($\mathcal{G}(A) = \mathcal{G}$) will be called an *SPN matrix realization* of G ($\mathcal{G}$).

Example 2.205. Small graphs, that is, graphs with at most 4 vertices, are SPN graphs by Theorem 2.151.

There exist (signed) graphs on 5 vertices that are not SPN.

Example 2.206. Let F_5 be the fan graph on 5 vertices (see Section 1.5). The matrix

$$A = \begin{pmatrix} 1 & -1 & 1 & 0 & 0 \\ -1 & 1 & -1 & 1 & 0 \\ 1 & -1 & 1 & -1 & 1 \\ 0 & 1 & -1 & 1 & -1 \\ 0 & 0 & 1 & -1 & 1 \end{pmatrix}$$

of Example 2.142 is a COP realization of F_5, which is not SPN. Hence the graph F_5 and the signed F_5 shown below are not SPN (throughout this section we draw signed graphs with solid lines representing positive edges, and dashed lines representing negative edges).

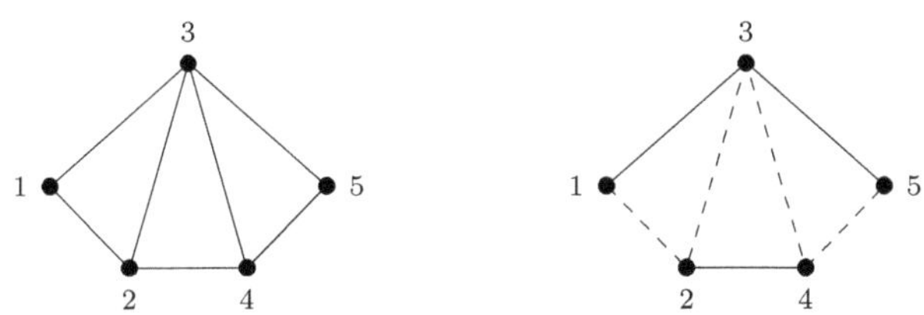

Proposition 2.207. *If G is an SPN graph, then every subgraph of G is an SPN graph.*

Proof. Let H be a subgraph of G, and let A be a COP realization of H. Without loss of generality we may assume that G has vertices $\{1, \ldots, n\}$, the vertices of H are $\{1, \ldots, k\}$, $k \leq n$, and the positive diagonal entries of A are all equal to 1. For every $\varepsilon > 0$ extend $A \oplus I_{n-k}$ to a COP realization A_ε of G by $(A_\varepsilon)_{ij} = \varepsilon$ for every $\{i, j\} \in E(G) \backslash E(H)$. As G is an SPN graph, the copositive A_ε is SPN. Let $A_\varepsilon \geq P_\varepsilon$, where P_ε is a positive semidefinite matrix with the same diagonal as A_ε. The sequence $P_{1/k}$ has a converging subsequence ($-J \leq P_{1/k} \leq J$ is bounded), say P_{1/k_l}. The limit matrix P is positive semidefinite. As $\lim_{l \to \infty} A_{1/k_l} = A \oplus I$ and $A_{1/k_l} \geq P_{1/k_l}$, $A \oplus I \geq P$. Thus $A \oplus I$ is SPN, and so is its principal submatrix A. □

By Proposition 2.207 we may characterize SPN graphs by identifying a list of all forbidden subgraphs.

Remark 2.208. Suppose every subgraph H of G such that $V(H)$ is a proper subset of $V(G)$ is SPN. If A is a COP realization of G that is not SPN, then $G_-(A)$ is a connected spanning subgraph of G. We leave the proof of this observation as an exercise.

By the next lemma, we may characterize SPN graphs by identifying the possible blocks of such graphs.

Proposition 2.209. *A graph G is an SPN graph if and only if every block of G is an SPN graph.*

Proof. The "only if" follows from Proposition 2.207. For the "if" part recall that if $G = G_1 \cup G_2$ where $G_1 \cap G_2$ is a cut vertex of G, and A is a COP realization of G, then $A = (A_1 \oplus 0) + (0 \oplus A_2)$, where A_i is a COP realization of G_i, $i = 1, 2$ (see Theorem 2.106). If both G_1 and G_2 are SPN graphs, then A_1 and A_2 are SPN matrices, and so is A. The result follows by induction on the number of blocks in G. □

As an example, we may deduce the following (see also Exercise 2.160).

Corollary 2.210. *Any acyclic graph is SPN.*

Proof. This follows from Example 2.205 and Proposition 2.209, since in an acyclic graph every block has at most 2 vertices. □

A key tool is the following result (compare it to Corollary 2.90 regarding copositive matrices).

Proposition 2.211. *Let*

$$A = \begin{pmatrix} B & C \\ C^T & E \end{pmatrix}$$

be an $n \times n$ symmetric matrix, such that E is a nonsingular M-matrix and $C \leq 0$. Then A is SPN if and only if A/E is SPN.

Proof. If A/E is SPN, then

$$A = \begin{pmatrix} CE^{-1}C^T & C \\ C^T & E \end{pmatrix} + \begin{pmatrix} A/E & 0 \\ 0 & 0 \end{pmatrix}$$

is a sum of a positive semidefinite matrix (see Remark 1.76) and an SPN matrix, and is therefore SPN.

Conversely, suppose A is SPN. Let

$$P = \begin{pmatrix} \widehat{B} & \widehat{C} \\ \widehat{C}^T & \widehat{E} \end{pmatrix}$$

be a positive semidefinite matrix such that $P \leq A$ and P has the same diagonal as A. Then $\widehat{E} \leq E$ is an M-matrix, and $\widehat{C} \leq C \leq 0$. By permuting rows and columns we may assume that $\widehat{E} = \widehat{E}_1 \oplus \widehat{E}_2$, where

$\widehat{E}_2$ is a direct sum of all the singular irreducible blocks of $\widehat{E}$, and $\widehat{E}_1$ is nonsingular. Then

$$P = \begin{pmatrix} \widehat{B} & \widehat{C}_1 & \widehat{C}_2 \\ \widehat{C}_1^T & \widehat{E}_1 & 0 \\ \widehat{C}_2^T & 0 & \widehat{E}_2 \end{pmatrix}.$$

Let $\mathbf{v}$ be a positive eigenvector of $\widehat{E}_2$ corresponding to the eigenvalue 0. Then $\mathbf{u} = \begin{pmatrix} \mathbf{0} \\ \mathbf{v} \end{pmatrix}$ is in the null space of the positive semidefinite P, $P\mathbf{u} = \mathbf{0}$. Thus $\widehat{C}_2\mathbf{v} = \mathbf{0}$, and since $\widehat{C}_2 \leq 0$ and $\mathbf{v} > \mathbf{0}$, this implies that $\widehat{C}_2 = 0$ and

$$P = \begin{pmatrix} \widehat{B} & \widehat{C}_1 & 0 \\ \widehat{C}_1^T & \widehat{E}_1 & 0 \\ 0 & 0 & \widehat{E}_2 \end{pmatrix}.$$

As $\widehat{C} \leq C \leq 0$, $\widehat{E} \leq E$ and E is an M-matrix,

$$A = \begin{pmatrix} B & C_1 & 0 \\ C_1^T & E_1 & 0 \\ 0 & 0 & E_2 \end{pmatrix},$$

and $CE^{-1}C^T = C_1E_1^{-1}C_1^T \oplus 0$. Since $\widehat{E}_1 \leq E_1$ are nonsingular M-matrices, $\widehat{E}_1^{-1} \geq E_1^{-1}$, and therefore

$$A/E = B - C_1E_1^{-1}C_1^T \geq \widehat{B} - \widehat{C}_1\widehat{E}_1^{-1}\widehat{C}_1^T.$$

The matrix on the right-hand side is a principal submatrix of $P/\widehat{E}_1$, which is positive semidefinite, and therefore A/E is SPN. □

Particularly useful is the following special case (for the case that $a_{ii} = 0$ see Exercise 2.162).

Corollary 2.212. *Let A be an $n \times n$ symmetric matrix with $a_{ii} > 0$, such that all off-diagonal elements in row i are nonpositive. Then A is SPN if and only if $A/[i]$ is SPN.*

Using these results we may show more graphs to be SPN.

Corollary 2.213. *Every cycle C_n, $n \geq 3$, is an SPN graph.*

Proof. By induction on n. For $n \leq 4$ the claim is clear (see Example 2.205).

Suppose an n-cycle is SPN, and let A be a copositive matrix whose graph is the $(n+1)$-cycle. By Remark 2.208 it suffices to consider the case that

$G_-(A)$ is connected and spans the $(n+1)$-cycle. If $G_-(A)$ is the whole cycle, the copositive A has all off-diagonal entries nonpositive, and is therefore positive semidefinite (Theorem 2.102). Otherwise, by Remark 2.208 we may assume that $a_{12} > 0$ and all other off-diagonal entries are nonpositive, in particular all the entries in the $(n+1)$-th row. By Corollary 2.91, $A/[n+1]$ is copositive, and it is easy to see that $G(A/[n+1])$ is the n-cycle. By the induction hypothesis, $A/[n+1]$ is SPN, and, by Corollary 2.212, so is A. □

Corollary 2.214. *Any subdivision of T_4, obtained by replacing none, any, some, or all of the edges of T_4 by independent paths, is an SPN graph.*

Proof. Let G be graph on n vertices that is a subdivision of T_4. Then G consists of three paths sharing common end vertices, and the number of edges in G is $n+1$. Each proper subgraph of G is either acyclic, or has one cycle block and all its other blocks have at most 2 vertices each. Thus every proper subgraph of G is SPN. To show that G is SPN it suffices, by Remark 2.208, to consider a copositive matrix A with $G(A) = G$ and $G_-(A)$ connected and spanning. As $G_-(A)$ has at least $n-1$ edges, at most two edges in $\mathcal{G}(A)$ are positive. Let α be the set of indices of rows in A that have all off-diagonal element nonpositive. Then $|\alpha| \geq n-4$, $A[\alpha]$ is a nonsingular M-matrix (see Exercise 2.82). The matrix A is SPN since $A/A[\alpha]$ is a copositive matrix of order at most 4, and is therefore SPN. □

Example 2.215. Let DR_5 denote the graph obtained from the complete graph K_4 by replacing one edge by an independent path of length 2.

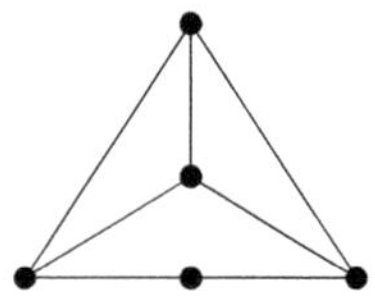

We will show that graph DR_5 is an SPN graph. By Remark 2.208, it suffices to consider copositive matrix realizations A of DR_5 such that $G_-(A)$ is connected and spans DR_5. If in row i of such A all the off-diagonal entries are nonpositive, then $A/[i]$ is a 4×4 copositive matrix, and therefore SPN, hence so is A. Thus we consider the case that each row of A has at least one positive off-diagonal entry. Up to isomorphism, $\mathcal{G}(A)$ has to be one of the two signed graphs shown in the figure below. We consider the two cases.

Case 1 Case 2

Case 1. In this case, the sign pattern of A is

$$\begin{pmatrix} + & - & 0 & + & 0 \\ - & + & - & 0 & + \\ 0 & - & + & - & + \\ + & 0 & - & + & - \\ 0 & + & + & - & + \end{pmatrix}.$$

Reduce the $1,4$ entry of A to get a $\{1,4\}$-irreducible copositive matrix $B \le A$. If $b_{14} \le 0$, then the first row of B has all off-diagonal entries nonpositive and B, and therefore A, is SPN. If $b_{14} > 0$, $\mathcal{G}(B) = \mathcal{G}(A)$. There exists a zero $\mathbf{u}$ of B such that $u_1 + u_4 > 0$ and $(B\mathbf{u})_1 = (B\mathbf{u})_4 = 0$. From $(B\mathbf{u})_1 = 0$ and $u_1 + u_4 > 0$ we get that $u_2 > 0$, and from $(B\mathbf{u})_4 = 0$ and $u_1 + u_4 > 0$ we get that $u_3 + u_5 > 0$. Hence $|\operatorname{supp} \mathbf{u}| \ge 3$ and $|\{i \mid (B\mathbf{u})_i = 0\}| \ge 4$. Therefore B is SPN, and so is A. (If $|\operatorname{supp} \mathbf{u}| = 4$, by Corollary 2.70; if $|\operatorname{supp} \mathbf{u}| = 3$ by Corollary 2.101.)

Case 2. In this case, the sign pattern of A is

$$\begin{pmatrix} + & - & 0 & + & 0 \\ - & + & - & 0 & + \\ 0 & - & + & + & - \\ + & 0 & + & + & - \\ 0 & + & - & - & + \end{pmatrix}.$$

Let $B \le A$ be a $\{1,4\}$-irreducible copositive matrix. The remainder of the proof that A is SPN follows the same ideas as in Case 1, and is left as an exercise.

Example 2.216. The graph T_5 is SPN. To see that, let A be a 5×5 copositive matrix with $G(A) = T_5$ and $G_-(A)$ connected and spanning. In particular, all diagonal entries are positive. By scaling A, we may assume that the diagonal of A is $\mathbf{e}$. It suffices to consider the case that each row i in A has also a positive off-diagonal entry. (For otherwise, $A/[i]$ is a 4×4

copositive matrix, and therefore $A/[i]$ is SPN, and so is A.) Hence we may assume that A has the following sign pattern

$$\begin{pmatrix} 1 & 0 & 0 & - & + \\ 0 & 1 & 0 & - & + \\ 0 & 0 & 1 & + & - \\ - & - & + & 1 & - \\ + & + & - & - & 1 \end{pmatrix},$$

and is therefore SPN by Exercise 2.143.

Proposition 2.217. *A graph G on 5 vertices is SPN if and only if G does not contain the fan graph F_5 as a subgraph.*

Proof. As F_5 is not SPN by Example 2.206, a graph containing F_5 is not SPN, by Proposition 2.207.

For the converse, consider a graph G that does not contain an F_5 subgraph, and let A be a copositive matrix with $G(A) = G$. The complement of G has at least 3 edges, since otherwise the complement of G would be a subgraph of $P_4 \cup K_1$, which is the complement of F_5, and G would contain F_5. Thus G has at most 7 edges. Any subgraph of G on at most 4 vertices is SPN, hence by Remark 2.208 it suffices to consider the case that $G_-(A)$ is connected and spans G. In particular, $G_-(A)$ has at least 4 edges. If $G_+(A)$ has at most 2 edges, then at most 4 rows have positive off-diagonal elements. In this case, there is a row i with all off-diagonal entries nonpositive, the matrix $A/[i]$ is is a 4×4 copositive matrix, and is therefore SPN, and thus A is SPN. Hence it suffices to consider the case that $G_+(A)$ has at least three edges. Combining all the assumptions on the number of edges in G, $G_-(A)$ and $G_+(A)$, the proof is reduced to the case that $G(A) = G \neq F_5$ has exactly 7 edges. Since the complement of G has 3 edges, and it is not $P_4 \cup K_1$, it has to be either $K_3 \cup 2K_1$, or $P_2 \cup P_3$. In the first case $G = T_5$ and in the second $G = DR_5$, and in both cases A is SPN (by Examples 2.216 and 2.215). □

Here are three examples of non-SPN graphs on 6 vertices. The first is a graph, sometimes called the *co-domino graph*, which we denote by CD_6.

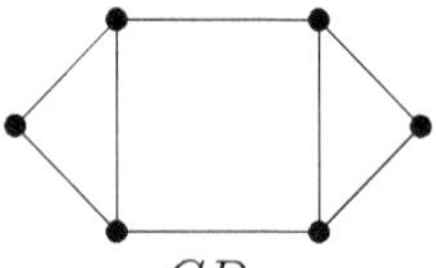

CD_6

Example 2.218. The graph CD_6 is non-SPN. To see that, let $t = \sqrt{0.5}$, and consider the matrix

$$A = \begin{pmatrix} 1 & -1 & t & 0 & 0 & 0 \\ -1 & 1 & -t & 0 & 1 & 0 \\ t & -t & 1 & -t & 0 & 0 \\ 0 & 0 & -t & 1 & -t & t \\ 0 & 1 & 0 & -t & 1 & -1 \\ 0 & 0 & 0 & t & -1 & 1 \end{pmatrix},$$

whose graph is CD_6. The submatrix $A(5)$ is copositive, since it is the positive semidefinite matrix $A(5,6)$ bordered by nonnegative column and row. Similarly, $A(2)$ is copositive. Every principal submatrix of A whose entries are less than 1 is a principal submatrix of one of these two 5×5 matrices, and is therefore copositive. By Proposition 2.16, A is copositive.

We now show that A is not SPN. Each of the submatrices $A[i, i+1, i+2]$, $i = 1, \ldots, 4$, is an $\mathcal{HN}$-irreducible positive semidefinite matrix. Thus if $A \geq P$, where P is positive semidefinite and the diagonal of P is $\mathbf{e}$, then

$$P = \begin{pmatrix} 1 & -1 & t & * & * & * \\ -1 & 1 & -t & 0 & * & * \\ t & -t & 1 & -t & 0 & * \\ * & 0 & -t & 1 & -t & t \\ * & * & 0 & -t & 1 & -1 \\ * & * & * & t & -1 & 1 \end{pmatrix}.$$

Since $(1,1,0,0,0)^T$ is a zero of the positive semidefinite P, we have that $p_{1j} = -p_{2j}$ (and $p_{j1} = -p_{j2}$) for every $j = 2, \ldots, 6$. Similarly, since $(0,0,0,1,1)^T$ is a zero of P, $p_{6i} = -p_{5i}$ (and $p_{i6} = -p_{i5}$) for every $i = 2, \ldots, 6$. The vector $(0,0,t,1,t,0)^T$ is also a zero of P, which implies that $p_{52} \cdot t + 0 \cdot 1 + (-t) \cdot t = 0$, that is, $p_{25} = p_{52} = t > 0$. By the equalities above, $p_{15} = p_{26} = -t$ and then $p_{16} = t$. This contradicts the assumption that $A \geq P$. Hence A is not SPN.

The next example of a non-SPN graph is T_6.

Example 2.219. To see that T_6 is not SPN, let $0 < \theta < \pi/6$, and consider the matrix

$$A = \begin{pmatrix} 1 & -\cos\theta & \cos 2\theta & -\cos\theta & -\sin\theta & 1 \\ -\cos\theta & 1 & -\cos\theta & \cos 2\theta & 1 & -\sin\theta \\ \cos 2\theta & -\cos\theta & 1 & 0 & 0 & 0 \\ -\cos\theta & \cos 2\theta & 0 & 1 & 0 & 0 \\ -\sin\theta & 1 & 0 & 0 & 1 & 0 \\ 1 & -\sin\theta & 0 & 0 & 0 & 1 \end{pmatrix},$$

whose graph is T_6. Note that

$$\mathbf{u}_1 = (1, 2\cos\theta, 1, 0, 0, 0)^T, \quad \mathbf{u}_2 = (2\cos\theta, 1, 0, 1, 0, 0)^T$$
$$\mathbf{u}_3 = (0, 1, \cos\theta, 0, 0, \sin\theta)^T, \quad \mathbf{u}_4 = (1, 0, 0, \cos\theta, \sin\theta, 0)^T$$

are zeros of A. As in the previous example, the matrix A is not SPN: If $A \geq P$ and P is positive semidefinite, then $P\mathbf{u}_i = \mathbf{0}$, $i = 1, \ldots, 4$. This yields a unique P. However, $-p_{34} = p_{56} = \cos 3\theta > 0$ contradicts the assumption that $A \geq P$, as $a_{56} = 0$. To show that A is copositive, it suffices to check that $A[\alpha]$ is copositive for the following four α's:

$$\{1, 2, 3, 4\}, \{1, 3, 4, 5\}, \{2, 3, 4, 6\}, \{3, 4, 5, 6\}.$$

We leave the details as an exercise.

Certain graph transformations preserve SPN-ness (or non-SPN-ness) of a graph or a signed graph.

Proposition 2.220. *Let $\mathcal{G}$ be a signed graph, and let $\widehat{\mathcal{G}}$ be the signed graph obtained from $\mathcal{G}$ by replacing a negative edge by a path consisting of two negative edges. Then $\mathcal{G}$ is SPN if and only if $\widehat{\mathcal{G}}$ is.*

Proof. Suppose $\mathcal{G}$ is not SPN. Let a copositive non-SPN matrix A have signed graph $\mathcal{G}$. Suppose, without loss of generality, that the subdivided edge is $\{n-1, n\}$, and that $a_{n-1\,n} = -1$. Let $\widehat{A} = (A \oplus 0) + (0_{n-2} \oplus E)$, where

$$E = \begin{pmatrix} 1 & 1 & -1 \\ 1 & 1 & -1 \\ -1 & -1 & 1 \end{pmatrix}.$$

Then $\mathcal{G}(\widehat{A}) = \widehat{\mathcal{G}}$, $\widehat{A}$ is copositive as a sum of a copositive matrix and a positive semidefinite one. All the off-diagonal entries in the last row and column of $\widehat{A}$ are nonpositive. The matrix $\widehat{A}$ is not SPN, since $\widehat{A}/\widehat{A}[n+1] = A$ is not SPN.

Now suppose $\mathcal{G}$ is SPN. Let a copositive A have $\mathcal{G}(A) = \widehat{\mathcal{G}}$, and let k be the vertex added in the subdivision the negative edge $\{i, j\}$ in $\mathcal{G}$. Then $A/[k]$ is copositive and $\mathcal{G}(A/[k]) = \mathcal{G}$. Hence $A/[k]$ is SPN and, by Corollary 2.212, so is A. □

A repeated use of the previous proposition leads to the following corollary.

Corollary 2.221. *Let $\mathcal{G}$ be a signed graph. Let $\widehat{\mathcal{G}}$ by obtained by subdividing a negative edge in $\mathcal{G}$, replacing it by a path consisting of negative edges. Then $\widehat{\mathcal{G}}$ is SPN if and only if $\mathcal{G}$ is SPN.*

Another implication of the previous results is the following.

Corollary 2.222. *Any subdivision of the graph $K_{3,2}$ is SPN.*

Proof. Let a graph G with $n \geq 5$ vertices be a subdivision of $K_{3,2}$. Then G consists of 4 independent paths, each of length at least two, with the same end vertices x and y. In every subgraph of G with at most $n-1$ vertices, each block is either an edge, or a subdivision of K_4, and therefore such a subgraph is SPN. To check that G is SPN it suffices therefore to consider a copositive A with $G(A) = G$ such that $G_-(A)$ is connected and spanning. In particular, at least one of the independent paths in $\mathcal{G}(A)$ consists of negative edges. It is then easy to see that $\mathcal{G}(A)$ is obtained from a signed T_5 by subdividing negative edges. Hence A is SPN. □

The next simple result is left as an exercise.

Proposition 2.223. *Let $\mathcal{G}$ be a signed graph. Let a signed graph $\widehat{\mathcal{G}}$ be obtained by adding to $\mathcal{G}$ a path of positive edges, whose internal vertices are not in $\mathcal{G}$. Then $\widehat{\mathcal{G}}$ is SPN if and only if $\mathcal{G}$ is SPN.*

In the non-signed case, it is not yet clear whether a subdivision of an SPN graph can yield a non-SPN graph. However, we can deduce the following result.

Proposition 2.224. *Let a graph G contain a path P of length $\ell \geq 3$, whose internal vertices have degree 2 in G. Then the graph $\widehat{G}$ obtained from G by subdividing one of the edges of P is an SPN graph if and only if G is an SPN graph.*

Proof. Let $\widehat{P}$ be the path in $\widehat{G}$ obtained from subdividing an edge of P. For any edge e of the path $\widehat{P}$, $\widehat{G} - e$ is SPN if and only if G is, since in $\widehat{G} - e$ each block is a subgraph of G, or an edge.

Suppose first that G is an SPN graph, and let a copositive A have graph $\widehat{G}$. For any internal vertex i of $\widehat{P}$, $G(A(i))$ is SPN, and $G(A/[i]) = G$ is also SPN. Thus if both edges incident with such i in $\mathcal{G}(A)$ are positive or both are negative, then A is SPN. Otherwise, the signs of the edges of $\widehat{P}$ in $\widehat{\mathcal{G}}(A)$ alternate. As $\widehat{P}$ has at least three inner vertices, it has a negative edge e between two positive edges. In that case, $G_-(A)$ is not connected, and as $\widehat{G} - e$ is SPN, the matrix A is SPN by Proposition 2.105.

If G is not SPN, there is a sign assignment to its edges that yields a non-SPN signed graph $\mathcal{G}$. If one of the edges of P is negative in $\mathcal{G}$, it can be subdivided to obtain a non-SPN graph whose underlying graph is $\widehat{G}$.

If P is a positive path in $\mathcal{G}$, then the signed graph $\mathcal{G}'$ obtained from $\mathcal{G}$ by omitting the edges of P is not SPN by Proposition 2.223. By the same proposition, the signed graph $\mathcal{G}$ obtained by adding to $\mathcal{G}'$ the path $\widehat{P}$ with all positive edges is also non-SPN, and therefore its underlying graph $\widehat{G}$ is non-SPN. □

As an example, consider the graph DR_n, the graph on $n \geq 4$ vertices obtained from K_4 by replacing one edge of K_4 by an independent path of length $n-3$.

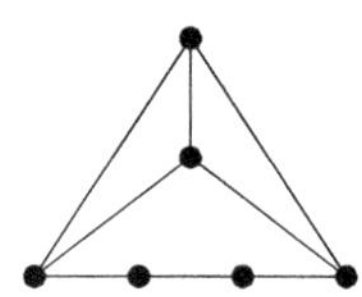

DR_6

Proposition 2.225. *For every $n \geq 5$, the graph DR_n is an SPN graph.*

Proof. For $n = 5$ this was proved in Example 2.215. We now prove that DR_6 is SPN. As any DR_n, $n \geq 7$, is obtained by further subdividing the length 3 path in DR_6, this will complete the proof by Proposition 2.224.

Every subgraph of DR_6 on 5 vertices is SPN, since it is either a cycle, or each of its blocks is a subgraph of the diamond T_4. So consider a copositive A with graph DR_6 and a connected $G_-(A)$ spanning DR_6. For every row i the graph of $A/[i]$ is either a subgraph of DR_5 or a subgraph of a subdivision of K_4, and therefore if all the off-diagonal entries in row i are nonpositive, the matrix $A/[i]$ is SPN and so is A. We therefore suppose that in each row of A there is a positive off-diagonal element. Up to isomorphism, we may assume that $\mathcal{G}(A)$ is the signed graph below.

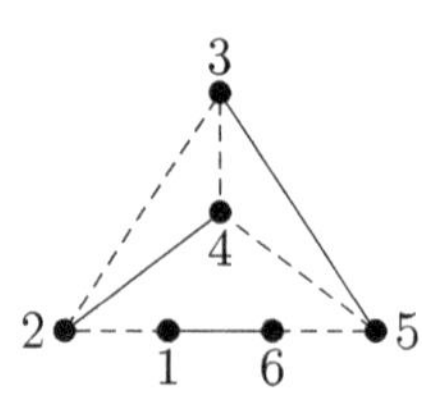

So A has the following sign pattern,

$$A = \begin{pmatrix} + & - & 0 & 0 & 0 & + \\ - & + & - & + & 0 & 0 \\ 0 & - & + & - & + & 0 \\ 0 & + & - & + & - & 0 \\ 0 & 0 & + & - & + & - \\ + & 0 & 0 & 0 & - & + \end{pmatrix}.$$

We may assume that A is $\{1,6\}$-irreducible (if replacing the $1,6$ entry by 0 yields a copositive matrix, then this matrix is SPN, since every proper subgraph of DR_6 is SPN, and so is A). Then there exists a zero $\mathbf{u}$ of A such that $u_1+u_6>0$ and $(A\mathbf{u})_1=(A\mathbf{u})_6=0$. From $(A\mathbf{u})_1=0$ and $u_1+u_6>0$ we deduce that $u_2>0$, and therefore $(A\mathbf{u})_2=0$. Denote $\mathbf{v}=\mathbf{u}[1,2]$. Then $(A\mathbf{u})_1=0$ and $\mathbf{u}\geq\mathbf{0}$ imply that $(A[1,2]\mathbf{v})_1\leq 0$. Since $A[1,2]$ is positive definite (as a principal submatrix of the irreducible M-matrix $A[1,2,3]$) and $\mathbf{v}\neq\mathbf{0}$, we have $\mathbf{v}^TA[1,2]\mathbf{v}>0$, and $(A[1,2]\mathbf{v})_2>0$. Combined with $(A\mathbf{u})_2=0$ this implies that $u_3>0$. By the same argument we get from $u_1+u_6>0$ and $(A\mathbf{u})_6=0$ that $u_5>0$ and $u_4>0$. That is, $|\operatorname{supp}\mathbf{u}|\geq 5$, and therefore A is SPN by Corollary 2.70 (in fact $A\mathbf{u}=\mathbf{0}$, hence A is positive semidefinite by Corollary 2.100). □

Next we consider another transformation on signed graphs. A *paw graph* has two blocks, a triangle and an edge. We call the following transformation on a signed graph a *Λ-paw transformation* on $x-y-z$: If $x-y-z$ is a negative path in $\mathcal{G}$, where the end vertices x and z may or may not be adjacent (by an edge of any sign), erase the negative path $x-y-z$, add a new vertex w to $\mathcal{G}$, connected to each of x, y and z by a negative edge, and add (or replace the existing edge by) a positive edge $\{x,z\}$. The figure below describes this transformation (with a dashed line denoting a negative edge, a solid line a positive edge, and a double dotted line denoting a possible, but not mandatory, edge of either sign).

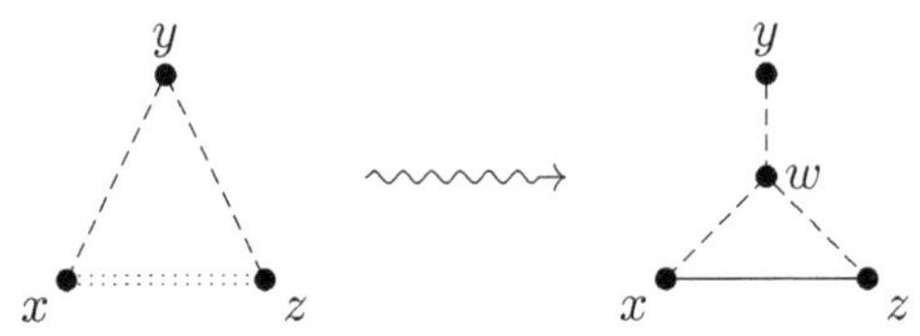

Proposition 2.226. *Let $\mathcal{G}$ be a non-SPN signed graph, and let $x-y-z$ be a negative path in $\mathcal{G}$. Then the graph $\widehat{\mathcal{G}}$ obtained by a Λ-paw operation on $x-y-z$ is a non-SPN signed graph.*

Proof. Let A be a non-SPN copositive matrix with signed graph $\mathcal{G}$, and suppose that $x = n-2$, $y = n-1$ and $z = n$. Let $a_{n-2\,n-1} = -a$ and $a_{n-1\,n} = -b$, where $a, b > 0$. Choose $c > 0$ such that $a_{n-2\,n} + c > 0$. Let

$$E = \begin{pmatrix} \frac{ac}{b} & a & c & -\sqrt{\frac{ac}{b}} \\ a & \frac{ab}{c} & b & -\sqrt{\frac{ab}{c}} \\ c & b & \frac{bc}{a} & -\sqrt{\frac{bc}{a}} \\ -\sqrt{\frac{ac}{b}} & -\sqrt{\frac{ab}{c}} & -\sqrt{\frac{bc}{a}} & 1 \end{pmatrix}.$$

The matrix E is a rank 1 positive semidefinite matrix. Then $\widehat{A} = (A \oplus 0) + (0_{n-2} \oplus E)$ is copositive, has signed graph $\widehat{\mathcal{G}}$, and all the off-diagonal entries in its $(n+1)$-th row and column are nonpositive. Since $\widehat{A}/\widehat{A}[n+1] = A$ is not SPN, $\widehat{A}$ is also non-SPN. □

The next proposition considers a Δ-Y *transformation* on a signed graph. Suppose there is a triangle with negative edges on vertices x, y, z in $\mathcal{G}$. In a Δ-Y transformation the negative edges of the triangle are erased, and a vertex w is added, connected by a negative edge to each of x, y, z. The proof of the proposition is practically the same as the proof of Proposition 2.226, and is left as an exercise.

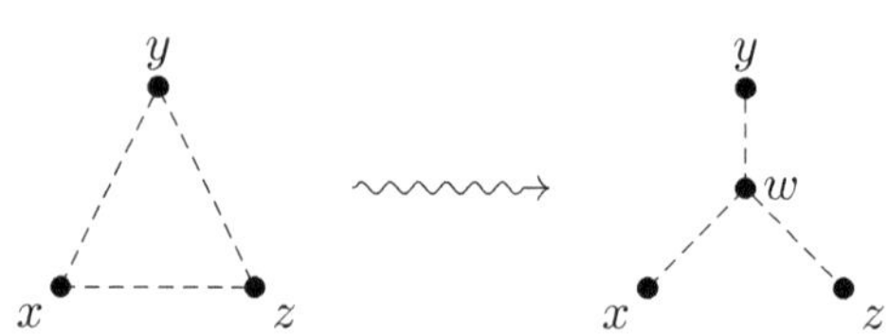

Proposition 2.227. *Let $\mathcal{G}$ be a non-SPN signed graph on n vertices. Then the signed graph $\widehat{\mathcal{G}}$ obtained by a Δ-Y transformation on a triangle of negative edges in $\mathcal{G}$ is a non-SPN signed graph.*

Corollary 2.228. *Any subdivision of F_5, obtained by replacing none, any, some, or all of the edges in F_5 by independent paths, is non-SPN.*

Proof. By Example 2.206, the signed F_5 shown on the left in the figure below is non-SPN. Therefore the subgraph obtained from it by Λ-paw transformation on $1-2-3$ (shown on the right in this figure) is also non-SPN.

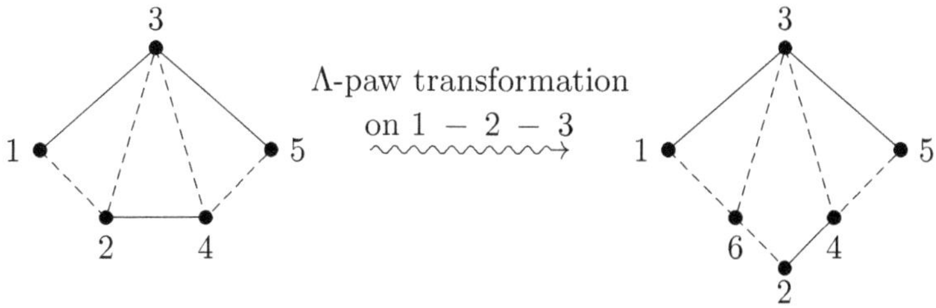

Let a graph G be a subdivision of F_5. Then there is a sign assignment to G's edges that yields a signed graph, which may be obtained from one of the graphs in the figure above by subdividing negative edges, replacing them by paths of negative edges. Thus G is non-SPN. □

Corollary 2.229. *Any subdivision of CD_6, obtained by replacing none, any, some, or all of the edges in CD_6 by independent paths, is non-SPN.*

Proof. In the figure below, the signed graph of the copositive non-SPN realization of CD_6 (Example 2.218) is shown on the left. On the right is the graph obtained from this graph by Λ-paw transformation on $4 - 5 - 6$. Thus the graph on the right is also non-SPN by Proposition 2.226.

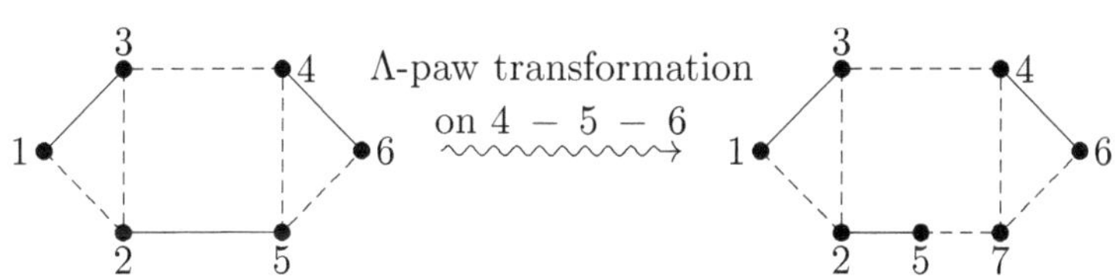

If G is a subdivision of CD_6, signs can be assigned to its edges so that the resulting signed graph $\mathcal{G}$ is obtained from one of the two signed graphs shown above, by replacing negative edges by paths of negative edges. Hence such G is not SPN. □

Next we consider which subdivisions of K_4 are SPN. We already know that subdividing one edge, repeatedly, yields the SPN graphs DR_n. We now show that if two or more edges of K_4 are subdivided, then the resulting graph is non-SPN. Consider first subdivision of two edges in K_4.

Proposition 2.230. *Let G be a subdivision of K_4, obtained by subdividing two edges, once each. Then G is not SPN.*

Proof. There are two possible cases: either the two subdivided edges are not adjacent (Case 1), or they are adjacent (Case 2).

Case 1 Case 2

Case 1. Let $t = \sqrt{0.5}$, and

$$A = \begin{pmatrix} 1 & -t & 0 & 1 & 0 & 0 \\ -t & 1 & -t & 0 & 1 & 0 \\ 0 & -t & 1 & -t & 0 & 1 \\ 1 & 0 & -t & 1 & -t & 0 \\ 0 & 1 & 0 & -t & 1 & -t \\ 0 & 0 & 1 & 0 & -t & 1 \end{pmatrix}.$$

The matrix A is copositive, since every principal submatrix in which the off-diagonal entries are less than 1 is copositive (actually, positive semidefinite). For every $i = 1, \ldots, 4$ the submatrix $A[i, i+1, i+2]$ is a singular M-matrix, with a zero $(t, 1, t)^T$. Hence if $A \geq P$, where P is positive semidefinite and $\text{diag}P = \text{diag}A$, then

$$P = \begin{pmatrix} 1 & -t & 0 & * & * & * \\ -t & 1 & -t & 0 & * & * \\ 0 & -t & 1 & -t & 0 & * \\ * & 0 & -t & 1 & -t & 0 \\ * & * & 0 & -t & 1 & -t \\ * & * & * & 0 & -t & 1 \end{pmatrix}.$$

The vector $\mathbf{u} = (t, 1, t, 0, 0)^T$ is a zero of P, hence $P\mathbf{u} = \mathbf{0}$. Hence $p_{41} = p_{14} = t$. Similarly, the vectors $\mathbf{v} = (0, t, 1, t, 0)^T$ and $\mathbf{w} = (0, 0, t, 1, t)^T$ are zeros of P, and we deduce that $p_{52} = p_{25} = t$ and $p_{63} = p_{36} = t$. But then $P\mathbf{u} = \mathbf{0}$ and $P\mathbf{v} = \mathbf{0}$ imply that $p_{51} = p_{15} = -1$, and $p_{62} = p_{26} = -1$. Then $P\mathbf{w} = \mathbf{0}$ implies that $p_{61} = p_{16} = t^2 > 0$, contradicting the assumption that $A \geq P$. Thus A is not SPN.

Case 2. A signed subdivision of K_4 in which two adjacent vertices were subdivided (once each) can be obtained from a signed F_5 by a Λ-paw transformation:

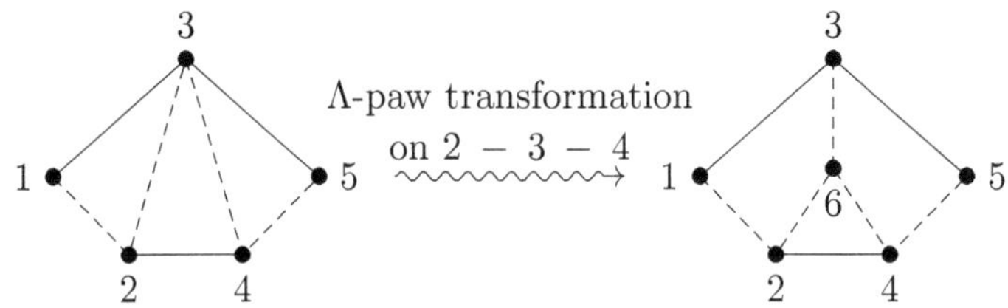

This signed F_5 is non-SPN by Example 2.206. Thus the graph obtained by subdividing two adjacent edges of K_4, once each, is not SPN. □

Corollary 2.231. *Any graph obtained by subdividing at least two edges and at most five edges of K_4, is not SPN.*

Proof. By the proof of Proposition 2.230, the following two signed graphs are non-SPN.

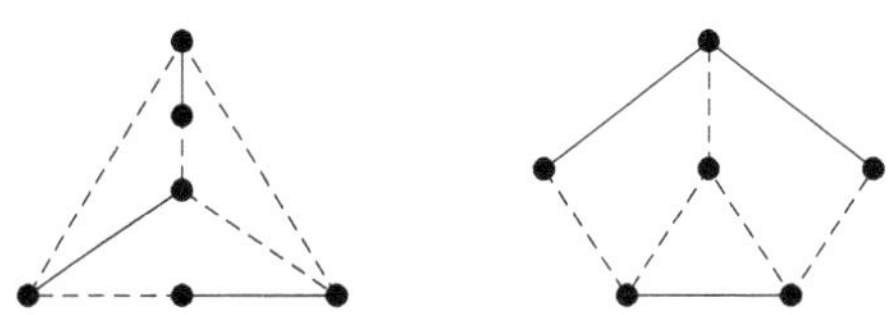

By further subdividing negative edges in these graphs, replacing them by paths of negative edges, one obtains more non-SPN signed graphs. Each graph obtained by subdividing between two to five edges of K_4 is an underlying graph of such a non-SPN signed graph, and is therefore non-SPN. □

Finally, the case that each edge of K_4 is subdivided at least once.

Proposition 2.232. *The graph obtained by subdividing each edge of K_4 at least once is non-SPN.*

Proof. Let $\theta_i > 0$, $1 \le i \le 4$, be such that

$$\sin(\theta_1) < \frac{\cos(\theta_2)\cos(\theta_4)}{\cos(\theta_3)}$$

and $\sum_{i=1}^{4} \theta_i < \frac{\pi}{2}$. Let

$$A = \begin{pmatrix} 1 & -\cos\theta_1 & \cos(\theta_1+\theta_2) & 0 & 0 \\ -\cos\theta_1 & 1 & -\cos\theta_2 & \cos(\theta_2+\theta_3) & \cos(\pi/2+\theta_1) \\ \cos(\theta_1+\theta_2) & -\cos\theta_2 & 1 & -\cos\theta_3 & \cos(\theta_3+\theta_4) \\ 0 & \cos(\theta_2+\theta_3) & -\cos\theta_3 & 1 & -\cos\theta_4 \\ 0 & \cos(\pi/2+\theta_1) & \cos(\theta_3+\theta_4) & -\cos\theta_4 & 1 \end{pmatrix}.$$

By Exercise 2.142, the matrix A is an exceptional copositive matrix, and therefore the signed graph $\mathcal{G}_1 = \mathcal{G}(A)$ below is non-SPN.

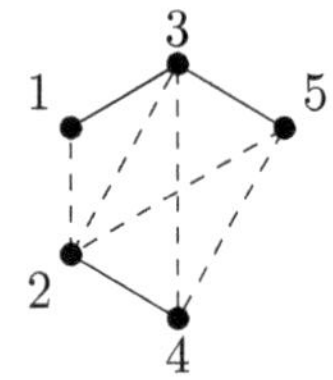

$\mathcal{G}_1$

Let

$$\mathbf{a}^T = \left(0, -\frac{\sin\theta_1}{\cos\theta_4}, -\cos\theta_3, -1, -\cos\theta_4\right)$$

and

$$B = \begin{pmatrix} A + \mathbf{a}\mathbf{a}^T & \mathbf{a} \\ \mathbf{a}^T & 1 \end{pmatrix}.$$

The matrix B is copositive (as a sum of the copositive $A \oplus 0$ and a rank-1 positive semidefinite matrix). The off-diagonal entries in the 6th column of B are nonpositive, and $A = B/B[6]$ is exceptional, hence B is also exceptional, by Proposition 2.211. Denoting $\mathcal{G}(B) = \mathcal{G}_2$, we see that the signed graph $\mathcal{G}_2$, shown below, is non-SPN.

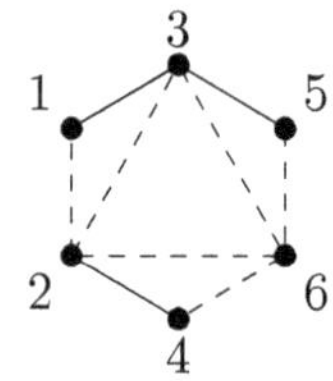

$\mathcal{G}_2$

By Proposition 2.227, the signed graph $\mathcal{G}_3$ obtained from $\mathcal{G}_2$ by a Δ-Y transformation is non-SPN.

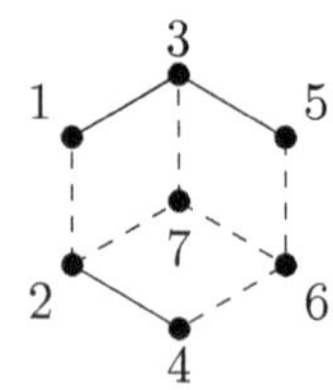

$\mathcal{G}_3$

Let $\mathcal{G}_4$ be the signed graph shown below. It is obtained from $\mathcal{G}_3$ by subdivision of three negative edges, and is therefore also non-SPN, by Proposition 2.223.

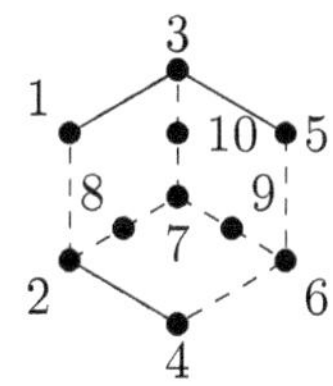

$\mathcal{G}_4$

This shows that a subdivision of K_4, where each edge of K_4 was subdivided exactly once, is non-SPN. By further replacing negative edges in $\mathcal{G}_4$ by longer paths of negative edges, it follows that any subdivision of K_4 with all edges subdivided, each at least once, is non-SPN. □

We combine the results to get a list of possible blocks of an SPN graph, and a list of forbidden subgraphs of SPN graphs.

Theorem 2.233. *Let G be a graph, in which each block is one of the following:*

(1) *an edge,*
(2) *a cycle,*
(3) *a subdivision of T_4,*
(4) *a subdivision of K_4, where at most one edge was subdivided,*
(5) *a T_5,*
(6) *a subdivision of $K_{3,2}$.*

Then G is SPN.

Note that any cycle is a subdivision of T_3, hence the possible blocks of an SPN graph include all subdivision of T_n, $n = 3, 4$. It is still not known whether subdivisions of T_5 (other than $K_{3,2}$ and its subdivisions) are SPN.

Theorem 2.234. *Let G be an SPN graph. Then G does not contain the following subgraphs:*

(1) *a subdivision of the fan F_5,*
(2) *a subdivision of CD_6,*
(3) *a subdivision of T_6,*

(4) *a subdivision of K_4 in which at least two edges were subdivided at least once each.*

In these theorems, any graph is considered to be a subdivision of itself.

Remark 2.235. Suppose G is a 2-connected graph, which contains no subdivision of F_5, no subdivision of CD_6, no subdivision of T_6, and no subdivision of K_4 in which at least two edges were subdivided at least once each. Then either G is a subdivision of T_n, $3 \leq n \leq 5$, or $G = DR_n$, $n \geq 4$.

To see that, recall that if H_0 is a 2-connected subgraph of a 2-connected graph G, then G can be generated from H_0 by successively adding H-paths to graphs H already constructed, beginning with H_0 (see Proposition 1.84). Let k be maximal such that G contains a subdivision of T_k. Then $3 \leq k \leq 5$ by the assumption that G does not contain a subdivision of T_6.

If $k = 3$, G contains a cycle H_0, and since it does not contain a subdivision of T_4, it contains no H_0-path, so $G = H_0$ is a cycle (i.e., a subdivision of T_3).

If $k = 5$, let H_0 be a subdivision of T_5 contained in G. The subgraph H_0 consists of 4 independent paths, which share the same two end vertices x and y. There is no H_0-path in G whose ends are x and y, since G does not contain a subdivision of T_6; there is no H_0-path in G whose ends both lie on the same xy-paths in H_0, since G does not contain a subdivision of F_5 or of CD_6. There is no H_0-path in G one of whose ends lie on two independent xy-paths in H_0 since G does not contain a subdivision of F_5. Hence there is no H_0-path in G, and $G = H_0$ is a subdivision of T_5.

Finally consider the case $k = 4$. Let H_0 be a subdivision of T_4 contained in G. Then H_0 consists of 3 independent paths sharing the same end vertices x and y. If $G \neq H_0$, then G contains an H_0-path. As in the case $k = 5$, the vertices x and y cannot both be ends of the H_0-path, and the ends of the path cannot both lie on the same xy-path in H_0. The only option is that the H_0-path has one end on one of the xy-paths, and the other end on another. That is, G contains a subdivision H_1 of K_4. By our assumptions on G, necessarily $H_1 = DR_n$, for some $n \geq 4$. As above, G cannot contain an H_1-path, since it does not contain a subdivision of F_5 or a subdivision of CD_6. Thus in this case $G = H_1 = DR_n$.

The one question that remains is whether (or which) subdivisions of T_5 are SPN graphs. Note that subdivisions of T_5 that are also subdivisions of $K_{3,2}$ are SPN, and only subdivisions where the base edges of T_5 is not subdivided need to be checked. If all subdivisions of T_5 are SPN, then

Theorem 2.234 gives the complete list of forbidden graphs, and in Theorem 2.233 replacing (5) and (6) by "subdivisions of T_5" would make the list of possible blocks of SPN graphs complete.

Remark 2.236. It is an interesting question whether the family of SPN graphs is minor-closed. Determining whether all subdivisions of T_5 are SPN would give an answer to this question too.

Exercises

2.160. Let G be an acyclic graph, and let A be a symmetric matrix with graph G. Let $N(A)$ be defined by

$$(N(A))_{ij} = \begin{cases} a_{ij}, & i = j \text{ or } a_{ij} < 0 \\ 0, & \text{otherwise} \end{cases} .$$

Show that the following are equivalent.

(a) A is copositive.
(b) A is SPN.
(c) $N(A)$ is positive semidefinite.

2.161. Prove the claim in Remark 2.208.
Hint: Note that Proposition 2.105 holds also if "copositive" is replaced by "SPN".

2.162. Let an $n \times n$ symmetric A have $a_{ii} = 0$. Prove that A is SPN if and only if the i-th row and column are nonnegative and $A(i)$ is SPN.

2.163. Complete the proof of Example 2.215.

2.164. Complete the missing details in the proof in Example 2.219.

2.165. Prove Corollary 2.221.

2.166. Prove Proposition 2.223.

2.167. Prove Proposition 2.227.

2.168. Complete the following missing detail the proof of Proposition 2.232. Show that $\mathcal{G}_3$ is not SPN since $\mathcal{G}_2$ is not SPN.

2.169. For a signed graph $\mathcal{G}$, and a subset S of its vertices, let $G_-[S]$ denote the graph on vertices S whose edges are the negative edges of $\mathcal{G}$ with both ends in S. Prove the following generalization of Propositions 2.220 and

2.226: Let $\mathcal{G}$ be a non-SPN signed graph and S a subset of its vertex set. Let $\widehat{\mathcal{G}}$ be a signed graph obtained by adding a vertex v adjacent to each of the vertices in S by a negative edge, (possibly) deleting some of the negative edges of $\mathcal{G}[S]$, (possibly) changing some negative edges between vertices in S to positive edges, and adding a positive edge between each pair of vertices in S that are not adjacent in $\mathcal{G}$. If every positive $G_-[S]$-partial matrix has a completion to a rank-1 positive symmetric matrix, then $\widehat{\mathcal{G}}$ is non-SPN.

Notes. The problem of characterizing SPN graphs was presented in [Shaked-Monderer (2016)], and most of this section is based on that paper and the corrigendum [Shaked-Monderer (2018)]. Proposition 2.232 and Exercise 2.169 are based on [Hogben and Shaked-Monderer (2019)]. In [Hogben and Shaked-Monderer (2019)] a graph G is defined to be *rank-1 completable* if every positive G-partial matrix can be completed to a rank-1 symmetric positive matrix. It is shown there that a connected graph G is rank-1 completable if and only if each component of G is either a tree or unicyclic with an odd cycle. Example 2.219 is due to [Drury (2020)].

Still open: *Are subdivisions of T_5 SPN? Is the family of SPN graphs minor-closed?*

Chapter 3

Complete positivity

3.1 Definition and basic properties

Consider the following question:

Question 3.1. *Let* $\mathbf{v}_1, \ldots, \mathbf{v}_n$ *be vectors in an* m*-dimensional Euclidean space* V. *Can the vectors be embedded in a nonnegative orthant of some Euclidean space? In other words, are there a natural number* k *and an isometry* $T : V \to \mathbb{R}^k$ *such that* $T\mathbf{v}_1, T\mathbf{v}_2, ..., T\mathbf{v}_n$ *belong to* $\mathbb{R}^k_+$*?*

Let A be the Gram matrix of $\mathbf{v}_1, \ldots, \mathbf{v}_n$, i.e., $a_{ij} = \langle \mathbf{v}_i, \mathbf{v}_j \rangle$, $i, j = 1, \ldots, n$. Then the answer to Question 3.1 is affirmative if and only if A is (also) the Gram matrix of nonnegative vectors.

Definition 3.2. A matrix is *completely positive* if it can be decomposed as $A = BB^T$, where B is a (not necessarily square) nonnegative matrix.

The answer to Question 3.1 is "yes" if and only if $\mathrm{Gram}(\mathbf{v}_1, \ldots, \mathbf{v}_n)$ is completely positive. Here k is the number of columns of B and for each $i = 1, 2, \ldots, n$, $T\mathbf{v}_i$ is the i-th row of B.

Completely positive matrices appear in the study of block designs in combinatorics, in probability, and in various applications of statistics, including a Markovian model for DNA evolution and a model for energy demands. The most prominent application is apparently in mathematical optimization. For more details, see the references in the notes at the end of this section.

It is very easy to construct completely positive matrices, but it is generally hard to determine whether a given square matrix is completely positive or not. An obvious necessary condition for a matrix to be completely positive is that it is doubly nonnegative, that is, both positive semidefinite and entrywise nonnegative. In some cases, this condition is also sufficient, but not always. So, how can one identify completely positive matrices? This is one of the major problems in the theory of completely positive matrices, and a significant part of this chapter is devoted to it. The next chapter deals with the following, related, problem: Suppose $\mathbf{v}_1, \ldots, \mathbf{v}_n \in V$ can be embedded in the nonnegative orthant of $\mathbb{R}^k$ for some k. What is the minimal such k? In other words, suppose an $n \times n$ matrix A is a completely positive. What is the minimal k for which there exists an $n \times k$ nonnegative matrix B such that $A = BB^T$? In the study of these problems both the geometric approach and the algebraic approach are helpful, and we will use both.

We start with simple examples.

Example 3.3. Any positive diagonal matrix is completely positive. If $d_1, \ldots, d_n \geq 0$, then $\mathrm{Diag}(d_1, \ldots, d_n) = \left(\mathrm{Diag}(\sqrt{d_1}, \ldots, \sqrt{d_n})\right)^2$.

Example 3.4. A rank 1 doubly nonnegative matrix is necessarily completely positive. If $A \in \mathcal{S}_n$ is doubly nonnegative matrix and has rank 1, then $A = \mathbf{b}\mathbf{b}^T$ for some $\mathbf{b} \in \mathbb{R}^n$ (by the proof of Theorem 1.25 and Remark 1.26). Since $A \geq 0$, we may take $\mathbf{b}$ to be nonnegative.

Theorem 3.5. *A rank* 2 *doubly nonnegative matrix is necessarily completely positive.*

Proof. Let A be an $n \times n$ doubly nonnegative matrix of rank 2. Then A is the Gram matrix of vectors $\mathbf{v}_1, \ldots, \mathbf{v}_n$ in $\mathbb{R}^2$. Since A is nonnegative, the angle between any two of these vectors is at most $\pi/2$. The two vectors which form the widest angle may therefore be rotated into the first quadrant of $\mathbb{R}^2$. In this rotation the other vectors also end up in $\mathbb{R}^2_+$. □

Remark 3.6. The discussion of Question 3.1 suggests one method for showing that a doubly nonnegative matrix A is completely positive. Being positive semidefinite, A is the Gram matrix of some vectors $\mathbf{v}_1, \ldots, \mathbf{v}_n$ in some k-dimensional Euclidean space V. If an orthonormal basis E of V can be found such that the coordinate vectors $[\mathbf{v}_1]_E, [\mathbf{v}_2]_E, \ldots, [\mathbf{v}_n]_E$ are nonnegative, then $T\mathbf{v} = [\mathbf{v}]_E$ is an isometric embedding of $\mathbf{v}_1, \ldots, \mathbf{v}_n$ into $\mathbb{R}^k_+$, and A is completely positive. This is basically the idea in the last proof: take E to be the basis obtained by the Gram-Schmidt procedure from the two vectors which form the widest angle (that is, the pair of nonzero vectors where $\min\{\langle \mathbf{v}_i/\|\mathbf{v}_i\|, \mathbf{v}_j/\|\mathbf{v}_j\|\rangle \mid \mathbf{v}_i, \mathbf{v}_j \neq \mathbf{0}\}$ is obtained).

By Example 3.4 and Theorem 3.5, every 2×2 doubly nonnegative matrix is completely positive. In Example 3.7 we show this fact directly, this time using the algebraic approach.

Example 3.7. Let

$$A = \begin{pmatrix} a & b \\ b & c \end{pmatrix},$$

be a 2×2 doubly nonnegative matrix. Then A is completely positive, for if $c = 0$, then $A = BB^T$, where

$$B = \begin{pmatrix} \sqrt{a} & 0 \\ 0 & 0 \end{pmatrix},$$

and if $c > 0$, then $A = BB^T$ where

$$B = \begin{pmatrix} \sqrt{a - b^2/c} & b/\sqrt{c} \\ 0 & \sqrt{c} \end{pmatrix}.$$

(This decomposition may be obtained by using Schur complements — see the end of Section 1.4.)

Not every doubly nonnegative matrix is completely positive.

Example 3.8. The doubly nonnegative matrix

$$A = \begin{pmatrix} 1 & 1 & 0 & 0 & 1 \\ 1 & 2 & 1 & 0 & 0 \\ 0 & 1 & 2 & 1 & 0 \\ 0 & 0 & 1 & 2 & 1 \\ 1 & 0 & 0 & 1 & 3 \end{pmatrix}$$

is not completely positive. The reader is invited to try verifying that. We will prove it in Section 3.4.

We call a factorization $A = BB^T$, $B \geq 0$, a *CP factorization.* Note that a completely positive matrix may have many different CP factorizations. In particular, consider the following addition to Theorem 3.5.

Proposition 3.9. *Let* $A = BB^T$, *where* $B = \begin{pmatrix} \mathbf{b}_1 & \mathbf{b}_2 \end{pmatrix} \geq 0$, *and* $\operatorname{supp} \mathbf{b}_2 \subseteq \operatorname{supp} \mathbf{b}_1$. *Then* $A = CC^T$, *where* $C = \begin{pmatrix} \mathbf{c}_1 & \mathbf{c}_2 \end{pmatrix} \geq 0$, $\operatorname{supp} \mathbf{c}_1 = \operatorname{supp} \mathbf{b}_1$, $\operatorname{supp} \mathbf{c}_2 \supseteq \operatorname{supp} \mathbf{b}_1 \setminus \operatorname{supp} \mathbf{b}_2$, *and for some* i *such that* $(\mathbf{b}_2)_i \neq 0$, $(\mathbf{c}_2)_i = 0$.

Proof. Let $\mu = \max\{\lambda \mid \mathbf{b}_1 - \lambda\mathbf{b}_2 \geq \mathbf{0}\}$. Then set

$$\mathbf{c}_1 = \frac{1}{\sqrt{1+\mu^2}}(\mathbf{b}_2 + \mu\mathbf{b}_1) \quad \text{and} \quad \mathbf{c}_2 = \frac{1}{\sqrt{1+\mu^2}}(\mathbf{b}_1 - \mu\mathbf{b}_2).$$

(If $\mathbf{v}$ is the column of B^T which forms the largest angle with $\mathbf{e}_1$, then $C = BR$, where the 2×2 orthogonal matrix R is a reflection about the bisector of the angle between $\mathbf{v}$ and $\mathbf{e}_1$.) □

If we start with a CP factorization $A = BB^T$, and apply Proposition 3.9 repeatedly, replacing at each step two columns of the CP factor that have the same support by two with one support strictly contained in the other, we get the following observation.

Corollary 3.10. *Let A be a completely positive matrix with a CP factorization $A = BB^T$ where B has m columns. Then it has a CP factorization $A = CC^T$, where $C \geq 0$ has m columns, no two of which have the same support.*

The following proposition contains an equivalent definition of complete positivity.

Proposition 3.11. *A symmetric matrix A is completely positive if and only if A can be represented as a sum*

$$A = \sum_{i=1}^{k} \mathbf{b}_i\mathbf{b}_i^T, \quad \mathbf{b}_i \geq 0,\ i = 1, \ldots, k. \tag{3.1}$$

Proof. $A = BB^T$ is equivalent to $A = \sum_{i=1}^{k} \mathbf{b}_i\mathbf{b}_i^T$, where $\mathbf{b}_i$ is the i-th column of B. □

In (3.1) the completely positive matrix A is represented a sum of rank 1 doubly nonnegative (completely positive) matrices. We call (3.1) a *CP representation* of A.

Example 3.12. Suppose that A is as in Example 3.7, with $c = 1$. Then a CP representation of A is

$$\begin{pmatrix} b^2 & b \\ b & 1 \end{pmatrix} + \begin{pmatrix} a - b^2 & 0 \\ 0 & 0 \end{pmatrix} = \begin{pmatrix} b \\ 1 \end{pmatrix}\begin{pmatrix} b \\ 1 \end{pmatrix}^T + \begin{pmatrix} \sqrt{a-b^2} \\ 0 \end{pmatrix}\begin{pmatrix} \sqrt{a-b^2} \\ 0 \end{pmatrix}^T.$$

Corollary 3.13. *The sum of completely positive matrices is completely positive.*

Proof. Let $A = \sum_{i=1}^{k} \mathbf{b}_i \mathbf{b}_i^T$ and $B = \sum_{i=k+1}^{k+l} \mathbf{b}_i \mathbf{b}_i^T$ be CP representations of A and B. Then $A + B = \sum_{i=1}^{k+l} \mathbf{b}_i \mathbf{b}_i^T$. □

The proofs of the next corollaries are similar to the corresponding proofs in Section 1.2.

Corollary 3.14. *The Hadamard product of completely positive matrices is completely positive.*

Proof. Let $A = \sum_{i=1}^{k} \mathbf{b}_i \mathbf{b}_i^T$ and $B = \sum_{j=1}^{l} \mathbf{c}_j \mathbf{c}_j^T$ be CP representations of A and B. Then

$$A \circ B = \sum_{i=1}^{k} \sum_{j=1}^{l} (\mathbf{b}_i \circ \mathbf{c}_j)(\mathbf{b}_i \circ \mathbf{c}_j)^T. \qquad \square$$

Corollary 3.15. *The Hadamard powers $A^{(k)} = A \circ A \circ \cdots \circ A$ of a completely positive matrix A are also completely positive.*

A similar result holds for ordinary powers. To prove it, we need the following proposition.

Proposition 3.16. *If A is an $n \times n$ completely positive matrix and S is an $m \times n$ nonnegative matrix, then SAS^T is also completely positive.*

Proof. Let $A = BB^T$, $B \geq 0$. Then $SAS^T = (SB)(SB)^T$. □

Corollary 3.17. *If A is completely positive and k is a natural number then A^k is completely positive.*

Proof. If k is even, $k = 2l$, then $A^k = (A^l)^2 = A^l(A^l)^T$. If $k = 2l + 1$, this follows from Proposition 3.16, since $A^k = (A^l)A(A^l)^T$. □

Corollary 3.18. *If $f(x)$ is a real polynomial with nonnegative coefficients and A is completely positive, then $f(A)$ is also completely positive.*

As in the case of positive semidefinite matrices, Corollary 3.14 also follows from the corresponding result for Kronecker products.

Proposition 3.19. *The Kronecker product $A \otimes C$ of completely positive matrices A and C is completely positive.*

Proof. Let $A = BB^T$, $B \geq 0$, $C = FF^T$, $F \geq 0$. The matrix $B \otimes F$ is nonnegative, and $A \otimes C = (B \otimes F)(B \otimes F)^T$. □

The observation in the next proposition may be used to deduce Corollary 3.14 from Proposition 3.19.

Proposition 3.20. *Principal submatrices of completely positive matrices are completely positive.*

Proof. Let $A[\alpha]$ be a principal submatrix of a completely positive matrix A, and let $A = BB^T$ where $B \geq 0$. Then $A[\alpha]$ is a product of the submatrix of B based on the rows indexed by α and the transpose of this submatrix.

(Alternatively, if A is the Gram matrix of $\mathbf{v}_1, \ldots, \mathbf{v}_n$, then $A[\alpha]$ is the Gram matrix of $\{\mathbf{v}_i \mid i \in \alpha\}$. Since the whole set $\{\mathbf{v}_1, \ldots, \mathbf{v}_n\}$ may be embedded in the nonnegative orthant of some $\mathbb{R}^k$, so does any subset of it.) □

The next proposition contains two simple results of Proposition 3.16, which are very useful in the study of completely positive matrices. We leave the proof as an exercise.

Proposition 3.21. *Let A be a symmetric $n \times n$ matrix. Then*

(a) *If P is an $n \times n$ permutation matrix, then A is completely positive if and only if $P^T AP$ is completely positive.*
(b) *If D is an $n \times n$ positive diagonal matrix, then A is completely positive if and only if DAD is completely positive.*

The two parts of Proposition 3.21 are algebraic statements of facts which are geometrically quite obvious: When checking whether $\mathbf{v}_1, \ldots, \mathbf{v}_n$ can be embedded in a nonnegative orthant of some $\mathbb{R}^k$, neither the order of the vectors nor their lengths matter. Here is another simple but useful observation.

Proposition 3.22. *If a symmetric matrix A is reducible, say*

$$A = A_1 \oplus A_2 \oplus \cdots \oplus A_m,$$

then A is completely positive if and only if each A_i is.

Remark 3.23. We conclude this section with a warning. The notion of complete positivity should not be confused with the notion of total positivity. However, the two positivity concepts are not totally unrelated. In Section 3.8 we will show that a symmetric totally nonnegative matrix is completely positive.

Exercises

3.1. Show that

$$A = \begin{pmatrix} 8 & 12 & 16 & 4 & 6 & 8 \\ 12 & 20 & 28 & 6 & 10 & 14 \\ 16 & 28 & 40 & 8 & 14 & 20 \\ 4 & 6 & 8 & 2 & 3 & 4 \\ 6 & 10 & 14 & 3 & 5 & 7 \\ 8 & 14 & 20 & 4 & 7 & 10 \end{pmatrix}$$

is completely positive and find a nonnegative matrix B such that $BB^T = A$.

3.2. Prove Corollary 3.10.

3.3. Let A be an $n \times n$ completely positive matrix, which is positive and has rank $A = 2$. Show that

(a) A has a CP representation $A = \mathbf{b}_1\mathbf{b}_1^T + \mathbf{b}_2\mathbf{b}_2^T$, where $\mathbf{b}_1$ and $\mathbf{b}_2$ are both positive.
(b) A has a CP representation $A = \mathbf{b}_1\mathbf{b}_1^T + \mathbf{b}_2\mathbf{b}_2^T$, where $\mathbf{b}_1$ is positive, and $\mathbf{b}_2$ has a zero entry.

Hint: Let $A = CC^T$, where $C = \begin{pmatrix} \mathbf{c}_1 & \mathbf{c}_2 \end{pmatrix} \geq 0$, and consider an appropriate rotation or reflection of the rows of C, as in Theorem 3.5.

3.4. Prove Proposition 3.21.

3.5. Show that if A is an $n \times n$ completely positive matrix, $E \geq 0$ is an $n \times m$ matrix, and

$$A' = \begin{pmatrix} A & AE \\ E^TA & E^TAE \end{pmatrix},$$

then A' is completely positive.

3.6. Prove Proposition 3.22 geometrically.

3.7. Suppose A is an $n \times n$ doubly nonnegative matrix of rank r. Then $A = \text{Gram}(\mathbf{v}_1, \ldots, \mathbf{v}_n)$, where $\mathbf{v}_1, \ldots, \mathbf{v}_n \in \mathbb{R}^r$ (see Remark 1.26). Let $\mathcal{C} = \text{cone}\{\mathbf{v}_1, \ldots, \mathbf{v}_n\} \subseteq \mathbb{R}^r$. Show that A is completely positive if and only if there exist $\mathbf{q}_1, \ldots, \mathbf{q}_m \in \mathcal{C}^*$ such that

$$\sum_{i=1}^{m} \mathbf{q}_i\mathbf{q}_i^T = I_r.$$

3.8. Let $A = \mathrm{Gram}(\mathbf{v}_1, \ldots, \mathbf{v}_n)$, where $\mathbf{v}_1, \ldots, \mathbf{v}_n \in \mathbb{R}^r$. Show that if A is a rank r completely positive matrix, then $\mathcal{C} = \mathrm{cone}\{\mathbf{v}_1, \ldots, \mathbf{v}_n\} \subseteq \mathbb{R}^r$ is a pointed solid cone.

3.9. Let A be a doubly nonnegative matrix whose entries are all zeros and ones. Show that A is permutationally similar to a direct sum $A_1 \oplus \cdots \oplus A_k$, where each A_i is either a matrix of all ones or a zero matrix. In particular, every doubly nonnegative $\{0,1\}$-matrix is completely positive.

3.10. Show that a matrix $A = (a_{ij})_{i,j=1}^n$ is completely positive if and only if there exist a measure space (Ω, μ) and nonnegative functions $f_i \in L^2(\Omega, \mu)$, $i = 1, \ldots, n$, such that

$$a_{ij} = \int_\Omega f_i(\omega) f_j(\omega)\, d\mu(\omega) \qquad i, j = 1, \ldots, n.$$

3.11. We say that an $n \times n$ completely positive matrix A is *critical* if for every $1 \le i \le n$ and every $\varepsilon > 0$, $A - \varepsilon E_{ii}$ is not completely positive. Show that

(a) Every $n \times n$ completely positive matrix, $n \ge 2$, whose rank is 1, is critical.
(b) Every nonsingular 2×2 completely positive matrix is not critical.

3.12. Consider the following definitions. Let A be an $n \times n$ positive semidefinite matrix, and $1 \le i \le n$. Let ε_0 be the maximal $\varepsilon \ge 0$ such that $A - \varepsilon E_{ii}$ is positive semidefinite. The *i-minimized matrix* of A is the matrix $A_{(i)} = A - \varepsilon_0 E_{ii}$. We say that the matrix A is *i-minimized* if $A = A_{(i)}$. A completely positive matrix A is *i-minimizable* if $A_{(i)}$ is also completely positive.

(a) Show that if

$$A = \begin{pmatrix} a & \mathbf{c}^T \\ \mathbf{c} & A_1 \end{pmatrix}$$

is positive semidefinite, then

$$A_{(1)} = \begin{pmatrix} \mu & \mathbf{c}^T \\ \mathbf{c} & A_1 \end{pmatrix},$$

where $\mu = \mathbf{c}^T A_1^\dagger \mathbf{c}$, and $A = A_{(1)}$ if and only if $\operatorname{rank} A = \operatorname{rank} A_1$.
(b) Every i-minimized positive semidefinite matrix is singular.
(c) Every singular positive semidefinite matrix is i-minimized for some i.

(d) If a completely positive matrix A is i-minimized for some i, then it is i-minimizable.
(e) If A is a singular completely positive matrix, then it is i-minimizable for some i.
(f) Every rank 1 completely positive matrix is i-minimizable for every i.
(g) If a completely positive matrix A is i-minimized for every i, then it is critical.

Hint: For part (a), see Theorems 1.75 and 1.77.

Notes. The discussion of Question 3.1 is based on [Gray and Wilson (1980)] and [Hannah and Laffey (1983)]. The idea in Proposition 3.9, Corollary 3.10 and Exercises 3.2–3.3 dates back to [Hall (1998)], see [Loewy and Tam (2003)]. The proof here follows [Dickinson (2013)] and [Dickinson and Dür (2012)]. The idea in Exercise 3.7 will be used extensively in the next chapter about the cp-rank. It was first presented in [Hannah and Laffey (1983)]. Complete positivity of $(0, 1)$-matrices (see Exercise 3.9) is discussed in [Xiang and Xiang (1998)]. Exercise 3.10 is taken from [Ando (1991)], where it is used to prove a nontrivial result on Hankel matrices. A matrix $A = (a_{ij})$ is a *Hankel matrix* if $a_{i+1\,j} = a_{i\,j+1}$ for $i, j = 1, \ldots, n-1$ (the Hilbert matrices are Hankel matrices). A positive semidefinite Hankel matrix is necessarily doubly nonnegative, because of the nonnegativity of the diagonal elements. In [Ando (1991)] it is shown that if A is a positive semidefinite Hankel matrix and if the matrix obtained from A by deleting the first column and the last row is also positive semidefinite, then A is completely positive. The notion of a critical completely positive matrix was introduced by [Zhang and Li (2002)], and that of an i-minimizable completely positive matrix by [Barioli (1998b)].

Previous surveys of the topic of completely positive matrices are [Berman (1993)], [Ando (1991)] and [Berman (1988)], and the book [Berman and Shaked-Monderer (2003)], which this book extends.

While it is easy to construct completely positive matrices, determining whether a given doubly nonnegative matrix is completely positive is hard. Computationally it was shown, in [Dickinson and Gijben (2014)], to be NP-hard. In [Berman and Rothblum (2006)] a symbolic algorithm for determining complete positivity and finding a CP factorization $A = BB^T$ with a minimal number of columns in B was presented. This algorithm, Renegar's algorithm of quantifier elimination, is shown to be highly exponential in this case, hence very expensive to implement. Various algorithms for deciding whether a given matrix is completely positive and finding a CP matrix factorization have been suggested. We do not go into these in the book, but mention here a selection of these results. A quadratic factorization heuristic algorithm generating completely positive matrices that converge to the given matrix, if it is completely positive, is suggested in [Jarre and Schmallowsky (2009)]. In [Kalofolias and Gallopoulos (2012)] an algorithm for the case that $G(A)$ has an independent set of size rank A is presented. [Brandts and Křížek (2016)] suggest a finite and exact algorithm of geometric nature for determining whether $A = BB^T$ for some $B \in \mathbb{R}_+^{n\times 3}$, and finding such B if it exists. In [Dickinson and Dür (2012)] a linear-time algorithm is presented to determine complete positivity and find a CP factorization for any matrix whose graph has only blocks that are either edges or cycles. In [Nie (2014)] the problem is considered as an $\mathcal{A}$-truncated moments problem, and an algorithm for solving such problems is applied. The algorithm consists of solving a sequence of semidefinite optimization problems, and may be used for determining complete positivity and finding a CP factorization. Another algorithm is developed in [Sponsel and Dür (2014)]. A heuristic algorithm is suggested in [Groetzner and Duer (2020)]. Not surprisingly, the non-heuristic of these algorithms, that are not limited to special matrices, tend to work mainly for matrices of rather limited order.

As far as we know, the first appearance of the term "completely positive matrix" is in [Markham (1972)]. Complete positivity was first defined for quadratic forms (see [Hall and Newman (1963); Hall (1958)]).

Definition 3.24. Let A be an $n \times n$ real symmetric matrix. A real quadratic form $Q = \sum_{i,j=1}^n a_{ij}x_ix_j$ is called *completely positive* if it can be written as $Q = L_1^2 + L_2^2 + \ldots + L_k^2$, where $L_i = b_{i1}x_1 + b_{i2}x_2 + \ldots + b_{in}x_n$, $b_{ij} \geq 0$, $i = 1, 2, \ldots, k$, $j = 1, 2, \ldots, n$.

Clearly the quadratic form $Q = \sum_{i,j=1}^n a_{ij}x_ix_j$ is completely positive if and only if the matrix $A = (a_{ij})_{i,j=1}^n$ is completely positive.

In [Gantmacher and Krein (2002)] the name "completely positive matrices" refers to what is now called "totally nonnegative" (see Section 1.3).

The complete positivity studied in this book is not related to the concept of completely positive maps known in operator theory and used in physics (see, e.g., [Benatti *et al.* (2002)]). For the sake of (positive) completeness we bring here the definition of a completely positive map (a finite dimensional version). Let V be a subspace of $\mathbb{C}^{n\times n}$ which contains the identity and such that if A belongs to V then A^*, the Hermitian adjoint of A, also belongs to V. A linear map $\varphi : V \to \mathbb{C}^{n\times n}$ is completely positive if for any positive semidefinite block matrix (X_{ij}), all of whose blocks belong to V, the block matrix $(\varphi(X_{ij}))$ is also positive semidefinite.

Completely positive matrices appear in several applications. One of the basic papers in the theory of completely positive matrices is [Gray and Wilson (1980)]. The motivation for this paper is "a proposed mathematical model of energy demand". [Drew and Johnson (1996)] mention a talk by Persi Diaconis at the Institute for Mathematics and its Applications (Minneapolis, 1993), in which completely positive matrices were related to exchangeable probability distributions on finite sample spaces. Here are references to some more applications.

Block designs. As far as we know the first application of complete positivity was to block designs. The following description is based on [Hall (1998, 1958)]. A *balanced incomplete block design* $D(v, b, r, k, \lambda)$ is an arrangement of v distinct objects $a_1, \ldots, a_v$ into b blocks $B_1, \ldots, B_b$ such that each block contains exactly k distinct objects, each object occurs in exactly r different blocks, and every pair of distinct objects occurs together in exactly λ blocks. The term balanced incomplete block design comes from the theory of design of experiments in statistics. (A *complete* block design is the set of all subsets of size k of a set of v elements. A design is *balanced* if every pair of distinct elements occurs in the same number of blocks.) A block design is described by its *incidence matrix* $B = (b_{ij})$, where $b_{ij} = 1$ if a_i belongs to B_j and $b_{ij} = 0$ if a_i does not belong to B_j. Each column of B describes a block and contains k 1's. Each row corresponds to an element of the set and contains r 1's. If B is the incidence matrix of $D(v, b, r, k, \lambda)$, then $A = BB^T = (r - \lambda)I_v + J_v$. Clearly A is completely positive. Complete positivity also appears in connection with the following question: Given parameters v, b, r, k, λ and blocks $B_1, \ldots, B_t$, is there a block design $D(v, b, r, k, \lambda)$ with $B_1, \ldots, B_t$ as initial blocks? Let $A = (r-\lambda)I_v + J_v$, and let $\mathbf{b}_1, \ldots, \mathbf{b}_t$ be $(0,1)$-vectors in R^v which describe the blocks $B_1, \ldots, B_t$. Let C be a $v \times t$ matrix with columns $\mathbf{b}_1, \ldots, \mathbf{b}_t$. Then clearly a necessary condition for the existence of the block design $D(v, b, r, k, \lambda)$ with initial blocks $B_1, \ldots, B_t$ is that $A - CC^T$ is also completely positive.

Maximin efficiency-robust tests. The maximin efficiency-robust test (MERT) idea of Gastwirth (1966) is to maximize the minimum asymptotic power versus special local families of alternatives over some specially chosen families of score statistics. In the case that the number of the alternatives is finite, the MERT algorithm reduces to a maximization problem involving a correlation matrix C (a positive semidefinite matrix C with all diagonal elements equal to 1) and vectors $\mathbf{b} \in \operatorname{cs} C$ satisfying $|b_i| \geq 1$. The associated problem in which $b_i \geq 1$ has a unique solution that can be found by quadratic programming techniques. [Bose and Slud (1995)] show that if C is completely positive, then the optimal solution of the associated problem is optimal for the MERT problem. They also show that if $\rho_i \in L^2(\mathbb{R}, \mu)$, $i = 1, \ldots, m$, where μ is a Borel probability measure on the real line, and the functions ρ_i all have the same signs μ-almost everywhere, i.e.

$$\mu(\{t|\rho_i(t) \geq 0 \text{ for all } i, \text{ or } \rho_i(t) \leq 0 \text{ for all } i\}) = 1$$

and if c_{ij} is the inner product of ρ_i and ρ_j, then $C = (c_{ij})$ is completely positive. These hypotheses are satisfied in many versions of MERT, so in these versions the MERT problem can be solved by solving the associate problem.

Markovian model of DNA evolution. [Kelly (1994)] describes Markov models for the evolution of two species from a common ancestor. Using these models one can calculate the probability of observing two particular DNA sequences from present-day species that share a common ancestor. The data is summarized in a 4×4 matrix

$$N = \begin{pmatrix} N_{AA} & N_{AC} & N_{AG} & N_{AT} \\ N_{CA} & N_{CC} & N_{CG} & N_{CT} \\ N_{GA} & N_{GC} & N_{GG} & N_{GT} \\ N_{TA} & N_{TC} & N_{TG} & N_{TT} \end{pmatrix}$$

where A, C, G, T denote the four bases of the DNA, and N_{jk} is the number of sites in which base j is observed in the first species and base k is observed in the second. The different models are based on some simplifying assumptions and it is of interest to determine if a model assumptions are met in certain data sets. In one of the models it is assumed that the rates of evolution for all positions in the DNA sequence are equal, that the positions are independent and that the transition matrices for the two species are the same. Kelly observed that a necessary and sufficient condition for the data to fit this model is that the expected value of $N + N^T$ is completely positive and since N is a 4×4 matrix, it is enough to check that it is doubly nonnegative (see Theorem 3.35 in Section 3.3).

Data Clustering. A clustering problem is the problem of partitioning a given data into k clusters based on some similarity measures of the data points. In the "hard" version, the input consists of n points $\mathbf{x}_1, \ldots, \mathbf{x}_n \in \mathbb{R}^d$, which we wish to assign to k mutually disjoint clusters $C_1, \ldots, C_k$. That is, with each $\mathbf{x}_i$ we wish to associate a label $y_i \in \{1, \ldots, k\}$. In a probabilistic ("soft") version, the goal is to assign probability $\text{Prob}(y_i = j|\mathbf{x}_i)$ to a point $\mathbf{x}_i$ being assigned to cluster C_j.

Suppose we are given pairwise "affinity" functions $\kappa(\mathbf{x}_i, \mathbf{x}_j)$, interpreted as the probability that the pair of points belong to the same cluster. These assign a high values for pairs of points likely to be in the same cluster (e.g., $\kappa(\mathbf{x}_i, \mathbf{x}_j)$ may be a distance-based function such as $\exp\left(-\|\mathbf{x}_i - \mathbf{x}_j\|^2/\sigma\right)$). Suppose the matrix K with $K_{ij} = \kappa(\mathbf{x}_i, \mathbf{x}_j)$ is symmetric positive semidefinite. Then K is the Gram matrix of $\phi(\mathbf{x}_i)$, $i = 1, \ldots, n$, viewed as high dimension representations of $\mathbf{x}_i$, $i = 1, \ldots, n$. A crucial ingredient for a good clustering outcome is the quality of this affinity matrix K. In [Ding *et al.* (2005)] and [Zass and Shashua (2005)] it is demonstrated that a key principal in known clustering methods such as K-means (for both the "hard" and probabilistic clustering) is the approximate complete positivity of K. In these methods, clustering amounts to finding $H \geq 0$ such that $\|K - HH^T\|^2$ is minimal (for hard clustering one requires also that $H^T H = I$).

Machine learning. Online services, such as sites trying to match job seekers and companies, or dating sites trying to match people, recommend preferred matches to the users. The service provides recommendations of potentially preferred partners to new/unobserved users, based on the observed information. The quality of the service depends on the accuracy of these prediction. This is a *reciprocal recommendation* problem, which can be presented as edge prediction of random graphs, where the graphs are bipartite graphs, in which the two independent vertex sets represent the two groups (job seekere/companies or male/female) and a directed edge from one set to the other represents preference. In such recommendation problems, it is often assumed that similar people have similar preferences. One approach is to use various similarity measures based on the topological structure of the graph, via a similarity matrix derived from the similarity measure. Another approach uses statistical models. In [Kanamori and Osugi (2019)] it is shown that most of the similarity matrices used in recommendation problems are completely positive. The complete positivity provides a statistical interpretation of these similarity-based methods.

Separability of diagonal symmetric quantum states. Quantum entanglement is a feature of quantum physics, where the quantum states of (groups of) particles, possibly very distant from each other, are correlated.

(The quantum state is a probability distribution for each possible measurement on the system of particles.) This phenomenon of quantum physics is important in quantum information. In [Tura *et al.* (2018)] a small class of the symmetric states, called Mixtures of Dicke states or Diagonal Symmetric states, is considered. This class arises naturally in certain dissipative systems. Specifically, [Tura *et al.* (2018)] consider states ρ acting on $\mathbb{C}^d \times \mathbb{C}^d$, of the form (in quantum physics notations)

$$\rho = \sum_{0 \leq i \leq j < d} p_{ij} \left| D_{ij} \right\rangle \left\langle D_{ij} \right|,$$

where each $|D_{ij}\rangle$ is a symmetric (Dicke) state (density operator/matrix), $\langle D_{ij}|$ is its Hermitian adjoint, and $\sum_{0 \leq i \leq j < d} p_{ij} = 1$, with $p_{ij} \geq 0$ for every i, j. The state ρ is separable (i.e., not entangled) if it has a decomposition as a convex combination of states of the form $\rho^1 \otimes \rho^2$, where ρ^1, ρ^2 are quantum states, each acting on $\mathbb{C}^d$ (each a positive semidefinite operator of trace 1). It is shown in [Tura *et al.* (2018)] that separability of ρ above is equivalent to complete positivity of the $d \times d$ matrix

$$M(\rho) = \begin{pmatrix} p_{00} & \frac{1}{2}p_{01} & \cdots & \frac{1}{2}p_{0\,d-1} \\ \frac{1}{2}p_{01} & p_{11} & \cdots & \frac{1}{2}p_{1\,d-1} \\ \vdots & \vdots & \ddots & \vdots \\ \frac{1}{2}p_{0\,d-1} & \frac{1}{2}p_{0\,d-1} & \cdots & p_{d-1\,d-1} \end{pmatrix}.$$

Results on complete positivity may then be used to identify such separable states.

Mathematical Optimization. As mentioned in the notes for Section 2.1, many nonconvex and combinatorial optimization problems may be reformulated as linear problems in matrix variables, over the copositive cone or the completely positive one. The dual program of the copositive program

$$\begin{aligned} &\min \langle C, X \rangle \\ &\text{s.t. } \langle A_i, X \rangle = b_i, \;\; i = 1, \ldots, m \\ &X \in \mathcal{COP}_n \end{aligned}$$

(see page 87) is the following linear program in matrix variable, over the completely positive cone.

$$\begin{aligned} &\max \textstyle\sum_{i=1}^m b_i y_i \\ &\text{s.t. } X - \textstyle\sum_{i=1}^m y_i A_i \in \mathcal{CP}_n, \;\; y_i \in \mathbb{R}. \end{aligned}$$

Some common conditions on the constraints guarantee the equality of the optimal values of these dual programs. In [Burer (2009)] it was shown that every optimization problem with quadratic objective function, linear constraints and some binary variables can be written as a linear problem over the completely positive cone. For example, the problem

$$\begin{aligned}&\min \mathbf{x}^T Q\mathbf{x}\\&\text{s.t. } \mathbf{e}^T\mathbf{x} = 1\,, \quad \mathbf{x} \in \mathbb{R}^n_+\end{aligned}$$

is equivalent to the problem

$$\begin{aligned}&\min\langle X, Q\rangle\\&\text{s.t. } \langle J, X\rangle = 1\,, \quad X \in \mathcal{CP}_n\,.\end{aligned}$$

Of course, this shifts the difficulty of the problem to the cone. As mentioned above, checking that a matrix is completely positive is NP-hard. Again we refer to [Dür (2010)] and [Bomze (2012)] for details on co(mpletely)positive programming.

For applications in dynamical systems and control see [Berman *et al.* (2012)].

We mention that the notion of complete positivity was also extended to tensors, see [Qi *et al.* (2014)], but we do not go into the theory of completely positive tensors here. Also, the notion of completely positive matrices was generalized to consider matrices over other algebraic structures (instead of the reals). But in this book we consider almost exclusively completely positive matrices over the reals. We discuss briefly, in Sections 3.11 and 4.9, complete positivity over the rationals or the integers,

3.2 Cones of completely positive matrices

Let $\mathcal{CP}_n$ denote the set of $n \times n$ completely positive matrices. In this section we observe that $\mathcal{CP}_n$ is a full cone, and that the cone $\mathcal{CP}_n$ and the cone $\mathcal{COP}_n$ of copositive matrices defined in Section 2.3 are duals of each other with respect to the inner product $\langle A, B\rangle = \operatorname{trace} AB$ on the space $\mathcal{S}_n$ of symmetric $n \times n$ matrices. We also describe the interior of this cone and its extreme rays. For a discussion of the faces of this cone see Chapter 5.

Theorem 3.25. *$\mathcal{CP}_n$ is a full cone.*

Proof. The sum of completely positive matrices is completely positive by Corollary 3.13. Let $A = BB^T$, $B \geq 0$ and let $t \geq 0$. Then

$$tA = (\sqrt{t}B)(\sqrt{t}B)^T,$$

so tA is also completely positive. Hence $\mathcal{CP}_n$ is a convex cone in $\mathcal{S}_n$.

To prove that $\mathcal{CP}_n$ is closed, suppose $\{A_i\}_{i=1}^{\infty}$ is a sequence of matrices in $\mathcal{CP}_n$ converging to A. We show that $A \in \mathcal{CP}_n$. By Carathéodory's Theorem, each matrix in $\mathcal{CP}_n$ is a sum of at most $d = n(n+1)/2$ rank 1 completely positive matrices. Hence each A_i is the Gram matrix of n vectors in $\mathbb{R}^d_+$ (some of which may be zero), $\mathbf{v}_1^{(i)}, \ldots, \mathbf{v}_n^{(i)}$. For every j, the sequence $\left\|\mathbf{v}_j^{(i)}\right\|^2 = (A_i)_{jj}$ converges to a_{jj} as i tends to infinity, hence it is a bounded sequence. There is therefore a converging subsequence of $\mathbf{v}_n^{(1)}$. The corresponding subsequence of $\mathbf{v}_n^{(2)}$ has a converging subsequence, and continuing similarly we get that there exists a sequence of indices $\{i_k\}$ such that $\lim \mathbf{v}_1^{(i_k)} = \mathbf{v}_1 \geq \mathbf{0}$, $\lim \mathbf{v}_2^{(i_k)} = \mathbf{v}_2 \geq \mathbf{0}, \ldots, \lim \mathbf{v}_n^{(i_k)} = \mathbf{v}_n \geq \mathbf{0}$. The limit of the subsequence A_{i_k} is $A = \mathrm{Gram}(\mathbf{v}_1, \ldots, \mathbf{v}_n)$, i.e., A is completely positive.

To prove that the cone $\mathcal{CP}_n$ is solid it suffices, by Proposition 1.152, to show that $\mathcal{CP}_n - \mathcal{CP}_n = \mathcal{S}_n$. Indeed, $E_{ii} = \mathbf{e}_i\mathbf{e}_i^T$ for every $1 \leq i \leq n$, and $E_{ij} = \frac{1}{2}\left((\mathbf{e}_i + \mathbf{e}_j)(\mathbf{e}_i + \mathbf{e}_j)^T - \mathbf{e}_i\mathbf{e}_i^T - \mathbf{e}_j\mathbf{e}_j^T\right)$ for every $1 \leq i < j \leq n$. As these matrices span $\mathcal{S}_n$, $\mathcal{CP}_n - \mathcal{CP}_n = \mathcal{S}_n$. □

Remark 3.26. By Proposition 3.11, Proposition 1.130, and the extremality of rank 1 matrices in $\mathcal{PSD}_n$, it is easy to see that the extreme rays of $\mathcal{CP}_n$ are the rays generated by rank 1 completely positive matrices — matrices of the form $\mathbf{x}\mathbf{x}^T$, where $\mathbf{x}$ is a nonnegative vector.

Remark 3.27. By Proposition 3.21, the transformations $A \mapsto P^TAP$ and $A \mapsto DAD$, where P is a permutation matrix and D is a positive diagonal matrix, preserve complete positivity. In fact, it can be shown that all the automorphisms of the cone $\mathcal{CP}_n$ are generated by these two types of transformations — see Section 5.2.

It is also easy to see that $A \in \mathcal{CP}_n$ generates an extreme ray of $\mathcal{CP}_n$ if and only if P^TAP does, and if and only if DAD does.

Theorem 3.28. *$\mathcal{CP}_n$ and $\mathcal{COP}_n$ are dual cones in the space $\mathcal{S}_n$. That is, $\mathcal{CP}_n^* = \mathcal{COP}_n$ and $\mathcal{COP}_n^* = \mathcal{CP}_n$.*

Proof. Let M be an $n \times n$ symmetric matrix. Then $M \in \mathcal{CP}_n^*$ if and only if $\mathrm{trace}\,(MA) \geq 0$ for every $n \times n$ completely positive matrix A; That is, if and only if $\mathrm{trace}\,(MBB^T) \geq 0$ for every nonnegative matrix B with n rows; If and only if $\mathrm{trace}\,(B^TMB) \geq 0$ for every nonnegative matrix B with n rows; If and only if $\mathbf{b}^TM\mathbf{b} \geq 0$ for every $\mathbf{b} \in \mathbb{R}^n_+$, i.e., M is copositive. The dual claim that $\mathcal{COP}_n^* = \mathcal{CP}_n$ follows from Theorem 1.147, since both cones are closed. □

We now take a closer look at the interior of $\mathcal{CP}_n$. As $\mathcal{CP}_n \subseteq \mathcal{DNN}_n$, every matrix in the interior of $\mathcal{CP}_n$ is necessarily positive definite and positive. However, these conditions are not sufficient. The following theorem contains several descriptions of $\text{int}\mathcal{CP}_n$.

Theorem 3.29. *Let $A \in \mathcal{CP}_n$. Then the following are equivalent.*

(a) *The matrix A is in the interior of $\mathcal{CP}_n$.*
(b) *There exists $B \geq 0$ with a positive column such that* $\text{rank}\, B = n$ *and* $A = BB^T$.
(c) *There exists $B > 0$ such that* $\text{rank}\, B = n$ *and* $A = BB^T$.
(d) *There exists $B = \begin{pmatrix} C & D \end{pmatrix}$, where C is an $n \times n$ positive nonsingular matrix and $D \geq 0$, such that $A = BB^T$.*

Proof. Let A be in the interior of $\mathcal{CP}_n$. As $\text{rank}\, A = n$, any matrix $B \geq 0$ such that $A = BB^T$ has rank n. Let $\mathbf{b}$ be a positive vector in $\mathbb{R}^n$ of sufficiently small norm, then $A - \mathbf{b}\mathbf{b}^T$ is completely positive, so

$$A - \mathbf{b}\mathbf{b}^T = CC^T, \quad C \geq 0,$$

and therefore

$$A = CC^T + \mathbf{b}\mathbf{b}^T = \begin{pmatrix} C & \mathbf{b} \end{pmatrix} \begin{pmatrix} C & \mathbf{b} \end{pmatrix}^T$$

is a CP factorization as described in (b). By Exercise 3.3 we may replace $\mathbf{b}$ and one column of C by two positive columns, and we may repeat this process until all columns of $\begin{pmatrix} C & \mathbf{b} \end{pmatrix}$ are replaced by positive columns, getting a CP factorization as described in (c). Clearly, (c) implies (d).

It remains to show that (d) implies (a). Indeed, suppose $A = BB^T$, where $B = \begin{pmatrix} \mathbf{b}_1 & \dots & \mathbf{b}_m \end{pmatrix} \geq 0$ is an $n \times m$ matrix, $\text{rank}\, B = n$ and $\mathbf{b}_1 > \mathbf{0}$. Suppose on the contrary that A lies on the boundary of $\mathcal{CP}_n$. Then there exists a nonzero copositive matrix M such that $\langle A, M \rangle = 0$ (see Exercise 1.59). That is $\sum_{i=1}^m \mathbf{b}_i^T M \mathbf{b}_i = 0$, and therefore $\mathbf{b}_1, \dots, \mathbf{b}_m$ are all zeros of M. Since $\mathbf{b}_1 > 0$, M is positive semidefinite (Proposition 2.47). Hence the zeros of M are all in the null space of M. As the zeros $\mathbf{b}_1, \dots, \mathbf{b}_m$ span the whole $\mathbb{R}^n$, $M = 0$. A contradiction. □

Remark 3.30. If A has a known CP factorization $A = BB^T$ with at least one positive column, one may easily find alternative representations in either of the forms described in parts (b), (c) or (d) of the theorem — by replacing two columns at a time, one of them positive, as in Proposition 3.9.

Checking whether a matrix is in the interior of $\mathcal{CP}_n$ requires finding a CP factorization BB^T, where B has full row rank and at least one positive column. Like checking complete positivity, this is generally not easy. For example, let

$$B = \frac{1}{\sqrt{3}} \begin{pmatrix} 1 & 0 & 0 & 1 & 1 & 1 & 0 & 1 & 1 & 0 \\ 1 & 1 & 0 & 0 & 1 & 0 & 1 & 0 & 1 & 1 \\ 1 & 1 & 1 & 0 & 0 & 1 & 0 & 1 & 0 & 1 \\ 0 & 1 & 1 & 1 & 0 & 1 & 1 & 0 & 1 & 0 \\ 0 & 0 & 1 & 1 & 1 & 0 & 1 & 1 & 0 & 1 \end{pmatrix}.$$

Then

$$A = BB^T = \begin{pmatrix} 2 & 1 & 1 & 1 & 1 \\ 1 & 2 & 1 & 1 & 1 \\ 1 & 1 & 2 & 1 & 1 \\ 1 & 1 & 1 & 2 & 1 \\ 1 & 1 & 1 & 1 & 2 \end{pmatrix}$$

is obviously completely positive. But is it in the interior? In this particular case the answer is positive. The following, slightly more general, example of a matrix in the interior of $\mathcal{CP}_n$ turns out to be very useful.

Proposition 3.31. *Let* $\mathbf{b}$ *be a positive vector in* $\mathbb{R}^n$, *and let* $\varepsilon > 0$. *Then* $\mathbf{b}\mathbf{b}^T + \varepsilon I$ *is in the interior of* $\mathcal{CP}_n$.

Proof. This follows from the equivalence of (a) and (b) in Theorem 3.29, as

$$\mathbf{b}\mathbf{b}^T + \varepsilon I = \mathbf{b}\mathbf{b}^T + \sum_{i=1}^{n} (\sqrt{\varepsilon}\mathbf{e}_i)(\sqrt{\varepsilon}\mathbf{e}_i)^T$$

is a CP representation of the nonsingular matrix $\mathbf{b}\mathbf{b}^T + \varepsilon I$, in which one of the rank 1 summands is positive. □

Remark 3.32. The CP representation used in the proof has $n+1$ summands. Note, however, that there are other factorizations with one less summand. Indeed, if $\mathbf{x} \in \mathbb{R}^n$ is a positive vector and $A = I + \mathbf{x}\mathbf{x}^T$, then the square root of A is $B = I + \mu\mathbf{x}\mathbf{x}^T$, where

$$\mu = \frac{\sqrt{\mathbf{x}^T\mathbf{x} + 1} - 1}{\mathbf{x}^T\mathbf{x}} > 0.$$

Hence $A = B^2 = BB^T$ is a CP factorization in which B has n positive columns.

A matrix $A \in \mathcal{CP}_n$ is on the boundary of $\mathcal{CP}_n$ if and only if it is orthogonal to a matrix $M \in \mathcal{COP}_n$ (and then M is necessarily on the boundary of $\mathcal{COP}_n$). We make an observation regarding such a pair of orthogonal matrices.

Proposition 3.33. *Let $A \in \mathcal{CP}_n$ and $M \in \mathcal{COP}_n$ be orthogonal to each other. Then for every $1 \le i \le n$ the i-th column of A is orthogonal to the i-th column of M.*

Proof. Let $A = \sum_{j=1}^m \mathbf{b}_j \mathbf{b}_j^T$ be a CP representation of A. Let $\mathbf{x} = A\mathbf{e}_l$ be the l-th column of A, and $\mathbf{y} = M\mathbf{e}_i$ be the i-th column of M. Then

$$\mathbf{x}^T\mathbf{y} = \mathbf{e}_l^T A M \mathbf{e}_i = \sum_{j=1}^m \mathbf{e}_l^T \mathbf{b}_j \mathbf{b}_j^T M \mathbf{e}_i = \sum_{\substack{j \text{ s.t.} \\ l \in \operatorname{supp} \mathbf{b}_j}} \mathbf{e}_l^T \mathbf{b}_j (M\mathbf{b}_j)_i. \tag{3.2}$$

In particular, for $l = i$, the scalar product of the i-th column of M and the corresponding column of A is

$$\mathbf{e}_i^T A M \mathbf{e}_i = \sum_{\substack{j \text{ s.t.} \\ i \in \operatorname{supp} \mathbf{b}_j}} \mathbf{e}_i^T \mathbf{b}_j (M\mathbf{b}_j)_i = 0,$$

as each $\mathbf{b}_j$ is a zero of the copositive matrix M, and thus $(M\mathbf{b}_j)_i = 0$ whenever $i \in \operatorname{supp} \mathbf{b}_j$ (by Proposition 2.45). □

Remark 3.34. In Section 3.1 we observed that a necessary condition for A to be completely positive is that it is doubly nonnegative. This means that $\mathcal{DNN}_n = \mathcal{N}_n \cap \mathcal{PSD}_n$ contains $\mathcal{CP}_n$. Since $\mathcal{N}_n$ and $\mathcal{PSD}_n$ are both self dual, and in view of Theorem 1.146(e), this is consistent with the fact that $\mathcal{DNN}_n^* = \mathcal{SPN}_n = \mathcal{N}_n + \mathcal{PSD}_n \subseteq \mathcal{COP}_n$ (see Remark 2.32). For $n \le 4$, $\mathcal{DNN}_n$ and $\mathcal{CP}_n$ have the same extreme rays (see Proposition 1.170 and Remark 3.26). The fact that $\mathcal{SPN}_5 \ne \mathcal{COP}_5$ implies that $\mathcal{DNN}_5 \ne \mathcal{CP}_5$. We will see later specific examples of 5×5 doubly nonnegative matrices which are not completely positive.

Exercises

3.13. Prove the statement in Remark 3.26.

3.14. Prove the final statements in Remark 3.27, regarding the effect of diagonal scaling and permutation similarity on the extremality of a matrix.

3.15. Use Theorem 3.29 to show that if $A \in \operatorname{int}\mathcal{CP}_n$, and B is an $n \times n$ completely positive matrix, then $A + B \in \operatorname{int}\mathcal{CP}_n$.

3.16. Let A be an $n \times n$ symmetric matrix. Prove that the following are equivalent

(a) $A \in \mathcal{CP}_n$.
(b) $K \circ A \in \mathcal{COP}_n$ for every $K \in \mathcal{COP}_n$.
(c) $K \circ A \in \mathcal{COP}_n$ for every extreme copositive matrix $K \in \mathcal{COP}_n$.
(d) $K \circ A \in \mathcal{COP}_n$ for every extreme copositive matrix $K \in \mathcal{COP}_n$ whose diagonal entries are either zero or one.

Hint: For (a)⇒(b) use the fact that $K \circ \mathbf{u}\mathbf{u}^T = DKD$, where $D = \mathrm{Diag}(\mathbf{u})$. For (d)⇒(a) note that if $K = DK'D$, where D is a positive diagonal matrix, then $K \circ A = D(K' \circ A)D$.

3.17. Let A be an $n \times n$ doubly nonnegative matrix. Prove that the following are equivalent

(a) $A \in \mathcal{CP}_n$.
(b) $K \circ A \in \mathcal{COP}_n$ for every $K \in \mathcal{COP}_n \setminus \mathcal{SPN}_n$.
(c) $K \circ A \in \mathcal{COP}_n$ for every extreme copositive matrix $K \in \mathcal{COP}_n \setminus \mathcal{SPN}_n$.
(d) $K \circ A \in \mathcal{COP}_n$ for every extreme copositive matrix $K \in \mathcal{COP}_n \setminus \mathcal{SPN}_n$ whose diagonal entries are either zero or one.

3.18. Let $A \in \mathcal{CP}_n$ and $M \in \mathcal{COP}_n$ be orthogonal to each other, and let $A = \sum_{j=1}^{m} \mathbf{b}_j \mathbf{b}_j^T$ be a CP representation of A. Show that if $i \in \operatorname{supp} \mathbf{b}_j$ for every $1 \le j \le m$, then the i-th column of M is in the null space of A.
Hint: Let $\mathbf{z} = M\mathbf{e}_i$ and use (3.2) to show that the $(A\mathbf{z})_l = 0$ for every l.

3.19. Try to find an example of a 5×5 matrix which is doubly nonnegative but not completely positive.

Notes. Properties of the cone $\mathcal{CP}_n$, its extreme rays and its dual were first studied in [Hall and Newman (1963)]. The first description of the interior of $\mathcal{CP}_n$, however, is due to [Dür and Still (2008)]. It was then reexamined in [Dickinson (2010)]. Proposition 3.29 combines results of these two papers. Remark 3.32 is based on [Shaked-Monderer (2009)], see also [Shaked-Monderer *et al.* (2013)]. Exercises 3.16 and 3.17 are based on [Berman *et al.* (2016)]. Proposition 3.33 and Exercise 3.18 are due to [Shaked-Monderer *et al.* (2015)].

3.3 Small completely positive matrices

For small matrices the necessary condition that A is doubly nonnegative is also sufficient for A to be completely positive. How small is small? At most 4×4.

Theorem 3.35. *If A is a doubly nonnegative matrix of order n, $n \leq 4$, then there exists a nonnegative matrix B such that $A = BB^T$.*

This theorem follows from the dualities $\mathcal{DNN}_n = \mathcal{SPN}_n^*$ and $\mathcal{CP}_n = \mathcal{COP}_n^*$, and the fact that $\mathcal{COP}_n = \mathcal{SPN}_n$ for $n \leq 4$ (see Theorem 2.151). In this section we provide a direct proof (which in turn could be used to give an alternative proof of Theorem 2.151). The following lemma will be used in this direct proof, and also later on.

Lemma 3.36. *Let A and B be $n \times n$ matrices. Suppose A is completely positive, B is positive semidefinite and all its entries are zero except for a 2×2 principal submatrix, and $A + B$ is nonnegative. Then $A + B$ is completely positive.*

Proof. We may assume that the nonzero 2×2 submatrix of B is $B[1, 2]$. We may also assume that B is not diagonal, otherwise B is completely positive, and so is $A + B$. So $B = \mathbf{b}\mathbf{b}^T + \mathbf{c}\mathbf{c}^T$, where $\operatorname{supp} \mathbf{b} = \{1, 2\}$ and $\operatorname{supp} \mathbf{c} \subseteq \{1\}$ (by the same argument as in Example 3.7). Since $\mathbf{c}\mathbf{c}^T$ is completely positive, it suffices to show that $A + \mathbf{b}\mathbf{b}^T$ is completely positive. If $\mathbf{b} \geq \mathbf{0}$ this is clear, so we consider the case that $b_1 b_2 < 0$.

Let

$$A = \sum_{i=1}^{m} \mathbf{b}_i \mathbf{b}_i^T$$

be a CP representation of A. Let

$$\mathbf{d}_i = \sqrt{\frac{(\mathbf{b}_i)_1 (\mathbf{b}_i)_2}{a_{12}}}\, \mathbf{b}, \quad i = 1, \ldots, m.$$

Then for each i, $\mathbf{b}_i \mathbf{b}_i^T + \mathbf{d}_i \mathbf{d}_i^T \geq 0$, since

$$\begin{aligned} (\mathbf{b}_i)_1 (\mathbf{b}_i)_2 + (\mathbf{d}_i)_1 (\mathbf{d}_i)_2 &= (\mathbf{b}_i)_1 (\mathbf{b}_i)_2 + \frac{(\mathbf{b}_i)_1 (\mathbf{b}_i)_2}{a_{12}} b_1 b_2 \\ &= (\mathbf{b}_i)_1 (\mathbf{b}_i)_2 \left(1 + \frac{b_1 b_2}{a_{12}}\right), \end{aligned}$$

and $a_{12}+b_1b_2 = (A+B)_{12} \geq 0$. Hence $\mathbf{b}_i\mathbf{b}_i^T+\mathbf{d}_i\mathbf{d}_i^T$ is a doubly nonnegative matrix of rank ≤ 2, and is therefore completely positive. Now

$$\sum_{i=1}^{m} \mathbf{d}_i\mathbf{d}_i^T = \frac{1}{a_{12}}\left(\sum_{i=1}^{m}(\mathbf{b}_i)_1(\mathbf{b}_i)_2\right)\mathbf{b}\mathbf{b}^T = \mathbf{b}\mathbf{b}^T,$$

so $A + \mathbf{b}\mathbf{b}^T = \sum_{i=1}^{m}(\mathbf{b}_i\mathbf{b}_i^T + \mathbf{d}_i\mathbf{d}_i^T)$. Hence $A + \mathbf{b}\mathbf{b}^T$ is completely positive, as a sum of m completely positive matrices. □

Proof of Theorem 3.35. The claim is clear for $n \leq 2$, by Examples 3.4 and 3.7.

Now for $n = 3$. Let A be a 3×3 doubly nonnegative matrix. If $a_{ii} = 0$, then the i-th row and the i-th column of A are equal to zero, and the result follows from the case $n \leq 2$. Thus we may assume that the diagonal entries of A are all positive. By choosing $D = \mathrm{Diag}(1/\sqrt{a_{11}}, 1/\sqrt{a_{22}}, 1/\sqrt{a_{33}})$ in Proposition 3.21(b), we see that we may assume that the diagonal entries of A are equal to 1. By Proposition 3.21(a), we may permute the rows and columns of A simultaneously. So assume that

$$A = \begin{pmatrix} 1 & a_{12} & a_{13} \\ a_{12} & 1 & a_{23} \\ a_{13} & a_{23} & 1 \end{pmatrix}$$

with

$$a_{12} \geq a_{13}. \tag{3.3}$$

We have to show that if A is doubly nonnegative, then it is completely positive. We do it by writing A as a sum of two matrices:

$$A = \mathbf{b}\mathbf{b}^T + \begin{pmatrix} \begin{matrix} & C & \end{matrix} & \begin{matrix} 0 \\ 0 \end{matrix} \\ \begin{matrix} 0 & 0 \end{matrix} & 0 \end{pmatrix},$$

where

$$\mathbf{b} = \begin{pmatrix} a_{13} \\ a_{23} \\ 1 \end{pmatrix} \quad \text{and } C = \begin{pmatrix} 1 - a_{13}^2 & a_{12} - a_{13}a_{23} \\ a_{12} - a_{13}a_{23} & 1 - a_{23}^2 \end{pmatrix}.$$

Since $\mathbf{b}\mathbf{b}^T$ is obviously completely positive, we complete the proof by showing that C is completely positive. It is positive semidefinite, being the Schur complement $A/[3]$. It is nonnegative by (3.3), and the fact that $1 \geq a_{13}, a_{23}$ since A is positive semidefinite. Thus C is a 2×2 doubly nonnegative matrix and is therefore completely positive.

For $n = 4$: Let A be a 4×4 doubly nonnegative matrix. We have to show that A is completely positive. If $a_{ii} = 0$, then the i-th row and the i-th column of A are zero, and the problem reduces to the case $n = 3$. If the diagonal entries of A are all positive, we may assume, by Proposition 3.21, that they are all equal to 1, and that a_{34} is the smallest entry of A. Let

$$S = \begin{pmatrix} 1 & 0 & 0 & 0 \\ 0 & 1 & 0 & 0 \\ 0 & 0 & 1 & a_{34} \\ 0 & 0 & 0 & 1 \end{pmatrix}.$$

The matrix S is nonnegative, so by Proposition 3.16 it suffices to prove that the matrix $B = S^{-1}A(S^{-1})^T$ is completely positive.

The matrix B is equal to

$$\begin{pmatrix} 1 & a_{12} & a_{13} - a_{14}a_{34} & a_{14} \\ a_{12} & 1 & a_{23} - a_{24}a_{34} & a_{24} \\ a_{13} - a_{14}a_{34} & a_{23} - a_{24}a_{34} & 1 - a_{34}^2 & 0 \\ a_{14} & a_{24} & 0 & 1 \end{pmatrix}.$$

It is positive semidefinite because A is. The nonnegativity of B follows from the positive semidefiniteness of A, which implies that $a_{ij} \leq 1$ for every i, j, and the fact that a_{34} is the smallest entry of A. If $a_{34} = 1$, the problem reduces to the case $n = 3$. If $a_{34} < 1$, we can divide the third row and the third column of B by $\sqrt{1 - a_{34}^2}$ to obtain a doubly nonnegative matrix C with all diagonal entries equal to 1, and, again by Proposition 3.21, it suffices to prove that C is completely positive.

So let

$$C = \begin{pmatrix} 1 & c_{12} & c_{13} & c_{14} \\ c_{12} & 1 & c_{23} & c_{24} \\ c_{13} & c_{23} & 1 & 0 \\ c_{14} & c_{24} & 0 & 1 \end{pmatrix}$$

be doubly nonnegative. To show that C is completely positive, we split C as $C = EE^T + F$, where

$$E = \begin{pmatrix} c_{13} & c_{14} \\ c_{23} & c_{24} \\ 1 & 0 \\ 0 & 1 \end{pmatrix}.$$

Then

$$F = \begin{pmatrix} f_{11} & f_{12} \\ f_{12} & f_{22} \end{pmatrix} \oplus 0_2,$$

where

$$\begin{pmatrix} f_{11} & f_{12} \\ f_{12} & f_{22} \end{pmatrix} = C/C[3,4].$$

F is therefore positive semidefinite, and EE^T completely positive. By Lemma 3.36, C is also completely positive. □

Remark 3.37. In fact, for each doubly nonnegative matrix A of order $n \leq 4$ there exists a *square* nonnegative matrix B such that $A = BB^T$. For $n \leq 3$ this follows easily from the proof of Theorem 3.35. The proof for $n = 4$ will be given in Section 4.1 (Theorem 4.10).

Theorem 3.35 can also be proved using the geometric approach. This was done for $n = 2$ in Theorem 3.5. We now demonstrate this approach in the case $n = 3$ and $n = 4$.

Geometric proof for $n = 3$. Let $\mathbf{v}_1, \mathbf{v}_2, \mathbf{v}_3$ be three vectors in $\mathbb{R}^3$ such that $\langle \mathbf{v}_i, \mathbf{v}_j \rangle \geq 0$ for $i, j = 1, 2, 3$. We may assume that all three vectors are nonzero, otherwise the problem is reduced to the case $n = 2$. We will show that there is an isometry $T : \mathbb{R}^3 \to \mathbb{R}^3$ such that $T\mathbf{v}_1, T\mathbf{v}_2$ and $T\mathbf{v}_3$ lie in $\mathbb{R}^3_+$. If the three vectors are in the same plane, then, as in the proof of Theorem 3.5, all three vectors can be embedded in $\mathbb{R}^2_+$, and thus in $\mathbb{R}^3_+$. If $\mathbf{v}_1, \mathbf{v}_2$ and $\mathbf{v}_3$ are linearly independent, let $S = \operatorname{span}\{\mathbf{v}_1, \mathbf{v}_2\}$ and project $\mathbf{v}_3$ onto S to get $\mathbf{v}_3 = \mathbf{v}_3' + \mathbf{v}_3''$, where $\mathbf{v}_3' \in S$ and $\mathbf{v}_3'' \in S^\perp$. The vectors $\mathbf{v}_1, \mathbf{v}_2$ and $\mathbf{v}_3'$ can be embedded in $\mathbb{R}^2_+$: We may assume that the widest angle between pairs of these vectors is formed between $\mathbf{v}_1$ and one of the other vectors. Then, by applying the Gram-Schmidt procedure, we obtain an orthonormal basis $E = (\mathbf{b}_1, \mathbf{b}_2)$ of S such that the coordinate vectors $[\mathbf{v}_1]_E, [\mathbf{v}_2]_E$ and $[\mathbf{v}_3']_E$ are nonnegative, and $\mathbf{b}_1 = \mathbf{v}_1/\|\mathbf{v}_1\|$. Let $E' = (\mathbf{b}_1, \mathbf{b}_2, \mathbf{b}_3)$, where $\mathbf{b}_3 = \mathbf{v}_3''/\|\mathbf{v}_3''\|$. Then E' is an orthonormal basis of $\mathbb{R}^3$ satisfying $[\mathbf{v}_i]_{E'} \geq 0$ for $i = 1, 2, 3$, which is what we were looking for. □

Geometric proof for $n = 4$. As in the algebraic proof, the first step is to show that it is suffices to consider matrices with at least one zero entry. The geometric reasoning is this: Suppose A is a positive 4×4 doubly nonnegative matrix, and let $S = \{\mathbf{v}_1, \mathbf{v}_2, \mathbf{v}_3, \mathbf{v}_4\}$ be a set of vectors such that $A = \operatorname{Gram}(S)$. By Proposition 3.21 we may assume that $\|\mathbf{v}_i\| = 1$. We also have $0 < \langle \mathbf{v}_i, \mathbf{v}_j \rangle$, $1 \leq i, j \leq 4$. Suppose the maximal angle between the vectors is the one between $\mathbf{v}_p$ and $\mathbf{v}_q$, and project $\mathbf{v}_p$ onto $\operatorname{span}\{\mathbf{v}_q\}$. Then $\mathbf{v}_p = \mathbf{v}_p' + \langle \mathbf{v}_p, \mathbf{v}_q \rangle \mathbf{v}_q$, where $\mathbf{v}_p'$ is orthogonal to $\mathbf{v}_q$. Let S' be the set obtained from S by replacing $\mathbf{v}_p$ by $\mathbf{v}'_p$, and let $A' = \mathrm{Gram}(S')$. If A' is completely positive, S' can be embedded in $\mathbb{R}^k_+$ for some k. And since $\mathbf{v}_p = \mathbf{v}'_p + \langle \mathbf{v}_p, \mathbf{v}_q \rangle \mathbf{v}_q$ and $\langle \mathbf{v}_p, \mathbf{v}_q \rangle \geq 0$, this implies that S can also be embedded in $\mathbb{R}^k_+$.

By Proposition 3.22, we may consider only irreducible matrices, that is, matrices whose graphs are connected. If A is a 4×4 irreducible doubly nonnegative matrix with a zero entry, its graph is one of the following.

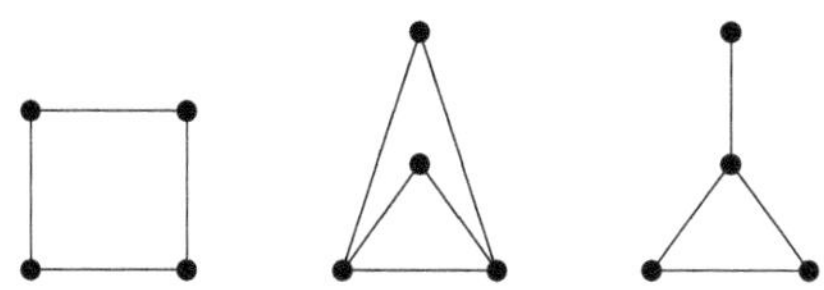

Doubly nonnegative matrices with these graphs will be treated in Section 3.3 (algebraically and geometrically), so we skip this part of the proof. The fact that such matrices are completely positive will follow from Theorems 3.54, 3.57, 3.55, and the case $n = 3$. □

Exercises

3.20. Show that if A is a 2×2 completely positive matrix, then there exist a nonnegative lower triangular matrix L such that $A = LL^T$ and a nonnegative upper triangular matrix U such that $A = UU^T$.

3.21. Let

$$A = \begin{pmatrix} 1 & a & b \\ a & 1 & c \\ b & c & 1 \end{pmatrix},$$

where $0 \leq a < 1$, $0 \leq b \leq 1$, $ab \leq c \leq 1$, and $\det A \geq 0$. Check that $A = BB^T$ where

$$B = \begin{pmatrix} 1 & 0 & 0 \\ a & \sqrt{1-a^2} & 0 \\ b & \frac{c-ab}{\sqrt{1-a^2}} & \frac{\sqrt{\det A}}{\sqrt{1-a^2}} \end{pmatrix}.$$

Explain how this proves that a 3×3 matrix is completely positive if and only if it is doubly nonnegative.

3.22. Show that for $n \leq 4$, an $n \times n$ completely positive matrix A with $\operatorname{rank} A = n$ is not critical (see definition in Exercise 3.11).

Notes. Theorem 3.35 is due to [Maxfield and Minc (1962/1963)]. The proof here is based on [Ando (1991)], but we make use of Lemma 3.36, which is due to [Barioli (2001a)]. A geometric proof is given in [Gray and Wilson (1980)]. The geometric proofs here are taken from this paper ($n = 3$) and [Kogan (1989)] ($n = 4$). As was mentioned in Section 2.9, the dual result, that for $n \leq 4$ every $n \times n$ copositive matrix can be written as a sum of a positive semidefinite matrix and a nonnegative one is proved in [Diananda (1962)]. Exercise 3.21 is based on the original proof of Maxfield and Minc.

3.4 Complete positivity and the comparison matrix

There are two main results in this section. First, if $A \in \mathbb{R}^{n\times n}$ is nonnegative and symmetric and its comparison matrix is positive semidefinite, then A is completely positive. Second, this sufficient condition is also necessary when the graph of A contains no triangles.

We start with an easy-to-check sufficient condition for complete positivity.

Theorem 3.38. *Nonnegative symmetric diagonally dominant matrices are completely positive.*

Proof. Let A be a nonnegative symmetric diagonally dominant matrix. Let

$$a_i = a_{ii} - \sum_{\substack{j=1\\ j\neq i}}^{n} a_{ij},$$

then $a_i \geq 0$. Let $F_{ij} \in \mathbb{R}^{n\times n}$ have 1 in positions i,i, i,j, ji and j,j, and 0 elsewhere. Then

$$A = \sum_{1\leq j<i\leq n} a_{ij}F_{ij} + \mathrm{Diag}(a_1,\ldots,a_n)$$

so A is completely positive. □

Example 3.39. Let

$$\begin{pmatrix} 5 & 2 & 2 \\ 2 & 4 & 1 \\ 2 & 1 & 3 \end{pmatrix}.$$

Then

$$A = \begin{pmatrix} 2 & 2 & 0 \\ 2 & 2 & 0 \\ 0 & 0 & 0 \end{pmatrix} + \begin{pmatrix} 2 & 0 & 2 \\ 0 & 0 & 0 \\ 2 & 0 & 2 \end{pmatrix} + \begin{pmatrix} 0 & 0 & 0 \\ 0 & 1 & 1 \\ 0 & 1 & 1 \end{pmatrix} + \begin{pmatrix} 1 & 0 & 0 \\ 0 & 1 & 0 \\ 0 & 0 & 0 \end{pmatrix}.$$

Hence $A = BB^T$, where

$$B = \begin{pmatrix} \sqrt{2} & \sqrt{2} & 0 & 1 & 0 \\ \sqrt{2} & 0 & 1 & 0 & 1 \\ 0 & \sqrt{2} & 1 & 0 & 0 \end{pmatrix}.$$

Theorem 3.40. *If A is symmetric and nonnegative and its comparison matrix $M(A)$ is positive semidefinite, then A is completely positive.*

Proof. $M(A)$ is an M-matrix, so by Theorem 1.66 there exists a positive diagonal matrix D such that $DM(A)D$ is diagonally dominant. Since the entries of $DM(A)D$ and DAD are equal in absolute value, DAD is also diagonally dominant. By the previous theorem, there exists a nonnegative matrix B such that $DAD = BB^T$. But then $A = (D^{-1}B)(D^{-1}B)^T$ and since $D^{-1}B \geq 0$, A is completely positive. □

Example 3.41. The matrix

$$\begin{pmatrix} 1 & 1 & 0 & 0 & 1 \\ 1 & 2 & 1 & 0 & 0 \\ 0 & 1 & 3 & 1 & 0 \\ 0 & 0 & 1 & 4 & 1 \\ 1 & 0 & 0 & 1 & 5 \end{pmatrix}$$

is completely positive, since its comparison matrix,

$$\begin{pmatrix} 1 & -1 & 0 & 0 & -1 \\ -1 & 2 & -1 & 0 & 0 \\ 0 & -1 & 3 & -1 & 0 \\ 0 & 0 & -1 & 4 & -1 \\ -1 & 0 & 0 & -1 & 5 \end{pmatrix},$$

is positive semidefinite.

The sufficient condition in Theorem 3.40 is not necessary.

Example 3.42. J_3 is completely positive but

$$M(J_3) = \begin{pmatrix} 1 & -1 & -1 \\ -1 & 1 & -1 \\ -1 & -1 & 1 \end{pmatrix}$$

is not positive semidefinite.

Observe that the graph of J_3 is a triangle. This is not a coincidence, as the following theorem shows.

Theorem 3.43. *If A is completely positive and $G(A)$ is triangle free, then $M(A)$ is positive semidefinite.*

Proof. Let $A = \sum_{i=1}^{m} \mathbf{b}_i \mathbf{b}_i^T$ be a CP representation of A. The support of each $\mathbf{b}_i$ is a clique in $G(A)$, and since $G(A)$ contains no triangles, the size of each of these supports is at most 2. Change each nonnegative vector $\mathbf{b}_i$ of support 2 to a vector $\mathbf{d}_i$ by multiplying one of its two positive entries by -1. Let $\mathbf{d}_i = \mathbf{b}_i$ if $|\operatorname{supp} \mathbf{b}_i| = 1$. Then $M(A) = \sum_{i=1}^{m} \mathbf{d}_i \mathbf{d}_i^T$, so $M(A)$ is positive semidefinite. □

Combining Theorems 3.40 and 3.43 we obtain

Theorem 3.44. *If a graph G contains no triangles and A is a nonnegative symmetric matrix realization of G, then A is completely positive if and only if $M(A)$ is positive semidefinite.*

Example 3.45. The doubly nonnegative matrix

$$\begin{pmatrix} 1 & 1 & 0 & 0 & 1 \\ 1 & 2 & 1 & 0 & 0 \\ 0 & 1 & 2 & 1 & 0 \\ 0 & 0 & 1 & 2 & 1 \\ 1 & 0 & 0 & 1 & 3 \end{pmatrix}$$

is not completely positive, since its graph is the triangle free C_5, and its comparison matrix,

$$\begin{pmatrix} 1 & -1 & 0 & 0 & -1 \\ -1 & 2 & -1 & 0 & 0 \\ 0 & -1 & 2 & -1 & 0 \\ 0 & 0 & -1 & 2 & -1 \\ -1 & 0 & 0 & -1 & 3 \end{pmatrix},$$

is not positive semidefinite.

The property of G in Theorem 3.44 is, in fact, a characterization of being triangle free.

Theorem 3.46. *A graph G has the property*

for every nonnegative symmetric matrix realization A of G,

A is completely positive iff $M(A)$ is positive semidefinite

if and only if G is triangle free.

Proof. The "if part" is Theorem 3.44. For the "only if part", suppose G has this property, $V(G) = \{1, 2, \ldots, n\}$, and the subgraph of G induced by $\{1, 2, 3\}$ is a triangle. Define B and C in $\mathbb{R}^{n\times n}$ by

$$b_{ij} = \begin{cases} n-1\,, & \text{if } i = j \\ 1\,, & \text{if } i \neq j \text{ and } \{i,j\} \in E(G) \\ 0\,, & \text{if } i \neq j \text{ and } \{i,j\} \notin E(G) \end{cases}$$

and $C = (n-1)J_3 \oplus 0_{n-3}$. The matrix C is clearly completely positive. The matrix B, being diagonally dominant, is also completely positive. Hence $A = B + C$ is completely positive. But $M(A)$ is not positive semidefinite since its 3×3 left upper principal submatrix is

$$\begin{pmatrix} 2n-2 & -n & -n \\ -n & 2n-2 & -n \\ -n & -n & 2n-2 \end{pmatrix},$$

which has -2 as an eigenvalue. □

The proof of Theorem 3.43 relied on the sizes of the supports of the vectors in a CP representation of the matrix. It seems natural to make the following definition.

Definition 3.47. A CP representation $A = \sum_{i=1}^{m} \mathbf{b}_i \mathbf{b}_i^T$, $\mathbf{b}_i \geq 0$, is *of support k* if the size of the support of each $\mathbf{b}_i$ is at most k, and at least one $\mathbf{b}_i$ has support of size exactly k.

Theorem 3.43 is therefore a special case of the following theorem, whose proof is outlined in Exercises 3.26–3.29.

Theorem 3.48. *Let A be a completely positive matrix, which has a CP representation of support k. Let $A = D + F$, where $D = \mathrm{Diag}(a_{11}, \ldots, a_{nn})$. Then the matrix $M = (k-1)D - F$ is an M-matrix.*

In the case that $k = 2$ the converse is also true.

Proposition 3.49. *A completely positive matrix A has a CP representation of support 2 if and only if A is nondiagonal and $M(A)$ is positive semidefinite.*

The proof of this proposition is also left as an exercise.

Exercises

3.23. Show that the following statement is false: "For every positive semidefinite matrix A, there exists a positive diagonal matrix D such that DAD is diagonally dominant".

3.24. Show that if A is an $n \times n$ symmetric diagonally dominant matrix, and A is positive, then $A \in \text{int}\mathcal{CP}_n$. Deduce that every $n \times n$ positive symmetric matrix with a positive semidefinite comparison matrix is in the interior of $\mathcal{CP}_n$.
Hint: For a positive diagonally dominant A, there exists $\mu > 0$ such that $A - \mu J$ is strictly diagonally dominant. Use Theorem 3.29.

3.25. Let A be a completely positive matrix with a triangle free graph. Let C be a real symmetric matrix with nonnegative diagonal elements such that $|C| = A$. Prove that C is positive semidefinite.

3.26. Prove the following special case of Theorem 3.48. Let $B = \mathbf{b}\mathbf{b}^T$ be a rank 1 doubly nonnegative matrix. Let $B = D + F$ where $D = \text{Diag}(b_{11}, ..., b_{nn})$, and let $k = |\text{supp}\,\mathbf{b}|$. Then $(k-1)\mathbf{e}^T D\mathbf{e} \geq \mathbf{e}^T F\mathbf{e}$.

3.27. Deduce from Exercise 3.26: Under the assumptions of Theorem 3.48, $\mathbf{e}^T M\mathbf{e} \geq 0$.

3.28. Deduce from Exercise 3.27: Under the assumptions of Theorem 3.48, $\mathbf{x}^T M\mathbf{x} \geq 0$ for every positive vector $\mathbf{x}$.

3.29. Prove: If $M = sI - B$, where $B \geq 0$ is irreducible, and $\mathbf{x}^T M\mathbf{x} \geq 0$ for every positive vector $\mathbf{x}$, then M is an M-matrix. Use this general result to complete the proof of Theorem 3.48.

3.30. Prove Proposition 3.49.

3.31. Let X be a symmetric matrix such that $\text{diag}(X) = \mathbf{e}$ and $G(X)$ is triangle free. Prove that X is completely positive if and only if $\rho(X) \leq 2$.
Hint: $M(X) = 2I - X$.

3.32. Show that the completely matrix

$$\begin{pmatrix} 2 & 1 & 0 & 0 & 1 \\ 1 & 2 & 1 & 0 & 0 \\ 0 & 1 & 2 & 1 & 0 \\ 0 & 0 & 1 & 2 & 1 \\ 1 & 0 & 0 & 1 & 2 \end{pmatrix}$$

is not i-minimizable for any $1 \leq i \leq 5$. (See definition of i-minimizability in Exercise 3.12. This example shows that the singularity requirement in part (e) of that exercise cannot be dismissed.)

Notes. Most of the section is based on [Drew *et al.* (1994)]. Theorem 3.38 was stated and proved in [Kaykobad (1987)] and Theorem 3.46, the proof of Theorem 3.43 and Exercise 3.25 are taken from [Berman and Shaked-Monderer (1998)]. A variation of Proposition 3.49 may be found in [Watkins (1993)]. Exercise 3.24 is based on [Shaked-Monderer *et al.* (2015)]. Exercise 3.31 is based on [Berman *et al.* (2016)]. It is used there for the solution of a completely optimization program. The idea is this: as a first step replace $\mathcal{CP}_n$ in the original program by $\mathcal{DNN}_n$. The resulting relaxation of the original program is exact only if $n \leq 4$. For $n \geq 5$ the solution may not be completely positive. Hence it is desirable to add a cut: add a linear constraints, which separates the obtained doubly nonnegative solution from the completely positive cone, resulting in an improved relaxation yielding a better bound on the optimal solution. Such a cut may be defined by a copositive matrix, and is called a *copositive cut.* Copositive cuts were introduced in [Bomze *et al.* (2010)]. In [Berman *et al.* (2016)] the result of Exercise 3.24 was used to generate a copositive cut that separates a doubly nonnegative X from the completely positive cone when the graph of X is triangle free.

3.5 Completely positive graphs

As mentioned in Section 3.1, an obvious necessary condition for a matrix to be completely positive is that it is doubly nonnegative. When is this necessary condition sufficient? In this section we give a qualitative answer to this question.

Definition 3.50. A graph G is *completely positive* if every doubly nonnegative matrix A whose graph is G is completely positive.

A doubly nonnegative matrix A such that $G(A) = G$ will be called a *DNN matrix realization* of the graph G. So a graph G is completely positive if every DNN matrix realization of G is completely positive. A completely positive matrix A such that $G(A) = G$ will be called a *CP matrix realization* of the graph G.

Example 3.51. Small graphs, with up to 4 vertices, are completely positive, as we saw in Section 3.3.

The main result of this section is the following.

Theorem 3.52. *A graph G is completely positive if and only if it does not contain an odd cycle of length greater than 4.*

We shall refer to odd cycles of length > 4 as "long odd cycles". (The triangles that played a major role in the previous section may be regarded as "small odd cycles".)

Theorem 3.52 is proved through a series of results.

Theorem 3.53. *A graph that contains a long odd cycle is not completely positive.*

Proof. We first show that a long odd cycle is not completely positive. Let B be the $k \times (k-1)$ matrix in which $b_{ii} = 1$, $i = 1, \ldots, k-1$, $b_{j+1\,j} = 1$ and $b_{kj} = (-1)^{j+1}$, $j = 1, \ldots, k-2$, and all other entries are zero:

$$B = \begin{pmatrix} 1 & 0 & \ldots & \ldots & 0 \\ 1 & 1 & \ldots & \ldots & 0 \\ 0 & 1 & 1 & \ldots & 0 \\ & & \ddots & \ddots & \\ & & & 1 & 1 \\ 1 & -1 & \ldots & \pm 1 & 0 \end{pmatrix}.$$

Let $A = BB^T$. Then

$$A = \begin{pmatrix} 1 & 1 & 0 & \ldots & \ldots & 1 \\ 1 & 2 & 1 & 0 & \ldots & 0 \\ 0 & 1 & 2 & 1 & & \vdots \\ \vdots & & \ddots & \ddots & \ddots & 0 \\ 0 & & & 1 & 2 & (-1)^{k-1} \\ 1 & 0 & \ldots & 0 & (-1)^{k-1} & k-2 \end{pmatrix}. \tag{3.4}$$

The matrix A is positive semidefinite, and for odd k it is also nonnegative. The graph of A is a k-cycle, so for $k > 3$ it does not contain a triangle.

$$M(A) = \begin{pmatrix} 1 & -1 & 0 & \ldots & \ldots & -1 \\ -1 & 2 & -1 & 0 & \ldots & 0 \\ 0 & -1 & 2 & -1 & & \vdots \\ \vdots & & \ddots & \ddots & \ddots & 0 \\ 0 & & & -1 & 2 & -1 \\ -1 & 0 & \ldots & 0 & -1 & k-2 \end{pmatrix},$$

and $\det M(A) = k - 2 - 1 - 2 - (k-1) = -4$, so $M(A)$ is not positive semidefinite. Thus it follows from Theorem 3.43 that an odd cycle of length $k > 4$ is not completely positive.

Now let G be a graph on n vertices, which contains a k-cycle, $k > 4$ odd. We have to show that G is not completely positive. Let S be the adjacency matrix of G and let C be an $n \times n$ matrix whose principal submatrix corresponding to the cycle is A above, and all its other entries are equal to 0. Then for every $a > 0$ there exists $0 < b \leq a$ such that $C_a = C + aI_n + bS$ is doubly nonnegative and $G(C_a) = G$. But $\lim_{a\to 0+} C_a = C$, and C is not completely positive. Since the cone $\mathcal{CP}_n$ is closed (Theorem 3.25), there exists some $a > 0$ for which C_a is doubly nonnegative but not completely positive. □

Theorem 3.54. *Bipartite graphs are completely positive.*

Proof. Let A be a doubly nonnegative matrix realization of a bipartite graph G. In particular, A is permutationally similar to a block matrix

$$\begin{pmatrix} D_1 & C \\ C^T & D_2 \end{pmatrix},$$

where D_1 and D_2 are diagonal matrices.

$$\begin{pmatrix} I & 0 \\ 0 & -I \end{pmatrix}\begin{pmatrix} D_1 & C \\ C^T & D_2 \end{pmatrix}\begin{pmatrix} I & 0 \\ 0 & -I \end{pmatrix} = \begin{pmatrix} D_1 & -C \\ -C^T & D_2 \end{pmatrix}.$$

That is,

$$\begin{pmatrix} I & 0 \\ 0 & -I \end{pmatrix} A \begin{pmatrix} I & 0 \\ 0 & -I \end{pmatrix} = M(A).$$

That is, $M(A)$ similar to A, and is therefore also positive semidefinite. By Theorem 3.40, A is completely positive. This proves that G is completely positive. □

Theorem 3.55. *Let G be a graph with a cut vertex, i.e., $G = G_1 \cup G_2$ where G_1 and G_2 have only one vertex in common. Then if both G_1 and G_2 are completely positive, so is G.*

Proof. Without loss of generality, assume that $V(G_1) = \{1, \ldots, k\}$ and $V(G_2) = \{k, k+1, \ldots, n\}$. Let A be a doubly nonnegative matrix realization of G, and suppose G_1 and G_2 are completely positive. The matrix A is the Gram matrix of some (not necessarily nonnegative) vectors $\mathbf{c}_1, \ldots, \mathbf{c}_n$. We have to show that there exist a Euclidean space V

(not necessarily n-dimensional) and nonnegative vectors $\mathbf{b}_1, \ldots, \mathbf{b}_n \in V$ such that $\mathrm{Gram}(\mathbf{b}_1, \ldots, \mathbf{b}_n) = A$. The subspaces $\mathrm{span}\{\mathbf{c}_1, \ldots, \mathbf{c}_{k-1}\}$ and $\mathrm{span}\{\mathbf{c}_{k+1}, \ldots, \mathbf{c}_n\}$ are orthogonal. Let $\mathbf{c}'_k$ be the orthogonal projection of $\mathbf{c}_k$ on $\mathrm{span}\{\mathbf{c}_1, \ldots, \mathbf{c}_{k-1}\}$ and let $\mathbf{c}''_k$ be the orthogonal projection of $\mathbf{c}_k$ on $\mathrm{span}\{\mathbf{c}_{k+1}, \ldots, \mathbf{c}_n\}$. Also let $\mathbf{c}'''_k = \mathbf{c}_k - \mathbf{c}'_k - \mathbf{c}''_k$ (it may be equal to $\mathbf{0}$). Define $A' = \mathrm{Gram}(\mathbf{c}_1, \ldots, \mathbf{c}_{k-1}, \mathbf{c}'_k)$. The fact that $\langle \mathbf{c}_i, \mathbf{c}'_k \rangle = \langle \mathbf{c}_i, \mathbf{c}_k \rangle$, $i = 1, \ldots, k-1$, implies that A' is equal to $A[1, \ldots, k]$, except in the lower right entry. This entry of A' is $\langle \mathbf{c}'_k, \mathbf{c}'_k \rangle$, which is positive. Thus A' is doubly nonnegative, and since the graph G_1 is completely positive, there exist a Euclidean space V' and nonnegative vectors $\mathbf{d}_1, \ldots, \mathbf{d}_{k-1}, \mathbf{d}'_k \in V'$ such that $A' = \mathrm{Gram}(\mathbf{d}_1, \ldots, \mathbf{d}_{k-1}, \mathbf{d}'_k)$. Similarly, for $A'' = \mathrm{Gram}(\mathbf{c}''_k, \mathbf{c}_{k+1}, \ldots, \mathbf{c}_n)$ there exists a Euclidean space V'' (of possibly different dimension) and nonnegative vectors $\mathbf{d}''_k, \mathbf{d}_{k+1}, \ldots, \mathbf{d}_n \in V''$ such that the matrix A'' is equal to $\mathrm{Gram}(\mathbf{d}''_k, \mathbf{d}_{k+1}, \ldots, \mathbf{d}_n)$. Finally, let $d'''_k = \|\mathbf{c}'''_k\|$, so that $d'''_k \geq 0$. Now let $V = V' \oplus V'' \oplus \mathbb{R}$, and let $\mathbf{b}_i = \mathbf{d}_i \oplus \mathbf{0} \oplus 0$ for $i = 1, \ldots, k-1$, $\mathbf{b}_k = \mathbf{d}'_k \oplus \mathbf{d}''_k \oplus d'''_k$, $\mathbf{b}_i = \mathbf{0} \oplus \mathbf{d}_i \oplus 0$ for $i = k+1, \ldots, n$ be vectors in V. (The third summand in each of these vectors is redundant if $\mathbf{c}'''_k = \mathbf{0}$.) The vectors $\mathbf{b}_1, \ldots, \mathbf{b}_n$ belong to the nonnegative orthant of V, and A is their Gram matrix, which proves that A is completely positive. $\square$

Example 3.56. The “butterfly” graph

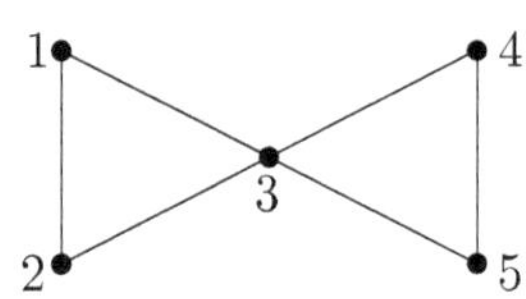

is completely positive, since 3 is a cut vertex and the triangles on vertices $\{1, 2, 3\}$ and $\{3, 4, 5\}$ are completely positive.

Theorem 3.57. *The graph T_n on n vertices, consisting of $n-2$ triangles with a common base, is completely positive.*

Proof. Let A be a DNN matrix realization of T_n. As in the previous proofs, we may assume that A is nonsingular, and that its diagonal entries are all equal to 1. If the common base of T_n is $\{1, 2\}$, then A is of the form

$$A = \begin{pmatrix} B & C \\ C^T & I_{n-2} \end{pmatrix},$$

where B is a 2×2 doubly nonnegative matrix, and C is a positive $2 \times (n-2)$

matrix. So

$$A = \begin{pmatrix} A/I_{n-2} & 0 \\ 0 & 0_{n-2} \end{pmatrix} + \begin{pmatrix} C \\ I_{n-2} \end{pmatrix} \begin{pmatrix} C^T & I_{n-2} \end{pmatrix}.$$

By Theorem 1.75, A/I_{n-2} is positive semidefinite. Since A/I_{n-2} is 2×2, and A and C are nonnegative, A is completely positive by Lemma 3.36. □

We can now prove Theorem 3.52.

Proof of Theorem 3.52. If G contains a long odd cycle, then, by Theorem 3.53, it is not completely positive. To prove the reverse implication we show that if a graph G does not contain a long odd cycle, then every block of G is either bipartite, or the complete graph on 4 vertices, or a T_k. This would imply that each block is completely positive and, by Theorem 3.55, that G is also completely positive.

So suppose G is a block which is not bipartite and does not contain a long odd cycle. Then G must contain a triangle H_0, say on vertices $1, 2, 3$. If $G = H_0$ is a triangle, then $G = T_3$ and we are done. Otherwise, there exists an H_0-path P in G, since G is 2-connected (see Proposition 1.84). Let the length of P be s. Then $s \geq 2$ and we may assume that 4 is an internal vertex of P. We may also assume that the end vertices of P are 1 and 2. So G contains two cycles: $P \cup \{1, 2\}$ of length $s+1$, and $P \cup \{1, 3\} \cup \{3, 2\}$ of length $s+2$. If $s > 2$ then either $s+1$ or $s+2$ is an odd number greater than 4, hence $s = 2$.

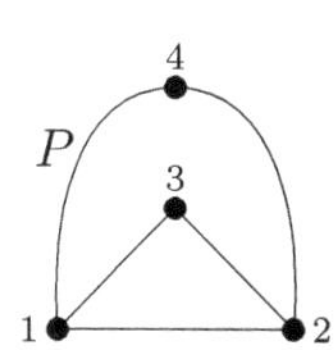

We have shown that G contains a T_4 subgraph H. If G has more than 4 vertices, then there is an H-path in G. The path P cannot connect 1 and 4 (or 1 and 3, or 2 and 4, or 2 and 3), since then G would contain cycles $P \cup \{1, 2\} \cup \{2, 4\}$ of length $s+2$ and $P \cup \{1, 3\} \cup \{3, 2\} \cup \{2, 4\}$ of length $s+3$, and one of these cycles would be of odd length greater than 4.

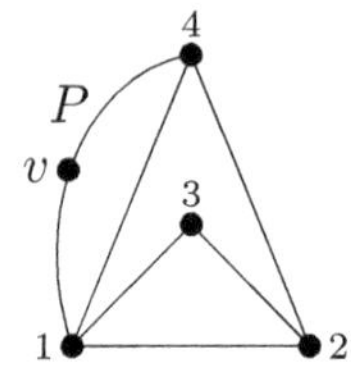

Hence P connects 1 and 2, and by the same argument as above, the length of P is 2. Continuing in this fashion we see that if G has n vertices, then $G \supseteq T_n$. So if $n = 4$, G is either K_4 or T_4. But if $n \geq 5$, G must be a T_n, because otherwise it would have an additional edge, say $\{3, 4\}$.

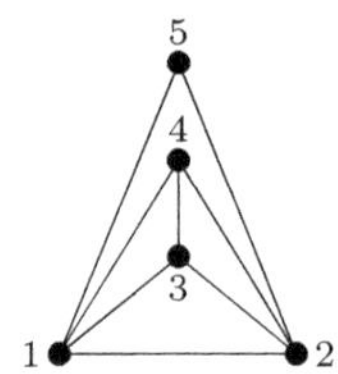

But then G contains a 5-cycle, $1 - 3 - 4 - 2 - 5 - 1$, contrary to our assumption. This completes the proof. □

Graphs that do not contain long odd cycles are known to be the line perfect graphs. Thus Theorem 3.52 can be rephrased as

Corollary 3.58. *The following properties of a graph G are equivalent.*

(a) *G is completely positive.*
(b) *Each block of G is completely positive.*
(c) *Each block of G is either bipartite, or a K_4, or a T_k.*
(d) *G does not contain a long odd cycle.*
(e) *G is a line perfect graph.*

Another corollary is obtained by combining Theorems 3.52 and 3.46. It is left as an exercise.

Corollary 3.59. *A graph G has the property*

for every nonnegative symmetric matrix realization A of G,
$M(A)$ is positive semidefinite iff A is positive semidefinite

if and only if G is bipartite.

Alternative proofs

There are other proofs of the results presented in this section. The preceding proofs were chosen partly for being the shortest, or the most elegant, we know. But different proofs provide different insights, so in the remainder of this section we present alternative proofs for the results of this section. We do it in the hope that the ideas in these proofs may be useful in studying other questions concerning completely positive matrices. The additional proof for the fact that bipartite graphs are completely positive (Theorem 3.54) has also the advantage of being algorithmic.

We start with the proof of Theorem 3.53. The original proof that the matrix A in (3.4) is not completely positive used the following lemma.

Lemma 3.60. *Let A be doubly nonnegative, and let l be a vertex of $G(A)$ that does not lie on a triangle. Then A is completely positive if and only if for each neighbour j of l there exists a positive number d_j such that*

$$\sum_{a_{lj}>0} \frac{a_{lj}^2}{d_j} \leq a_{ll}$$

and the matrix A' obtained from A by subtracting every d_j from a_{jj} and deleting the l-th row and column is completely positive.

Proof. We prove one direction, and leave the other as an exercise. Without loss of generality, assume that $l = 1$. Let N be the set of neighbours of 1 in $G(A)$. Suppose A is completely positive. Let

$$A = \sum_{i=1}^{m} \mathbf{b}_i \mathbf{b}_i^T$$

be a CP representation of A. For every $j \in N$ let

$$\Omega_j = \{i \mid \operatorname{supp} \mathbf{b}_i = \{1, j\}\},$$

and

$$B_j = \sum_{i \in \Omega_j} \mathbf{b}_i \mathbf{b}_i^T.$$

Each B_j is zero except for its 2×2 principal submatrix based on rows and columns 1 and j. Hence by Example 3.7

$$B_j = a_j E_{11} + F_j,$$

where $a_j \geq 0$, and F_j is zero except for its $\{1, j\}$-principal submatrix, and

$$F_j[1, j] = \begin{pmatrix} a_{1j}^2/d_j & a_{1j} \\ a_{1j} & d_j \end{pmatrix}$$

for some $d_j > 0$. For these d_j's, define A' as in the statement of the lemma. Then A' is a principal submatrix of the completely positive matrix $A - \sum_{j \in N} F_j$, and is therefore completely positive. □

We use the lemma in the following alternative proof of Theorem 3.53.

Another proof of Theorem 3.53. We only demonstrate here a different way to prove that (for odd k) the doubly nonnegative matrix

$$A = \begin{pmatrix} 1 & 1 & 0 & \dots & \dots & 1 \\ 1 & 2 & 1 & 0 & \dots & 0 \\ 0 & 1 & 2 & 1 & & \vdots \\ \vdots & & \ddots & \ddots & \ddots & 0 \\ 0 & & & 1 & 2 & 1 \\ 1 & 0 & \dots & 0 & 1 & k-2 \end{pmatrix},$$

is not completely positive. The rest of the proof remains unchanged.

The graph of A is the k-cycle. None of the vertices of $G(A)$ lies on a triangle, and in particular k does not. So by the lemma, if A is completely positive then there exist positive numbers d and e such that

$$\frac{1}{d} + \frac{1}{e} \leq k - 2 \tag{3.5}$$

and the $(k-1) \times (k-1)$ matrix

$$A' = \begin{pmatrix} 1-d & 1 & 0 & \dots & & 0 \\ 1 & 2 & 1 & \dots & & 0 \\ 0 & 1 & 2 & 1 & & \vdots \\ \vdots & & \ddots & \ddots & \ddots & 1 \\ 0 & & & & 1 & 2-e \end{pmatrix}$$

is completely positive. But $\det A'[1, 2, \dots, k-2] = 1 - (k-2)d \geq 0$, hence $1/d \geq k - 2$. Since this contradicts (3.5), A is not completely positive. □

Lemma 3.60 can be used to show also the following corollary, which in turn can be used to show that every tree is a completely positive graph.

Corollary 3.61. *If A is a doubly nonnegative matrix and l is a pendant vertex of $G(A)$, then A is completely positive if and only if $A/[l]$ is completely positive.*

We leave the proof of the corollary as an exercise. For the coming proofs, we point out that in order to show that G is a completely positive graph it is enough to consider only nonsingular DNN realizations of G, or consider only singular DNN realizations of G, according to the next lemma.

Lemma 3.62. *The following three conditions are equivalent.*

(a) *Every nonsingular DNN realization of G is completely positive.*
(b) *Every singular DNN realization of G is completely positive.*
(c) *G is completely positive.*

Proof. Since (c) is (a) and (b) combined, it suffices to show that (a) implies (b) and vice versa. We prove that (a)$\Rightarrow$(b), and leave the reverse implication as an exercise (Exercise 3.36).

Suppose (a) holds, and let A be a singular DNN realization of G. Then for every $\varepsilon > 0$, $A + \varepsilon I$ is also doubly nonnegative, it is nonsingular, and $G(A + \varepsilon I) = G(A)$. Now $A = \lim_{m\to\infty}(A + 1/mI)$. By the assumption, each $A + 1/mI$ is completely positive, hence by Theorem 3.25 so is A. $\square$

We now turn to the original proof of Theorem 3.54. This is a much longer proof than the one presented above. It has, however, the advantage of yielding a very simple algorithm for finding a CP representation of an irreducible doubly nonnegative matrix whose graph is bipartite. Here we sketch the proof and describe the algorithm, leaving the following proposition, which is part of the proof, as an exercise.

Proposition 3.63. *Let*

$$A = \begin{pmatrix} D_1 & C \\ C^T & D_2 \end{pmatrix},$$

where C is $k \times (n-k)$ and D_1 and D_2 are diagonal. If A is doubly nonnegative, singular and irreducible, then the null space of A is one dimensional, and is spanned by

$$\begin{pmatrix} \mathbf{x} \\ -\mathbf{y} \end{pmatrix}, \tag{3.6}$$

where $\mathbf{x} \in \mathbb{R}^k$ and $\mathbf{y} \in \mathbb{R}^{n-k}$ are positive.

Another proof of Theorem 3.54. Let A be a DNN matrix realization of a bipartite graph. As usual, we may permute the rows and columns of A. So let

$$A = \begin{pmatrix} D_1 & C \\ C^T & D_2 \end{pmatrix}.$$

By Lemma 3.62 we may assume that A is singular. We may also assume that it is irreducible, for if it is reducible, it is a direct sum of irreducible matrices and the theorem can be proved for each direct summand.

The proof is by induction on the number of edges in $G(A)$. The theorem is true if $G(A)$ has one edge. Assume that it holds for graphs that have less than k edges and that $G(A)$ has k edges. By Proposition 3.63, $\operatorname{ns} A$ is spanned by $\mathbf{z} = (x_1, \ldots, x_k, -y_{k+1}, \ldots, -y_n)^T$, where $x_i > 0$, $i = 1, \ldots, k$ and $y_i > 0$, $i = k+1, \ldots, n$. Choose any $i < j$ such that $a_{ij} > 0$, and let $\mathbf{b}_{ij} = y_j \mathbf{e}_i + x_i \mathbf{e}_j$. This is a nonnegative vector in $\mathbb{R}^n$, orthogonal to $\mathbf{z}$. Hence $\mathbf{z}$ is in the null space of $B_{ij} = \mathbf{b}_{ij}\mathbf{b}_{ij}^T$, which implies that $\mathbf{z}$ is in the null space of every matrix of the form $A_\alpha = A - \alpha B_{ij}$. Let $\alpha_0 \geq 0$ be the largest value of α for which A_{α_0} is still doubly nonnegative. There are two possible cases: either $\alpha_0 = a_{ij}/x_i y_j$, in which case the i, j entry of A_{α_0} is equal to zero, or $\alpha_0 < a_{ij}/x_i y_j$ and A_{α_0} still has positive i, j and j, i entries, but for $\alpha > \alpha_0$ the matrix A_α is not positive semidefinite. By the continuity of the eigenvalues and since $\mathbf{z}$ is in $\operatorname{ns} A_\alpha$ for every α, the second case is, in fact, impossible. Hence $G(A_{\alpha_0})$ has fewer edges than $G(A)$, and the result follows from the induction hypothesis. □

As mentioned above, this proof suggests a simple algorithm. If A_{α_0} is still irreducible, its null space is spanned by $\mathbf{z}$ and we may repeat the procedure to eliminate another edge from the graph. If A_{α_0} is reducible, then it is permutationally similar to a direct sum of irreducible matrices. In this case a permutation of $\mathbf{z}$ will be a null vector for the direct sum and since $\mathbf{z}$ has no zero entries, each of the summands is singular. The nullity of each summand is then exactly 1, and we can proceed to eliminate another edge using the newly labeled version of G. At each stage we obtain a matrix that is permutationally similar to a direct sum of doubly nonnegative singular matrices, so when we finish eliminating the off-diagonal entries the diagonal entries also vanish, since 1×1 singular matrices are equal to zero. Observe that the permutation of rows and columns at each stage so that each irreducible summand has the special form

$$\begin{pmatrix} D_1 & C \\ C^T & D_2 \end{pmatrix}$$

was needed only to simplify the description of $\mathbf{z}$, but is not really necessary in the actual factorization. The algorithm is therefore simpler.

Let A be an irreducible doubly nonnegative matrix whose graph is bipartite.

1. Obtain from A a row equivalent upper triangular matrix U by forward elimination (that is, using only operations of subtracting from one row scalar multiples of preceding rows — this is possible since all the leading principal minors, except possibly $\det A$ itself, are positive).
2. Let $U_1 = U - u_{nn}E_{nn}$. Find a vector $\mathbf{z}$ that spans $\operatorname{ns} U_1$.
3. For every $i < j$ such that $a_{ij} > 0$, if $z_j\mathbf{e}_i - z_i\mathbf{e}_j \geq \mathbf{0}$, let $b_{ij} = z_j\mathbf{e}_i - z_i\mathbf{e}_j$, otherwise let $b_{ij} = z_i\mathbf{e}_j - z_j\mathbf{e}_i$.
4. $$A = u_{nn}E_{nn} + \sum_{\substack{i<j \\ a_{ij}>0}} \frac{a_{ij}}{-z_iz_j}\mathbf{b}_{ij}\mathbf{b}_{ij}^T.$$

For $n \times n$ matrices, the above algorithm requires about $n^3/3$ multiplications.

Algebraic proof of Theorem 3.55. Let $G = G_1 \cup G_2$, where G_1 and G_2 are completely positive graphs which share exactly one vertex. Suppose that $V(G_1) = \{1, 2, \ldots, k\}$, $V(G_2) = \{k, k+1, \ldots, n\}$. To prove that G is completely positive, we have to show that every DNN matrix realization A of G is completely positive. Suppose then that A is a doubly nonnegative matrix of the block form

$$A = \begin{pmatrix} A_{11} & A_{21}^T \\ A_{21} & A_{22} \end{pmatrix},$$

where A_{11} is a $k \times k$ matrix and A_{21} is an $(n-k) \times k$ matrix in a block form $\begin{pmatrix} 0_{(n-k)\times(k-1)} & \mathbf{y} \end{pmatrix}$. By Theorem 1.75 the (generalized) Schur complement A/A_{22} is positive semidefinite. Since the first $k-1$ columns in A_{21} are zero,

$$A/A_{22} = A_{11} - \mathbf{y}^T A_{22}^\dagger \mathbf{y} E_{kk},$$

where E_{kk} is a $k \times k$ matrix whose last right entry is 1 and all other entries are 0. So the entries of A/A_{22} are the same as the entries of A_{11} except for the k, k entry which is positive, as a diagonal entry of a positive definite matrix. Thus A/A_{22} is doubly nonnegative and has the same graph as A_{11}, which is the completely positive G_1. Hence A/A_{22} is completely positive. Consider now the $(n-k+1) \times (n-k+1)$ matrix

$$A' = \begin{pmatrix} \mathbf{y}^T A_{22}^\dagger \mathbf{y} & \mathbf{y}^T \\ \mathbf{y} & A_{22} \end{pmatrix}.$$

Since A_{22} is positive semidefinite, so is $A_{22}^\dagger$, and $\mathbf{y}^T A_{22}^\dagger \mathbf{y} \geq 0$. The matrix A' is also positive semidefinite by Theorem 1.75, so it is doubly nonnegative.

$G(A') = G_2$, so A' is completely positive. Finally, the decomposition

$$A = \begin{pmatrix} A/A_{22} & 0 \\ 0^T & 0_{n-k} \end{pmatrix} + \begin{pmatrix} 0_{k-1} & 0 \\ 0 & A' \end{pmatrix}$$

shows that A is completely positive. □

Geometric proof of Theorem 3.57. The proof is by induction on n. For $n = 3$ the graph is a triangle, which is a small graph and is thus completely positive by Theorem 3.35. Suppose that every T_k, $3 \leq k < n$, is completely positive, and consider T_n. Let A be a DNN matrix realization of T_n. The matrix A is the Gram matrix of vectors $\mathbf{c}_1, \ldots, \mathbf{c}_n$. We want to show that $\mathbf{c}_1, \ldots, \mathbf{c}_n$ may be embedded in some m-dimensional vector space V (m may be different than n), with an orthonormal basis E such that $[\mathbf{c}_1]_E, [\mathbf{c}_2]_E, \ldots, [\mathbf{c}_n]_E$ belong to $\mathbb{R}^m_+$. We may assume that $\{1, 2\}$ is the common base of T_n, and since complete positivity is not affected by positive diagonal congruence, we may assume that $a_{ii} = 1$ for every $i = 1, \ldots, n$. By this assumption, $\|\mathbf{c}_i\| = 1$ and $\langle \mathbf{c}_i, \mathbf{c}_j \rangle \leq 1$, $1 \leq i \neq j \leq n$. Let $\langle \mathbf{c}_p, \mathbf{c}_q \rangle$ be the minimal inner product of all the positive $\langle \mathbf{c}_i, \mathbf{c}_j \rangle$'s. Consider two cases.

Case 1. $\{p, q\} = \{1, 2\}$. Define $\mathbf{c}_1' = \mathbf{c}_1 - \langle \mathbf{c}_1, \mathbf{c}_2 \rangle \mathbf{c}_2$, and consider the vectors $\mathbf{c}_1', \mathbf{c}_2, \ldots, \mathbf{c}_n$. Then $\langle \mathbf{c}_1', \mathbf{c}_2 \rangle = 0$, and for $i > 2$,

$$\langle \mathbf{c}_1', \mathbf{c}_i \rangle = \langle \mathbf{c}_1, \mathbf{c}_i \rangle - \langle \mathbf{c}_1, \mathbf{c}_2 \rangle \langle \mathbf{c}_2, \mathbf{c}_i \rangle \geq \langle \mathbf{c}_1, \mathbf{c}_i \rangle - \langle \mathbf{c}_1, \mathbf{c}_2 \rangle \geq 0.$$

This shows that the positive semidefinite matrix $A' = \text{Gram}(\mathbf{c}_1', \mathbf{c}_2, \ldots, \mathbf{c}_n)$ is doubly nonnegative. The graph of A' is obtained from $G(A)$ by erasing the base (and possibly other edges), hence it is bipartite, and as such completely positive. Hence we may assume that $\mathbf{c}_1', \mathbf{c}_2, \ldots, \mathbf{c}_n \in V$, for some vector space V with an orthonormal basis E such that $[\mathbf{c}_1']_E, [\mathbf{c}_2]_E, \ldots, [\mathbf{c}_n]_E$ are nonnegative. But $[\mathbf{c}_1]_E = [\mathbf{c}_1']_E + \langle \mathbf{c}_1, \mathbf{c}_2 \rangle [\mathbf{c}_2]_E$, so $[\mathbf{c}_1]_E$ is also nonnegative.

Case 2. $\{p, q\} \neq \{1, 2\}$. Without loss of generality we may assume that $\{p, q\} = \{1, n\}$. Defining $\mathbf{c}_1' = \mathbf{c}_1 - \langle \mathbf{c}_1, \mathbf{c}_n \rangle \mathbf{c}_n$, we get (as in Case 1) vectors $\{\mathbf{c}_1', \mathbf{c}_2, \ldots, \mathbf{c}_n\}$ with a nonnegative Gram matrix A', whose graph is obtained from $G(A)$ by erasing the edge $\{1, n\}$ and possibly some other edges incident with the vertex 1. The vertex 2 is a cut vertex of the graph $G(A')$, connecting a graph T_k with $\{1, 2\}$ for a base and $k < n$, the edge $\{2, n\}$, and possibly some other edges.

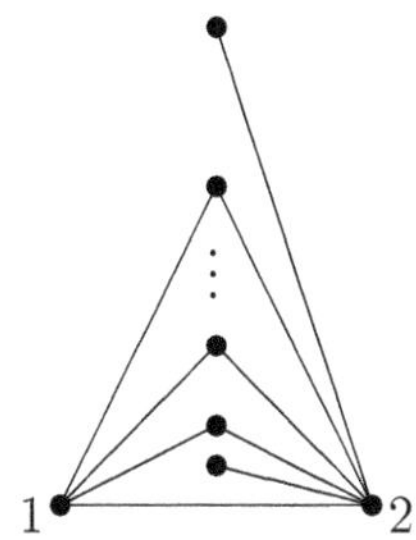

By the induction hypothesis and Theorem 3.55, $G(A')$ is completely positive and so is A'. Thus there is a vector space V with an orthonormal basis E such that $[\mathbf{c}_1']_E, [\mathbf{c}_2]_E, \ldots, [\mathbf{c}_n]_E$ are nonnegative. But $\mathbf{c}_1 = \mathbf{c}_1' + \langle \mathbf{c}_1, \mathbf{c}_n \rangle \mathbf{c}_n$, so $[\mathbf{c}_1]_E$ is also nonnegative and A is completely positive. □

We conclude with a proof of the "if" part of Theorem 3.52 characterizing completely positive graphs. It uses the duality of the cones $\mathcal{CP}_n$ and $\mathcal{COP}_n$, and the characterization of SPN completable graphs.

Another proof of the "if" part of Theorem 3.52. Let A be a doubly nonnegative matrix realization of a graph G that has no odd cycle of length greater than 4. Given any copositive matrix M, let B be the G-partial COP matrix whose specified entries are equal to those of M. Since G contains no long odd cycle of length greater than 4, any fully specified submatrix of B is at most 4×4, and is therefore SPN. Hence B is a G-partial SPN matrix. By Theorem 2.203, the partial B may be completed to an SPN matrix B. Since A is doubly nonnegative and B is SPN, $\langle A, B \rangle \geq 0$. And as $a_{ij} = 0$ whenever $b_{ij} \neq m_{ij}$,

$$\langle A, M \rangle = \langle A, B \rangle \geq 0.$$

This holds for every copositive M, hence $A \in \mathcal{COP}_n^* = \mathcal{CP}_n$. □

Exercises

3.33. Let

$$A = \begin{pmatrix} 1 & a & b & 0 & 0 \\ a & 1 & c & 0 & 0 \\ b & c & 1 & d & e \\ 0 & 0 & d & 1 & f \\ 0 & 0 & e & f & 1 \end{pmatrix}$$

be doubly nonnegative. Find a nonnegative matrix B such that $BB^T = A$.

3.34. Complete the proof of Lemma 3.60.

3.35. Prove Corollary 3.61. Then use the corollary to prove by induction that every tree is a completely positive graph.

3.36. Let A be a nonsingular DNN realization of a graph G. Use Proposition 1.47 to prove that $A - (\det A/\det A(n))E_{nn}$ is a singular DNN realization of G. Use that to prove that in Lemma 3.62 (b)⇒(a).

3.37. Prove Proposition 3.63 using the following steps.

(a) Show that if Q is an $k \times (n-k)$ matrix,
$$S = \begin{pmatrix} 0 & Q \\ Q^T & 0 \end{pmatrix}$$
is nonnegative and irreducible, and $\mathbf{x} \in \mathbb{R}^k$ and $\mathbf{y} \in \mathbb{R}^{n-k}$, then
$$S\begin{pmatrix} \mathbf{x} \\ \mathbf{y} \end{pmatrix} = \lambda \begin{pmatrix} \mathbf{x} \\ \mathbf{y} \end{pmatrix} \text{ iff } S\begin{pmatrix} \mathbf{x} \\ -\mathbf{y} \end{pmatrix} = -\lambda \begin{pmatrix} \mathbf{x} \\ -\mathbf{y} \end{pmatrix}.$$
(b) Show that if S is as in (a), then the smallest eigenvalue of S is simple and the associated eigenspace is spanned by a vector of the form (3.6).
(c) Prove Proposition 3.63.

3.38. Use the algorithm that follows from the original proof of Theorem 3.54 to find a CP representation of
$$A = \begin{pmatrix} 6 & 0 & 2 & 2 \\ 0 & 5 & 4 & 2 \\ 2 & 4 & 6 & 0 \\ 2 & 2 & 0 & 6 \end{pmatrix}.$$

3.39. By Proposition 3.63, if A is an $n \times n$ completely positive matrix whose graph is bipartite, then $\operatorname{rank} A \geq n-1$. Show that every bipartite graph has both a nonsingular CP matrix realization and a singular one.

3.40. Let G be a completely positive graph. Let A be a completely positive matrix with $G(A) = G$. Show that

(a) A is critical if and only if $A - \varepsilon E_{ii}$ is not positive semidefinite for every $1 \leq i \leq n$.
(b) If A is critical, then A is singular.
(c) A is i-minimizable for every i.

3.41. Show that if A is a doubly nonnegative matrix realization of an odd cycle, then A is completely positive if and only if $\det M(A) \geq 0$.

Notes. Theorem 3.52 is due to [Kogan and Berman (1993)]. See also [Ando (1991)], [Berman (1988)], [Berman (1993)], [Kogan (1989)] and [Drew and Johnson (1996)] (where graphs which contain no long odd cycles are called *NLOC graphs*). Theorem 3.53 is due to [Berman and Hershkowitz (1987)]. The first part of the proof of this theorem, which appears on page 270, is taken from [Berman and Shaked-Monderer (1998)]. The second part is taken from the original proof, and the rest of the original proof is given on page 276. The deletion of a pendant vertex in Corollary 3.61 is used in [Berman and Hershkowitz (1987)] to show that trees are completely positive. This idea is extended in Section 3.6 to a deletion of a vertex of degree 2. Theorem 3.54 is due to [Berman and Grone (1988)]. The original proof is given on page 277. Proposition 3.63 appears in [Berman and Grone (1988)], and the proof outlined in Exercise 3.37 is based on that paper. The equation in part (a) of that exercise, for the case that S is the adjacency matrix of a bipartite graph, is known in chemistry as the Coulson-Rushbrooke Pairing Theorem. Theorem 3.55 is proved in [Kogan and Berman (1993)]. The algebraic proof (page 279) is taken from [Ando (1991)]. The graphs T_n discussed in this section are called *crowns* in [Salce and Zanardo (1993)]. The geometric proof of Theorem 3.57 (page 280) is the original proof of [Kogan and Berman (1993)]. The proof that we give on page 272 is a simplification of the algebraic proof of [Ando (1991)], using Lemma 3.36, which is due to [Barioli (2001a)]. Corollary 3.59 is taken from [Berman and Shaked-Monderer (1998)]. The alternative proof to Theorem 3.52 on page 281 is a late addition, due to [Shaked-Monderer *et al.* (2016)]. Exercise 3.40 is due to [Zhang and Li (2002)] (parts (a) and (b)) and to [Barioli (1998b)] (part (c)). Exercise 3.41 is taken from [Zhang and Li (2000)].

3.6 Completely positive matrices whose graphs are not completely positive

In this section we collect miscellaneous results on complete positivity of matrices whose graphs are not completely positive. The section is divided into three subsections: positivity of least squares solution, book graphs and odd cycles, and deletion of vertices of degree 2.

Positivity of least squares solution

We describe an inductive sufficient condition for complete positivity. Consider an $n \times n$ matrix A in block form

$$A = \begin{pmatrix} a & \mathbf{c}^T \\ \mathbf{c} & A_1 \end{pmatrix} \tag{3.7}$$

where A_1 is a symmetric $(n-1) \times (n-1)$ matrix.

Theorem 3.64. *Let A be an $n \times n$ matrix in the block form* (3.7). *Then A is completely positive if and only if A_1 is completely positive and there exist a (not necessarily square) nonnegative matrix C and a nonnegative vector $\mathbf{x}$ such that $A_1 = CC^T$, $\mathbf{c} = C\mathbf{x}$, and $a = \mathbf{x}^T\mathbf{x}$.*

Proof. If A is completely positive, then $A = BB^T$, where B is nonnegative. Denote the first row of B by $\mathbf{x}^T$, and let C be the matrix consisting of rows $2, \ldots, n$ of B. Then $\mathbf{x}$ and C are nonnegative and satisfy the required equalities, since

$$BB^T = \begin{pmatrix} \mathbf{x}^T \\ C \end{pmatrix} \begin{pmatrix} \mathbf{x} & C^T \end{pmatrix} = \begin{pmatrix} \mathbf{x}^T\mathbf{x} & \mathbf{x}^T C^T \\ C\mathbf{x} & CC^T \end{pmatrix}.$$

Conversely, suppose there exist C and $\mathbf{x}$ satisfying the conditions in the theorem. Then

$$B = \begin{pmatrix} \mathbf{x}^T \\ C \end{pmatrix}$$

is a nonnegative matrix satisfying $A = BB^T$. □

The matrix C and the vector $\mathbf{x}$ in the statement of Theorem 3.64 are not unique.

Definition 3.65. A symmetric nonnegative matrix A in block form (3.7) has *property PLSS (positivity of least squares solution)* if there exists a nonnegative matrix C such that $CC^T = A_1$ and $C^\dagger \mathbf{c}$ is a nonnegative vector.

If A_1 is completely positive and the first row of A is zero, then A trivially has property PLSS. Here is another example.

Example 3.66. If A is a nonnegative symmetric matrix in block form (3.7) and A_1 is a nonnegative rank 1 symmetric matrix, then A has property PLSS. To see that, note that such A_1 is equal to $\mathbf{u}\mathbf{u}^T$ for some nonnegative vector $\mathbf{u}$. The matrix A has the PLSS property since $\mathbf{u}^\dagger = \|\mathbf{u}\|^{-2}\mathbf{u}^T$.

The next example is left as an exercise.

Example 3.67. If A is a nonnegative symmetric matrix in block form (3.7) and A_1 is a diagonal matrix, then A has property PLSS.

Observe that if A has property PLSS, then so does every nonnegative matrix which differs from A only in its $1,1$ entry.

Theorem 3.68. *A doubly nonnegative matrix that has the PLSS property is completely positive.*

Proof. Let $C \geq 0$ satisfy $CC^T = A_1$ and $\mathbf{x} = C^\dagger \mathbf{c} \geq \mathbf{0}$. Then the matrix

$$\begin{pmatrix} \mathbf{x}^T \\ C \end{pmatrix} \begin{pmatrix} \mathbf{x} & C^T \end{pmatrix} = \begin{pmatrix} \mathbf{x}^T\mathbf{x} & \mathbf{x}^T C^T \\ C\mathbf{x} & CC^T \end{pmatrix} = \begin{pmatrix} \mathbf{x}^T\mathbf{x} & \mathbf{c}^T \\ \mathbf{c} & CC^T \end{pmatrix}$$

is completely positive. By Theorem 1.75,

$$a - \mathbf{x}^T\mathbf{x} = a - \mathbf{c}^T (C^\dagger)^T C^\dagger \mathbf{c} = a - \mathbf{c}^T A_1^\dagger \mathbf{c} \geq 0.$$

Hence

$$A = \begin{pmatrix} a - \mathbf{x}^T\mathbf{x} & \mathbf{0}^T \\ \mathbf{0} & 0 \end{pmatrix} + \begin{pmatrix} \mathbf{x}^T\mathbf{x} & \mathbf{x}^T C^T \\ C\mathbf{x} & CC^T \end{pmatrix}$$

is completely positive. □

The converse of Theorem 3.68 is not true.

Example 3.69. Let

$$A = \begin{pmatrix} 2 & 1 & 0 & 0 & 1 \\ 1 & 2 & 1 & 0 & 0 \\ 0 & 1 & 2 & 1 & 0 \\ 0 & 0 & 1 & 2 & 1 \\ 1 & 0 & 0 & 1 & 2 \end{pmatrix}.$$

The matrix A is completely positive since it is diagonally dominant. However, A does not have property PLSS. If it had that property, then so would its 1-minimized matrix

$$\begin{pmatrix} 1.2 & 1 & 0 & 0 & 1 \\ 1 & 2 & 1 & 0 & 0 \\ 0 & 1 & 2 & 1 & 0 \\ 0 & 0 & 1 & 2 & 1 \\ 1 & 0 & 0 & 1 & 2 \end{pmatrix},$$

but the latter is not completely positive since its comparison matrix is not positive semidefinite.

Unlike double nonnegativity or complete positivity, the PLSS property is not preserved under permutation similarity.

Example 3.70. Let A be as in Example 3.69, B the 6×6 matrix in block form

$$B = \begin{pmatrix} 0 & \mathbf{0}^T \\ \mathbf{0} & A \end{pmatrix}$$

and

$$B' = \begin{pmatrix} A & \mathbf{0} \\ \mathbf{0}^T & 0 \end{pmatrix}.$$

The matrices B and B' are permutationally similar, B trivially has the PLSS property, but B' does not, since A does not have this property.

Since property PLSS is not preserved under permutation similarity, it is worthwhile to mention the following obvious corollary of Theorem 3.68.

Corollary 3.71. *If PAP^T has property PLSS for some permutation matrix P, then A is completely positive.*

The matrix A in Example 3.69 can be used to show that the converse of Corollary 3.71 also does not hold. We leave the proof as an exercise, together with the proof of the following proposition, which asserts that some permutation similarities do preserve the PLSS property.

Proposition 3.72. *If $P = 1 \oplus P'$, where P' is an $(n-1) \times (n-1)$ permutation matrix, then A has property PLSS if and only if PAP^T has this property.*

The following is a corollary of Theorem 3.68 and Examples 3.66 and 3.67.

Corollary 3.73. *If A is a doubly nonnegative matrix in block form* (3.7) *and* $\operatorname{rank} A_1 = 1$ *or A_1 is diagonal, then A is completely positive.*

Property PLSS can be used to verify complete positivity. We leave it as an exercise to show that $n \times n$ doubly nonnegative matrices, $n \leq 4$, are not only completely positive — they actually have the PLSS property. It is also possible to prove directly that a matrix whose graph is T_n, where 1 is one of the vertices of the common base, has property PLSS, thus providing an alternative proof to Theorem 3.57.

Book graphs, odd cycles and house graphs

Definition 3.74. A graph G is a *book graph* with r completely positive pages if $V(G) = V_1 \cup \ldots \cup V_r$, where

(a) There exist two distinct vertices u and v such that $V_i \cap V_j = \{u, v\}$ for every $1 \leq i \neq j \leq r$.
(b) The subgraph of G induced by V_i, G_i, is completely positive.
(c) $G = G_1 \cup \ldots \cup G_r$.

Note that $\{u, v\}$ may or may not be an edge of G, and that by condition (c) two vertices of G which belong to different V_i's are not adjacent, unless they are u and v and $\{u, v\} \in E(G)$. Several examples of book graphs are shown bellow.

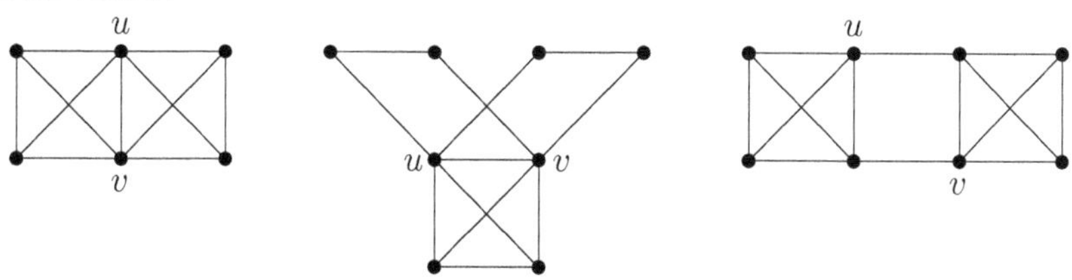

A matrix realization of a book graph is permutationally similar to a matrix A in the following block form

$$A = \begin{pmatrix} a & h & \mathbf{x}_1^T & \mathbf{x}_2^T & \ldots & \mathbf{x}_r^T \\ h & b & \mathbf{y}_1^T & \mathbf{y}_2^T & \ldots & \mathbf{y}_r^T \\ \mathbf{x}_1 & \mathbf{y}_1 & A_1 & 0 & \ldots & 0 \\ \mathbf{x}_2 & \mathbf{y}_2 & 0 & A_2 & \ldots & 0 \\ \vdots & \vdots & \vdots & \vdots & \ddots & \vdots \\ \mathbf{x}_r & \mathbf{y}_r & 0 & 0 & \ldots & A_r \end{pmatrix} \tag{3.8}$$

where for every $1 \leq i \leq r$ the graph of

$$\begin{pmatrix} a & h & \mathbf{x}_i^T \\ h & b & \mathbf{y}_i^T \\ \mathbf{x}_i & \mathbf{y}_i & A_i \end{pmatrix}$$

is the completely positive G_i. Let $\tilde{A}$ be the direct sum of $A_1, \ldots, A_r$, $\mathbf{x}^T = (\mathbf{x}_1^T, \mathbf{x}_2^T, \ldots, \mathbf{x}_r^T)$ and $\mathbf{y}^T = (\mathbf{y}_1^T, \mathbf{y}_2^T, \ldots, \mathbf{y}_r^T)$. Then

$$A = \begin{pmatrix} a & h & \mathbf{x}^T \\ h & b & \mathbf{y}^T \\ \mathbf{x} & \mathbf{y} & \tilde{A} \end{pmatrix}. \tag{3.9}$$

Denote

$$\Upsilon = \{1 \le i \le r \mid \mathbf{x}_i^T A_i^\dagger \mathbf{y}_i < 0\} \text{ and } \Upsilon' = \{1, \ldots, r\} \setminus \Upsilon .$$

Using these notations we state a necessary and sufficient condition for a DNN matrix realization of a book graph to be completely positive.

Theorem 3.75. *Let A be a doubly nonnegative matrix such that $G(A)$ is a book graph, A in block form (3.8), (3.9). Then A is completely positive if and only if $\Upsilon = \emptyset$, or $\Upsilon \ne \emptyset$ and*

$$\left(a - \mathbf{x}^T \tilde{A}^\dagger \mathbf{x}\right)^{1/2} \left(b - \mathbf{y}^T \tilde{A}^\dagger \mathbf{y}\right)^{1/2} \ge -2 \sum_{i \in \Upsilon} \mathbf{x}_i^T A_i^\dagger \mathbf{y}_i . \qquad (3.10)$$

For the proof we need the following lemma.

Lemma 3.76. *Let*

$$A = \begin{pmatrix} a & h \\ h & b \end{pmatrix}$$

be a real positive semidefinite 2×2 matrix and $y \in \mathbb{R}$. There exist $x, z \in \mathbb{R}$ such that

$$\begin{pmatrix} x & y \\ y & z \end{pmatrix} \quad \textit{and} \quad \begin{pmatrix} a - x & h - y \\ h - y & b - z \end{pmatrix}$$

are both positive semidefinite if and only if

$$\frac{h - \sqrt{ab}}{2} \le y \le \frac{h + \sqrt{ab}}{2} . \qquad (3.11)$$

Proof. If

$$\begin{pmatrix} x & y \\ y & z \end{pmatrix} \quad \text{and} \quad \begin{pmatrix} a - x & h - y \\ h - y & b - z \end{pmatrix}$$

are positive semidefinite, then so are

$$\begin{pmatrix} x & -y \\ -y & z \end{pmatrix}$$

and

$$\begin{pmatrix} a & h - 2y \\ h - 2y & b \end{pmatrix} = \begin{pmatrix} a - x & h - y \\ h - y & b - z \end{pmatrix} + \begin{pmatrix} x & -y \\ -y & z \end{pmatrix} .$$

The fact that the matrix on the left is positive semidefinite implies that $ab \ge (h - 2y)^2$, and the desired inequalities follow.

The converse is clearly true if either $a = 0$ or $b = 0$. If $ab > 0$, then $h = \sqrt{ab}\cos 2\theta$ for some $0 \le \theta \le \pi/2$. So $A = \cos^2\theta A_1 + \sin^2\theta A_2$, where

$$A_1 = \begin{pmatrix} a & \sqrt{ab} \\ \sqrt{ab} & b \end{pmatrix} \text{ and } A_2 = \begin{pmatrix} a & -\sqrt{ab} \\ -\sqrt{ab} & b \end{pmatrix}.$$

The matrices A_1 and A_2 are both positive semidefinite.

By (3.11),

$$-\sin^2\theta \le \frac{y}{\sqrt{ab}} \le \cos^2\theta.$$

If $0 \le (y/\sqrt{ab}) \le \cos^2\theta$, we may choose

$$\begin{pmatrix} x & y \\ y & z \end{pmatrix} = \frac{y}{\sqrt{ab}} A_1,$$

$$\begin{pmatrix} a-x & h-y \\ h-y & b-z \end{pmatrix} = \left(\cos^2\theta - \frac{y}{\sqrt{ab}}\right) A_1 + \sin^2\theta A_2.$$

If $-\sin^2\theta \le (y/\sqrt{ab}) \le 0$, we may choose

$$\begin{pmatrix} x & y \\ y & z \end{pmatrix} = \frac{-y}{\sqrt{ab}} A_1,$$

$$\begin{pmatrix} a-x & h-y \\ h-y & b-z \end{pmatrix} = \left(\sin^2\theta + \frac{y}{\sqrt{ab}}\right) A_1 + \cos^2\theta A_2. \qquad \square$$

In the proof of Theorem 3.75 we use a few more notations. For each $1 \le i \le r$, let

$$\begin{pmatrix} a_i & h_i \\ h_i & b_i \end{pmatrix} = \begin{pmatrix} \mathbf{x}_i^T \\ \mathbf{y}_i^T \end{pmatrix} A_i^\dagger \begin{pmatrix} \mathbf{x}_i & \mathbf{y}_i \end{pmatrix}$$

and let C_i be the $n \times n$ matrix with zero entries except for $C_i[V_i]$, which is

$$\begin{pmatrix} a_i & h_i & \mathbf{x}_i^T \\ h_i & b_i & \mathbf{y}_i^T \\ \mathbf{x}_i & \mathbf{y}_i & A_i \end{pmatrix}.$$

Let

$$\begin{pmatrix} a_\perp & h_\perp \\ h_\perp & b_\perp \end{pmatrix} = A/\tilde{A}.$$

By Theorem 1.75 the latter matrix is positive semidefinite, and since

$$\begin{aligned} \begin{pmatrix} \mathbf{x}^T \\ \mathbf{y}^T \end{pmatrix} \tilde{A}^\dagger \begin{pmatrix} \mathbf{x} & \mathbf{y} \end{pmatrix} &= \sum_{i=1}^r \begin{pmatrix} \mathbf{x}_i^T \\ \mathbf{y}_i^T \end{pmatrix} A_i^\dagger \begin{pmatrix} \mathbf{x}_i & \mathbf{y}_i \end{pmatrix} \\ &= \sum_{i=1}^r \begin{pmatrix} a_i & h_i \\ h_i & b_i \end{pmatrix}, \end{aligned}$$

$A = \sum_{i=1}^{r} C_i + C_\perp$, where

$$C_\perp = \begin{pmatrix} a_\perp & h_\perp \\ h_\perp & b_\perp \end{pmatrix} \oplus 0_{n-2}.$$

With these notations, $h_\perp + \sum_{i=1}^{r} h_i = h$, and condition (3.10) becomes $\sqrt{a_\perp b_\perp} + h_\perp \geq -2\sum_{i\in\Upsilon} h_i$.

Proof of Theorem 3.75. First we show that if A is completely positive, then condition (3.10) holds. Let

$$A = \sum_{j=1}^{m} \mathbf{b}_j \mathbf{b}_j^T$$

be a CP representation of A. For $1 \leq j \leq r-1$ let

$$\Omega_i = \{1 \leq j \leq m \mid \operatorname{supp} \mathbf{b}_j \cap (V_i \setminus \{1,2\}) \neq \emptyset\},$$

and let

$$\Omega_r = \{1,\ldots,m\} \setminus \cup_{i=1}^{r-1}\Omega_i\,.$$

Finally for $1 \leq i \leq r$ let

$$B_i = \sum_{j\in\Omega_i} \mathbf{b}_j \mathbf{b}_j^T.$$

Each B_i is a completely positive matrix and B_i is zero except for $B_i[V_i]$ which is

$$\begin{pmatrix} d_i & e_i & \mathbf{x}_i^T \\ e_i & f_i & \mathbf{y}_i^T \\ \mathbf{x}_i & \mathbf{y}_i & A_i \end{pmatrix}.$$

By Theorem 1.75,

$$B_i[V_i]/A_i = \begin{pmatrix} d_i & e_i \\ e_i & f_i \end{pmatrix} - \begin{pmatrix} a_i & h_i \\ h_i & b_i \end{pmatrix}$$

is positive semidefinite for every $1 \leq i \leq r$. Hence

$$\begin{pmatrix} a_\Upsilon & h_\Upsilon \\ h_\Upsilon & b_\Upsilon \end{pmatrix} = \sum_{i\in\Upsilon}\left(\begin{pmatrix} d_i & e_i \\ e_i & f_i \end{pmatrix} - \begin{pmatrix} a_i & h_i \\ h_i & b_i \end{pmatrix}\right)$$

and

$$\begin{pmatrix} a_{\Upsilon'} & h_{\Upsilon'} \\ h_{\Upsilon'} & b_{\Upsilon'} \end{pmatrix} = \sum_{i\in\Upsilon'}\left(\begin{pmatrix} d_i & e_i \\ e_i & f_i \end{pmatrix} - \begin{pmatrix} a_i & h_i \\ h_i & b_i \end{pmatrix}\right)$$

are both positive semidefinite. Since

$$\sum_{i=1}^{r} B_i = A = \sum_{i=1}^{r} C_i + C_\perp,$$

the equality

$$\sum_{i=1}^{r} \begin{pmatrix} d_i & e_i \\ e_i & f_i \end{pmatrix} = \sum_{i=1}^{r} \begin{pmatrix} a_i & h_i \\ h_i & b_i \end{pmatrix} + \begin{pmatrix} a_\perp & h_\perp \\ h_\perp & b_\perp \end{pmatrix}$$

holds. Hence

$$\begin{pmatrix} a_\Upsilon & h_\Upsilon \\ h_\Upsilon & b_\Upsilon \end{pmatrix} + \begin{pmatrix} a_{\Upsilon'} & h_{\Upsilon'} \\ h_{\Upsilon'} & b_{\Upsilon'} \end{pmatrix} = \begin{pmatrix} a_\perp & h_\perp \\ h_\perp & b_\perp \end{pmatrix}.$$

By Lemma 3.76 this implies that $\sqrt{a_\perp b_\perp} + h_\perp \geq 2h_\Upsilon$, which in turn implies (3.10), since $h_\Upsilon \geq -\sum_{i\in\Upsilon} h_i \geq 0$.

For the converse, we consider three possible cases.

Case 1. $\Upsilon = \emptyset$ and $h_\perp \geq 0$. In this case, $C_\perp$ is doubly nonnegative and therefore completely positive. Every C_i is doubly nonnegative and $G(C_i)$ is the completely positive graph G_i, hence every C_i is also completely positive. So A is completely positive as a sum of $r+1$ completely positive matrices.

Case 2. $\Upsilon = \emptyset$ and $h_\perp < 0$. In this case $\sum_{i=1}^{r} h_i > 0$ and we may define

$$C_i' = C_i + \frac{h_i}{\sum_{i=1}^{r} h_i} C_\perp.$$

It is easy to see that each C_i' is doubly nonnegative and $G(C_i') = G_i$. Hence it is completely positive and so is $A = \sum_{i=1}^{r} C_i'$.

Case 3. $\Upsilon \neq \emptyset$ and $\sqrt{a_\perp b_\perp} + h_\perp \geq -2\sum_{i\in\Upsilon} h_i$. In this case, we may apply Lemma 3.76 to

$$\begin{pmatrix} a_\perp & h_\perp \\ h_\perp & b_\perp \end{pmatrix} \quad \text{and} \quad y = -\sum_{i\in\Upsilon} h_i$$

and get $x, z \in \mathbb{R}$ such that

$$\begin{pmatrix} x & y \\ y & z \end{pmatrix} \quad \text{and} \quad \begin{pmatrix} a_\perp - x & h_\perp - y \\ h_\perp - y & b_\perp - z \end{pmatrix}$$

are both positive semidefinite. Let

$$C_i' = \begin{cases} C_i - \dfrac{h_i}{y} \begin{pmatrix} x & y & \mathbf{0}^T \\ y & z & \mathbf{0}^T \\ \mathbf{0} & \mathbf{0} & 0_{n-2} \end{pmatrix}, & i \in \Upsilon \\[2ex] C_i + \dfrac{h_i}{\sum_{j\in\Upsilon'} h_j} \begin{pmatrix} a_\perp - x & h_\perp - y & \mathbf{0}^T \\ h_\perp - y & b_\perp - z & \mathbf{0}^T \\ \mathbf{0} & \mathbf{0} & 0_{n-2} \end{pmatrix}, & i \in \Upsilon' \end{cases}.$$

Each C_i' is a doubly nonnegative matrices with $G(C_i') \subseteq G_i$, and is therefore completely positive. Hence $A = \sum_{i=1}^{r} C_i'$ is also completely positive. □

For the statement of the following corollary, we denote

$$A_0 = \begin{pmatrix} a & h \\ h & b \end{pmatrix} \tag{3.12}$$

and for $1 \leq i \leq r$

$$H_i^T = \begin{pmatrix} \mathbf{x}_i^T \\ \mathbf{y}_i^T \end{pmatrix}. \tag{3.13}$$

The proof of the corollary is left as an exercise.

Corollary 3.77. *Let A be a doubly nonnegative matrix such that $G(A)$ is a book graph. Suppose A is in block form (3.8), A_0 and H_i are as defined in (3.12) and (3.13). Then A is completely positive if and only if the matrix $A_0 - \sum_{i=1}^r \left| H_i^T A_i^\dagger H_i \right|$ is positive semidefinite.*

Next we consider a special type of book graph.

Definition 3.78. A graph G is a $\widehat{CP}$ *graph* if there exists a vertex $v \in V(G)$ such that $d(v) = 2$, the two vertices adjacent to v are not adjacent to each other, and $G - v$ is completely positive. The vertex v is the *hat vertex* in this $\widehat{\text{CP}}$ graph.

So a $\widehat{\text{CP}}$ graph is created by adding a vertex v to a completely positive graph, along with two edges connecting it to two nonadjacent vertices u and w. Here are two examples.

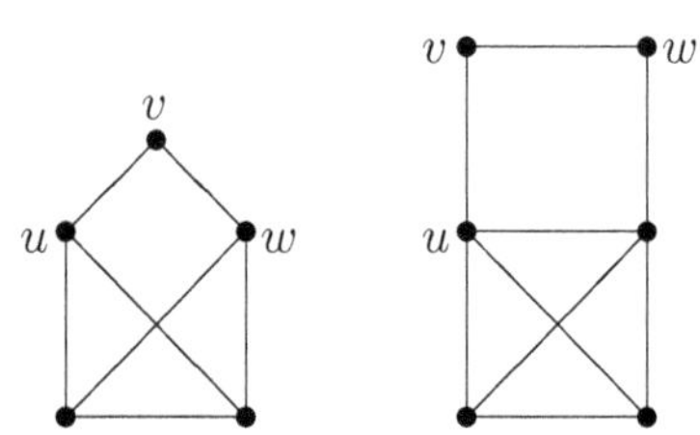

If G is a $\widehat{\text{CP}}$ graph on n vertices, we may assume that $v = 1$ and the two nonadjacent vertices which are neighbours of v are $u = 2$ and $w = n$. A symmetric matrix realization of G then has the form

$$A = \begin{pmatrix} a & \mathbf{c}^T \\ \mathbf{c} & \tilde{A} \end{pmatrix} \tag{3.14}$$

where $\mathbf{c} = \mathbf{c}_1 + \mathbf{c}_2 \in \mathbb{R}^{n-1}$,

$$\mathbf{c}_1^T = (s_2, 0, \ldots, 0) \text{ and } \mathbf{c}_2^T = (0, \ldots, 0, s_n),$$

and $a_{2n} = a_{n2} = 0$.

We denote $\overline{\mathbf{c}} = \mathbf{c}_1 - \mathbf{c}_2$ and

$$A^- = \begin{pmatrix} a & \overline{\mathbf{c}}^T \\ \overline{\mathbf{c}} & \tilde{A} \end{pmatrix}. \tag{3.15}$$

Using these notations we now state a necessary and sufficient condition for a matrix whose graph is $\widehat{\mathrm{CP}}$ to be completely positive.

Theorem 3.79. *Let A be a doubly nonnegative matrix whose graph is $\widehat{\mathrm{CP}}$. Suppose A be in block form* (3.14) *and let A^- be as in* (3.15). *Then A is completely positive if and only if A^- is positive semidefinite.*

Proof. If A is completely positive, let $A = \sum_{i=1}^m \mathbf{b}_i\mathbf{b}_i^T$ be a CP representation of A, and let

$$\Omega_1 = \{1 \le i \le m \mid \{1, n\} = \operatorname{supp} \mathbf{b}_i\} \text{ and } \Omega_2 = \{1, \ldots, m\} \setminus \Omega_1.$$

For every $i \in \Omega_1$ let $\mathbf{d}_i$ be the vector obtained from $\mathbf{b}_i$ by multiplying the first entry by -1. Then

$$A^- = \sum_{i\in\Omega_1} \mathbf{d}_i\mathbf{d}_i^T + \sum_{i\in\Omega_2} \mathbf{b}_i\mathbf{b}_i^T$$

and is therefore positive semidefinite.

We now prove the sufficiency of this condition. Suppose A and A^- are both positive semidefinite. Then there exist $n \times n$ matrices W and $\overline{W}$ such that $A = WW^T$ and $A^- = \overline{W}\,\overline{W}^T$. Let

$$W = \begin{pmatrix} \mathbf{w}^T \\ W_0 \end{pmatrix} \text{ and } \overline{W} = \begin{pmatrix} \overline{\mathbf{w}}^T \\ \overline{W}_0 \end{pmatrix}.$$

Since $W_0W_0^T = \overline{W}_0\overline{W}_0^T$, there exists an orthogonal $n \times n$ matrix U such that $UW_0^T = \overline{W}_0^T$, and we may replace $\overline{W}$ by $\overline{W}U^T$. So assume

$$W = \begin{pmatrix} \mathbf{w}^T \\ W_0 \end{pmatrix} \text{ and } \overline{W} = \begin{pmatrix} \overline{\mathbf{w}}^T \\ W_0 \end{pmatrix}.$$

Then

$$A = \begin{pmatrix} \mathbf{w}^T\mathbf{w} & \mathbf{w}^TW_0^T \\ W_0\mathbf{w} & W_0W_0^T \end{pmatrix} \text{ and } A^- = \begin{pmatrix} \overline{\mathbf{w}}^T\overline{\mathbf{w}} & \overline{\mathbf{w}}^TW_0^T \\ W_0\overline{\mathbf{w}} & W_0W_0^T \end{pmatrix},$$

and in particular

$$W_0\mathbf{w} = \mathbf{c} \text{ and } W_0\overline{\mathbf{w}} = \overline{\mathbf{c}}. \tag{3.16}$$

Let

$$\mathbf{u} = \frac{1}{2}(\mathbf{w} + \overline{\mathbf{w}}) \text{ and } \mathbf{v} = \frac{1}{2}(\mathbf{w} - \overline{\mathbf{w}}).$$

Obviously, $\mathbf{w} = \mathbf{u}+\mathbf{v}$ and since $\|\mathbf{w}\|^2 = \|\overline{\mathbf{w}}\|^2 = a$, $\mathbf{u}$ and $\mathbf{v}$ are orthogonal. By (3.16),

$$W_0\mathbf{u} = \frac{1}{2}(\mathbf{c}+\overline{\mathbf{c}}) = \mathbf{c}_1 \text{ and } W_0\mathbf{v} = \frac{1}{2}(\mathbf{c}-\overline{\mathbf{c}}) = \mathbf{c}_2.$$

Let

$$\hat{A} = \begin{pmatrix} \mathbf{u}^T \\ \mathbf{v}^T \\ W_0 \end{pmatrix} \begin{pmatrix} \mathbf{u} & \mathbf{v} & W_0^T \end{pmatrix}.$$

Then $\hat{A}$ is an $(n+1)\times(n+1)$ positive semidefinite matrix, and since $\mathbf{u}$ and $\mathbf{v}$ are orthogonal,

$$\hat{A} = \begin{pmatrix} \mathbf{u}^T\mathbf{u} & 0 & \mathbf{c}_1^T \\ 0 & \mathbf{v}^T\mathbf{v} & \mathbf{c}_2^T \\ \mathbf{c}_1 & \mathbf{c}_2 & \tilde{A} \end{pmatrix}.$$

Thus the matrix $\hat{A}$ is nonnegative, and its graph is the graph of $\tilde{A}$ with two additional vertices, each joined by an edge to a vertex of $G(\tilde{A})$. The graph $G(\hat{A})$ is therefore completely positive, and so is the matrix $\hat{A}$. Let

$$E = \begin{pmatrix} 1 & 1 & 0 & 0 & \dots & 0 \\ 0 & 0 & 1 & 0 & \dots & 0 \\ 0 & 0 & 0 & 1 & \dots & 0 \\ \vdots & \vdots & \vdots & & \ddots & \vdots \\ 0 & 0 & 0 & \dots & 0 & 1 \end{pmatrix} \in \mathbb{R}^{n\times(n+1)}.$$

Then $A = E\hat{A}E^T$. This proves that A is completely positive. □

Remark 3.80. Observe that the proof of the sufficiency part in the last theorem did not rely on the fact that $a_{2n} = a_{n2} = 0$. So if a graph G has a vertex v such that $d(v) = 2$ and $G - v$ is completely positive, and A is a doubly nonnegative realization of G, we may assume that A is in block form (3.14) and define A^- as in (3.15). If A^- is positive semidefinite, then A is completely positive.

Odd cycles of length ≥ 5 are $\widehat{\text{CP}}$ graphs which are not completely positive. We use Theorem 3.79 to obtain a qualitative necessary and sufficient condition for a DNN matrix realization of such a graph to be completely positive. If A is a doubly nonnegative matrix whose graph is the n-cycle, we may assume that A has the form

$$\begin{pmatrix} d_1 & h_1 & 0 & & \dots & 0 & h_n \\ h_1 & d_2 & h_2 & 0 & & & 0 \\ 0 & h_2 & d_3 & \ddots & \ddots & & \\ \vdots & \ddots & & \ddots & \ddots & & \vdots \\ & & & \ddots & d_{n-2} & h_{n-2} & 0 \\ 0 & & & 0 & h_{n-2} & d_{n-1} & h_{n-1} \\ h_n & 0 & \dots & & 0 & h_{n-1} & d_n \end{pmatrix}. \tag{3.17}$$

Theorem 3.81. *Let A be a doubly nonnegative matrix realization of the n-cycle, $n \geq 5$ odd. Suppose A is in the form* (3.17). *Then A is completely positive if and only if*

$$\det A \geq 4 \prod_{i=1}^{n} h_i \,. \tag{3.18}$$

Proof. By Theorem 3.79 A is completely positive if and only if A^- is positive semidefinite. But since n is odd,

$$\begin{aligned} \det A - \det A^- &= 4h_1 h_n \det \begin{pmatrix} h_2 & d_3 & h_3 & \dots & 0 & 0 \\ 0 & h_3 & d_4 & \ddots & \dots & 0 \\ \vdots & 0 & \ddots & \ddots & \ddots & \vdots \\ & \vdots & \ddots & \ddots & d_{n-2} & h_{n-2} \\ 0 & & & \ddots & h_{n-2} & d_{n-1} \\ 0 & 0 & & \dots & 0 & h_{n-1} \end{pmatrix} \\ &= 4 \prod_{i=1}^{n} h_i \,. \end{aligned}$$

Hence $\det A \geq 4\prod_{i=1}^n h_i$ if and only if $\det A^- \geq 0$. If A is completely positive, then A^- is positive semidefinite, and therefore inequality (3.18) holds. Conversely, if (3.18) holds, then $\det A > 0$, so the doubly nonnegative A is nonsingular. But then $A^-(1) = A(1)$ is positive definite, and therefore $\det A^- \geq 0$ implies that A^- is positive semidefinite. □

Remark 3.82. Theorem 3.81 implies that a completely positive matrix whose graph is an odd cycle of length 5 or more is necessarily positive definite (but see also Exercise 3.50).

Now consider the case that the book graph has two "pages", one of them a triangle. That is, G is a graph obtained by adding to a completely positive graph a vertex v adjacent to both ends of an edge $\{u, w\}$. We may call such graph a *nearly-*$\widehat{CP}$ *graph.* An example of such graph is obtained by adding the edge $\{u, w\}$ to the $\widehat{\text{CP}}$ graph shown on the left in the figure on page 292. Then a doubly nonnegative matrix A with $G(A) = G$ has the following form, up to permutation similarity.

$$A = \begin{pmatrix} a & h & x_1 & \mathbf{x}_2^T \\ h & b & y_1 & \mathbf{y}_2^T \\ x_1 & y_1 & a_1 & \mathbf{0}^T \\ \mathbf{x}_2 & \mathbf{y}_2 & \mathbf{0} & A_2 \end{pmatrix}, \tag{3.19}$$

with $a_1, x_1, y_1 > 0$.

Theorem 3.83. *Let A be a doubly nonnegative matrix of the form* (3.19). *Then the following are equivalent.*

(a) *A is completely positive.*
(b) *Either* $\det A[1, 3|2, 3] \geq 0$, *or* $\mathbf{x}_2^T A_2^\dagger \mathbf{y}_2 \geq 0$, *or*

$$\det(A/\tilde{A}) \geq 4\frac{1}{a_1} \det A[1, 3|2, 3] \cdot \mathbf{x}_2^T A_2^\dagger \mathbf{y}_2,$$

where $\tilde{A} = a_1 \oplus A_2$.

Proof. First note that $(A/[3])_{12} = h - \frac{1}{a_1} x_1 y_1 = \frac{1}{a_1} \det A[1, 3|2, 3]$. Hence if $\det A[1, 3|2, 3] \geq 0$ the matrix $A/[3]$ is a doubly nonnegative matrix, with a completely positive graph. As $\frac{1}{a_1} x_1 y_1 > 0$, we get that in the case that $\mathbf{x}_2^T A_2^\dagger \mathbf{y}_2 \geq 0$, $\Upsilon = \emptyset$ in Theorem 3.75. Thus in both cases A is completely positive.

Now suppose $\mathbf{x}_2^T A_2^\dagger \mathbf{y}_2 < 0$. In the notations of Theorem 3.75, in this case $\Upsilon = \{2\}$, and A is completely positive if and only if

$$\sqrt{a_\perp b_\perp} + h_\perp \geq -2\mathbf{x}_2^T A_2^\dagger \mathbf{y}_2 \tag{3.20}$$

(see page 290). As $A/\tilde{A}$ is positive semidefinite

$$-\sqrt{a_\perp b_\perp} \leq h_\perp \leq \sqrt{a_\perp b_\perp}.$$

In particular, we have in this case

$$-\sqrt{a_\perp b_\perp} \leq -2\mathbf{x}_2^T A_2^\dagger \mathbf{y}_2 - h_\perp.$$

Hence (3.20) holds if and only if

$$-\sqrt{a_\perp b_\perp} \leq -2\mathbf{x}_2^T A_2^\dagger \mathbf{y}_2 - h_\perp \leq \sqrt{a_\perp b_\perp}.$$

That is, A is completely positive if and only if

$$a_\perp b_\perp \geq (2\mathbf{x}_2^T A_2^\dagger \mathbf{y}_2 + h_\perp)^2,$$

if and only if

$$\det\left(A/\tilde{A}\right) = a_\perp b_\perp - h_\perp^2 \geq 4(\mathbf{x}_2^T A_2^\dagger \mathbf{y}_2 + h_\perp)\mathbf{x}_2^T A_2^\dagger \mathbf{y}_2.$$

Finally, note that $h_\perp + \mathbf{x}_2^T A_2^\dagger \mathbf{y}_2 = h - x_1 y_1 \frac{1}{a_1} = \frac{1}{a_1}\det A[1,3|2,3]$. □

Remark 3.84. If in Theorem 3.83 the matrix A_2 is invertible,

$$\det\left(A/\tilde{A}\right) = \det A/\det\tilde{A} = \frac{\det A}{a_1 \det A_2},$$

and $\det A_2 > 0$. Hence in this case A is completely positive if an only if either $\det A[1,3|2,3] \geq 0$, or $\mathbf{x}_2^T \operatorname{adj} A_2\, \mathbf{y}_2 \geq 0$, or

$$\det A \geq 4 \det A[1,3|2,3] \cdot \mathbf{x}_2^T \operatorname{adj} A_2\, \mathbf{y}_2.$$

A graph, which is both a special type of a $\widehat{\text{CP}}$ graph and a nearly-$\widehat{\text{CP}}$ graph is a *house graph*. A house graph consists of an n-cycle C, $n \geq 5$, plus one chord connecting two vertices u and v such that $d_C(u,v) = 2$. Suppose that the edges of C are $\{1,2\},\{2,3\},\ldots,\{n-1,n\}$ and $\{1,n\}$, and that $u = 1$, $v = 3$. Let A be a DNN matrix realization of G. Then

$$A = \begin{pmatrix} d_1 & h_1 & p & 0 & & \ldots & 0 & h_n \\ h_1 & d_2 & h_2 & 0 & & \ldots & 0 & 0 \\ p & h_2 & d_3 & h_3 & & \ldots & 0 & 0 \\ 0 & 0 & h_3 & d_4 & \ddots & & \vdots & \vdots \\ \vdots & \vdots & & \ddots & \ddots & \ddots & & \vdots \\ 0 & 0 & \ldots & & \ddots & a_{n-2} & h_{n-2} & 0 \\ 0 & 0 & \ldots & \ldots & & h_{n-2} & d_{n-1} & h_{n-1} \\ h_n & 0 & \ldots & \ldots & & 0 & h_{n-1} & d_n \end{pmatrix}. \quad (3.21)$$

That is, $a_{31} = a_{13} = p > 0$, $a_{ii} = d_i > 0$ and $a_{i\,i\dotplus 1} = a_{i\dotplus 1\,i} = h_i > 0$, $i = 1,\ldots,n$, where $\dotplus$ is addition modulo n on $\{1,\ldots,n\}$.

Theorem 3.85. *Let A be an $n\times n$ doubly nonnegative matrix with a house graph in the form (3.21), and suppose A^- is obtained from A by replacing the $1,n$ and $n,1$ entries by $-h_n$. If n is odd, then the following are equivalent.*

(a) *A is completely positive.*
(b) $\det A^- \geq 0$.

(c) $\det A \geq \det A[1,2|2,3] \cdot 4\prod_{i=3}^{n} h_i$.

In particular, if A is singular, then it is completely positive if and only if $\det A[1,2|2,3] \leq 0$.

Proof. By Theorem 3.79, (a) implies (b).

Computing $\det A$ and $\det A^-$ along the last column, and then the minors along the last row, we get

$$\begin{aligned}\det A - \det A^- &= (-1)^{n-1} 4h_{n-1}h_n \det A(n-1,n|1,n) \\ &= \det A[1,2|2,3] \cdot 4\prod_{i=3}^{n} h_i.\end{aligned}$$

Hence (b) is equivalent to (c).

To show that (c) or (b) implies (a), consider three possible cases. If A is positive definite, so is $A^-(1) = A(1)$. In that case, $\det A^- \geq 0$ implies that A^- is positive semidefinite, and A is completely positive by Theorem 3.79.

If A is singular, and $\det A[1,2|2,3] < 0$, then $\det A^- > 0$. As $A^-(1) = A(1)$ is positive semidefinite, at most one eigenvalue of A^- may be negative, but since $\det A^- > 0$, all the eigenvalues of A^- are necessarily positive. Hence A^- is positive definite, and again Theorem 3.79 implies that A is completely positive.

Finally, if $\det A = \det A[1,2|2,3] = 0$, then $A_\varepsilon = A + \varepsilon I$ is a doubly nonnegative matrix with the same house graph, which is positive definite for every $\varepsilon > 0$. Then $\det A_\varepsilon[1,2|2,3] = \det A[1,2|2,3] - \varepsilon a_{13} < 0$, and by the previous case A_ε is completely positive. Hence $A = \lim_{\varepsilon \to 0+} A_\varepsilon$ is also completely positive. □

Theorem 3.86. *Let A be an $n \times n$ doubly nonnegative with a house graph in the form* (3.21), *and suppose A^- is obtained from A by replacing the $1,n$ and $n,1$ entries by $-h_n$. If n is even, then the following are equivalent.*

(a) *A is completely positive.*
(b) *$A[1,2|2,3]$ is nonsingular and* $\det A \geq -\det A[1,2|2,3] \cdot 4\prod_{i=3}^{n} h_i$, *or* $A[1,2|2,3]$ *is singular and* $\det A(1) > 0$.

We leave the proof as an exercise.

Deletion of a vertex of degree 2

In this subsection we extend the idea of deletion of a pendant vertex (Corollary 3.61) to deletion of a vertex of degree 2. Suppose then that a graph G has a vertex v such that $d(v) = 2$ and u and w are the vertices adjacent to v. Given a doubly nonnegative matrix A whose graph is G, we construct a smaller doubly nonnegative matrix A' whose graph is obtained from G be deletion of the vertex v and possibly also the edge $\{u, w\}$. We may assume that $V(G) = \{1, \ldots, n\}$, $v = 1$, $u = 2$ and $w = 3$. Then if A is a DNN matrix realization of G,

$$A = \begin{pmatrix} a_{11} & a_{12} & a_{13} & 0 & \ldots & 0 \\ a_{12} & a_{22} & a_{23} & \ldots & \ldots & a_{2n} \\ a_{13} & a_{23} & a_{33} & & & \vdots \\ 0 & \vdots & & \ddots & & \\ \vdots & \vdots & & & \ddots & \vdots \\ 0 & a_{2n} & \ldots & & \ldots & a_{nn} \end{pmatrix}. \tag{3.22}$$

We write

$$A = \begin{pmatrix} \tilde{A} & S^T \\ S & M \end{pmatrix} \tag{3.23}$$

where $M = A[3, \ldots, n]$, and define

$$A^S = \begin{pmatrix} S^T M^\dagger S & S^T \\ S & M \end{pmatrix}$$

and

$$A^R = A - A^S = (A/M) \oplus 0_{n-2}.$$

By Theorem 1.75 the matrices A^S and A^R are positive semidefinite. Let

$$A_t = tA^S + (1 - t)A.$$

For every $0 \leq t \leq 1$ the matrix A_t is positive semidefinite, $A^S \preceq A_t \preceq A$, and all the entries in the last $n - 2$ rows and columns of A_t are equal to those of A. In particular, most of the entries of A_t are nonnegative — the $1, 2$ entry is the only one which may be negative.

Let

$$\delta(A) = \det A[1, 2|1, 3] = a_{11}a_{23} - a_{12}a_{13}.$$

For every t,

$$\delta(A_t) = t\delta(A^S) + (1 - t)\delta(A) = t(\delta(A^S) - \delta(A)) + \delta(A). \tag{3.24}$$

Now let

$$d = \begin{cases} 0 & \text{if } \delta(A) \geq 0 \\ \dfrac{-\delta(A)}{\delta(A^S) - \delta(A)} & \text{if } \delta(A) < 0 \leq \delta(A^S) \\ 1 & \text{otherwise} \end{cases}$$

and

$$A^{[1/2]} = A_d = A - dA^R.$$

That is, if $\delta(A) \geq 0$, $A^{[1/2]} = A$; If $\delta(A) < 0 \leq \delta(A^S)$, $A^{[1/2]} = A_t$ for the unique t such that $\delta(A_t) = 0$; and if both $\delta(A)$ and $\delta(A^S)$ are negative, $A^{[1/2]} = A^S$. Finally, let

$$A^{[1]} = A^{[1/2]}/[1].$$

Remark 3.87. The graph $G(A^{[1]})$ is obtained from $G(A)$ by deleting the vertex 1 and possibly also the edge $\{2, 3\}$.

Proposition 3.88. *Let A be a doubly nonnegative matrix in the form (3.22), then $A^{[1/2]}$ is also doubly nonnegative, and $A^{[1]}$ is positive semidefinite. The matrix $A^{[1]}$ is doubly nonnegative if and only if* $\max(\delta(A), \delta(A^S)) \geq 0$.

Proof. To show that $A^{[1/2]}$ is doubly nonnegative we only need to show that its 1, 2 entry is nonnegative. This is clear if $\delta(A) \geq 0$, since in that case $A^{[1/2]} = A$. If $\delta(A) < 0 \leq \delta(A^S)$, then $\delta(A^{[1/2]}) = 0$, so that

$$(A^{[1/2]})_{12} = \frac{(A^{[1/2]})_{11}(A^{[1/2]})_{23}}{(A^{[1/2]})_{13}} \geq 0.$$

Finally, if $\delta(A)$ and $\delta(A^S)$ are both negative, then $\delta(A^{[1/2]}) = \delta(A^S) < 0$, and this implies that

$$(A^{[1/2]})_{12} > \frac{(A^{[1/2]})_{11}(A^{[1/2]})_{23}}{(A^{[1/2]})_{13}} \geq 0.$$

The matrix $A^{[1]}$ is positive semidefinite by Theorem 1.75, and only its 1, 2 entry may be negative. But

$$(A^{[1]})_{12} = (A^{[1/2]})_{23} - \frac{(A^{[1/2]})_{12}(A^{[1/2]})_{13}}{(A^{[1/2]})_{11}} = \frac{\delta(A^{[1/2]})}{(A^{[1/2]})_{11}},$$

so $A^{[1]}$ is nonnegative if and only if $\delta(A^{[1/2]}) \geq 0$. By the definition of $A^{[1/2]}$ and (3.24), this occurs if and only if $\max(\delta(A), \delta(A^S)) \geq 0$. □

Remark 3.89. If $A^{[1]}$ is completely positive, then so is $A^{[1/2]}$, since

$$A^{[1/2]} - (0_1 \oplus A^{[1]}) = \frac{1}{(A^{[1/2]})_{11}} \mathbf{u}\mathbf{u}^T,$$

where $\mathbf{u}$ is the first column of $A^{[1/2]}$.

If $A^{[1/2]}$ is completely positive, then by Lemma 3.36 the matrix A is also completely positive, since $A - A^{[1/2]}$ is a positive semidefinite matrix which is zero except for a 2×2 principal submatrix.

So if $A^{[1]}$ is completely positive, A is also completely positive.

If $\delta(A) \geq 0$, then the converse is also true.

Theorem 3.90. *Let A be a doubly nonnegative matrix in the form* (3.22) *with $\delta(A) \geq 0$. Then A is completely positive if and only if $A^{[1]}$ is completely positive.*

Proof. Suppose A is completely positive. Let $A = \sum_{i=1}^m \mathbf{b}_i \mathbf{b}_i^T$ be any CP representation of A, and let

$$\Omega_1 = \{1 \leq i \leq m \mid 1 \in \operatorname{supp} \mathbf{b}_i\} \text{ and } \Omega_2 = \{1, \ldots, m\} \setminus \Omega_1.$$

Let

$$B = \sum_{i \in \Omega_1} \mathbf{b}_i \mathbf{b}_i^T \text{ and } C = \sum_{i \in \Omega_2} \mathbf{b}_i \mathbf{b}_i^T.$$

Then B and C are completely positive matrices and $A = B + C$. The matrix B is zero except for its 3×3 leading principal submatrix, so the matrix $H = 0_1 \oplus (B/[1])$ is zero except for its principal submatrix based on rows 2 and 3.

Observe that

$$A - (0_1 \oplus (A/[1])) = B - H = \begin{pmatrix} a_{11} & a_{12} & a_{13} \\ a_{12} & a_{12}^2/a_{11} & a_{12}a_{13}/a_{11} \\ a_{13} & a_{12}a_{13}/a_{11} & a_{13}^2/a_{11} \end{pmatrix} \oplus 0_{n-3}.$$

We may therefore write

$$0_1 \oplus A^{[1]} = A - (B - H) = C + H.$$

Since C is completely positive and $C+H$ is nonnegative, $C+H$ is completely positive by Lemma 3.36. Hence $0_1 \oplus A^{[1]}$ is completely positive, and so is $A^{[1]}$. □

If $\delta(A) < 0$, it is possible for A to be completely positive when $A^{[1]}$ is not.

Example 3.91. Let

$$A = \begin{pmatrix} 2 & 1 & 1 & 0 & 0 \\ 1 & 2 & 0 & 0 & 1 \\ 1 & 0 & 2 & 1 & 0 \\ 0 & 0 & 1 & 2 & 1 \\ 0 & 1 & 0 & 1 & 2 \end{pmatrix}.$$

Then A is diagonally dominant and therefore completely positive, and

$$A^S = \begin{pmatrix} 5/4 & 3/4 & 1 & 0 & 0 \\ 3/4 & 5/4 & 0 & 0 & 1 \\ 1 & 0 & 2 & 1 & 0 \\ 0 & 0 & 1 & 2 & 1 \\ 0 & 1 & 0 & 1 & 2 \end{pmatrix}.$$

Hence in this case $\max(\delta(A), \delta(A^S)) < 0$, and $A^{[1]}$ is not even doubly nonnegative.

In the case that $\delta(A) < 0$, an additional condition is required to ensure that complete positivity of A is equivalent to that of $A^{[1]}$.

Definition 3.92. Let A be a doubly nonnegative matrix in the form (3.23) and let 1 be a vertex of degree 2 in $G(A)$. We say that A is *diminishable* if $\operatorname{rank} A \leq \operatorname{rank} M + 1$.

Remark 3.93. Clearly, every diminishable matrix is singular. The converse is generally not true. But if M is nonsingular and A is singular then A is diminishable.

The diminishability of A implies that A_t, $0 \leq t \leq 1$, are the only matrices B satisfying $A^S \preceq B \preceq A$.

Proposition 3.94. *Let A be a diminishable doubly nonnegative matrix in block form* (3.23). *Then every matrix B such that $A^S \preceq B \preceq A$ is a convex combination of A^S and A.*

Proof. By Theorem 1.77, $\operatorname{rank} A = \operatorname{rank} A^R + \operatorname{rank} M$. Since A is diminishable, this implies that $\operatorname{rank} A^R \leq 1$. But then if $A^S \preceq B \preceq A$, the equality

$$A^R = (A - B) + (B - A^S)$$

implies that each of the positive semidefinite matrices $A - B$ and $B - A^S$ is a nonnegative multiple of A^R. $\square$

We can now prove the following.

Theorem 3.95. *If A is a doubly nonnegative diminishable matrix of the form (3.22), then A is completely positive if and only if $A^{[1]}$ is completely positive.*

Proof. We already know that if $A^{[1]}$ is completely positive, then so are $A^{[1/2]}$ and A (see Remark 3.89). By Theorem 3.90 we only need to show that if A is completely positive and $\delta(A) < 0$ then $A^{[1]}$ is completely positive. We first show that under these conditions $A^{[1/2]}$ is completely positive.

So suppose A is completely positive. Let $A = \sum_{i=1}^{m} \mathbf{b}_i \mathbf{b}_i^T$ be any CP representation of A, and let

$$\Omega_1 = \{1 \leq i \leq m \mid 1 \in \operatorname{supp} \mathbf{b}_i\} \text{ and } \Omega_2 = \{1, \ldots, m\} \setminus \Omega_1.$$

Let

$$B = \sum_{i \in \Omega_1} \mathbf{b}_i \mathbf{b}_i^T \text{ and } C = \sum_{i \in \Omega_2} \mathbf{b}_i \mathbf{b}_i^T.$$

Then $B = B' \oplus 0_{n-3}$ and $C = 0_1 \oplus C'$ are completely positive matrices such that $A = B + C$, and

$$\delta(B') = b_{11}b_{23} - b_{12}b_{13} = a_{11}b_{23} - a_{12}a_{13} \leq \delta(A) < 0.$$

Hence

$$b_{12}b_{13} > b_{11}b_{23}.$$

Multiply this inequality by the inequality

$$b_{11}b_{33} \geq b_{13}^2$$

to get that

$$b_{33}b_{12} > b_{13}b_{23}. \tag{3.25}$$

Inequality (3.25) implies that $H = (B'/[3]) \oplus 0_{n-2}$ is nonnegative. The matrix $B - H$ is equal to $\mathbf{u}\mathbf{u}^T$ where $\mathbf{u}$ is the third column of B divided by $\sqrt{b_{33}}$, so $\mathbf{u}$ is a nonnegative vector with $\operatorname{supp} \mathbf{u} = \{1, 2, 3\}$. Hence $B - H$ is also completely positive, and $F = A - H = C + \mathbf{u}\mathbf{u}^T$ is completely positive. Since $F[3, \ldots, n] = M$, the matrix F^S is equal to A^S, which implies that $A^S \preceq F$. Since H is positive semidefinite, we also have $F \preceq A$. By Proposition 3.94, $F = A_{t_0}$ for some $0 \leq t_0 \leq 1$. Now

$$\delta(A_{t_0}) = f_{11}f_{23} - f_{12}f_{13} = u_1^2(c_{23} + u_2u_3) - u_1^2u_2u_3 = u_1^2c_{23} \geq 0. \tag{3.26}$$

Combining (3.24) and (3.26) and the fact that $\delta(A) < 0$, we get that $\delta(A^S) \geq 0$. But then (again by (3.24)) $\delta(A_t)$ is an increasing function. Hence $\delta(A^{[1/2]}) = 0$ implies that $d \leq t_0$. We may write

$$A^{[1/2]} = A_{t_0} + (t_0 - d)A^R.$$

So $A^{[1/2]}$ is completely positive, as a sum of a completely positive matrix and a positive semidefinite matrix which is zero except for a 2×2 principal submatrix. Since $\delta(A^{[1/2]}) = 0$, Theorem 3.90 implies that $(A^{[1/2]})^{[1]} = A^{[1]}$ is also completely positive. □

Remark 3.96. Observe that Theorem 3.95 also holds when 1 is a vertex of degree 1. In that case $\delta(A) = 0$, and $A^{[1]} = A/[1]$. By Corollary 3.61, the theorem holds in this case too.

Successive deletion of pendant vertices may be used to show that every DNN matrix realization of a tree is completely positive. A similar use of the technic of deletion of a vertex of degree two is possible, if we start with a diminishable matrix. To be able to use the technic on $A^{[1]}$, the matrix $A^{[1]}$ needs to be diminishable with respect to some vertex of degree 2 in $G(A^{[1]})$, or $G(A^{[1]})$ has to have a vertex of degree 1. We describe a family of graphs for which such repeated deletions are possible.

Definition 3.97. We say that a graph G on n vertices is an *NCC graph* if G has two vertices of degree 2, u and v, and $E(G)$ consists of two internally disjoint paths P and Q from u to v,

$$P = u - p_1 - p_2 - \ldots - p_{m-1} - v$$
$$Q = u - q_1 - q_2 - \ldots - q_{k-1} - v\,,$$

and edges $\{p_i, q_j\}$ such that

$$\text{If for } i \leq l \text{ both } \{p_i, q_j\} \text{ and } \{p_l, q_r\} \text{ are edges of } G, \text{ then } j \leq r. \quad (3.27)$$

The two paths form a cycle, and the additional edges are noncrossing chords of the cycle, with one end an internal vertex of one path and the other end an internal vertex of the other path. Here is one example of an NCC graph.

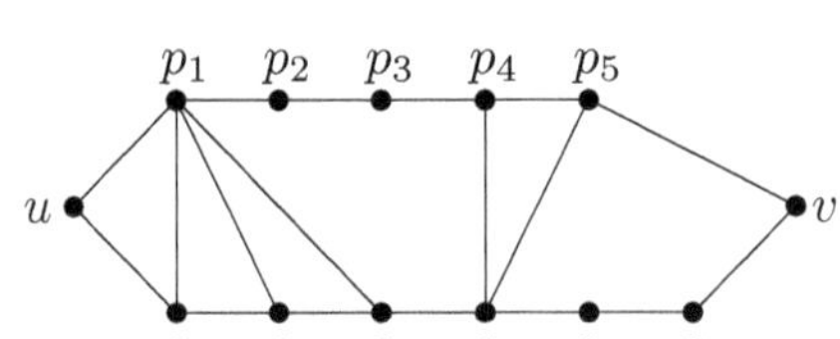

While every NCC graph is a cycle with noncrossing chords, not every cycle with noncrossing chords is an NCC graph. For example,

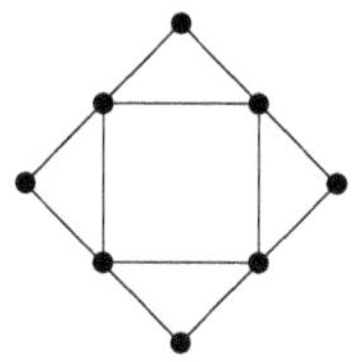

is not an NCC graph.

Given an NCC graph G on n vertices as in Definition 3.97, we order its vertices. For that we define $\nu : V(G) \to \{1, \ldots, n\}$ as follows: $\nu(u) = 1$ and $\nu(v) = n$. Next, we consider p_1 and q_1. If there exists $l > 1$ such that $\{p_1, q_l\}$ is an edge of G, we let $\nu(q_1) = 2$. Otherwise, we let $\nu(p_1) = 2$. We continue the same way. If the r-th vertex was already chosen, we consider the first p_i and the first q_j that are not yet assigned a number. If p_i is adjacent in G to some q_l, $l > j$, we let $\nu(q_j) = r + 1$. Otherwise, we let $\nu(p_i) = r + 1$. When $\nu(p_i)$ is defined for all the vertices of the path P, there may still be some vertices on the path Q which were not yet assigned a number. We complete the numbering by keeping these vertices in their natural order on Q. For the NCC graph drawn above, the new order is

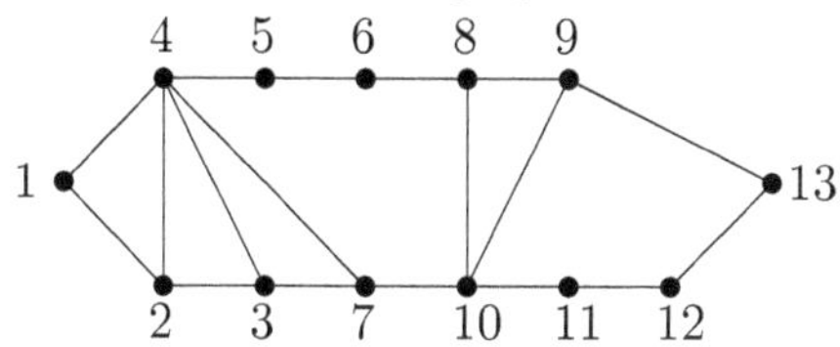

Given this order, we denote by G_i the subgraph induced by $\{i, \ldots, n\}$. Then G_i is either an NCC graph, or an NCC graph with one or two "tails" attached to it. In this subgraph, the degree of i is at most 2, and if i has degree 2 in G_i, then $i + 1$ is one of the two vertices adjacent to i. Given a diminishable DNN matrix realization of G, we may therefore try deleting successively vertices of degree 2 or 1. The process will work, due to the next two results.

Theorem 3.98. *If A is doubly nonnegative and $G(A)$ is an NCC graph, then A is diminishable if and only if A is singular.*

Proof. We already know that every diminishable matrix is singular, so we only have to prove the reverse implication. Suppose A is a doubly nonnegative singular matrix and $G(A)$ is an NCC graph whose vertices are ordered as described before the statement of the theorem. The vertex 1 is adjacent to 2 and some $l \geq 3$. The matrix A therefore is in the form (3.23), where S has one nonzero entry in its first row. We will show that $A[1,\ldots,n-2|3,\ldots,n]$ is necessarily nonsingular, and since A is positive semidefinite, M is nonsingular. This, together with the singularity of A would imply that A is diminishable.

To prove that $A[1,\ldots,n-2|3,\ldots,n]$ is nonsingular, we reorder the columns of A. Let $\pi : \{1,\ldots,n-2\} \to \{3,\ldots,n\}$ be defined as follows.

$$\begin{aligned} &\pi(1) = l && \\ &\pi(i) = \nu(p_{h+1}) && \text{if } i = \nu(p_h), h < m-1 \\ &\pi(i) = \nu(q_{r+1}) && \text{if } i = \nu(q_r), r < k-1 \\ &\pi(i) = n && \text{if } i = \nu(p_{m-1}) \end{aligned}$$

Now let A^π be the matrix obtained from A by leaving the first two columns as they are, and for $1 \leq i \leq n-2$ letting column $i+2$ of A^π be column $\pi(i)$ of A. We claim that $A^\pi[1,\ldots,n-2|3,\ldots,n]$ is a lower triangular matrix with all its diagonal elements nonzero. Indeed, for $1 \leq i \leq n-2$

$$A^\pi_{i\,i+2} = a_{i\,\pi(i)} > 0,$$

since if i is on the path P, $\pi(i)$ is the next vertex in the natural order of P, and if i is on Q, $\pi(i)$ is the next vertex in the natural order of Q. If $j > i+2$, then $\{i, \pi(j)\}$ is not an edge of G. This is clear if both i and j are internal vertices of P, or both are internal vertices of Q, and it is clear also if $i = 1$. If $i = \nu(p_h)$ for some h and $j = \nu(q_r)$ for some r, then $\pi(j) = \nu(q_{r+1})$, and by the definition of ν, p_h is not adjacent to q_{r+1}. If $i = \nu(q_r)$ for some r and $j = \nu(p_k)$ for some k, then by the definition of ν, some vertex p_s, $s \leq k$ is adjacent to some q_t, $t > r$. By (3.27), p_k cannot be adjacent to q_r. □

Proposition 3.99. *If A is an $n \times n$ diminishable doubly nonnegative matrix in block form* (3.23) *and $d_{G(A)}(1) \leq 2$, then $A^{[1]}$ is singular.*

Proof. Since $A = A^{[1/2]} + dA^R$ and A^R is positive semidefinite,

$$\operatorname{rank} A^{[1/2]} \leq \operatorname{rank} A \leq n-1.$$

By Theorem 1.77,

$$\operatorname{rank} A^{[1]} = \operatorname{rank} A^{[1/2]} - 1 \leq n-2,$$

so $A^{[1]}$ is singular. □

Given a singular doubly nonnegative matrix A whose graph is an NCC graph, we may define for $r \geq 2$,

$$A^{[r]} = (A^{[r-1]})^{[1]}.$$

By Theorems 3.95 and 3.98 and Proposition 3.99, the matrix $A^{[r-1]}$ is completely positive if and only if $A^{[r]}$ is completely positive. We therefore have the following result, which is an easy consequence of the previous discussion.

Corollary 3.100. *Let A be a singular $n \times n$ doubly nonnegative matrix whose graph is an NCC graph. Then A is completely positive if and only if for every $1 \leq r \leq n-4$ the matrix $A^{[r]}$ is nonnegative.*

Exercises

3.42. Prove the statement in Example 3.67.

3.43. Prove Proposition 3.71, and use it to show that a matrix that is permutationally similar to the matrix A of Example 3.69 does not have property PLSS.

3.44. Use the fact that every $n \times n$ doubly nonnegative matrix, $n \leq 4$, is completely positive to show that every such matrix has property PLSS.

3.45. Prove (without using Theorem 3.57) that if a graph G is T_n and the vertex 1 is incident with the common base of the $n-2$ triangles, then every DNN matrix realization of G has property PLSS.

3.46. Check which of the following doubly nonnegative matrices is completely positive.

$$\text{(a)} \quad \begin{pmatrix} 12 & 2 & 2 & 6 & 4 & 2 \\ 2 & 23 & 7 & 9 & 6 & 4 \\ 2 & 7 & 5 & 7 & 0 & 0 \\ 6 & 9 & 7 & 13 & 0 & 0 \\ 4 & 6 & 0 & 0 & 8 & 6 \\ 2 & 4 & 0 & 0 & 6 & 5 \end{pmatrix}$$

(b) $\begin{pmatrix} 5 & 0 & 1 & 1 & 1 & 1 \\ 0 & 13 & 1 & 3 & 3 & 5 \\ 1 & 1 & 2 & 3 & 0 & 0 \\ 1 & 3 & 3 & 5 & 0 & 0 \\ 1 & 3 & 0 & 0 & 2 & 4 \\ 1 & 5 & 0 & 0 & 4 & 10 \end{pmatrix}$

(c) $\begin{pmatrix} 7 & 1 & 2 & 2 & 1 & 1 \\ 1 & 12 & 1 & 3 & 3 & 5 \\ 2 & 1 & 2 & 3 & 0 & 0 \\ 2 & 3 & 3 & 5 & 0 & 0 \\ 1 & 3 & 0 & 0 & 2 & 4 \\ 1 & 5 & 0 & 0 & 4 & 10 \end{pmatrix}$

3.47. Prove Corollary 3.77.

3.48. Construct an example of a nonsingular critical CP matrix realization of C_5.

3.49. Show that every CP realization of an odd cycle is not i-minimizable for every i.

3.50. Show that if n is odd, an $n \times n$ matrix with the sign pattern

$$\begin{pmatrix} + & 0 & \dots & 0 & + \\ + & + & \ddots & & 0 \\ 0 & + & \ddots & \ddots & \vdots \\ \vdots & \ddots & \ddots & + & 0 \\ 0 & \dots & 0 & + & + \end{pmatrix}$$

has a positive determinant. Use this observation to deduce that a completely positive matrix whose graph is an odd cycle is nonsingular. (This is an alternative argument to prove the claim in Remark 3.82.)

3.51. Prove Theorem 3.86.
Hint: For the case that A and $A[1,2|2,3]$ are singular, show that if A in (3.21) is completely positive and singular, then $\det A(1) > 0$, by considering the possible pattern of a CP factorization.

3.52. Let

$$A = \begin{pmatrix} 3 & 3 & 3 & 0 & 0 & 0 \\ 3 & 11 & 1 & 4 & 1 & 0 \\ 3 & 1 & 7 & 2 & 1 & 0 \\ 0 & 4 & 2 & 6 & 4 & 2 \\ 0 & 1 & 1 & 4 & 5 & 4 \\ 0 & 0 & 0 & 2 & 4 & 4 \end{pmatrix}.$$

Show that

$$A^{[1]} = \begin{pmatrix} 6 & 0 & 4 & 1 & 0 \\ 0 & 2 & 2 & 1 & 0 \\ 4 & 2 & 6 & 4 & 2 \\ 1 & 1 & 4 & 5 & 4 \\ 0 & 0 & 2 & 4 & 4 \end{pmatrix},$$

and conclude that A is completely positive.

3.53. Check whether the matrix

$$A = \begin{pmatrix} 2 & 2 & 2 & 0 & 0 & 0 \\ 2 & 7 & 1 & 4 & 1 & 0 \\ 2 & 1 & 7 & 1 & 1 & 0 \\ 0 & 4 & 1 & 6 & 4 & 2 \\ 0 & 1 & 1 & 4 & 5 & 4 \\ 0 & 0 & 0 & 2 & 4 & 4 \end{pmatrix}$$

is completely positive.

3.54. Check whether the matrix

$$A = \begin{pmatrix} 2 & 1 & 2 & 0 & 0 & 0 & 0 \\ 1 & 2 & 3 & 1 & 0 & 0 & 0 \\ 2 & 3 & 7 & 3 & 2 & 0 & 0 \\ 0 & 1 & 3 & 4 & 1 & 1 & 0 \\ 0 & 0 & 2 & 1 & 5 & 1 & 2 \\ 0 & 0 & 0 & 1 & 1 & 3 & 2 \\ 0 & 0 & 0 & 0 & 2 & 2 & 2 \end{pmatrix}$$

is completely positive.

Notes. Property PLSS was defined in [Salce and Zanardo (1993)], and the relevant subsection is based on that paper. Book graphs are defined and discussed in [Barioli (1998a)]. $\widehat{\text{CP}}$ graphs are called there *excellent graphs*, and DNN matrix realizations of such graphs are called *excellent matrices*. Corollary 3.77 was proved, along with other related results, in [Drew *et al.* (2001)]. The proof given there is direct, and does not depend on Theorem 3.75. Theorem 3.81 was proved in [Barioli (1998a)], and also in [Xu (2002)], using a different approach and relying on a theorem proved in [Xu and Li (2000)]. Theorem 3.83 is based on [Di Virgilio (2019)]. Matrices with a house graph were studied in [Berman and Shasha (2012)], where the term house graph was coined. A result for the odd case appeared previously in [Xu (2002)]. Theorems 3.85 and 3.86 and Exercise 3.51 combine and extend the results of these papers, while the proofs use also [Di Virgilio (2019)], where the case $n = 5$ is dealt with, and Barioli's Theorem 3.75. The subsection on deletion of a vertex of degree 2 is based on [Barioli (2001a)], where a triple consisting of a vertex of degree 2 and the two vertices adjacent to it is called a *dog-ear*, and NCC graphs are called *non crossing cycles*. Exercises 3.46 and 3.52–3.54 are taken from [Barioli (1998a)] and [Barioli (1998b)].

3.7 $\mathcal{CP}_5$

The cone $\mathcal{CP}_5$ is the first to differ from $\mathcal{DNN}_5$. It is time to check whether the results amassed so far suffice to characterize the 5×5 completely positive matrices. The answer is, unfortunately, no. To date, there is no complete characterization of 5×5 completely positive matrices. We present here existing results in cases where a characterization is known. We also consider the difference between $\mathcal{DNN}_5$ and $\mathcal{CP}_5$. For a description of the faces of $\mathcal{CP}_5$ see Chapter 5 on the structure of $\mathcal{COP}_n$ and $\mathcal{CP}_n$.

Determining complete positivity of a 5×5 matrix

If A is a 5×5 doubly nonnegative matrix such that $G(A)$ has no odd cycle, then A is completely positive. We therefore need to consider only doubly nonnegative matrices whose graph contains an odd cycle, which in the case of a 5×5 matrix is simply a cycle on all 5 vertices. Note that in such matrices all the diagonal entries are necessarily positive. Up to isomorphism, there are eight graphs on 5 vertices containing the 5-cycle. These are shown in the following figure.

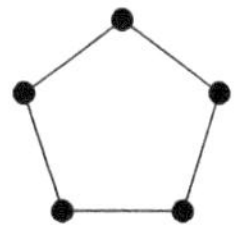
G_1

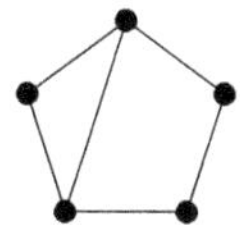
G_2

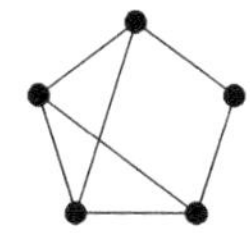
G_3

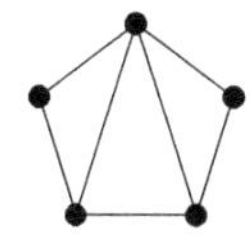
G_4

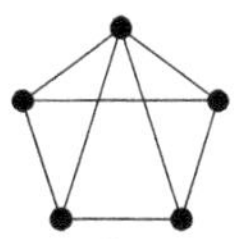
G_5

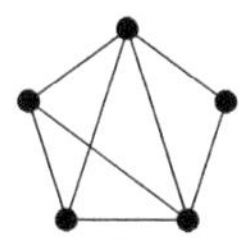
G_6

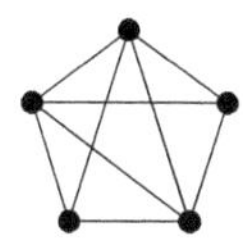
G_7

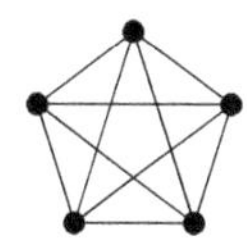
G_8

The results presented in this section address only the cases that $G(A)$ is not the complete graph G_8. A complete characterization in terms of the entries of the matrix is only given in the case that $G(A)$ has a vertex of degree 2.

The graphs G_1, G_2 and G_3 are $\widehat{\mathrm{CP}}$ graphs, and the graphs G_4 and G_6 are nearly-$\widehat{\mathrm{CP}}$ graphs. For these five graphs the results of Section 3.6 apply. We present an interpretation of these results for the five graphs in terms of the weights of cycles in the graphs, defined as follows.

Definition 3.101. Let C be a cycle of length $k \geq 3$ in the graph $G = G(A)$ of a symmetric matrix A, with edges

$$\{i_1, i_2\}, \{i_2, i_3\}, \ldots, \{i_{k-1}, i_k\}, \{i_k, i_1\}.$$

The *weight* of the cycle C in A is

$$\mathrm{w}_A(C) = (-1)^{k-1} \prod_{j=1}^{k} a_{i_j\, i_{j \oplus 1}},$$

with $\oplus$ denoting addition modulo k on $\{1, \ldots, k\}$.

The *corrected weight* of C is

$$\mathrm{cw}_A(C) = (-1)^{k-1} \prod_{j=1}^{k} a_{i_j\, i_{j \oplus 1}} \prod_{i \notin V(C)} a_{ii}.$$

If U is a set of vertices of $G = G(A)$, the *total weight* of U in A is the sum of all corrected weights of cycles that include all the vertices of U,

$$\mathrm{tw}_A(U) = \sum_{U \subseteq V(C)} \mathrm{cw}_A(C).$$

Example 3.102. Consider a matrix $A \in \mathcal{DNN}_5$, whose graph is one of the graphs G_1, G_2, G_3, where the vertices are labeled in counter-clockwise order, starting from the top vertex, labeled 1. In all three cases, consider the vertex 5 as the hat vertex in these $\widehat{\text{CP}}$ graphs. Then

$$A = \begin{pmatrix} a_{11} & a_{12} & a_{13} & 0 & a_{15} \\ a_{12} & a_{22} & a_{23} & a_{24} & 0 \\ a_{13} & a_{23} & a_{33} & a_{34} & 0 \\ 0 & a_{24} & a_{34} & a_{44} & a_{45} \\ a_{15} & 0 & 0 & a_{45} & a_{55} \end{pmatrix}. \tag{3.28}$$

If $G(A) = G_1$, $a_{13} = a_{24} = 0$; if $G(A) = G_2$, only $a_{24} = 0$.

If $G(A) = G_3$, there are four cycles through the hat vertex 5, two 5-cycles and two 4-cycles:

$$\begin{aligned} C_1 &= 1 - 2 - 3 - 4 - 5 - 1 \\ C_2 &= 1 - 3 - 2 - 4 - 5 - 1 \\ C_3 &= 1 - 2 - 4 - 5 - 1 \\ C_4 &= 1 - 3 - 4 - 5 - 1. \end{aligned}$$

Their corrected weights are

$$\begin{aligned} \text{cw}_A(C_1) &= a_{12}a_{23}a_{34}a_{45}a_{15}, \\ \text{cw}_A(C_2) &= a_{13}a_{23}a_{24}a_{45}a_{15}, \\ \text{cw}_A(C_3) &= -a_{12}a_{24}a_{45}a_{15}a_{33}, \\ \text{cw}_A(C_4) &= -a_{13}a_{34}a_{45}a_{15}a_{22}. \end{aligned}$$

The total weight of the hat vertex $\{5\}$ is

$$\begin{aligned} \text{tw}_A(\{5\}) = a_{12}a_{23}a_{34}a_{45}a_{15} &+ a_{13}a_{23}a_{24}a_{45}a_{15} \\ &- a_{12}a_{24}a_{45}a_{15}a_{33} - a_{13}a_{34}a_{45}a_{15}a_{22}. \end{aligned}$$

If $G(A) = G_2$, only the cycles C_1 and C_4 pass through 5 in $G(A)$, and

$$\text{tw}_A(\{5\}) = a_{12}a_{23}a_{34}a_{45}a_{15} - a_{13}a_{34}a_{45}a_{15}a_{22}.$$

If $G(A) = G_1$, only the cycle C_1 passes through 5 in $G(A)$, and

$$\text{tw}_A(\{5\}) = a_{12}a_{23}a_{34}a_{45}a_{15}.$$

In all three cases

$$\text{tw}_A(\{5\}) = a_{45}a_{15} \det A(4, 5|1, 5).$$

Consider a 5×5 doubly nonnegative matrix A with a $\widehat{\text{CP}}$ graph, where i is a hat vertex in $G(A)$ (i.e., i is a vertex of degree 2 in $G(A)$, and the two neighbours of i are not adjacent to each other). As in Section 3.6, we denote by A^- a matrix obtained from A by replacing one of the off-diagonal nonzero entries in row i (and column i) by its negative.

Theorem 3.103. *Let A be a 5×5 doubly nonnegative matrix with a $\widehat{\text{CP}}$ graph, and let i be a hat vertex in $G(A)$. Then the following are equivalent:*

(a) *A is completely positive.*
(b) $\det A \geq 4\text{tw}_A(\{i\})$.
(c) *A^- is positive semidefinite.*
(d) $\det A^- \geq 0$.

Proof. Suppose without loss of generality that A is as in (3.28), the graphs are labeled as in Example 3.101, and $i = 5$. Let

$$A^- = \begin{pmatrix} a_{11} & a_{12} & a_{13} & 0 & -a_{15} \\ a_{12} & a_{22} & a_{23} & a_{24} & 0 \\ a_{13} & a_{23} & a_{33} & a_{34} & 0 \\ 0 & a_{24} & a_{34} & a_{44} & a_{45} \\ -a_{15} & 0 & 0 & a_{45} & a_{55} \end{pmatrix}.$$

Proceed exactly as in the proof of Theorem 3.85 (using the calculation of $\text{tw}_A(\{5\})$ in Example 3.101). □

Now suppose A is a doubly nonnegative matrix whose graph is one of the nearly-$\widehat{\text{CP}}$ graphs G_4 or G_6. Again we interpret the results of Section 3.6 in terms of cycle weights. First, consider a doubly nonnegative A with $G(A) = G_4$, with the vertices labeled in counter-clockwise order, starting with the top vertex labeled 1. In this case,

$$A = \begin{pmatrix} a_{11} & a_{12} & a_{13} & a_{14} & a_{15} \\ a_{12} & a_{22} & a_{23} & 0 & 0 \\ a_{13} & a_{23} & a_{33} & a_{34} & 0 \\ a_{14} & 0 & a_{34} & a_{44} & a_{45} \\ a_{15} & 0 & 0 & a_{45} & a_{55} \end{pmatrix}. \tag{3.29}$$

Theorem 3.104. *Let A be a 5×5 doubly nonnegative matrix in the form (3.29). Then the following are equivalent.*

(a) *A is completely positive.*
(b) *At least one of the following holds:*
$\det A[1,2|2,3] \leq 0$ *or* $\det A[1,4|4,5] \geq 0$ *or* $\det A \geq 4\text{tw}_A(\{3,4\})$.

Proof. Consider $G(A)$ to be the cycle 1-3-4-5-1, to which the vertex 2 was added, connected to the vertices 1 and 3 by edges. That is, in Theorem 3.83,

$$\begin{pmatrix} a & h \\ h & b \end{pmatrix} = A[1,3], \quad a_1 = a_{22}, \quad A_2 = A[4,5],$$

$$x_1 = a_{12}, \quad y_1 = a_{23}, \quad \mathbf{x}_2 = \begin{pmatrix} a_{14} \\ a_{15} \end{pmatrix}, \quad \mathbf{y}_2 = \begin{pmatrix} a_{34} \\ 0 \end{pmatrix}.$$

Note that in this case, as columns 4 and 5 are linearly independent by their sign pattern, A_2 is invertible, and

$$\mathbf{x}_2^T A_2^{-1} \mathbf{y}_2 = \frac{1}{\det A_2}(a_{14}a_{55} - a_{15}a_{45})a_{34} = \frac{a_{34}}{\det A_2} \det A[1,5|4,5].$$

Hence $\mathbf{x}_2^T \operatorname{adj} A_2 \, \mathbf{y}_2 \geq 0$ if and only if $\det A[1,5|4,5] \geq 0$. Applying Remark 3.84 we get that A is completely positive if and only if $\det A[1,2|3,2] = -\det A[1,2|2,3] \geq 0$ or $\det A[1,5|4,5] \geq 0$ or

$$\det A \geq -4 \det A[1,2|2,3] a_{34} \det A[1,5|4,5].$$

A simple calculation shows that

$$-\det A[1,2|2,3] a_{34} \det A[1,5|4,5] = \operatorname{tw}(\{3,4\}). \qquad \square$$

The plot thickens as we turn to consider G_6. Here, the characterization is stated in terms of the total weight of the degree 2 vertex in a matrix related to A, rather than A itself. Labeling the vertices of G_6 counter-clockwise, starting from the top vertex, we have

$$A = \begin{pmatrix} a_{11} & a_{12} & a_{13} & a_{14} & a_{15} \\ a_{12} & a_{22} & a_{23} & a_{24} & 0 \\ a_{13} & a_{23} & a_{33} & a_{34} & 0 \\ a_{14} & a_{24} & a_{34} & a_{44} & a_{45} \\ a_{15} & 0 & 0 & a_{45} & a_{55} \end{pmatrix}. \tag{3.30}$$

Let $S = I - \frac{a_{14}}{a_{15}} E_{45}$, and

$$\hat{A} = SAS^T = \begin{pmatrix} a_{11} & a_{12} & a_{13} & 0 & a_{15} \\ a_{12} & a_{22} & a_{23} & a_{24} & 0 \\ a_{13} & a_{23} & a_{33} & a_{34} & 0 \\ 0 & a_{24} & a_{34} & \hat{a}_{44} & \hat{a}_{45} \\ a_{15} & 0 & 0 & \hat{a}_{45} & a_{55} \end{pmatrix}. \tag{3.31}$$

The matrix $\hat{A}$ is positive semidefinite with $\det \hat{A} = \det A$, and in particular $\hat{A}_{44} = a_{44} - 2\frac{a_{14}}{a_{15}} a_{45} + \frac{a_{14}^2}{a_{15}^2} a_{55} > 0$. Also, $\hat{a}_{45} = a_{45} - \frac{a_{14}}{a_{15}} a_{55} > 0$ if and only if $-\det A[1,5|4,5] > 0$. In the latter case, $G(\hat{A})$ is the graph G_3. We get the following characterization.

Theorem 3.105. *Let A be a 5×5 doubly nonnegative matrix in the form (3.30). Then the following are equivalent.*

(a) *A is completely positive.*
(b) *At least one of the following holds:*
$\det A[1,5|4,5] \geq 0$ *or* $\det \hat{A}(4,5|1,5) \leq 0$ *or* $\det A \geq 4\mathrm{tw}_{\hat{A}}(\{5\})$.

Proof. In this case, A is as in Theorem 3.83, with

$$\begin{pmatrix} a & h \\ h & b \end{pmatrix} = A[1,4], \quad a_1 = a_{55}, \quad A_2 = A[2,3],$$

$$x_1 = a_{15}, \quad y_1 = a_{45}, \quad \mathbf{x}_2 = \begin{pmatrix} a_{12} \\ a_{13} \end{pmatrix}, \quad \mathbf{y}_2 = \begin{pmatrix} a_{24} \\ a_{34} \end{pmatrix}.$$

A simple computation shows that $\mathbf{x}_2^T \operatorname{adj} A_2 \, \mathbf{y}_2 = -\det \hat{A}(4,5|1,5)$.

If $A_2 = A[2,3]$ is nonsingular, then by Remark 3.84, A is completely positive if and only if $\det A[1,5|4,5] \geq 0$ or $-\det \hat{A}(4,5|1,5) \geq 0$ or

$$\det A \geq -4 \det A[1,5|4,5] \det \hat{A}(4,5|1,5).$$

If $\det A[1,5|4,5] < 0$, $\hat{A}$ is a doubly nonnegative matrix with graph G_3, and

$$\begin{aligned} \mathrm{tw}_{\hat{A}}(\{5\}) &= \hat{a}_{45} a_{15} \det \hat{A}(4,5|1,5) \\ &= (a_{45}a_{15} - a_{14}a_{55}) \det \hat{A}(4,5|1,5) \\ &= -\det A[4,5|1,5] \det \hat{A}(4,5|1,5), \end{aligned}$$

which completes the proof for a nonsingular $A[2,3]$.

If $A_2 = A[2,3]$ is singular, then the third column of A is a multiple of the second column of A. As $A(3)$ is completely positive, so is A (see Exercise 3.5). On the other hand, in this case $\det A = 0$ and also $\det \hat{A}(4,5|1,5) = 0$ (and $\mathrm{tw}_{\hat{A}}(\{5\}) = 0$). That is, in this case both (a) and (b) hold, and are thus equivalent. □

The next few results give conditions, most of which are only sufficient, for complete positivity of a 5×5 doubly nonnegative matrix which is not positive. To simplify the discussion, let us utilize the fact that both double nonnegativity and complete positivity are preserved under diagonal scaling by a positive diagonal matrix, and by permutation similarity. We therefore consider 5×5 doubly nonnegative matrices in the form

$$A = \begin{pmatrix} a_{11} & a_{12} & a_{13} & a_{14} & a_{15} \\ a_{12} & a_{22} & a_{23} & a_{24} & a_{25} \\ a_{13} & a_{23} & a_{33} & a_{34} & a_{35} \\ a_{14} & a_{24} & a_{34} & 1 & 0 \\ a_{15} & a_{25} & a_{35} & 0 & 1 \end{pmatrix} = \begin{pmatrix} A_0 & \mathbf{x}_1 & \mathbf{x}_2 \\ \mathbf{x}_1^T & 1 & 0 \\ \mathbf{x}_2^T & 0 & 1 \end{pmatrix}, \tag{3.32}$$

where $A_0 = A[1,2,3]$. We also denote

$$X_i = \begin{pmatrix} \mathbf{x}_i \\ 0 \\ 1 \end{pmatrix} \begin{pmatrix} \mathbf{x}_i \\ 0 \\ 1 \end{pmatrix}^T, \quad i = 1, 2. \tag{3.33}$$

The matrices X_1 and X_2 are rank-1 completely positive matrices.

The results are stated in terms of the number of negative elements in certain Schur complements of A. First, some obvious sufficient conditions for complete positivity.

Theorem 3.106. *Let A be a 5×5 doubly nonnegative matrix of the form (3.32). If at least one of the matrices $A/[5]$ or $A/[4]$ or $A/A[4,5]$ is nonnegative, then A is completely positive.*

Proof. If the matrix $A/[5]$ is nonnegative, then it is a 4×4 doubly nonnegative matrix, and therefore completely positive. Then $A = ((A/[5]) \oplus 0_1) + X_2$ is also completely positive. The case of $A/[4] \geq 0$ is similar. And if $A/A[4,5] \geq 0$, the result follows from $A = ((A/A[4,5]) \oplus 0_2) + X_1 + X_2$. □

Now consider the case that none of the conditions in Theorem 3.106 holds. Let

$$C = A/A[4,5] = A_0 - \mathbf{x}_1 \mathbf{x}_1^T - \mathbf{x}_2 \mathbf{x}_2^T$$

have at least one negative entry above the diagonal.

Theorem 3.107. *Let A be a 5×5 doubly nonnegative matrix of the form (3.32). Suppose one of the following holds.*

(a) *Exactly one off-diagonal entry above the diagonal in $C = A/A[4,5]$ is negative.*
(b) *All the off-diagonal entries in $C = A/A[4,5]$ are nonpositive.*

Then A is completely positive.

Proof. In case (a), without loss of generality suppose $c_{23} < 0$. Then

$$C = \mathbf{c}\mathbf{c}^T + (0_1 \oplus (C/[1])),$$

where $\mathbf{c} \geq 0$ is the first column of C multiplied by $1/\sqrt{c_{11}}$. Then

$$A = (X_1 + X_2 + \mathbf{c}\mathbf{c}^T) + (0_1 \oplus (C/[1]) \oplus 0_2).$$

The matrix $X_1 + X_2 + \mathbf{c}\mathbf{c}^T$ is completely positive, and $0_1 \oplus (C/[1]) \oplus 0_2$ is positive semidefinite, and zero except for a 2×2 principal submatrix. Their sum is the doubly nonnegative A. By Lemma 3.36, A is completely positive.

In case (b), C is an M-matrix and thus may be diagonally scaled to a diagonally dominant matrix. Hence (see Theorem 3.38)

$$C = \sum_{1 \leq i < j \leq 3} F_{ij} + D,$$

where each F_{ij} is a 3×3 positive semidefinite matrix of rank at most 1, which is zero except, possibly, for the 2×2 principal submatrix on rows and columns $\{i, j\}$, and D is a 3×3 nonnegative diagonal matrix. The matrix

$$A' = X_1 + X_2 + (D \oplus 0_2)$$

is a completely positive matrix. Adding to it successively the matrices $F'_{12} = F_{12} \oplus 0_2$, $F'_{13} = F_{13} \oplus 0_2$, $F'_{23} = F_{23} \oplus 0_2$ results at each step in a nonnegative matrix, finally ending with A. By successive use of Lemma 3.36 we get that A is completely positive. □

It remains to consider the case where C has exactly two negative entries above the diagonal, and one positive entry. For every $1 \leq i < j \leq 3$,

$$c_{ij} = a_{ij} - (X_1)_{ij} - (X_2)_{ij} = a_{ij} - a_{i4}a_{j4} - a_{i5}a_{j5}.$$

Theorem 3.108. *Let A be a 5×5 doubly nonnegative matrix of the form (3.32). Let $C = A/A[4,5]$ have exactly two negative entries above the diagonal. If $\operatorname{rank} C = 1$, then A is completely positive if and only if*

$$\min_{1 \leq i < j \leq 3} \left\{ \frac{a_{i4}a_{j4}}{-c_{ij}} \,\middle|\, c_{ij} < 0 \right\} + \min_{1 \leq i < j \leq 3} \left\{ \frac{a_{i5}a_{j5}}{-c_{ij}} \,\middle|\, c_{ij} < 0 \right\} \geq 1.$$

Proof. Without loss of generality we may assume that

$$c_{12} > 0, \quad c_{13} < 0, \quad c_{23} < 0.$$

Let

$$t_1 = \min\left\{ \frac{(X_1)_{13}}{-c_{13}}, \frac{(X_1)_{23}}{-c_{23}} \right\}, \qquad t_2 = \min\left\{ \frac{(X_2)_{13}}{-c_{13}}, \frac{(X_2)_{23}}{-c_{23}} \right\}.$$

That is, t_i is the largest (nonnegative) t such that $X_i + t(C \oplus 0_2)$ is nonnegative. For every $0 \le t \le t_i$, $X_i + t(C \oplus 0_2)$ is a 5×5 doubly nonnegative matrix with a zero row and column, and is therefore completely positive.

If $t_1 + t_2 \ge 1$, we may choose $0 \le t \le t_1$ such that $0 \le 1 - t \le t_2$. Then $A = A_1 + A_2$, where

$$A_1 = X_1 + t(C \oplus 0_2) \quad \text{and} \quad A_2 = X_2 + (1-t)(C \oplus 0_2),$$

and is therefore completely positive.

Conversely, if A is completely positive, let $A = \sum_{i=1}^m \mathbf{b}_i \mathbf{b}_i^T$ be a CP representation of A. Let

$$\Omega_1 = \{1 \le i \le m \mid 4 \in \operatorname{supp} \mathbf{b}_i\} \quad \text{and} \quad \Omega_2 = \{1, \ldots, m\} \setminus \Omega_1.$$

Let $A_1 = \sum_{i \in \Omega_1} \mathbf{b}_i \mathbf{b}_i^T$, and $A_2 = \sum_{i \in \Omega_2} \mathbf{b}_i \mathbf{b}_i^T$. Then

$$A_1 = \begin{pmatrix} A' & \mathbf{x}_1 & \mathbf{0} \\ \mathbf{x}_1^T & 1 & 0 \\ \mathbf{0}^T & 0 & 0 \end{pmatrix}, \quad A_2 = \begin{pmatrix} A'' & \mathbf{0} & \mathbf{x}_2 \\ \mathbf{0}^T & 0 & 0 \\ \mathbf{x}_2^T & 0 & 1 \end{pmatrix}$$

are both completely positive. In particular, $C' = A_1/[4,5] = A' - \mathbf{x}_1 \mathbf{x}_1^T$ and $C'' = A_2/[4,5] = A'' - \mathbf{x}_2 \mathbf{x}_2^T$ are positive semidefinite. Since $A = A_1 + A_2$,

$$C = A/A[4,5] = A' + A'' - \mathbf{x}_1 \mathbf{x}_1^T - \mathbf{x}_2 \mathbf{x}_2^T = C' + C''.$$

As the positive semidefinite matrix C has rank 1, this implies that $C' = tC$ and $C'' = (1-t)C$ for some $0 \le t \le 1$. Thus $\mathbf{x}_1 \mathbf{x}_1^T + tC_1 = A' \ge 0$ and $\mathbf{x}_2 \mathbf{x}_2^T + (1-t)C = A'' \ge 0$, $0 \le t \le t_1$ and $0 \le 1 - t \le t_2$, implying that $t_1 + t_2 \ge 1$. □

Note that the "if" part of the proof did not use the fact that $\operatorname{rank} C = 1$. Hence we have proved the following.

Corollary 3.109. *Let A be a 5×5 doubly nonnegative matrix of the form (3.32), and $C = A/A[4,5]$. If*

$$\min_{1 \le i < j \le 3} \left\{ \frac{a_{i4} a_{j4}}{-c_{ij}} \,\middle|\, c_{ij} < 0 \right\} + \min_{1 \le i < j \le 3} \left\{ \frac{a_{i5} a_{j5}}{-c_{ij}} \,\middle|\, c_{ij} < 0 \right\} \ge 1,$$

then A is completely positive.

The reverse implication is not true in general when $\operatorname{rank} C \ge 2$.

Example 3.110. Let

$$A = \begin{pmatrix} 8 & 3 & 0 & 2 & 1 \\ 3 & 4 & 0 & 1 & 1 \\ 0 & 0 & 7 & 1 & 1 \\ 2 & 1 & 1 & 1 & 0 \\ 1 & 1 & 1 & 0 & 1 \end{pmatrix}.$$

The matrix A is completely positive, since the doubly nonnegative A is obtained by adding to the completely positive

$$\begin{pmatrix} 2 \\ 1 \\ 1 \\ 1 \\ 0 \end{pmatrix}\begin{pmatrix} 2 \\ 1 \\ 1 \\ 1 \\ 0 \end{pmatrix}^T + \begin{pmatrix} 1 \\ 1 \\ 1 \\ 0 \\ 1 \end{pmatrix}\begin{pmatrix} 1 \\ 1 \\ 1 \\ 0 \\ 1 \end{pmatrix}^T = \begin{pmatrix} 5 & 3 & 3 & 2 & 1 \\ 3 & 2 & 2 & 1 & 1 \\ 3 & 2 & 2 & 1 & 1 \\ 2 & 1 & 1 & 1 & 0 \\ 1 & 1 & 1 & 0 & 1 \end{pmatrix}$$

two positive semidefinite matrices, each zero except for a 2×2 principal submatrix,

$$F_1 = \begin{pmatrix} 3 & 0 & -3 & 0 & 0 \\ 0 & 0 & 0 & 0 & 0 \\ -3 & 0 & 3 & 0 & 0 \\ 0 & 0 & 0 & 0 & 0 \\ 0 & 0 & 0 & 0 & 0 \end{pmatrix} \quad \text{and} \quad F_2 = \begin{pmatrix} 0 & 0 & 0 & 0 & 0 \\ 0 & 2 & -2 & 0 & 0 \\ 0 & -2 & 2 & 0 & 0 \\ 0 & 0 & 0 & 0 & 0 \\ 0 & 0 & 0 & 0 & 0 \end{pmatrix}.$$

Yet

$$C = A/A[4,5] = \begin{pmatrix} 3 & 0 & -3 \\ 0 & 2 & -2 \\ -3 & -2 & 5 \end{pmatrix}.$$

Hence

$$t_1 = \min\left\{\frac{2}{3}, \frac{1}{2}\right\}, \quad t_2 = \min\left\{\frac{1}{2}, \frac{1}{2}\right\}$$

and $t_1 + t_2 = \frac{1}{2} + \frac{1}{3} < 1$. Note that $\operatorname{rank} C = 2$.

For the following cases there is currently no complete positivity characterization in terms of the entries of the matrix: If A is a 5×5 doubly nonnegative matrix, whose graph contains a 5-cycle, and A is either positive, or has exactly one zero above the diagonal at position i, j, $\operatorname{rank} A/A[i,j] \geq 2$ and $A/A[i,j]$ has at exactly two negative entries and one positive entry above the diagonal.

The difference between $\mathcal{DNN}_5$ and $\mathcal{CP}_5$

While a complete characterization of the extreme rays of $\mathcal{DNN}_n$ is not known for a general n, it is known that are exactly two types of extremal matrices in $\mathcal{DNN}_5$: rank 1 matrices and matrices of rank 3 whose graph is a 5-cycle (Proposition 1.170). Each rank 1 extremal doubly nonnegative matrix has the form $\mathbf{b}\mathbf{b}^T$, where $\mathbf{b} \in \mathbb{R}^5_+$, and is completely positive. Every matrix $A \in \mathcal{DNN}_5$ is a sum of extremal matrices. If the matrix is not completely positive, each such representation of A has at least one rank 3 summand. In this subsection it is shown that every 5×5 doubly nonnegative matrix, which is not completely positive, may be represented as a sum of extremal matrices in which exactly one of the matrices is a rank 3 extremal.

Though we are currently considering matrices in $\mathcal{DNN}_5$ and $\mathcal{CP}_5$, let us define a more general notion of irreducibility/reducibility of an $n\times n$ doubly nonnegative matrix with respect to $\mathcal{CP}_n$ (similar to Definition 2.36 for copositive matrices).

Definition 3.111. A matrix $A \in \mathcal{DNN}_n$ is *$\mathcal{CP}$-irreducible* if for every completely positive matrix $X \in \mathcal{CP}$ the matrix $A - X \notin \mathcal{DNN}_n$. The matrix A is *$\mathcal{CP}$-reducible* if there exists a nonzero completely positive matrix $X \in \mathcal{CP}$ such that the matrix $A - X$ is doubly nonnegative.

Since both cones, $\mathcal{DNN}_n$ and $\mathcal{CP}_n$, are invariant under permutation similarity and diagonal scaling, we have the following.

Proposition 3.112. *Let A, P and D be an $n\times n$ matrices, where A is doubly nonnegative, P is a permutation matrix, and D a positive diagonal matrix. Then the following are equivalent.*

(a) *A is $\mathcal{CP}$-reducible.*
(b) *P^TAP is $\mathcal{CP}$-reducible.*
(c) *DAD is $\mathcal{CP}$-reducible.*

Proposition 3.113. *An $n\times n$ doubly nonnegative matrix A is $\mathcal{CP}$-reducible if and only if there exists $\mathbf{x} \in (\operatorname{cs} A) \cap \mathbb{R}^n_+$ such that $\operatorname{supp}\mathbf{x}$ is a clique in $G(A)$.*

Proof. Suppose A is $\mathcal{CP}$-reducible. Then there exists a nonzero completely positive X such that $A - X$ is doubly nonnegative. Let $X = \sum_{i=1}^m \mathbf{x}_i\mathbf{x}_i^T$, $m \geq 1$, where $\mathbf{x}_1,\ldots,\mathbf{x}_m$ are nonzero nonnegative vectors in $\mathbb{R}^n$. Denote $\mathbf{x} = \mathbf{x}_1$. The matrix $A - \mathbf{x}\mathbf{x}^T = (A-X)+\sum_{i=2}^m \mathbf{x}_i\mathbf{x}_i^T$ is a doubly nonnegative matrix as a sum of two such matrices. Moreover, if $\sigma = \operatorname{supp}\mathbf{x}$, then $A[\sigma] \geq \mathbf{x}\mathbf{x}^T[\sigma] > 0$ implies that σ is a clique in $G(A)$.

Conversely, suppose $\mathbf{x} \in (\operatorname{cs} A) \cap \mathbb{R}^n_+$ and $\operatorname{supp}\mathbf{x}$ is a clique in $G(A)$. As $\mathbf{x} \in \operatorname{cs} A$, $\operatorname{cs}\mathbf{x}\mathbf{x}^T \subseteq \operatorname{cs} A$, and therefore there exists $\varepsilon_0 > 0$ such that $A - \varepsilon\mathbf{x}\mathbf{x}^T$ is positive semidefinite for every $0 < \varepsilon \leq \varepsilon_0$ (see Exercise 1.22). Let $\sigma = \operatorname{supp}\mathbf{x}$. Then $(\mathbf{x}\mathbf{x}^T)_{ij} = 0$ whenever either $i \notin \sigma$ or $j \notin \sigma$, and as σ is a clique in $G(A)$, $A[\sigma] > 0$. Hence there exists $\varepsilon_1 > 0$ such that $A - \varepsilon\mathbf{x}\mathbf{x}^T \geq 0$ for every $0 < \varepsilon \leq \varepsilon_1$. For small enough $\varepsilon > 0$, $A - \varepsilon\mathbf{x}\mathbf{x}^T$ is doubly nonnegative, and clearly $\varepsilon\mathbf{x}\mathbf{x}^T$ is completely positive. □

Theorem 3.114. *A matrix $A \in \mathcal{DNN}_n$ that is not completely positive has a representation as $A = X + Y$, where $X \in \mathcal{CP}_n$, and Y is a $\mathcal{CP}$-reducible matrix.*

Proof. For the purpose of this proof we denote the number of zeros in a symmetric matrix A on and above its diagonal by $\zeta(A)$. We first show that if A is $\mathcal{CP}$-reducible, then there exists a vector $\mathbf{v} \in (\operatorname{cs} A) \cap \mathbb{R}^n_+$ such that $A' = A - \mathbf{v}\mathbf{v}^T$ is doubly nonnegative, and either $\operatorname{rank} A' = \operatorname{rank} A - 1$ or $\zeta(A') < \zeta(A)$. Indeed, if A is $\mathcal{CP}$-reducible and $\mathbf{x}$ is any nonnegative vector in $\operatorname{cs} A$, there exists a maximal $\varepsilon_0 > 0$ such that $A - \varepsilon \mathbf{x}\mathbf{x}^T$ is positive semidefinite for every $0 < \varepsilon \leq \varepsilon_0$ ($\mathcal{PSD}_n$ is a closed cone). Then $\mathbf{x}$ is not in $\operatorname{cs}(A - \varepsilon_0 \mathbf{x}\mathbf{x}^T)$, and thus $\operatorname{rank}(A - \varepsilon_0 \mathbf{x}\mathbf{x}^T) < \operatorname{rank} A$. Similarly, there exists a maximal $\varepsilon_1 > 0$ such that $A - \varepsilon \mathbf{x}\mathbf{x}^T \geq 0$ for every $0 < \varepsilon \leq \varepsilon_1$. By the maximality of ε_1, $A - \varepsilon_1 \mathbf{x}\mathbf{x}^T$ has at least one additional zero on or above the diagonal compared to A. Let $t = \min\{\sqrt{\varepsilon_0}, \sqrt{\varepsilon_1}\}$, and $\mathbf{v} = t\mathbf{x}$. Then $A' = A - \mathbf{v}\mathbf{v}^T$ is doubly nonnegative, and either $\operatorname{rank} A' = \operatorname{rank} A - 1$ or $\zeta(A') < \zeta(A)$.

For $A \in \mathcal{DNN}_n$, we have $\operatorname{rank} A \leq n$ and $\zeta(A) \leq \frac{n(n+1)}{2}$. Hence starting with A we get a $\mathcal{CP}$-irreducible matrix after at most $\frac{n(n+1)}{2}$ steps of such maximal rank 1 $\mathcal{CP}$-reductions. □

We now turn our attention to the 5×5 case.

Proposition 3.115. *If two 5×5 matrices A and B are extremal doubly nonnegative matrices of rank 3, and B is not a scalar multiple of A, then $A + B$ is $\mathcal{CP}$-reducible.*

Proof. Since each of $G(A)$ and $G(B)$ is a 5-cycle, the graph of $A + B$ is one of the following graphs, up to isomorphism.

H_1 H_2 H_3 H_4

By Proposition 3.112, permutation similarity does not change $\mathcal{CP}$-reducibility. Hence we may assume the vertices are labeled counter clockwise, with the top one labeled 1, and that $G(A)$ is the cycle 1–2–3–4–5–1.

We consider two cases: $G(A) \neq G(B)$ and $G(A) = G(B)$.

In the first case (where $G(A + B)$ is one of H_1, H_2 and H_3), one of the edges of $G(B)$ connects two edges in $G(A)$ whose distance in $G(A)$

is 2. Then $G(A+B)$ has a triangle on three subsequent vertices in the cycle $1-2-3-4-5-1$. We may assume without loss of generality that this triangle has vertices $\{1,2,3\}$. Let $\mathbf{x}$ be the second column of A. Then $\mathbf{x} \geq \mathbf{0}$, $\operatorname{supp}\mathbf{x} = \{1,2,3\}$ is a clique in $G(A+B)$, and as $\mathbf{x} \in \operatorname{cs} A$ and $\operatorname{cs}(A+B) = \operatorname{cs} A + \operatorname{cs} B$ (see Proposition 1.21), $\mathbf{x} \in \operatorname{cs}(A+B)$. By Proposition 3.113, $A+B$ is $\mathcal{CP}$-reducible.

Now consider the case that $G(A+B) = G(A) = G(B)$ is the 5-cycle. Since the matrices A and B are symmetric, B is not a scalar multiple of A, and $G(B) = G(A)$ is connected, at least one column of B is not a scalar multiple of the corresponding column of A. Without loss of generality suppose this is the second column. Let $\mathbf{x}$ be the second column of A, and $\mathbf{y}$ the second column of B. These two linearly independent vectors are nonnegative, and $\operatorname{supp}\mathbf{x} = \operatorname{supp}\mathbf{y} = \{1,2,3\}$. Then there exists a nonzero $\mathbf{z} \in \operatorname{span}\{\mathbf{x},\mathbf{y}\} \subseteq \operatorname{cs} A + \operatorname{cs} B = \operatorname{cs}(A+B)$, such that $\mathbf{z} \geq 0$ and either $\operatorname{supp}\mathbf{z} \subseteq \{1,2\}$, or $\operatorname{supp}\mathbf{z} \subseteq \{2,3\}$. By Proposition 3.113 the matrix $A+B$ is $\mathcal{CP}$-reducible in this case too. □

Clearly, an extremal doubly nonnegative matrix that is not completely positive is $\mathcal{CP}$-irreducible. Any other matrix is a sum of extremal matrices. Either one of these extremal matrices is rank 1, and then the matrix is $\mathcal{CP}$-reducible, or at least two are rank 3, and again the matrix is $\mathcal{CP}$-reducible by Proposition 3.115. This proves the following.

Theorem 3.116. *A matrix $A \in \mathcal{DNN}_5$ is $\mathcal{CP}$-irreducible if and only if it is an extremal rank 3 matrix.*

Combined with Theorem 3.114 we get the following description for matrices that are in $\mathcal{DNN}_5$, but not in $\mathcal{CP}_5$.

Theorem 3.117. *A matrix $A \in \mathcal{DNN}_5$ that is not completely positive has a representation as $A = X + Y$, where $X \in \mathcal{CP}_5$, and Y is an extremal rank 3 matrix.*

Exercises

3.55. Use Theorem 3.103 to check that the matrix A in Example 3.110 is completely positive.

Verify that A does not satisfy the condition of Theorem 3.107.

3.56. Use Theorem 3.103 to check that the matrix

$$A = \begin{pmatrix} 2 & 1 & 0 & 0 & 1 \\ 1 & 2 & 0 & 1 & 0 \\ 0 & 0 & 4 & 1 & 1 \\ 0 & 1 & 1 & 1 & 0 \\ 1 & 0 & 1 & 0 & 1 \end{pmatrix}$$

is not completely positive.

3.57. Show that A is a rank 3 extremal matrix in $\mathcal{DNN}_5$ if and only if there exists a permutation matrix P, a positive diagonal matrix D, and a 5×3 matrix

$$R = \begin{pmatrix} 1 & 0 & 0 \\ a & b & 1 \\ 0 & 1 & 0 \\ 0 & 1 & -b \\ 1 & 0 & -a \end{pmatrix}, \qquad a > 0, b > 0,$$

such that $P^T AP = DRR^T D$.

3.58. Show that if $\mathbf{x}, \mathbf{y} \in \mathbb{R}^3$ are linearly independent positive vectors, then there exists a nonnegative $\mathbf{z} \in \operatorname{span}\{\mathbf{x}, \mathbf{y}\}$ such that $\operatorname{supp} \mathbf{z} \subseteq \{1, 2\}$ or $\operatorname{supp} \mathbf{z} \subseteq \{2, 3\}$.

Notes. Theorem 3.103 is based on [Cedolin and Salce (2016)] and the version of its proof in [Di Virgilio (2019)]. Theorems 3.104 and 3.105 are based on the latter. In these two papers, the weight notions in Definition 3.101 are defined and used to correct results in this spirit that appeared in [Xu (2001)]. Theorems 3.106–3.108 and Corollary 3.109 appear in [Berman and Xu (2004)]. Note that the final section in the paper [Berman and Xu (2004)] has an error, pointed out in [Dong and Anstreicher (2010)]. Another error in [Berman and Xu (2004)] is Theorem 4.3, and Exercise 3.56 is a counter example to this theorem. The discussion of the difference between $\mathcal{DNN}_5$ and $\mathcal{CP}_5$, including the notions of $\mathcal{CP}$-reducibility/irreducibility, is based on [Burer *et al.* (2009)]. Exercise 3.57 is the basis for the discussion in that paper, but the proofs presented here are slightly different. A doubly nonnegative matrix that is not completely positive is called in [Burer *et al.* (2009)] a *bad* matrix (B(urer)A(nstreicher)D(ür)...). In that paper, an algorithm to decompose any doubly nonnegative matrix into a sum of a completely positive matrix and a $\mathcal{CP}$-irreducible doubly nonnegative matrix, as in Theorem 3.114, is presented. The number of required iterations is polynomial, however at each iteration several (and not necessarily a polynomial number of) linear programming problems need to be solved.

3.8 Square and rank-revealing CP factorizations

By Remark 3.37, small (order ≤ 4) completely positive matrices have *square CP factorizations.* That is, such a matrix A can be represented as BB^T where B is a nonnegative square matrix. This is generally not true for larger completely positive matrices.

Example 3.118. The matrix

$$A = \begin{pmatrix} 2 & 0 & 0 & 1 & 1 \\ 0 & 2 & 0 & 1 & 1 \\ 0 & 0 & 2 & 1 & 1 \\ 1 & 1 & 1 & 3 & 0 \\ 1 & 1 & 1 & 0 & 3 \end{pmatrix}$$

is completely positive, but does not have a square nonnegative decomposition. If $A = BB^T$, $B \geq 0$, then the size of the support of each column of B is at most 2. Hence B has to have at least 6 columns.

In this section we discuss conditions which guarantee that the completely positive A does have a square CP factorization. We also consider conditions that guarantee the existence of a *rank-revealing CP factorization*, i.e., factorizations of the form $A = BB^T$, where $B \geq 0$ and has $r = \operatorname{rank} A$ columns. (Clearly, for a nonsingular completely positive matrix a rank-revealing CP factorization is a square one, and a rank-revealing CP factorization $A = BB^T$ of a singular A may be extended to a square CP factorization by adding zero columns to B.)

We begin with triangular CP factorizations.

Definition 3.119. A matrix A is a *UL-completely positive* matrix if there exists a nonnegative upper triangular matrix B such that $A = BB^T$. A is a *LU-completely positive matrix* if there exists a nonnegative lower triangular matrix B such that $A = BB^T$.

Example 3.120. By Example 3.7, every 2×2 doubly nonnegative matrix is *UL*-completely positive. By the same argument, every 2×2 doubly nonnegative matrix is also *LU*-completely positive: If

$$A = \begin{pmatrix} a & b \\ b & c \end{pmatrix}$$

is doubly nonnegative, then in the case that $a = 0$

$$A = \begin{pmatrix} 0 & 0 \\ 0 & \sqrt{c} \end{pmatrix}^2,$$

and in the case that $a > 0$,

$$A = \begin{pmatrix} \sqrt{a} & 0 \\ b/\sqrt{a} & \sqrt{c - b^2/a} \end{pmatrix} \begin{pmatrix} \sqrt{a} & 0 \\ b/\sqrt{a} & \sqrt{c - b^2/a} \end{pmatrix}^T.$$

In this section we will see that a 3×3 doubly nonnegative matrix is either *UL*-completely positive or *LU*-completely positive, but not necessarily both. But for $n \geq 4$, an $n \times n$ completely positive matrix may be neither *UL*- nor *LU*-completely positive.

The following proposition plays a crucial role in determining whether a given completely positive matrix with positive diagonal entries is *UL*-completely positive.

Proposition 3.121. *Let A be an $n\times n$ doubly nonnegative matrix such that a_{nn} is positive. Then A is a UL-completely positive if and only if $A/[n]$ is UL-completely positive.*

Proof. Let $A = UU^T$ where U is an upper triangular nonnegative matrix. Partition U as

$$U = \begin{pmatrix} K & \mathbf{u} \\ \mathbf{0}^T & u_{nn} \end{pmatrix},$$

where K is of order $n-1$. Then $A/[n] = KK^T$.

Conversely, if $A/[n] = KK^T$, where K is a nonnegative upper triangular matrix, partition A as

$$A = \begin{pmatrix} A_{11} & \mathbf{a} \\ \mathbf{a}^T & a_{nn} \end{pmatrix},$$

where $\mathbf{a} \in \mathbb{R}_+^{n-1}$, and A_{11} is $(n-1)\times(n-1)$. Let

$$U = \begin{pmatrix} K & \mathbf{a}/\sqrt{a_{nn}} \\ \mathbf{0}^T & \sqrt{a_{nn}} \end{pmatrix},$$

then $A = UU^T$. □

For a sufficient condition for *UL*-complete positivity we need the following terminology.

Definition 3.122. Let A be an $n \times n$ matrix. Let $i, j, i_2, ..., i_k$ be indices such that $1 \leq i \neq j < i_2 < \ldots < i_k$ and let $\alpha = \{i_2, \ldots, i_k\}$. Then the $k \times k$ minor $\det A[\{i\}\cup\alpha|\{j\}\cup\alpha]$ is called a *left almost principal minor of A*.

Definition 3.123. Let A be an $n \times n$ matrix. Let $i, j, i_1, ...i_{k-1}$ be indices such that $i_1 < \ldots < i_{k-1} < i \neq j \leq n$ and let $\alpha = \{i_1, ..., i_{k-1}\}$. Then the $k \times k$ minor $\det A[\alpha \cup \{i\}|\alpha \cup \{j\}]$ is called a *right almost principal minor* of A.

We can now prove the following.

Theorem 3.124. *If A is an $n \times n$ doubly nonnegative matrix, $n \geq 3$, and all its left almost principal minors are nonnegative, then A is UL-completely positive.*

Proof. The proof is by induction. Since any 2×2 doubly nonnegative matrix is *UL*-completely positive (Example 3.120), we may consider the claim to hold for $n = 2$.

Let $n \geq 3$ and suppose the theorem holds for $n - 1$. Let A be an $n \times n$ doubly nonnegative matrix with nonnegative left almost principal minors. Then

$$A = \begin{pmatrix} A_{11} & \mathbf{a} \\ \mathbf{a}^T & a_{nn} \end{pmatrix},$$

where $\mathbf{a} \in \mathbb{R}_+^{n-1}$. If $a_{nn} = 0$, then $\mathbf{a} = \mathbf{0}$, and the claim follows from the induction hypothesis. If $a_{nn} > 0$, it suffices to prove that $A/[n]$ is *UL*-completely positive. The matrix $A/[n]$ is doubly nonnegative: positive semidefinite by Theorem 1.75, and nonnegative since, by Proposition 1.71,

$$(A/[n])_{ij} = \frac{\det A[i, n|j, n]}{a_{nn}}.$$

If $A/[n]$ is at least 3×3, then all its left almost principal minors are nonnegative: For any $\alpha \subseteq \{1, \ldots, n-1\}$,

$$(A/[n])[\{i\} \cup \alpha|\{j\} \cup \alpha] = A[\alpha \cup \{i, n\}|\alpha \cup \{j, n\}]/[n].$$

So by Theorem 1.70,

$$\det\left((A/[n])[\{i\} \cup \alpha|\{j\} \cup \alpha]\right) = \frac{\det A[\alpha \cup \{i, n\}|\alpha \cup \{j, n\}]}{a_{nn}} \geq 0$$

for every $1 \leq i \neq j < \min \alpha$. Hence $A/[n]$ is an $(n-1) \times (n-1)$ doubly nonnegative matrix with nonnegative left almost principal minors, and by the induction hypothesis it is *UL*-completely positive. By Proposition 3.121, so is A. □

We leave it as an exercise to show that the converse of Theorem 3.124 is not true (see Exercise 3.60).

Analogous results hold for LU-completely positive matrices.

Theorem 3.125. *An $n \times n$ doubly nonnegative matrix, $n \geq 3$, whose right almost principal minors are all nonnegative is LU-completely positive.*

To prove the promised result concerning 3×3 doubly nonnegative matrices we need the following lemma.

Lemma 3.126. *If A is a 3×3 doubly nonnegative matrix, then at least two of the following three inequalities hold.*

$$\begin{aligned} a_{11}a_{23} &\geq a_{12}a_{13} \\ a_{22}a_{13} &\geq a_{12}a_{23} \\ a_{33}a_{12} &\geq a_{13}a_{23}. \end{aligned}$$

Proof. Suppose that two of these inequalities do not hold, say,

$$a_{11}a_{23} < a_{12}a_{13} \quad \text{and} \quad a_{22}a_{13} < a_{12}a_{23}.$$

Then

$$a_{11}a_{23}a_{22}a_{13} < a_{12}a_{13}a_{12}a_{23}.$$

But this implies that $a_{11}a_{22} < a_{12}^2$, which is impossible in a positive semidefinite A. □

Corollary 3.127. *If A is a 3×3 doubly nonnegative matrix, then A is UL-completely positive or LU-completely positive.*

Proof. If any diagonal entry of A is zero, the result is true by Example 3.120. So suppose $a_{ii} > 0$ for $i = 1, 2, 3$. The only left almost principal minor of A is

$$\det A[1,3|2,3] = \det A[2,3|1,3] = a_{33}a_{12} - a_{13}a_{23},$$

and the only right almost principal minor of A is

$$\det A[1,3|1,2] = \det A[1,2|1,3] = a_{11}a_{23} - a_{12}a_{13}.$$

By Lemma 3.126 at least one of these minors is nonnegative. The result therefore follows from Theorems 3.124 and 3.125. □

Example 3.128. A can be LU-completely positive and not UL-completely positive. For example, consider the matrix

$$A = \begin{pmatrix} 2 & 1 & 3 \\ 1 & 3 & 2 \\ 3 & 2 & 5 \end{pmatrix}.$$

It is LU-completely positive by Theorem 3.125, but it is not UL-completely positive since $A/[3]$ is not nonnegative.

Corollary 3.129. *Every symmetric totally nonnegative matrix is (LU- and UL-) completely positive.*

Corollary 3.130. *Symmetric nonnegative Cauchy matrices, and in particular Hilbert matrices, are completely positive.*

UL-complete positivity is related to the concept of an *inverse M-matrix*, the inverse of an M-matrix.

Theorem 3.131. *A nonsingular doubly nonnegative matrix is an inverse M-matrix if and only if PAP^T is UL-completely positive for every permutation matrix P.*

Proof. If A^{-1} is a symmetric M-matrix, then there exists a lower triangular M-matrix L such that $A^{-1} = LL^T$ (see Exercise 1.35). Hence $A = UU^T$ where $U = (L^T)^{-1}$ is a nonnegative upper triangular matrix. If A^{-1} is an M-matrix and P is a permutation matrix, then $(PAP^T)^{-1} = PA^{-1}P^T$ is also an M-matrix, so this proves the "only if" part.

For the converse, we show that if $A = UU^T$, where U is an upper triangular nonnegative matrix, then the $2,1$ entry of A^{-1} is nonpositive. Indeed, this entry is equal to $-\det A(1|2)/\det A$. As $\det A = (\prod_{i=1}^n u_{ii})^2$ is positive and $\det A(1|2) = (\prod_{i=2}^n u_{ii})\, u_{12}\, (\prod_{i=3}^n u_{ii})$ is nonnegative, the entry is nonpositive. So if PAP^T is UL-completely positive for every permutation matrix P, then all the off-diagonal entries of A^{-1} are nonpositive. □

The class of matrices with square CP factorizations includes, of course, matrices that are squares of symmetric nonnegative matrices. Square factorizations of this type have nothing to do with the triangular CP factorizations discussed so far. A matrix may be the square of a nonnegative symmetric matrix, but not be LU- or UL-completely positive, and vice versa (see Exercise 3.66). We present now a case in which it is possible to verify that A is a square of a symmetric nonnegative matrix.

Theorem 3.132. *Let A be an $n \times n$ irreducible doubly nonnegative matrix, with Perron root ρ and a corresponding positive eigenvector* $\mathbf{v}$, *and let*

$$\mu = \sum_{\substack{i=1 \\ i \neq k}}^{n} v_i^2,$$

where $v_k = \min_{1 \le i \le n} v_i$. If $a_{ii} \le \rho \frac{v_i^2}{\mu}$ for every $1 \le i \le n$, then A is completely positive and there exists a doubly nonnegative matrix B such that $A = B^2$.

Proof. By the irreducibility of A, $\rho > 0$ and $\mathbf{v}$ is a positive vector. Let B be the square root of A, which is a positive semidefinite matrix (see Remark 1.26). We now show that $B \ge 0$. Suppose on the contrary that $b_{i_0 j_0} < 0$. We have

$$a_{i_0 i_0} = \mathbf{e}_{i_0}^T B^2 \mathbf{e}_{i_0} = b_{1 i_0}^2 + \cdots + b_{n i_0}^2 > \sum_{\substack{j=1 \\ j \neq j_0}}^{n} b_{j i_0}^2.$$

By Cauchy-Schwarz inequality,

$$\left(\sum_{\substack{j=1 \\ j \neq j_0}}^{n} b_{j i_0} v_j \right)^2 \le \sum_{\substack{j=1 \\ j \neq j_0}}^{n} b_{j i_0}^2 \sum_{\substack{j=1 \\ j \neq j_0}}^{n} v_j^2.$$

As $B\mathbf{v} = \sqrt{\rho}\mathbf{v}$ and $b_{i_0 j_0} < 0$,

$$\sum_{\substack{j=1 \\ j \neq j_0}}^{n} b_{j i_0} v_j = \sqrt{\rho} v_{j_0} - b_{i_0 j_0} v_{j_0} > \sqrt{\rho} v_{j_0}.$$

And by the definition of μ,

$$\sum_{\substack{j=1 \\ j \neq j_0}}^{n} v_j^2 \le \mu.$$

Combining these last three inequalities we get that

$$a_{i_0 i_0} > \sum_{\substack{j=1 \\ j \neq j_0}}^{n} b_{j i_0}^2 \ge \frac{\left(\sum_{\substack{j=1 \\ j \neq j_0}}^{n} b_{j i_0} v_j \right)^2}{\sum_{\substack{j=1 \\ j \neq j_0}}^{n} v_j^2} > \frac{\rho v_{j_0}^2}{\mu},$$

a contradiction. □

Remark 3.133. From the proof it is clear that the result of Theorem 3.132 holds also for a reducible doubly nonnegative A with a positive eigenvector corresponding to $\rho(A)$. Moreover, as any doubly nonnegative A is a direct sum of irreducible doubly nonnegative matrices and a zero matrix, a sufficient condition for nonnegativity of the square root of such A may be obtained by applying the theorem to each irreducible summand.

Recall that a square nonnegative matrix A is *stochastic* if all its row sums are equal to 1. Such A is *doubly stochastic* if all its column sums are also equal to 1. That is, $A \geq 0$ is doubly stochastic if and only if

$$A\mathbf{e} = \mathbf{e} \quad \text{and} \quad A^T\mathbf{e} = \mathbf{e}.$$

In view of the last remark, in the case of a doubly stochastic matrix Theorem 3.132 translates to the following.

Corollary 3.134. *Let A be an $n \times n$ doubly nonnegative doubly stochastic matrix which has no diagonal entry greater than $1/(n-1)$. Then A is completely positive and there exists a doubly nonnegative doubly stochastic matrix B such that $A = B^2$.*

The requirement that the diagonal entries are not greater than $1/(n-1)$ (or, rather, not greater than $1/(k-1)$, where k is the size of the largest irreducible block of A) cannot be dismissed — see Exercise 3.66.

A more intricate sufficient condition for nonnegativity of the positive semidefinite square root of a doubly nonnegative matrix (and thus, for complete positivity of A and the existence of a square CP factorization) is the following.

Theorem 3.135. *Let A be an $n \times n$ irreducible doubly nonnegative and doubly stochastic matrix. Let $\lambda_2 < \rho(A)$ be the second largest eigenvalue of A, and let $\mathbf{w}$ be an eigenvector corresponding to λ_2. Let $w_p = \min_{1\leq i\leq n} w_i$, $w_q = \max_{1\leq i\leq n} w_i$, and*

$$\nu = \max_{1\leq l<k\leq n} \sum_{\substack{1\leq i<j\leq n \\ (i,j)\neq(l,k)}} (w_j - w_i)^2.$$

If for every $1 \leq i < j \leq n$

$$\det\begin{pmatrix} a_{ii} & a_{ij} \\ a_{ij} & a_{jj} \end{pmatrix} \leq \frac{\lambda_2(w_i - w_j)^2}{\nu}$$

and $a_{pp} \leq \frac{1}{n-1}$ and $a_{qq} \leq \frac{1}{n-1}$, then the positive semidefinite square root of A is nonnegative.

(Note that by the irreducibility of A, $\mathbf{w}$ is not a multiple of $\mathbf{e}$, and therefore $\nu > 0$.) We outline the proof here, and leave some of the details to Exercise 3.68. In the proof, $C_2(X)$ denotes the *second compound matrix* of an $m \times n$ matrix. That is, $C_2(X)$ is an $\binom{m}{2} \times \binom{n}{2}$ matrix, whose rows are the subsets of size 2 of $\{1, \ldots, m\}$, and the columns are the subsets of size 2 of $\{1, \ldots, n\}$, ordered lexicographically, and for every such α and β, $(C_2(X))_{\alpha,\beta} = \det X[\alpha|\beta]$. It is known that $C_2(XY) = C_2(X)C_2(Y)$.

Proof. Denote the positive semidefinite square root of A by B. If P is a permutation matrix, $P^TAP = (P^TBP)^2$, and $P^T\mathbf{w}$ is an eigenvector of P^TAP corresponding to the second largest eigenvalue λ_2. By permuting the rows and columns of A and B we may assume that the entries of $\mathbf{w}$ are non-decreasing, and therefore $p = 1$ and $q = n$. Since A is irreducible, so is B. In particular, B has no zero diagonal entry, so $b_{ii} > 0$ for every i.

Let $X = \begin{pmatrix} \mathbf{e} & \mathbf{w} \end{pmatrix}$. Then $AX = X\mathrm{Diag}(1, \lambda_2)$, and thus $C_2(A)\mathbf{x} = \lambda_2\mathbf{x}$, where $\mathbf{x} = C_2(X)$. Each entry of $\mathbf{x}$ has the form $w_j - w_i$ for $i < j$. Hence $\mathbf{x}$ is a nonzero nonnegative vector of length $\binom{n}{2}$.

Suppose on the contrary that B has a negative entry, necessarily an off-diagonal one. We leave it as an exercise to check that in that case $C_2(B)$ also has a negative entry. As $C_2(A) = (C_2(B))^2$, an argument similar to the one in the proof of Theorem 3.133 shows that if $C_2(B)$ has a negative entry in the row indexed by $\alpha = \{i_0, j_0\}$, then $C_2(A)_{\alpha,\alpha} > \frac{\lambda_2 x_\alpha^2}{\nu}$, which contradicts the assumption on the 2×2 principal minors of A. □

The next few results present conditions that guarantee the existence of a rank-revealing CP factorization. An $n \times n$ doubly nonnegative matrix A of rank r is the Gram matrix of n vectors $\mathbf{v}_1, \ldots, \mathbf{v}_n$ in $\mathbb{R}^r$, such that the angle between any two of them is sharp. The following proposition gives a sufficient condition for such vectors to be in a circular cone that may be embedded into the nonnegative orthant.

Proposition 3.136. *Let* $\mathbf{v}_i \in \mathbb{R}^r$, $i = 1, \ldots, n$. *If there exists a nonzero vector* $\mathbf{x} \in \mathbb{R}^r$ *such that*

$$\langle \mathbf{v}_i, \mathbf{x} \rangle \geq \sqrt{\frac{r-1}{r}} \|\mathbf{v}_i\| \, \|\mathbf{x}\|$$

for every i, *then there exists an* $r \times r$ *orthogonal matrix* Q *such that* $Q\mathbf{v}_i \geq \mathbf{0}$ *for every* i.

Proof. First consider the case that $\mathbf{x}$ is a positive multiple of $\mathbf{e}$. In this case it suffices to show that if $\langle \mathbf{z}, \mathbf{e} \rangle \geq \sqrt{\frac{r-1}{r}} \|\mathbf{z}\| \, \|\mathbf{e}\|$, then $\mathbf{z} \in \mathbb{R}^r_+$. (Then

$Q = I$ will satisfy the requirements.) Indeed, suppose on the contrary that $\mathbf{z}$ satisfies this inequality, and is not nonnegative. Then $\mathbf{z} \neq \mathbf{0}$ and the assumption on $\langle \mathbf{z}, \mathbf{e} \rangle$ implies that $\sum_{i=1}^{r} z_i > 0$. We may assume that for some $1 \leq k < r$

$$z_1 \geq \ldots z_k \geq 0 > z_{k+1} \geq \cdots \geq z_r.$$

As $\|\mathbf{e}\| = \sqrt{r}$, the assumption on $\mathbf{z}$ is

$$\sum_{i=1}^{r} z_i \geq \sqrt{(r-1)\sum_{i=1}^{r} z_i^2},$$

or, equivalently,

$$\left(\sum_{i=1}^{r} z_i\right)^2 \geq (r-1)\sum_{i=1}^{r} z_i^2.$$

Using the fact that $\sum_{i=1}^{k} z_i > 0$ and $\sum_{i=k+1}^{r} z_i \leq 0$, and applying Cauchy-Schwarz inequality, we get

$$\left(\sum_{i=1}^{r} z_i\right)^2 \leq \left(\sum_{i=1}^{k} z_i\right)^2 + \left(\sum_{i=k+1}^{r} z_i\right)^2 \leq k\sum_{i=1}^{k} z_i^2 + (r-k)\sum_{i=k+1}^{r} z_i^2.$$

Combined with the assumption on $\mathbf{z}$, this yields

$$(r-1)\sum_{i=1}^{r} z_i^2 \leq k\sum_{i=1}^{k} z_i^2 + (r-k)\sum_{i=k+1}^{r} z_i^2,$$

that is,

$$(r-1-k)\sum_{i=1}^{k} z_i^2 + (k-1)\sum_{i=k+1}^{r} z_i^2 \leq 0.$$

As $\sum_{i=1}^{r} z_i > 0$ and $k \geq 1$, we get that $r-1-k \leq 0$. That is, $r \leq k+1$. Thus $r = k+1$. Then the inequality $(k-1)\sum_{i=k+1}^{r} z_i^2 \leq 0$ implies $k = 1$, $r = 2$. That is, $\mathbf{z} \in \mathbb{R}^2$ has $z_1 > 0 > z_2$. But the assumption on $\mathbf{z}$ is $z_1 + z_2 \geq \sqrt{z_1^2 + z_2^2}$, i.e., $z_1 z_2 \geq 0$. A contradiction.

Now, if $\mathbf{x}$ is not a positive multiple of $\mathbf{e}$, let $\mathbf{v} = \mathbf{x} - \frac{\|\mathbf{x}\|}{\sqrt{r}}\mathbf{e} \neq \mathbf{0}$. It is easy to check that $\|\mathbf{v}\|^2 = 2\mathbf{v}^T\mathbf{x}$, the matrix $Q = I - \frac{2}{\mathbf{v}^T\mathbf{v}}\mathbf{v}\mathbf{v}^T$ is an orthogonal matrix, and $Q\mathbf{x} = \frac{\|\mathbf{x}\|}{\sqrt{r}}\mathbf{e}$. Then for every i,

$$\langle Q\mathbf{v}_i, \mathbf{e} \rangle = \frac{\sqrt{r}}{\|\mathbf{x}\|}\langle Q\mathbf{v}_i, Q\mathbf{x} \rangle = \frac{\sqrt{r}}{\|\mathbf{x}\|}\langle \mathbf{v}_i, \mathbf{x} \rangle \geq \sqrt{\frac{r-1}{r}}\|\mathbf{v}_i\|\,\|\mathbf{e}\|,$$

since $\langle \mathbf{v}_i, \mathbf{x} \rangle \geq \sqrt{\frac{r-1}{r}}\|\mathbf{v}_i\|\,\|\mathbf{x}\|$. By the first part of the proof, this shows that $Q\mathbf{v}_i \in \mathbb{R}_+^r$ for every i. $\square$

Proposition 3.136 may not be strengthened by decreasing the constant $\sqrt{\frac{r-1}{r}}$ — see Exercise 3.69. Of course, verifying the existence of a vector $\mathbf{x}$ as in the previous proposition, and finding it, is not a straight forward task. The next corollary presents one case, where such a vector $\mathbf{x}$ exists. The result is a condition that is easy to check, and guarantees both that the given doubly nonnegative matrix A is completely positive, and that it has a rank-revealing CP factorization.

Theorem 3.137. *Let A be an $n \times n$ doubly nonnegative matrix with* $\operatorname{rank} A = r$. *Let R_i be the sum of row i of A, $1 \leq i \leq n$. If*

$$rR_i^2 \geq (r-1)a_{ii}\mathbf{e}^T A\mathbf{e}$$

for every $1 \leq i \leq n$, then A is completely positive, and there exists a nonnegative matrix B with r columns such that $A = BB^T$.

Proof. Let A be the Gram matrix of $\mathbf{v}_1, \ldots, \mathbf{v}_n \in \mathbb{R}^r$. That is, $A = C^T C$, where $C = \begin{pmatrix} \mathbf{v}_1 & \mathbf{v}_2 & \ldots & \mathbf{v}_n \end{pmatrix}$.

Then for every i, $\mathbf{v}_i = C\mathbf{e}_i$ and $\|\mathbf{v}_i\|^2 = a_{ii}$. Let $\mathbf{x} = \sum_{i=1}^n \mathbf{v}_i = C\mathbf{e}$. For every i,

$$\langle \mathbf{v}_i, \mathbf{x} \rangle = \mathbf{v}_i^T \mathbf{x} = \mathbf{e}_i^T C^T C\mathbf{e} = \mathbf{e}_i^T A\mathbf{e} = R_i,$$

and

$$\|\mathbf{x}\|^2 = \mathbf{x}^T\mathbf{x} = \mathbf{e}^T C^T C\mathbf{e} = \mathbf{e}^T A\mathbf{e}.$$

Thus the condition in the theorem means that for every i

$$\langle \mathbf{v}_i, \mathbf{x} \rangle \geq \sqrt{\frac{r-1}{r}} \|\mathbf{v}_i\| \, \|\mathbf{x}\|,$$

and the result follows by Proposition 3.136. □

For a doubly stochastic matrix A this implies the following. (Compare to Corollary 3.134.)

Corollary 3.138. *Let A be a doubly nonnegative doubly stochastic matrix of rank r. If $a_{ii} \leq \frac{r}{(r-1)n}$, then A is completely positive, and $A = BB^T$, where $B \geq 0$ has r columns.*

We conclude this section with the class of Soules matrices. This is a class of doubly nonnegative matrices that turn out to be completely positive. Moreover, each Soules matrix A has both a a square CP factorization and a rank-revealing CP factorization. We now define and discuss this special class of completely positive matrices, and their CP factorizations. A matrix

R is a *Soules basis matrix* if R is an orthogonal matrix whose first column is a positive vector, and for every nonnegative diagonal matrix D with nonincreasing diagonal entries, the matrix RDR^T is nonnegative. A matrix A is called a *Soules matrix* if $A = RDR^T$, where R is a Soules basis matrix, and $D = \mathrm{Diag}(d_1, \ldots, d_n)$, $d_1 \geq \cdots \geq d_n \geq 0$.

Theorem 3.139. *Every Soules matrix completely positive and has a positive semidefinite square root (and thus a square CP factorization).*

Proof. Let $A = RDR^T$, where R is a Soules basis matrix and $D = \mathrm{Diag}(d_1, \ldots, d_n)$, $d_1 \geq \cdots \geq d_n \geq 0$. The square root of A is $B = R\sqrt{D}R^T$, $\sqrt{D} = \mathrm{Diag}(\sqrt{d_1}, \ldots, \sqrt{d_n})$. As $\sqrt{d_1} \geq \cdots \geq \sqrt{d_n} \geq 0$, B is also a Soules matrix. In particular, it is nonnegative. □

We should point out that Soules basis matrices, and therefore Soules matrices, of any order (≥ 2) do exist. In fact, given a positive unit vector $\mathbf{x} \in \mathbb{R}^n$, we may construct a Soules basis matrix R whose first column is $\mathbf{x}$ as follows. Let $\Pi_1, \ldots, \Pi_n$ be partitions of $\{1, \ldots, n\}$ such that Π_i is a partition of $\{1, \ldots, n\}$ into i subsets, $\pi_{i,1}, \ldots, \pi_{i,i}$, and for every $2 \leq i \leq n$, Π_i refines Π_{i-1} by splitting one of the partition sets π_{i-1,k_i} of Π_{i-1} into two disjoint subsets, π_{i,k_i} and π_{i,k_i+1}. We call such $\Pi_1, \ldots, \Pi_n$ a *sequence of S-partitions.* Let $\mathbf{x}_i \in \mathbb{R}^n$ be a unit vector supported by π_{i-1,k_i}, such that $\mathbf{x}_i[\pi_{i,k_i}]$ is a positive multiple of $\mathbf{x}[\pi_{i,k_i}]$, and $\mathbf{x}_i[\pi_{i,k_i+1}]$ is a negative multiple of $\mathbf{x}[\pi_{i,k_i+1}]$. That is

$$\mathbf{x}_i[\pi_{i,j}] = \begin{cases} s_i\mathbf{x}[\pi_{i,j}], & \text{if } j = k_i \\ -t_i\mathbf{x}[\pi_{i,j}], & \text{if } j = k_i + 1 \\ \mathbf{0}, & \text{otherwise} \end{cases}, \tag{3.34}$$

where s_i, t_i are the unique positive numbers such that

$$\begin{aligned} s_i^2\|\mathbf{x}[\pi_{i,k_i}]\|^2 + t_i^2\|\mathbf{x}[\pi_{i,k_i+1}]\|^2 &= 1, \\ s_i\|\mathbf{x}[\pi_{i,k_i}]\|^2 - t_i\|\mathbf{x}[\pi_{i,k_i+1}]\|^2 &= 0. \end{aligned}$$

More explicitly,

$$\begin{aligned} s_i &= \frac{1}{\sqrt{\|\mathbf{x}[\pi_{i,k_i}]\|^2 + \|\mathbf{x}[\pi_{i,k_i+1}]\|^2}} \frac{\|\mathbf{x}[\pi_{i,k_i+1}]\|}{\|\mathbf{x}[\pi_{i,k_i}]\|}, \\ t_i &= \frac{1}{\sqrt{\|\mathbf{x}[\pi_{i,k_i}]\|^2 + \|\mathbf{x}[\pi_{i,k_i+1}]\|^2}} \frac{\|\mathbf{x}[\pi_{i,k_i}]\|}{\|\mathbf{x}[\pi_{i,k_i+1}]\|}. \end{aligned} \tag{3.35}$$

Then $R = \begin{pmatrix} \mathbf{x}_1 & \mathbf{x}_2 & \ldots & \mathbf{x}_n \end{pmatrix}$ is a Soules basis matrix. This is one of the assertions in following lemma.

Lemma 3.140. *Let* $R = \begin{pmatrix} \mathbf{x}_1 & \mathbf{x}_2 & \ldots & \mathbf{x}_n \end{pmatrix}$ *be an* $n \times n$ *real matrix with* $\mathbf{x}_1 = \mathbf{x} > \mathbf{0}$ *a unit vector. Let* $E_m = \sum_{i=1}^m \mathbf{x}_i \mathbf{x}_i^T$, $m = 1, \ldots, n$. *Then the following are equivalent.*

(a) *R is a Soules basis matrix.*
(b) $E_m \geq 0$ *for every* $1 \leq m \leq n$ *and* $E_n = I$.
(c) *For every* $1 \leq m \leq n$, *the matrix* E_m *is a direct sum of* m *rank 1 matrices* $\mathbf{b}_{m,i}\mathbf{b}_{m,i}^T$, $i = 1, \ldots, m$, *where each* $\mathbf{b}_{m,i}$ *is a positive unit vector, and* $\Pi_1, \ldots, \Pi_n$, *where* Π_m *is the partition of* $\{1, \ldots, n\}$ *into the* m *supports of* $\mathbf{b}_{m,1}, \ldots, \mathbf{b}_{m,m}$, *is a sequence of S-partitions.*
(d) *There exists a sequence of S-partitions* $\Pi_1, \ldots, \Pi_n$ *of* $\{1, \ldots, n\}$ *such that for every* $2 \leq i \leq n$ *the vector* $\mathbf{x}_i$ *satisfies* (3.34) *with* (3.35).
(e) *R is an orthogonal matrix, and there exists a sequence of S-partitions* $\Pi_1, \ldots, \Pi_n$ *of* $\{1, \ldots, n\}$ *such that for every* $2 \leq i \leq n$ *the vector* $\mathbf{x}_i$ *satisfies*

$$\operatorname{supp} \mathbf{x}_i = \pi_{i-1,k_i}, \quad \operatorname{supp}(\mathbf{x}_i)_+ = \pi_{i,k_i}, \quad \textit{and} \quad \operatorname{supp}(\mathbf{x}_i)_- = \pi_{i,k_i+1}.$$

We outline the proof, and leave some of the details to the exercises.

Proof. (a) $\Rightarrow$ (b): If R is a Soules matrix, then in particular $E_m = R(I_m \oplus 0_{n-m})R^T \geq 0$, and $E_n = RI_nR^T = I_n$.

(b) $\Rightarrow$ (a): If $E_m \geq 0$ for every m, then for $D = \operatorname{Diag}(d_1, \ldots, d_n)$, where $d_1 \geq \cdots \geq d_n \geq 0$,

$$RDR^T = (d_1 - d_2)E_1 + (d_2 - d_3)E_2 + \cdots + (d_{n-1} - d_n)E_{n-1} + d_nE_n \geq 0.$$

The implications (c) $\Rightarrow$ (b) and (d) $\Rightarrow$ (e) are trivial. Hence it suffices to prove (b) $\Rightarrow$ (d) and (e) $\Rightarrow$ (c).

(b) $\Rightarrow$ (d): This may be proved by induction. Clearly, $E_1 = \mathbf{b}_{1,1}\mathbf{b}_{1,1}^T$, where $\mathbf{b}_{1,1} = \mathbf{x}$ is a positive unit vector. Suppose E_m is a direct sum of m rank 1 matrices $\mathbf{b}_{m,i}\mathbf{b}_{m,i}^T$, $i = 1, \ldots, m$, where each $\mathbf{b}_{m,i}$ is a positive unit vector (equivalently, $\mathbf{b}_{m,i}\mathbf{b}_{m,i}^T$ has trace 1). Let $\pi_{m,i} = \operatorname{supp} \mathbf{b}_{m,i}$. Then $E_{m+1} = E_m + \mathbf{x}_{m+1}\mathbf{x}_{m+1}^T \geq 0$ implies that every $i \in \operatorname{supp}(\mathbf{x}_{m+1})_+$ and $j \in \operatorname{supp}(\mathbf{x}_{m+1})_-$ have to be in the same support $\pi_{m,k}$. Hence $\operatorname{supp} \mathbf{x}_{m+1} \subseteq \pi_{m,k}$. It remains to show that if $\mathbf{u}$ is a positive vector and $\mathbf{v}$ is a unit vector orthogonal to $\mathbf{u}$ such that $\mathbf{u}\mathbf{u}^T + \mathbf{v}\mathbf{v}^T \geq 0$, $\sigma_+ = \operatorname{supp} \mathbf{v}_+$ and $\sigma_- = \operatorname{supp} \mathbf{v}_-$, then $\mathbf{v}[\sigma_+]$ is a positive multiple of $\mathbf{u}[\sigma_+]$, $\mathbf{v}[\sigma_-]$ is a negative multiple of $\mathbf{u}[\sigma_-]$, and $\mathbf{u}\mathbf{u}^T + \mathbf{v}\mathbf{v}^T$ is a direct sum of two positive rank 1 matrices with trace 1 each. See Exercise 3.72. In particular, this would show that Π_{m+1} is obtained from Π_m by splitting $\pi_{m,k}$ into two disjoint subsets. Hence

$\Pi_1, \ldots, \Pi_n$ form a sequence of S-partitions. As $E_n = I$, R is an orthogonal matrix. Its columns therefore satisfy (3.34) with (3.35).

(e) $\Rightarrow$ (c) may be proved by reverse induction. Clearly, $E_n = I$ is a direct sum of n positive rank 1 matrices, each with trace 1. In the induction step, assume E_{m+1} is a direct sum of $m+1$ rank 1 matrices with trace 1 each. Without loss of generality we may suppose that $E_{m+1} = \mathbf{b}_1\mathbf{b}_1^T \oplus \mathbf{b}_2\mathbf{b}_2^T \oplus B$, where $\mathbf{b}_1 \in \mathbb{R}^p$ and $\mathbf{b}_2 \in \mathbb{R}^q$ are positive vectors of norm 1 each, and B is a direct sum of $m-1$ such rank 1 matrices. We may also assume that

$$\mathbf{x}_{m+1} = \begin{pmatrix} \mathbf{a} \\ -\mathbf{c} \\ \mathbf{0} \end{pmatrix}, \quad \mathbf{a} \in \mathbb{R}^p_+, \mathbf{c} \in \mathbb{R}^q_+ .$$

To show that the positive semidefinite matrix E_m is a direct sum of m positive rank 1 matrices, it suffices to consider the leading principal submatrix of $E_m = E_{m+1} - \mathbf{x}_{m+1}\mathbf{x}_{m+1}^T$ of order $p+q$. That is, it suffices to show that if $\mathbf{b}_1, \mathbf{a} \in \mathbb{R}^p$ and $\mathbf{b}_2, \mathbf{c} \in \mathbb{R}^q$ are positive vectors, $\|\mathbf{b}_1\| = \|\mathbf{b}_2\| = 1$, $\|\mathbf{a}\|^2 + \|\mathbf{c}\|^2 = 1$, and

$$C = \begin{pmatrix} \mathbf{b}_1 \\ \mathbf{b}_2 \end{pmatrix} \begin{pmatrix} \mathbf{b}_1 \\ \mathbf{b}_2 \end{pmatrix}^T - \begin{pmatrix} \mathbf{a} \\ -\mathbf{c} \end{pmatrix} \begin{pmatrix} \mathbf{a} \\ -\mathbf{c} \end{pmatrix}^T$$

is doubly nonnegative, then C is positive and $\operatorname{rank} C = 1$. See Exercise 3.73. □

Based on this lemma we can prove the following.

Theorem 3.141. *Let A be a Soules matrix. Then $A = BB^T$, where $B \geq 0$ has* $\operatorname{rank} A$ *columns.*

Proof. We have $A = RDR^T$, where R is a Soules basis matrix, and $D = \operatorname{Diag}(d_1, \ldots, d_r, 0, \ldots, 0)$, with $d_1 \geq \cdots \geq d_r > 0$. Let $\widetilde{R}$ consist of the first r columns of R, and $\widetilde{D} = \operatorname{Diag}(d_1, \ldots, d_r)$. Then $A = \widetilde{R}\widetilde{D}\widetilde{R}^T$. By permutation of the rows and columns of A we may assume that each of the partitions sets $\pi_{r,i}$ consists of consecutive integers, and the smallest integer in $\pi_{r,i+1}$ is the successor of the largest one in $\pi_{r,i}$. Then, by the structure of a Soules basis matrix described in Lemma 3.140(d),

$$\widetilde{R} = \begin{bmatrix} \mathbf{v}_1 & \ell_{12}\mathbf{v}_1 & \ell_{13}\mathbf{v}_1 & \ldots & \ell_{1r}\mathbf{v}_1 \\ \mathbf{v}_2 & \ell_{22}\mathbf{v}_2 & \ell_{23}\mathbf{v}_2 & \ldots & \ell_{2r}\mathbf{v}_2 \\ \vdots & \vdots & \vdots & & \vdots \\ \mathbf{v}_r & \ell_{r2}\mathbf{v}_r & \ell_{r3}\mathbf{v}_r & \ldots & \ell_{rr}\mathbf{v}_r \end{bmatrix},$$

where for $i = 1, \dots, r$ the vector $\mathbf{v}_i$ is a positive vector supported on $\pi_{r,i}$, and for $i = 1, \dots, r$ and $j = 2, \dots, r$

$$\ell_{ij} = \begin{cases} s_j, & \text{if } \pi_{r,i} \subseteq \pi_{j,k_j} \\ -t_j, & \text{if } \pi_{r,i} \subseteq \pi_{j,k_j+1} \\ 0, & \text{otherwise} \end{cases},$$

s_j and t_j as in (3.35). Now let S be the $n \times r$ nonnegative matrix

$$S = \begin{pmatrix} \frac{\mathbf{v}_1}{\|\mathbf{v}_1\|} & 0 & 0 & 0 \\ 0 & \frac{\mathbf{v}_2}{\|\mathbf{v}_2\|} & 0 & 0 \\ 0 & 0 & \ddots & 0 \\ 0 & 0 & 0 & \frac{\mathbf{v}_r}{\|\mathbf{v}_r\|} \end{pmatrix}$$

and let C be the $r \times r$ matrix

$$C = \begin{pmatrix} \|\mathbf{v}_1\| & \ell_{12}\|\mathbf{v}_1\| & \ell_{13}\|\mathbf{v}_1\| & \dots & \ell_{1r}\|\mathbf{v}_1\| \\ \|\mathbf{v}_2\| & \ell_{22}\|\mathbf{v}_2\| & \ell_{23}\|\mathbf{v}_2\| & \dots & \ell_{2r}\|\mathbf{v}_2\| \\ \vdots & \vdots & \vdots & & \vdots \\ \|\mathbf{v}_r\| & \ell_{r2}\|\mathbf{v}_r\| & \ell_{r3}\|\mathbf{v}_r\| & \dots & \ell_{rr}\|\mathbf{v}_r\| \end{pmatrix}.$$

Then $\widetilde{R} = SC$. Since the columns of $\tilde{R}$ are orthonormal, C is an orthonormal matrix. By the sign pattern of C, it is a Soules basis matrix, with a positive first column and a sequence of S-partitions of $\{1, \dots, r\}$, where the p-th partition, $1 \le p \le r$, has the following partition sets

$$\omega_{p,j} = \{1 \le i \le r \mid \pi_{r,i} \subseteq \pi_{p,j}\}, \quad j = 1, \dots, p.$$

Hence by Theorem 3.139

$$C\widetilde{D}C^T = BB^T,$$

where B is an $r \times r$ nonnegative matrix, and

$$A = \widetilde{R}\widetilde{D}\widetilde{R}^T = S(C\widetilde{D}C^T)S^T = (SB)(SB)^T$$

is a CP factorization of A such that $SB \ge 0$ has r columns. □

In addition to having a square CP factorization and a rank-revealing CP factorization, a Soules matrix is also an inverse M-matrix, and is therefore both UL- and LU-completely positive, see Exercise 3.74.

Some more cases in which matrices with a certain graph and rank have a rank-revealing factorization are discussed in the next chapter, about the cp-rank.

Exercises

3.59. Prove the claim in Example 3.118.

3.60. Show that

$$A = \begin{pmatrix} 6 & 5 & 3 & 0 \\ 5 & 11 & 4 & 0 \\ 3 & 4 & 2 & 0 \\ 0 & 0 & 0 & 0 \end{pmatrix}$$

is *UL*-completely positive but has a negative left almost principal minor, or construct an example of your own to demonstrate that the converse of Theorem 3.124 is not true.

3.61. Find an example of a 4×4 doubly nonnegative matrix which is a square of a nonnegative matrix, and is neither *LU*- nor *UL*-completely positive.

3.62. Let $A = H(5)$, the 5×5 Hilbert matrix. Find a nonnegative matrix B such that $A = BB^T$.

3.63. Let A be an $n \times n$ *UL*-completely positive matrix, $n \geq 3$. Show that $\det A[\{i\} \cup \alpha | \{j\} \cup \alpha] \geq 0$ for every $1 \leq i \neq j < \min \alpha$, for all $\alpha = \{k, \ldots, n\}$, $k = 3, \ldots, n$.

3.64. Let A be an $n \times n$ nonsingular doubly nonnegative matrix, $n \geq 2$. Then A is *UL*-completely positive if and only if

$$\det A[i, k, \ldots, n | j, k, \ldots, n] \geq 0$$

for every $1 \leq i, j < k$ and $2 \leq k \leq n$.

3.65. Prove Corollary 3.134.

3.66. Let A be the doubly stochastic doubly nonnegative matrix

$$\begin{pmatrix} \frac{3}{4} & 0 & \frac{1}{4} \\ 0 & \frac{3}{4} & \frac{1}{4} \\ \frac{1}{4} & \frac{1}{4} & \frac{1}{2} \end{pmatrix}.$$

(a) Show that there is no square nonnegative matrix B such that $A = B^2$ (in particular, there is no doubly stochastic B such that $A = B^2$).
(b) Show that there is no upper triangular nonnegative matrix U such that $A = UU^T$.
(c) Find a lower triangular nonnegative matrix L such that $A = LL^T$.

3.67. Let $\mathbf{x} \in \mathbb{R}^n_+$, and $A = I + \mathbf{x}\mathbf{x}^T$.

(a) Show that the positive semidefinite square root of A is nonnegative.
(b) Show that there exists an upper triangular nonnegative matrix U such that $A = UU^T$.
(c) Show that there exists a lower triangular nonnegative matrix L such that $A = LL^T$.

Hint: For (a), see Remark 3.32. For (b) and (c), note that $(I + \mathbf{x}\mathbf{x}^T)^{-1} = I - \frac{1}{1+\|\mathbf{x}\|^2}\mathbf{x}\mathbf{x}^T$ is an M-matrix.

3.68. Complete the missing details in the proof of Theorem 3.135: Show that if the positive semidefinite root B of the doubly nonnegative matrix A has a negative entry, then the second compound matrix $C_2(B)$ also has a negative entry.

(a) Suppose B has a negative entry. Show by the argument in the proof of Theorem 3.132 that if $b_{ij} < 0$ for some $i < j$, then $(i, j) \neq (1, n)$.
(b) Let i_0 be the minimal index of a row in which B has a negative entry, and let j_0 be the minimal $j > i_0$ for which $b_{i_0 j_0} < 0$. Consider $m > j_0$. Denote $\alpha = \{i_0, j_0\}$, $\beta = \{i_0, m\}$ and $\gamma = \{j_0, m\}$. Show that if both $(C_2(B))_{\alpha,\beta} \geq 0$ and $(C_2(B))_{\alpha,\gamma} \geq 0$, then $b_{i_0 m} b_{j_0 m} \leq 0$. Similarly for $m < i_0$.
(c) Use the fact that $a_{i_0\, i_0+1}$ is the scalar product of rows i_0 and $i_0 + 1$ of B to deduce from (b) that $j_0 > i_0 + 1$.
(d) Show that if $i_0 = 1$ and $3 \leq j_0 \leq n-1$, then there is a negative element in the column of $C_2(B)$ indexed by $\{j_0, n\}$.
(e) To complete the proof, show that if $i_0 > 1$, then either $C_2(B)_{\{1,i_0\},\{i_0,j_0\}} < 0$ or $C_2(B)_{\{1,i_0\},\{1,j_0\}} < 0$.

3.69. Show that if $\mathbf{z} \in \mathbb{R}^r$ satisfies $\langle \mathbf{z}, \mathbf{e} \rangle \geq c\|\mathbf{z}\|\,\|\mathbf{e}\|$, where $c < \sqrt{\frac{r-1}{r}}$, then $\mathbf{z}$ need not be nonnegative.
Hint: Let $\varepsilon > 0$ sufficiently small so that $\sqrt{r-1}\sqrt{1-\varepsilon^2} - \varepsilon > c\sqrt{r}$, and consider $\mathbf{z}^T = \left(\sqrt{1-\varepsilon^2}, \ldots, \sqrt{1-\varepsilon^2}, -\varepsilon\sqrt{r-1}\right)$.

3.70. The matrix

$$A = \begin{pmatrix} 41 & 43 & 80 & 56 & 50 \\ 43 & 62 & 89 & 78 & 51 \\ 80 & 89 & 162 & 120 & 93 \\ 56 & 78 & 120 & 104 & 62 \\ 50 & 51 & 93 & 62 & 65 \end{pmatrix}$$

is doubly nonnegative and rank $A = 3$. Show that it is completely positive, and $A = BB^T$ for some $B \geq 0$ with 3 columns.

3.71. Prove Corollary 3.138.

3.72. Let $\mathbf{u}$ and $\mathbf{v}$ be vectors in $\mathbb{R}^{p+q}$, $\mathbf{u}$ positive and $\mathbf{v}$ a unit vector orthogonal to $\mathbf{u}$. Suppose

$$\mathbf{v} = \begin{pmatrix} \mathbf{v}_1 \\ -\mathbf{v}_2 \end{pmatrix}, \quad \mathbf{v}_1 \in \mathbb{R}^p_+ \text{ and } \mathbf{v}_2 \in \mathbb{R}^q_+,$$

and split $\mathbf{u}$ conformally:

$$\mathbf{u} = \begin{pmatrix} \mathbf{u}_1 \\ \mathbf{u}_2 \end{pmatrix}, \quad \mathbf{v}_1 \in \mathbb{R}^p_+ \text{ and } \mathbf{v}_2 \in \mathbb{R}^q_+$$

$\mathbf{u}_1, \mathbf{u}_2$ positive. Show that if $\mathbf{u}\mathbf{u}^T + \mathbf{v}\mathbf{v}^T \geq 0$, then $\mathbf{v}_1 = \alpha\mathbf{u}_1$, $\mathbf{v}_2 = t\mathbf{u}_2$, where $\alpha, t > 0$, and

$$\mathbf{u}\mathbf{u}^T + \mathbf{v}\mathbf{v}^T = \left(\frac{1}{\|\mathbf{u}_1\|^2}\mathbf{u}_1\mathbf{u}_1^T\right) \oplus \left(\frac{1}{\|\mathbf{u}_2\|^2}\mathbf{u}_2\mathbf{u}_2^T\right).$$

3.73. Complete the induction step in the proof that (e) $\Rightarrow$ (c) in Lemma 3.140: Let $\mathbf{b}_1, \mathbf{a} \in \mathbb{R}^p$ and $\mathbf{b}_2, \mathbf{c} \in \mathbb{R}^q$ are positive vectors, $\|\mathbf{b}_1\| = \|\mathbf{b}_2\| = 1$ and $\|\mathbf{a}\|^2 + \|\mathbf{c}\|^2 = 1$, and suppose

$$C = \begin{pmatrix} \mathbf{b}_1 \\ \mathbf{b}_2 \end{pmatrix}\begin{pmatrix} \mathbf{b}_1 \\ \mathbf{b}_2 \end{pmatrix}^T - \begin{pmatrix} \mathbf{a} \\ -\mathbf{c} \end{pmatrix}\begin{pmatrix} \mathbf{a} \\ -\mathbf{c} \end{pmatrix}^T$$

is doubly nonnegative. Show that C is necessarily positive and $\operatorname{rank} C = 1$.
Hint: C is positive semidefinite, and $\mathbf{b}_1\mathbf{b}_1^T - \mathbf{a}\mathbf{a}^T$ and $\mathbf{b}_2\mathbf{b}_2^T - \mathbf{c}\mathbf{c}^T$ are nonzero principal submatrices of C.

3.74. Show that any 2×2 orthogonal matrix with a positive first column is a Soules basis matrix, and any positive 2×2 doubly nonnegative matrix is a Soules matrix.

3.75. Show that if A is a nonsingular Soules matrix then A is an inverse M-matrix. (Hence P^TAP is UL-completely positive for every permutation matrix P.)
Hint: In the notations of Lemma 3.140,
$RD^{-1}R^T = \left(\frac{1}{d_1} - \frac{1}{d_2}\right)E_1 + \left(\frac{1}{d_2} - \frac{1}{d_3}\right)E_2 + \cdots + \left(\frac{1}{d_{n-1}} - \frac{1}{d_n}\right)E_{n-1} + \frac{1}{d_n}E_n.$

3.76. Show that for every positive $\mathbf{x} \in \mathbb{R}^n$ the matrix $A = I + \mathbf{x}\mathbf{x}^T$ is a Soules matrix.

3.77. Show that if A is a doubly nonnegative matrix with at most two different eigenvalues, then A has a nonnegative square root (in particular,

A is completely positive and has a square CP factorization). Conclude that for such A, $A^{1/2^k} \geq 0$ for every positive integer k.
Hint: $A = \lambda_1 E_1 + \lambda_2 E_2 = (\lambda_1 - \lambda_2)E_1 + \lambda_2 I$, where $\lambda_1 \geq \lambda_2$, E_1 and E_2 are orthogonal projection matrices such that $E_1 + E_2 = I$. Deduce that $E_2 \geq 0$.

3.78. Let A be a 3×3 completely positive matrix, which is irreducible and singular. Suppose A has three different eigenvalues, and there exists an eigenvector corresponding to 0 which has a zero entry. Show that A has a nonnegative square root. Conclude that for such A, $A^{1/2^k} \geq 0$ for every positive integer k.

Show that if for such A the eigenvector corresponding to 0 does not have a zero entry, then there exists a positive integer k such that $A^{1/2^k}$ is not nonnegative.
Hint: Consider the sign pattern of an eigenvector of 0 in each case.

Notes. Most of the discussion on the existence of LL^T and UU^T factorizations is based on [Markham (1972)]. Lemma 3.126 appeared in [Maxfield and Minc (1962/1963)] as part of the proof that 3×3 doubly nonnegative matrices are completely positive. The M-matrix results are taken from [Ando (1987)] and [Jacobson (1974)]. See also [Lau and Markham (1978)]. Theorem 3.132 was proved in [Reams (2011)], generalizing Corollary 3.134, which is due to [Marcus and Minc (1962)]. The latter is also the source for the example in Exercise 3.66. We point out that Theorem 3.132 may be deduced from Corollary 3.134, using the fact that every irreducible nonnegative matrix is diagonally congruent by a positive diagonal matrix to a doubly stochastic matrix. Theorem 3.135 and its proof, as outlined in Exercise 3.68, are due to [Reams (2011)]. Proposition 3.136, Theorem 3.137, Corollary 3.138 and Exercises 3.69–3.71 are due to [So and Xu (2015)]. The first Soules basis matrices were constructed in [Soules (1983)], the general notion of Soules basis matrices was introduced, and named after Soules, in [Elsner *et al.* (1998)]. The characterization in Lemma 3.140, the result in Exercise 3.75, and connections of Soules matrices to other special matrices also appeared there. Part (e) of Lemma 3.140 was added here to emphasize that the sign pattern of an orthogonal matrix R alone suffices to determine whether it is a Soules basis matrix. Theorems 3.139 and 3.140 were proved in [Shaked-Monderer (2004/05)]. (Note a slight change of terminology: in these two papers a Soules matrix is what we call here a Soules basis matrix, following some later works on the topic.) In [Salce (2016)], a positive semidefinite matrix A such that $A^{1/2^k} \geq 0$, where k is a positive integer, is called a *k-rooted* matrix, and a matrix which is k-rooted for every positive integer k is called an *∞-rooted* matrix. The study of completely positive rooted matrices was initiated in [Salce (2016)], and the results in Exercises 3.77–3.78 were proved there. (Note that Soules matrices are ∞-rooted.)

3.9 Functions of completely positive matrices

Given a completely positive matrix A it is natural to ask for which functions f is $f(A)$ also completely positive. But first we have to define what we mean by $f(A)$. The ordinary definition of a function of a matrix is given through its Jordan form. In the case of real symmetric matrices it amounts to the following definition.

Definition 3.142. Given an $n \times n$ real symmetric matrix A, let U be an orthogonal matrix and $D = \mathrm{Diag}(d_1, d_2, \ldots, d_n)$ a diagonal matrix such that $U^T AU = D$. If $d_1, d_2, \ldots, d_n$ belong to the domain of a function f, then $f(A)$ is defined by $f(A) = U\mathrm{Diag}(f(d_1), f(d_2), ..., f(d_n))U^T$.

In the case that $f(x) = \sum_{i=0}^n a_i x^i$ is a polynomial, this definition agrees with the natural definition $f(A) = \sum_{i=0}^n a_i A^i$. And if $f(x)$ admits a power series expansion, $f(x) = \sum_{i=0}^\infty a_i x^i$, and $\lim_{m\to\infty} \sum_{i=0}^m a_i A^i$ exists, then $f(A)$ equals that limit, that is, $f(A) = \sum_{i=0}^\infty a_i A^i$.

We also consider Hadamard functions of A.

Definition 3.143. Let $f(x)$ be a function whose domain includes the entries of a matrix A. The *Hadamard function of a matrix* A, $f_H(A)$, is defined entrywise by $(f_H(A))_{ij} = f(a_{ij})$.

Example 3.144. Let

$$A = \begin{pmatrix} 1 & 2 \\ 3 & 4 \end{pmatrix},$$

and let $f(x) = x^2 - 5x - 2$. Then

$$f_H(A) = \begin{pmatrix} -6 & -8 \\ -8 & -6 \end{pmatrix}.$$

The ordinary $f(A)$ is in this case the zero matrix.

Note that we use the natural convention $A^{(0)} = J$.

Theorem 3.145. *Suppose $f(x)$ is defined on $[0, c)$, and has a series expansion $f(x) = \sum_{k=0}^{\infty} a_k x^k$ for $0 \leq x < c$, where $a_k \geq 0$ for every k. Then for every completely positive matrix A with $a_{ij} < c$, $i, j = 1, \ldots, n$, the matrix $f_H(A)$ is also completely positive.*

Proof. $f_H(A) = \sum_{k=0}^{\infty} a_k A^{(k)}$. For every $k \geq 1$ the matrix $A^{(k)}$ is completely positive by Corollary 3.15, and $A^{(0)} = J$ is also completely positive. For every m, $\sum_{k=0}^{m} a_k A^{(k)}$ is completely positive by Theorem 3.25 and, by the same theorem,

$$f_H(A) = \lim_{m \to \infty} \left(\sum_{k=0}^{m} a_k A^{(k)} \right)$$

is completely positive. □

Corollary 3.146. *If A is completely positive, then so is* $\exp_H(A)$.

Proof. The function $\exp(x) = e^x$ satisfies the assumptions of the previous theorem for $c = \infty$. □

The last corollary can be used to obtain a stronger result.

Theorem 3.147. *If A is a real positive semidefinite matrix, then* $\exp_H(A)$ *is completely positive.*

Proof. Since A is positive semidefinite, it can be represented as sum of rank 1 positive semidefinite matrices, $\sum_{k=1}^{m} \mathbf{x}_k \mathbf{x}_k^T$, where every $\mathbf{x}_k$ is a (not necessarily nonnegative) vector in $\mathbb{R}^n$. Thus $\exp_H(A)$ is the Hadamard product of the Hadamard exponents $\exp_H(\mathbf{x}_k \mathbf{x}_k^T)$. By Corollary 3.14 it is therefore enough to prove that for every $\mathbf{x} \in \mathbb{R}^n$, $\exp_H(\mathbf{x}\mathbf{x}^T)$ is completely positive. The i, j entry of $\exp_H(\mathbf{x}\mathbf{x}^T)$ is $e^{x_i x_j}$. Let $M = \max_{1 \leq i \leq n}(|x_i|)$. Then

$$e^{x_i x_j} = e^{-M^2 + (M + x_i)(M + x_j) - M x_i - M x_j}.$$

Thus

$$\exp_H(\mathbf{x}\mathbf{x}^T) = e^{-M^2} \exp_H(\mathbf{y}\mathbf{y}^T) \circ \mathbf{z}\mathbf{z}^T$$

for $\mathbf{y}, \mathbf{z} \in \mathbb{R}^n$ with $y_i = M + x_i$, $z_i = e^{-M x_i}$. Since both $\mathbf{y}$ and $\mathbf{z}$ are nonnegative, $\mathbf{y}\mathbf{y}^T$ and $\mathbf{z}\mathbf{z}^T$ are completely positive. By Corollary 3.146, $\exp_H(\mathbf{y}\mathbf{y}^T)$ is completely positive, and since $e^{-M^2} > 0$, it follows from Corollary 3.14 that $\exp_H(\mathbf{x}\mathbf{x}^T)$ is completely positive. □

In the next definition we use $A^{(p)}$ to denote the matrix whose i, j entry is a_{ij}^p, where p is any positive number.

Definition 3.148. A matrix A is *infinitely divisible* if it is doubly nonnegative and the matrix $A^{(p)}$ is positive semidefinite for every real $p > 0$.

Theorem 3.149. *Infinitely divisible matrices are completely positive.*

Proof. The equality

$$a_{ij} = \lim_{p \to 0^+} \exp\left(\frac{1}{p}(a_{ij}^p - 1)\right)$$

implies that an infinitely divisible A satisfies

$$A = \lim_{p \to 0^+} e^{-1/p} \exp_H\left(\frac{1}{p}A^{(p)}\right).$$

As $\exp_H\left(\frac{1}{p}A^{(p)}\right)$ is completely positive for every $p > 0$ by Theorem 3.147, A is also completely positive by Theorem 3.25. □

The converse of Theorem 3.149 is not true.

Example 3.150. For every $\varepsilon > 0$,

$$A_\varepsilon = \begin{pmatrix} 1+\varepsilon & 1+2\varepsilon & 1 \\ 1+2\varepsilon & 1+4\varepsilon & 1 \\ 1 & 1 & 1 \end{pmatrix} = \begin{pmatrix} 1 & \sqrt{\varepsilon} \\ 1 & 2\sqrt{\varepsilon} \\ 1 & 0 \end{pmatrix} \begin{pmatrix} 1 & 1 & 1 \\ \sqrt{\varepsilon} & 2\sqrt{\varepsilon} & 0 \end{pmatrix}$$

is completely positive, but for small ε it is not infinitely divisible, since $\det A_\varepsilon^{(p)}$ is negative for $0 < p < 1$ and small ε.

Results similar to Theorem 3.145 and its corollaries hold for ordinary functions of completely positive matrices. Here we have to rely on the fact that natural powers of completely positive matrices are completely positive (Corollary 3.17), instead of the corresponding result for Hadamard powers.

The results are stated here, and left as exercises.

Theorem 3.151. *Suppose $f(x)$ has a power series expansion*

$$f(x) = \sum_{i=0}^{\infty} a_i x^i \ , \ 0 \le x \le c,$$

in which $a_i \ge 0$ for every i. If A is completely positive and $\|A\| \le c$, then $f(A)$ is completely positive.

Corollary 3.152. *If A is a nonnegative symmetric matrix, then* $\exp(A)$ *is completely positive.*

Exercises

3.79. If A is a real positive semidefinite matrix and $|a_{ij}| < 1$ for every i and j, then the matrix $(1/(1 - a_{ij}))_{i,j=1}^n$ is completely positive.
Hint: $1/(1 - x) = \exp(-\log(1 - x))$.

3.80. If A is nonnegative and symmetric, and if for every positive q the matrix $(a_{ij}/(q + a_{ij}))_{i,j=1}^n$ is positive semidefinite, then A is completely positive.
Hint: Use the identity

$$x^p = \frac{\sin(\pi p)}{\pi} \int_0^\infty \frac{x}{t + x}\, t^{p-1}\, dt,$$

which is valid for every $0 < p < 1$.

3.81. Prove Theorem 3.151.

3.82. Prove Corollary 3.152.
Hint: $\exp(A) = \left(\exp\left(\frac{1}{2}A\right)\right)^2$.

3.83. Show that the equivalent of Exercise 3.79 for functions of matrices does not hold. That is, find an example of a positive semidefinite matrix A such that $\|A\| < 1$ and $(I - A)^{-1}$ is not completely positive.

Show that if A is completely positive and $\|A\| < 1$, then $(I - A)^{-1}$ is completely positive.

Notes. This section is based on [Ando (1991)]. Infinitely divisible matrices were defined and characterized by R. A. Horn, see [Horn (1967)] and [Horn (1969)]. The name comes from the related concept of infinitely divisible function, see [FitzGerald and Horn (1977)]. The integral identity in the hint for Exercise 3.80 can be derived by methods of complex analysis — see, e.g., [Beardon (1979)]. It is also true that if A is a real positive semidefinite matrix, and if $A(pI + A)^{-1}$ is doubly nonnegative for every $p > 0$, then A is completely positive — see [Ando (1991)].

3.10 The CP completion problem

The PSD completion problem was described in Section 1.8, the COP and SPN completion problems in Section 2.13. Here we describe the analogous CP completion problem.

We will need the terms defined in Section 1.8, such as a partial symmetric matrix, a partial PSD matrix, and specification graph. Recall that in a partial symmetric matrix all the diagonal entries are specified. Here we make the following definitions.

Definition 3.153. A *partial CP matrix* is a partial symmetric matrix in which all the fully specified principal submatrices are completely positive. A *CP completion* of A is a completion of A which is completely positive. If A is a partial CP matrix with specification graph G, we say that A is a *G-partial CP matrix.*

The question is: which partial CP matrices have a CP completion? Our first aim in this section is to characterize the graphs G such that every G-partial CP matrix is CP completable. By the following remark it suffices to characterize the *connected* graphs with this property.

Remark 3.154. A graph G has the property that all its partial CP matrix realizations can be completed to a completely positive matrix if and only if all its components have this property.

The main theorem in this section is the following.

Theorem 3.155. *Every partial CP matrix realization of a connected graph G can be completed to a completely positive matrix if and only if G is a block-clique graph.*

Recall that a block-clique graph is a graph in which each block is a clique (see Section 1.5). Before we prove the theorem, we consider two examples of partial CP matrices that cannot be completed to a doubly nonnegative matrix (let alone to a completely positive matrix).

Example 3.156. Let

$$A = \begin{pmatrix} 1 & 1 & ? & & \dots & ? & 0 \\ 1 & 1 & 1 & ? & & ? & ? \\ ? & 1 & 1 & \ddots & \ddots & & \vdots \\ & ? & \ddots & \ddots & \ddots & ? & \\ \vdots & & \ddots & \ddots & 1 & 1 & ? \\ ? & ? & & ? & 1 & 1 & 1 \\ 0 & ? & \dots & \dots & ? & 1 & 1 \end{pmatrix}.$$

It is the $k \times k$ partial symmetric matrix of Example 1.183, where the diagonal, first superdiagonal and first subdiagonal entries are equal to 1, the $1, k$ and the $k, 1$ entries are equal to 0, and all other entries are unspecified. Clearly, the fully specified principal submatrices are the diagonal entries and the 2×2 principal submatrices, each of which is either J_2 or I_2, so A is a partial CP matrix. It was shown in Example 1.183 that A cannot be completed to a positive semidefinite matrix.

Example 3.157. Let A be the partial symmetric matrix

$$A = \begin{pmatrix} 2 & 3 & 0 & ? \\ 3 & 6 & 3 & 0 \\ 0 & 3 & 6 & 3 \\ ? & 0 & 3 & 2 \end{pmatrix}.$$

All fully specified submatrices are doubly nonnegative of order less than 4 so by Theorem 3.35 they are completely positive. Hence A is a partial CP matrix. However, when $a_{14} = a_{41}$ are specified, $\det A = -27(a_{14} + 1)^2$, so A cannot be completed to a doubly nonnegative matrix. Observe that the specification graph of A is T_4.

We can now prove Theorem 3.155.

Proof of Theorem 3.155. If G is not chordal, then it contains an induced cycle of length $k > 3$. The partial PSD matrix in the proof of Theorem 1.185 is actually partial CP, and since it cannot be completed to a positive semidefinite matrix, it cannot be completed to a completely positive one.

If G is chordal but not block-clique, then it contains an induced T_4. In this case we use Example 3.157 to specify the entries of the double triangle, set all remaining diagonal entries to be 1, and all remaining off-diagonal specified entries to be 0. By Example 3.157 the partial CP matrix obtained that way has no doubly nonnegative completion. Hence to be CP completable, the graph must be block-clique.

Suppose now that G is a block-clique graph, and let A be any partial CP matrix realization of G. We have to show that A can be completed to a completely positive matrix. If G is a clique, then any "partial" CP matrix realization of G is already a completely positive matrix. Suppose that G has just two blocks, and let A be a partial CP matrix realization of G. Since complete positivity is invariant under permutation similarity or positive diagonal congruence we may assume that

$$A = \begin{pmatrix} B & \mathbf{c} & ? \\ \mathbf{c}^T & 1 & \mathbf{d} \\ ? & \mathbf{d} & E \end{pmatrix},$$

where

$$A_1 = \begin{pmatrix} B & \mathbf{c} \\ \mathbf{c}^T & 1 \end{pmatrix} \text{ and } A_2 = \begin{pmatrix} 1 & \mathbf{d} \\ \mathbf{d} & E \end{pmatrix}$$

are completely positive, and each question mark stands for a block of unspecified entries. Suppose

$$A_1 = \sum_{i=1}^{m_1} \mathbf{f}_i \mathbf{f}_i^T \text{ and } A_2 = \sum_{j=1}^{m_2} \mathbf{g}_j \mathbf{g}_j^T,$$

$\mathbf{f}_i \in \mathbb{R}^p_+$, $\mathbf{g}_j \in \mathbb{R}^q_+$, and $p+q-1 = n$ is the order of A. Partition the vectors $\mathbf{f}_i$ and $\mathbf{g}_j$:

$$\mathbf{f}_i = \begin{pmatrix} \mathbf{v}_i \\ s_i \end{pmatrix}, \ \mathbf{g}_j = \begin{pmatrix} t_j \\ \mathbf{w}_j \end{pmatrix},$$

where $\mathbf{v}_i$ is obtained from $\mathbf{f}_i$ by deleting the last entry, and $\mathbf{w}_j$ is obtained from $\mathbf{g}_j$ by deleting the first entry.

Define $m_1 m_2$ vectors $\mathbf{h}_{ij}$ in $\mathbb{R}^n_+$.

$$\mathbf{h}_{ij} = \begin{pmatrix} \mathbf{v}_i t_j \\ s_i t_j \\ s_i \mathbf{w}_j \end{pmatrix}.$$

Then

$$\sum_{i,j} \mathbf{h}_{ij} \mathbf{h}_{ij}^T = \begin{pmatrix} B & \mathbf{c} & \mathbf{c}\mathbf{d}^T \\ \mathbf{c}^T & 1 & \mathbf{d} \\ \mathbf{d}\mathbf{c}^T & \mathbf{d} & E \end{pmatrix},$$

so A has a CP completion. For the general case, where G consists of k blocks, we proceed by induction on k. We use the fact that in a graph with k blocks, there is a block which contains only one cut vertex (see Exercise 1.38), and apply a construction similar to the one above. □

One may also consider a DNN completion problem.

Definition 3.158. A *partial DNN matrix* is a partial PSD matrix whose specified entries are nonnegative. A *G-partial DNN matrix* is a partial DNN matrix whose specification graph is G.

It is interesting that although a matrix (of order > 4) may be doubly nonnegative but not completely positive, the DNN completion problem has the same qualitative solution as the CP completion problem. The interested reader is invited to prove this fact (Exercise 3.86).

Since the existence of induced cycles indicates that a graph is not chordal, it is natural to ask what additional conditions are required on the specified entries of a partial CP matrix whose specification graph contains cycles, to assure that it has a CP completion.

Let A be a C_n-partial CP matrix and let $D = \mathrm{Diag}(a_{11}^{-1/2}, \ldots, a_{11}^{-1/2})$. Then

$$DAD = \begin{pmatrix} 1 & \cos\theta_2 & ? & \ldots & ? & \cos\theta_1 \\ \cos\theta_2 & 1 & \cos\theta_3 & \ddots & ? & ? \\ ? & \ddots & \ddots & \ddots & \ddots & \vdots \\ \vdots & \ddots & \ddots & \ddots & \ddots & \vdots \\ ? & ? & \ddots & \cos\theta_{n-1} & 1 & \cos\theta_n \\ \cos\theta_1 & ? & \ldots & ? & \cos\theta_n & 1 \end{pmatrix}$$

is also partial CP. We refer to DAD as the *normalization* of A.

Definition 3.159. A partial DNN (CP) matrix *meets the cycle conditions* if in the normalization of each of its partial submatrices that correspond to a cycle (of length n),

$$2 \max_{1\le i\le n} \theta_i \le \sum_{i=1}^{n} \theta_i.$$

We conclude this section by bringing, without proof, the following generalization of Theorem 3.155.

Theorem 3.160. *Let G be a graph. Then the following are equivalent.*

(a) *Every G-partial CP matrix which meets the cycle conditions has a completely positive completion.*
(b) *Every G-partial DNN matrix which meets the cycle conditions has a doubly nonnegative completion.*
(c) *All the blocks of G are cliques or cycles.*

Exercises

3.84. Prove the claim in Remark 3.154.

3.85. The partial symmetric matrix

$$\begin{pmatrix} a & 3 & 0 & ? \\ 3 & 6 & 3 & 0 \\ 0 & 3 & 6 & 3 \\ ? & 0 & 3 & a \end{pmatrix}$$

is partial CP for $a \geq 2$. Show that it is CP completable if and only if $a \geq 3$, and find a CP completion for it when $a \geq 3$.

3.86. Prove: A graph G has the property that every partial DNN matrix realization of G can be completed to a doubly nonnegative matrix if and only if G is a block-clique graph.

Notes. Theorem 3.155 and the theorem in Exercise 3.86 are due to [Drew and Johnson (1998)]. Theorem 3.160 was proved in [Drew *et al.* (2000)]. Block-clique graphs were used in [Johnson and Smith (1996)], in a study of completion problems for M-matrices and inverse M-matrices. In [Drew *et al.* (2000)] they are called 1-chordal graphs.

3.11 Rational and integral completely positive matrices

Consider the following question regarding rational completely positive matrices.

Question 3.161. *Let A be a completely positive matrix with rational entries. Does A have a factorization $A = BB^T$ in which the entries of B are nonnegative rationals?*

If a matrix A has such a factorization, we say that A is *completely positive over the rationals.* One may ask similar questions over other algebraic structures: Does a completely positive matrix with integer entries have a factorization $A = BB^T$ in which B is nonnegative and has integral entries? Which integral completely positive matrices have such a factorization with B a $\{0,1\}$-matrix? The notion of complete positivity was generalized to complete positivity over Boolean rings, over semirings, and over min-max rings. We do not go into these generalizations (but see the notes at the end of the section for some references). We consider here only complete positivity over the rationals, and, as a side, also over the integers. Complete positivity over the rationals is of particular interest due to its relevance to determining whether the membership problem for $\mathcal{CP}_n$ is NP-complete.

It is not yet known whether any rational completely positive matrix is completely positive over the rationals, but there is strong evidence in favour of a positive answer, and no counter example yet. Hence we make the following conjecture.

Conjecture 3.162. *Every rational completely positive matrix is completely positive over the rationals.*

Before we go into the results on this question, let us take another look at the definition. By the definition above, a matrix A is completely positive over the rationals if $A = \sum_{i=1}^{k} \mathbf{b}_i \mathbf{b}_i^T$, where the vectors $\mathbf{b}_i$ are rational and nonnegative. An alternative approach is to say that A is completely positive over the rationals if it can be decomposed in the form

$$A = \sum_{i=1}^{m} d_i \mathbf{x}_i \mathbf{x}_i^T, \tag{3.36}$$

where each d_i is a nonnegative rational number and each $\mathbf{x}_i$ is a nonnegative rational vector. Such CP representation of A can also be written as

$$A = XDX^T, \tag{3.37}$$

where the matrix $X = \begin{pmatrix} \mathbf{x}_1 & \dots & \mathbf{x}_m \end{pmatrix}$ is nonnegative and has rational entries, and $D = \mathrm{Diag}(d_1, \dots, d_m)$ is a nonnegative diagonal rational matrix. Note that for real matrices, these two definitions of complete positivity are identical: just let $\mathbf{b}_i = \sqrt{d_i}\mathbf{x}_i$. Although this simple argument does not apply in the rational case, the two approaches are equivalent in this case also.

Proposition 3.163. *An $n \times n$ symmetric matrix A is completely positive over the rationals if and only if* (3.36) *or* (3.37) *hold.*

Proof. Clearly if $A = BB^T$, where B is a nonnegative rational matrix, then $A = XDX^T$, where $X = B$ and $D = I$.

For the converse, suppose $A = \sum_{i=1}^{m} d_i \mathbf{x}_i \mathbf{x}_i^T$, where each d_i is nonnegative and rational, and each $\mathbf{x}_i$ is a nonnegative vector with rational entries. We may assume all d_i's are positive. Let $d_i = \frac{p_i}{q_i}$, where p_i and q_i are positive integers. Then

$$\sum_{i=1}^{m} d_i \mathbf{x}_i \mathbf{x}_i^T = \sum_{i=1}^{m} \frac{p_i}{q_i} \mathbf{x}_i \mathbf{x}_i^T = \sum_{i=1}^{m} \frac{p_i q_i}{q_i^2} \mathbf{x}_i \mathbf{x}_i^T.$$

According to Lagrange's four-square theorem there exist nonnegative integers $s_{i_1}, s_{i_2}, s_{i_3}, s_{i_4}$ such that $p_i q_i = s_{i_1}^2 + s_{i_2}^2 + s_{i_3}^2 + s_{i_4}^2$. Hence

$$\sum_{i=1}^{m} d_i \mathbf{x}_i \mathbf{x}_i^T = \sum_{i=1}^{m} \frac{s_{i_1}^2 + s_{i_2}^2 + s_{i_3}^2 + s_{i_4}^2}{q_i^2} \mathbf{x}_i \mathbf{x}_i^T.$$

Let $\mathbf{b}_{ij} = \frac{s_{ij}}{q_i}\mathbf{x}_i$, $i = 1, \ldots, m$, $j = 1, \ldots, 4$. Then

$$A = \sum_{i=1}^{m} \sum_{j=1}^{4} \mathbf{b}_{ij} \mathbf{b}_{ij}^T . \qquad \square$$

As the proof of Proposition 3.163 shows, the number of summands in the rational rank 1 representations $A = \sum \mathbf{b}_i \mathbf{b}_i^T$ and $A = \sum d_i \mathbf{x}_i \mathbf{x}_i^T$ may differ by a factor of up to 4. For now we will use both approaches. The (minimal) number of rank-one matrices in a CP representation will be discussed, for the real case and also for the rational case, in the next chapter.

We denote by $\mathbb{Z}_+^n$ the set of nonnegative integral vectors in $\mathbb{R}_+^n$, and by $\mathbb{Q}_+^n$ the set of nonnegative rational vectors. The set of $n \times n$ matrices that are completely positive over the rationals is denoted by $\mathcal{CP}_n^{\mathbb{Q}}$. Note that $\mathcal{CP}_n^{\mathbb{Q}}$ is a subset of $\mathcal{CP}_n$ that is not a cone (over $\mathbb{R}$). We collect in Proposition 3.165 some observations regarding this set. For the proof of the last part of the proposition we recall Dirichlet's Theorem (see reference to a proof of this theorem in the notes at the end of the section).

Lemma 3.164. **[Dirichlet's Theorem]** *Let $a_1, \ldots, a_n$ be real numbers, and let ε be a real number such that $0 < \varepsilon < 1$. Then there exist integers $p_1, \ldots, p_n$ and a positive integer q such that $1 \le q \le \frac{1}{\varepsilon^n}$ and*

$$\left| a_i - \frac{p_i}{q} \right| \le \frac{\varepsilon}{q} \quad \textit{for every } i = 1, \ldots, n.$$

Proposition 3.165. *The following holds for the set $\mathcal{CP}_n^{\mathbb{Q}}$ of matrices that are completely positive over the rationals, and its dual in $\mathcal{S}_n$.*

(a) $\mathrm{cl}\mathcal{CP}_n^{\mathbb{Q}} = \mathrm{cl}\left(\mathrm{cone}\mathcal{CP}_n^{\mathbb{Q}}\right) = \mathcal{CP}_n$.

(b) $\left(\mathcal{CP}_n^{\mathbb{Q}}\right)^* = \mathcal{COP}_n$.

(c) $\{X \in \mathcal{S}_n \mid \langle X, \mathbf{v}\mathbf{v}^T \rangle \ge 0 \ \textit{for every} \ \mathbf{v} \in \mathbb{Z}_+^n\} = \mathcal{COP}_n$.

(d) $\{X \in \mathcal{S}_n \mid \langle X, \mathbf{v}\mathbf{v}^T \rangle \ge 1 \ \textit{for every nonzero} \ \mathbf{v} \in \mathbb{Z}_+^n\} \subseteq \mathrm{int}\mathcal{COP}_n$.

Proof. (a): As $\mathcal{CP}_n^{\mathbb{Q}} \subseteq \mathcal{CP}_n$, we have $\mathcal{CP}_n^{\mathbb{Q}} \subseteq \mathrm{cone}\mathcal{CP}_n^{\mathbb{Q}} \subseteq \mathcal{CP}_n$. And since $\mathbb{Q}_+^n$ is dense in $\mathbb{R}_+^n$, the set $\mathcal{CP}_n^{\mathbb{Q}}$ is dense in $\mathcal{CP}_n$.

(b): By (a) and Theorem 1.146(d), the dual of $\mathcal{CP}_n^{\mathbb{Q}}$ is $\mathcal{CP}_n^* = \mathcal{COP}_n$.

(c): It is obvious from (b) that $X \in \mathcal{COP}_n$ satisfies $\langle X, \mathbf{v}\mathbf{v}^T \rangle \ge 0$ for every $\mathbf{v} \in \mathbb{Z}_+^n$. For the converse, note that if $\mathbf{v} \in \mathbb{Q}_+^n$, then $\mathbf{v} = \frac{1}{q}\mathbf{p}$, where q is a positive integer and the entries of $\mathbf{p}$ are nonnegative integers. Hence if a symmetric matrix X satisfies

$$\langle X, \mathbf{v}\mathbf{v}^T \rangle \ge 0 \ \text{ for every } \ \mathbf{v} \in \mathbb{Z}_+^n,$$

it also satisfies

$$\langle X, \mathbf{v}\mathbf{v}^T \rangle \geq 0 \quad \text{for every} \quad \mathbf{v} \in \mathbb{Q}_+^n.$$

As matrices in $\mathcal{CP}_n^{\mathbb{Q}}$ are sums of rank 1 matrices of the form $\mathbf{v}\mathbf{v}^T$, $\mathbf{v} \in \mathbb{Q}_+^n$, such X satisfies also

$$\langle X, A \rangle \geq 0 \quad \text{for every} \quad A \in \mathcal{CP}_n^{\mathbb{Q}}.$$

That is, $X \in \left(\mathcal{CP}_n^{\mathbb{Q}}\right)^* = \mathcal{COP}_n$.

(d): From (c) we already know that if $\langle X, \mathbf{v}\mathbf{v}^T \rangle \geq 1$ for every nonzero $\mathbf{v} \in \mathbb{Z}_+^n$, then X is copositive. Suppose on the contrary that a certain X in this set is not strictly copositive. Then there exists a nonnegative vector $\mathbf{u} \in \mathbb{R}_+^n$ such that $\langle X, \mathbf{u}\mathbf{u}^T \rangle = \mathbf{u}^T X \mathbf{u} = 0$. As principal submatrices of X also have the same property, we may assume that $\mathbf{u}$ is a positive vector. Then, by Proposition 2.47, X is positive semidefinite and $X\mathbf{u} = \mathbf{0}$.

Let $0 < \varepsilon < 1$ be sufficiently small, so that it satisfies also

$$\varepsilon < \min_{1 \leq i \leq n} u_i \quad \text{and} \quad \varepsilon < \frac{1}{\sqrt{\|X\|_1}},$$

where $\|X\|_1 = \sum_{i,j=1}^n |x_{ij}|$ (note that the assumption on X implies $X \neq 0$).

By Lemma 3.164 there exist a positive integer q and integers $p_1, \ldots, p_n$ that satisfy

$$\left| u_i - \frac{p_i}{q} \right| \leq \frac{\varepsilon}{q} \qquad \text{for every } i = 1, \ldots, n.$$

Let $\mathbf{p} = (p_1, \ldots, p_n)^T$. By the choice of ε we get that $\frac{1}{q}\mathbf{p}$ is positive, hence $\mathbf{p} \in \mathbb{Z}_+^n$ is also positive. Let $\mathbf{y} = q\mathbf{u} - \mathbf{p}$. Then $|y_i| \leq \varepsilon$ for every i, and therefore $\mathbf{y}^T X \mathbf{y} \leq \|X\|_1 \varepsilon^2$. We get that

$$1 \leq \langle X, \mathbf{p}\mathbf{p}^T \rangle = (q\mathbf{u} - \mathbf{y})^T X (q\mathbf{u} - \mathbf{y}) = \mathbf{y}^T X \mathbf{y} \leq \|X\|_1 \varepsilon^2,$$

which contradicts our choice of ε. □

The following theorem is the main evidence in support of Conjecture 3.162.

Theorem 3.166. *Every rational matrix that lies in the interior of $\mathcal{CP}_n$ is completely positive over the rationals.*

Proof. Let A be a rational matrix in $\operatorname{int}\mathcal{CP}_n$. Then for some $c > 1$, the matrix $cA + X$ has minimal eigenvalue greater or equal to 1 for every $X \in \mathcal{S}_n$ with $\|X\|_F \leq 1$. In particular, $\langle cA + X, \mathbf{v}\mathbf{v}^T \rangle \geq \|\mathbf{v}\|^2 \geq 1$ for every nonzero $\mathbf{v} \in \mathbb{Z}_+^n$ and every such X. For $\mu > c\|A\|_F^2 + \|A\|_F$ let

$$\mathcal{P}(A, \mu) = \left\{ X \in \mathcal{S}_n \mid \langle X, A \rangle \leq \mu \text{ and } \langle X, \mathbf{v}\mathbf{v}^T \rangle \geq 1 \text{ for every } \mathbf{0} \neq \mathbf{v} \in \mathbb{Z}_+^n \right\}.$$

The set $\mathcal{P}(A,\mu)$ has a nonempty interior: $cA + X \in \mathcal{P}(A,\mu)$ for every X in the unit ball of the Frobenius norm, by the choice of c and μ. By Proposition 3.165(d), $\mathcal{P}(A,\mu) \subseteq \text{int}\mathcal{COP}_n$, and it is clearly convex, as the intersection of (infinitely many) halfspaces.

We now show that $\mathcal{P}(A,\mu)$ is bounded. Suppose on the contrary that the set is unbounded. Then there exists $X_0 \in \text{int}\mathcal{P}(A,\mu)$ and a nonzero $X_1 \in \mathcal{S}_n$ such that $X_0 + tX_1 \in \mathcal{P}(A,\mu)$ for every $t \geq 0$ (see Exercise 1.45). But then for every $\mathbf{v} \in \mathbb{Z}_+^n$

$$\langle X_1, \mathbf{v}\mathbf{v}^T \rangle \geq \frac{1}{t}\left(1 - \langle X_0, \mathbf{v}\mathbf{v}^T \rangle\right)$$

holds for every $t > 0$. Hence $\langle X_1, \mathbf{v}\mathbf{v}^T \rangle \geq 0$ for every $\mathbf{v} \in \mathbb{Z}_+^n$, which implies that $X_1 \in \mathcal{COP}_n$ by Proposition 3.165(c). In particular, $\langle X_1, A \rangle \geq 0$. On the other hand, for every $t > 0$

$$\langle X_1, A \rangle \leq \frac{1}{t}\left(\mu - \langle X_0, A \rangle\right),$$

hence $\langle X_1, A \rangle \leq 0$. We therefore have $\langle X_1, A \rangle = 0$. But as A is in the interior of $\mathcal{CP}_n$, this implies that $X_1 = 0$, a contradiction.

Next we show that $\mathcal{P}(A,\mu)$ is in fact an intersection of finite number of halfspaces (i.e., it is a *polytope*). Suppose on the contrary that this is not the case. Then there exist an infinite sequence $\mathbf{v}_k$ of mutually different nonzero vectors in $\mathbb{Z}_+^n$, and a sequence $B_k \in \mathcal{P}(A,\mu)$ such that $\langle B_k, \mathbf{v}_k\mathbf{v}_k^T \rangle = 1$. Note that the norms sequence $\|\mathbf{v}_k\|$ tends to infinity, since there is only a finite number of vectors from $\mathbb{Z}_+^n$ in any ball about $\mathbf{0}$. Replacing the bounded B_k by a converging subsequence, we may assume $\lim_{k\to\infty} B_k = B_0$. Let $\mathbf{u}_k = \mathbf{v}_k/\|\mathbf{v}_k\| \in \Gamma$ (the nonnegative part of the unit sphere). As Γ is compact, we may similarly assume that $\mathbf{u}_k$ converges to some $\mathbf{u}_0 \in \Gamma$ (in particular, $\mathbf{u}_0 \neq \mathbf{0}$). Then

$$\langle B_0, \mathbf{u}_0\mathbf{u}_0^T \rangle = \lim_{k\to\infty} \langle B_k, \mathbf{u}_k\mathbf{u}_k^T \rangle = \lim_{k\to\infty} \frac{\langle B_k, \mathbf{v}_k\mathbf{v}_k^T \rangle}{\|\mathbf{v}_k\|^2} = \lim_{k\to\infty} \frac{1}{\|\mathbf{v}_k\|^2} = 0,$$

contradicting the fact that $B_0 \in \mathcal{P}(A,\mu)$ is in the interior of $\mathcal{COP}_n$. Hence there exist m nonzero mutually different nonzero vectors $\mathbf{v}_i$, $i = 1,\ldots,m$, in $\mathbb{Z}_+^n$ such that

$$\mathcal{P}(A,\mu) = \left\{X \in \mathcal{S}_n \mid \langle X, A \rangle \leq \mu \text{ and } \langle X, \mathbf{v}_i\mathbf{v}_i^T \rangle \geq 1 \text{ for every } i = 1,\ldots,m\right\}.$$

To prove that A is completely positive over the rationals, we consider the minimum of the linear functional $B \mapsto \langle B, A \rangle$ over $\mathcal{P}(A,\mu)$. As $\mathcal{P}(A,\mu)$ is a polytope, the minimum is attained at a vertex B_0 of the polytope. Since this minimum is clearly smaller than μ, we may assume without loss of generality that

$$\langle B_0, \mathbf{v}_i\mathbf{v}_i^T\rangle = 1, \quad i = 1, \ldots, p$$

for some $p \leq m$, $\langle B_0, \mathbf{v}_i\mathbf{v}_i^T\rangle > 1$ for $p < i \leq m$, and $\langle B_0, A\rangle < \mu$. We claim that A is in $\text{cone}\{\mathbf{v}_i\mathbf{v}_i^T \mid i = 1, \ldots, p\}$. Suppose on the contrary that $A \notin \text{cone}\{\mathbf{v}_i\mathbf{v}_i^T \mid i = 1, \ldots, p\}$. Then by Theorem 1.124 there exists $X \in \mathcal{S}_n$ such that

$$\langle X, A\rangle < 0 \quad \text{and} \quad \langle X, \mathbf{v}_i\mathbf{v}_i^T\rangle \geq 0, \quad i = 1, \ldots, p.$$

But then for sufficiently small $t > 0$ we get that

$$B_0 + tX \in \mathcal{P}(A, \mu) \quad \text{and} \quad \langle B_0 + tX, A\rangle < \langle B_0, A\rangle,$$

which contradicts the choice of B_0. Hence

$$A \in \text{cone}\{\mathbf{v}_i\mathbf{v}_i^T \mid i = 1, \ldots, p\},$$

and by Carathéodory's Theorem (see Remark 1.133) there exists a linearly independent subset $I \subseteq \{1, \ldots, p\}$ such that A is a positive combination of $\{\mathbf{v}_i\mathbf{v}_i^T \mid i \in I\}$. Without loss of generality, $I = \{1, \ldots, q\}$ for some $q \leq p$. By the linear independence of $\mathbf{v}_1\mathbf{v}_1^T, \ldots, \mathbf{v}_q\mathbf{v}_q^T$, the linear system

$$\sum_{i=1}^{q} t_i\mathbf{v}_i\mathbf{v}_i^T = A$$

has a unique solution (by the above, a positive one). As this system of $n(n+1)/2$ equations in the q variables $t_1, \ldots, t_q$ has all rational coefficients, this unique solution consists of rational numbers, yielding a rational decomposition of A of the form (3.36). □

To settle the conjecture it remains to consider the rational matrices on the boundary of $\mathcal{CP}_n$. For that, it is useful to consider the tools used in determining complete positivity, and check which of them may be used in checking complete positivity over the rationals. The following results contain such rational variations of previous results. Proposition 3.167 is the first on the list. All the claims in this proposition may be easily verified, similarly to their real counterparts, and are left as an exercise.

Proposition 3.167. *The following hold.*

(a) *Let matrices A_1 and A_2 of the same order be completely positive over the rationals, and let d_1, d_2 be nonnegative rational numbers. Then $d_1A_1 + d_2A_2$ is completely positive over the rationals.*

(b) *Let* $A = A_1 \oplus A_2$ *(*A_1 *and* A_2 *not necessarily of the same order). Then* A *is completely positive over the rationals if and only if both* A_1 *and* A_2 *are completely positive over the rationals.*

(c) *Let* P *be a permutation matrix,* D *a rational positive diagonal matrix of the same order. Then the following are equivalent for a symmetric* A *(also of the same order):*

- A *is completely positive over the rationals.*
- P^TAP *is completely positive over the rationals.*
- DAD *is completely positive over the rationals.*

(d) *Let* S *be a nonsingular rational matrix with* $S^{-1} \geq 0$*. If* SAS^T *is completely positive over the rationals, then* A *is completely positive over the rationals.*

The Schur complement was an important tool in the study of complete positivity. For rational matrices we have the following.

Proposition 3.168. *Let*

$$A = \begin{pmatrix} B & C \\ C^T & E \end{pmatrix}$$

be positive semidefinite and rational. Then both the (generalized) Schur complement $A/E = B - CE^\dagger C^T$ *and the matrix*

$$\begin{pmatrix} CE^\dagger C^T & C \\ C^T & E \end{pmatrix}$$

are rational positive semidefinite matrices. If both matrices are completely positive over the rationals, then A *is also completely positive over the rationals.*

Proof. Since A is positive semidefinite, there exist a matrix X such that $C^T = EX$. As the columns of X are solutions to linear systems with rational coefficients, we may take X to be rational. Then we have that $CE^\dagger C^T = X^TEE^\dagger EX = X^TEX$ is rational, and therefore the two matrices

$$A/E = B - CE^\dagger C^T \quad \text{and} \quad \begin{pmatrix} CE^\dagger C^T & C \\ C^T & E \end{pmatrix},$$

which are positive semidefinite by Theorem 1.75, are rational. The final assertion is a special case of Proposition 3.167(a). □

As a result, we may also consider a variation of the Cholesky decomposition described in Section 1.4, which yields a decomposition of a positive semidefinite into rational factors. For a positive semidefinite A of rank r we may write

$$A = XDX^T$$

where X and D are rational matrices, D is a rational positive diagonal matrix. We start by taking $\mathbf{a}_1$ to be the first nonzero column of A, say it is column i. Let $\mathbf{x}_1 = \frac{1}{a_{ii}}\mathbf{a}_1 \in \mathbb{Q}_+^n$ and $d_1 = a_{ii}$. Then

$$A = d_1\mathbf{x}_1\mathbf{x}_1^T + A^{(1)},$$

where $A^{(1)}$ is $A/[i]$ with added row and column of zeros in position i, and is therefore rational and positive semidefinite. In $A^{(1)}$ any column that in A was a scalar multiple of the i-th column is zero. We then repeat the process, choosing at the $(k+1)$-th step $\mathbf{x}_{k+1}$ to be the first nonzero column in $A^{(k)} = A - \sum_{i=1}^{k} d_i\mathbf{x}_i\mathbf{x}_i^T$, divided by the corresponding diagonal entry. We have $\operatorname{rank} A^{(k)} = r - k$, and after r steps we get $A = \sum_{i=1}^{r} d_i\mathbf{x}_i\mathbf{x}_i^T$, where the d_i's are positive rationals. Let $X = \begin{pmatrix} \mathbf{x}_1 & \mathbf{x}_2 & \dots & \mathbf{x}_r \end{pmatrix}$, $D = \operatorname{Diag}(d_1, \dots, d_r)$. Then X and D are rational, X has the property that $\min\operatorname{supp}\mathbf{x}_{i+1} > \min\operatorname{supp}\mathbf{x}_i$ for every i, and the first nonzero entry in each column of X is equal to 1.

Corollary 3.169. *Let A be a rational doubly nonnegative matrix, and let $A = XDX^T$ be its Cholesky decomposition with X and D rational, D a positive diagonal matrix. If X is nonnegative, then A is completely positive over the rationals*

Example 3.170. Every rational completely positive matrix of rank at most 2 is completely positive over the rationals.

If $\operatorname{rank} A = 1$, without loss of generality $a_{11} > 0$, and then $A = \frac{1}{a_{11}}\mathbf{a}\mathbf{a}^T$, where $\mathbf{a}$ is the first column of A.

If $\operatorname{rank} A = 2$, then by the argument in Theorem 3.5, A is the Gram matrix of some vectors in $\mathbb{R}_+^2$, that may be rotated in the nonnegative orthant until one of them lies on the horizontal axis. We get that $A = BB^T$, where $B \in \mathbb{R}_+^{n\times 2}$ has a row with one positive entry in the first position. By using a permutation similarity on A, if necessary, we may assume that the second column of B has first entry equal to zero. If $B = \begin{pmatrix} \mathbf{y}_1 & \mathbf{y}_2 \end{pmatrix}$, then the first column of A is a multiple of $\mathbf{y}_1$, and in the Cholesky decomposition of A, $A = XDX^T$, we have $d_1\mathbf{x}_1\mathbf{x}_1^T = \mathbf{y}_1\mathbf{y}_1^T$, hence $d_2\mathbf{x}_2\mathbf{x}_2^T = A - d_2\mathbf{x}_1\mathbf{x}_1^T = A - \mathbf{y}_1\mathbf{y}_1^T = \mathbf{y}_2\mathbf{y}_2^T$. That is, $X = B\sqrt{D}^{-1} \geq 0$.

And now the following variation of Lemma 3.36 may be proved.

Corollary 3.171. *Let A be completely positive over the rationals, and let B be a rational positive semidefinite matrix of the same order, which is zero except for a 2×2 principal submatrix. If $A+B$ is nonnegative, then $A+B$ is completely positive over the rationals.*

Proof. Without loss of generality suppose $B[1,2]$ is the nonzero principal submatrix of B. If $b_{12} \geq 0$ the result is trivial by the previous example. Hence we consider the case that $b_{12} < 0$. By Example 3.170,

$$B = t\mathbf{b}\mathbf{b}^T + s\mathbf{c}\mathbf{c}^T,$$

where $\mathbf{b}, \mathbf{c}$ are rational nonnegative vectors, $\operatorname{supp}\mathbf{b} = \{1,2\}$ and $\mathbf{c} = \mathbf{e}_2$, and $t > 0$, $s \geq 0$ are rational numbers. Clearly, it suffices to show that $A + t\mathbf{b}\mathbf{b}^T$ is completely positive over the rationals.

Suppose $A = \sum_{i=1}^m d_i\mathbf{x}_i\mathbf{x}_i^T$, where each d_i is nonnegative and rational, and $\mathbf{x}_i \in \mathbb{Q}_+^n$. Let

$$t_i = \frac{td_i(\mathbf{x}_i)_1(\mathbf{x}_i)_2}{a_{12}}, \quad i = 1, \ldots, m.$$

(Note that each t_i is rational and $\sum_{i=1}^m t_i = 1$.) Then for each i we have $d_i\mathbf{x}_i\mathbf{x}_i^T + t_i\mathbf{b}\mathbf{b}^T \geq 0$, as

$$\begin{aligned} d_i(\mathbf{x}_i)_1(\mathbf{x}_i)_2 + \frac{t(\mathbf{x}_i)_1(\mathbf{x}_i)_2}{a_{12}}b_1b_2 &= d_i(\mathbf{x}_i)_1(\mathbf{x}_i)_2\left(1 + \frac{tb_1b_2}{a_{12}}\right) \\ &= \frac{d_i(\mathbf{x}_i)_1(\mathbf{x}_i)_2}{a_{12}}(A+B)_{12} \geq 0. \end{aligned}$$

Thus for every i the matrix $d_i\mathbf{x}_i\mathbf{x}_i^T + t_i\mathbf{b}\mathbf{b}^T$ is doubly nonnegative of rank 2, and is therefore completely positive over the rationals, and so is

$$A + B = \sum_{i=1}^m (d_i\mathbf{x}_i\mathbf{x}_i^T + t_i\mathbf{b}\mathbf{b}^T). \qquad \square$$

With these tools, we may verify that in each of the cases where complete positivity of a rational matrix could be verified by one of the main results in this section, the matrix is completely positive over the rationals.

Theorem 3.172. *Let A be an $n \times n$ rational completely positive matrix. Then A is completely positive over the rationals in any of the following cases.*

(a) $\operatorname{rank} A \leq 2$.

(b) *A is diagonally dominant.*
(c) *$M(A)$ is positive semidefinite.*
(d) *$G(A)$ is triangle free.*
(e) *$G(A)$ is completely positive.*

Proof. For (a), see Example 3.170.

For (b), note that if A is a rational diagonally dominant matrix, the decomposition of A obtained in the proof of Theorem 3.38 translates to a decomposition of the form (3.36).

To prove (c), we show that if $M(A)$ is positive semidefinite, then there exists a rational positive diagonal matrix D such that DAD (equivalently, $DM(A)D$) is diagonally dominant. We can then apply (b).

If $M(A)$ is nonsingular, then there exists a real positive diagonal matrix D such that $M(A)D\mathbf{e} > \mathbf{0}$ (e.g., take $D = \mathrm{Diag}(\mathbf{d})$, where $\mathbf{d}$ is the positive eigenvector corresponding to the minimal eigenvalue of $M(A)$). By slightly perturbing the entries of D if necessary, we may choose D to be rational. This makes $DM(A)D$ a rational strictly diagonally dominant matrix.

If $M(A)$ is singular, then there exists a positive diagonal matrix D such that $DM(A)D\mathbf{e} = \mathbf{0}$. Since $D\mathbf{e}$ is a solution to the rational linear system $M(A)\mathbf{x} = \mathbf{0}$, D may be chosen to be rational. This makes $DM(A)D$ a rational diagonally dominant singular matrix.

Now (d) follows from (c), by Theorem 3.43.

Finally, we prove (e). First, we consider blocks that are completely positive. If $G(A)$ is bipartite, A is completely positive over the rationals by (d).

If $G(A) = T_k$, $k \geq 3$, we may apply a suitable permutation similarity to assume that A has the form

$$A = \begin{pmatrix} B & C \\ C^T & D \end{pmatrix},$$

D is an $(k-2) \times (k-2)$ rational positive diagonal matrix. Let

$$A_1 = \begin{pmatrix} C^T D^{-1} C & C^T \\ C & D \end{pmatrix} \quad \text{and} \quad A_2 = \begin{pmatrix} A/D & 0 \\ 0 & 0 \end{pmatrix}.$$

Then $A = A_1 + A_2$. The matrix A_1 is completely positive over the rationals by the following decomposition:

$$A_1 = \begin{pmatrix} C^T D^{-1} \\ I \end{pmatrix} D \begin{pmatrix} C^T D^{-1} \\ I \end{pmatrix}^T.$$

The matrix A_2 is rational and positive semidefinite, and it is zero except for a 2×2 principal submatrix. Corollary 3.171 therefore implies that A is completely positive over the rationals. (Note that the argument works also if B and C have some zero entries, i.e., when $G(A)$ is a subgraph of T_4, as long as D is a positive diagonal matrix.)

If $G(A) = K_4$, let

$$\mu = \min_{1 \le j \le n} \frac{a_{1j}}{a_{2j}},$$

and let S be a 4×4 rational matrix whose diagonal entries are all equal to 1, and whose off-diagonal entries all equal zero except for $s_{12} = -\mu$. Then SAS^T is a 4×4 matrix which is rational, nonnegative, positive semidefinite and has an off-diagonal zero entry. That is, $G(SAS^T)$ is a subgraph of T_4. By the above argument SAS^T is completely positive over the rationals. Applying Proposition 3.167(d) we get that A is completely positive over the rationals.

In a general completely positive graph, each block is either bipartite, or a T_k, $k \ge 3$, or a K_4. Hence to complete the proof it suffices to show that if $G = G_1 \cup G_2$, where G_1 and G_2 share exactly one vertex, and G_1 and G_2 have the property that every rational doubly nonnegative matrix with one of these graphs is completely positive over the rationals, then every rational doubly nonnegative matrix with graph G is completely positive over the rationals.

So suppose A is doubly nonnegative with $G(A) = G$, where G_1 and G_2 have this property, G_1 has k vertices and G_2 has m vertices. We may assume that

$$A = \begin{pmatrix} B & \mathbf{u} & 0 \\ \mathbf{u}^T & c & \mathbf{w}^T \\ 0 & \mathbf{w} & E \end{pmatrix}.$$

By Proposition 3.168 the matrices

$$A_1 = A/E = \begin{pmatrix} B & \mathbf{u} \\ \mathbf{u}^T & c - \mathbf{w}^T E \mathbf{w} \end{pmatrix} \quad \text{and} \quad A_2 = \begin{pmatrix} \mathbf{w}^T E \mathbf{w} & \mathbf{w}^T \\ \mathbf{w} & E \end{pmatrix}$$

are rational positive semidefinite, and they are clearly nonnegative. The graph of A_1 is G_1, and the graph of A_2 is G_2. By our assumption on G_1 and G_2 these two matrices are completely positive over the rationals. Hence $A = (A_1 \oplus 0_{m-1}) + (0_{k-1} \oplus A_2)$ is completely positive over the rationals. □

Theorem 3.172 does not cover all rational matrices on the boundary of $\mathcal{CP}_n$. It does not deal even with all such matrices that have zeros, let alone the singular ones, or those that are orthogonal to an exceptional copositive matrix. See, however, the following example.

Example 3.173. A 5×5 rational completely positive matrix A, which is orthogonal to the Horn matrix, is completely positive over the rationals.

To see that, let

$$W = \begin{pmatrix} 1 & 0 & 0 & 0 & 1 \\ 1 & 1 & 0 & 0 & 0 \\ 0 & 1 & 1 & 0 & 0 \\ 0 & 0 & 1 & 1 & 0 \\ 0 & 0 & 0 & 1 & 1 \end{pmatrix}$$

be the matrix whose columns are all the minimal zeros of the Horn matrix H, up to scalar multiplication. Each zero of H is a nonnegative combination of two consecutive columns in W (considering the first column to be the successor of the fifth). If $A = BB^T$, where $B \geq 0$, then each column of B is a zero of H. That is,

$$A = (WX)(WX)^T,$$

where X is a nonnegative matrix, with at most two positive entries in each column. Let $S = W^{-1}$. Then $SAS^T = XX^T$ is rational (as A and S are rational) and completely positive (as $X \geq 0$). Since each column of X has at most two positive entries, the matrix $M(XX^T)$ is positive semidefinite by Proposition 3.49, and Theorem 3.172(b) implies that $SAS^T = XX^T$ is completely positive over the rationals. Since $S^{-1} = W \geq 0$, A is also completely positive over the rationals by Proposition 3.167.

Integral completely positive matrices

Integral completely positive matrices are rational matrices, and Question 3.161 applies to them. But one may also ask the following question.

Question 3.174. *Let A be a completely positive matrix with integral entries. Does A have a factorization $A = BB^T$ where the entries of B are nonnegative integers?*

If an integral matrix A has such a factorization, we say that A is *completely positive over the integers.*

In general, the answer to this question is negative.

Example 3.175. Let

$$A(a,b) = \begin{pmatrix} 1 & 1 & 1 \\ 1 & a & 0 \\ 1 & 0 & b \end{pmatrix},$$

where $a \geq 2$ and $b \geq 2$ are integers. The matrix $A(a,b)$ is completely positive (check that it is doubly nonnegative), but is not completely positive over the integers. Indeed, if $A(a,b) = BB^T$, where $B \geq 0$ has integral entries, then $(A(a,b))_{12} = 1$ and $(A(a,b))_{23} = 0$ imply that B has a column of the form

$$\begin{pmatrix} 1 \\ 1 \\ 0 \end{pmatrix}.$$

Similarly, $(A(a,b))_{13} = 1$ and $(A(a,b))_{23} = 0$ imply that B has a column of the form

$$\begin{pmatrix} 1 \\ 0 \\ 1 \end{pmatrix}.$$

But then $(A(a,b))_{11} \geq 2$, a contradiction.

Example 3.176. Let

$$A(c) = \begin{pmatrix} 2 & 3 & 1 \\ 3 & 5 & 3 \\ 1 & 3 & c \end{pmatrix}, \qquad \text{where } c \geq 5 \text{ is an integer.}$$

We leave it as an exercise to show that the matrix $A(c)$ is completely positive, but is not completely positive over the integers.

More generally, we prove the following.

Theorem 3.177. *Let G be a graph on n vertices, which has at least one vertex of degree 2 or more. Then there exists an integral CP realization of G that is not completely positive over the integers.*

Proof. Note that if a $k \times k$ matrix B is an integral completely positive matrix that is not completely positive over the integers, then any matrix with B as a principal submatrix is not completely positive over the integers. So if $G = K_n$, let

$$A = \begin{pmatrix} I_3 \\ E \end{pmatrix} B \begin{pmatrix} I_3 & E^T \end{pmatrix},$$

where $B = A(c)$, $c \geq 5$, is the matrix of Example 3.176, and $E = J_{(n-3)\times 3}$. Then A is an integral completely positive matrix with $G(A) = K_n$ that is not completely positive over the integers.

If $G \neq K_n$, it has an induced $K_{1,2}$. Suppose without loss of generality 1 is a vertex of degree $k \geq 2$, whose neighbors are $2, \ldots, k+1$, and whose neighbors 2 and 3 are not adjacent to each other. Let

$$A = \sum_{i=2}^{k+1} \left(\frac{1}{n}\mathbf{e}_1 + n\mathbf{e}_i\right)\left(\frac{1}{n}\mathbf{e}_1 + n\mathbf{e}_i\right)^T + \sum_{\substack{2 \leq i < j \leq n \\ \{i,j\} \in E(G)}} (\mathbf{e}_i + \mathbf{e}_j)(\mathbf{e}_i + \mathbf{e}_j)^T$$
$$+ \operatorname{Diag}\left(1 - \frac{k}{n^2}, 0, \ldots, 0\right).$$

Then A is an integral completely positive matrix, with $G(A) = G$. It is not completely positive over the integers since $A[1, 2, 3]$ has the form $A(a, b)$ of Example 3.175. □

For very small matrices, namely 2×2, the answer to Question 3.161 is positive. The proof is based on the fact that Proposition 3.167 holds if the words "rational" and "rationals" is replaced by "integral" and "integers", respectively. In particular this is true for part (d) of the proposition.

Theorem 3.178. *Integral* 2×2 *completely positive matrices are completely positive over the integers.*

Proof. We consider nonzero completely positive matrices of the form

$$A = \begin{pmatrix} a & b \\ b & c \end{pmatrix}, \quad a, b, c \in \mathbb{Z}_+ . \tag{3.38}$$

First suppose $a, c \geq b$. In this case

$$A = \begin{pmatrix} b & b \\ b & b \end{pmatrix} + \begin{pmatrix} a-b & 0 \\ 0 & c-b \end{pmatrix}$$
$$= b\begin{pmatrix} 1 \\ 1 \end{pmatrix}\begin{pmatrix} 1 \\ 1 \end{pmatrix}^T + (a-b)\begin{pmatrix} 1 \\ 0 \end{pmatrix}\begin{pmatrix} 1 \\ 0 \end{pmatrix}^T + (c-b)\begin{pmatrix} 0 \\ 1 \end{pmatrix}\begin{pmatrix} 0 \\ 1 \end{pmatrix}^T.$$

As each of b, $a-b$ and $a-c$ is a sum of at most 4 square integers by Lagrange's four-square Theorem, this yields an integral CP representation of A with at most 12 summands.

Now suppose A is of the form (3.38) and there exists an integral matrix S with $\det S = 1$, $S^{-1} \geq 0$ such that $A' = SAS^T$ is an integral completely positive matrix with both diagonal entries greater or equal to the off-diagonal entry. Then A' has an integral CP representation with at most 12 summands by the previous case. That is, $A' = BB^T$, where $B \geq 0$ has integer entries. Then $A = (S^{-1}B)(S^{-1}B)^T$ is a CP decomposition of A

over the integers (and $S^{-1}B$ has the same number of columns as B, i.e., at most 12).

To complete the proof we show that such S exists for every completely positive A of the form (3.38). If $a, c \geq b$, then $S = I$. Otherwise, note that as $\det A \geq 0$, $b < c$ and $b < a$ cannot both hold. So suppose without loss of generality that $a \geq b > c$. Then $b = kc + b_0$, where the integers k and b_0 satisfy $k \geq 1$ and $0 \leq b_0 \leq c - 1$. Let

$$S_0 = \begin{pmatrix} 1 & -k \\ 0 & 1 \end{pmatrix}.$$

The matrix S_0 is has determinant 1 and a nonnegative integral inverse. Let

$$A_0 = S_0 A S_0^T = \begin{pmatrix} a - 2kb + k^2 c & b - kc \\ b - kc & c \end{pmatrix} = \begin{pmatrix} a_0 & b_0 \\ b_0 & c_0 \end{pmatrix}.$$

Then $b_0 \geq 0$, $c_0 = c > 0$, and $\det A_0 = \det A \geq 0$. That is, $a_0 c_0 \geq b_0^2$, and as $c_0 = c > 0$ we get that $a_0 \geq \frac{b_0^2}{c_0} > 0$. Hence A_0 is an integral completely positive matrix and $b_0 < c_0$. If also $b_0 \leq a_0$, we are done, taking $S = S_0$. Otherwise, $b_0 > a_0$ and we may repeat the process, dividing b_0 by a_0, $b_0 = k_1 a_0 + b_1$, and constructing

$$S_1 = \begin{pmatrix} 1 & 0 \\ -k_1 & 1 \end{pmatrix}, \qquad A_1 = S_1 A_1 S_1^{-1}.$$

Continue similarly, until we get a matrix A_m with both diagonal entries greater or equal to the off-diagonal one, and let $S = S_m \cdots S_0$. Such A_m necessarily will be reached, as in each step the nonnegative off-diagonal entry strictly decreases. □

Exercises

3.87. Prove Proposition 3.167.

3.88. Prove the claims in Example 3.175.

3.89. Show that if A is an $n \times n$ integral completely positive matrix of rank 1, then A is completely positive over the integers.

3.90. Follow the proof of Theorem 3.178 to find an integral CP factorization of $A = \begin{pmatrix} 56 & 40 \\ 40 & 30 \end{pmatrix}$.

Notes. For generalizations of complete positivity to other algebraic structures see [Mohindru (2014, 2015a,b); Mohindru and Pereira (2016); Shitov (2016); Dolžan and Oblak (2018)]. In [Berman and Xu (2005, 2007)] a matrix A is defined to be *S-completely positive* if $A = BB^T$ where the entries of B belong to S. The case that $S = \{0, 1\}$ is studied there and in [Dahl and Haufmann (2016)]. In the latter it is shown, in analogy with Theorem 3.46, that triangle free graphs are characterized by the following property: every integral symmetric nonnegative matrix realization A of the graph G is a $\{0, 1\}$-completely positive matrix iff A is diagonally dominant.

A proof of Dirichlet's Theorem (Lemma 3.164) may be found in [Grötschel *et al.* (1988)]. Theorem 3.166 was proved by [Dutour Sikirić *et al.* (2017)], and the remainder of the discussion of complete positivity over the rationals is based on [Shaked-Monderer *et al.* (2018)]. The proof of Theorem 3.177 is based on [Dickinson (2017a)], and Theorem 3.178 was first proved in [Laffey and Šmigoc (2018)], but the proof here is based on the same authors' paper [Laffey and Šmigoc (2019)]. The result may also be derived from an algorithm for finding rational CP factorizations of matrices that are completely positive over the rationals, which is presented in [Dutour Sikirić *et al.* (2020)].

Still open: *Is every rational matrix on the boundary of $\mathcal{CP}_n$ completely positive over the rationals?*

Chapter 4

CP-Rank

4.1 Definition and basic results

A completely positive matrix may typically have different CP factorizations, with different number of columns.

Example 4.1. Let $A = BB^T$, where

$$B = \frac{1}{\sqrt{3}} \begin{pmatrix} 1 & 0 & 0 & 1 & 1 & 1 & 0 & 1 & 1 & 0 \\ 1 & 1 & 0 & 0 & 1 & 0 & 1 & 0 & 1 & 1 \\ 1 & 1 & 1 & 0 & 0 & 1 & 0 & 1 & 0 & 1 \\ 0 & 1 & 1 & 1 & 0 & 1 & 1 & 0 & 1 & 0 \\ 0 & 0 & 1 & 1 & 1 & 0 & 1 & 1 & 0 & 1 \end{pmatrix} \in \mathbb{R}^{5\times 10}.$$

Then

$$A = \begin{pmatrix} 2 & 1 & 1 & 1 & 1 \\ 1 & 2 & 1 & 1 & 1 \\ 1 & 1 & 2 & 1 & 1 \\ 1 & 1 & 1 & 2 & 1 \\ 1 & 1 & 1 & 1 & 2 \end{pmatrix} = I_5 + J_5.$$

This A has also an LU-factorization obtained by the Cholesky decomposition, and a nonnegative square root — see Section 3.8, and in particular Exercise 3.67.

Definition 4.2. Let $A \neq 0$ be an $n \times n$ completely positive matrix. The minimal k such that $A = BB^T$ for some nonnegative $n \times k$ matrix B is the *cp-rank* of A. We denote the cp-rank of A by $\operatorname{cpr} A$. For completeness we define $\operatorname{cpr} 0 = 0$.

A CP factorization $A = BB^T$ where the nonnegative matrix B has $k = \operatorname{cpr} A$ columns is called a *minimal CP factorization* of A.

Geometrically, if A is a completely positive matrix with $\operatorname{rank} A = r$, and $A = \operatorname{Gram}(\mathbf{w}_1, \ldots, \mathbf{w}_n)$ where $\mathbf{w}_1, \ldots, \mathbf{w}_n$ are vectors in $\mathbb{R}^r$, then the cp-rank of A is the minimal dimension k such that $\mathbf{w}_1, \ldots, \mathbf{w}_n$ can be isometrically embedded into $\mathbb{R}^k_+$.

Another way to look at it: $\operatorname{cpr} A$ is the minimal number of summands in a CP representation of A, $A = \sum_{i=1}^k \mathbf{b}_i \mathbf{b}_i^T$, $\mathbf{b}_i \geq \mathbf{0}$. A CP representation with $k = \operatorname{cpr} A$ summands will be called a *minimal CP representation.* Using this latter approach, it is easy to see the following.

Proposition 4.3. *If A and B are $n \times n$ completely positive matrices, then*

$$\operatorname{cpr}(A + B) \leq \operatorname{cpr} A + \operatorname{cpr} B. \tag{4.1}$$

Here is another simple observation, which we leave as an exercise.

Proposition 4.4. *Let $A = A_1 \oplus A_2$ where A_1 and A_2 are completely positive. Then* $\operatorname{cpr} A = \operatorname{cpr} A_1 + \operatorname{cpr} A_2$.

Propositions 4.3 and 4.4 point to similarities between the properties of the rank parameter on positive semidefinite matrices, and the cp-rank parameter on completely positive matrices. However, this is where the similarities end. The next example shows at least one difference between these two parameters. If A and B are positive semidefinite matrices of the same order, then $\operatorname{rank}(A + B) \geq \max\{\operatorname{rank} A, \operatorname{rank} B\}$. The cp-rank does not satisfy such inequality.

Example 4.5. Let

$$A_1 = \begin{pmatrix} kI_k & J_k \\ J_k & kI_k \end{pmatrix} \quad \text{and} \quad A_2 = \begin{pmatrix} J_k & 0_k \\ 0_k & J_k \end{pmatrix}.$$

We leave it as an exercise to check that $\operatorname{cpr} A_1 = k^2$ and $\operatorname{cpr} A_2 = 2$.

Now consider

$$A = \begin{pmatrix} A_1 & 0 \\ 0 & A_2 \end{pmatrix} \quad \text{and} \quad B = \begin{pmatrix} A_2 & 0 \\ 0 & A_1 \end{pmatrix}.$$

By Proposition 4.4, $\operatorname{cpr} A = \operatorname{cpr} B = k^2 + 2$. As

$$A_1 + A_2 = (k-1)I_{2k} + J_{2k}$$

is a nonsingular matrix whose square root is nonnegative (see, e.g., Exercise 3.67), $\operatorname{cpr}(A_1 + A_2) = 2k$. Hence

$$A + B = (A_1 + A_2) \oplus (A_1 + A_2)$$

satisfies $\operatorname{cpr}(A + B) = 4k$. For $k \gg 4$,

$$\operatorname{cpr}(A + B) = 4k \ll k^2 + 2 = \min\{\operatorname{cpr} A, \operatorname{cpr} B\}.$$

The following simple observation follows from the definition of the cp-rank.

Proposition 4.6. *For every completely positive matrix A,*

$$\operatorname{cpr} A \geq \operatorname{rank} A. \tag{4.2}$$

Proof. Let $A = BB^T$, $B \in \mathbb{R}_+^{n\times k}$, be a minimal CP factorization of A. Then $\operatorname{cpr} A = k \geq \operatorname{rank} B = \operatorname{rank} A$. □

Equality may occur in (4.2).

Theorem 4.7. *If A is a completely positive matrix and* $\operatorname{rank} A \leq 2$, *then*

$$\operatorname{cpr} A = \operatorname{rank} A.$$

Proof. This follows from Example 3.4 and the proof of Theorem 3.5. □

Theorem 4.8. *Let A be an $n \times n$ completely positive matrix, $n \leq 3$. Then*

$$\operatorname{cpr} A = \operatorname{rank} A.$$

Proof. For $n \leq 2$ and for singular 3×3 completely positive matrices this follows from the previous theorem. So assume A is a 3×3 nonsingular completely positive matrix. Let δ_0 be the maximal $\delta > 0$ for which $A - \delta E_{33}$ is positive semidefinite ($\delta_0 = \det A / \det A(3)$). Then

$$A = \delta_0 E_{33} + (A - \delta_0 E_{33}),$$

and $A - \delta_0 E_{33}$ is a singular 3×3 doubly nonnegative matrix. Hence

$$\operatorname{cpr}(A - \delta_0 E_{33}) = \operatorname{rank}(A - \delta_0 E_{33}). \tag{4.3}$$

Combining (4.1), (4.2) and (4.3), we get that

$$3 = \operatorname{rank} A \leq \operatorname{cpr} A \leq 1 + \operatorname{cpr}(A - \delta_0 E_{33}) = 1 + \operatorname{rank}(A - \delta_0 E_{33}) \leq 1 + 2,$$

which implies that $\operatorname{cpr} A = \operatorname{rank} A = 3$. □

For $n \times n$ matrices of rank r, where $r \geq 3$ and $n \geq 4$, the cp-rank may be greater than the rank.

Example 4.9. The matrix

$$\begin{pmatrix} 6 & 3 & 3 & 0 \\ 3 & 5 & 1 & 3 \\ 3 & 1 & 5 & 3 \\ 0 & 3 & 3 & 6 \end{pmatrix}$$

satisfies $\operatorname{rank} A = 3$, $\operatorname{cpr} A = 4$. To see that $\operatorname{cpr} A = 4$, note that if $A = \sum_{i=1}^{m} \mathbf{b}_i \mathbf{b}_i^T$ is a minimal CP representation of A, then $1 \in \operatorname{supp} \mathbf{b}_i$ for at least one i, say $1 \in \operatorname{supp} \mathbf{b}_1$. If $1 \notin \operatorname{supp} \mathbf{b}_i$ for every $i \neq 1$, then

$$\mathbf{b}_1 \mathbf{b}_1^T = \begin{pmatrix} 6 & 3 & 3 & 0 \\ 3 & \frac{3}{2} & \frac{3}{2} & 0 \\ 3 & \frac{3}{2} & \frac{3}{2} & 0 \\ 0 & 0 & 0 & 0 \end{pmatrix}.$$

This implies that $(A - \mathbf{b}_1 \mathbf{b}_1^T)_{23} < 0$. But $A - \mathbf{b}_1 \mathbf{b}_1^T = \sum_{i=2}^{m} \mathbf{b}_i \mathbf{b}_i^T$ is completely positive, a contradiction. Hence 1 must belong to two of the supports $\operatorname{supp} \mathbf{b}_i$, $i = 1, \ldots, m$. By symmetry, 4 must also belong to two of the supports. Add the fact that 1 and 4 cannot belong to the same support, to conclude that we necessarily have $m \geq 4$. That is,

$$\operatorname{cpr} A \geq 4 > \operatorname{rank} A = 3.$$

The next theorem shows that we actually have here $\operatorname{cpr} A = 4$.

Theorem 4.10. *Let A be an $n \times n$ completely positive matrix, $n \leq 4$. Then* $\operatorname{cpr} A \leq n$.

A few simple observations will help in the proof (and in future proofs as well). We state them here and leave the proofs as exercises.

Proposition 4.11. *If S is a nonsingular matrix with a nonnegative inverse, and SAS^T is completely positive, then A is also completely positive, and* $\operatorname{cpr} A \leq \operatorname{cpr}(SAS^T)$.

Corollary 4.12. *Let A be a completely positive matrix, D a positive diagonal matrix, and P a permutation matrix. Then*

$$\operatorname{cpr} A = \operatorname{cpr} DAD = \operatorname{cpr} P^T AP.$$

We also use the following lemma.

Lemma 4.13. *If $x, y, z \geq 0$ and $0 \leq x + y - z \leq 1$, then there exist $0 \leq a, b \leq 1$ such that*

$$\sqrt{ab} \leq x, \quad \sqrt{(1-a)(1-b)} \leq y,$$

and

$$\sqrt{ab} + \sqrt{(1-a)(1-b)} = x + y - z.$$

Proof. Since $0 \le x + y - z \le 1$,

$$x + y - z = \cos\theta$$

for some $0 \le \theta \le \pi/2$. For every $0 \le t \le \pi/2 - \theta$,

$$\cos t \cos(t+\theta) + \sin t \sin(t+\theta) = \cos\theta.$$

We show that there exists $0 \le t \le \pi/2 - \theta$ such that

$$\cos t \cos(t+\theta) \le x, \tag{4.4}$$

and

$$\sin t \sin(t+\theta) \le y. \tag{4.5}$$

If (4.5) holds for $t = \pi/2 - \theta$, then we are done, since (4.4) certainly holds for this value of t. Hence we consider the case that (4.5), which holds for $t = 0$, does not hold for $t = \pi/2 - \theta$. In that case, let t_0 be the maximal $0 \le t \le \pi/2 - \theta$ for which (4.5) holds. Then

$$\sin t_0 \sin(t_0 + \theta) = y.$$

Since

$$\cos t_0 \cos(t_0+\theta) + \sin t_0 \sin(t_0+\theta) = \cos\theta \le x + y,$$

(4.4) also holds for t_0. Take $a = \cos^2 t_0$ and $b = \sin^2 t_0$. □

Proof of Theorem 4.10. For $n \le 3$, the theorem follows from Theorem 4.8. (It also follows from the proof of Theorem 3.35 in Section 3.3, or Exercise 3.21, or Corollary 3.127.)

Let A is a 4×4 completely positive matrix. As in the proof of Theorem 3.35, we generate a doubly nonnegative matrix

$$C = \begin{pmatrix} 1 & c_{12} & c_{13} & c_{14} \\ c_{12} & 1 & c_{23} & c_{24} \\ c_{13} & c_{23} & 1 & 0 \\ c_{14} & c_{24} & 0 & 1 \end{pmatrix}.$$

That is, $C = D_2 S D_1 A D_1 S^T D_2$, where D_1 and D_2 are positive diagonal matrices, and S has a nonnegative inverse. In view of Proposition 4.11 and Corollary 4.12, it suffices to prove that $\operatorname{cpr} C \le 4$. As in the proof of Theorem 3.35, split C into $C = EE^T + F$, where

$$E = \begin{pmatrix} c_{13} & c_{14} \\ c_{23} & c_{24} \\ 1 & 0 \\ 0 & 1 \end{pmatrix}.$$

Then

$$F = \begin{pmatrix} f_{11} & f_{12} \\ f_{12} & f_{22} \end{pmatrix} \oplus 0_2.$$

Clearly EE^T is completely positive with cp-rank $= 2$, and

$$\begin{pmatrix} f_{11} & f_{12} \\ f_{12} & f_{22} \end{pmatrix} = C/C[3,4]$$

is positive semidefinite. Hence

$$f_{11}f_{22} - f_{12}^2 \geq 0. \tag{4.6}$$

If $f_{12} \geq 0$, then $F[1,2]$ is a 2×2 completely positive matrix. In this case, $\operatorname{cpr} F = \operatorname{cpr} F[1,2]$ is at most 2 and

$$\operatorname{cpr} C \leq \operatorname{cpr} EE^T + \operatorname{cpr} F \leq 2 + 2 = 4.$$

So consider the case that $f_{12} < 0$. We write $EE^T = G_1 + G_2$, where

$$G_1 = \begin{pmatrix} c_{13} \\ c_{23} \\ 1 \\ 0 \end{pmatrix} \begin{pmatrix} c_{13} \\ c_{23} \\ 1 \\ 0 \end{pmatrix}^T \quad \text{and } G_2 = \begin{pmatrix} c_{14} \\ c_{24} \\ 0 \\ 1 \end{pmatrix} \begin{pmatrix} c_{14} \\ c_{24} \\ 0 \\ 1 \end{pmatrix}^T.$$

G_1 and G_2 are rank 1 doubly nonnegative matrices, and

$$G_1 + G_2 + F = C.$$

In particular,

$$c_{13}c_{23} + c_{14}c_{24} + f_{12} = c_{12}. \tag{4.7}$$

By (4.6) and (4.7) we may apply Lemma 4.13 to

$$x = c_{13}c_{23}/\sqrt{f_{11}f_{22}}\,, \quad y = c_{14}c_{24}/\sqrt{f_{11}f_{22}}\,, \quad \text{and} \quad z = c_{12}/\sqrt{f_{11}f_{22}},$$

to get a and b such that

$$\sqrt{ab} \leq c_{13}c_{23}/\sqrt{f_{11}f_{22}}\,, \quad \sqrt{(1-a)(1-b)} \leq c_{14}c_{24}/\sqrt{f_{11}f_{22}},$$

and

$$\sqrt{ab} + \sqrt{(1-a)(1-b)} = -f_{12}/\sqrt{f_{11}f_{22}}.$$

For these a and b, let

$$\mathbf{x} = \begin{pmatrix} \sqrt{af_{11}} \\ -\sqrt{bf_{22}} \\ 0 \\ 0 \end{pmatrix} \quad \text{and } \mathbf{y} = \begin{pmatrix} \sqrt{(1-a)f_{11}} \\ -\sqrt{(1-b)f_{22}} \\ 0 \\ 0 \end{pmatrix}.$$

Then each of the matrices $G_1 + \mathbf{x}\mathbf{x}^T$ and $G_2 + \mathbf{y}\mathbf{y}^T$ is a doubly nonnegative matrix of rank ≤ 2 and hence completely positive of cp-rank ≤ 2. Since $C = (G_1 + \mathbf{x}\mathbf{x}^T) + (G_2 + \mathbf{y}\mathbf{y}^T)$, this implies that

$$\operatorname{cpr} C \leq \operatorname{cpr}(G_1 + \mathbf{x}\mathbf{x}^T) + \operatorname{cpr}(G_2 + \mathbf{y}\mathbf{y}^T) \leq 4.$$

□

For $n \geq 5$, there exist $n \times n$ completely positive matrices A such that $\operatorname{cpr} A > n$.

Example 4.14. Let A be any 5×5 completely positive matrix whose graph is $K_{2,3}$. Let $A = \sum_{i=1}^m \mathbf{b}_i \mathbf{b}_i^T$ be any CP representation of A. The support of each $\mathbf{b}_i$ is a clique in $G(A) = K_{2,3}$, and the corresponding complete graphs necessarily cover all the edges of $G(A)$. Since the maximal cliques in $K_{2,3}$ are of size 2, and there are 6 such cliques (edges), we get that $\operatorname{cpr} A \geq 6$.

Remark 4.15. The argument in Example 4.14 suggests a general lower bound on the cp-rank of a matrix, given in terms of its graph.

If $A = \sum_{i=1}^m \mathbf{b}_i \mathbf{b}_i^T$ is any CP representation of a completely positive matrix A, then the supports of the vectors $\mathbf{b}_i$ are cliques in $G(A)$, and the complete subgraphs induced on these cliques cover all the edges of $G(A)$ (that is, every edge of $G(A)$ is an edge of one of these complete subgraphs). Recall that the edge-clique covering number of a graph G is the minimum number of complete subgraphs of G needed to cover all of G's edges, and it is denoted by $\operatorname{cc}(G)$. That is, we have observed that

$$\operatorname{cpr} A \geq \operatorname{cc}(G(A)). \tag{4.8}$$

Strict inequality may occur — see Example 4.9. (A strict inequality will also occur if $G(A)$ has an isolated vertex $\{i\}$ such that $a_{ii} > 0$.)

It is useful to define also the cp-rank of a graph.

Definition 4.16. For any graph G,

$$\operatorname{cpr} G = \max\{\operatorname{cpr} A \mid A \text{ is a CP matrix realization of } G\}.$$

The inequality $\operatorname{cpr} A \geq \operatorname{rank} A$ implies that $\operatorname{cpr} G \geq |V(G)|$, and by Remark 4.15 we also have $\operatorname{cpr} G \geq \operatorname{cc}(G)$. We state some properties of the parameter $\operatorname{cpr} G$, leaving the proofs as exercises.

Proposition 4.17. *For every graph G,*

(a) *If H is a subgraph of G, then* $\operatorname{cpr} H \leq \operatorname{cpr} G$.
(b) *If $G = G_1 \cup G_2$, where G_1 and G_2 are induced subgraphs of G, then* $\operatorname{cpr} G \leq \operatorname{cpr} G_1 + \operatorname{cpr} G_2$.
(c) *If $G = G_1 \cup G_2$, where the union is disjoint, then* $\operatorname{cpr} G = \operatorname{cpr} G_1 + \operatorname{cpr} G_2$.

Proposition 4.18. *For every graph G,*

$$\min\{\operatorname{cpr} A \mid A \text{ is a CP matrix realization of } G\} = \operatorname{cc}(G). \tag{4.9}$$

Proposition 4.17(a) suggests other lower bounds on $\operatorname{cpr} G$. We will define the parameter $\operatorname{tf}(G)$ to be the maximum size (i.e., number of edges) of a triangle free subgraph of G.

Proposition 4.19. *For every graph G,*

$$\operatorname{cpr} G \geq \operatorname{tf}(G). \tag{4.10}$$

Proof. Let H be a triangle free subgraph of G of maximum size. Then $\operatorname{cpr} G \geq \operatorname{cpr} H \geq \operatorname{cc}(H) = |E(H)| = \operatorname{tf}(G)$. □

Strict inequality may occur in (4.10). For example, if G is a triangle, $\operatorname{cpr} G = 3$, but $\operatorname{tf}(G) = 2$, and if G consists of n isolated vertices, $\operatorname{cpr} G = n$ and $\operatorname{tf}(G) = 0$.

We note that it is known that $\operatorname{tc}(G) \leq \operatorname{tf}(G)$, where $\operatorname{tc}(G)$ is the minimum number of triangles and edges needed to cover all the edges of the graph G, and that equality holds if and only if G is triangle free (see the notes at the end of the section). Therefore we get the following result.

Proposition 4.20. *For every graph G,*

$$\operatorname{cc}(G) \leq \operatorname{tc}(G) \leq \operatorname{tf}(G).$$

This result means that $\operatorname{tf}(G)$ is a tighter lower bound to $\operatorname{cpr} G$, compared to $\operatorname{cc}(G)$.

There is no known method of computing the cp-rank of a general completely positive matrix, or computing $\operatorname{cpr} G$ for a graph G. The next four sections contain results on related problems:

(A) For $r \geq 3$, what is the maximal possible cp-rank of a rank r completely positive matrix?
(B) For $n \geq 5$, what is the maximal possible cp-rank of an $n \times n$ completely positive matrix?
(C) Which graphs have the property that $\operatorname{cpr} A = \operatorname{rank} A$ holds for every CP realization A of the graph?
(D) For which graphs G is $\operatorname{cpr} G$ equal to the lower bound $|V(G)|$? to $\operatorname{cc}(G)$? to $\operatorname{tf}(G)$?

Exercises

4.1. Prove Proposition 4.4.

4.2. Complete the details of Example 4.5: Show that $\operatorname{cpr} A_1 = k^2$ and $\operatorname{cpr} A_2 = 2$.
Hint: $A_1 = V_1 V_1^T$, where the columns of V_1 are $\mathbf{e}_i + \mathbf{e}_j$, $1 \le i \le k$, $k+1 \le j \le 2k$.

4.3. Let $A = \sum_{i=1}^k \mathbf{b}_i \mathbf{b}_i^T$ be a minimal CP representation of a completely positive matrix. Let $\Omega \subseteq \{1, \ldots, k\}$, and $A' = \sum_{i \in \Omega} \mathbf{b}_i \mathbf{b}_i^T$. Show that $\operatorname{cpr} A' = |\Omega|$.

4.4. Let $A = \sum_{i=1}^k \mathbf{b}_i \mathbf{b}_i^T$ be a minimal CP representation of a completely positive matrix. Show that any three of the vectors $\mathbf{b}_1, \ldots, \mathbf{b}_k$ are linearly independent.
Hint: Use Exercise 4.3 and Theorem 4.7.

4.5. Let $A = \sum_{i=1}^k \mathbf{b}_i \mathbf{b}_i^T$ be a minimal CP representation of a completely positive matrix. Show that $\{\mathbf{b}_1 \mathbf{b}_1^T, \ldots, \mathbf{b}_k \mathbf{b}_k^T\}$ is a linearly independent set in $\mathcal{S}_n$.
Hint: If $\sum_{i=1}^k \mu_i \mathbf{b}_i \mathbf{b}_i^T = 0$ and $1 + t\mu_i \ge 0$ for every i, then $A = \sum_{i=1}^k (1 + t\mu_i) \mathbf{b}_i \mathbf{b}_i^T$ is a CP representation of A.

4.6. Prove Proposition 4.3.

4.7. Prove Proposition 4.11.

4.8. Prove Corollary 4.12.

4.9. Suppose that A_k is completely positive for every k, and that

$$A = \lim_{k \to \infty} A_k.$$

Show that

$$\operatorname{cpr} A \le \liminf_{k \to \infty} \operatorname{cpr} A_k.$$

4.10. Show that for every $1 \le k \le 4$ there exists a critical 4×4 completely positive matrix A such that $G(A) = K_4$ and $\operatorname{cpr} A = k$.
Hint: For $k = 4$, let

$$A_\varepsilon = \begin{pmatrix} 2 & 0 & 1 & 1 \\ 0 & 2 & 1 & 1 \\ 1 & 1 & 2 & 0 \\ 1 & 1 & 0 & 2 \end{pmatrix} + \varepsilon J_4.$$

Show that $\operatorname{cpr} A_\varepsilon$ must be 4 for some $\varepsilon > 0$.

4.11. Let A be an $n \times n$ completely positive matrix, $\mathbf{b} \in \mathbb{R}^n_+$, and

$$A' = \begin{pmatrix} \mathbf{b}^T A \mathbf{b} & \mathbf{b}^T A \\ A\mathbf{b} & A \end{pmatrix}.$$

Show that $\operatorname{cpr} A' = \operatorname{cpr} A$, and if A is critical, so is A'.

4.12. Prove Proposition 4.17.

Hint: For (a), use Exercise 4.9. For (b), use Exercise 4.3.

4.13. Prove equality (4.9) (Proposition 4.18).

4.14. Prove that

$$\operatorname{cpr} G = \max\{\operatorname{cpr} A \mid A \text{ is a nonsingular CP matrix realization of } G\}.$$

Hint: Exercise 4.9.

Notes. The cp-rank of a completely positive matrix A is called the *factorization index* of A in some papers, and denoted by $\varphi(A)$. In other papers it is denoted by $\#(A)$. Example 4.5 and Exercise 4.4 are due to [Shaked-Monderer *et al.* (2013)]. The fact that small matrices have square factorizations was proved by [Maxfield and Minc (1962/1963)]. The proof we give here for 4×4 matrices is based on [Ando (1991)]. Exercise 4.5 is due to [Shaked-Monderer (2009)], and Exercise 4.10 is based on [Zhang and Li (2002)]. The observations regarding the lower bound $\operatorname{tf}(G)$ on $\operatorname{cpr} G$, and its relation to $\operatorname{cc}(G)$, were made in [Shaked-Monderer (2015)]. The fact that $\operatorname{tc}(G) \leq \operatorname{tf}(G)$, with equality if and only if G is triangle free, was proved in [Lehel and Tuza (1982)].

A notion related to the cp-rank is the *nonnegative rank* of a nonnegative matrix A. It is the smallest number of summands in a representation of A as a sum of rank 1 nonnegative matrices. For completely positive matrices, the nonnegative rank is a lower bound for the cp-rank. The two can be different, since $\min\{m, n\}$ is an upper bound on the nonnegative rank of an $m \times n$ matrix (see [Cohen and Rothblum (1993)]).

4.2 Completely positive matrices of a given rank

This section deals with problem (A) mentioned in Section 4.1: Find an upper bound on the cp-ranks of completely positive matrices of rank r. A tight such upper bound is known, and we present it here in Theorem 4.23, with the original proof. The second part of this section presents a different proof that, unlike the first proof of Theorem 4.23, requires no limit argument, and a slight refinement of the bound for the case that the rank r completely positive matrix is not positive.

If A is a rank r completely positive matrix, then $A = W^T W$ where W is an $r \times n$ matrix of rank r. Since A is completely positive, there exists a matrix Q such that $Q^T W \geq 0$ and $QQ^T = I_r$. The cp-rank of A is the minimal number of columns in such Q. Throughout this section we denote the columns of such W by $\mathbf{w}_1, \ldots, \mathbf{w}_n$, and the columns of such Q by $\mathbf{q}_1, \ldots, \mathbf{q}_k$. The cone generated by $\mathbf{w}_1, \ldots, \mathbf{w}_n$ in $\mathbb{R}^r$ is denoted by C_W,

$$C_W = \operatorname{cone}\{\mathbf{w}_1, \ldots, \mathbf{w}_n\}. \tag{4.11}$$

Since $A = \operatorname{Gram}(\mathbf{w}_1, \ldots, \mathbf{w}_n)$ is completely positive and $\operatorname{rank} W = r$, the polyhedral cone C_W is pointed and solid. The nonnegativity of $Q^T W$ means that

$$\mathbf{q}_1, \ldots, \mathbf{q}_k \in C_W^* \tag{4.12}$$

and the equality $QQ^T = I_r$ translates to

$$\sum_{i=1}^{k} \mathbf{q}_i \mathbf{q}_i^T = I_r. \tag{4.13}$$

Hence the cp-rank of A is the minimal number k of vectors in C_W^* which satisfy (4.13). We use this approach and these notations throughout the first part of the section.

Theorem 4.21. *Let A be a rank r completely positive matrix, $r \geq 2$, and let N be the maximal number of zeroes above the diagonal in a nonsingular principal submatrix of A. Then*

$$\operatorname{cpr} A \leq \frac{r(r+1)}{2} - N.$$

Proof. Let A be a $n \times n$ completely positive matrix of rank r. There exist an $r \times n$ matrix W such that $A = W^T W$ and an $r \times k$ matrix Q whose columns $\mathbf{q}_1, \ldots, \mathbf{q}_k$ satisfy (4.12) and (4.13). Let $\mathcal{C}$ be the cone generated in $\mathcal{S}_r$ by $\{\mathbf{q}_i \mathbf{q}_i^T \mid i = 1, \ldots, k\}$, and

$$V = \operatorname{span}\left\{\mathbf{q}_i \mathbf{q}_i^T \mid i = 1, \ldots, k\right\} = \operatorname{span} C \subseteq \mathcal{S}_r.$$

Let $d = \dim V$. By (4.13) $I_r \in \mathcal{C}$, and since $\dim(\operatorname{span}\mathcal{C}) = d$, Theorem 1.132 (Carathéodory's Theorem) implies that I_r can be represented as a nonnegative combination of at most d of the $\mathbf{q}_i \mathbf{q}_i^T$'s. Without loss of generality,

$$I_r = \sum_{i=1}^{d} \alpha_i \mathbf{q}_i \mathbf{q}_i^T \quad , \; \alpha_i \geq 0, \; i = 1, \ldots, d.$$

Then the matrix $P = \left(\sqrt{\alpha_1}\mathbf{q}_1 \dots \sqrt{\alpha_d}\mathbf{q}_d\right)$ is an $r \times d$ matrix such that $PP^T = I_r$ and $P^T W \geq 0$. Hence $\operatorname{cpr} A \leq d$. To complete the proof we show that $d \leq \frac{r(r+1)}{2} - N$.

Let $A[\alpha]$, $\alpha \subseteq \{1, \dots, n\}$, be a nonsingular principal submatrix of A with N zeroes above the diagonal. Without loss of generality assume that $\alpha = \{1, \dots, h\}$. There are N pairs (p, q), $p, q \in \alpha$ and $p < q$, for which $a_{pq} = 0$, i.e., $\mathbf{w}_p^T \mathbf{w}_q = 0$. For each of these N pairs define a linear functional $f_{p,q} : \mathcal{S}_r \to \mathbb{R}$,

$$f_{p,q}(X) = \mathbf{w}_p^T X \mathbf{w}_q.$$

Obviously, $I_r \in \ker(f_{p,q})$ for each such pair. By (4.13)

$$\sum_{i=1}^{k} \mathbf{w}_p^T \mathbf{q}_i \mathbf{q}_i^T \mathbf{w}_q = \mathbf{w}_p^T I_r \mathbf{w}_q = 0, \tag{4.14}$$

and since $\mathbf{w}_p^T \mathbf{q}_i \mathbf{q}_i^T \mathbf{w}_q \geq 0$ for every i, (4.14) implies that $\mathbf{w}_p^T \mathbf{q}_i \mathbf{q}_i^T \mathbf{w}_q = 0$ for each i. That is,

$$\mathbf{q}_i \mathbf{q}_i^T \in \ker(f_{p,q}) \quad , i = 1, \dots, k.$$

Hence

$$V = \operatorname{span}\{\mathbf{q}_i \mathbf{q}_i^T \mid i = 1, \dots, k\} \subseteq \ker(f_{p,q})$$

for each such pair (p, q). It follows that $V \subseteq K$, where K is the intersection of all these kernels.

Let U be an $r \times r$ nonsingular matrix satisfying $U\mathbf{e}_i = \mathbf{w}_i$, $, i = 1, \dots, h$. Then $X \in K$ if and only if $\mathbf{e}_p^T U^T X U \mathbf{e}_q$ is zero for each of the N pairs (p, q). That is, $(U^T X U)_{pq} = 0$ for each such pair. Since $X \mapsto U^T X U$ is an isomorphism of $\mathcal{S}_r$ onto itself, this means that K is isomorphic to a subspace of $\mathcal{S}_r$ consisting of all $r \times r$ symmetric matrices with zeroes in N specified entries above the diagonal. In particular, $\dim K = \frac{r(r+1)}{2} - N$. And since $V \subseteq K$,

$$d = \dim V \leq \dim K = \frac{r(r+1)}{2} - N. \qquad \square$$

The next theorem provides a slight improvement of the bound in Theorem 4.21 in the case that A is a *positive* completely positive matrix. For the proof, we need one more observation.

Remark 4.22. If A is a rank r completely positive matrix, we may assume that $A = W^T W$ where W is $r \times n$ and every nonzero vector in C_W^* has a positive last entry. To see that, suppose $A = W^T W$, $W \in \mathbb{R}^{r \times n}$. Let

α_i, $i = 1, \ldots, n$, be positive real numbers for which $\mathbf{w}_0 = \sum_{i=1}^n \alpha_i \mathbf{w}_i$ is a nonzero unit vector in C_W, and let U be an orthogonal $r \times r$ matrix satisfying $U\mathbf{w}_0 = \mathbf{e}_r$. Then $A = (UW)^T(UW)$. If $0 \neq \mathbf{q} \in C_{UW}^*$, then $\mathbf{q}_r = \mathbf{q}^T\mathbf{e}_r = \mathbf{q}^T U\mathbf{w}_0 = \sum_{i=1}^n \alpha_i \mathbf{q}^T U\mathbf{w}_i$. For every $i = 1, \ldots, n$, $\mathbf{q}^T U\mathbf{w}_i \geq 0$, and since $U\mathbf{w}_1, \ldots, U\mathbf{w}_n$ span $\mathbb{R}^r$ and $\mathbf{q} \neq 0$, at least one of the products $\mathbf{q}^T U\mathbf{w}_i$ is positive. Thus $\mathbf{q}_r > 0$.

Theorem 4.23. *For every rank r completely positive matrix A, $r \geq 2$,*

$$\operatorname{cpr} A \leq \frac{r(r+1)}{2} - 1.$$

Proof. Suppose that A is an $n \times n$ rank r completely positive matrix. Let W be an $r \times n$ matrix such that $A = W^T W$, and let $\mathbf{q}_1, \ldots, \mathbf{q}_k \in C_W^*$ satisfy (4.13). We may assume that the last component of each $\mathbf{q}_i$ is positive (by Remark 4.22), that $k \leq r(r+1)/2$ (by Theorem 4.21), and that the k rank 1 matrices $\mathbf{q}_1\mathbf{q}_1^T, \ldots, \mathbf{q}_k\mathbf{q}_k^T$ are linearly independent (by Remark 1.133). If $k < r(r+1)/2$, there is nothing to prove, since $\operatorname{cpr} A \leq k$. We therefore consider only the case $k = r(r+1)/2$.

For $i = 1, \ldots, k$, let $\mu_i = (\mathbf{q}_i)_r > 0$ and $\mathbf{p}_i = 1/\mu_i \mathbf{q}_i$. Then $(\mathbf{p}_i)_r = 1$ for each i, $\mathbf{p}_1\mathbf{p}_1^T, \ldots, \mathbf{p}_k\mathbf{p}_k^T$ are linearly independent, and

$$\sum_{i=1}^k \mu_i^2 \mathbf{p}_i \mathbf{p}_i^T = I_r.$$

Define $\mathbf{p}_1(\beta) = (1-\beta)\mathbf{p}_1 + \beta\mathbf{p}_2$, and consider the matrix equation

$$x_1 \mathbf{p}_1(\beta)\mathbf{p}_1^T(\beta) + \sum_{i=2}^k x_i \mathbf{p}_i \mathbf{p}_i^T = I_r. \tag{4.15}$$

Equation (4.15) translates into a system of k linear equations in the k unknowns $x_1, \ldots, x_k$. For $\beta = 0$ there is a unique solution $x_i = \mu_i^2$, $i = 1, \ldots, k$, and therefore the coefficients matrix of the system is nonsingular. For $\beta = 1$, the first two columns of the coefficients matrix are identical, so the matrix is singular. Let $0 < \beta_0 \leq 1$ be the smallest positive β for which the coefficients matrix is singular. The determinant $\Delta(\beta)$ of the coefficients matrix is continuous, hence for every $0 \leq \beta < \beta_0$ the coefficients matrix is nonsingular and its determinant $\Delta(\beta)$ has the same sign as $\Delta(0)$. For each such β the system has a unique solution, which may be represented by Cramer's rule:

$$\lambda_i(\beta) = \frac{\Delta_i(\beta)}{\Delta(\beta)}, \quad i = 1, \ldots, k.$$

$\Delta_1(\beta)$ does not depend on β. The sign of $\lambda_1(\beta)$ is therefore fixed for every $0 \le \beta < \beta_0$, i.e., $\lambda_1(\beta) > 0$. Since $\lim_{\beta\to\beta_0-} \Delta(\beta) = \Delta(\beta_0) = 0$, this implies

$$\lim_{\beta\to\beta_0-} \lambda_1(\beta) = \lim_{\beta\to\beta_0-} \frac{\Delta_1(\beta)}{\Delta(\beta)} = \infty.$$

Thus there exists $0 < \beta_1 < \beta_0$ for which $\lambda_1(\beta_1) > 1$. In particular

$$\lambda_1(\beta_1)(\mathbf{p}_1(\beta_1)\mathbf{p}_1^T(\beta_1))_{rr} + \sum_{i=2}^{k} \lambda_i(\beta_1)(\mathbf{p}_i\mathbf{p}_i^T)_{rr} = 1.$$

But $(\mathbf{p}_1(\beta_1))_r = (\mathbf{p}_2)_r = \ldots = (\mathbf{p}_k)_r = 1$, hence

$$\sum_{i=1}^{k} \lambda_i(\beta_1) = 1.$$

Since $\lambda_1(\beta_1) > 1$, $\lambda_i(\beta_1) < 0$ for at least one of the other i's. Let β_2 be the largest $0 \le \beta < \beta_1$ such that $\lambda_i(\beta) \ge 0$ for every $i = 1, \ldots, k$. By the continuity of all the $\lambda_i(\beta)$'s, we have $\lambda_i(\beta_2) = 0$ for at least one i. So

$$\lambda_1(\beta_2)\mathbf{p}_1(\beta_2)\mathbf{p}_1^T(\beta_2) + \sum_{i=2}^{k} \lambda_i(\beta_2)\mathbf{p}_i\mathbf{p}_i^T = I_r$$

is a representation of I_r as a sum of less than $k = r(r+1)/2$ rank 1 matrices. Thus

$$\operatorname{cpr} A \le \frac{r(r+1)}{2} - 1. \qquad \square$$

The next theorem shows that the upper bound in Theorem 4.23 is tight. The proof uses the following lemma.

Lemma 4.24. *Suppose there exist vectors $\mathbf{p}_1, \ldots, \mathbf{p}_k \in \mathbb{R}^r$ and positive real numbers $\mu_1, \ldots, \mu_k$ such that the following three conditions are satisfied.*

(a) *For each $1 \le i \le k$, $\mathbf{p}_i^T = (\mathbf{u}_i^T, 1)$, where each $\mathbf{u}_i \in \mathbb{R}^{r-1}$ satisfies $\|\mathbf{u}_i\| = \sqrt{r-1}$.*
(b) *$\{\mathbf{p}_1\mathbf{p}_1^T, \ldots, \mathbf{p}_k\mathbf{p}_k^T \mid i = 1, \ldots, k\}$ is a linearly independent set of matrices in $\mathcal{S}_r$.*
(c) *$\sum_{i=1}^k \mu_i\mathbf{p}_i\mathbf{p}_i^T = I_r$.*

Then there exists a completely positive matrix A such that $\operatorname{rank} A = r$ *and* $\operatorname{cpr} A = k$.

Proof. Let C be the convex cone generated in $\mathbb{R}^r$ by $\mathbf{p}_1, \ldots, \mathbf{p}_k$. Then

$$x_r \geq 0 \text{ for every } \mathbf{x} \in C, \text{ and } C \cap \{\mathbf{x} \in \mathbb{R}^r \mid x_r = 0\} = \{\mathbf{0}\}. \qquad (4.16)$$

To see that the second part of (4.16) holds, note that if

$$\sum_{i=1}^{k} \alpha_i \mathbf{p}_i = \begin{pmatrix} \mathbf{y} \\ 0 \end{pmatrix}, \quad \mathbf{y} \in \mathbb{R}^{r-1}, \ \alpha_1, \ldots, \alpha_k \geq 0,$$

then equating the r-th component on both sides yields $\sum_{i=1}^{k} \alpha_i = 0$, and since the α_i's are all nonnegative, this implies that $\alpha_1 = \ldots = \alpha_k = 0$, hence $\mathbf{y} = 0$.

Another property of C is the following.

$$\begin{aligned} &\text{If } \begin{pmatrix} \mathbf{y} \\ 1 \end{pmatrix} \in C, \ \mathbf{y} \in \mathbb{R}^{r-1}, \text{ then } \|\mathbf{y}\| \leq \sqrt{r-1}, \\ &\text{and equality occurs iff } \mathbf{y} = \mathbf{u}_i \text{ for some } i = 1, \ldots, k. \end{aligned} \qquad (4.17)$$

To see that (4.17) holds, assume

$$\begin{pmatrix} \mathbf{y} \\ 1 \end{pmatrix} = \sum_{i=1}^{k} \alpha_i \mathbf{p}_i, \quad \alpha_1, \ldots, \alpha_k \geq 0.$$

Then $\sum_{i=1}^{k} \alpha_i = 1$ and $\sum_{i=1}^{k} \alpha_i \mathbf{u}_i = \mathbf{y}$, and therefore

$$\|\mathbf{y}\| \leq \sum_{i=1}^{k} \alpha_i \|\mathbf{u}_i\| = \sum_{i=1}^{k} \alpha_i \sqrt{r-1} = \sqrt{r-1}.$$

Since the $\mathbf{u}_i$'s are k different vectors with equal norm, equality occurs if and only if one of the α_i's is 1 and the others are all zero.

C is a polyhedral cone in $\mathbb{R}^r$ and (c) implies that it is reproducing. Hence C^* is also a reproducing polyhedral cone in $\mathbb{R}^r$. Let W be an $r \times n$ matrix whose columns generate C^*, and $A = W^T W$. Let Q be the matrix whose columns are $\sqrt{\mu_1}\mathbf{p}_1, \ldots, \sqrt{\mu_k}\mathbf{p}_k$. Then $Q^T W$ is $k \times n$, $Q^T W \geq 0$, and since $QQ^T = \sum_{i=1}^{k} \mu_i \mathbf{p}_i \mathbf{p}_i^T = I_r$,

$$A = W^T W = (Q^T W)^T (Q^T W).$$

That is, A is a completely positive matrix, $\operatorname{rank} A = r$ and $\operatorname{cpr} A \leq k$.

To see that $\operatorname{cpr} A = k$, suppose $\hat{Q}$ is an $r \times t$ matrix satisfying $\hat{Q}^T W \geq 0$ and $\hat{Q}\hat{Q}^T = I_r$, where $t = \operatorname{cpr} A$. Let $\hat{\mathbf{q}}_1, \ldots, \hat{\mathbf{q}}_t$ be $\hat{Q}$'s columns (necessarily, all nonzero). Let $\lambda_i = (\hat{\mathbf{q}}_i)_r$, $i = 1, \ldots, t$. By (4.16), $\lambda_i > 0$ for every i. Thus $\hat{\mathbf{q}}_i = \lambda_i \mathbf{b}_i$, where $\mathbf{b}_i^T = (\mathbf{v}_i^T, 1) \in C$. The equality

$$\sum_{i=1}^{t} \hat{\mathbf{q}}_i \hat{\mathbf{q}}_i^T = \sum_{i=1}^{t} \lambda_i^2 \mathbf{b}_i \mathbf{b}_i^T = I_r$$

implies that $\sum_{i=1}^{t} \lambda_i^2 = 1$, and since

$$\operatorname{trace}(I_r) = \sum_{i=1}^{t} \lambda_i^2 \operatorname{trace}(\mathbf{b}_i \mathbf{b}_i^T),$$

we have

$$r = \sum_{i=1}^{t} \lambda_i^2 \|\mathbf{v}_i\|^2 + \sum_{i=1}^{t} \lambda_i^2 = \sum_{i=1}^{t} \lambda_i^2 \|\mathbf{v}_i\|^2 + 1. \tag{4.18}$$

By (4.17), $\|\mathbf{v}_i\|^2 \le r-1$ for each i, and thus from (4.18) it follows that $\|\mathbf{v}_i\|^2 = r-1$ for each i. This in turn implies (again by (4.17)) that each $\mathbf{v}_i$ is one of $\mathbf{u}_1, \ldots, \mathbf{u}_k$. Hence

$$\{\mathbf{b}_1, \ldots, \mathbf{b}_t\} \subseteq \{\mathbf{p}_1, \ldots, \mathbf{p}_k\}.$$

Since $\{\mathbf{p}_1\mathbf{p}_1^T, \ldots, \mathbf{p}_k\mathbf{p}_k^T \mid i = 1, \ldots, k\}$ is linearly independent, the two representations of I_r,

$$I_r = \sum_{i=1}^{k} \mu_i \mathbf{p}_i \mathbf{p}_i^T \quad , \quad I_r = \sum_{i=1}^{t} \lambda_i^2 \mathbf{b}_i \mathbf{b}_i^T$$

are the same. In particular, $t = k$. □

We can now state and prove

Theorem 4.25. *For every $r \ge 2$ there exists a completely positive matrix A with* $\operatorname{rank} A = r$ *and* $\operatorname{cpr} A = r(r+1)/2 - 1$.

Proof. For $r = 2$ this follows from Theorem 4.7. We therefore consider the case $r \ge 3$. Let $k = r(r+1)/2 - 1$. We prove the theorem by constructing k vectors in $\mathbb{R}^r$ and k positive real numbers such that conditions (a), (b), and (c) of Lemma 4.24 are satisfied.

For every $2 \le h \le r$ and every $1 \le l \le h$ define $\mathbf{p}_{hl} \in \mathbb{R}^r$. For $l = 1$ let

$$(\mathbf{p}_{h1})_i = \begin{cases} -ab^{h-2}, & i = 1 \\ -ab^{h-i}, & 2 \le i \le h-1 \\ 0, & h \le i \le r-1 \\ 1, & i = r \end{cases}$$

and for $2 \le l \le h$ let

$$(\mathbf{p}_{hl})_i = \begin{cases} 0, & i \le l-2 \\ ab^{h-l}, & i = l-1 \\ -ab^{h-i}, & l \le i \le h-1 \\ 0, & h \le i \le r-1 \\ 1, & i = r \end{cases},$$

where $a = \sqrt{r-1}$ and $b = 1/\sqrt{2}$.

We first show that these vectors satisfy (a) and (b). Clearly, each $\mathbf{p}_{hl}^T$ is of the form $(\mathbf{u}_{hl}^T, 1)$, and it is easy to check that $\|\mathbf{u}_{hl}\| = \sqrt{r-1}$. That is, (a) is satisfied. We now show that $\{\mathbf{p}_{hl}\mathbf{p}_{hl}^T \mid h = 2, \ldots, r, l = 1, \ldots, h\}$ is a linearly independent set. First, we show that for each $2 \le h \le r$ the set $\{\mathbf{p}_{hl} \mid 1 \le l \le h\} \subseteq \mathbb{R}^r$ is linearly independent. Consider the matrix

$$M_h = \begin{pmatrix} -ab^{h-2} & ab^{h-2} & 0 & 0 & \cdots & 0 & 0 \\ -ab^{h-2} & -ab^{h-2} & ab^{h-3} & 0 & \cdots & 0 & 0 \\ -ab^{h-3} & -ab^{h-3} & -ab^{h-3} & ab^{h-4} & \cdots & 0 & 0 \\ \vdots & \vdots & \vdots & \vdots & \ddots & \vdots & \vdots \\ -ab^2 & -ab^2 & -ab^2 & -ab^2 & \cdots & ab & 0 \\ -ab & -ab & -ab & -ab & \cdots & -ab & a \\ 1 & 1 & 1 & 1 & \cdots & 1 & 1 \end{pmatrix}.$$

M_h consists of rows $1, \ldots, h-1, r$ of the matrix $(\mathbf{p}_{h1} \ldots \mathbf{p}_{hh})$. Divide each row by the absolute value of the first element in the row, to get

$$\begin{pmatrix} -1 & 1 & 0 & 0 & \cdots & 0 & 0 \\ -1 & -1 & 1/b & 0 & \cdots & 0 & 0 \\ -1 & -1 & -1 & 1/b & \cdots & 0 & 0 \\ \vdots & \vdots & \vdots & \vdots & \ddots & \vdots & \vdots \\ -1 & -1 & -1 & -1 & \cdots & 1/b & 0 \\ -1 & -1 & -1 & -1 & \cdots & -1 & 1/b \\ 1 & 1 & 1 & 1 & \cdots & 1 & 1 \end{pmatrix}.$$

The determinant of this $h \times h$ matrix is $-2(-1/b-1)^{h-2}$ (prove by induction on h). Hence

$$\det(M_h) = -2a^{h-1}b^{\frac{h^2-h-2}{2}}(-1/b-1)^{h-2} \ne 0.$$

We can now prove, by induction on m, that for each $2 \le m \le r$ the set $\{\mathbf{p}_{hl}\mathbf{p}_{hl}^T \mid h = 2, \ldots, m, l = 1, \ldots, h\}$ is linearly independent. For $m = 2$ this follows from the fact that $\{\mathbf{p}_{21}, \mathbf{p}_{22}\}$ are linearly independent. Let $2 < m \le r$, and suppose $\{\mathbf{p}_{hl}\mathbf{p}_{hl}^T \mid h = 2, \ldots, m-1, l = 1, \ldots, h\}$ is linearly independent. Assume

$$\sum_{h=2}^{m}\sum_{l=1}^{h} \alpha_{hl}\mathbf{p}_{hl}\mathbf{p}_{hl}^T = 0. \tag{4.19}$$

For every $2 \le h \le m-1$ and $1 \le l \le h$

$$\operatorname{supp}(\mathbf{p}_{hl}) \subseteq \{1, \ldots, m-2, r\}.$$

Hence the $(m-1)$-th column of $\sum_{h=2}^{m-1}\sum_{l=1}^{h}\alpha_{hl}\mathbf{p}_{hl}\mathbf{p}_{hl}^T$ is zero. Equating the $(m-1)$-th columns on both sides of (4.19) we therefore get that

$$\sum_{l=1}^{m}\alpha_{ml}(\mathbf{p}_{ml})_{m-1}\mathbf{p}_{ml}=0.$$

Since $\{\mathbf{p}_{m1},\ldots,\mathbf{p}_{mm}\}$ is linearly independent and $(\mathbf{p}_{ml})_{m-1}\neq 0$ for every $l=1,\ldots,m$, we get that $\alpha_{ml}=0$ for every l. Hence (4.19) means

$$\sum_{h=2}^{m-1}\sum_{l=1}^{h}\alpha_{hl}\mathbf{p}_{hl}\mathbf{p}_{hl}^T=0,$$

and by the induction hypothesis this implies that $\alpha_{hl}=0$ for every $h=1,\ldots,m-1$, $l=1,\ldots,h$. This completes the proof that (b) is satisfied.

Finally, we define μ_{hl}, $h=2,\ldots,r, l=1,\ldots,h$, and show that (c) is satisfied. The numbers μ_{hl} are defined in the following table, where in addition to a and b we have $p=2-\sqrt{2}$.

	$l=1,2$	$l=3,\ldots,h$
$h=2$	$\mu_{hl}=\frac{b^2p}{a^2}$	—
$h=3,\ldots,r-1$	$\mu_{hl}=\frac{b^2p^{h-2}}{a^2}$	$\mu_{hl}=\frac{bp^{h-l+1}}{a^2}$
$h=r$	$\mu_{hl}=\frac{bp^{h-2}}{a^2}$	$\mu_{hl}=\frac{p^{h-l+1}}{a^2}$

To show that (c) holds, let $D_h=\sum_{l=1}^{h}\mu_{hl}\mathbf{p}_{hl}\mathbf{p}_{hl}^T$ for $h=2,\ldots,r$. Then

$$D_2=\operatorname{Diag}\left(2(1-b),0,\ldots,0,2(1-b)/a^2\right),$$

$$D_h=\operatorname{Diag}\left((1-b)^{h-2},b(1-b)^{h-3},b(1-b)^{h-4},\ldots,b(1-b)^0,0,\ldots,1/a^2\right)$$

for $h=3,\ldots,r-1$, and

$$D_r=\operatorname{Diag}\left((1-b)^{r-2}/b,(1-b)^{r-3},(1-b)^{r-4},\ldots,(1-b)^0,1/a^2\right).$$

From this it is easy to deduce that $\sum_{h=2}^{r}D_h=I_r$. The straightforward computations are left to the reader. □

Observe that the proof of Theorem 4.25 provides no information on the order of this rank r completely positive matrix whose cp-rank is $\frac{r(r+1)}{2}-1$.

By the approach used in this section, if A is a rank r completely positive matrix, $A=W^TW$ where W is an $r\times n$ matrix, then $\operatorname{cpr}A$ is the minimal number of vectors $\mathbf{q}_1,\ldots,\mathbf{q}_k$ in C_W^* that satisfy (4.13). The next proposition shows that in one special case it is enough to consider only vectors

$\mathbf{q}_1, \ldots, \mathbf{q}_k$ which lie in a certain finite set of rays. The proof uses arguments similar to the ones used in Lemma 4.24, and is left as an exercise.

Proposition 4.26. *Let A be a rank r doubly nonnegative $n \times n$ matrix, $A = W^T W$ where W is $r \times n$ and every $\mathbf{0} \neq \mathbf{x} \in C_W^*$ has a positive last entry. Denote $K = C_W^* \cap \{\mathbf{x} \in \mathbb{R}^r \mid x_r = 1\}$, and suppose that $\|\mathbf{x}\| \leq r$ for every $\mathbf{x} \in K$. Then*

(a) *If A is completely positive, then $K \cap \{\mathbf{x} \mid \|\mathbf{x}\| = r\}$ is a nonempty finite set.*
(b) *If $K \cap \{\mathbf{x} \mid \|\mathbf{x}\| = r\} = \{\mathbf{p}_1, \ldots, \mathbf{p}_k\}$, then A is completely positive if and only if there exist nonnegative numbers λ_i such that*

$$\sum_{i=1}^{k} \lambda_i \mathbf{p}_i \mathbf{p}_i^T = I_r \tag{4.20}$$

and

$$\operatorname{cpr} A = \min_{(\lambda_1,\ldots,\lambda_k)\in\Lambda} \{\operatorname{supp}(\lambda_1, \ldots, \lambda_k)\},$$

where Λ is the set of all nonnegative solutions to (4.20).

Alternative proofs and a refinement

We present now alternative proofs to Theorems 4.21 and 4.23. The approach in these proofs is also used to show that the upper bounds in Theorems 4.21 and 4.23 may be slightly reduced for completely positive matrices that are not positive.

For an $n \times n$ symmetric matrix A, let

$$\operatorname{cs}_+ A = \left\{\mathbf{x} \in \operatorname{cs} A \mid \mathbf{x} \geq \mathbf{0} \text{ and } G(\mathbf{x}\mathbf{x}^T) \subseteq G(A)\right\}.$$

If A is completely positive and $A = \sum_{i=1}^k \mathbf{b}_i \mathbf{b}_i^T$ is a CP representation of A, then $\mathbf{b}_1, \ldots, \mathbf{b}_k$ are all in $\operatorname{cs}_+ A$, and span $\operatorname{cs} A$. Hence

$$A \in \operatorname{cone}\{\mathbf{b}\mathbf{b}^T \mid \mathbf{b} \in \operatorname{cs}_+ A\}.$$

Proposition 4.27. *Let A be an $n \times n$ completely positive matrix. Let L be any subspace of $\mathbb{R}^n$ containing* $\operatorname{cone}\{\mathbf{b}\mathbf{b}^T \mid \mathbf{b} \in \operatorname{cs}_+ A\}$. *Then*

$$\operatorname{cpr} A \leq \dim L.$$

Proof. By Carathéodory's Theorem

$$A = \sum_{i=1}^{k} \mu_i \mathbf{b}_i \mathbf{b}_i^T,$$

where $\mu_i > 0$, $\mathbf{b}_i \in \operatorname{cs}_+ A$ for every i, and $k \leq \dim L$. As

$$A = \sum_{i=1}^{k} \mu_i \mathbf{b}_i \mathbf{b}_i^T = \sum_{i=1}^{k} (\sqrt{\mu_i}\mathbf{b}_i)(\sqrt{\mu_i}\mathbf{b}_i)^T$$

is a CP representation of A, this implies that $\operatorname{cpr} A \leq \dim L$. □

Proof of Theorem 4.21. By Proposition 4.27, it suffices to find a subspace L of $\mathbb{R}^n$ that contains $\operatorname{cone}\{\mathbf{b}\mathbf{b}^T \mid \mathbf{b} \in \operatorname{cs}_+ A\}$ with $\dim L = \frac{r(r+1)}{2} - N$.

Without loss of generality, we may assume that

$$A = \begin{pmatrix} A_1 & A_1 E^T \\ EA_1 & EA_1E^T \end{pmatrix}, \tag{4.21}$$

where A_1 is $r \times r$, nonsingular, and has N zero entries above the diagonal. Consider the following subspaces of $\mathcal{S}_n$:

$$L_1 = \operatorname{span}\{\mathbf{b}\mathbf{b}^T \mid \mathbf{b} \in \operatorname{cs} A\},$$

and

$$L_2 = \{X \in \mathcal{S}_n \mid x_{ij} = 0 \text{ for every } (i,j) \in \mathcal{Z}\},$$

where $\mathcal{Z}$ is the set of positions of the zero entries above the diagonal in A_1, $|\mathcal{Z}| = N$.

Let $L = L_1 \cap L_2$. Clearly, $\operatorname{cone}\{\mathbf{b}\mathbf{b}^T \mid \mathbf{b} \in \operatorname{cs}_+ A\} \subseteq L$. We leave it as an exercise to show that $\dim L_1 = \frac{r(r+1)}{2}$ and $\dim L_2 = \frac{n(n+1)}{2} - N$. We now show that $L_1 + L_2 = \mathcal{S}_n$, and this will imply that

$$\dim L = \dim L_1 + \dim L_2 - \dim \mathcal{S}_n = \frac{r(r+1)}{2} - N,$$

and will conclude the proof.

To see that $L_1 + L_2 = \mathcal{S}_n$, note that by (4.21), $\mathbf{b} \in \operatorname{cs} A$ if and only if $\mathbf{b} = \begin{pmatrix} \mathbf{x} \\ E\mathbf{x} \end{pmatrix}$, where $\mathbf{x} \in \mathbb{R}^r$.

For every $X \in \mathcal{S}_n$,

$$X = \begin{pmatrix} X_1 & X_2^T \\ X_2 & X_3 \end{pmatrix},$$

where X_1 is $r \times r$. Then

$$X = \begin{pmatrix} X_1 & X_1E^T \\ EX_1 & EX_1E^T \end{pmatrix} + \begin{pmatrix} 0 & X_2^T - X_1E^T \\ X_2 - EX_1 & X_3 - EX_1E^T \end{pmatrix}. \tag{4.22}$$

If $X_1 = \sum_{i=1}^{r} \lambda_i \mathbf{u}_i \mathbf{u}_i^T$, where the λ_i's are the positive eigenvalues of X_1, and the $\mathbf{u}_i$'s are the corresponding eigenvectors. Then

$$\begin{pmatrix} X_1 & X_1E^T \\ EX_1 & EX_1E^T \end{pmatrix} = \sum_{i=1}^{r} \lambda_i \begin{pmatrix} \mathbf{u}_i \\ E\mathbf{u}_i \end{pmatrix} \begin{pmatrix} \mathbf{u}_i \\ E\mathbf{u}_i \end{pmatrix}^T,$$

so (4.22) show that $X \in L_1 + L_2$. □

For the proof of Theorem 4.23, and a further reduction of the bound in this theorem for matrices that are not positive, we need some more results.

Proposition 4.28. *Let $A = \sum_{j=1}^{k} \mathbf{b}_i \mathbf{b}_i^T$ be a CP representation of a completely positive matrix A. If $\mathbf{b}_k$ is a nonnegative linear combination of $\mathbf{b}_1, \ldots, \mathbf{b}_{k-1}$, then* $\operatorname{cpr} A \leq k - 1$.

Proof. Let $B = \begin{pmatrix} \mathbf{b}_1 & \ldots & \mathbf{b}_{k-1} \end{pmatrix}$. Then $A = BB^T + \mathbf{b}_k \mathbf{b}_k^T$, and $\mathbf{b}_k = B\mathbf{x}$ for some $\mathbf{x} \geq \mathbf{0}$, $\mathbf{x} \in \mathbb{R}^{k-1}$. Hence,

$$A = BB^T + (B\mathbf{x})(B\mathbf{x})^T = BB^T + B\mathbf{x}\mathbf{x}^T B^T = B(I + \mathbf{x}\mathbf{x}^T)B^T.$$

By Exercise 3.67, $I + \mathbf{x}\mathbf{x}^T = C^2$, where $C \in \mathcal{S}_{k-1}$ is nonnegative. Hence $A = (BC)(BC)^T$, where $BC \geq 0$ has $k - 1$ columns. □

Proposition 4.29. *Let A be an $n \times n$ completely positive matrix, and let L be a subspace of $\mathcal{S}_n$ of dimension d such that* $\operatorname{cone}\{\mathbf{b}\mathbf{b}^T \mid \mathbf{b} \in \operatorname{cs}_+ A\} \subseteq L$. *Let $\mathbf{a}$ be the i-th column of A for some i such that $a_{ii} > 0$ and the Schur complement $A/[i]$ is not completely positive. If $\mathbf{a}\mathbf{a}^T \in L$, then*

$$\operatorname{cpr} A \leq d - 1.$$

Proof. We may assume that $i = 1$, and $\mathbf{a}$ is the first column of A. Let $A = \sum_{i=1}^{k} \mathbf{b}_i \mathbf{b}_i^T$ be a minimal CP representation of A. By Proposition 4.27, $k \leq d$. Suppose on the contrary that $k = d$. That is,

$$A = \sum_{i=1}^{d} \mathbf{b}_i \mathbf{b}_i^T. \tag{4.23}$$

By the minimality of the CP representation, $\{\mathbf{b}_1 \mathbf{b}_1^T, \ldots, \mathbf{b}_d \mathbf{b}_d^T\}$ is a linearly independent set (see Exercise 4.20), and is therefore a basis for L. Since $\mathbf{a}\mathbf{a}^T \in L$, there exist scalars μ_i, at least one of which is positive, such that

$$\mathbf{a}\mathbf{a}^T = \sum_{i=1}^{d} \mu_i \mathbf{b}_i \mathbf{b}_i^T.$$

That is,

$$0 = \mathbf{a}\mathbf{a}^T - \sum_{i=1}^{d} \mu_i \mathbf{b}_i \mathbf{b}_i^T. \tag{4.24}$$

Adding a proper positive multiple ν of (4.24) to (4.23), we get that

$$A = \sum_{i=1}^{d} \nu_i \mathbf{b}_i \mathbf{b}_i^T + \nu \mathbf{a}\mathbf{a}^T,$$

where $\nu_i \geq 0$ for every i, and $\nu_i = 0$ for at least one i. By the assumption that $\operatorname{cpr} A = d$, exactly one of the ν_i's, say ν_d, is zero. Hence

$$A = \sum_{j=1}^{d-1} \nu_i \mathbf{b}_i \mathbf{b}_i^T + \nu \mathbf{a}\mathbf{a}^T. \tag{4.25}$$

But then

$$A\mathbf{e}_1 = \sum_{j=1}^{d-1} \nu_i \mathbf{b}_i \mathbf{b}_i^T \mathbf{e}_1 + \nu \mathbf{a}\mathbf{a}^T \mathbf{e}_1.$$

As $A\mathbf{e}_1 = \mathbf{a}$, we get that

$$(1 - \nu \mathbf{a}^T \mathbf{e}_1)\mathbf{a} = \sum_{j=1}^{d-1} \nu_i \mathbf{b}_i \mathbf{b}_i^T \mathbf{e}_1. \tag{4.26}$$

Since $A/[1]$ is not completely positive, $|\{1 \leq i \leq d \mid 1 \in \operatorname{supp} \mathbf{b}_i\}| \geq 2$. Hence there exists $1 \leq i \leq d-1$ such that $\mathbf{b}_i^T \mathbf{e}_1 \neq 0$, and the right-hand side of (4.26) is not zero. Therefore $1 - \nu \mathbf{a}^T \mathbf{e}_1 > 0$ and (4.26) implies that $\mathbf{a}$ is a nonnegative linear combination of $\mathbf{b}_1, \ldots, \mathbf{b}_{d-1}$. This yields, by (4.25) and Proposition 4.28, that $\operatorname{cpr} A \leq d-1$, contradicting our assumption. □

Using Proposition 4.29, we can now add the following.

Proof of Theorem 4.23. Assume without loss of generality that the leading principal $r \times r$ submatrix of A is nonsingular, and let $\mathbf{a} = A\mathbf{e}_1$.

If $A/[1]$ is completely positive, then from

$$A = \frac{1}{a_{11}} \mathbf{a}\mathbf{a}^T + (0 \oplus A/[1])$$

we get that

$$\operatorname{cpr} A \leq 1 + \operatorname{cpr}(A/[1]) \leq 1 + \frac{(r-1)r}{2} \leq \frac{r(r+1)}{2} - 1,$$

where the middle inequality follows from Theorem 4.21, since $\operatorname{rank}(A/[1]) = r - 1 \geq 1$, and the right inequality holds for $r \geq 2$.

If $A/[1]$ is not completely positive, let $L = \operatorname{span}\{\mathbf{b}\mathbf{b}^T \mid \mathbf{b} \in \operatorname{cs} A\}$. As $\dim L = \frac{r(r+1)}{2}$ (see Exercise 4.17) and $\mathbf{a}\mathbf{a}^T \in L$, Proposition 4.29 implies that $\operatorname{cpr} A \leq \dim L - 1 = \frac{r(r+1)}{2} - 1$. □

We may also improve slightly on Theorems 4.21 and 4.23 in the case that $r \geq 3$ and $N = 1$.

Theorem 4.30. *Let $r \geq 3$, and let A be a completely positive matrix with* $\operatorname{rank} A = r$. *If in some $r \times r$ nonsingular principal submatrix of A there is a zero above the diagonal, then*

$$\operatorname{cpr} A \leq \frac{r(r+1)}{2} - 2.$$

Proof. Assume again without loss of generality that the $r \times r$ leading principal submatrix A_1 of A is nonsingular. If there are two or more zeros above the diagonal in this submatrix, the result follows from Theorem 4.23. Consider therefore the case of a unique zero above the diagonal in A_1. We may assume that $a_{12} = 0$. Let $L = L_1 \cap L_2$, where L_1 and L_2 are as in the alternative proof of Theorem 4.21, with $\mathcal{Z} = \{(1,2)\}$ and $N = 1$. Then $\dim L = \frac{r(r+1)}{2} - 1$, and the vector $\mathbf{a} = A\mathbf{e}_1$ satisfies $\mathbf{a}\mathbf{a}^T \in L$.

If $A/[1]$ is completely positive, then

$$\operatorname{cpr} A \leq 1 + \operatorname{cpr}(A/[1]) \leq 1 + \frac{(r-1)r}{2} - 1 < \frac{r(r+1)}{2} - 2,$$

where the middle inequality follows from Theorem 4.23, as $\operatorname{rank}(A/[1]) = r - 1 \geq 2$, and the right inequality holds since $r \geq 3$.

If $A/[1]$ is not completely positive, the result follows from Proposition 4.29. □

Remark 4.31. By Theorem 4.25, there exists a completely positive matrix with rank r and cp-rank $\frac{r(r+1)}{2} - 1$. By Theorem 4.30, any such matrix has to be positive.

Remark 4.32. The bound in Theorem 4.21 may not be reduced by 1 for every N. For example, if a copositive matrix has a nonsingular $r \times r$ principal submatrix with $N = \frac{r(r-1)}{2}$ zeros above the diagonal, this means that A has an $r \times r$ positive diagonal principal submatrix. In this case $\operatorname{cpr} A \leq \frac{r(r+1)}{2} - \frac{(r-1)r}{2} = r$ by Theorem 4.21. This upper bound cannot be reduced, since $\operatorname{cpr} A \geq \operatorname{rank} A = r$ as always, hence $\operatorname{cpr} A = r$. See also Exercise 4.18.

Exercises

4.15. Use Theorem 4.23 to find an upper bound on the cp-ranks of all $n \times n$ completely positive matrices, and Theorem 4.21 to find an upper bound on the cp-ranks of all $n \times n$ completely positive matrices whose graph is triangle free (in both cases, for $n \geq 5$).

4.16. Prove Proposition 4.26.

4.17. Complete the missing details in the alternative proof of Theorem 4.21. Let A be an $n \times n$ completely positive matrix of rank r in the form (4.21), with $\operatorname{rank} A_1 = r$, and let $\mathcal{Z}$ be the set of positions of the zero entries above the diagonal in A_1, $|\mathcal{Z}| = N$. Let

$$L_1 = \operatorname{span}\{\mathbf{b}\mathbf{b}^T \mid \mathbf{b} \in \operatorname{cs} A\} \quad \text{and} \quad L_2 = \{X \in \mathcal{S}_n \mid x_{ij} = 0 \ \ \forall\,(i,j) \in \mathcal{Z}\}.$$

(a) Let V be an r dimensional subspace of $\mathbb{R}^n$ and $L = \{\mathbf{b}\mathbf{b}^T \mid \mathbf{b} \in V\}$. Show that if $\{\mathbf{a}_1, \ldots, \mathbf{a}_r\}$ be a basis for V, then $\{(\mathbf{a}_i + \mathbf{a}_j)(\mathbf{a}_i + \mathbf{a}_j)^T \mid 1 \le i \le j \le r\}$ is a basis for L.
Deduce that $\dim L_1 = \frac{r(r+1)}{2}$.

(b) Show that $\dim L_2 = \frac{n(n+1)}{2} - N$.

Hint: For (a): Let $U : \operatorname{cs} A \to \mathbb{R}^r$ be the isomorphism $U\mathbf{a}_i = \mathbf{e}_i$. Show that the mapping $X \mapsto UXU^T$ is an isomorphism of L onto $\mathcal{S}_r$. (Note that $\mathbf{a}_i\mathbf{a}_j^T + \mathbf{a}_j\mathbf{a}_i^T = (\mathbf{a}_i + \mathbf{a}_j)(\mathbf{a}_i + \mathbf{a}_j)^T - \mathbf{a}_i\mathbf{a}_i^T - \mathbf{a}_j\mathbf{a}_j^T$.)

4.18. Show that if A is a doubly nonnegative matrix with $\operatorname{rank} A = r$, and A has an $r \times r$ positive diagonal principal submatrix D, then A is completely positive and $\operatorname{cpr} A = \operatorname{rank} A = r$.

Notes. Theorem 4.21 is due to [Hannah and Laffey (1983)] and Theorems 4.23 and 4.25 are due to [Barioli and Berman (2003)]. The alternative proofs of Theorem 4.21 and Theorem 4.23 are due to [Shaked-Monderer (2009)], and so are Theorem 4.30 and Exercise 4.18. Regarding the latter, see [Kalofolias and Gallopoulos (2012)] for a suggested algorithm for computing a minimal CP factorization in this case. One advantage of the alternative proofs is that they remain valid if we consider matrices over the field of rational numbers, and rational CP factorizations, instead of real completely positive matrices (see Section 4.9). The case $r = 3$ of Theorem 4.30 was previously proved in [Barioli (2001b)]. Both proofs of Theorem 4.23 do not give an indication on what is the order of a rank r matrix that attains the maximum cp-rank.

Still open: *What is the minimal n for which there exists a matrix $A \in \mathcal{CP}_n$ such that* $\operatorname{rank} A = r$ *and* $\operatorname{cpr} A = r(r+1)/2 - 1$? *And what is the maximum cp-rank of a singular matrix in $\mathcal{CP}_n$?*

4.3 The cp-ranks and minimal CP factorizations in $\mathcal{CP}_n$

In the next section we go into problem (B), of bounding the cp-ranks of $n \times n$ completely positive matrices. But before that, we consider here where in $\mathcal{CP}_n$ the various cp-ranks are attained. First, let us denote

$$p_n = \max \{\operatorname{cpr} A \mid A \in \mathcal{CP}_n\}$$

and consider where in the cone $\mathcal{CP}_n$ matrices attaining this maximum cp-rank may be found.

Proposition 4.33. *For every $n > 1$, and every $0 \leq p \leq p_n$, the set*

$$\{X \in \mathcal{CP}_n \mid \operatorname{cpr} X \leq p\}$$

is closed.

Proof. For $p = 0$ the set consists of the zero matrix, and is trivially closed. For $p \geq 1$, let A_k be a sequence of matrices in $\mathcal{CP}_n$, each with $\operatorname{cpr} A_k \leq p$, such that A_k converges to $A \in \mathcal{CP}_n$. For each k, $A_k = B_k B_k^T$, where $B_k \geq 0$ has p columns. As B_k is a bounded sequence in $\mathbb{R}^{n \times p}$, it has a convergent subsequence. Without loss of generality, $\lim_{k\to\infty} B_k = B$. Then $B \geq 0$ and $A = BB^T$, and therefore $\operatorname{cpr} A \leq p$. □

Corollary 4.34. *For every $n > 1$, and every $1 \leq p \leq p_n$, the set*

$$\{X \in \mathcal{CP}_n \mid \operatorname{cpr} X \geq p\}$$

is open in $\mathcal{CP}_n$.

The corollary implies, in particular, that the set

$$\{X \in \mathcal{CP}_n \mid \operatorname{cpr} X = p_n\}$$

of matrices in $\mathcal{CP}_n$ attaining the maximum cp-rank is open in $\mathcal{CP}_n$, and thus intersects $\operatorname{int}\mathcal{CP}_n$.

Corollary 4.35. *The maximum cp-rank of completely positive $n \times n$ matrices, p_n, is attained at a matrix in the interior of $\mathcal{CP}_n$.*

However, the next result tells us that the maximum cp-rank p_n is also attained on the boundary of $\mathcal{CP}_n$.

Theorem 4.36. *For $n > 1$, the maximum cp-rank of completely positive $n \times n$ matrices, p_n, is attained at a positive definite matrix on the boundary of $\mathcal{CP}_n$.*

Proof. Let $A \in \text{int}\mathcal{CP}_n$ have $\text{cpr}\, A = p_n$. Then A is positive definite and positive. Let $\rho = \lambda_1 > \lambda_2 \geq \cdots \geq \lambda_n$ be the eigenvalues of A, and $\{\mathbf{v}_1, \mathbf{v}_2, \ldots, \mathbf{v}_n\}$ an orthonormal basis of $\mathbb{R}^n$ consisting of the corresponding eigenvectors, with $\mathbf{v}_1 = \mathbf{v} > \mathbf{0}$ the Perron vector. There exists $\delta > 0$ such that $A' = A - \delta \mathbf{v}\mathbf{v}^T$ is on the boundary of $\mathcal{CP}_n$. Then $\mathbf{v}_1, \mathbf{v}_2, \ldots, \mathbf{v}_n$ are eigenvectors of A' corresponding to the eigenvalues $\rho - \delta, \lambda_2, \ldots, \lambda_n$ of A'. Since $A' \in \mathcal{CP}_n$ is nonnegative, the positive eigenvector $\mathbf{v}_1 = \mathbf{v}$ is necessarily a Perron vector of A', hence $\rho - \delta \geq \lambda_2 > 0$. Let

$$A' = \sum_{i=1}^{p} \mathbf{b}_i \mathbf{b}_i^T$$

be a minimal CP representation of A', $p = \text{cpr}\, A' \leq p_n$. Then

$$A'\mathbf{v} = (\rho - \delta)\mathbf{v},$$

and also

$$A'\mathbf{v} = \sum_{i=1}^{p} (\mathbf{b}_i^T \mathbf{v})\mathbf{b}_i.$$

That is, $\mathbf{v}$ is a nonnegative combination of $\mathbf{b}_1, \ldots, \mathbf{b}_p$. By Proposition 4.28, as $A = \sum_{i=1}^{p} \mathbf{b}_i \mathbf{b}_i^T + \delta \mathbf{v}\mathbf{v}^T$ is a CP representation of A, this implies that $p_n = \text{cpr}\, A \leq p$. Hence $\text{cpr}\, A' = p_n$. □

Corollary 4.37. *For every $1 \leq p \leq p_n$, there exists a matrix on the boundary of $\mathcal{CP}_n$, which has cp-rank p.*

Proof. Let A be a matrix on the boundary of $\mathcal{CP}_n$ with $\text{cpr}\, A = p_n$. Then A is orthogonal to some copositive matrix M. Let $A = \sum_{i=1}^{p_n} \mathbf{b}_i \mathbf{b}_i^T$ be a minimal CP representation of A. As A is orthogonal to M, so is each $\mathbf{b}_i \mathbf{b}_i^T$. For $1 \leq p \leq p_n$ let

$$A_p = \sum_{i=1}^{p} \mathbf{b}_i \mathbf{b}_i^T.$$

The matrix A_p has cp-rank p (see Exercise 4.3), and it is orthogonal to M, since each $\mathbf{b}_i \mathbf{b}_i^T$ is (see also Exercise 3.15). Hence this is a matrix on the boundary of $\mathcal{CP}_n$. □

The next proposition is proved similarly, and we leave the proof as an exercise.

Proposition 4.38. *There exists a matrix in the interior of $\mathcal{CP}_n$ with cp-rank equal to p if and only if $n \leq p \leq p_n$.*

CP$^+$ factorizations and p_n^+

Recall that every matrix in the interior has a CP factorization BB^T with B positive. Some matrices on the boundary of $\mathcal{CP}_n$ also have a factorization BB^T with a positive B (for example, $\mathbf{x}\mathbf{x}^T$, where $\mathbf{x} \in \mathbb{R}^n$ is a positive vector). We now consider the question: Do such matrices have a *minimal* CP factorization with B positive?

If a matrix A has a factorization $A = BB^T$ with $B > 0$, we say that A has a *CP$^+$ factorization.* The corresponding CP representation $A = \sum_{i=1}^{p} \mathbf{b}_i\mathbf{b}_i^T$ with $\mathbf{b}_i > \mathbf{0}$ for every i, is called a *CP$^+$ representation.* By Theorem 3.29, a nonsingular matrix has a CP$^+$ factorization if and only if it is in the interior of $\mathcal{CP}_n$; a completely positive matrix on the boundary that has a CP$^+$ factorization is necessarily positive and singular.

For A with a CP$^+$ factorization we define the *cp$^+$-rank* to be the minimum number of columns in a positive B such that $A = BB^T$, and denote it by $\mathrm{cp^+r}\, A$. A factorization in which B has $\mathrm{cp^+r}\, A$ columns is a *minimal CP$^+$ factorization.* Alternatively, $\mathrm{cp^+r}\, A$ is the minimum number of summands in a CP$^+$ representation $A = \sum_{i=1}^{n} \mathbf{b}_i\mathbf{b}_i^T$. We call such a representation a *minimal CP$^+$ representation.* Using Proposition 3.9 repeatedly, one sees that $\mathrm{cp^+r}\, A$ is also the minimum number of summands in a CP representation $A = \sum_{i=1}^{n} \mathbf{b}_i\mathbf{b}_i^T$ such that $\mathbf{b}_i > \mathbf{0}$ for at least one i.

Proposition 4.39. *Let $A \in \mathrm{int}\mathcal{CP}_n$ such that $\mathrm{cp^+r}\, A = p$. Then there exists a neighbourhood of A such that for every X in this neighbourhood $\mathrm{cp^+r}\, X \le p$.*

Proof. First consider the case $p = n$. Let $A = BB^T$, where B is an $n \times n$ positive nonsingular matrix. If S is any symmetric matrix of norm 1 and $\varepsilon > 0$, then

$$A + \varepsilon S = B(I + \varepsilon B^{-1}SB^{-T})B^T.$$

For small enough ε, $\mathrm{rank}\,(A + \varepsilon S) = n$. In the Cholesky decomposition $I + \varepsilon B^{-1}SB^{-T} = Q(\varepsilon)Q(\varepsilon)^T$, the matrix $Q(\varepsilon)$ depends continuously on ε, and $Q(0) = I$. As $B > 0$, we get that $BQ(\varepsilon) > 0$ for $\varepsilon > 0$ sufficiently small. For such ε,

$$A + \varepsilon S = (BQ(\varepsilon))(BQ(\varepsilon))^T$$

is a CP$^+$ factorization of $A + \varepsilon S$, hence $\mathrm{cp^+r}\,(A + \varepsilon S) \le n$. Since

$$\mathrm{cp^+r}\,(A + \varepsilon S) \ge \mathrm{rank}\,(A + \varepsilon S) = n,$$

we get that $\mathrm{cp^+r}\,(A + \varepsilon S) = n$.

Now if $\operatorname{cp^+r} A = p > n$, then $A = B_1B_1^T + B_2B_2^T$, where B_1 is an $n \times n$ positive nonsingular matrix, and B_2 is an $n \times (p-n)$ positive matrix. For every symmetric S of norm 1 and sufficiently small $\varepsilon > 0$ we get from the first part of the proof that $B_1B_1^T + \varepsilon S = C(\varepsilon)C(\varepsilon)^T$, where $C(\varepsilon)$ is a positive $n \times n$ matrix. Hence

$$A + \varepsilon S = C(\varepsilon)C(\varepsilon)^T + B_2B_2^T,$$

and $\operatorname{cp^+r}(A + \varepsilon S) \leq p$. □

Combining the last proposition with Corollary 4.34 we get the following.

Corollary 4.40. *If $A \in \operatorname{int}\mathcal{CP}_n$ has $\operatorname{cp^+r} A = \operatorname{cpr} A = p$, then there exists a neighbourhood of A in which every matrix X has*

$$\operatorname{cp^+r} X = \operatorname{cpr} X = p.$$

Corollary 4.41. *The set $\{A \in \operatorname{int}\mathcal{CP}_n \mid \operatorname{cp^+r} A = \operatorname{cpr} A\}$ is open.*

In the case that A in the interior of $\mathcal{CP}_n$ has $\operatorname{cpr} A < \operatorname{cp^+r} A$, the next proposition guarantees that every neighbourhood of A includes matrices with both cp-rank and cp$^+$-rank equal to $\operatorname{cpr} A$, and other matrices with both cp-rank and cp$^+$-rank equal to $\operatorname{cp^+r} A$.

Proposition 4.42. *Let $A \in \operatorname{int}\mathcal{CP}_n$ have $\operatorname{cpr} A = p$ and $\operatorname{cp^+r} A = q$, $p \leq q$. Let $\mathbf{v}$ be the unit Perron vector of A, and let*

$$A(t) = A + t\mathbf{v}\mathbf{v}^T.$$

Then there exists $\varepsilon_0 > 0$ such that for every $0 < \varepsilon < \varepsilon_0$,

$$\operatorname{cpr} A(\varepsilon) = \operatorname{cp^+r} A(\varepsilon) = p$$

and

$$\operatorname{cpr} A(-\varepsilon) = \operatorname{cp^+r} A(-\varepsilon) = q.$$

Proof. Clearly $A(\varepsilon) \in \operatorname{int}\mathcal{CP}_n$ for every $\varepsilon > 0$, and for small $\varepsilon > 0$ also $A(-\varepsilon) \in \operatorname{int}\mathcal{CP}_n$. By an argument similar to the proof of Theorem 4.36, for every $\varepsilon > 0$

$$\operatorname{cp^+r} A(\varepsilon) \leq \operatorname{cpr} A = p. \tag{4.27}$$

And as $\mathbf{v}$ is the unit Perron vector of $A(-\varepsilon)$ and $A = A(-\varepsilon) + \varepsilon\mathbf{v}\mathbf{v}^T$, we also have

$$q = \operatorname{cp^+r} A \leq \operatorname{cpr} A(-\varepsilon). \tag{4.28}$$

Combine (4.27) with $\operatorname{cpr} A(\varepsilon) \leq \operatorname{cp^+r} A(\varepsilon)$ to get that

$$\operatorname{cpr} A(\varepsilon) \leq \operatorname{cpr} A = p.$$

By Corollary 4.34, for sufficiently small $\varepsilon > 0$,

$$p = \operatorname{cpr} A \leq \operatorname{cpr} A(\varepsilon),$$

and therefore for such ε

$$\operatorname{cpr} A(\varepsilon) = \operatorname{cp^+r} A(\varepsilon) = p.$$

Similarly, from (4.28) and $\operatorname{cpr} A(-\varepsilon) \leq \operatorname{cp^+r} A(-\varepsilon)$ we get that

$$q \leq \operatorname{cp^+r} A(-\varepsilon).$$

By Proposition 4.39, for sufficiently small $\varepsilon > 0$

$$\operatorname{cp^+r} A(-\varepsilon) \leq \operatorname{cp^+r} A = q,$$

which completes the proof that

$$\operatorname{cpr} A(-\varepsilon) = \operatorname{cp^+r} A(-\varepsilon) = q. \qquad \square$$

Proposition 4.42 implies the following.

Corollary 4.43. *The set* $\{A \in \operatorname{int}\mathcal{CP}_n \mid \operatorname{cpr} A = \operatorname{cp^+r} A\}$ *is dense in* $\operatorname{int}\mathcal{CP}_n$.

Remark 4.44. By Corollaries 4.40 and 4.43 the set

$$\mathcal{A} = \{A \in \operatorname{int}\mathcal{CP}_n \mid \operatorname{cpr} A = \operatorname{cp^+r} A\}$$

is open and dense in $\operatorname{int}\mathcal{CP}_n$. In fact, the set $(\operatorname{int}\mathcal{CP}_n) \setminus \mathcal{A}$ also has a zero Lebesque measure, hence membership in the open set $\mathcal{A}$ is *generic*. We will not go into that in this book. For details, see the reference in the notes at the end of this section.

Now to another corollary of Proposition 4.42.

Corollary 4.45. *Let* $1 \leq p \leq p_n$. *Then there exists* $A \in \mathcal{CP}_n$ *such that* $\operatorname{cp^+r} A = \operatorname{cpr} A = p$ *(on the boundary if* $1 \leq p \leq n-1$*, in the interior if* $n \leq p \leq p_n$*).*

Proof. By Proposition 4.42, there exists a matrix A in $\operatorname{int}\mathcal{CP}_n$ such that $\operatorname{cpr} A = \operatorname{cp^+r} A = p_n$. Let

$$A = \sum_{i=1}^{p_n} \mathbf{b}_i \mathbf{b}_i^T$$

be a minimal CP$^+$ representation of A, with $\mathbf{b}_1, \ldots, \mathbf{b}_n$ linearly independent. Then for $1 \leq p \leq p_n$ the matrix $A_p = \sum_{i=1}^{p} \mathbf{b}_i \mathbf{b}_i^T$ has $\operatorname{cp^+r} A_p = \operatorname{cpr} A_p = p$. $\square$

Remark 4.46. In fact, not only there exists a matrix $A \in \text{int}\mathcal{CP}_n$ with $\text{cpr}\, A = \text{cp}^+\text{r}\, A = p_n$, but every A in the interior of $\mathcal{CP}_n$ that has $\text{cpr}\, A = p_n$ also has $\text{cp}^+\text{r}\, A = p_n$. This follows from (4.28) in the proof of Proposition 4.42: For such A we get that for some $\varepsilon > 0$,

$$p_n = \text{cpr}\, A \le \text{cp}^+\text{r}\, A \le \text{cpr}\, A(-\varepsilon) \le p_n.$$

Still, not every matrix in $\text{int}\mathcal{CP}_n$ has cp-rank equal to the cp$^+$-rank.

Proposition 4.47. *Let $n + 1 \le p \le p_n$. Then there exists $A \in \text{int}\mathcal{CP}_n$ such that* $\text{cp}^+\text{r}\, A = p$ *and* $\text{cpr}\, A = p - 1$.

Proof. Let $A \in \text{int}\mathcal{CP}_n$ have $\text{cpr}\, A = \text{cp}^+\text{r}\, A = p_n$, and let

$$A = \sum_{i=1}^{p_n} \mathbf{b}_i \mathbf{b}_i^T$$

be a minimal CP$^+$ representation of A, with $\{\mathbf{b}_1, \ldots, \mathbf{b}_n\}$ linearly independent. Let $n + 1 \le p \le p_n$. For every $0 \le \theta \le 1$ let

$$A_\theta = \sum_{i=1}^{p-1} \mathbf{b}_i \mathbf{b}_i^T + \theta \mathbf{b}_p \mathbf{b}_p^T.$$

Clearly, $\text{cp}^+\text{r}\, A_\theta \le p$.

Let $0 < \theta \le 1$. Then $A_\theta \in \text{int}\mathcal{CP}_n$ and

$$A = A_\theta + (1 - \theta)\mathbf{b}_p \mathbf{b}_p^T + \sum_{i=p+1}^{p_n} \mathbf{b}_i \mathbf{b}_i^T$$

implies that

$$p_n = \text{cpr}\, A \le \text{cpr}\, A_\theta + 1 + p_n - p,$$

hence $\text{cpr}\, A_\theta \ge p - 1$. Therefore,

$$p - 1 \le \text{cpr}\, A_\theta \le \text{cp}^+\text{r}\, A_\theta \le p. \tag{4.29}$$

Let

$$\theta_p = \sup_{0 \le \theta \le 1} \{\theta \mid \text{cpr}\, A_\theta = p - 1\}.$$

As $\text{cpr}\, A_0 = \text{cp}^+\text{r}\, A_0 = p - 1$ and $\text{cpr}\, A_1 = \text{cp}^+\text{r}\, A_1 = p$, by Corollary 4.40 we have $0 < \theta_p < 1$. For $\theta_p < \theta \le 1$ we have $\text{cpr}\, A_\theta = \text{cp}^+\text{r}\, A_\theta = p$. Hence, by Proposition 4.39, $\text{cp}^+\text{r}\, A_{\theta_p} \ge p$. Combined with (4.29) this implies that $\text{cp}^+\text{r}\, A_{\theta_p} = p$. By the definition of θ_p and Proposition 4.33, $\text{cpr}\, A_{\theta_p} = p - 1$. □

We define

$$p_n^+ = \max\left\{\mathrm{cp}^+\mathrm{r}\, A \mid A \in \mathcal{CP}_n\right\}.$$

Note that by Proposition 4.42, the maximum cp$^+$-rank on int$\mathcal{CP}_n$ is equal to p_n. However, the following proposition states all we can say with certainty on the relation between p_n^+ and p_n. The proof is left as an exercise.

Proposition 4.48.

(a) $p_n \le p_n^+ \le p_n + 1$.
(b) *For* $n \le 4$, $p_n^+ = p_n = n$.

It is not known whether $p_n^+ = p_n$ or $p_n^+ = p_n + 1$.

Exercises

4.19. Prove Proposition 4.38.

4.20. Show that for every n, $p_n < p_{n+1}$.
Hint: Consider $A \oplus 1$ for $A \in \mathcal{CP}_n$ with $\mathrm{cpr}\, A = p_n$.

4.21. Prove that if A and B are $n \times n$ completely positive matrices, and A has a CP$^+$ factorization, then $A + B$ has a CP$^+$ factorization, and

$$\mathrm{cp}^+\mathrm{r}\,(A + B) \le \mathrm{cp}^+\mathrm{r}\, A + \mathrm{cpr}\, B.$$

4.22. Prove Corollaries 4.40 and 4.41.

4.23. Complete the proof of Proposition 4.42. Show that if $A \in \mathrm{int}\mathcal{CP}_n$, and $\mathbf{v}$ is its unit Perron vector, then (4.27) holds. Deduce that (4.28) holds for sufficiently small $\varepsilon > 0$.

4.24. Prove Proposition 4.48.
Hint: For (a): Let A have $\mathrm{cp}^+\mathrm{r}\, A = p_n^+$. For some positive $\mathbf{v}$, $A - \mathbf{v}\mathbf{v}^T \in \mathcal{CP}_n$. For (b): for $n \le 4$, and a positive $A \in \mathcal{CP}_n$ with a unit Perron vector $\mathbf{v}$, $A - \varepsilon\mathbf{v}\mathbf{v}^T$ is positive and in $\mathcal{CP}_n$ for sufficiently small ε.

Notes. The results in this section concerning the cp-ranks and p_n are based on [Shaked-Monderer *et al.* (2013)]. The results concerning cp$^+$-rank, p_n^+ were proved in [Bomze *et al.* (2015a)]. In the latter, the concept of genericity is explained, and genericity of the set $\{A \in \mathcal{S}_n \mid \mathrm{cpr}\, A = \mathrm{cp}^+\mathrm{r}\, A\}$ is proved, where $\mathrm{cpr}\, A$ is considered to be ∞ if $A \in \mathcal{S}_n$ is not completely positive, and $\mathrm{cp}^+\mathrm{r}\, A$ is considered to be ∞ if $A \in \mathcal{S}_n$ does not have a CP$^+$ factorization.

For the search of an explicit example of a completely positive A where $\mathrm{cp^+r}\, A > \mathrm{cpr}\, A$ the results of [Shader *et al.* (2014)] may be useful.

Still open: *It is not known whether* $p_n^+ = p_n$ *or* $p_n^+ = p_n + 1$. *An explicit example of a completely positive* A *with* $\mathrm{cp^+r}\, A > \mathrm{cpr}\, A$ *is also not yet known.*

4.4 Completely positive matrices of a given order, with a given graph

In this section we begin to consider problem (B) of Section 4.1, the problem of finding a tight upper bound on the cp-ranks of completely positive matrices of order n. By Theorem 4.23, if A is an $n \times n$ completely positive matrix, then

$$\mathrm{cpr}\, A \leq \frac{n(n+1)}{2} - 1. \tag{4.30}$$

For $n = 3$ this bound is 5, and for $n = 4$ this bound is 9. In both cases we have a better bound in Theorem 4.10, $\mathrm{cpr}\, A \leq n$. So is there a better upper bound then (4.30) on the cp-ranks of $n \times n$ matrices? And if there is, what is the best bound?

We start searching for this bound "under the lamppost", by considering the cp-ranks of $n \times n$ matrices whose complete positivity may be deduced based on their graph. It turns out that for such matrices the cp-rank is at most $n^2/4$. This led Drew, Johnson and Loewy to conjecture the following.

Conjecture 4.49. [The DJL Conjecture] *If* A *is an* $n \times n$ *completely positive matrix,* $n \geq 4$, *then*

$$\mathrm{cpr}\, A \leq \frac{n^2}{4}.$$

That is, $p_n \leq \frac{n^2}{4}$.

The DJL conjecture may be rephrased in terms of graphs as follows. For each graph G on $n \geq 4$ vertices,

$$\mathrm{cpr}\, G \leq \frac{n^2}{4}.$$

We now know that the conjecture is wrong, but it remained open for about twenty years, before examples were found that disproved it, and these will be given in the next section. In this section we consider matrices for which the DJL bound does hold. Mainly, matrices with certain graphs, but also matrices with a positive semidefinite comparison matrix.

We start with the cp-ranks of irreducible acyclic matrices.

Theorem 4.50. *If A is a completely positive matrix, and $G(A)$ is a tree, then* $\operatorname{cpr} A = \operatorname{rank} A$.

Proof. If $G(A)$ has 2 vertices, this is already known by Theorem 4.7. Suppose the theorem holds for every matrix whose graph is a tree on $n-1$ vertices. If $G(A)$ is a tree on n vertices, we may assume that 1 is a vertex of degree 1 in $G(A)$. Then

$$A = \begin{pmatrix} a_{11} & a_{12} & 0 & \cdots & 0 \\ a_{12} & a_{22} & \cdots & \cdots & a_{2n} \\ 0 & \vdots & & & \vdots \\ \vdots & \vdots & & & \vdots \\ 0 & a_{2n} & \cdots & \cdots & a_{nn} \end{pmatrix},$$

and we may write

$$A = \left(\begin{pmatrix} a_{11} & a_{12} \\ a_{12} & a_{12}^2/a_{11} \end{pmatrix} \oplus 0_{n-2} \right) + (0 \oplus \hat{A}),$$

where $\hat{A}$ is the Schur complement $A/[1]$. The matrix $\hat{A}$ is an $(n-1)\times(n-1)$ completely positive matrix whose graph is a tree, and $\operatorname{rank}\hat{A} = \operatorname{rank} A - 1$. By induction $\operatorname{cpr}\hat{A} = \operatorname{rank}\hat{A}$. Thus,

$$\operatorname{rank} A \le \operatorname{cpr} A \le 1 + \operatorname{cpr}\hat{A} = 1 + \operatorname{rank}\hat{A} = \operatorname{rank} A,$$

so $\operatorname{cpr} A = \operatorname{rank} A$. □

Since every graph has a nonsingular CP matrix realization, Theorem 4.50 implies the following.

Corollary 4.51. *If G is a tree on n vertices, then* $\operatorname{cpr} G = n$.

Trees are not the only graphs with the property that every CP matrix realization of the graph has cp-rank equal to the rank. The question which graphs have this property will be discussed in Section 4.6. For now, here is one such example.

Example 4.52. If

$$A = \begin{pmatrix} a_{11} & a_{12} & 0 & 0 \\ a_{12} & a_{22} & a_{23} & a_{24} \\ 0 & a_{23} & a_{33} & a_{34} \\ 0 & a_{24} & a_{34} & a_{44} \end{pmatrix}$$

is completely positive and $\operatorname{rank} A = 3$, then $\operatorname{cpr} A = \operatorname{rank} A = 3$. This implies that the graph G consisting of two blocks, a triangle and an edge, has the property that every completely positive matrix A with $G(A) = G$ satisfies $\operatorname{cpr} A = \operatorname{rank} A$. We leave the details of the proof as an exercise.

Next we consider CP matrix realizations of triangle free graphs.

Theorem 4.53. *If A is an $n \times n$ completely positive matrix and $G(A)$ is triangle free, then*

$$\operatorname{cpr} A \leq \max\{n, |E(G(A))|\}.$$

Proof. It suffices to prove the theorem for irreducible matrices.

Consider first the case that the irreducible A is singular. By Theorem 3.43, $M(A)$ is positive semidefinite and, as in the proof of Theorem 3.40, this implies that for some positive diagonal D the matrix DAD is diagonally dominant. Hence we may assume that A itself is diagonally dominant, and since it is singular, this means that

$$a_{ii} = \sum_{\substack{j=1 \\ j \neq i}}^{n} a_{ij} \quad \text{for } i = 1, \ldots, n.$$

Let $B_{ij} = a_{ij}(\mathbf{e}_i + \mathbf{e}_j)(\mathbf{e}_i + \mathbf{e}_j)^T$, $1 \leq i < j \leq n$. Then $A = \sum_{1 \leq i < j \leq n} B_{ij}$ is a CP representation of A with $|E(G(A))|$ nonzero summands, hence

$$\operatorname{cpr} A \leq |E(G(A))| \leq \max\{n, |E(G(A))|\}.$$

Next we consider the case that A is nonsingular and the connected graph $G(A)$ is not a tree. By applying a positive diagonal congruence we may assume that A is strictly diagonally dominant (see Theorem 1.66 and Remark 1.67). Since $G(A)$ is connected and not a tree, there exists an edge which may be removed without disconnecting $G(A)$. Using a permutation similarity we may assume that the entry corresponding to this edge is a_{12}. For each $x \geq 1$ let $\mathbf{b}(x)^T = \sqrt{a_{12}}(x, 1/x, 0, \ldots, 0)$, and $B(x) = \mathbf{b}(x)\mathbf{b}(x)^T$. The matrix $M(A - B(1))$ is strictly diagonally dominant, and hence nonsingular. The matrix $M(A - B(\sqrt{a_{11}/a_{12}}))$ is not positive semidefinite. Let x_0 be the smallest $x \geq 1$ such that $M(A - B(x))$ is singular. The matrix $\hat{A} = A - B(x_0)$ is a singular completely positive matrix whose graph is triangle free and connected. By the first part of the proof, $\operatorname{cpr} \hat{A} \leq |E(G(\hat{A}))| = |E(G(A))| - 1$. Since $A = \hat{A} + B(x_0)$, we have

$$\operatorname{cpr} A \leq |E(G(A))|.$$

Note that in this case $|E(G(A))| = \max\{n, |E(G(A))|\}$.

Finally, consider the case that A is nonsingular and $G(A)$ is a tree. In that case $\max\{n, |E(G(A))|\} = n = |E(G(A))| + 1$, and by Theorem 4.50, $\operatorname{cpr} A = \operatorname{rank} A = n$. □

Remark 4.54. The maximal cliques in a triangle free graph G are the edges. Hence, by Remark 4.15, $\operatorname{cpr} A \geq |E(G)|$ for every CP matrix realization of G. This, combined with the proof of Theorem 4.53, implies the next corollary.

Corollary 4.55. *If G is a connected triangle free graph which is not a tree, then* $\operatorname{cpr} G = |E(G)|$.

In the coming results we need the following lemma.

Lemma 4.56. *If v is a vertex of degree* 1 *or* 2 *in a graph G, then*

$$\operatorname{cpr} G \leq d(v) + \operatorname{cpr}(G - v).$$

Proof. Suppose that G is a graph on vertices $\{1, \ldots, n\}$ and $v = 1$. We consider first the case that $d(v) = 2$, and assume that 2 and 3 are the vertices adjacent to $v = 1$.

Let A be a CP matrix realization of G, and let $A = \sum_{i=1}^{m} \mathbf{b}_i \mathbf{b}_i^T$ be any CP representation of A. Denote

$$\Omega_1 = \{1 \leq i \leq m \mid 1 \in \operatorname{supp} \mathbf{b}_i\}, \qquad \Omega_2 = \{1, \ldots, m\} \setminus \Omega_1$$
$$B = \sum_{i \in \Omega_1} \mathbf{b}_i \mathbf{b}_i^T, \qquad C = \sum_{i \in \Omega_2} \mathbf{b}_i \mathbf{b}_i^T.$$

Then $A = B + C$, $B = B' \oplus 0_{n-3}$ and $C = 0_1 \oplus C'$ are both completely positive, and $G(C') \subseteq G - v$. We may assume that B' is singular. Otherwise, let δ_0 be the maximal δ for which $B - \delta E_{33}$ is positive semidefinite. The matrix $B_1 = B - \delta_0 E_{33}$ is a completely positive matrix, and it is zero except for its 3×3 leading principal submatrix, which is singular. The matrix $C_1 = C + \delta_0 E_{33}$ is obviously also completely positive, and has the same graph as C. Hence we may replace B and C by B_1 and C_1, respectively. Since the 3×3 matrix B' is singular,

$$\operatorname{cpr} A \leq \operatorname{cpr} B' + \operatorname{cpr} C' \leq 2 + \operatorname{cpr} C' \leq 2 + \operatorname{cpr}(G - v).$$

This completes the proof in the case that $d(v) = 2$.

Suppose then that $d(v) = 1$ and $v = 1$ is adjacent to 2. Let A be a CP realization of G. By the same argument as in the proof of Theorem 4.50,

$$A = \left(\begin{pmatrix} a_{11} & a_{12} \\ a_{12} & a_{12}^2/a_{11} \end{pmatrix} \oplus 0_{n-2} \right) + (0 \oplus \hat{A}),$$

where $\hat{A}$ a completely positive matrix whose graph is $G - v$. The desired inequality easily follows. □

Theorem 4.57. *For every* $n \geq 4$, $\operatorname{cpr} T_n = 2n - 4$, *and there exists a CP matrix realization* A *of* T_n *such that* $\operatorname{cpr} A = k$ *if and only if* $n - 2 \leq k \leq 2n - 4$.

Proof. By Example 4.9 and Theorem 4.10, $\operatorname{cpr} T_4 = 4$. The graph T_n has $n-2$ vertices of degree 2. Using Lemma 4.56 successively $n - 4$ times, we find that for $n \geq 4$

$$\operatorname{cpr} T_n \leq \operatorname{cpr} T_4 + 2(n-4) = 4 + 2(n-4) = 2n - 4. \tag{4.31}$$

This implies that $\operatorname{cpr} A \leq 2n - 4$ for every CP matrix realization A of T_n. The inequality $n - 2 \leq \operatorname{cpr} A$ holds for every such A since $\operatorname{cc}(T_n) = n - 2$.

We need to show that for every $n-2 \leq k \leq 2n-4$ there exists a matrix A such that $G(A) = T_n$ and $\operatorname{cpr} A = k$. This will also imply that equality holds in (4.31). So let

$$A_{n,0} = \begin{pmatrix} (n-2)J_2 & J_{2\times(n-2)} \\ J^T_{2\times(n-2)} & I_{n-2} \end{pmatrix} = \begin{pmatrix} n-2 & n-2 & 1 & 1 & \dots & 1 \\ n-2 & n-2 & 1 & 1 & \dots & 1 \\ 1 & 1 & 1 & 0 & \dots & 0 \\ 1 & 1 & 0 & 1 & \ddots & \vdots \\ \vdots & \vdots & \vdots & \ddots & \ddots & 0 \\ 1 & 1 & 0 & \dots & 0 & 1 \end{pmatrix}.$$

Then $A_{n,0} = BB^T$, where B is the $n \times (n-2)$ matrix whose j-th column is zero, except for 1 in positions $1, 2, j+2$,

$$B = \begin{pmatrix} 1 & 1 & 1 & \dots & 1 \\ 1 & 1 & 1 & \dots & 1 \\ 1 & 0 & 0 & \dots & 0 \\ 0 & 1 & 0 & & \vdots \\ \vdots & 0 & 1 & \ddots & \\ \vdots & & \ddots & \ddots & 0 \\ 0 & \dots & & 0 & 1 \end{pmatrix}.$$

Hence A is completely positive and $\operatorname{cpr} A_{n,0} = n - 2$. By Lemma 3.36, for every $0 < t < n - 2$ the matrix $A_{n,t} = A_{n,0} + \left(\begin{pmatrix} t & -t \\ -t & t \end{pmatrix} \oplus 0_{n-2} \right)$ is also completely positive. We leave it as an exercise to show that for $r = 1, \dots, n-2$ and $r - 1 < t < r$ the cp-rank of $A_{n,t}$ is $r + (n-2)$ (Exercise 4.26). □

Looking for an upper bound on the cp-ranks of the graphs discussed so far, we get the following.

Corollary 4.58. *If G is a graph on $n \geq 4$ vertices which is either triangle free or T_n, then*

$$\operatorname{cpr} G \leq \frac{n^2}{4}.$$

Proof. If G is triangle free, then $|E(G)| \leq n^2/4$ by Theorem 1.88, and since $n \leq n^2/4$ for $n \geq 4$, the inequality follows from Theorem 4.53.

Since $2n - 4 \leq n^2/4$ for every n, we also have $\operatorname{cpr} T_n \leq n^2/4$. □

To prove that the same bound holds for all completely positive graphs on n vertices, we need the following lemma.

Lemma 4.59. *Let $G = G_1 \cup G_2$, where G_1 is a completely positive graph on vertices $\{1, \ldots, k\}$, and G_2 is a graph on vertices $\{k, \ldots, n\}$. If A is a completely positive matrix with $G(A) = G$, then*

$$A = (A_1 \oplus 0_{n-k}) + (0_{k-1} \oplus A_2),$$

where A_1 and A_2 are completely positive, $G(A_1) = G_1$, $G(A_2) = G_2$, A_1 is singular, and $\operatorname{rank} A = \operatorname{rank} A_1 + \operatorname{rank} A_2$.

Proof. Let $A = \sum_{i=1}^m \mathbf{b}_i \mathbf{b}_i^T$ be any CP representation of A, and denote

$$\Omega_1 = \{1 \leq i \leq m \mid \operatorname{supp} \mathbf{b}_i \subseteq \{1, \ldots, k\}\}, \qquad \Omega_2 = \{1, \ldots, m\} \setminus \Omega_1$$
$$B = \sum_{i \in \Omega_1} \mathbf{b}_i \mathbf{b}_i^T, \qquad C = \sum_{i \in \Omega_2} \mathbf{b}_i \mathbf{b}_i^T.$$

Then $B = B' \oplus 0_{n-k}$, $C = 0_{k-1} \oplus C'$, B' and C' are completely positive, $G(B') = G_1$ and $G(C') = G_2$. If $\mathbf{e}_k \notin \operatorname{cs}(B)$, let $A_1 = B'$ and $A_2 = C'$. If $\mathbf{e}_k \in \operatorname{cs}(B)$, let δ_0 be the maximal $\delta > 0$ for which $B - \delta \mathbf{e}_k \mathbf{e}_k^T$ is positive semidefinite. Then $\mathbf{e}_k \notin \operatorname{cs}(B - \delta_0 \mathbf{e}_k \mathbf{e}_k^T)$ (see Exercise 1.22). $B - \delta_0 \mathbf{e}_k \mathbf{e}_k^T = A_1 \oplus 0_{n-k}$ is doubly nonnegative, and since $G(A_1) = G_1$ is completely positive graph, A_1 is completely positive. Clearly the matrix $C + \delta_0 \mathbf{e}_k \mathbf{e}_k^T = 0_{k-1} \oplus A_2$ is also completely positive, and

$$A = (A_1 \oplus 0_{n-k}) + (0_{k-1} \oplus A_2).$$

Since $\mathbf{e}_k \notin \operatorname{cs}(A_1 \oplus 0_{n-k}) \cap \operatorname{cs}(0_{k-1} \oplus A_2)$, $\operatorname{rank} A = \operatorname{rank} A_1 + \operatorname{rank} A_2$. □

We can now prove that the cp-rank of any completely positive graph is bounded by $n^2/4$.

Theorem 4.60. *If G is a completely positive graph on n vertices, then for every CP matrix realization A of G,*

$$\operatorname{cpr} A \leq \max\{|E(G)|, \operatorname{rank} A\}.$$

If $n \geq 4$, then also

$$\operatorname{cpr} G \leq \frac{n^2}{4}.$$

Proof. If $n \leq 3$, the first claim holds by Theorem 4.8. If G is tree on $n \geq 4$ vertices and A is a completely positive matrix with $G(A) = G$, then by Theorem 4.50

$$\operatorname{cpr} A = \operatorname{rank} A \leq \max\{|E(G)|, \operatorname{rank} A\} \leq n \leq \frac{n^2}{4}.$$

Hence both claims are true in this case.

So we need to prove the two claims for any completely positive graph G on $n \geq 4$ vertices which is not a tree. In this case $\max\{|E(G)|, \operatorname{rank} A\} = |E(G)|$. We therefore need to prove that if G is such a graph, then

$$\operatorname{cpr} G \leq |E(G)| \text{ and } \operatorname{cpr} G \leq n^2/4.$$

It suffices to consider connected graphs (Exercise 4.28). We prove both claims by induction on the number of blocks.

If G has no cut vertex, then G is a K_4, or a T_n, or it is bipartite. The two desired inequalities follow from Theorems 4.10 and 4.57, Corollary 4.55 and Corollary 4.58.

Let G be a connected completely positive graph which is not a tree and has k blocks, and suppose both claims hold for any such graph with less than k blocks. Then $G = G_1 \cup G_2$, where G_1 and G_2 are connected completely positive graphs, and $G_1 \cap G_2$ consists of a single vertex. Since G is not a tree, at least one of the graphs G_1 and G_2 is not a tree. If both are not trees, then for every $i = 1, 2$, either G_i is a triangle, in which case $\operatorname{cpr} G_i = 3 = |E(G_i)|$, or G_i has at least 4 vertices, in which case $\operatorname{cpr} G_i \leq |E(G_i)|$ by the induction hypothesis. In any case,

$$\operatorname{cpr} G \leq \operatorname{cpr} G_1 + \operatorname{cpr} G_2 \leq |E(G_1)| + |E(G_2)| = |E(G)|.$$

Now suppose one of these subgraphs, say G_1, is a tree. We may assume that the vertices of G_1 are $\{1, \ldots, k\}$. Let A be a completely positive matrix with $G(A) = G$. By Lemma 4.59, $A = (A_1 \oplus 0_{n-k}) + (0_{k-1} \oplus A_2)$, where A_1 and A_2 are completely positive, $G(A_1) = G_1$ and $G(A_2) = G_2$, and A_1 is singular. Hence

$$\operatorname{cpr} A \leq \operatorname{cpr} A_1 + \operatorname{cpr} A_2 = \operatorname{rank} A_1 + \operatorname{cpr} A_2.$$

Since A_1 is singular, $\operatorname{rank} A_1 \le |E(G_1)|$, and by the induction hypothesis (or the case $n = 3$), $\operatorname{cpr} A_2 \le |E(G_2)|$. Hence

$$\operatorname{cpr} A \le |E(G_1)| + |E(G_2)| = |E(G)|.$$

We now show that $\operatorname{cpr} G \le n^2/4$. Suppose $|V(G_i)| = n_i$, $i = 1, 2$. If both graphs have at least 4 vertices each, then

$$\operatorname{cpr} G \le \frac{n_1^2}{4} + \frac{n_2^2}{4} \le \frac{(n_1 + n_2 - 1)^2}{4} = \frac{n^2}{4} \tag{4.32}$$

by the induction hypothesis. If one of the graphs, say G_1, has less than 4 vertices, then by Lemma 4.59 and Theorem 4.8,

$$\operatorname{cpr} G \le (n_1 - 1) + \operatorname{cpr} G_2.$$

Since $n_1 - 1 \le n_1^2/4$, we get again (4.32) in the case that G_2 has 4 vertices or more. If G_2 also has less than 4 vertices, than $\operatorname{cpr} G_2 \le n_2$ and we get that $\operatorname{cpr} G \le (n_1 - 1) + n_2 = n \le n^2/4$. □

Recall that for every graph G,

$$\min\{\operatorname{cpr} A \mid A \text{ is a CP matrix realization of } G\} = \operatorname{cc}(G).$$

It is natural to ask whether every $\operatorname{cc}(G) \le k \le \operatorname{cpr} G$ is the cp-rank of some CP matrix realization of G. In the case that G is completely positive the answer is yes. In other words, the set of cp-ranks of CP matrix realizations of G has no gaps.

Theorem 4.61. *If G is a completely positive graph and*

$$\operatorname{cc}(G) \le k \le \operatorname{cpr} G,$$

then k is the cp-rank of some CP realization of G.

Observe that the theorem holds trivially for any graph G which is triangle free and not a tree, since $\operatorname{cc}(G) = \operatorname{cpr} G = |E(G)|$ for such graphs. The proof for general completely positive graphs is outlined in Exercises 4.29–4.33. It is not known whether the a similar holds for other types of graphs.

The final theorem in the section shows that the same bound of $n^2/4$ on the cp-rank holds also for completely positive matrices with a positive semidefinite comparison matrix, regardless of their graph. The proof relies on the following technical lemma.

Lemma 4.62. *Let A be an $n \times n$ diagonally dominant symmetric nonnegative matrix. Let r, s, t be distinct indices, $1 \le r, s, t \le n$. For $x > 0$ define $\mathbf{b}(x) \in \mathbb{R}^n$ by*

$$\mathbf{b}(x)_r = \sqrt{x}, \quad \mathbf{b}(x)_s = \frac{a_{rs}}{\sqrt{x}}, \quad \mathbf{b}(x)_t = \frac{a_{rt}}{\sqrt{x}}, \quad \mathbf{b}(x)_i = 0, \quad i \neq r, s, t.$$

If

$$a_{rs}, a_{rt} \le x \le a_{rr} \tag{4.33}$$

and

$$a_{st}x \ge a_{rs}a_{rt}, \tag{4.34}$$

then

(a) $A(x) = A - \mathbf{b}(x)\mathbf{b}(x)^T$ *is symmetric and nonnegative, and has diagonal dominance in all rows except possibly the r-th row.*

(b) *If in addition to* (4.33) *and* (4.34),

$$a_{rs}a_{rt} > 0, \tag{4.35}$$

then the diagonal dominance in rows s and t is strict.

(c) *If*

$$x = a_{rs} + a_{rt} \text{ and (4.34) holds,} \tag{4.36}$$

then there is also a diagonal dominance in the r-th row, that is, $A(x)$ is a diagonally dominant symmetric nonnegative matrix.

Proof. We may assume that $r = 1$, $s = 2$, $t = 3$. Then the matrix $A(x)$ coincides with A in rows $4, \ldots, n$, and

$$A(x)[1,2,3] = \begin{pmatrix} a_{11} - x & 0 & 0 \\ 0 & (a_{22}x - a_{12}^2)/x & (a_{23}x - a_{12}a_{13})/x \\ 0 & (a_{23}x - a_{12}a_{13})/x & (a_{33}x - a_{13}^2)/x \end{pmatrix}.$$

The nonnegativity of $A(x)$ in rows $4, \ldots, n$ and the diagonal dominance in these rows follow from the properties of A. The nonnegativity in rows $1, 2, 3$ is guaranteed by (4.33) and (4.34). If $i = 2$ or $i = 3$, diagonal dominance in row i is equivalent to

$$x\left(a_{ii} - \sum_{j \neq i} a_{ij}\right) + (x - a_{1i})a_{1i} + a_{12}a_{13} \ge 0,$$

and this inequality follows from (4.33) and the diagonal dominance of A. If (4.35) holds, the last inequality is actually a strict inequality and there is a strict diagonal dominance in rows $i = 2, 3$. Finally, if x is as in (4.36), then

$$(a_{11} - x) - \sum_{j=4}^{n} a_{1j} = a_{11} - \sum_{j=2}^{n} a_{1j} \geq 0,$$

and there is diagonal dominance in row 1 also. □

Theorem 4.63. *If A is an $n \times n$ doubly nonnegative matrix, $n \geq 4$, with a positive semidefinite comparison matrix, then*

$$\operatorname{cpr} A \leq \frac{n^2}{4}.$$

Moreover, A has a CP representation of support 3 with at most $\lfloor n^2/4 \rfloor$ summands.

Proof. We may assume that A is a diagonally dominant symmetric nonnegative matrix (see the proof of Theorem 3.40). We prove the theorem by induction on n.

First, the case $n = 4$. Let A be a diagonally dominant completely positive matrix. By Theorem 4.10, $\operatorname{cpr} A \leq 4$, so $A = \sum_{i=1}^{4} \mathbf{b}_i \mathbf{b}_i^T$, where $\mathbf{b}_i$ are nonnegative vectors. We want to show that there exists such a representation in which no vector is positive. If A has a zero entry this is clear. So we need only consider the case that A is positive. Without loss of generality we may assume that $a_{12} = \min a_{ij}$. Every $a_{12}, a_{13} \leq x \leq a_{11}$ satisfies (4.33), (4.34) and (4.35) (with $r = 1, s = 2, t = 3$). For such x, the matrix $A(x)$ is nonnegative, symmetric and has strict diagonal dominance in rows 2 and 3, and since $a_{14} > 0$, $A(x)[2, 3, 4]$ is strictly diagonally dominant. The matrix $A(a_{11})$ is not positive semidefinite, hence $\det(A(a_{11})) < 0$. On the other hand, by Lemma 4.62(c), $A(a_{12} + a_{13})$ is diagonally dominant. It is therefore positive semidefinite, and $\det(A(a_{12} + a_{13})) \geq 0$. Let x_0 be the smallest $x \geq a_{12} + a_{13}$ for which $\det(A(x)) = 0$. Then $A(x_0)$ is a rank 3 doubly nonnegative matrix, and $A(x_0)_{12} = A(x_0)_{13} = 0$. By Example 4.52, $\operatorname{cpr} A(x_0) = \operatorname{rank} A(x_0) = 3$. Because of the zero entries of $A(x_0)$, if $A(x_0) = \sum_{i=1}^{3} \mathbf{b}_i \mathbf{b}_i^T$ is a CP representation of $A(x_0)$, then $|\operatorname{supp}(\mathbf{b}_i)| \leq 3$ for each $1 \leq i \leq 3$. Hence $A = \sum_{i=1}^{3} \mathbf{b}_i \mathbf{b}_i^T + \mathbf{b}(x_0)\mathbf{b}(x_0)^T$ is a CP representation of A in which each of the four vectors has support of size at most 3.

Now let $n \geq 5$, and assume that the theorem is true for $4 \leq m < n$. Let A be an $n \times n$ diagonally dominant completely positive matrix. By continuity we may assume that A is positive. We will show that such A can be represented as

$$A = \mathbf{b}_1\mathbf{b}_1^T + \ldots + \mathbf{b}_k\mathbf{b}_k^T + \tilde{A},$$

where $k = \lfloor n/2 \rfloor$, the vectors $\mathbf{b}_i$ are all nonnegative with $|\mathrm{supp}\,(\mathbf{b}_i)| \leq 3$, and the matrix $\tilde{A}$ is a diagonally dominant completely positive matrix with a zero row. By the induction hypothesis, $\tilde{A}$ has a CP representation with $\lfloor (n-1)^2/4 \rfloor$ summands, each with support of size at most 3. This will imply that A has such a representation with $\lfloor (n-1)^2/4 \rfloor + \lfloor n/2 \rfloor = \lfloor n^2/4 \rfloor$ summands.

Consider two cases: n is even, or n is odd.

If $n = 2k$ is even, we may assume by permutation similarity that

$$\begin{aligned} \max_{1 \leq i \neq j \leq n} a_{ij} &= a_{n-1\,n} \\ \max_{1 \leq i \neq j \leq n-2} a_{ij} &= a_{n-3\,n-2} \\ &\vdots \\ \max_{1 \leq i \neq j \leq 4} a_{ij} &= a_{34}. \end{aligned}$$

Apply Lemma 4.62 successively $k-1$ times, each time with $x = a_{rs} + a_{rt}$. The first time apply it to A, with $r = 1$, $s = n-1$, $t = n$, $x_1 = a_{1\,n-1} + a_{1n}$. Let $\mathbf{b}_1 = \mathbf{b}(x_1)$ and $A_1 = A - \mathbf{b}_1\mathbf{b}_1^T$. By the lemma, A_1 is a diagonally dominant completely positive matrix. Now apply the lemma to A_1, with $r = 1$, $s = n-3$, $t = n-2$, $x_2 = a_{1\,n-3} + a_{1\,n-2}$. Let $\mathbf{b}_2 = \mathbf{b}(x_2)$, and $A_2 = A_1 - \mathbf{b}_2\mathbf{b}_2^T$. Continue until the lemma is applied to A_{k-2}, with $r = 1$, $s = 3$, $t = 4$ and $x_{k-1} = a_{13} + a_{14}$. Denote $\mathbf{b}_{k-1} = \mathbf{b}(x_{k-1})$, and $A_{k-1} = A - \sum_{i=1}^{k-1} \mathbf{b}_i\mathbf{b}_i^T$. Finally, denote $a = (A_{k-1})_{11}$, and $\mathbf{b}_k^T = (\sqrt{a}, a_{12}/\sqrt{a}, 0, \ldots, 0)$. The matrix $\tilde{A} = A_{k-1} - \mathbf{b}_k\mathbf{b}_k^T$ is a diagonally dominant completely positive matrix with a zero first row, and $A = \mathbf{b}_1\mathbf{b}_1^T + \ldots + \mathbf{b}_k\mathbf{b}_k^T + \tilde{A}$. We have $k = n/2$ and the vectors $\mathbf{b}_i$ are all nonnegative with $|\mathrm{supp}\,(\mathbf{b}_i)| \leq 3$.

If $n = 2k+1$ is odd, we may assume by permutation similarity that

$$\begin{aligned} \max_{1 \leq i \neq j \leq n} a_{ij} &= a_{n-1\,n} \\ \max_{1 \leq i \neq j \leq n-2} a_{ij} &= a_{n-3\,n-2} \\ &\vdots \\ \max_{1 \leq i \neq j \leq 5} a_{ij} &= a_{45}. \end{aligned}$$

The matrix

$$\begin{pmatrix} a_{11} - \sum_{j=4}^n a_{1j} & a_{12} & a_{13} \\ a_{12} & a_{22} - \sum_{j=4}^n a_{2j} & a_{23} \\ a_{13} & a_{23} & a_{33} \end{pmatrix}$$

is completely positive, so by Lemma 3.126 either

$$\left(a_{11} - \sum_{j=4}^{n} a_{1j}\right) a_{23} \geq a_{12}a_{13}, \tag{4.37}$$

or

$$\left(a_{22} - \sum_{j=4}^{n} a_{2j}\right) a_{13} \geq a_{12}a_{23}. \tag{4.38}$$

Without loss of generality we may assume that (4.37) holds. As in the previous case, define $\mathbf{b}_1, \ldots, \mathbf{b}_{k-1}$, $A_1, \ldots, A_{k-1}$, using repeatedly Lemma 4.62. Then

$$A_{k-1}[1,2,3] = \begin{pmatrix} a_{11} - \sum_{j=4}^{n} a_{1j} & a_{12} & a_{13} \\ a_{12} & a_{22} & a_{23} \\ a_{13} & a_{23} & a_{33} \end{pmatrix}.$$

By (4.37) we may use Lemma 4.62 once more, with $r = 1$, $s = 2$, $t = 3$ and $x = a_{11} - \sum_{j=4}^{n} a_{1j}$. Denote $\mathbf{b}_k = \mathbf{b}(x)$, and $\tilde{A} = A - \mathbf{b}_k\mathbf{b}_k^T$. By the lemma, $\tilde{A}$ is a symmetric nonnegative matrix that has diagonal dominance in rows 2 to n. Since its first row is zero, it is a diagonally dominant completely positive matrix. We have $A = \mathbf{b}_1\mathbf{b}_1^T + \ldots + \mathbf{b}_k\mathbf{b}_k^T + \tilde{A}$, $k = (n-1)/2$, and the vectors $\mathbf{b}_i$ are all nonnegative with $|\mathrm{supp}\,(\mathbf{b}_i)| \leq 3$. □

Observe that there exist $n \times n$ matrices that have a CP representation with at most $\lfloor n^2/4 \rfloor$ summands, but do not have a CP representation of support 3 (see Exercise 4.35). Moreover, the following example shows that there exist matrices that have both a CP representation with at most $\lfloor n^2/4 \rfloor$ summands, and a CP representation of support 3, but not a CP representation which satisfies both these properties.

Example 4.64. The matrix $A = J_5 + I_5$ has a CP representation of support 3, and $\mathrm{cpr}\, A = 5 < 5^2/4$. But A does not have a CP representation with at most $\lfloor 5^2/4 \rfloor = 6$ summands which is at the same time also of support 3.

There are 10 different matrices obtained from $(1/3J_3) \oplus 0_2$ by permutation similarity. The matrix A is the sum of these 10 matrices. The fact that $\mathrm{cpr}\, A = 5$ follows from Exercise 3.67.

Now suppose $A = \sum_{i=1}^{6} \mathbf{b}_i\mathbf{b}_i^T$ is a CP representation of A of support 3. The matrix A has 20 nonzero off-diagonal elements, and each of the 6 matrices $\mathbf{b}_i\mathbf{b}_i^T$ has at most 6 nonzero off-diagonal entries. Therefore, there must exist $1 \leq r < s \leq 5$, and $1 \leq i \leq 6$ such that $\left(\mathbf{b}_i\mathbf{b}_i^T\right)_{rs} = a_{rs}$. We may assume that $r = 1$, $s = 2$, $i = 1$. Then $\mathbf{b}_1^T = (x, y, z, 0, 0)$ for some nonnegative x, y, z. The matrix

$$A' = A - \mathbf{b}_1\mathbf{b}_1^T = \begin{pmatrix} 2-x^2 & 1-xy & 1-xz & 1 & 1 \\ 1-xy & 2-y^2 & 1-yz & 1 & 1 \\ 1-xz & 1-yz & 2-z^2 & 1 & 1 \\ 1 & 1 & 1 & 2 & 1 \\ 1 & 1 & 1 & 1 & 2 \end{pmatrix}$$

is completely positive, and $xy = 1$. Then $\det(A'[1,2,4]) \le 0$, with equality only if $x = y = 1$. But if $x = y = 1$, $\det(A'[1,2,4,5]) = -1$, a contradiction.

Exercises

4.25. Prove the claims in Example 4.52.

4.26. Complete the proof of Theorem 4.57 by showing that for $r = 1, \ldots, n-2$, if $r - 1 < t < r$, then $\operatorname{cpr} A_{n,t} = r + n - 2$.
Hint: Show that if $A_{n,t} = \sum \mathbf{b}_i \mathbf{b}_i^T$ is a CP representation of $A_{n,t}$, $r - 1 < t < r$, then at most $n - 2 - r$ of the vertices $3, \ldots, n$ may be in exactly one of the supports $\operatorname{supp} \mathbf{b}_i$. Conclude that $\operatorname{cpr} A_{n,t} \ge 2r + (n - 2 - r)$. To prove the equality, consider first the case $r = n - 2$.

4.27. Show that the completely positive matrix $A_{n,t}$ in the proof of Theorem 4.57 is critical for every $0 < t < n - 2$.

4.28. Prove Theorem 4.60 for completely positive graphs which are not connected.

4.29. Say that a completely positive graph G is CP^1 if for every $\operatorname{cc}(G) \le k \le \operatorname{cpr} G$ there exists a critical completely positive matrix A such that $G(A) = G$ and $\operatorname{cpr} A = k$. By Exercise 4.10, K_4 is CP^1, and by Theorem 4.57 and Exercise 4.27 so is every T_n, $n \ge 4$. Show that every bipartite graph which is not a tree is also CP^1.

4.30. Say that a completely positive graph G is CP^2 if for every $\operatorname{cc}(G) \le k \le \operatorname{cpr} G - 1$ there exists a critical completely positive matrix A such that $G(A) = G$ and $\operatorname{cpr} A = k$, and there exists no singular completely positive matrix such that $G(A) = G$ and $\operatorname{cpr} A = \operatorname{cpr} G$. Show that K_3 is CP^2 and so is every tree.

4.31. Let $G = G_1 \cup G_2$, where G_1 and G_2 are connected subgraphs sharing exactly one vertex. Suppose $V(G_1) = \{1, \ldots, k\}$ and $V(G_2) = \{k, \ldots, n\}$.

Show that if A_1 is a critical CP realization of G_1 and A_2 is a critical CP realization of G_2, then $A = (A_1 \oplus 0_{n-k}) + (0_{k-1} \oplus A_2)$ is a critical CP realization of G.

4.32. Suppose a connected completely positive graph G has a cut vertex, say $G = G_1 \cup G_2$, where G_1 and G_2 share exactly one vertex. Show that

(a) $\operatorname{cc}(G) = \operatorname{cc}(G_1) + \operatorname{cc}(G_2)$.
(b) If G_1 and G_2 are both CP^1, then

$$\operatorname{cpr} G = \operatorname{cpr} G_1 + \operatorname{cpr} G_2$$

and G is CP^1.
(c) If G_1 is CP^2 and G_2 is CP^1, then

$$\operatorname{cpr} G = \operatorname{cpr} G_1 + \operatorname{cpr} G_2 - 1$$

and G is CP^1.
(d) If G_1 is either K_2 or K_3 and G_2 is CP^2, then

$$\operatorname{cpr} G = \operatorname{cpr} G_1 + \operatorname{cpr} G_2 - 1$$

and G is CP^2.

4.33. Prove Theorem 4.61. Show also that every completely positive graph is either CP^1 or CP^2.

4.34. Construct a completely positive matrix A which has a CP representation of support 3 with no more than $\lfloor n^2/4 \rfloor$ summands (where n is the order of A), such that $M(A)$ is not positive semidefinite.

4.35. Find an $n \times n$ completely positive matrix which has a CP representation with no more than $\lfloor n^2/4 \rfloor$ summands, but does not have a CP representation of support 3.

Notes. Theorem 4.53 and the conclusion that $\operatorname{cpr} A \leq \lfloor n^2/4 \rfloor$ for every completely positive matrix of order $n \geq 4$ whose graph is triangle free were proved in [Drew *et al.* (1994)]. The fact that the bound was valid also in all other known cases led the authors to wonder whether this holds for every completely positive matrix of order $n \geq 4$. Ten years later there was still no counter example, and in the book that preceded this one, "Completely Positive Matrices", we included Conjecture 4.49 as The DJL Conjecture. [Barioli (2002)] announced an example of a 7×7 completely positive matrix with $\operatorname{cpr} A = 14$, but his result was never published. It took until 2014 for the conjecture to be disproved, in [Bomze *et al.* (2014)] — see next section.

Theorem 4.50 was proved before, in [Berman and Hershkowitz (1987)]. The bound in Theorem 4.53 improved known bounds. By [Kaykobad (1987)], $\operatorname{cpr} A \leq |E(G(A))| + |V(G(A))|$ when A is diagonally dominant (see the proof of Theorem 3.38). In [Berman and Grone (1988)] it was shown that $|E(G(A))| \leq \operatorname{cpr} A \leq |E(G(A))| + 1$ when $G(A)$ is bipartite (see the algorithm in Section 3.5 for finding a CP representation in this case).

It is shown in [Goldberg (2003)] that if G is a completely positive graph on $n \geq 6$ vertices, then $|E(G)| \leq n^2/4$. Thus the second part of Theorem 4.60 follows from the first for $n \geq 6$.

Theorem 4.57 was proved in [Drew and Johnson (1996)]. The proof presented here is different. Prior to that paper, it was known that $\operatorname{cpr} T_n \leq 2n-3$ by [Kogan and Berman (1993)]. The fact that every $n-2 \leq k \leq 2n-4$ is the cp-rank of some CP matrix realization of T_n is mentioned in [Drew and Johnson (1996)] and proved in [Zhang and Li (2002)], where it is part of the proof that there are no gaps in the set of cp-ranks of CP matrix realizations of a completely positive graph. The matrices $A_{n,t}$ constructed here for the proof are different from the ones used in the original proof of [Zhang and Li (2002)].

Theorem 4.60 was proved in [Drew and Johnson (1996)]. Theorem 4.61 is taken from [Zhang and Li (2002)], and its proof as outlined in the exercises is based on the proof there. Theorem 4.63 and Example 4.64 are taken from [Berman and Shaked-Monderer (1998)].

4.5 Bounding p_n

We now consider problem (B), the problem of finding a tight upper bound on the maximum cp-rank of $n \times n$ completely positive matrices. While $p_n = n$ for $n \leq 4$, $p_n > n$ for $n \geq 5$. The exact value of p_n, $n \geq 6$, is not yet known. In this section we compute p_5 exactly and present upper and lower bounds on p_n for every n. In particular we will see that the conjectured DJL bound $p_n \leq \frac{n^2}{4}$ does not hold for $n \geq 7$.

By Theorem 4.36, to find p_n, $n \geq 5$, it suffices to consider the cp-ranks of positive definite matrices on the boundary of $\mathcal{CP}_n$. Recall that every matrix A on the boundary of $\mathcal{CP}_n$ is orthogonal to some extremal matrix in $\mathcal{COP}_n$. If A is positive definite, the extremal matrix cannot be of the form $\mathbf{x}\mathbf{x}^T$, since otherwise $\mathbf{x}$ is in the null space of A. Hence either A is orthogonal to some extremal nonnegative matrix, in which case A is not positive, or A is orthogonal to an exceptional extremal matrix M in $\mathcal{COP}_n$.

Such M has a finite number of minimal zeros, and if $A = BB^T$ is a CP factorization of A, then each column of B is a zero of M. We therefore get the following observation.

Lemma 4.65. *Let A be a completely positive matrix on the boundary of $\mathcal{CP}_n$, orthogonal to an exceptional extremal $M \in \mathcal{COP}_n$. Let $W \in \mathbb{R}^{n\times k}$ be a matrix whose columns are all the different minimal zeros of M in Δ_n (or any other representing set of all the minimal zeros). Let $A = BB^T$ be a CP factorization of A, where $B \in \mathbb{R}^{n\times m}$. Then there exists a nonnegative matrix $X \in \mathbb{R}^{k\times m}$ such that $B = WX$. For any such X,* $\operatorname{cpr} A \leq \operatorname{cpr}(XX^T)$.

Proof. As each column of B is a zero of M, each such column is a nonnegative combination of minimal zeros. Hence there exists $X \geq 0$ such that $B = WX$. Then

$$A = BB^T = (WX)(WX)^T = W(XX^T)W^T.$$

If YY^T is a minimal CP factorization of XX^T, then $A = (WY)(WY)^T$ is a CP factorization of A, hence $\operatorname{cpr} A \leq \operatorname{cpr}(XX^T)$. □

Finding p_5

We now prove that $p_5 = 6$. As $\lfloor 5^2/4 \rfloor = 6$, this shows that the DJL bound applies for $n = 5$. For the computation, it suffices to consider positive definite matrices A on the boundary of $\mathcal{CP}_5$. We first show that if A is orthogonal to an exceptional extremal matrix in $\mathcal{COP}_5$ (i.e., to a matrix in the orbit of either the Horn matrix or a Hildebrand matrix), then $\operatorname{cpr} A = \operatorname{rank} A$. This, of course, implies that for such matrices $\operatorname{cpr} A \leq 5$.

Proposition 4.66. *Let a 5×5 completely positive matrix A be orthogonal to a matrix in the orbit of the Horn matrix. Then* $\operatorname{cpr} A = \operatorname{rank} A$.

Proof. Let H be the Horn matrix. Note that if P is a permutation matrix and D is a positive diagonal matrix then

$$\langle A, DPHP^TD\rangle = \langle P^TDADP, H\rangle.$$

Since $\operatorname{cpr} A = \operatorname{cpr}(P^TDADP)$ by Proposition 4.12, it suffices to prove the claim for a 5×5 matrix A orthogonal to H.

Suppose A is orthogonal to H and $A = BB^T$ is a minimal CP factorization of A. Then each column of B is a zero of H. Recall that the minimal zeros of H are

$$\mathbf{u}_i = \mathbf{e}_i + \mathbf{e}_{i+1}, \quad i = 1, \ldots, 5,$$

where $\dotplus$ denotes addition modulo 5 on $\{1,\dots,5\}$, and each zero of H is $s\mathbf{u}_i + t\mathbf{u}_{i\dotplus 1}$, $s,t \geq 0$ and $s+t>0$. Let

$$W = \begin{pmatrix} \mathbf{u}_1 & \mathbf{u}_2 & \mathbf{u}_3 & \mathbf{u}_4 & \mathbf{u}_5 \end{pmatrix}.$$

Then $B = WX$, where $X \geq 0$ and the support of each column of X is contained in $\{i, i \dotplus 1\}$ for some $1 \leq i \leq 5$. Then $G(XX^T)$ is a subgraph of the 5-cycle, and therefore $\operatorname{cpr}(XX^T) = \operatorname{rank}(XX^T)$ (if $G(XX^T)$ is a proper subgraph of the cycle, this follows from Theorem 4.50, and if $G(XX^T)$ is the whole cycle this follows from Theorem 4.53 and Remark 3.82). As W is nonsingular (see Exercise 3.50) and

$$A = (WX)(WX)^T = W(XX^T)W^T,$$

$\operatorname{rank} A = \operatorname{rank}(XX^T)$. We also have $\operatorname{cpr} A \leq \operatorname{cpr}(XX^T)$ by Lemma 4.65, hence

$$\operatorname{rank} A \leq \operatorname{cpr} A \leq \operatorname{cpr}(XX^T) = \operatorname{rank}(XX^T) = \operatorname{rank} A. \qquad \square$$

Proposition 4.67. *Let a 5×5 completely positive matrix A be orthogonal to a matrix in the orbit of a Hildebrand matrix. Then* $\operatorname{cpr} A = \operatorname{rank} A$.

Proof. As in the previous proof, we may assume that A is orthogonal to a Hildebrand matrix

$$S(\boldsymbol{\theta}) = \begin{pmatrix} 1 & -\cos\theta_1 & \cos(\theta_1+\theta_2) & \cos(\theta_4+\theta_5) & -\cos\theta_5 \\ -\cos\theta_1 & 1 & -\cos\theta_2 & \cos(\theta_2+\theta_3) & \cos(\theta_1+\theta_5) \\ \cos(\theta_1+\theta_2) & -\cos\theta_2 & 1 & -\cos\theta_3 & \cos(\theta_3+\theta_4) \\ \cos(\theta_4+\theta_5) & \cos(\theta_2+\theta_3) & -\cos\theta_3 & 1 & -\cos\theta_4 \\ -\cos\theta_5 & \cos(\theta_1+\theta_5) & \cos(\theta_3+\theta_4) & -\cos\theta_4 & 1 \end{pmatrix},$$

where $\boldsymbol{\theta} = (\theta_1,\theta_2,\theta_3,\theta_4,\theta_5)^T$ has entries $\theta_i \in (0,\pi)$ and $\sum_{i=1}^5 \theta_i < \pi$. The matrix $S(\boldsymbol{\theta})$ has exactly 5 zeros up to multiplication by a positive scalar, which are the columns of the matrix

$$W = \begin{pmatrix} \sin\theta_2 & 0 & 0 & \sin\theta_4 & \sin(\theta_1+\theta_5) \\ \sin(\theta_1+\theta_2) & \sin\theta_3 & 0 & 0 & \sin\theta_5 \\ \sin\theta_1 & \sin(\theta_2+\theta_3) & \sin\theta_4 & 0 & 0 \\ 0 & \sin\theta_2 & \sin(\theta_3+\theta_4) & \sin\theta_5 & 0 \\ 0 & 0 & \sin\theta_3 & \sin(\theta_4+\theta_5) & \sin\theta_1 \end{pmatrix}.$$

As the 5×5 matrix $S(\boldsymbol{\theta})$ is extremal, it has at least 5 linearly independent minimal zeros (by Corollary 2.68). Hence the matrix W is nonsingular.

Let $A = BB^T$, $B \geq 0$, be a minimal CP factorization of A, i.e., B has $p = \operatorname{cpr} A$ columns. Then each column of B is a zero of $S(\boldsymbol{\theta})$, i.e., a positive multiple of a column of W. Since $A = BB^T$ is a *minimal* CP factorization, each column of B is a multiple of a different zero of A. The p columns of B are therefore linearly independent, and $\operatorname{rank} B = p$. This implies

$$\operatorname{rank} A = \operatorname{rank} B = p = \operatorname{cpr} A. \qquad \square$$

As there are matrices in $\mathcal{CP}_5$ with cp-rank 6, this means that matrices in $\mathcal{CP}_5$ orthogonal to an exceptional matrix do not attain p_5. To show that $p_5 = 6$, it remains to show that the cp-rank of a 5×5 copositive matrix, which is not positive, is at most 6. The following lemma is used repeatedly in the proof.

Lemma 4.68. *Let A be an $n \times n$ doubly nonnegative matrix. Suppose that for some $1 \leq r \neq s \leq n$ the support of row r is a subset of the support of row s. Let*

$$\alpha = \min\left\{ \frac{a_{sj}}{a_{rj}} \;\middle|\; a_{rj} > 0 \right\},$$

$S = I_n - \alpha E_{sr}$, and $\hat{A} = SAS^T$. Then

(a) *$\hat{A}$ is doubly nonnegative and* $\operatorname{rank} \hat{A} = \operatorname{rank} A$.
(b) *If $a_{ij} = 0$ for some $1 \leq i, j \leq n$, then $\hat{a}_{ij} = 0$.*
(c) *There is at least one more zero off-diagonal entry in row s of $\hat{A}$, compared to row s of A.*

Proof. We may assume without loss of generality that $r = 1$ and $s = 2$, and

$$\alpha = \min\left\{ \frac{a_{2j}}{a_{1j}} \;\middle|\; a_{1j} > 0 \right\}.$$

Then $\alpha > 0$ and

$$\hat{A} = SAS^T = \begin{pmatrix} a_{11} & a_{12} - \alpha a_{11} & a_{13} & \cdots & a_{1n} \\ a_{12} - \alpha a_{11} & \beta & a_{23} - \alpha a_{13} & \cdots & a_{2n} - \alpha a_{1n} \\ a_{13} & a_{23} - \alpha a_{13} & a_{33} & \cdots & a_{3n} \\ \vdots & \vdots & \vdots & & \vdots \\ a_{1n} & a_{2n} - \alpha a_{1n} & a_{3n} & \cdots & a_{nn} \end{pmatrix}$$

where $\beta = a_{22} - \alpha a_{12} - \alpha(a_{12} - \alpha a_{11})$. From this (and the assumption on the supports of rows 1 and 2) it is obvious that (b) holds. The matrix $\hat{A}$ is positive semidefinite, hence $\beta \geq 0$. By the choice of α, all other entries of

$\hat{A}$ are also nonnegative, hence $\hat{A}$ is doubly nonnegative. Since A is positive semidefinite, $a_{22}/a_{12} \geq a_{12}/a_{11}$, hence the minimum α is attained at some $j \neq 2$. For this j we have $a_{2j} > 0$ and $\hat{a}_{2j} = 0$. □

Proposition 4.69. *For every* 5×5 *completely positive matrix* A*, which is not positive,* $\operatorname{cpr} A \leq 6$.

Proof. Without loss of generality, suppose $a_{15} = 0$. If there is an additional zero entry in the first row of A, then the degree of the vertex $v = 1$ in $G(A)$ is at most 2. In that case, by Lemma 4.56,

$$\operatorname{cpr} A \leq 2 + \operatorname{cpr}(G(A) - v) \leq 2 + 4 = 6.$$

Similarly, $\operatorname{cpr} A \leq 6$ if there is an additional zero entry in the fifth row of A. We therefore assume that all entries in the first and fifth row are nonzero, except for a_{15} and a_{51}. Let $A = \sum_{i=1}^{m} \mathbf{b}_i \mathbf{b}_i^T$ be any CP representation of A. Let

$$\Omega_1 = \{1 \leq i \leq m \mid 1 \in \operatorname{supp} \mathbf{b}_i\}, \qquad \Omega_2 = \{1, \ldots, m\} \setminus \Omega_1$$
$$B = \sum_{i \in \Omega_1} \mathbf{b}_i \mathbf{b}_i^T, \qquad C = \sum_{i \in \Omega_2} \mathbf{b}_i \mathbf{b}_i^T.$$

Then $B = B' \oplus 0$, and $C = 0 \oplus C'$, where B' and C' are both 4×4 completely positive matrices. All the entries in the first row of B' and all the entries in the last row of C' are positive. So

$$B = \begin{pmatrix} + & + & + & + & 0 \\ + & + & * & * & 0 \\ + & * & + & * & 0 \\ + & * & * & + & 0 \\ 0 & 0 & 0 & 0 & 0 \end{pmatrix}, \qquad C = \begin{pmatrix} 0 & 0 & 0 & 0 & 0 \\ 0 & + & * & * & + \\ 0 & * & + & * & + \\ 0 & * & * & + & + \\ 0 & + & + & + & + \end{pmatrix},$$

where each $+$ denotes a positive entry and each $*$ a nonnegative one. We show that it is enough to consider the case that $B[2,3,4]$ and $C[2,3,4]$ each have at least two zero entries above the diagonal. If, say, $B[2,3,4]$ is positive or has just one zero entry above the diagonal, then by permuting rows (and columns) 2, 3, 4 of A, we may assume that the second row of B' is positive. The support of row 1 in B is then contained in that of row 2. Let

$$\alpha = \min \left\{ \frac{b_{2j}}{b_{1j}} \,\middle|\, 1 \leq j \leq 4 \right\}$$

and $S = I_5 - \alpha E_{21}$. By Lemma 4.68, $\hat{B} = SBS^T$ is a doubly nonnegative matrix. Since it has at most 4 nonzero rows, it is completely positive.

At least one of the entries $\hat{b}_{12}, \hat{b}_{23}, \hat{b}_{24}$ is zero. Since row 1 in C is zero, $SCS^T = C$ and

$$\hat{A} = SAS^T = SBS^T + SCS^T = \hat{B} + C.$$

In particular, $\hat{A}$ is completely positive and $\operatorname{cpr} A \leq \operatorname{cpr} \hat{A}$ (see Proposition 4.11). If $\hat{b}_{12} = 0$, then $\hat{a}_{12} = 0$, and by Lemma 4.56 we deduce that $\operatorname{cpr} \hat{A} \leq 6$. So we need to consider the case that (at least) one of the entries $\hat{b}_{23}$, $\hat{b}_{24}$ is zero. If only one of them is zero, we may assume by permutation similarity that it is $\hat{b}_{23}$. But then the support of row 1 in $\hat{B}$ is contained in that of row 4, and we may repeat the above argument with

$$\beta = \min \left\{ \frac{\hat{b}_{4j}}{\hat{b}_{1j}} \,\middle|\, 1 \leq j \leq 4 \right\}$$

and $\hat{S} = I_5 - \beta E_{41}$. We get a matrix $\tilde{B} = \hat{S}\hat{B}\hat{S}^T$. Again we may restrict our attention to the case that the additional zero entry in the fourth row of $\tilde{B}$ is not in the first column. That is, $\tilde{B}[2,3,4]$ has at least two zero nondiagonal entries, and $\tilde{A} = \hat{S}\hat{A}\hat{S}^T = \tilde{B} + C$ is a completely positive matrix, $\operatorname{cpr} A \leq \operatorname{cpr} \hat{A} \leq \operatorname{cpr} \tilde{A}$. A similar argument applies to C. We may use the fifth row of C to generate a matrix $\hat{C}$ which has two zeros in its $\{2,3,4\}$ principal submatrix, without changing $\tilde{B}$, except possibly permuting rows and columns $2,3,4$.

So assume $A = B + C$, $B = B' \oplus 0$, $C = 0 \oplus C'$, and each of the matrices B and C has at least two zero entries above the diagonal in its principal submatrix on rows $2,3,4$. If the two zeros of B are in the same positions as those of C, then A has two zeros in one of the rows 2, 3 or 4. In this case we deduce from Lemma 4.56 that $\operatorname{cpr} A \leq 6$. Thus there is one case left to consider, the case that only one of the zeros of B and one of the zeros of C coincide. By permuting rows and columns $2,3,4$ we may assume that

$$B = \begin{pmatrix} + & + & + & + & 0 \\ + & + & 0 & 0 & 0 \\ + & 0 & + & + & 0 \\ + & 0 & + & + & 0 \\ 0 & 0 & 0 & 0 & 0 \end{pmatrix}, \qquad C = \begin{pmatrix} 0 & 0 & 0 & 0 & 0 \\ 0 & + & + & 0 & + \\ 0 & + & + & 0 & + \\ 0 & 0 & 0 & + & + \\ 0 & + & + & + & + \end{pmatrix},$$

where each $+$ denotes a positive entry. If both B' and C' are singular, then by Example 4.52, $\operatorname{cpr} B' \leq 3$ and $\operatorname{cpr} C' \leq 3$, and $\operatorname{cpr} A \leq 6$. Hence we assume that at least one of the matrices B' and C' is nonsingular. By symmetry we may assume B' is. By applying a positive diagonal congruence on $A = B + C$ we may assume that $b_{34} = b_{33}$. Then $b_{44} > b_{33}$, and for each $0 \le \alpha < b_{33}$ the matrix $\begin{pmatrix} b_{33}-\alpha & b_{33}-\alpha \\ b_{33}-\alpha & b_{44}-\alpha \end{pmatrix}$ is positive definite and nonnegative. For each such α and nonnegative β let

$$B(\alpha,\beta) = \begin{pmatrix} b_{11} & b_{12} & b_{13} & b_{14} \\ b_{12} & b_{22}+\beta & 0 & 0 \\ b_{13} & 0 & b_{33}-\alpha & b_{33}-\alpha \\ b_{14} & 0 & b_{33}-\alpha & b_{44}-\alpha \end{pmatrix}$$

$$C(\alpha,\beta) = \begin{pmatrix} c_{22}-\beta & c_{23} & 0 & c_{25} \\ c_{23} & c_{33}+\alpha & \alpha & c_{35} \\ 0 & \alpha & c_{44}+\alpha & c_{45} \\ c_{25} & c_{35} & c_{45} & c_{55} \end{pmatrix}.$$

Note that $B(0,0) = B'$ and $C(0,0) = C'$, and for every (α,β)

$$A = (B(\alpha,\beta) \oplus 0) + (0 \oplus C(\alpha,\beta)).$$

Define

$$T = \left\{ (\alpha,\beta) \;\middle|\; \begin{array}{c} \alpha > 0, \ \beta \ge 0, \ B(\alpha,\beta) \text{ is singular, and} \\ B(\alpha,\beta) \text{ and } C(\alpha,\beta) \text{ are doubly nonnegative} \end{array} \right\}.$$

The set T is not empty, as the matrix $B(0,0)$ is nonnegative and positive definite, while $B(b_{33},0)$ is nonnegative and not positive semidefinite. Hence there exists $0 < \alpha_0 < b_{33}$ such that $B(\alpha_0,0)$ is a singular doubly nonnegative matrix. Since $C(\alpha_0,0)$ is clearly doubly nonnegative, $(\alpha_0,0) \in T$.

Let

$$T' = \{\alpha \mid \alpha > 0 \text{ and there exists } \beta \ge 0 \text{ such that } (\alpha,\beta) \in T\}.$$

T' is a bounded nonempty subset of $\mathbb{R}$. Let

$$\varphi = \sup_{\alpha \in T'} \alpha.$$

Then $\varphi \ge \alpha_0$. Let $\{\alpha_j\}$ be a nondecreasing sequence of elements of T' which converges to φ, and $\{\beta_j\}$ a sequence of nonnegative numbers such that $(\alpha_j,\beta_j) \in T$. Since $\{\beta_j\}$ is bounded (*e.g.*, by c_{22}), it has a converging subsequence. We may assume that $\{\beta_j\}$ itself converges to a limit ψ. Then the sequences $\{B(\alpha_j,\beta_j)\}$ and $\{C(\alpha_j,\beta_j)\}$ both converge, to $B(\varphi,\psi)$ and $C(\varphi,\psi)$ respectively. The matrices $B(\varphi,\psi)$ and $C(\varphi,\psi)$ are clearly doubly nonnegative, and $B(\varphi,\psi)$ is singular.

We will now show that $C(\varphi,\psi)$ is also singular. If this is not the case, $C(\varphi,\psi)$ is positive definite, and therefore there exists $\delta > 0$ such that

$C(\varphi, \psi + \delta)$ is positive semidefinite. Consider the equality

$$\begin{aligned} 0 &= \det B(\varphi, \psi) \\ &= -b_{12} \det \begin{pmatrix} b_{12} & b_{13} & b_{14} \\ 0 & b_{33} - \varphi & b_{33} - \varphi \\ 0 & b_{33} - \varphi & b_{44} - \varphi \end{pmatrix} + (b_{22} + \psi) \det(B(\varphi, \psi)[1, 3, 4]) \\ &= -b_{12}^2(b_{33} - \varphi)(b_{44} - b_{33}) + (b_{22} + \psi) \det(B(\varphi, \psi)[1, 3, 4]). \end{aligned}$$

It implies that $B(\varphi, \psi)[1, 3, 4]$ has positive determinant, hence

$$\det B(\varphi, \psi + \delta) = \det B(\varphi, \psi) + \delta \det(B(\varphi, \psi)[1, 3, 4]) > 0,$$

and the doubly nonnegative matrix $B(\varphi, \psi + \delta)$ is positive definite. On the other hand, $B(b_{33}, \psi+\delta)$ is nonnegative but clearly not positive semidefinite. Thus there exists an $\varepsilon > 0$ such that $B(\varphi + \varepsilon, \psi + \delta)$ is a singular doubly nonnegative matrix. Since $C(\varphi, \psi+\delta)$ is doubly nonnegative, $C(\varphi+\varepsilon, \psi+\delta)$ is a doubly nonnegative matrix too. But this implies that $\varphi+\varepsilon \in T'$, which contradicts our choice of φ. Hence $C(\varphi, \psi)$ is a singular matrix. We may therefore assume that $A = B + C$, where $B = B' \oplus 0$ and $C = 0 \oplus C'$, B' and C' are singular completely positive matrices, and

$$B = \begin{pmatrix} + & + & + & + & 0 \\ + & + & 0 & 0 & 0 \\ + & 0 & + & + & 0 \\ + & 0 & + & + & 0 \\ 0 & 0 & 0 & 0 & 0 \end{pmatrix}, \qquad C = \begin{pmatrix} 0 & 0 & 0 & 0 & 0 \\ 0 & + & + & 0 & + \\ 0 & + & + & + & + \\ 0 & 0 & + & + & + \\ 0 & + & + & + & + \end{pmatrix}.$$

We use once more Lemma 4.68, this time for

$$\alpha = \min \left\{ \frac{c_{3j}}{c_{5j}} \,\middle|\, 2 \le j \le 5 \right\},$$

$S = I_5 - \alpha E_{35}$. Then $\hat{A} = SAS^T = B + SCS^T$ is a completely positive matrix and it suffices to show that $\operatorname{cpr} \hat{A} \le 6$. Let $\hat{C} = SCS^T$. If $\hat{c}_{23} = 0$ or $\hat{c}_{35} = 0$, then $G(\hat{A})$ has a vertex of degree at most 2, and we deduce by Lemma 4.56 that $\operatorname{cpr} \hat{A} \le 6$. Otherwise, we have $\hat{A} = B + \hat{C}$, $B = B' \oplus 0$, $\hat{C} = 0 \oplus \hat{C}'$, B' and $\hat{C}'$ are both singular 4×4 completely positive matrices, and the graph of each of them has two blocks, a triangle and a single edge. By Example 4.52 we have $\operatorname{cpr} B = \operatorname{cpr} B' = \operatorname{rank} B' \le 3$ and $\operatorname{cpr} \hat{C} = \operatorname{cpr} \hat{C}' = \operatorname{rank} \hat{C}' \le 3$, which implies

$$\operatorname{cpr} A \le \operatorname{cpr} \hat{A} \le \operatorname{cpr} B + \operatorname{cpr} \hat{C} \le 3 + 3 = 6. \qquad \square$$

Matrices in $\mathcal{CP}_5$ with cp-rank 6 do exist. For example, any matrix whose graph is $K_{2,3}$. Together with Propositions 4.66, 4.67 and 4.69, this completes the computation of p_5.

Theorem 4.70. $p_5 = 6$.

Remark 4.71. Note that the maximum cp-rank p_5 is attained not only at nonsingular matrices on the boundary, but also at some singular matrices, e.g., singular matrices A with graph $K_{2,3}$, and positive singular matrices close enough to such A (e.g., $A + \varepsilon \mathbf{v}\mathbf{v}^T$, where $\mathbf{v}$ is the unit Perron vector of A and $\varepsilon > 0$ is sufficiently small).

Bounding p_n for $n \geq 6$

For $n = 6$, information on the minimal supports of exceptional extremal matrices, and arguments similar to those in Propositions 4.66 and 4.67 lead to the following result, which we do not prove here (see notes at the end of this section, and Exercises 4.53 and 4.54 in Section 4.7).

Proposition 4.72. *Let $A \in \mathcal{CP}_6$, be orthogonal to an exceptional extremal $M \in \mathcal{COP}_6$. Then*

$$\operatorname{cpr} A \leq 9.$$

Since there exist matrices in $\mathcal{CP}_6$ with cp-rank 9 (e.g., matrices with graph $K_{3,3}$), Proposition 4.72 and Theorem 4.36 imply that $p_6 \geq 9$ is attained at a nonsingular matrix in $\mathcal{CP}_6$, which is not positive. However, p_6 is not yet known, and it is not clear whether the DJL bound holds in this case. But for $n \geq 7$ there are examples of matrices $M \in \mathcal{CP}_n$ with $\operatorname{cpr} M > n^2/4$. Before we present these, let us state and prove the best upper bound on p_n known at this point.

Theorem 4.73. *For every $n \geq 6$,*

$$p_n \leq \frac{n(n+1)}{2} - 4.$$

The proof uses the following result.

Proposition 4.74. *Let $A \in \mathcal{CP}_n$, $n \geq 5$, be orthogonal to a matrix $M \in \mathcal{COP}_n$ that has $k \geq 1$ nonzero diagonal elements. Then*

$$\operatorname{cpr} A \leq \frac{n(n+1)}{2} - k.$$

Proof. Without loss of generality, suppose $m_{ii} > 0$ for $1 \le i \le k$. Consider the subspace of $\mathcal{S}_n$

$$L = \left\{ X \in \mathcal{S}_n \mid \mathbf{e}_i^T X M \mathbf{e}_i = 0 \text{ for every } 1 \le i \le k \right\}.$$

This subspace is isomorphic to the solution space of the following homogenous system of k linear equations in the $n(n+1)/2$ variables x_{ij}, $1 \le i \le j \le n$,

$$m_{ii}x_{ii} + \sum_{j<i} m_{ij}x_{ji} + \sum_{j>i} m_{ij}x_{ij} = 0.$$

As $\mathrm{Diag}(m_{11}, \ldots, m_{kk})$ is a submatrix of the coefficients matrix of this system, the rank of this coefficients matrix is k, and therefore

$$\dim L = \frac{n(n+1)}{2} - k.$$

For each zero $\mathbf{v}$ of M, $\mathbf{v}\mathbf{v}^T \in \mathcal{CP}_n$ is orthogonal to M, hence by Proposition 3.33,

$$\left\{ \mathbf{v}\mathbf{v}^T \mid \mathbf{v} \in \mathcal{V}^M \right\} \subseteq L.$$

As $A \in \mathrm{cone}\left\{ \mathbf{v}\mathbf{v}^T \mid \mathbf{v} \in \mathcal{V}^M \right\}$, by Caratheodory's Theorem

$$\mathrm{cpr}\, A \le \dim L = \frac{n(n+1)}{2} - k. \qquad \square$$

Proof of Theorem 4.73. Let A be a matrix in the interior of $\mathcal{CP}_n$ with $\mathrm{cpr}\, A = p_n$. Let $\delta > 0$ be maximal such that $A_0 = A - \delta E_{nn} \in \mathcal{CP}_n$. Then A_0 is a matrix on the boundary of $\mathcal{CP}_n$, and since all its off-diagonal entries are positive, all its diagonal entries are also positive. As $A = A_0 + \delta E_{nn}$,

$$p_n = \mathrm{cpr}\, A \le \mathrm{cpr}\, A_0 + 1.$$

If A_0 is singular, $\mathrm{cpr}\, A_0 \le \frac{(n-1)n}{2} - 1$ by Theorem 4.23 and, as $n \ge 5$,

$$p_n = \mathrm{cpr}\, A \le \frac{(n-1)n}{2} < \frac{n(n+1)}{2} - 4.$$

If the positive A_0 is nonsingular, then it is orthogonal to an exceptional extremal matrix M in $\mathcal{COP}_n$. Such M has at least 5 positive entries on the diagonal. Therefore $\mathrm{cpr}\, A_0 \le \frac{n(n+1)}{2} - 5$ by Proposition 4.74, and

$$p_n = \mathrm{cpr}\, A \le \mathrm{cpr}\, A_0 + 1 \le \frac{n(n+1)}{2} - 4. \qquad \square$$

Remark 4.75. The bound in Theorem 4.73 is definitely not tight in the case $n = 6$. While we conjecture that for $n = 6$ the DJL bound applies, for now this is only a conjecture. However we can tell for sure that $p_6 \leq 12$, which is a smaller bound than the bound $17 = \frac{6(6+1)}{2} - 4$ of Theorem 4.73. Indeed, by Proposition 4.72 we know that if $A \in \mathcal{CP}_6$ is orthogonal to an exceptional copositive matrix, then $\operatorname{cpr} A \leq 9$. If $A \in \mathcal{CP}_6$ has a zero entry, we may suppose $a_{16} = 0$. Then

$$A = (B \oplus 0) + (0 \oplus C),$$

where B and C are 5×5 completely positive matrices, and therefore

$$\operatorname{cpr} A \leq \operatorname{cpr} B + \operatorname{cpr} C \leq 12.$$

As p_6 is attained at a nonsingular completely positive matrix on the boundary of $\mathcal{CP}_6$, this yields the inequality $p_6 \leq 12$.

We now consider $n \geq 7$. To bound p_n from below we will consider completely positive matrices that are orthogonal to certain copositive matrices, which have only a finite number of zeros up to scalar multiplication. In the following example, the copositivity of the matrix M was verified by the method described in Exercise 2.90, using exact arithmetic computations to avoid numerical errors. The fact that the matrix M in the example is circulant helps to cut down some of the computations, by considering only representatives of the σ sets due to the circulant nature of M (for example, it suffices to consider σ such that $1 \in \sigma$).

Example 4.76. Consider the matrix

$$M = \begin{pmatrix} 162 & -153 & 127 & -27 & -27 & 127 & -153 \\ -153 & 162 & -153 & 127 & -27 & -27 & 127 \\ 127 & -153 & 162 & -153 & 127 & -27 & -27 \\ -27 & 127 & -153 & 162 & -153 & 127 & -27 \\ -27 & -27 & 127 & -153 & 162 & -153 & 127 \\ 127 & -27 & -27 & 127 & -153 & 162 & -153 \\ -153 & 127 & -27 & -27 & 127 & -153 & 162 \end{pmatrix}.$$

The matrix M is a copositive circulant matrix (see Exercise 2.91).

Since M is circulant, for each zero $\mathbf{x} \in \Delta$ of M and every $1 \le i \le 6$ the vector $P^i\mathbf{x} \in \Delta$ is also a zero of M, where

$$P = \begin{pmatrix} 0 & 0 & 0 & 0 & 0 & 0 & 1 \\ 1 & 0 & 0 & 0 & 0 & 0 & 0 \\ 0 & 1 & 0 & 0 & 0 & 0 & 0 \\ 0 & 0 & 1 & 0 & 0 & 0 & 0 \\ 0 & 0 & 0 & 1 & 0 & 0 & 0 \\ 0 & 0 & 0 & 0 & 1 & 0 & 0 \\ 0 & 0 & 0 & 0 & 0 & 1 & 0 \end{pmatrix}.$$

Let

$$\mathbf{u} = \frac{1}{7}(3,3,0,0,1,0,0)^T \quad \text{and} \quad \mathbf{v} = \frac{1}{35}(9,17,9,0,0,0,0)^T.$$

The zeros of M are

$$\mathbf{u}_i = P^{i-1}\mathbf{v}, \quad i = 1, \dots, 7$$

and

$$\mathbf{u}_i = P^{i-8}\mathbf{u}, \quad i = 8, \dots, 14.$$

Since these form a complete representing set of the zeros of M, any completely positive matrix A orthogonal to M satisfies

$$A \in \operatorname{cone}\left\{\mathbf{u}_i\mathbf{u}_i^T \mid 1 \le i \le 14\right\}.$$

We will now show that $\mathbf{u}_i\mathbf{u}_i^T$, $i = 1, \dots, 14$, are linearly independent.

Suppose

$$\sum_{i=1}^{14} x_i \mathbf{u}_i\mathbf{u}_i^T = 0.$$

For $j = 1, \dots, 7$, $(\mathbf{u}_i\mathbf{u}_i^T)_{j\,j\dotplus 2} = 0$ for every $i \ne j$, and $(\mathbf{u}_j\mathbf{u}_j^T)_{j\,j\dotplus 2} = \frac{81}{1225}$, where $\dotplus$ denotes addition modulo 7 on $\{1, \dots, 7\}$. Hence

$$x_i = 0, \quad i = 1, \dots, 7,$$

and

$$\sum_{i=8}^{14} x_i \mathbf{u}_i\mathbf{u}_i^T = 0.$$

For $j = 1, \dots, 7$, $(\mathbf{u}_i\mathbf{u}_i^T)_{j\,j\dotplus 1} = 0$ if $8 \le i \le 14$ and $i \ne j+7$, and $(\mathbf{u}_{j+7}\mathbf{u}_{j+7}^T)_{j\,j\dotplus 1} = \frac{9}{49}$. Hence

$$x_i = 0, \quad i = 8, \dots, 14,$$

and the proof of linear independence is complete.

By the linear independence of $\mathbf{u}_i\mathbf{u}_i^T$, $i = 1, \ldots, 14$, and the fact that $\mathbf{u}_1, \ldots, \mathbf{u}_{14}$ are the only zeros of M in Δ, any matrix

$$A = \sum_{i=1}^{14} x_i \mathbf{u}_i \mathbf{u}_i^T$$

with all $x_i > 0$ is an example of a completely positive matrix orthogonal to M, with a unique CP representation and $\operatorname{cpr} A = 14$. This shows that $p_7 \geq 14$, while the DJL bound for $n = 7$ is $\lfloor 7^2/4 \rfloor = 12$. (See also Proposition 1.135.)

The same idea, with some extensions, can be used to show that

$$p_8 \geq 18, \quad p_9 \geq 26, \quad p_{10} \geq 27, \quad p_{11} \geq 32, \quad p_{15} \geq 95$$

— see Exercises 4.36–4.41. Example 4.84 below uses another construction to show that $p_{12} \geq 37$. These lower bounds on p_n, $n \in \{8, 9, 10, 11, 12, 15\}$, are all greater than the DJL bounds for these orders.

The best known lower bounds on p_n for general values of n are given in the following theorem.

Theorem 4.77. *For $n \geq 5$,*

$$p_n \geq \frac{n^2}{2} - (n+1)\sqrt{2n} + \frac{9}{4}n + \frac{1}{16}. \tag{4.39}$$

Also, for $n \geq 1$,

$$p_n \geq \frac{n^2}{2} - n(2\log_2 \log n + e^2). \tag{4.40}$$

The lower bound (4.39) is better than the DJL bound for $n \geq 13$, the bound (4.40) is better than the DJL bound for $n \geq 45$. Starting at $n \geq 97$ the bound (4.40) is the best known general lower bound. Both bounds are asymptotically closer to the known upper bound of Theorem 4.73, than to the originally conjectured DJL bound, and clearly (4.40) is asymptotically the better of the two. It makes the gap between the best known upper bound and the best known lower bound to be of order $O(n \log \log n)$.

We now outline some ideas behind the proofs of the two lower bounds in Theorem 4.77. The complete proofs use some elaborate computations, which we will not fully go into — see the references in the notes to this section. The idea is to start with some matrices, such that the rows of each form a complete representing set of zeros of some copositive matrix, for which there exists a corresponding completely positive matrix of high cp-rank (for example, $U = \begin{pmatrix} \mathbf{u}_1 & \ldots & \mathbf{u}_{14} \end{pmatrix}^T$, where $\mathbf{u}_1, \ldots, \mathbf{u}_{14}$ are as in

Example 4.76, is such a matrix). Several such matrices are then combined, using some matrix constructions described below, to create a matrix whose rows form a complete representing set of zeros of some copositive matrix of higher order. By carefully choosing the basic building blocks, a corresponding completely positive matrix with high cp-rank is generated.

We now describe these matrix constructions, and state their properties. Since several all-ones vectors are used in the same construction, it would be convenient to denote here the vector $\mathbf{e} \in \mathbb{R}^n$ by $\mathbf{1}_n$.

First, for $A \in \mathbb{R}^{m\times n}$,

$$A = \begin{pmatrix} \mathbf{a}_2 & \mathbf{a}_2 & \ldots & \mathbf{a}_m \end{pmatrix}^T, \qquad \mathbf{a}_i \in \mathbb{R}^n,\ i = 1, \ldots, n, \tag{4.41}$$

we define $A^{\langle 2\rangle} = \begin{pmatrix} \mathbf{a}_1 \otimes \mathbf{a}_1 & \mathbf{a}_2 \otimes \mathbf{a}_2 & \ldots & \mathbf{a}_n \otimes \mathbf{a}_n \end{pmatrix}^T$. The matrix $A^{\langle 2\rangle}$ is of order $m \times n^2$, and its rows are vectorizations of $\mathbf{a}_i\mathbf{a}_i^T$, $i = 1, \ldots, m$. Hence $\operatorname{rank} A^{\langle 2\rangle}$ is the dimension of the cone generated in $\mathcal{S}_n$ by the m matrices $\mathbf{a}_i\mathbf{a}_i^T$.

Next, for any pair of matrices $A \in \mathbb{R}^{m\times n}$ and $B \in \mathbb{R}^{k\times \ell}$ we define

$$A \circledast B = \begin{pmatrix} A \otimes \mathbf{1}_k & \mathbf{1}_m \otimes B \end{pmatrix}.$$

The matrix $A \circledast B$ has order $mk \times (n+\ell)$, and its rows are all the row vectors of the form $\begin{pmatrix} \mathbf{a}_i^T & \mathbf{b}_j^T \end{pmatrix}$, where $\mathbf{a}_1^T, \ldots, \mathbf{a}_m^T$ and $\mathbf{b}_1^T, \ldots, \mathbf{b}_\ell^T$ are the rows of A and B, respectively. We use $A \circledast B \circledast C$ to denote $(A \circledast B) \circledast C$, and interpret $A_1 \circledast \cdots \circledast A_k$ accordingly. (Note that while $\circledast$ is not associative, $(A \circledast B) \circledast C$ and $A \circledast (B \circledast C)$ differ only by the order of rows.) We will show that if the rows of A are all the zeros in Δ_n of some copositive matrix, and the rows of B are all the zeros in Δ_ℓ of some copositive matrix, then the rows of $tA \circledast (1-t)B$, $0 \le t \le 1$, form the set of all zeros in $\Delta_{n+\ell}$ of some copositive matrix.

Remark 4.78. We can now make the idea outlined above a bit more specific. The lower bound on p_n is obtained by constructing an $n \times n$ copositive matrix M with a finite, and rather large, number of zeros in Δ_n. If U an $m \times n$ matrix whose rows are these m zeros of M, every completely positive matrix orthogonal to M is in cone $\{\mathbf{u}\mathbf{u}^T \mid \mathbf{u}^T \text{ is a row of } U\} \subseteq \mathcal{S}_n$. The dimension of this cone is $d = \operatorname{rank} U^{\langle 2\rangle}$. By Carathéodory's Theorem (Theorem 1.132) and Proposition 1.135, the maximum cp-rank of completely positive matrices orthogonal to M is d.

Hence the idea is to start with matrices U_i, each of whose rows is a complete set of zeros in Δ_{n_i} of some copositive matrix. Combine these matrices "convexly" via $\circledast$, to obtain a matrix U whose rows are all the zeros in Δ_n, $n = \sum n_i$, of some copositive matrix M.

The next couple of results will tell us, among other details, how $\operatorname{rank} U^{\langle 2\rangle}$ relates to the ranks of $U_i^{\langle 2\rangle}$.

The following lemma lists properties of the matrices generated by the two operations. We leave most of the proof as an exercise.

Lemma 4.79. *Let $A \in \mathbb{R}^{m\times n}$ and $\mathbf{a}_1, \ldots, \mathbf{a}_n$ be as in (4.41), and let $B \in \mathbb{R}^{k\times \ell}$.*

(a) $\operatorname{rank} A^{\langle 2\rangle} = \dim \operatorname{span}\{\mathbf{a}_i\mathbf{a}_i^T \mid i = 1, \ldots, n\}$.
(b) $\operatorname{rank} A^{\langle 2\rangle} \geq \operatorname{rank} A$, *and if A has full row rank, equality holds. If $0 < \operatorname{rank} A < m$ and no two rows in A are linearly dependent, then* $\operatorname{rank} A^{\langle 2\rangle} > \operatorname{rank} A$.
(c) *If*

$$A = \begin{pmatrix} A_1 \\ A_2 \\ \vdots \\ A_p \end{pmatrix},$$

where each A_i is a block of rows of A, then

$$A^{\langle 2\rangle} = \begin{pmatrix} A_1^{\langle 2\rangle} \\ A_2^{\langle 2\rangle} \\ \vdots \\ A_p^{\langle 2\rangle} \end{pmatrix}.$$

(d) $\operatorname{rank}(A \circledast B) \geq \operatorname{rank} A + \operatorname{rank} B - 1$, *and equality holds if and only if* $\mathbf{1}_m \in \operatorname{cs} A$ *and* $\mathbf{1}_k \in \operatorname{cs} B$.
(e) *If*

$$A = \begin{pmatrix} A_1 \\ A_2 \\ \vdots \\ A_p \end{pmatrix} \quad \text{and} \quad B = \begin{pmatrix} B_1 \\ B_2 \\ \vdots \\ B_q \end{pmatrix},$$

where each A_i a block of rows of A, and each B_i a block of rows of B, then $A \circledast B$ is a row permutation of

$$\begin{pmatrix} A_1 \circledast B_1 \\ A_1 \circledast B_2 \\ \vdots \\ A_p \circledast B_q \end{pmatrix}.$$

(f) $A \otimes B$ *is a submatrix of* $(A \circledast B)^{\langle 2 \rangle}$.
(g) $A^{\langle 2 \rangle} \circledast B^{\langle 2 \rangle}$ *is a submatrix of* $(A \circledast B)^{\langle 2 \rangle}$.

Proof. We only prove here parts (a), (b) and (d), and leave the rest as an exercise. For (a), just note that $\mathbf{a}_i \otimes \mathbf{a}_i$ is a vectorization of the matrix $\mathbf{a}_i \mathbf{a}_i^T$.

For (b), use the fact that if $\mathbf{a}_1, \ldots, \mathbf{a}_k$ are linearly independent, then $\mathbf{a}_1 \mathbf{a}_1^T, \ldots, \mathbf{a}_k \mathbf{a}_k^T$ are also linearly independent to obtain the first assertion. For the second assertion in (b), note that under the conditions in this part, $n \geq 2$, $m \geq 3$. Also, if D is a nonsingular $m \times m$ diagonal matrix and R is a nonsingular $n \times n$ matrix, then

$$(DA)^{\langle 2 \rangle} = D^2 A^{\langle 2 \rangle} \quad \text{and} \quad (AR)^{\langle 2 \rangle} = A^{\langle 2 \rangle}(R \otimes R),$$

where both D^2 and $R \otimes R$ are nonsingular. Hence for such D and R,

$$\operatorname{rank} A = \operatorname{rank} DAR \qquad \text{and} \qquad \operatorname{rank} A^{\langle 2 \rangle} = \operatorname{rank} (DAR)^{\langle 2 \rangle}, \tag{4.42}$$

and every two rows in DAR are linearly independent. Now let $\mathbf{u} \in \mathbb{R}^n$ be such that $\mathbf{a}_i^T \mathbf{u} \neq 0$ for every $1 \leq i \leq m$, and let D be the diagonal matrix that satisfies $DA\mathbf{u} = \mathbf{e}$. Every two rows in DA are linearly independent, so $n \geq 2$ and there exists $\mathbf{v}$ such that $\mathbf{x} = DA\mathbf{v}$ has m different entries. Let R be an $n \times n$ nonsingular matrix whose first columns are $\mathbf{u}$ and $\mathbf{v}$. Then in DAR the first two columns are $\mathbf{e}$ and $\mathbf{x}$, and by (4.42), it suffices to prove the claimed rank inequality for DAR. So suppose without loss of generality the first column of A is $\mathbf{e}$, and the second is some $\mathbf{x}$ with m different entries. Since the first column is $\mathbf{e}$, A is a submatrix of $A^{\langle 2 \rangle}$, and therefore $\operatorname{cs} A \subseteq \operatorname{cs} A^{\langle 2 \rangle}$. To prove that $\operatorname{rank} A < \operatorname{rank} A^{\langle 2 \rangle}$ we show that the inclusion is strict. Let X be a diagonal matrix whose diagonal is $\mathbf{x}$. As $\mathbf{x}$ is a column of A, XA is also a submatrix of $A^{\langle 2 \rangle}$. Hence if $\operatorname{cs} A = \operatorname{cs} A^{\langle 2 \rangle}$, we get that $\operatorname{cs}(XA) \subseteq \operatorname{cs} A$, and therefore $\operatorname{cs}(X^k A) \subseteq \operatorname{cs} A$ for every k. As $\mathbf{e}$ and $\mathbf{x}$ are columns of A, this means that $\operatorname{cs} A$ contains the columns of the nonsingular $m \times m$ Vandermonde matrix

$$\begin{pmatrix} 1 & x_1 & \ldots & x_1^{m-1} \\ 1 & x_2 & & x_2^{m-1} \\ \vdots & \vdots & & \vdots \\ 1 & x_m & \ldots & x_m^{m-1} \end{pmatrix},$$

and therefore $\operatorname{rank} A = m$, contrary to the assumption.

To prove (d), note that

$$\operatorname{rank}(A \circledast B) = \operatorname{rank}(A \otimes \mathbf{1}_k) + \operatorname{rank}(\mathbf{1}_m \otimes B) - \dim L,$$

where $L = \operatorname{cs}(A \otimes \mathbf{1}_k) \cap \operatorname{cs}(\mathbf{1}_m \otimes B)$. As $\operatorname{rank}(A \otimes \mathbf{1}_k) = \operatorname{rank} A$, and $\operatorname{rank}(\mathbf{1}_m \otimes B) = \operatorname{rank} B$, it remains to show that $\dim L \leq 1$, and $\dim L = 1$ if and only if $\mathbf{1}_m \in \operatorname{cs} A$ and $\mathbf{1}_k \in \operatorname{cs} B$.

Suppose $\mathbf{w} \in \operatorname{cs}(A \otimes \mathbf{1}_k) \cap \operatorname{cs}(\mathbf{1}_m \otimes B)$. Then for some $\mathbf{x} \in \mathbb{R}^n$ and some $\mathbf{y} \in \mathbb{R}^\ell$,

$$\mathbf{w} = (A \otimes \mathbf{1}_k)\mathbf{x} = (A \otimes \mathbf{1}_k)(\mathbf{x} \otimes 1) = A\mathbf{x} \otimes \mathbf{1}_k$$

and

$$\mathbf{w} = (\mathbf{1}_m \otimes B)\mathbf{y} = (\mathbf{1}_m \otimes B)(\mathbf{1}_m \otimes \mathbf{y}) = \mathbf{1}_m \otimes B\mathbf{y},$$

and $A\mathbf{x} \otimes \mathbf{1}_k = \mathbf{1}_m \otimes B\mathbf{y}$. This implies that $B\mathbf{y} = (A\mathbf{x})_i \mathbf{1}_k$ for every $i = 1, \ldots, m$. Denoting the common value of all $(A\mathbf{x})_i$'s by c, we get that $\mathbf{w} = c\mathbf{1}_{mk}$. That is, $\dim L \leq 1$. Equality holds if and only if there exists such $\mathbf{x}$, $\mathbf{y}$ and $c \neq 0$, that is, if and only if $A\mathbf{x} = c\mathbf{1}_m$ and $B\mathbf{y} = c\mathbf{1}_k$ for some $c \neq 0$. □

Proposition 4.80. *Let $A \in \mathbb{R}^{m\times n}$, $B \in \mathbb{R}^{k\times \ell}$. Then*
$\operatorname{rank}(A \circledast B)^{\langle 2\rangle} \geq \operatorname{rank} A \cdot \operatorname{rank} B + \operatorname{rank} A^{\langle 2\rangle} - \operatorname{rank} A + \operatorname{rank} B^{\langle 2\rangle} - \operatorname{rank} B.$
In the case that $\mathbf{1}_m \in \operatorname{cs} A$ and $\mathbf{1}_k \in \operatorname{cs} B$, equality holds.

Proof. We prove the inequality, and leave the rest of the proof to the exercises (Exercises 4.43–4.45). Denote

$$\operatorname{rank} A = r_A, \quad \operatorname{rank} B = r_B, \quad \operatorname{rank} A^{\langle 2\rangle} = r_A^{\langle 2\rangle} \quad \text{and} \quad \operatorname{rank} B^{\langle 2\rangle} = r_B^{\langle 2\rangle}.$$

Since permuting rows of A and B does not change the ranks in the formula, we may assume that

$$A = \begin{pmatrix} A_1 \\ A_2 \\ A_3 \end{pmatrix} \quad \text{and} \quad B = \begin{pmatrix} B_1 \\ B_2 \\ B_3 \end{pmatrix},$$

where A_1 consists of r_A rows of A and has full row rank, A_2 consists of $r_A^{\langle 2\rangle} - r_A$ rows, such that

$$\operatorname{rank} \begin{pmatrix} A_1 \\ A_2 \end{pmatrix}^{\langle 2\rangle} = \operatorname{rank} \begin{pmatrix} A_1^{\langle 2\rangle} \\ A_2^{\langle 2\rangle} \end{pmatrix} = r_A^{\langle 2\rangle},$$

and A_3 consists of the remaining $m - r_A^{\langle 2\rangle}$ rows of A. Similarly, B_1 consists of r_B rows of B and has full row rank, B_2 consists of $r_B^{\langle 2\rangle} - r_B$ rows, such that

$$\operatorname{rank} \begin{pmatrix} B_1 \\ B_2 \end{pmatrix}^{\langle 2\rangle} = \operatorname{rank} \begin{pmatrix} B_1^{\langle 2\rangle} \\ B_2^{\langle 2\rangle} \end{pmatrix} = r_B^{\langle 2\rangle},$$

and B_3 consists of the remaining $k - r_B^{\langle 2\rangle}$ rows of B.

As $A_1 \otimes B_1$ is a submatrix of $(A_1 \circledast B_1)^{\langle 2\rangle}$, which has $r_A r_B$ rows,

$$r_A r_B = \operatorname{rank} A_1 \operatorname{rank} B_1 = \operatorname{rank}(A_1 \otimes B_1) \le \operatorname{rank}(A_1 \circledast B_1)^{\langle 2\rangle} \le r_A r_B.$$

That is, $\operatorname{rank}(A_1 \circledast B_1)^{\langle 2\rangle} = r_A r_B$. Let $\mathbf{a}_1^T$ and $\mathbf{b}_1^T$ be the first row of A and the first row of B, respectively. By Lemma 4.79(e),

$$T = \begin{pmatrix} A_1 \circledast B_1 \\ \mathbf{a}_1^T \circledast B_2 \\ A_2 \circledast \mathbf{b}_1^T \end{pmatrix} \tag{4.43}$$

(or, rather, a row permutation thereof) is a submatrix of $A \circledast B$, hence by parts (g) and (c) of the same lemma, a row permutation of

$$S = \begin{pmatrix} A_1^{\langle 2\rangle} \circledast B_1^{\langle 2\rangle} \\ (\mathbf{a}_1^T)^{\langle 2\rangle} \circledast B_2^{\langle 2\rangle} \\ A_2^{\langle 2\rangle} \circledast (\mathbf{b}_1^T)^{\langle 2\rangle} \end{pmatrix}$$

is a submatrix of $(A \circledast B)^{\langle 2\rangle}$. The matrix S has $r_A r_B + r_A^{\langle 2\rangle} - r_A + r_B^{\langle 2\rangle} - r_B$ rows, and we now show it has full row rank. Indeed, suppose $\mathbf{x} \in \mathbb{R}^{r_A r_B}$, $\mathbf{y} \in \mathbb{R}^{r_B^{\langle 2\rangle} - r_B}$ and $\mathbf{z} \in \mathbb{R}^{r_A^{\langle 2\rangle} - r_A}$ satisfy $\begin{pmatrix} \mathbf{x}^T & \mathbf{y}^T & \mathbf{z}^T \end{pmatrix} S = \mathbf{0}^T$. Then

$$\mathbf{x}^T\left(A_1^{\langle 2\rangle} \circledast B_1^{\langle 2\rangle}\right) + \mathbf{y}^T\left((\mathbf{a}_1^T)^{\langle 2\rangle} \circledast B_2^{\langle 2\rangle}\right) + \mathbf{z}^T\left(A_2^{\langle 2\rangle} \circledast (\mathbf{b}_1^T)^{\langle 2\rangle}\right) = \mathbf{0}^T.$$

Considering the first n^2 columns in S, we get that

$$\mathbf{x}^T(A_1^{\langle 2\rangle} \otimes \mathbf{1}_{r_B}) + \mathbf{y}^T(\mathbf{1}_{r_B^{\langle 2\rangle} - r_B} \otimes (\mathbf{a}_1^T)^{\langle 2\rangle}) + \mathbf{z}^T A_2^{\langle 2\rangle} = \mathbf{0}^T.$$

As the first two summands are in the row space of $A_1^{\langle 2\rangle}$, we get that $\mathbf{z}^T A_2^{\langle 2\rangle}$ is in the intersection of $\operatorname{rs} A_1^{\langle 2\rangle}$ and $\operatorname{rs} A_2^{\langle 2\rangle}$. But the rows of

$$\begin{pmatrix} A_1^{\langle 2\rangle} \\ A_2^{\langle 2\rangle} \end{pmatrix}$$

are linearly independent, implying that $\mathbf{z}^T A_2^{\langle 2\rangle} = \mathbf{0}^T$, and therefore $\mathbf{z} = \mathbf{0}$. Similarly, by considering the last ℓ^2 columns of S we find that $\mathbf{y} = \mathbf{0}$. But then

$$\mathbf{x}^T\left(A_1^{\langle 2\rangle} \circledast B_1^{\langle 2\rangle}\right) = \mathbf{0}^T,$$

and this implies $\mathbf{x} = \mathbf{0}$, since $A_1^{\langle 2\rangle} \circledast B_1^{\langle 2\rangle}$ has full row rank. Hence we get that

$$\operatorname{rank}(A \circledast B)^{\langle 2\rangle} \ge \operatorname{rank} S = r_A r_B + r_A^{\langle 2\rangle} - r_A + r_B^{\langle 2\rangle} - r_B. \qquad \square$$

If the set of rows of U and the set of rows of V are each a complete set of zeros in Δ of some copositive matrix, we may use $\circledast$ to generate a matrix whose rows are a complete set of representing zeros of some copositive matrix of greater order.

Proposition 4.81. *Let $U \in \mathbb{R}^{m\times n}$ be a matrix whose rows are all the zeros in Δ_n of an $n \times n$ copositive matrix M_U. Let $V \in \mathbb{R}^{k\times \ell}$ be a matrix whose rows are all the zeros in Δ_ℓ of a $k \times k$ copositive matrix M_V. Then for every $0 < t, s < 1$ such that $t + s = 1$ there exists an $(n+\ell) \times (n+\ell)$ copositive matrix M, such that the rows of $(tU \circledast sV)$ are all the zeros of M in $\Delta_{n+\ell}$.*

Proof. Let

$$M = \begin{pmatrix} M_U & 0 \\ 0 & M_V \end{pmatrix} + \begin{pmatrix} s\mathbf{1}_n \\ -t\mathbf{1}_\ell \end{pmatrix} \begin{pmatrix} s\mathbf{1}_n \\ -t\mathbf{1}_\ell \end{pmatrix}^T.$$

The matrix M is a copositive matrix, as a sum of a copositive matrix and a positive semidefinite one. A vector $\mathbf{w} = \begin{pmatrix} \mathbf{x} \\ \mathbf{y} \end{pmatrix} \in \Delta_{n+\ell}$, where $\mathbf{x} \in \mathbb{R}^n_+$ and $\mathbf{y} \in \mathbb{R}^\ell_+$, is a zero of M if and only if it is a zero of both $M_U \oplus M_V$ and of $\begin{pmatrix} s\mathbf{1}_n \\ -t\mathbf{1}_\ell \end{pmatrix} \begin{pmatrix} \mathbf{1}_n \\ -\mathbf{1}_\ell \end{pmatrix}^T$. That is, if and only if $\mathbf{x}$ is a zero of M_U, $\mathbf{y}$ is a zero of M_V, and $s\mathbf{1}_n^T\mathbf{x} = t\mathbf{1}_\ell^T\mathbf{y}$, $\mathbf{1}_n^T\mathbf{x} + \mathbf{1}_\ell^T\mathbf{y} = 1$. As the rows of $(tU \circledast sV)$ consist of all the vectors of the form $\begin{pmatrix} t\mathbf{u}^T & s\mathbf{v}^T \end{pmatrix}$, where $\mathbf{u} \in \Delta_n$ is a zero of M_U and $\mathbf{v} \in \Delta_\ell$ is a zero of M_V, this concludes the proof. $\square$

Note that if U and V, t and s, are as in Proposition 4.22, then U, V (and thus also $tU \circledast sV$) are row stochastic. By Lemma 4.79(d) we deduce that $\operatorname{rank}(tU \circledast sV) = \operatorname{rank} U + \operatorname{rank} V - 1$. Every completely positive matrix that is orthogonal to the matrix M constructed in the proposition is in cone $\{\mathbf{w}\mathbf{w}^T \mid \mathbf{w}^T \text{ is a row of } tU \circledast sV\}$. By Lemma 4.79(a), this cone spans a subspace of $\mathcal{S}_{n+\ell}$ of dimension $d = \operatorname{rank}(U \circledast V)^{\langle 2\rangle}$. As in Remark 4.78, the maximum cp-rank of a matrix orthogonal to M is d, where by Proposition 4.80

$$d = \operatorname{rank} U \operatorname{rank} V + \operatorname{rank} U^{\langle 2\rangle} - \operatorname{rank} U + \operatorname{rank} V^{\langle 2\rangle} - \operatorname{rank} V.$$

For $A \in \mathbb{R}^{m\times n}$, $A^{\circledast 1} = A$, and for $p \geq 2$ define recursively $A^{\circledast p} = (A^{\circledast(p-1)}) \circledast A$. Then if $U \in \mathbb{R}^{m\times n}$ is as in Proposition 4.22, $U^{\circledast p}$ is of order $m^p \times np$, and it is a complete representing set of zeros of some $np \times np$ copositive matrix. Hence we may deduce the following.

Corollary 4.82. *Let U be a matrix with row sums equal to 1, whose rows are all the zeros in Δ_n of some $n \times n$ copositive matrix. Suppose* $\operatorname{rank} U = r$ *and* $\operatorname{rank} U^{\langle 2\rangle} = q$. *Then for any positive integer p,*

$$\operatorname{rank}\left(U^{\circledast p}\right)^{\langle 2\rangle} = \frac{1}{2}(r-1)^2 p(p-1) + (q-1)p + 1,$$

and there exists a completely positive matrix A of order $np \times np$ such that

$$\operatorname{cpr} A = \frac{1}{2}(r-1)^2 p(p-1) + (q-1)p + 1.$$

Proof. As explained above, there exists a completely positive matrix in cone $\left\{\mathbf{w}\mathbf{w}^T \mid \mathbf{w}^T \text{ is a row of } U^{\circledast p}\right\}$, whose cp-rank is the dimension of the cone, which is $\operatorname{rank}\left(U^{\circledast p}\right)^{\langle 2\rangle}$. Using induction, we get that this rank is $\frac{1}{2}(r-1)^2 p(p-1) + (q-1)p + 1$. (Note that $\operatorname{rank} U^{\circledast p} = pr - p + 1$.) □

For the bound (4.39) in Theorem 4.77 consider $U = I_m$, which is a row stochastic matrix whose rows are all the zeros in Δ_m of the copositive matrix $J_m - I_m$. As $\operatorname{rank} I_m = m$, we get, by Corollary 4.82, that $\operatorname{rank}\left(I_m^{\circledast i}\right)^{\langle 2\rangle} = \rho(m,i)$, where

$$\rho(m,i) = \frac{1}{2}(m-1)^2 i(i-1) + (m-1)i + 1,$$

which may be rewritten as

$$\rho(m,i) = \frac{(mi)^2}{2} - mi\left(i + \frac{m}{2}\right) + 2mi + \frac{i(i-3)}{3} + 1. \tag{4.44}$$

Example 4.83. For $n = 6$, the matrix $U = I_3^{\circledast 2}$ is a 9×6 matrix, whose rows form a representative set of zeros of a 6×6 copositive matrix, and $\operatorname{rank} U = 5$, as $\operatorname{rank} I_3^{\langle 2\rangle} = 3$. Hence this the already known lower bound

$$p_6 \geq \operatorname{rank} U^{\langle 2\rangle} \geq \rho(3,2) = 9.$$

Example 4.84. A more interesting use of the above construction is the following. Let $U = I_4$. Then $\operatorname{rank} U = 4$ and $U^{\langle 2\rangle} = \begin{pmatrix} E_{11} & E_{22} & E_{33} & E_{44} \end{pmatrix} \in \mathbb{R}^{4\times 16}$ also has rank 4. By Corollary 4.23, $p_{12} \geq \frac{1}{2}(4-1)^2 \cdot 6 + 3\cdot 3 + 1 = 37$. This implies that there exists a completely positive matrix of order 12 with cp-rank at least 37. Note that the DJL bound for $n = 12$ is $144/4 = 36$.

Proof of the first bound in Theorem 4.77. For $n \geq 5$, if m, k and $1 \leq i \leq k$ are positive integers such that $n = mi + (m+1)(k-i) = mk + k - i$, the matrix $U = I_m^{\circledast i} \circledast I_{m+1}^{\circledast(k-i)}$ is, by the above discussion, a $\{0,1\}$-matrix, whose rows are a representing set of all the zeros of some $n \times n$ completely positive matrix, and thus $p_n \geq \operatorname{rank} U^{\langle 2\rangle}$. As $\operatorname{rank} I_m = m$, it follows from Lemma 4.79 and Proposition 4.80 and that rank $I_m^{\circledast i} = (m-1)i+1$, rank $I_{m+1}^{\circledast k-i} = m(k-i)i+1$, rank $\left(I_m^{\circledast i}\right)^{\langle 2\rangle} = \rho(m,i)$, and rank $\left(I_{m+1}^{\circledast k-i}\right)^{\langle 2\rangle} = \rho(m+1,k-i)$, where $\rho(\cdot,\cdot)$ is as in (4.44). By Proposition 4.80, after some tidying up, we get that rank $U^{\langle 2\rangle} \geq f_n(m,k)$ where

$$f_n(m,k) = \frac{1}{2}(n-k)^2 + \frac{3}{2}(n-k) - mn + \frac{1}{2}m(m+1)k + 1.$$

That is, $p_n \geq f_n(m,k)$ for every m,k such that $mk \leq n < (m+1)k$.

Let k and m be positive integers such that

$$\sqrt{\frac{n}{2}} - \frac{1}{4} \leq k \leq \sqrt{\frac{n}{2}} + \frac{3}{4}, \qquad \frac{n}{k} - 1 < m \leq \frac{n}{k}.$$

For such k and m it can be shown that

$$f(m,k) \geq \frac{n^2}{2} - n\sqrt{2n} + \frac{9}{4}n - \sqrt{2n} + \frac{1}{16}. \qquad \square$$

To derive the bound (4.40), further matrix operations are employed: For an $m \times n$ matrix A and a real t, define

$$t \star A = \begin{cases} tA, & t \neq 0, \\ \mathbf{0}_n^T, & t = 0. \end{cases}$$

If U is an $m \times n$ matrix, and for $1 \leq i \leq n$ the matrix V_i is of order $k_i \times \ell_i$, denote

$$\circledast_U(V_1,\ldots,V_k) = \begin{pmatrix} u_{11} \star V_1) \circledast \cdots \circledast (u_{1n} \star V_n) \\ \vdots \\ (u_{m1} \star V_1) \circledast \cdots \circledast (u_{mn} \star V_n) \end{pmatrix}.$$

(This generalizes the $sU \circledast tV$ construction of Proposition 4.81.) It can be seen that if each of the matrices U and $V_1,\ldots,V_n$ is the complete set of zeros in Δ of a copositive matrix, then so is $\circledast_U(V_1,\ldots,V_n)$. We leave the proof as an exercise. The bound (4.40) is obtained by constructing sets $\mathcal{J}^{(k)}$ of matrices, each containing one matrix of each order n, whose rows are zeros of copositive matrices. Starting with $\mathcal{J}^{(1)} = \{I_n \mid n \geq 1\}$, the sets $\mathcal{J}^{(k)}$, $k \geq 2$, are constructed recursively from matrices in $\mathcal{J}^{(m)}$, $m < k$, with the aid of $\circledast_U$ for matrices U whose rows are the zeros of some circulant matrices. The matrix of order n in $\mathcal{J}^{(k)}$ is chosen so that the corresponding completely positive matrix has a high cp-rank. The lower bound on p_n is obtained by bounding from below the cp-ranks of the matrices corresponding to the order n matrices in $\left(\mathcal{J}^{(k)}\right)_{k\geq 1}$. See a reference in the notes to this section for the details.

Remark 4.85. Both lower bounds on p_n in Theorem 4.77 may be improved for specific values of n. The first bound may be improved by starting with a matrix whose rows are zeros of a certain copositive matrix other than I_m. For example, we may use the matrices U_k of zeros of the copositive matrices M of order k for $k = 7, 9, 11, 15$ that appear in Example 4.76 and in Exercises 4.38, 4.40 and 4.41 — see Remark 4.78. In these cases, the exact value of rank $U_k^{\langle 2 \rangle}$ is known. Also the second bound may be improved for sufficiently small values of n by choosing the members in $\mathcal{J}^{(k)}$ optimally.

Exercises

4.36. Let M be an $n \times n$ symmetric circulant copositive matrix with first column equal to $\mathbf{a}$. That is,

$$M = \begin{pmatrix} \mathbf{a} & P\mathbf{a} & P^2\mathbf{a} & \dots & P^{n-1}\mathbf{a} \end{pmatrix},$$

where $P = \begin{pmatrix} \mathbf{e}_2 & \mathbf{e}_3 & \dots & \mathbf{e}_n & \mathbf{e}_1 \end{pmatrix}$, and $a_i = a_{n+2-i}$ for every $2 \le i \le n$. Let

$$R = \begin{pmatrix} \mathbf{e}_n & \mathbf{e}_{n-1} & \dots & \mathbf{e}_2 & \mathbf{e}_1 \end{pmatrix}$$

be the reversal matrix. Denote by $\oplus$ and $\ominus$ addition and substraction modulo n on $\{1, \dots, n\}$. For each zero $\mathbf{u} \in \Delta$ of M let

$$\mathcal{D}_\mathbf{u} = \{d \in \{1, \dots, n-1\} \mid d = i \ominus j \text{ for at least one pair } \{i, j\} \subseteq \operatorname{supp} \mathbf{u}\},$$
$$\mathcal{U}_\mathbf{u} = \{d \in \{1, \dots, n-1\} \mid d = i \ominus j \text{ for exactly one pair } \{i, j\} \subseteq \operatorname{supp} \mathbf{u}\}.$$

Suppose $\mathbf{u} \in \Delta$ is a zero of M.

(a) Show that $P^i\mathbf{u}$ is a zero of M for every i, and $R\mathbf{u}$ is a zero of M.
(b) Show that $\mathcal{D}_{P^i\mathbf{u}} = \mathcal{D}_\mathbf{u}$ and $\mathcal{U}_{P^i\mathbf{u}} = \mathcal{U}_\mathbf{u}$ for every i, and also $\mathcal{D}_{R\mathbf{u}} = \mathcal{D}_\mathbf{u}$ and $\mathcal{U}_{R\mathbf{u}} = \mathcal{U}_\mathbf{u}$.
(c) Show that for an even n, $\frac{n}{2} \notin \mathcal{U}_\mathbf{u}$.
(d) Suppose that $Q = \{P^i\mathbf{u} \mid i = 1, \dots, n\}$ has n different elements, $R\mathbf{u} \notin Q$, and there exist two different $d, d' \in \mathcal{U}_\mathbf{u}$ such that $d = r \ominus s, d' = r' \ominus s'$, where $r \oplus s \ominus r' \ominus s'$ and n are coprime. Consider the set of symmetric matrices

$$\mathcal{Q} = \{\mathbf{v}\mathbf{v}^T \mid \mathbf{v} \in Q\} \cup \{(R\mathbf{v})(R\mathbf{v})^T \mid \mathbf{v} \in Q\}.$$

Show that $\mathcal{Q}$ spans a subspace of dimension $2n-1$ in $\mathcal{S}_n$, and every $2n-1$ elements in $\mathcal{Q}$ are linearly independent.

Hint: A matrix $A \in \mathbb{R}^{n\times n}$ is circulant if and only if $PAP^T = A$, and if and only if $P^TAP = A$. If a circulant A is also symmetric, then $RAR = A$.

In (d), to show that $\mathcal{Q}$ is not linearly independent, show that $\sum_{\mathbf{v}\in Q} \mathbf{v}\mathbf{v}^T$ is a symmetric circulant matrix, hence

$$\sum_{\mathbf{v}\in Q} \mathbf{v}\mathbf{v}^T = R\left(\sum_{\mathbf{v}\in Q} \mathbf{v}\mathbf{v}^T\right)R = \sum_{\mathbf{v}\in Q} (R\mathbf{v})(R\mathbf{v})^T.$$

To show that every $2n-1$ elements in $\mathcal{Q}$ are linearly independent, let $d = r \ominus s$ and $d' = r' \ominus s'$, and $\mathbf{u}_i = P^i\mathbf{u}$, $\mathbf{v}_i = R\mathbf{u}_i$ for $i = 1, \ldots, n$. Suppose

$$\sum_{i=1}^{n} x_i \mathbf{u}_i \mathbf{u}_i^T + \sum_{i=1}^{n} y_i \mathbf{v}_i \mathbf{v}_i^T = 0,$$

and consider the $r \oplus l, s \oplus l$ and $r' \oplus l, s' \oplus l$ entries of this sum for $l = 1, \ldots, n$, to conclude that $x_l + y_{r \oplus s \ominus 1 \ominus l} = 0$ and $x_l + y_{r' \oplus s' \ominus 1 \ominus l} = 0$ for each l. Deduce that there exists t such that $y_i = -x_i = -t$ for every $i = 1, \ldots, n$.

4.37. Let M, P and R, $\mathbf{u}$, Q and $\mathcal{Q}$ satisfy the conditions of Exercise 4.36(d). Let $Q' = \{R\mathbf{v} \mid \mathbf{v} \in Q\}$, and suppose also that $Q \cup Q'$ is the set of all the zeros of M in Δ. Show that if $A \in \operatorname{cone}\mathcal{Q}$, then $\operatorname{cpr} A \leq 2n-1$, and if A has a representation the form

$$A = \sum_{i=1}^{n} x_i \mathbf{u}_i \mathbf{u}_i^T + \sum_{i=1}^{n} y_i \mathbf{v}_i \mathbf{v}_i^T \tag{4.45}$$

with $x_i, y_i \geq 0$, in which the number of positive coefficients among $x_1, \ldots, x_n, y_1, \ldots, y_n$ is $p < 2n$, then $\operatorname{cpr} A = p$ and A has a unique minimal CP representation.
Hint: For the uniqueness note that if A has two minimal representations in the form (4.45), then the two coefficients vectors of these representations differ by a vector $t\mathbf{u}$, where $\mathbf{u}^T = \begin{pmatrix} \mathbf{e}^T & -\mathbf{e}^T \end{pmatrix}$, and the minimality implies $t = 0$.

4.38. Let M be the symmetric circulant matrix in $\mathcal{S}_9$ with first column

$$\mathbf{a} = (1089, -1056, 959, -484, 231, 231, -484, 959, -1056)^T.$$

Using exact arithmetic as in Example 4.76, check copositivity of M and show that the zeros of M in Δ_9 are

$$\mathbf{u}_i = \begin{cases} P^i\mathbf{u}, & i = 1, \ldots, 9 \\ P^i\mathbf{v}, & i = 10, \ldots, 18 \\ P^i\mathbf{w}, & i = 19, \ldots, 27 \end{cases},$$

where $P = \begin{pmatrix} \mathbf{e}_2 & \mathbf{e}_3 & \mathbf{e}_4 & \mathbf{e}_5 & \mathbf{e}_6 & \mathbf{e}_7 & \mathbf{e}_8 & \mathbf{e}_9 & \mathbf{e}_1 \end{pmatrix}$ and

$$\mathbf{u} = \frac{1}{26}(11, 12, 0, 0, 3, 0, 0, 0, 0)^T,$$
$$\mathbf{v} = \frac{1}{26}(12, 11, 0, 0, 0, 0, 3, 0, 0)^T,$$
$$\mathbf{w} = \frac{1}{320}(33, 64, 33, 0, 0, 0, 0, 0, 0)^T.$$

Show that for any $1 \le k \le 18$ the set $\{\mathbf{u}_i\mathbf{u}_i^T \mid 1 \le i \le 27, i \ne k\}$ in $\mathcal{S}_9$ is linearly independent, and hence for any $x_i > 0$, $1 \le i \le 27$, $i \ne k$, the matrix

$$A = \sum_{\substack{i=1 \\ i \ne k}}^{27} x_i \mathbf{u}_i \mathbf{u}_i^T$$

is a completely positive matrix with $\operatorname{cpr} A = 26$. Thus $p_9 \ge 26$.
Hint: If $\sum_{\substack{i=1 \\ i \ne k}}^{27} x_i \mathbf{u}_i \mathbf{u}_i^T = B$, then for $i = 19, \ldots 27$, $x_i = \frac{102400}{1089} b_{i+1\, i+3}$. Then use Exercise 4.36(d).

4.39. Find lower bounds for p_8 and p_{10} using the example in Exercise 4.38. (We use here the same notation as in that exercise.)

(1) Let $\Omega = \{1 \le i \le 27 \mid i \ne 4, 7, 8, 11, 17, 18, 24, 25, 26\}$. Then if $x_i > 0$, for every $i \in \Omega$, the matrix $\sum_{i \in \Omega} x_i \mathbf{u}_i \mathbf{u}_i^T$ has the form $A \oplus 0_1$, where A is an 8×8 completely positive matrix with $\operatorname{cpr} A = |\Omega| = 18$, hence $p_8 \ge 18$.
(2) Let A be a completely positive matrix of order 9 with $\operatorname{cpr} A = 26$. Then $A \oplus 1$ is a completely positive matrix of order 10 with cp-rank 27. Hence $p_{10} \ge 27$.

4.40. Show that $p_{11} \ge 32$, by considering, as in Exercise 4.38, the zeros of the 11×11 symmetric circulant copositive matrix M, whose first column is

$$\mathbf{a} = (32, 32, 18, 4, -24, -31, -31, -24, 4, 18, 32)^T.$$

4.41. Show that $p_{15} \ge 95$, by considering, as in Exercise 4.38, the zeros of the 15×15 symmetric circulant copositive matrix M, whose first column is

$$\mathbf{a} = (d, a, a, b, a, c, b, a, a, b, c, a, b, a, a)^T,$$

where $(a, b, c, d) = (-2609, 1803, 4009, 7318)$.

4.42. Complete the proof of Lemma 4.79.
Hint: The rows of $A \circledast B$ are $\mathbf{c}_{(i,j)}^T = \begin{pmatrix} \mathbf{a}_i^T & \mathbf{b}_j^T \end{pmatrix}$ ordered lexicographically, where $\mathbf{a}_i^T$, $\mathbf{b}_j^T$ are the rows of A and B.

4.43. Let A be an $m \times n$ matrix such that $A\mathbf{u} = \mathbf{1}_m$ for some $\mathbf{u} \in \mathbb{R}^n$. Let M be a submatrix of A consisting of p rows, and let $\mathbf{a}^T$ be a row of A.

(a) Show that if $\mathbf{a}^T = \mathbf{x}^T M$, then $\mathbf{x}^T \mathbf{1}_p = 1$.
(b) Show that if $(\mathbf{a}^T)^{\langle 2 \rangle} = \mathbf{x}^T M^{\langle 2 \rangle}$, then also $\mathbf{a}^T = \mathbf{x}^T M$.

Hint: For (b): $\mathbf{a}^T = (\mathbf{a}^T \otimes \mathbf{a}^T)(1 \otimes \mathbf{u}) = \sum_{i=1}^p x_i \left((\mathbf{e}_i^T M) \otimes (\mathbf{e}_i^T M) \right) (1 \otimes \mathbf{u})$.

4.44. Let A be an $m \times n$ matrix such that $A\mathbf{u} = \mathbf{1}_m$ for some $\mathbf{u} \in \mathbb{R}^n$, B a $k \times \ell$ matrix such that $B\mathbf{v} = \mathbf{1}_k$ for some $\mathbf{v} \in \mathbb{R}^\ell$. Let M be a submatrix of A consisting of p rows, and N a submatrix of B consisting of q rows. Let $\mathbf{a}^T$ be a row of A, $\mathbf{b}^T$ and $\mathbf{c}^T$ rows of B.

(a) Show that if $(\mathbf{a}^T)^{\langle 2 \rangle} = \mathbf{x}^T M^{\langle 2 \rangle}$, and $\mathbf{b}^T = \mathbf{y}^T N$, then $(\mathbf{a}^T \circledast \mathbf{b}^T)^{\langle 2 \rangle} = \mathbf{x}^T (M \circledast \mathbf{b}^T)^{\langle 2 \rangle}$.
(b) Show that if $\mathbf{a}^T = \mathbf{x}^T M$ and $\mathbf{b}^T = \mathbf{y}^T N$, then
$$(\mathbf{a}^T \circledast \mathbf{b}^T)^{\langle 2 \rangle} = \mathbf{x}^T (M \circledast \mathbf{b}^T)^{\langle 2 \rangle} + \mathbf{y}^T \left(\mathbf{a}^T \circledast N\right)^{\langle 2 \rangle} - (\mathbf{x} \otimes \mathbf{y})^T \left(M \circledast N\right)^{\langle 2 \rangle}.$$
(c) Show that if $\mathbf{a}^T = \mathbf{x}^T M$, then
$$(\mathbf{a}^T \circledast \mathbf{b}^T)^{\langle 2 \rangle} = \mathbf{x}^T (M \circledast \mathbf{b}^T)^{\langle 2 \rangle} + (\mathbf{a}^T \circledast \mathbf{c}^T)^{\langle 2 \rangle} - \mathbf{x}^T (M \circledast \mathbf{c}^T)^{\langle 2 \rangle}.$$

Note that similar results hold when the conditions on $\mathbf{a}$ and $\mathbf{b}$, M and N, are reversed.
Hint: $(\mathbf{a}^T \circledast \mathbf{b}^T)^{\langle 2 \rangle} = \begin{pmatrix} (\mathbf{a}^T)^{\langle 2 \rangle} & \mathbf{a}^T \otimes \mathbf{b}^T & \mathbf{b}^T \otimes \mathbf{a}^T & (\mathbf{b}^T)^{\langle 2 \rangle} \end{pmatrix}$. Prove the equalities for each of the four row sections separately.

4.45. Let A, B, A_i and B_i, $i = 1, \ldots, 3$, $\mathbf{a}_1^T$, $\mathbf{b}_1^T$ and T as in Proposition 4.80 and its proof. Suppose that $A\mathbf{u} = \mathbf{1}_m$ and $B\mathbf{v} = \mathbf{1}_k$ for some $\mathbf{u} \in \mathbb{R}^n$, $\mathbf{v} \in \mathbb{R}^\ell$. Prove the following.

(a) Let $\mathbf{a}^T$ be a row of A. Then $(\mathbf{a}^T \circledast \mathbf{b}_1^T)^{\langle 2 \rangle} \in \operatorname{rs} T^{\langle 2 \rangle}$.
(b) Let $\mathbf{a}^T$ be a row of A, $\mathbf{b}^T$ a row of B_1. Then $(\mathbf{a}^T \circledast \mathbf{b}^T)^{\langle 2 \rangle} \in \operatorname{rs} T^{\langle 2 \rangle}$. Conclude that $\operatorname{rs}(A \circledast B_1)^{\langle 2 \rangle} \subseteq \operatorname{rs} T^{\langle 2 \rangle}$, and similarly $\operatorname{rs}(A_1 \circledast B)^{\langle 2 \rangle} \subseteq \operatorname{rs} T^{\langle 2 \rangle}$.
(c) Let $\mathbf{a}^T$ be a row of A_2, $\mathbf{b}^T$ a row of $\begin{pmatrix} B_1 \\ B_2 \end{pmatrix}$. Then $(\mathbf{a}^T \circledast \mathbf{b}^T)^{\langle 2 \rangle} \in \operatorname{rs} T^{\langle 2 \rangle}$.

  Conclude that $\left(\begin{pmatrix} A_1 \\ A_2 \end{pmatrix} \circledast \begin{pmatrix} B_1 \\ B_2 \end{pmatrix} \right)^{\langle 2 \rangle} \subseteq \operatorname{rs} T^{\langle 2 \rangle}$.
(d) Let $\mathbf{a}^T$ be a row of A, $\mathbf{b}^T$ a row of B. Then $(\mathbf{a}^T \circledast \mathbf{b}^T)^{\langle 2 \rangle} \in \operatorname{rs} T^{\langle 2 \rangle}$. This completes the proof of the equality claim in Proposition 4.80.

Hint: Part (a) follows from Exercise 4.44(a) for $M = \begin{pmatrix} A_1 \\ A_2 \end{pmatrix}$. For (b) use part (a) and Exercise 4.44(c) for $M = A_1$, $N = B_1$, $\mathbf{c}^T = \mathbf{b}_1^T$. For (c), part (b) and Exercise 4.44(b) with $M = A_1$, $N = \begin{pmatrix} B_1 \\ B_2 \end{pmatrix}$. For (d) part (c) and Exercise 4.44(b) with $(\mathbf{a}^T)^{\langle 2\rangle} = \mathbf{x}^T \begin{pmatrix} A_1 \\ A_2 \end{pmatrix}^{\langle 2\rangle}$, $(\mathbf{b}^T)^{\langle 2\rangle} = \mathbf{y}^T \begin{pmatrix} B_1 \\ B_2 \end{pmatrix}^{\langle 2\rangle}$.

4.46. Let the rows of an $m \times n$ matrix U be all the zeros in Δ_n of a copositive M. Let the rows of V_i of order $k_i \times \ell_i$ be all the zeros in Δ_{ℓ_i} of a copositive M_i, $i = 1, \ldots, n$. Then the rows of $\circledast_U(V_1, \ldots, V_n)$ form a complete set of zeros in Δ_ℓ of some copositive matrix K, where $\ell = \sum_{i=1}^n \ell_i$.
Hint: Consider $K = (M_1 \oplus \cdots \oplus M_n) + C$, where C is the block matrix whose i,j block is $m_{ij} J_{\ell_i \times \ell_j}$, $1 \leq i,j \leq n$.

Notes. The computation of p_5 was initiated in [Loewy and Tam (2003)], where Proposition 4.69 was proved (including the preceding lemma). It was completed in [Shaked-Monderer *et al.* (2013)], where Propositions 4.66 and 4.67 were proved. The idea of Lemma 4.65 was used there, but the lemma was stated in this general form in [Shaked-Monderer (2017)]. This is where the proof of Proposition 4.72 may be found.

Theorem 4.73 and Proposition 4.74 were stated and proved in [Shaked-Monderer *et al.* (2015)].

The first examples of matrices disproving the DJL conjecture, matrices of orders $7 \leq n \leq 11$ with cp-rank greater than the DJL bound, are due to [Bomze *et al.* (2014)]. These are the matrices in Example 4.76 and Exercises 4.38–4.40, and the proof ideas of [Bomze *et al.* (2014)] are outlined before the example, and in Exercises 4.36–4.37.

The lower bound (4.39) of Theorem 4.77, and all the constructions and propositions that are part of its proof, are due to [Bomze *et al.* (2015b)], as are the examples for $n = 12$ and $n = 15$. The matrix product $\circledast$ is related the Khatri-Rao product. The lower bound (4.40) is due to [Schachinger (2020)]. The construction of $\circledast_U(V_1, \ldots, V_n)$ and the proof of the equality in Proposition 4.80 in Exercises 4.43–4.46 is based on that article, where a more general approach was applied, and used also to derive asymptotically matching upper and lower bounds on the cp-ranks of completely positive tensors.

Still open: *Does p_6 equal 9? (what is the maximum cp-rank of a 6×6 completely positive matrix that has a zero off-diagonal entry?) For $n \geq 6$,* *finding p_n, or bounding it more tightly from above and below, is still an open problem.*

4.6 When is the cp-rank equal to the rank?

There is no complete answer to the question in the title of this section. We have already encountered some examples of matrices that have cp-rank equal to the rank. We know that $\operatorname{cpr} A = \operatorname{rank} A$ if the completely positive matrix A is of rank 2 or of order 3 or less, and if $G(A)$ is a tree. Other examples appeared in Section 3.8: nonsingular matrices that have square factorizations, Soules matrices, and matrices whose row sums satisfy certain inequalities (Theorem 3.137). One more case in which $\operatorname{cpr} A = \operatorname{rank} A$ is when a matrix A has a chordal graph, and it rank is the minimum possible rank, which is the clique covering number of the chordal graph. See Theorem 4.117 in Section 4.8.

In this section we present a qualitative answer to the question in the title. We describe graphs with the property that every completely positive matrix realization of the graph has cp-rank equal to the rank.

Here is an example of such graph.

Example 4.86. By Corollary 4.55, if A is a CP matrix realization of C_n, $n \geq 4$, then $\operatorname{cpr} A = n$. If $n \geq 5$ is odd, then every CP realization of C_n is nonsingular by Remark 3.82. Hence $\operatorname{cpr} A = \operatorname{rank} A$ for every completely positive matrix A whose graph is an odd cycle of length 5 or more.

If $n \geq 4$ is even, there exist singular CP matrix realizations of C_n (see Exercise 3.39). If A is a such a matrix, then $\operatorname{cpr} A = n > \operatorname{rank} A$.

Remark 4.87. A graph G on n vertices has the property that every nonsingular matrix A with $G(A) = G$ satisfies $\operatorname{cpr} A = \operatorname{rank} A$, if and only if $\operatorname{cpr} G = n$. To see that, observe that if $\operatorname{cpr} G = n$, then for every nonsingular matrix A with $G(A) = G$

$$n = \operatorname{rank} A \leq \operatorname{cpr} A \leq \operatorname{cpr} G = n.$$

For the reverse implication, use the fact that every completely positive matrix A satisfying $G(A) = G$ is a limit of nonsingular completely positive matrices with the same graph, $A = \lim_{\varepsilon \to 0+}(A + \varepsilon I)$. If for each $\varepsilon > 0$ $\operatorname{cpr}(A + \varepsilon I) = n$, then $\operatorname{cpr} A \leq n$. Hence $\operatorname{cpr} G = n$.

Of course, for every graph G on n vertices, $\operatorname{cpr} G \geq n$, so the graphs which satisfy $\operatorname{cpr} G = n$ attain the minimal possible graph cp-rank.

In this section we characterize all graphs G which satisfy

$$\text{For any nonsingular CP matrix realization } A \text{ of } G, \quad \operatorname{cpr} A = \operatorname{rank} A. \tag{4.46}$$

(By Remark 4.87 these are the graphs G for which $\operatorname{cpr} G = |V(G)|$.) We will also characterize all graphs G which satisfy

$$\text{For any CP matrix realization } A \text{ of } G, \quad \operatorname{cpr} A = \operatorname{rank} A. \tag{4.47}$$

Trees, odd cycles and graphs on 3 vertices or less satisfy (4.47). Even cycles satisfy (4.46) but not (4.47). Every graph on 4 vertices satisfies (4.46). The graphs T_4 and K_4 do not satisfy (4.47) (see Example 4.9 and Exercise 4.10). Of course, every graph which satisfies (4.47), satisfies also (4.46). Some graphs satisfy neither.

Remark 4.88. A connected triangle free graph which has more edges than vertices does not satisfy (4.46). Such graph G is not a tree, hence

$$\operatorname{cpr} G = |E(G)| > |V(G)|.$$

A graph G satisfies (4.46) if and only if every component of G satisfies (4.46). The same holds for (4.47). We will therefore consider mostly connected graphs. In the next theorem we describe a class of connected graphs which satisfy (4.47).

Theorem 4.89. *If G is a block-clique graph, and each block of G is either an edge or a triangle, then G satisfies* (4.47).

Proof. We consider connected graphs, and use induction on the number of blocks in G. If G has no cut vertex, then G is either a K_2 or a K_3 and it satisfies (4.47). Suppose the claim holds for every block-clique graph which has less than m blocks, and let G be a graph with m blocks, $m \geq 2$, each of them a triangle or an edge. Then $G = G_1 \cup G_2$, where $G_1 \cap G_2$ is a single vertex, and G_1 and G_2 are connected. We may assume that $V(G_1) = \{1, \ldots, k\}$, and $V(G_2) = \{k, \ldots, n\}$. Let A be any CP matrix realization of G. Then, by Lemma 4.59 (since G_1 is a completely positive graph),

$$A = (A_1 \oplus 0_{n-k}) + (0_{k-1} \oplus A_2),$$

where A_1 and A_2 are completely positive, $G(A_1) = G_1$ and $G(A_2) = G_2$, A_1 is singular, and $\operatorname{rank} A = \operatorname{rank} A_1 + \operatorname{rank} A_2$. By the induction hypothesis,

$\operatorname{cpr} A_1 = \operatorname{rank} A_1$ and $\operatorname{cpr} A_2 = \operatorname{rank} A_2$. Therefore

$$\begin{aligned}\operatorname{rank} A &\leq \operatorname{cpr} A \\ &\leq \operatorname{cpr} A_1 + \operatorname{cpr} A_2 \\ &= \operatorname{rank} A_1 + \operatorname{rank} A_2 = \operatorname{rank} A\end{aligned}$$

and this yields the desired equality. □

Theorem 4.89 implies the following generalization.

Corollary 4.90. *If H is a block of a connected graph G, and every other block of G is either an edge or a triangle, then G satisfies* (4.46) ((4.47)) if and only if H satisfies (4.46) (respectively, (4.47)).

Proof. Again we use induction on the number of blocks in G. If G has one block, than $G = H$ and the claim is obviously true. Suppose the claim holds for every connected graph which has less than m blocks, and let G be a graph with m blocks, $m \geq 2$, one of them H, and each of the others is either a triangle or an edge. Then $G = G_1 \cup G_2$, where $G_1 \cap G_2$ is a single vertex, G_1 and G_2 are connected. We may assume that $V(G_1) = \{1, \dots, k\}$, and $V(G_2) = \{k, \dots, n\}$. One of these subgraphs, say G_2, has H as one of its blocks. Then each of G_1's blocks is a K_2 or a K_3. In particular, G_1 is completely positive. Let A be a CP matrix realization of G. Then, by Lemma 4.59,

$$A = (A_1 \oplus 0_{n-k}) + (0_{k-1} \oplus A_2),$$

where A_1 and A_2 are completely positive, $G(A_1) = G_1$ and $G(A_2) = G_2$, A_1 is singular, and $\operatorname{rank} A = \operatorname{rank} A_1 + \operatorname{rank} A_2$.

Now if H satisfies (4.47), then by the induction hypothesis G_2 also satisfies (4.47). Therefore

$$\begin{aligned}\operatorname{rank} A &\leq \operatorname{cpr} A \\ &\leq \operatorname{cpr} A_1 + \operatorname{cpr} A_2 \\ &= \operatorname{rank} A_1 + \operatorname{rank} A_2 = \operatorname{rank} A.\end{aligned}$$

That is, $\operatorname{cpr} A = \operatorname{rank} A$.

If H satisfies (4.46), then by the induction hypothesis G_2 satisfies (4.46). If $\operatorname{rank} A = n$, the inequality $\operatorname{rank} A_1 \leq k - 1$, together with the equality $\operatorname{rank} A_1 + \operatorname{rank} A_2 = n$ implies that $\operatorname{rank} A_2 \geq n - k + 1$, and therefore $\operatorname{rank} A_2 = n - k + 1$. But $G(A_2) = G_2$, and G_2 satisfies (4.46), hence $\operatorname{cpr} A_2 = \operatorname{rank} A_2$, and we deduce as above that $\operatorname{cpr} A = \operatorname{rank} A = n$. □

Theorem 4.91. *If G satisfies* (4.46), *then every subgraph of G also satisfies* (4.46).

Proof. First suppose that $V(H) = V(G) = \{1, \ldots, n\}$ and $E(H)$ is a proper subset of $E(G)$. Let A be a CP matrix realization of H that has rank n. For each edge $e \in E(G) \setminus E(H)$, denote by $\mathbf{1}_e$ the $(0,1)$-vector of support e. For every $\varepsilon > 0$ let

$$A_\varepsilon = A + \sum_{e \in E(G)\setminus E(H)} \varepsilon \mathbf{1}_e \mathbf{1}_e^T.$$

Then A_ε is clearly completely positive, $G(A_\varepsilon) = G$ and $\operatorname{rank} A_\varepsilon = n$. Hence for each $\varepsilon > 0$ $\operatorname{cpr} A_\varepsilon = n$. Since $A = \lim_{\varepsilon \to 0+} A_\varepsilon$, this implies that $\operatorname{cpr} A \leq n$, and therefore $\operatorname{cpr} A = n$.

Next suppose that $V(H)$ is a proper subset of $V(G)$. Without loss of generality assume that $V(H) = \{1, \ldots, k\}$ for some $k < n$. If A is a completely positive matrix such that $G(A) = H$ and $\operatorname{rank} A = k$, let

$$A_1 = A \oplus I_{n-k}.$$

Clearly A_1 is a completely positive matrix, $\operatorname{rank} A_1 = n$, $V(G(A_1)) = \{1, \ldots, n\}$, and $E(G(A_1))$ is a subset of $E(G)$. By the first part of the proof, $\operatorname{cpr} A_1 = n$. It is easy to see that every minimal CP representation of A_1 is of the form

$$\sum_{i=1}^{k} \mathbf{b}_i \mathbf{b}_i^T + \sum_{i=k+1}^{n} \mathbf{e}_i \mathbf{e}_i^T,$$

where $\sum_{i=1}^{k} \mathbf{b}_i \mathbf{b}_i^T$ is a CP representation of $A \oplus 0_{n-k}$. Hence $\operatorname{cpr} A = k$. □

Theorem 4.91 together with Remark 4.88 imply the following.

Corollary 4.92. *A graph G which has a triangle free subgraph with more edges than vertices does not satisfy* (4.46).

We are going to show that the converse of Corollary 4.92 also holds. The proof requires a series of propositions. First we present some 2-connected graphs which satisfy (4.46), then we describe all the graphs which have no triangle free subgraphs with more edges than vertices. Using this description and Corollary 4.90 we will be able to show that all these graphs satisfy (4.46).

First — some additional 2-connected graphs which satisfy (4.46). For $n \geq 3$ denote by S_{2n} the graph obtained from C_{2n} by adding chords connecting each even vertex to the next even vertex (assuming the cycle vertices are numbered consecutively).

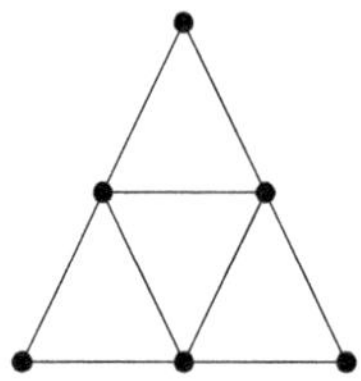

S_6

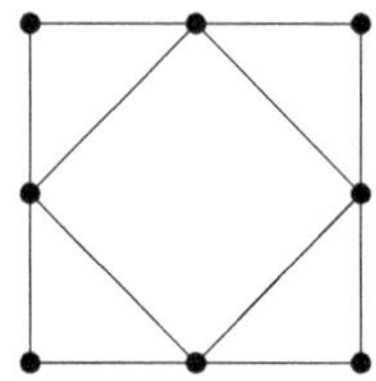

S_8

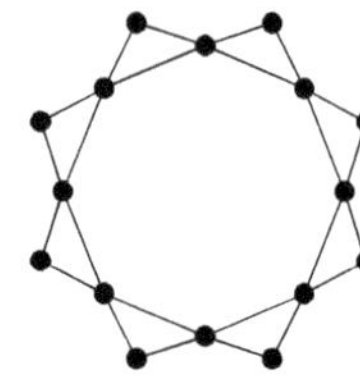

S_{16}

The chords generate an n-cycle (on the n even vertices). We will call this cycle the *inner cycle* of S_{2n}.

We denote by S_5 the graph

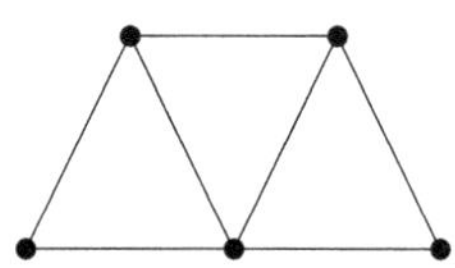

S_5

Proposition 4.93. *For every $n \geq 3$ the graph S_{2n} satisfies* (4.46).

Proof. We consider first the case $n \geq 4$. Renumber the vertices of S_{2n} so that 1 is a vertex of degree 2, adjacent to vertices 2 and 3. Let A be a completely positive matrix with $G(A) = S_{2n}$ and $\operatorname{rank} A = 2n$. Consider any CP representation of A, $A = \sum_{i=1}^{m} \mathbf{b}_i \mathbf{b}_i^T$, and let

$$\Omega_1 = \{1 \leq i \leq m \mid \operatorname{supp} \mathbf{b}_i \subseteq \{1,2,3\}\}, \quad \Omega_2 = \{1,2,\ldots,m\} \setminus \Omega_1,$$

$$B = \sum_{i \in \Omega_1} \mathbf{b}_i \mathbf{b}_i^T, \quad C = \sum_{i \in \Omega_2} \mathbf{b}_i \mathbf{b}_i^T.$$

Then B and C are completely positive, $B = B' \oplus 0_{2n-3}$, $C = 0_1 \oplus C'$. The graph $G(B')$ is a triangle, and $G(C')$ is a graph on $2n-1$ vertices which is a "chain" of $n-1$ triangles — every block of $G(C')$ is a K_3. Of course, $A = B + C$. Note that

$$B' = \begin{pmatrix} a_{11} & a_{12} & a_{13} \\ a_{12} & t_0 & a_{23} \\ a_{13} & a_{23} & s_0 \end{pmatrix},$$

where $a_{12}^2/a_{11} \leq \alpha_0 \leq a_{22}$. If $t_0 = a_{12}^2/a_{11}$, then $a_{23} = (a_{12}a_{13})/a_{11}$ and

$$B' = \begin{pmatrix} a_{11} & a_{12} & a_{13} \\ a_{12} & t_0 & a_{23} \\ a_{13} & a_{23} & a_{13}^2/a_{11} \end{pmatrix} + \delta E_{33},$$

for some $\delta \geq 0$. In this case, denote $B'' = B' - \delta E_{33}$ and $C'' = C' + \delta E_{33}$. B'' is completely positive, and clearly so is C''; $A = (B'' \oplus 0_{2n-3}) + (0_1 \oplus C'')$ and

$$2n = \operatorname{rank} A \leq \operatorname{cpr} A \leq \operatorname{cpr} B'' + \operatorname{cpr} C'' \leq 1 + (2n-1) = 2n.$$

Now consider the case $a_{12}^2/a_{11} < t_0 \leq a_{22}$. For every $a_{12}^2/a_{11} < t \leq a_{22}$ denote

$$B'(t) = \begin{pmatrix} a_{11} & a_{12} & a_{13} \\ a_{12} & t & a_{23} \\ a_{13} & a_{23} & f(t) \end{pmatrix},$$

where $f(t)$ is the unique real number for which $B'(t)$ is singular, i.e., $f(t) = [a_{23}(a_{11}a_{23} - a_{12}a_{13}) - a_{13}(a_{12}a_{23} - ta_{13})]/(a_{11}t - a_{12}^2)$. Since $a_{11} > 0$ and $a_{11}t - {a_{12}}^2 > 0$, $B'(t)$ is a positive semidefinite matrix. In particular $f(t) \geq 0$, so $B'(t)$ is nonnegative, and therefore completely positive (Theorem 3.35). Let $B(t) = B'(t) \oplus 0_{2n-3}$, and $C(t) = A - B(t)$; $C(t) = 0_1 \oplus C'(t)$. Now if B', B, C' and C are as above, then $s_0 \geq f(t_0)$, $B(t_0) = B - (s_0 - f(t_0))E_{33}$, and $C(t_0) = C + (s_0 - f(t_0))E_{33}$. Hence $C(t_0)$ is completely positive. On the other hand, it is clear that $C(a_{22})$ is not positive semidefinite. By the continuity of the eigenvalues of $C(t)$, it follows that there exists $t_0 < t_1 < a_{22}$ such that $C(t_1)$ is a singular positive semidefinite matrix. In particular, $C(t_1)_{22} > 0$, $C(t_1)_{33} > 0$, so $C(t_1)$ is doubly nonnegative. The graph $G(C(t_1))$ is completely positive, hence $C(t_1)$ is a completely positive matrix. Thus we have

$$\begin{aligned} 2n = \operatorname{rank} A &\leq \operatorname{cpr} A \\ &\leq \operatorname{cpr} B'(t_1) + \operatorname{cpr} C'(t_1) \\ &\leq 2 + (2n-2) = 2n. \end{aligned}$$

(The last inequality holds since both $B'(t_1)$ and $C'(t_1)$ are singular and their graphs satisfy (4.47) by Theorem 4.89.)

For the proof that S_6 also satisfies (4.46) we need to show first that S_5 satisfies (4.46). Number the vertices of S_5 as in the figure below.

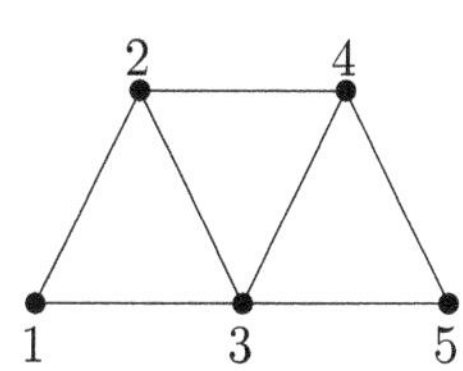

Let A be a completely positive matrix with $G(A) = S_5$ and $\operatorname{rank} A = 5$. We show that $\operatorname{cpr} A \leq 5$. Let $A = \sum_{i=1}^{m} \mathbf{b}_i \mathbf{b}_i^T$ be a CP representation of A, and denote

$$\Omega_1 = \{1 \leq i \leq m \mid 1 \in \operatorname{supp} \mathbf{b}_i\}, \quad \Omega_2 = \{1, \ldots, m\} \setminus \Omega_1$$
$$A_1 = \sum_{i \in \Omega_1} \mathbf{b}_i \mathbf{b}_i^T, \qquad A_2 = \sum_{i \in \Omega_2} \mathbf{b}_i \mathbf{b}_i^T.$$

Both matrices are completely positive and $A = A_1 + A_2$. Rows 4 and 5 of A_1 are zero, and $A_2 = 0_1 \oplus A_2'$. The support of row 5 in A_2 is contained in that of row 4. Let

$$a = \min \left\{ \frac{(A_2)_{4j}}{(A_2)_{5j}} \,\middle|\, j = 3, 4, 5 \right\}$$

and $S = I_5 - aE_{45}$. By Lemma 4.68, SA_2S^T is a doubly nonnegative matrix and at least one of its entries in positions 4, 3 and 4, 5 is zero. Since it has only four nonzero rows, it is completely positive. Row 4 of A_1 is zero, hence $SA_1S^T = A_1$ and $SAS^T = SA_1S^T + SA_2S^T = A_1 + SA_2S^T$ is completely positive. Also, $\operatorname{rank}(SAS^T) = \operatorname{rank} A = 5$, and $\operatorname{cpr} A \leq \operatorname{cpr}(SAS^T)$ (see Proposition 4.11). It therefore suffices to show that $\operatorname{cpr}(SAS^T) = 5$. But the graph of SAS^T is contained in one of the following graphs.

The graph on the left is a subgraph of S_8. The one on the right has two blocks, a K_2 and a T_4. By the beginning of this proof and Theorem 4.91, Theorem 4.10 and Corollary 4.90, both graphs satisfy (4.46), hence $\operatorname{cpr}(SAS^T) \leq 5$.

We now show by similar arguments that S_6 also satisfies (4.46). Renumber the vertices.

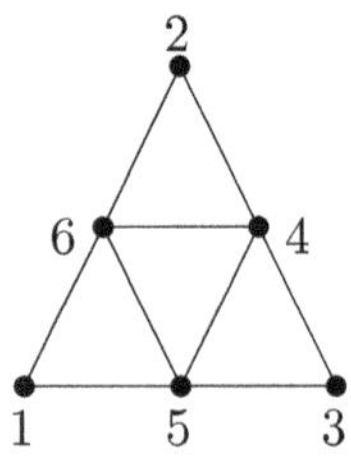

Let A be a CP matrix realization of S_6 that has rank 6. Let $A = \sum_{i=1}^{m} \mathbf{b}_i \mathbf{b}_i^T$ be a CP representation of A and denote

$$\Omega_1 = \{1 \leq i \leq m \mid 1 \in \operatorname{supp} \mathbf{b}_i\}$$
$$\Omega_2 = \{1 \leq i \leq m \mid 2 \in \operatorname{supp} \mathbf{b}_i\}$$
$$\Omega_3 = \{1, \ldots, m\} \setminus (\Omega_1 \cup \Omega_2).$$

(Note that $\Omega_1 \cap \Omega_2 = \emptyset$.) Let

$$A_1 = \sum_{i \in \Omega_1} \mathbf{b}_i \mathbf{b}_i^T, \qquad A_2 = \sum_{i \in \Omega_2} \mathbf{b}_i \mathbf{b}_i^T, \qquad A_3 = \sum_{i \in \Omega_3} \mathbf{b}_i \mathbf{b}_i^T.$$

These are three completely positive matrices, and $A = A_1 + A_2 + A_3$. Rows $2, 3, 4$ of A_1 are zero, rows $1, 3, 5$ of A_2 are zero, and rows 1 and 2 of A_3 are zero. In A_3, the support of row 3 is contained in that of row 4. Let

$$a = \min \left\{ \frac{(A_3)_{4j}}{(A_3)_{3j}} \,\middle|\, j = 3, 4, 5 \right\},$$

and $S = I_6 - aE_{43}$. Then SA_3S^T is a completely positive matrix, and at least one of its entries in positions $4, 3$ or $4, 5$ is zero.

$$SAS^T = SA_1S^T + SA_2S^T + SA_3S^T = A_1 + A_2 + SA_3S^T$$

is completely positive, and it suffices to prove that $\operatorname{cpr}(SAS^T) \leq 6$. The graph of SAS^T is contained in one of the following graphs

The graph on the left is a subgraph of S_8, and the graph on the right has two blocks, a K_2 and an S_5. By the beginning of this proof and Corollary 4.90 both graphs satisfy (4.46). Hence $G(SAS^T)$ satisfies (4.46) and $\operatorname{cpr}(SAS^T) \leq 6$. □

We turn to describe all blocks which contain no triangle free graph with more edges than vertices.

Proposition 4.94. *If G is a block which has no triangle free subgraph with more edges than vertices, then G is either a K_4 or a subgraph of S_{2n} for some $n \geq 3$.*

Proof. A block on 2 vertices is a K_2. A block on 3 vertices is a K_3. The only blocks on 4 vertices are C_4, T_4 and K_4. All these fit the description in the statement of the proposition. Hence we consider the case that G is a block on $n \geq 5$ vertices. Since G has no cut vertex, it contains a cycle. Let C be a cycle subgraph of G of maximal length. We first show that C is a spanning cycle of G. Suppose C is not a spanning cycle of G. Then C is a cycle on k vertices, $k < n$, and there exists a vertex v of G which is not a vertex of C. Since G is 2-connected, v lies on a path P connecting two vertices of C, v_1 and v_2, such that all the internal vertices of P are not vertices of C (see Exercise 1.37). Denote the length of P by s, then $s \geq 2$.

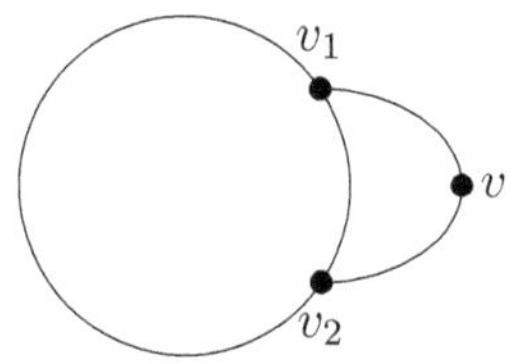

If v_1 and v_2 are adjacent in C, then the graph consisting of all edges of C other than $\{v_1, v_2\}$ and the edges of P is a cycle on $k - 1 + s \geq k + 1$ vertices, in contradiction to the choice of C. If v_1 and v_2 are not adjacent in C, then $k \geq 4$ and $C \cup P$ is a triangle free graph on $k + s - 1$ vertices, which has $k + s$ edges, and this contradicts the initial assumption regarding G. Hence there is no such vertex v, and the maximal cycle C is a spanning cycle of G.

We now consider which chords of C may be edges of G. There is no edge in G which is a chord between two vertices u, v such that the $d_C(u, v) \geq 3$, since the graph consisting of C and such a chord would be a triangle free subgraph of G which has n vertices and $n+1$ edges. Hence, if there exists an edge of G which is a chord of C, it would be some $\{u, v\}$ where $d_C(u, v) = 2$. If w is the vertex such that $d_C(u, w) = 1$ and $d_C(v, w) = 1$, then there is no edge $\{w, x\}$ of G which is a chord of C. Otherwise, $d_C(w, x) = 2$, so either u or v is halfway between w and x. Assume without loss of generality v is. Then the graph consisting of the two chords, and all of C's edges except for $\{w, v\}$, is a triangle free graph on n vertices with $2 + (n - 1) = n + 1$ edges. (The fact that $n \geq 5$ guarantees an additional vertex on the cycle C, between u and x.)

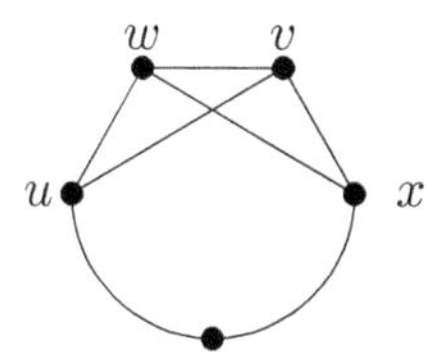

This completes the proof that G is a subgraph of some S_{2m}. □

Theorem 4.95. *If a connected graph G has no triangle free subgraph with more edges than vertices, then each block of G is a K_4 or a subgraph of S_{2n} for some $n \geq 3$, and at most one of the blocks of G has more than 3 vertices.*

Proof. If G has no triangle free subgraph with more edges than vertices, then no block of G has such subgraph. By Proposition 4.94 each block of G is either a K_4 or a subgraph of S_{2n} for some $n \geq 3$. Suppose two of these blocks have 4 vertices or more. By the proof of Proposition 4.94, each of these large blocks has a spanning cycle. Let C_1 and C_2 be spanning cycles of these two blocks. The cycles may share a vertex, or there is a path in G from a vertex of C_1 to a vertex of C_2, whose internal vertices are not vertices of $C_1 \cup C_2$. A graph consisting of two cycles on 4 or more vertices sharing a vertex, or two such cycles and a path connecting them, is a triangle free graph with more edges than vertices. But this contradicts the assumption on G, hence there cannot be two blocks on 4 or more vertices in G. □

With Theorem 4.95, the characterization of all graphs which satisfy (4.46) is complete. The results are summarized in Theorem 4.96.

Theorem 4.96. *Let G be a connected graph, then the following are equivalent.*

(a) *For every nonsingular completely positive matrix realization A of G,* $\operatorname{cpr} A = \operatorname{rank} A$.
(b) $\operatorname{cpr} G = |V(G)|$.
(c) *G contains no triangle free graph with more edges than vertices.*
(d) *Each block of G is either a K_4 or a subgraph of S_{2m} for some $m \geq 3$, and at most one of the blocks of G has more than 3 vertices.*

We now turn to characterize the graphs which satisfy (4.47). Since every such graph satisfies also (4.46), it suffices to check which of the graphs described in Theorem 4.96 satisfy (4.47).

Consider first 2-connected graphs which are subgraphs of S_{2m}, for some $m \geq 4$. Any such graph is a subgraph of some S_{2k}, $k \geq 4$, which contains the inner cycle C of S_{2k}. If it is not C itself, then it contains also $1 \leq r \leq k$ triangles, each consisting of an edge e of C, and a vertex not in C, which is joined by edges to e's ends. We start with one such example.

Example 4.97. Consider the house graph G, which is a subgraph of $S_{2(n-1)}$, $n > 4$ even, containing the inner cycle and exactly one triangle. Let $V(G) = \{1, \ldots, n\}$ and suppose the edges are $\{i, i+1\}$, $i = 1, \ldots, n-1$, $\{1, n-1\}$ and $\{1, n\}$.

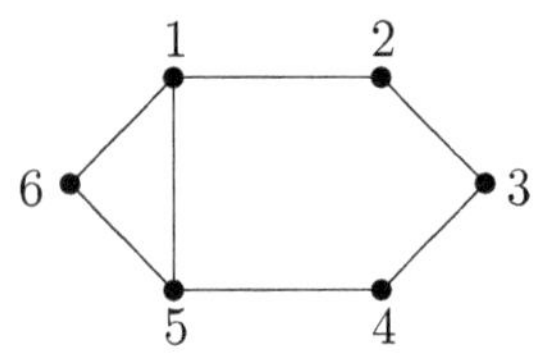

(for $n = 6$)

We show that G does not satisfy (4.47). Specifically, we construct a completely positive matrix A such that $G(A) = G$, $\operatorname{rank} A = n - 1$ and $\operatorname{cpr} A = n$. Let R be the following $n \times n$ matrix

$$R = \begin{pmatrix} 1 & 0 & 0 & \cdots & & \cdots & 1 & 2 \\ 1 & 1 & 0 & \cdots & & \cdots & 0 & 0 \\ 0 & 1 & 1 & & & & \vdots & \vdots \\ 0 & 0 & 1 & \ddots & & & \vdots & \vdots \\ & & & \ddots & \ddots & & & \\ 0 & 0 & 0 & & & 1 & 1 & 0 \\ 0 & 0 & 0 & \cdots & & 0 & 1 & 1 \end{pmatrix},$$

and let $A = RR^T$. Then

$$A = \begin{pmatrix} 6 & 1 & 0 & & \cdots & & \cdots & 1 & 3 \\ 1 & 2 & 1 & & & & \cdots & 0 & 0 \\ 0 & 1 & 2 & 1 & & & & \vdots & \vdots \\ \vdots & & \ddots & & \ddots & \ddots & & & \vdots \\ & & & & \ddots & 2 & & 1 & 0 \\ 1 & 0 & \cdots & & & 1 & & 2 & 1 \\ 3 & 0 & \cdots & & \cdots & 0 & & 1 & 2 \end{pmatrix}.$$

Clearly, A is completely positive, $G(A) = G$, and $\operatorname{cpr} A \leq n$. It is also easy to see that $\operatorname{rank} A = n-1$. Denote the i-th column of R by R_i, then the first $n-1$ columns of R are linearly independent and $R_n = \sum_{i=1}^{n-1}(-1)^{i+1}R_i$. We show that $\operatorname{cpr} A \geq n$. Let

$$A = \sum_{i=1}^{m} \mathbf{b}_i \mathbf{b}_i^T$$

be a minimal CP representation of A. Let

$$\Omega_1 = \{1 \leq i \leq m \mid n \in \operatorname{supp} \mathbf{b}_i\}, \quad \Omega_2 = \{1, \ldots, m\} \setminus \Omega_1$$
$$B = \textstyle\sum_{i \in \Omega_1} \mathbf{b}_i \mathbf{b}_i^T, \qquad C = \sum_{i \in \Omega_2} \mathbf{b}_i \mathbf{b}_i^T.$$

Then B and C are completely positive. In B only rows $1, n-1, n$ are nonzero, and the matrix C has the form $C' \oplus 0_1$. By the minimality of the representation, $\operatorname{cpr} B = |\Omega_1|$, $\operatorname{cpr} C = \operatorname{cpr} C' = |\Omega_2|$, and $\operatorname{cpr} A = \operatorname{cpr} B + \operatorname{cpr} C$. If $|\Omega_1| = 1$, then

$$B[1, n-1, n] = \begin{pmatrix} \frac{9}{2} & \frac{3}{2} & 3 \\ \frac{3}{2} & \frac{1}{2} & 1 \\ 3 & 1 & 2 \end{pmatrix},$$

but this would imply that the $1\,(n-1)$ element of $C = A - B$ is $-1/2$, which is impossible since C is completely positive. Hence $|\Omega_1| \geq 2$, which implies that $\operatorname{cpr} B \geq 2$. Since $G(C')$ is either the cycle on $n-1$ vertices or the path from vertex 1 to vertex $n-1$, $\operatorname{cc}(G') \geq n-2$. Thus $\operatorname{cpr} C' \geq n-2$, and we get

$$\operatorname{cpr} A = \operatorname{cpr} B + \operatorname{cpr} C \geq 2 + (n-2) = n.$$

We use Example 4.97 to prove that any block which is a subgraph of S_{2k}, $k \geq 4$, and is not an edge or a triangle, does not satisfy (4.47).

Proposition 4.98. *If $G \subseteq S_{2k}$, $k \geq 4$, is 2-connected which is not an odd cycle, then G does not satisfy* (4.47).

Proof. We may assume that G consists of the inner cycle of S_{2k} and r triangles, $r \geq 0$; G has $n = k + r$ vertices. We prove by induction on r that there exists a completely positive matrix A such that $G(A) = G$, $\operatorname{rank} A = n-1$, and $\operatorname{cpr} A = n$. By Example 4.86, if k is even and $r = 0$, there exists a completely positive matrix A such that $G(A) = G$, $\operatorname{rank} A = n-1$, $\operatorname{cpr} A = n$. By Example 4.97, this is true also in the case that k is odd and $r = 1$. For the induction step we assume that the claim holds for any 2-connected subgraphs of S_{2k} which contains the inner cycle and $r-1$ triangles. Let $G \subseteq S_{2k}$ be a 2-connected graph with r triangles. Then G

has $n = k+r$ vertices, and we may assume that the vertex 1 is not a vertex of the inner cycle and that it is adjacent to the vertices 2 and 3 which are adjacent vertices of the inner cycle. Denote by G' the graph obtained from G by deleting vertex 1 and the two edges incident with it. By the induction hypothesis there exists a completely positive matrix A' such that $G(A') = G'$, $\operatorname{rank} A' = n-2$ and $\operatorname{cpr} A' = n-1$. Let

$$A = (J_3 \oplus 0_{n-2}) + (0_1 \oplus A').$$

The matrix A is clearly completely positive, $G(A) = G$, and $\operatorname{rank} A = \operatorname{rank} A' + 1 = n-1$. We show that $\operatorname{cpr} A = \operatorname{cpr} A' + 1 = n$. Let

$$A = \sum_{i=1}^{m} \mathbf{b}_i \mathbf{b}_i^T$$

be a minimal CP representation of A. Let

$$\Omega_1 = \{1 \le i \le m \mid \operatorname{supp} \mathbf{b}_i \subseteq \{1,2,3\}\}, \qquad \Omega_2 = \{1,\ldots,m\} \setminus \Omega_1,$$
$$B = \textstyle\sum_{i\in\Omega_1} \mathbf{b}_i \mathbf{b}_i^T, \qquad C = \textstyle\sum_{i\in\Omega_2} \mathbf{b}_i \mathbf{b}_i^T.$$

Then $B = B' \oplus 0_{n-3}$, $C = 0_1 \oplus C'$, $A = B + C$, and by the minimality of the representation $\operatorname{cpr} A = \operatorname{cpr} B + \operatorname{cpr} C$. Note that

$$B' = \begin{pmatrix} 1 & 1 & 1 \\ 1 & + & t+1 \\ 1 & t+1 & + \end{pmatrix}, \tag{4.48}$$

where $+$ denotes a positive entry and $t > 0$ is the 23 entry of $0_1 \oplus A'$. We may write $B' = J_3 + (0_1 \oplus B'')$, where B'' is the Schur complement $B'/[1]$. Then $\operatorname{rank} B' = \operatorname{rank} B'' + 1$, and by (4.48) B'' is nonnegative. Since both B' and B'' are of order ≤ 3, the rank equality implies also that $\operatorname{cpr} B' = \operatorname{cpr} B'' + 1$. The equality

$$A = (J_3 \oplus 0_{n-3}) + [C + (0_1 \oplus B'' \oplus 0_{n-3})]$$

implies that $C + (0_1 \oplus B'' \oplus 0_{n-3}) = 0_1 \oplus A'$, and therefore

$$\operatorname{cpr} A' \le \operatorname{cpr} C + \operatorname{cpr} B''.$$

But then

$$\begin{aligned} \operatorname{cpr} C + \operatorname{cpr} B'' + 1 &= \operatorname{cpr} A \\ &\le 1 + \operatorname{cpr} A' \\ &\le 1 + \operatorname{cpr} C + \operatorname{cpr} B''. \end{aligned}$$

This implies that $\operatorname{cpr} A = \operatorname{cpr} A' + 1 = n$. □

Blocks which are subgraphs of S_6, and are neither an edge nor a triangle, also do not satisfy (4.47).

Proposition 4.99. *If $G \subseteq S_6$ is 2-connected, and is not an odd cycle, then G does not satisfy* (4.47).

Proof. Each such proper subgraph of S_6 is either a T_4, or a C_4, or a subgraph of S_8, or S_5. By Examples 4.9 and 4.86, and Proposition 4.98, we only have to show that S_5 and S_6 do not satisfy (4.47).

Recall that by Example 4.9,

$$A_1 = \begin{pmatrix} 6 & 3 & 3 & 0 \\ 3 & 5 & 1 & 3 \\ 3 & 1 & 5 & 3 \\ 0 & 3 & 3 & 6 \end{pmatrix}$$

is a CP matrix realization of T_4, $\operatorname{rank} A_1 = 3$, and $\operatorname{cpr} A_1 = 4$. Let $A_2 = (J_3 \oplus 0_2) + (0 \oplus A_1)$. The matrix A_2 is a CP matrix realization of S_5 and $\operatorname{rank} A_2 = 4$. Let $A_2 = \sum_{i=1}^m \mathbf{b}_i \mathbf{b}_i^T$ be a minimal CP representation of A_2. If $1 \in \operatorname{supp} \mathbf{b}_i$ for exactly one i, say $i = 1$, then necessarily $\mathbf{b}_1 \mathbf{b}_1^T = J_3 \oplus 0_2$, and $\operatorname{cpr} A_2 = \operatorname{cpr} \mathbf{b}_1 \mathbf{b}_1^T + \operatorname{cpr} A_1 = 1 + 4 = 5$. Suppose 1 belongs to two of the supports, say $\operatorname{supp} \mathbf{b}_1$ and $\operatorname{supp} \mathbf{b}_2$. We argue as in Example 4.9 that 5 also belongs to two supports, say $\operatorname{supp} \mathbf{b}_3, \operatorname{supp} \mathbf{b}_4$. But if $1 \in \operatorname{supp} \mathbf{b}_i$ or $5 \in \operatorname{supp} \mathbf{b}_i$, then $\{2, 4\} \not\subseteq \operatorname{supp} \mathbf{b}_i$. Thus there is a fifth vector $\mathbf{b}_5$ such that $\{2, 4\} \subseteq \operatorname{supp} \mathbf{b}_5$. Hence $\operatorname{cpr} A_2 \geq 5$ and S_5 does not satisfy (4.47). By a similar argument,

$$A_3 = (A_1 \oplus 0_1) + \begin{pmatrix} 1 & 0 & 1 & 0 & 1 \\ 0 & 0 & 0 & 0 & 0 \\ 1 & 0 & 1 & 0 & 1 \\ 0 & 0 & 0 & 0 & 0 \\ 1 & 0 & 1 & 0 & 1 \end{pmatrix} = \begin{pmatrix} 7 & 3 & 4 & 0 & 1 \\ 3 & 5 & 1 & 3 & 0 \\ 4 & 1 & 6 & 3 & 1 \\ 0 & 3 & 3 & 6 & 0 \\ 1 & 0 & 1 & 0 & 1 \end{pmatrix}$$

is also completely positive, $G(A_3) = S_5$, $\operatorname{rank} A_3 = 4$, and $\operatorname{cpr} A_3 = 5$. Using these results, and the same arguments, it is easy to see that

$$A_4 = (J_3 \oplus 0_3) + (0_1 \oplus A_3)$$

is a CP matrix realization of S_6, $\operatorname{rank} A_4 = 5$ and $\operatorname{cpr} A_4 = 6$. □

Combining Theorem 4.96 with Corollary 4.90 and Propositions 4.98 and 4.99 we obtain a characterization of the graphs which satisfy (4.47).

Theorem 4.100. *Let G be a connected graph, then the following are equivalent.*

(a) *For every CP matrix realization A of G,* $\operatorname{cpr} A = \operatorname{rank} A$.
(b) *G contains no even cycle and no triangle free graph with more edges than vertices.*
(c) *Each block of G is a K_2 or an odd cycle, and at most one of the blocks of G has more than 3 vertices.*

Exercises

4.47. By Proposition 4.99 there exists a CP matrix realization of S_6 of rank 5 and cp-rank 6. Show that there exists also a CP matrix realization of S_6 with rank and cp-rank both equal to 5.

4.48. Show that if A is a CP matrix realization of S_6 and $\operatorname{rank} A \neq 5$, then $\operatorname{cpr} A = \operatorname{rank} A$.

Notes. This section is based on [Shaked-Monderer (2001)].

4.7 Graphs attaining minimal cp-rank

As mentioned in Section 4.1, for every graph G, the following three lower bounds on $\operatorname{cpr} G$ hold: $\operatorname{cpr} G \geq |V(G)|$, $\operatorname{cpr} G \geq \operatorname{cc}(G)$ and $\operatorname{cpr} G \geq \operatorname{tf}(G)$. In the previous section we found all graphs such that $\operatorname{cpr} G$ is equal to its lower bound $|V(G)|$. In this section we consider graphs G attaining the two other lower bounds. We completely characterize graphs that satisfy $\operatorname{cpr} G = \operatorname{cc}(G)$, and find families of graphs that satisfy $\operatorname{cpr} G = \operatorname{tf}(G)$. Note that if a graph G on $n \geq 4$ vertices satisfies either of these equalities, then for any CP matrix realization A of G the DJL bound $\operatorname{cpr} A \leq n^2/4$ holds, since

$$\operatorname{cc}(G) \leq \operatorname{tf}(G) \leq \frac{n^2}{4}$$

(see Proposition 4.20 and Theorem 1.88).

Results on the cp-ranks of graphs may be useful also for the study the cp-ranks of positive completely positive matrices, as in the proof of Proposition 4.66. Some results in this section may be used in the proof of Proposition 4.72 — see Exercises 4.53 and 4.54 at the end of this section.

We begin with the parameter $\operatorname{cc}(G)$.

Theorem 4.101. *A connected graph G satisfies* $\operatorname{cpr} G = \operatorname{cc}(G)$ *if and only if G is triangle free, but not a tree.*

Proof. Let G have n vertices. If G is a tree, then the maximal cliques of G are G's edges, and $\text{cc}(G) = |E(G)|$. By Corollary 4.51, in this case $\text{cpr}\, G = n$. Thus $\text{cpr}\, G = n > n-1 = |E(G)| = \text{cc}(G)$. If G is triangle free and not a tree, then $n \geq 4$, and $\text{cpr}\, G = |E(G)|$ by Corollary 4.55. Hence in this case $\text{cpr}\, G = \text{cc}(G)$.

To show that equality does not hold for graphs which are not triangle free, recall that by Proposition 4.20

$$\text{cc}(G) \leq \text{tc}(G) \leq \text{tf}(G).$$

If G is not triangle free, the right inequality is strict. Hence if G is not triangle free, there exists a triangle free subgraph H of G such that $\text{tf}(G) = |E(H)| > \text{cc}(G)$, implying that

$$\text{cpr}\, G \geq \text{cpr}\, H = |E(H)| > \text{cc}(G).$$ □

If G is a disjoint union of G_1 and G_2, both $\text{cc}(G) = \text{cc}(G_1) + \text{cc}(G_2)$ and $\text{cpr}\, G = \text{cpr}\, G_1 + \text{cpr}\, G_2$. We therefore deduce from Theorem 4.101 the general characterization.

Corollary 4.102. *A graph G satisfies* $\text{cpr}\, G = \text{cc}(G)$ *if and only if G is triangle free, and none of its connected components is a tree.*

While the class of graphs satisfying $\text{cpr}\, G = \text{cc}(G)$ is quite limited, we already know several families of graphs satisfying $\text{cpr}\, G = \text{tf}(G)$.

Example 4.103. Here are some graphs G that satisfy $\text{cpr}\, G = \text{tf}(G)$.

- A connected triangle free graph G which is not a tree (by Corollary 4.55).
- $G = K_4$ satisfies $\text{tf}(G) = \text{cpr}\, G = 4$.
- $G = T_k$, $k \geq 4$. In this case $\text{cpr}\, T_k = 2k-4$ by Theorem 4.57, and also $\text{tf}(T_k) = 2k-4$, since the maximal size triangle free subgraph of T_k is $K_{2,k-2}$.
- Any graph G that is a 2-connected subgraph on $n \geq 3$ vertices of some S_{2k}. It is easy to see that $\text{tf}(G) = n$, and $\text{cpr}\, G = n$ by Theorem 4.96.

To extend the list we need some preliminary results. But first, we list some easy to prove facts about the parameter $\text{tf}(G)$ that are left as an exercise.

Proposition 4.104. *Let G be a graph, $G = G_1 \cup G_2$.*

(a) *If the union is disjoint, then* $\text{tf}(G) = \text{tf}(G_1) + \text{tf}(G_2)$.

(b) *If* $V(G_1) \cap V(G_2)$ *is an independent set of* G, *then* $\mathrm{tf}(G) = \mathrm{tf}(G_1) + \mathrm{tf}(G_2)$.
(c) *If* $V(G_1) \cap V(G_2)$ *is a single vertex and* G_1 *is a* K_m, $m = 2$ *or* 3, *then* $\mathrm{tf}(G) = m - 1 + \mathrm{tf}(G_2)$.

Proposition 4.105. *If* G' *is a triangle free subgraph of a connected graph* G, *then there exists a spanning triangle free subgraph of* G *which has* $|E(G')| + |V(G)| - |V(G')|$ *edges.*

Proposition 4.105 implies that any triangle free subgraph of G of maximal size is necessarily spanning. Since every connected graph on n vertices has a spanning tree, $\mathrm{tf}(G) \geq n - 1$ for every connected graph G on n vertices.

Proposition 4.106. *A connected graph* G *on* n *vertices has* $\mathrm{tf}(G) = n - 1$ *if and only if each block of* G *is either an edge or a triangle.*

Proof. By Proposition 4.105, if $\mathrm{tf}(G) = |V(G)| - 1$, then every connected triangle free subgraph G' of G is a tree. In particular, no block of G has more than 3 vertices. Conversely, suppose G is a connected graph on n vertices with blocks $G_1, \ldots, G_k$, and $|V(G_i)| = n_i$, $1 \leq i \leq k$, $n_i \in \{2, 3\}$. Then

$$n = 1 + \sum_{i=1}^{k}(n_i - 1) = 1 + \sum_{i=1}^{k} \mathrm{tf}(G_i) = 1 + \mathrm{tf}(G),$$

since $\mathrm{tf}(K_m) = m - 1$ if $m = 2$ or $m = 3$. □

Corollary 4.107. *A connected graph* G *on* n *vertices has* $\mathrm{tf}(G) = n$ *if and only if* G *has exactly one block on more than* 3 *vertices, and this block is a* K_4 *or a subgraph of* S_{2k} *for some* $k \geq 3$.

Proof. Suppose $\mathrm{tf}(G) = |V(G)| = n$. By Proposition 4.106, G cannot contain a triangle free subgraph G' with more edges than vertices, otherwise G' could be extended to a spanning triangle free subgraph of size greater than n by adding $|V(G)| - |V(G')|$ edges, which would imply that $\mathrm{tf}(G) > |V(G)|$. Thus, by the equivalence of part (c) and (d) in Proposition 4.96, each block of G is either a K_4, or a subgraph of S_{2k} for some $k \geq 3$, with at most one block on more than 3 vertices. On the other hand, Proposition 4.106 guarantees the existence of at least one such block.

Now suppose G has the block structure described in the theorem, then by Theorem 4.96, any spanning triangle free subgraph G' of G of maximal size has at most $|V(G')| = n$ edges. Thus $\operatorname{tf}(G) \le n$. On the other hand, $\operatorname{tf}(G) > n-1$ by Proposition 4.106. $\square$

Regarding the cp-rank of a graph, recall that in general, if $G = G_1 \cup G_2$, then $\operatorname{cpr} G \le \operatorname{cpr} G_1 + \operatorname{cpr} G_2$. In the following case, we have a slightly improved bound.

Proposition 4.108. *Let $G = G_1 \cup G_2$, where $V(G_1) \cap V(G_2)$ is a single vertex v, where G_1 is connected and has m vertices, $m = 2$ or 3. Then $\operatorname{cpr} G = m - 1 + \operatorname{cpr} G_2$.*

Proof. Without loss of generality assume $V(G_1) = \{1, \dots, m\}$ and $V(G_2) = \{m, \dots, n\}$.

First observe that if A is a CP matrix realization of G, then we may represent A as a sum $A = A_1 + A_2$, where A_1 and A_2 are completely positive and $\operatorname{cpr} A = \operatorname{cpr} A_1 + \operatorname{cpr} A_2$, $A_1 = B_1 \oplus 0_{n-m}$, $A_2 = 0_{m-1} \oplus B_2$, where $G(B_1) = G_1$, $G(B_2) = G_2$, and B_1 is singular.

Indeed, let $A = \sum_{i=1}^{p} \mathbf{b}_i \mathbf{b}_i^T$ be a minimal CP representation of A. Let

$$\Omega_1 = \{i \in \{1, \dots, p\} \mid \operatorname{supp} \mathbf{b}_i \subseteq \{1, \dots, m\}\}, \qquad \Omega_2 = \{1, \dots, p\} \setminus \Omega_1,$$

and

$$A_1 = \sum_{i \in \Omega_1} \mathbf{b}_i \mathbf{b}_i^T, \qquad A_2 = \sum_{i \in \Omega_2} \mathbf{b}_i \mathbf{b}_i^T.$$

Then $A_1 = B_1 \oplus 0_{n-m}$ and $A_2 = 0_{m-1} \oplus B_2$ satisfy the requirements if B_1 is singular. If B_1 is nonsingular, then (since it is either 2×2 or 3×3), it has a minimal CP representation where one of the vectors is supported by $\{m\}$ — see Example 3.7 and Lemma 3.126. Hence $B_1 = B_1' + c\mathbf{e}_m \mathbf{e}_m^T$ for some $c > 0$, and we may replace A_1 by $B_1' \oplus 0$ and A_2 by $A_2 + cE_{mm}$. And obtain the desired decomposition.

We now show that $\operatorname{cpr} G \le m - 1 + \operatorname{cpr}(G_2)$. Let A be a CP matrix realization of G such that $\operatorname{cpr} A = \operatorname{cpr} G$. Let $A = A_1 + A_2$ be a decomposition as described above, with B_1 singular. Then

$$\operatorname{cpr} G = \operatorname{cpr} A \le \operatorname{cpr} A_1 + \operatorname{cpr} A_2 \le m - 1 + \operatorname{cpr} G_2.$$

To prove the reverse inequality, let Q_1 be zero except for its leading $m \times m$ principal submatrix. If $m = 2$ let that leading 2×2 submatrix be J_2, and if $m = 3$ let the leading 3×3 submatrix be

$$\begin{pmatrix} 1 & 1 & 1 \\ 1 & 2 & 1 \\ 1 & 1 & 1 \end{pmatrix} \text{ (if } G_1 \text{ is a triangle) or } \begin{pmatrix} 1 & 1 & 0 \\ 1 & 2 & 1 \\ 0 & 1 & 1 \end{pmatrix} \text{ (if } G_1 \text{ is a path).}$$

In all these cases, Q_1 is completely positive and $\operatorname{cpr} Q_1 = \operatorname{rank} Q_1 = m-1$. Let Q_2 have zero $m-1$ first rows and columns with $Q_2[m,\ldots,n]$ a CP matrix realization of G_2 such that $\operatorname{cpr} Q_2 = \operatorname{cpr} G_2$. Let $A = Q_1 + Q_2$. Decompose the matrix A as described at the beginning of the proof: $A = A_1 + A_2$, where $\operatorname{cpr} A = \operatorname{cpr} A_1 + \operatorname{cpr} A_2$ and $A_1 = B_1 \oplus 0_{n-m}$, B_1 singular. Then A_1 is equal to Q_1 in all entries except possibly the m,m entry. Since B_1 is singular, $\operatorname{rank} B_1 \leq m-1$. Since the first $m-1$ rows of A_1 are equal to the corresponding rows of Q_1, and are linearly independent, $\operatorname{rank} A_1 = m-1$ and $A_1 = Q_1$. Thus $A_2 = Q_2$. We therefore have $\operatorname{cpr} A_1 = m-1$, $\operatorname{cpr} A_2 = \operatorname{cpr} G_2$, yielding that

$$\operatorname{cpr} G \geq \operatorname{cpr} A = \operatorname{cpr} A_1 + \operatorname{cpr} A_2 = m - 1 + \operatorname{cpr} G_2. \qquad \square$$

Combining the previous results we get the following.

Proposition 4.109. *Let G be a connected graph. If each block G' of G that has more than 3 vertices satisfies $\operatorname{cpr} G' = \operatorname{tf}(G')$, and G has at least one such block, then $\operatorname{cpr} G = \operatorname{tf}(G)$.*

Proof. Use induction on the number of blocks. For G consisting of one block there is nothing to prove. Now assume that the claim holds for graphs with less than k blocks, $k \geq 2$, and let G be a graph with k blocks that satisfies the assumptions.

First suppose $G = G_1 \cup G_2$, where G_1 and G_2 are connected and $G_1 \cap G_2$ is a single vertex, G_1 is either a K_2 or a K_3, and G_2 has $k-1$ blocks, at least one of which has more than 3 vertices. In that case,

$$\operatorname{tf}(G) = m - 1 + \operatorname{tf}(G_2) = m - 1 + \operatorname{cpr} G_2 = \operatorname{cpr} G$$

by Proposition 4.104(c), the induction hypothesis, and Proposition 4.108.

Otherwise, $G = G_1 \cup G_2$, where G_1 and G_2 are connected and $G_1 \cap G_2$ is a single vertex, G_1 has more than 3 vertices, and G_2 has $k-1$ blocks, at least one of which on more than 3 vertices. By (4.10), the induction hypothesis, and Proposition 4.104(b),

$$\operatorname{tf}(G) \leq \operatorname{cpr} G \leq \operatorname{cpr} G_1 + \operatorname{cpr} G_2 = \operatorname{tf}(G_1) + \operatorname{tf}(G_2) = \operatorname{tf}(G),$$

implying that $\operatorname{cpr} G = \operatorname{tf}(G)$. $\square$

We can now describe a family of graphs in which every graph G has $\operatorname{cpr} G = \operatorname{tf}(G)$.

Theorem 4.110. *Let G be a graph in which each block on more than 3 vertices is either triangle free, or a K_4, or a subgraph of some S_{2k}, $k \geq 3$, and has at least one such block. Then $\operatorname{cpr} G = \operatorname{tf}(G)$.*

The graphs described in this theorem include all triangle free graphs that are not trees, all completely positive graphs which contain a cycle on more than 3 vertices, and all graphs whose cp-rank is equal to the number of their vertices (satisfying (4.46)) that have a block on more than 3 vertices.

Proof. A 2-connected triangle free graph is not a tree, and satisfies $\operatorname{cpr} G = |E(G)| = \operatorname{tf}(G)$ by Corollary 4.55. From this and Example 4.103 the result follows by Proposition 4.109. □

Some outerplanar graphs also satisfy $\operatorname{cpr} G = \operatorname{tf}(G)$. A graph is *outerplanar* if it can be drawn in the plane so that no two edges cross, and all the vertices lie on the boundary of the outer face. Some well known facts are: Every subgraph of an outerplanar graph is outerplanar, every outerplanar graph has a vertex of degree at most 2, and every 2-connected outerplanar graph has a unique Hamiltonian cycle.

Proposition 4.111. *Let G be a connected outerplanar graph on n vertices, in which at least one block has more than 3 vertices. Then* $\operatorname{cpr} G = \operatorname{tf}(G)$.

Proof. We prove the theorem by induction on n. By the assumptions, $n \geq 4$. If $n = 4$, then $\operatorname{cpr} G = \operatorname{tf}(G)$ by Theorem 4.110. Let $n > 4$, assume the claim holds for graphs with fewer than n vertices, and let G be a connected outerplanar graph on n vertices with at least one block on more than 3 vertices. If G is not 2-connected, each of its blocks on more than 3 vertices satisfies the equality by the induction hypothesis, and Proposition 4.109 completes the proof. Hence we consider the case that G is 2-connected. The graph G has a Hamiltonian cycle, and a vertex v of degree 2. We consider two cases.

Case 1. The two vertices adjacent to v are not adjacent to each other.

If $G - v$ has only blocks that are edges and triangles, then G itself is a subgraph of S_{2k} for some $k \geq 3$ and, by Theorem 4.110, $\operatorname{cpr} G = \operatorname{tf}(G)$ $(= n)$. Otherwise, $G - v$ satisfies the induction hypothesis, and thus

$$\operatorname{tf}(G) \leq \operatorname{cpr} G \leq 2 + \operatorname{cpr}(G - v) = 2 + \operatorname{tf}(G - v) = \operatorname{tf}(G),$$

where the second inequality follows from Lemma 4.56, and the last equality from Proposition 4.104(c).

Case 2. The vertices u and w that are adjacent to v are also adjacent to each other.

Without loss of generality let $u = 1$, $v = 2$ and $w = 3$. Since the graph is Hamiltonian, there exists an additional vertex which is adjacent to 3 other than 1 and 2. Label this vertex 4. (The vertex 4 may or may not be adjacent to 1, and there is at least one more vertex on the Hamiltonian cycle.)

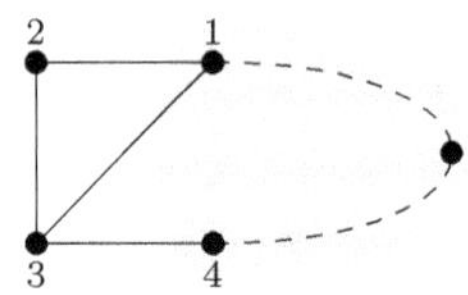

Let $\alpha_1 = \{1,2,3,4\}$, $\alpha_2 = \{1,4,\ldots,n\}$ and $G_i = G[\alpha_i]$, $i = 1,2$. Let A be a completely positive matrix such that $G(A) = G$ and $\operatorname{cpr} A = \operatorname{cpr} G$. As in previous proofs, $A = A_1 + A_2$, where $G(A_1[\alpha_1]) \subseteq G_1$, $G(A_2[\alpha_2]) \subseteq G_2$, and $\operatorname{cpr}(A) = \operatorname{cpr}(A_1) + \operatorname{cpr}(A_2)$.

Now set

$$t = \min_{1 \le j \le 3} \frac{a_{1j}}{a_{2j}}.$$

Suppose t is attained at $j = m$. Since A is positive semidefinite and its first two rows are linearly independent by their sign pattern, m cannot be equal to 1, so $m = 2$ or $m = 3$. Let $S = I - tE_{12}$. Then SA_1S^T is doubly nonnegative and, compared to A_1, has one or two additional zero entries in the first row, in position 1, 2 or in position 1, 3. Because row 2 of A_2 is zero, $SA_2S^T = A_2$. Every 4×4 doubly nonnegative matrix is completely positive, hence SA_1S^T is completely positive, and so is $SAS^T = SA_1S^T + A_2$. The graph of SAS^T is a subgraph of G, and thus $\operatorname{cpr}(SAS^T) \le \operatorname{cpr} G$. But since $S^{-1} \ge 0$, we have that $\operatorname{cpr} G = \operatorname{cpr} A \le \operatorname{cpr}(SAS^T)$ by Proposition 4.11. Thus $\operatorname{cpr}(SAS^T) = \operatorname{cpr} G$.

Denote the edge $\{1,2\}$ by e_1 and the edge $\{1,3\}$ by e_2. By Proposition 4.109 and the induction hypothesis $\operatorname{cpr}(G - e_1) = \operatorname{tf}(G - e_1)$, since the block of $G - e_1$ on vertices $\{1,3,4,\ldots,n\}$ is 2-connected and has at least 4 vertices. Thus if $G(SAS^T) = G - e_1$ we

$$\begin{aligned} \operatorname{tf}(G) \le \operatorname{cpr} G &= \operatorname{cpr}(SAS^T) \\ &\le \operatorname{cpr}(G - e_1) = \operatorname{tf}(G - e_1) \le \operatorname{tf}(G), \end{aligned}$$

and the equality $\operatorname{cpr}(G) = \operatorname{tf}(G)$ follows.

Since $\operatorname{cpr}(G - e_2) = \operatorname{tf}(G - e_2)$ by Case 1, a similar argument shows that if $G(SAS^T) = G - e_2$ then $\operatorname{cpr} G = \operatorname{tf}(G)$.

Finally, suppose $G(SAS^T) = G - \{e_1, e_2\}$. If $G - \{e_1, e_2\}$ has only K_2 and K_3 blocks, then G is a 2-connected subgraph of S_{2k} for some $k \ge 3$, and thus $\operatorname{cpr} G = n = \operatorname{tf}(G)$. Otherwise, $\operatorname{cpr}(G - \{e_1, e_2\}) = \operatorname{tf}(G - \{e_1, e_2\})$ by the induction hypothesis and Proposition 4.109. □

A particular example of an outerplanar graph is the fan F_n (a graph on n vertices consisting of a path on $n-1$ vertices and a vertex adjacent to all vertices of the path).

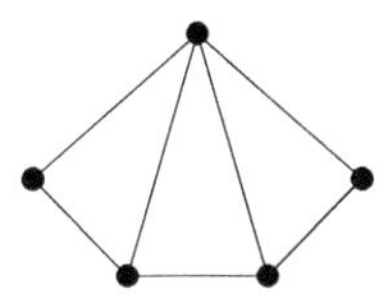

Corollary 4.112. *For $n \geq 4$,*

$$\operatorname{cpr} F_n = \begin{cases} \dfrac{3n-5}{2} & n \text{ is odd} \\ \\ \dfrac{3n-4}{2} & n \text{ is even} \end{cases}.$$

Proof. Since $n \geq 4$, F_n is a 2-connected outerplanar graph on more than 3 vertices, so $\operatorname{cpr} F_n = \operatorname{tf}(F_n)$. It is easy to see that the triangle free subgraph consisting of the spanning n-cycle and every other chord is a maximum size triangle free subgraph of F_n. (More explicitly, label the path vertices consecutively by $1, \ldots, n-1$ and the vertex of degree $n-1$ by n. The triangle free subgraph consists of the Hamiltonian cycle on all n vertices, and the chords $\{j, n\}$, $j = 2k+1$, $k = 2, \ldots, \lfloor (n-3)/2 \rfloor$.) $\square$

There are other graphs that have the property that $\operatorname{cpr} G = \operatorname{tf}(G)$ and do not belong to any of the families mentioned so far. We conclude the section with one such example. Let W_n denote the *wheel* on n vertices, a graph consisting of the cycle on $n-1$ vertices, and a vertex adjacent to all the cycle vertices. The wheel is a planar graph, but for $n \geq 4$ it is not an outerplanar one, and for $n \geq 5$ it does not belong to the graph family described in Theorem 4.110. The graph W_7 is shown here.

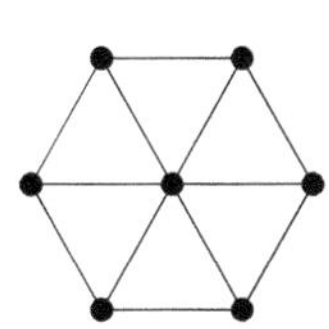

We leave the proof of the next proposition as an exercise.

Proposition 4.113. *For $n \geq 4$,*

$$\operatorname{cpr} W_n = \operatorname{tf}(W_n) = \begin{cases} \dfrac{3n-3}{2} & n \text{ is odd} \\ \dfrac{3n-4}{2} & n \text{ is even} \end{cases}.$$

Exercises

4.49. Prove Proposition 4.104.

4.50. Prove Proposition 4.105.

4.51. Let A be a 3×3 completely positive matrix. Prove:

(a) For every $1 \leq i < j \leq 3$ such that $a_{ij} > 0$, there exists a minimal CP factorization of A, $A = BB^T$, in which there is exactly one column of B whose support contains $\{i, j\}$.
(b) For every $1 \leq i \leq 3$ such that $a_{ii} > 0$, there exists a minimal CP factorization of A, $A = BB^T$, in which at least one column and at most two columns have i in their support.

Hint: Use Corollary 3.127 and Exercise 3.2.

4.52. Prove Proposition 4.113.
Hint: For $n \geq 5$, find $\operatorname{tf}(W_n)$. A matrix A with $\operatorname{cpr} A = \operatorname{cpr} W_n$ is a sum of completely positive matrices, each supported by one of the triangles of W_n. Use the previous exercise to show that $\operatorname{cpr} W_n \leq \operatorname{cpr} F_n + 1$.

4.53. Show that if $A \in \mathcal{CP}_6$ is orthogonal to $H \oplus 0$, where H is the Horn matrix, then $\operatorname{cpr} A \leq 7$.
Hint: Use Lemma 4.65 to show that $\operatorname{cpr} A \leq \operatorname{cpr} W_6$.

4.54. Show that if $A \in \mathcal{CP}_6$ is orthogonal to $H \oplus 0$, where H is a Hildebrand matrix, then $\operatorname{cpr} A \leq 6$.
Hint: Lemma 4.65.

Notes. This section is based on [Shaked-Monderer (2015)]. Exercises 4.53 and 4.54 are based on [Shaked-Monderer (2017)].

Still open: *Identify more families of graphs for which* $\operatorname{cpr}(G) = \operatorname{tf}(G)$.

4.8 The number of (minimal) CP factorizations

A completely positive matrix may have many different minimal CP factorizations. (We consider two minimal CP factorizations $A = BB^T$ and $A = CC^T$ to be the same if the columns of C are a permutation of the columns of B, i.e., $C = BP$, where P is a permutation matrix. This is equivalent to changing the order of summands in a minimal CP representation.) In particular, this is true for "most" matrices in the interior of $\mathcal{CP}_n$, namely those that have cp$^+$-rank equal to the rank. (Recall that these form an open dense subset of the interior of $\mathcal{CP}_n$.)

Proposition 4.114. *Let A be an $n \times n$ matrix in* $\operatorname{int}\mathcal{CP}_n$*, $n \geq 2$, with* $\operatorname{cp^+r} A = \operatorname{cpr} A$*. Then A has infinitely many minimal CP factorizations.*

Proof. Since $\operatorname{cp^+r} A = \operatorname{cpr} A = p$, there exists a minimal CP factorization $A = BB^T$, where B is a positive $n \times p$ matrix of rank n. Let $B = \begin{pmatrix} B_1 & B_2 \end{pmatrix}$, where B_1 consist of the first two columns of B. For any 2×2 orthogonal matrix $R \neq I$ such that $B_1 R > 0$, the matrix $C = \begin{pmatrix} B_1 R & B_2 \end{pmatrix}$ yields a different minimal CP factorization. As the n rows of B_1 are n positive vectors in $\mathbb{R}^2$, there are infinitely many rotations rotating them into the interior of $\mathbb{R}^2_+$. □

Note that by the same argument any matrix in the interior of $\mathcal{CP}_n$ has infinitely many CP factorizations. However, it is not clear whether a matrix in the interior with cp$^+$-rank greater than the cp-rank may have a unique *minimal* CP factorization.

Matrices with a unique minimal CP factorization do exist on the boundary of $\mathcal{CP}_n$.

Example 4.115. A rank 1 completely positive matrix A has a unique minimal CP representation as $A = \mathbf{b}\mathbf{b}^T$, $\mathbf{b} \geq \mathbf{0}$. Moreover, this is the only CP representation, apart from artificially extended representations of the form $A = \sum_{i=1}^{p} t_i \mathbf{b}\mathbf{b}^T$, where $t_i > 0$ and sum up to 1.

In general, when considering CP representations of a completely positive A that are not minimal, we will only consider representations without such redundancies. That is, representations $A = \sum_{i=1}^{k} \mathbf{b}_i \mathbf{b}_i^T$ where any two of the vectors $\mathbf{b}_1, \ldots, \mathbf{b}_k$ are linearly independent (i.e., the k vectors are *pairwise linearly independent*).

The graph of a rank 1 completely positive matrix is a clique and, possibly, some isolated vertices. Example 4.115 may be generalized to matrices with a chordal graph. We cite the following two theorems without proof — see references in the notes to this section. For Theorem 4.116 below, recall that $\mathrm{mr}_{\mathcal{M}}(G)$ denotes the minimum rank of matrices in the class $\mathcal{M}$ of symmetric matrices with graph G. The Cholesky decomposition of a positive semidefinite matrix A referred to in Theorem 4.117 is the decomposition $A = QQ^T$ described on page 32. The number of columns of Q is equal to $\operatorname{rank} A$.

Theorem 4.116. *Let G be a chordal graph. Then*

$$\mathrm{mr}_{\mathcal{CP}}(G) = \mathrm{mr}_{\mathcal{DNN}}(G) = \mathrm{mr}_{\mathcal{PSD}}(G) = \mathrm{cc}\,(G).$$

Theorem 4.117. *Let G be a chordal graph, and let A be a positive semidefinite A with $G(A) = G$ and $\operatorname{rank} A = r = \mathrm{cc}\,(G)$. Then the following are equivalent.*

(a) *A is completely positive.*
(b) *There exists a perfect elimination ordering of G's vertices such that for the corresponding permutation matrix P, the matrix Q in the Cholesky decomposition $P^TAP = QQ^T$ is nonnegative.*
(c) *For every perfect elimination ordering of G's vertices and a corresponding permutation matrix P, the matrix Q in the Cholesky decomposition $P^TAP = QQ^T$ is nonnegative.*

Moreover, if A is completely positive then $\operatorname{cpr} A = \operatorname{rank} A = r$. And if P and Q are as in (c), then $A = (PQ)(PQ)^T$ is the unique CP factorization of A.

One special case worth mentioning is that of trees. If G is a tree on n vertices, then

$$\mathrm{mr}_{\mathcal{CP}}(G) = \mathrm{mr}_{\mathcal{DNN}}(G) = \mathrm{mr}_{\mathcal{PSD}}(G) = n - 1,$$

and any singular completely positive matrix whose graph is the tree G has a unique minimal CP factorization. (This special case of Theorem 4.117 may be proved quite easily directly — see Exercise 4.56.)

For the n-cycle, $n > 3$, it is known that $\operatorname{cpr} A = n$ for every completely positive matrix A whose graph is the n-cycle (Corollary 4.55).

Theorem 4.118. *Let A be a completely positive matrix whose graph is the n-cycle. Then A has at most two different minimal CP factorizations.*

Proof. By considering an appropriate permutation similarity, we assume that

$$A = \begin{pmatrix} d_1 & h_1 & 0 & \dots & 0 & h_n \\ h_1 & d_2 & h_2 & \ddots & & 0 \\ 0 & h_2 & d_3 & h_3 & \ddots & \vdots \\ \vdots & \ddots & \ddots & \ddots & \ddots & 0 \\ 0 & & \ddots & h_{n-2} & d_{n-1} & h_{n-1} \\ h_n & 0 & \dots & 0 & h_{n-1} & d_n \end{pmatrix}$$

where h_i and d_i are positive, $i = 1, \dots, n$. In any minimal CP representation there are n summands, one of which has the form

$$\mathbf{b}_1\mathbf{b}_1^T = \begin{pmatrix} t & h_1 \\ h_1 & \frac{h_1^2}{t} \end{pmatrix} \oplus 0_{n-2}, \quad t > 0.$$

Then

$$A - \mathbf{b}_1\mathbf{b}_1^T = \begin{pmatrix} d_1 - t & 0 & 0 & \dots & 0 & h_n \\ 0 & d_2 - \frac{h_1^2}{t} & h_2 & \ddots & & 0 \\ 0 & h_2 & d_3 & h_3 & \ddots & \vdots \\ \vdots & \ddots & \ddots & \ddots & \ddots & 0 \\ 0 & & \ddots & h_{n-2} & d_{n-1} & h_{n-1} \\ h_n & 0 & \dots & 0 & h_{n-1} & d_n \end{pmatrix}$$

is a completely positive matrix which has a CP representation with $n-1$ summands, and its graph is a path on n vertices. In particular,

$$\operatorname{rank}(A - \mathbf{b}_1\mathbf{b}_1^T) = \operatorname{cpr}(A - \mathbf{b}_1\mathbf{b}_1^T) = n - 1.$$

By the previous result, $A - \mathbf{b}_1\mathbf{b}_1^T$ has a unique CP representation. Hence the number of rank one representations of A is equal to the number of positive t's for which $A - \mathbf{b}_1\mathbf{b}_1^T$ is doubly nonnegative and singular. Note that the last $n-2$ columns in $A' = A - \mathbf{b}_1\mathbf{b}_1^T$ are linearly independent by their sign pattern, hence

$$A' = \begin{pmatrix} d_1 - t & 0 & 0 & \dots & 0 & h_n \\ 0 & d_2 - \frac{h_1^2}{t} & h_2 & 0 & \dots & 0 \\ 0 & h_2 & & & & \\ \vdots & 0 & & & E & \\ 0 & \vdots & & & & \\ h_n & 0 & & & & \end{pmatrix},$$

where the matrix E of order $n-2$ is nonsingular. As $\operatorname{rank} A' = n-1$, the matrix A'/E is positive semidefinite and has rank 1. This latter matrix has the form

$$A'/E = \begin{pmatrix} a-t & b \\ b & c - \frac{h_1^2}{t} \end{pmatrix},$$

where a, b, c are some real numbers. Then $\det(A'/E) = 0$ translates into a quadratic equation in t, that has at most two positive solutions. $\square$

How can we tell if a given matrix with a cycle graph has one or two minimal CP representations? Here is a more general result (which generalizes the result on trees mentioned following Theorem 4.117).

Theorem 4.119. *Let A be an $n \times n$ irreducible completely positive matrix, $n \geq 2$, whose graph is triangle free. Then the following are equivalent*

(a) *$M(A)$ is singular.*
(b) *A has a unique CP representation.*
(c) *A has a unique minimal CP representation.*

Remark 4.120. If the triangle free $G(A)$ is bipartite, then A is singular if and only if $M(A)$ is singular: Let α and β be the independent sets of vertices in $G(A)$ such that every edge has one end in α and the other in β. Then $A = SM(A)S$, where S is the signature matrix with

$$s_{ii} = \begin{cases} 1, & i \in \alpha \\ -1, & i \in \beta \end{cases}.$$

Proof of Theorem 4.119. Only (a) $\Rightarrow$ (b) and (c) $\Rightarrow$ (a) need to be proved.

(a) $\Rightarrow$ (b): Let $A = \sum_{i=1}^k \mathbf{b}_i \mathbf{b}_i^T$ be any CP representation of A. Then $1 \leq |\operatorname{supp} \mathbf{b}_i| \leq 2$ for every i. As observed in the proof of Theorem 3.43,

$$M(A) = \sum_{i=1}^k M\left(\mathbf{b}_i \mathbf{b}_i^T\right),$$

where $M\left(\mathbf{b}_i \mathbf{b}_i^T\right) = \mathbf{d}_i \mathbf{d}_i^T$ is a rank one positive semidefinite matrix, with $\operatorname{supp} \mathbf{d}_i = \operatorname{supp} \mathbf{b}_i$, $|\mathbf{d}_i| = \mathbf{b}_i$.

The irreducible M-matrix $M(A)$ is singular, and therefore has a positive eigenvector $\mathbf{v}$ corresponding to the eigenvalue zero. As $M(A) = \sum_{i=1}^k \mathbf{d}_i \mathbf{d}_i^T$, we get that each $\mathbf{d}_i$ is in $\operatorname{cs} M(A)$, and is therefore orthogonal to $\mathbf{v}$. Since $1 \leq |\operatorname{supp} \mathbf{d}_i| \leq 2$, this forces $\sigma_i = \operatorname{supp} \mathbf{d}_i$ to have size 2, and hence

$\mathbf{d}_i[\sigma_i]$ spans the orthogonal complement of $\{\mathbf{v}[\sigma_i]\}$ in $\mathbb{R}^2$. As the vectors $\mathbf{b}_1, \ldots, \mathbf{b}_k$ are pairwise linearly independent, so are $\mathbf{d}_1, \ldots, \mathbf{d}_k$. Combined with the fact that the orthogonal complement of $\{\mathbf{v}[\sigma_i]\}$ in $\mathbb{R}^2$ is one dimensional, this implies that $\operatorname{supp} \mathbf{d}_j \neq \sigma_i$ for every $j \neq i$. Hence if $\sigma_i = \{p, q\}$, we have $\left(\mathbf{d}_i \mathbf{d}_i^T\right)_{pq} = (M(A))_{pq}$. This equality together with $\mathbf{d}_i^T \mathbf{v} = \mathbf{0}$ completely determines $\mathbf{d}_i$ up to sign, and therefore determines $\mathbf{b}_i = |\mathbf{d}_i|$. Hence the CP representation $A = \sum_{i=1}^k \mathbf{b}_i \mathbf{b}_i^T$ is unique (and necessarily $k = |E(G(A))|$).

(c) $\Rightarrow$ (a): We will show that if $M(A)$ is nonsingular, then A has at least two minimal CP factorizations. As the number of CP factorizations is clearly preserved by positive diagonal congruence, it suffices to prove the following claim: If $M(A)$ is diagonally dominant and nonsingular, then A has at least two different CP representations. We prove it by induction on the number m of edges of the connected $G(A)$. Recall that the number of summands in a minimal CP representation of such A is m.

Consider first the case that $G(A)$ is a tree, $m = n - 1$. As $M(A)$ is a nonsingular diagonally dominant matrix, A is also nonsingular by Remark 4.120. By Theorem 4.50, $\operatorname{cpr} A = \operatorname{rank} A$, hence any minimal CP representation of A has n summands. We can generate two different minimal CP representations as follows. Choose $\delta_1 > 0$ such that $A - \delta_1 E_{11}$ is positive semidefinite and singular, and $\delta_2 > 0$ such that $A - \delta_2 E_{22}$ is positive semidefinite and singular ($\delta_i = \det(A)/\det(A(i))$, $i = 1, 2$). Both matrices are doubly nonnegative, with the same tree graph as A, and are therefore completely positive. By Remark 4.120, both their comparison matrices are also singular. As (a) implies (b), there exist unique CP representations

$$A - \delta_1 E_{11} = \sum_{i=1}^{n-1} \mathbf{b}_i \mathbf{b}_i^T \quad \text{and} \quad A - \delta_2 E_{22} = \sum_{i=1}^{n-1} \mathbf{c}_i \mathbf{c}_i^T,$$

where for each i the nonnegative vectors $\mathbf{b}_i$ and $\mathbf{c}_i$ are supported by the same edge of the tree. Then

$$A = \delta_1 E_{11} + \sum_{i=1}^{n-1} \mathbf{b}_i \mathbf{b}_i^T \quad \text{and} \quad A = \delta_2 E_{22} + \sum_{i=1}^{n-1} \mathbf{c}_i \mathbf{c}_i^T$$

are two minimal rank 1 representations of A that are clearly different (by the only vector in each representation that is supported by a single vertex).

Now suppose the claim holds for $m \geq n - 1$, and let A be an irreducible completely positive matrix with a nonsingular diagonally dominant comparison matrix, whose triangle free graph $G(A)$ has $m + 1$ edges. As $M(A)$ is nonsingular, there exists $1 \leq i \leq n$ such that in row i there is strict diagonal dominance, i.e., $(M(A)\mathbf{e})_i > 0$. As $G(A)$ is not a tree, there exists an edge whose removal from $G(A)$ will keep the graph connected. Without loss of generality $\{1,2\}$ is such an edge. Let $F = \begin{pmatrix} a_{12} & -a_{12} \\ -a_{12} & a_{12} \end{pmatrix} \oplus 0_{n-2}$. Then $M(A) - F$ is diagonally dominant, and since $F\mathbf{e} = \mathbf{0}$,

$$((M(A) - F)\mathbf{e})_i = (M(A)\mathbf{e})_i > 0.$$

Hence $M(A)-F = M(A-|F|)$ is a nonsingular diagonally dominant matrix whose graph has m edges. By the induction hypothesis, $A-|F|$ has at least two different minimal CP representations,

$$A - |F| = \sum_{i=1}^{m-1} \mathbf{b}_i \mathbf{b}_i^T \quad \text{and} \quad A - |F| = \sum_{i=1}^{m-1} \mathbf{c}_i \mathbf{c}_i^T.$$

Then

$$A = |F| + \sum_{i=1}^{m-1} \mathbf{b}_i \mathbf{b}_i^T \quad \text{and} \quad A = |F| + \sum_{i=1}^{m-1} \mathbf{c}_i \mathbf{c}_i^T$$

are two different minimal CP representations of A. □

This result has implications also for some positive matrices on the boundary of $\mathcal{CP}_n$.

Proposition 4.121. *Let M be an $n \times n$ copositive matrix, which is exceptional and extremal, and has a set of n representing minimal zeros $\mathbf{w}_1, \ldots, \mathbf{w}_n$. Let $W = \begin{pmatrix} \mathbf{w}_1 & \mathbf{w}_2 & \ldots & \mathbf{w}_n \end{pmatrix}$. Then each completely positive matrix A, which is orthogonal to M in $\mathcal{S}_n$, has the form $A = WCW^T$ for some $n \times n$ completely positive C, $\operatorname{cpr} A = \operatorname{cpr} C$, and the number of (minimal) CP factorizations of A is equal to the number of (minimal) CP factorizations of C.*

Proof. As M is exceptional and extremal, W is nonsingular. Each zero of M is a nonnegative combination of W's columns. If $A = BB^T$ for some nonnegative B, each column of B is a zero of M, hence $B = WX$, where $X \geq 0$, and

$$A = (WX)(WX)^T = WCW^T, \tag{4.49}$$

where $C = XX^T$ is a completely positive matrix. By (4.49), $\operatorname{cpr} A \leq \operatorname{cpr} C$.

If $A = QQ^T$ is any CP factorization of A, then $Q = WY$, $Y \geq 0$. Since

$$A = QQ^T = W(YY^T)W^T$$

we get that $YY^T = W^{-1}AW^{-T} = C$ is a CP factorization of C. In particular, if $A = QQ^T$ is a minimal CP factorization, this shows that $\operatorname{cpr} C \leq \operatorname{cpr} A$, and we conclude that $\operatorname{cpr} A = \operatorname{cpr} C$. As W is nonsingular, $B \neq Q$ if and only if $X \neq Y$. Hence the number of CP factorizations/minimal CP factorizations of both matrices are equal. $\square$

We conclude this section with two examples in $\mathcal{CP}_5$.

Example 4.122. Let A be a 5×5 positive matrix, which is completely positive and orthogonal to the Horn matrix. Then A has at most two different minimal CP decompositions. To see that, recall that by Proposition 4.66, $\operatorname{cpr} A = \operatorname{rank} A$. A closer look at the proof shows that any minimal CP factorization of A has the form $A = (WX)(WX)^T$, where

$$W = \begin{pmatrix} \mathbf{u}_1 & \mathbf{u}_2 & \mathbf{u}_3 & \mathbf{u}_4 & \mathbf{u}_5 \end{pmatrix}$$

is the matrix of minimal zeros of the Horn matrix, and $X \geq 0$ has at most two consecutive positive entries in each column (where the first entry is considered the successor of the fifth). By the positivity of A, the graph of XX^T is the 5-cycle. Hence the completely positive XX^T may have at most two different minimal CP factorizations, and by Proposition 4.121, the number A is equal to the number of CP factorization of XX^T. By Theorem 4.120 there is a unique CP representation if and only if $M(XX^T)$ is singular.

A matrix orthogonal to a Hildebrand matrix has a unique CP representation.

Example 4.123. Let A be a 5×5 completely positive matrix orthogonal to a Hildebrand matrix. Then A has exactly one CP decompositions. To see that, recall that by Proposition 4.67, $\operatorname{cpr} A = \operatorname{rank} A$. Any CP factorization of A has the form $A = (WD)(WD)^T$, where

$$W = \begin{pmatrix} \mathbf{u}_1 & \mathbf{u}_2 & \mathbf{u}_3 & \mathbf{u}_4 & \mathbf{u}_5 \end{pmatrix}$$

is the matrix of minimal zeros of the Hildebrand matrix, and $D \geq 0$ has a single positive entry in each column. As DD^T is a diagonal matrix, it has a unique CP factorization (see Exercise 4.55) and so does A.

Exercises

4.55. Let $A_1, \ldots, A_k$ be completely positive matrices, each with a unique CP factorization. Show that $A_1 \oplus \cdots \oplus A_k$ has a unique rank CP factorization.

In particular, any nonzero nonnegative diagonal matrix has a unique CP factorization.

4.56. Prove directly that if A is an $n \times n$ singular completely positive matrix whose graph is a tree, then A has a unique CP factorization.
Hint: Use induction on n. If i is a pendant vertex in $G(A)$, and $\mathbf{a}_i$ is the i-th column of A, then the $\frac{1}{\sqrt{a_{ii}}}\mathbf{a}_i$ has to be one of the columns in the CP factorization.

4.57. Show that a completely positive matrix A of rank 2 has a unique minimal CP factorization if and only if A has a zero off-diagonal entry.

4.58. Let A be a completely positive matrix whose graph is the n-cycle $1 - 2 - \cdots - n - 1$. Show that the CP factorization $A = BB^T$, where

$$B = \begin{pmatrix} s_1 & 0 & 0 & \dots & t_n \\ t_1 & s_2 & 0 & & \vdots \\ 0 & t_2 & \ddots & & \\ \vdots & 0 & \ddots & s_{n-1} & 0 \\ 0 & 0 & 0 & t_{n-1} & s_n \end{pmatrix},$$

is the unique CP factorization of A if and only if $\Pi_{i=1}^n s_i = \Pi_{i=1}^n t_i$.
Hint: $M(A) = SS^T$, where S is obtained from B by replacing each t_i by $-t_i$.

4.59. Let A be an $n \times n$ irreducible completely positive matrix, whose graph is bipartite. Prove:

(a) If A is singular, it has a unique CP factorization.
(b) If A is nonsingular, it has at least two minimal CP factorizations, and infinitely many CP factorizations.

Notes. Theorems 4.116 and 4.117 were proved in [Shaked-Monderer (2013)]. Theorem 4.116 previously in [Shaked-Monderer (1993)], and in [Booth *et al.* (2008)] the result on $\mathrm{mr}_{\mathcal{PSD}}(G)$ was proved differently, for both the real and complex case. The number of CP factorizations and minimal CP factorizations for CP realizations of an acyclic or a unicylic graph was discussed in [Dickinson and Dür (2012)], where linear-time algorithms were presented to derive minimal CP factorizations of such matrices. Theorem 4.119 is due to [Shaked-Monderer (2020)], with the cyclic case previously proved (differently) in [Zhang and Li (2000)]. Examples 4.122 and

4.123 appeared in [Shaked-Monderer *et al.* (2013)]. Example 4.122 and Exercise 4.58 combine to a characterization of the uniqueness of a factorization of a positive matrix orthogonal to the Horn matrix, which was first proved (differently) in [Zhang (2019)].

Still open: *Does there exist a matrix in the interior of $\mathcal{COP}_n$ with a unique minimal CP factorization? Finding more examples of matrices with a unique CP factorization on the boundary is also of interest.*

4.9 Rational and integral cp-rank

Suppose A is completely positive over the rationals. Recall that this means that $A = BB^T$, where $B \geq 0$ has rational entries, or (if B is $n \times k$)

$$A = \sum_{i=1}^{p} \mathbf{b}_i \mathbf{b}_i^T, \quad \mathbf{b}_i \in \mathbb{Q}_+^n \quad \text{for } i = 1, \ldots, k\,.$$

Alternatively, $A = XDX^T$, where $X \geq 0$ and D is a rational positive diagonal matrix, or (if X is $n \times m$)

$$A = \sum_{i=1}^{m} d_i \mathbf{x}_i \mathbf{x}_i^T, \quad \mathbf{x}_i \in \mathbb{Q}_+^n,\ d_i \in \mathbb{Q}_+ \quad \text{for } i = 1, \ldots, m\,. \tag{4.50}$$

Although the two approaches are equivalent, it was explained in Section 3.11 that the number of columns in such B may be up to four times bigger than the number of columns of X.

Example 4.124. Let

$$A = \begin{pmatrix} 7 & 0 \\ 0 & 0 \end{pmatrix}.$$

Then $A = XDX^T$, where

$$X = \begin{pmatrix} 1 \\ 0 \end{pmatrix} \quad \text{and} \quad D = (7),$$

and $A = BB^T$, where

$$B = \begin{pmatrix} 2 & 1 & 1 & 1 \\ 0 & 0 & 0 & 0 \end{pmatrix}.$$

There is no decomposition $A = BB^T$, where $B \geq 0$ has less then 4 columns. Otherwise we would have

$$7 = \left(\frac{p_1}{q_1}\right)^2 + \left(\frac{p_2}{q_2}\right)^2 + \left(\frac{p_3}{q_3}\right)^2 = \frac{(p_1q_2q_3)^2 + (p_2q_1q_3)^2 + (p_3q_1q_2)^2}{q^2},$$

where $p_i \geq 0$ and $q_i > 0$ are integers, $q = q_1 q_2 q_3$. That is, $7q^2$ is a sum of less than four square integers, which is impossible by Lagrange's four-square theorem, which guarantees that any integer of the form $4^m(8k+7)$ cannot be represented as a sum of less than four squares.

For the definition of the rational cp-rank, we prefer to work with rational decompositions of the form XDX^T, or equivalently, with rational CP representations of the form (4.50).

Definition 4.125. If a matrix A is completely positive over the rationals, its *rational cp-rank*, denoted by $\operatorname{rcpr} A$, is the minimum number of summands in (4.50).

The following clearly holds.

Proposition 4.126. *For every matrix A, which is completely positive over the rationals,*

$$\operatorname{rcpr} A \geq \operatorname{cpr} A \geq \operatorname{rank} A.$$

To the basic results in Section 3.11 the following observations may be added.

Proposition 4.127. *The following hold.*

(a) *Let matrices A_1 and A_2 of the same order be completely positive over the rationals, and let d_1, d_2 be positive rational numbers. Then* $\operatorname{rcpr}(d_1 A_1 + d_2 A_2) \leq \operatorname{rcpr} A_1 + \operatorname{rcpr} A_2$.

(b) *Let $A = A_1 \oplus A_2$, where A_1 and A_2 are completely positive over the rationals (and not necessarily of the same order). Then* $\operatorname{rcpr} A = \operatorname{rcpr} A_1 + \operatorname{rcpr} A_2$.

(c) *Let A be completely positive over the rationals, let P be a permutation matrix, and let D be a rational positive diagonal matrix. Then*

$$\operatorname{rcpr} A = \operatorname{rcpr}(PAP^T) = \operatorname{rcpr}(DAD).$$

(d) *Let S be a nonsingular rational matrix with $S^{-1} \geq 0$. If SAS^T is completely positive over the rationals, then*

$$\operatorname{rcpr} A \leq \operatorname{rcpr}(SAS^T).$$

(e) *Let a A be a rational completely positive matrix of rank r, and suppose the matrix X in the Cholesky decomposition $A = XDX^T$ is nonnegative. Then* $\operatorname{rcpr} A = r$.

In particular, we get the following.

Corollary 4.128. *If A is a rational completely positive matrix of rank at most 2, then* $\operatorname{rcpr} A = \operatorname{cpr} A = \operatorname{rank} A$.

For matrices that are completely positive over the rationals, the best upper bounds on the rational cp-rank in terms of the rank are the same as the ones for the cp-rank of real completely positive matrices, given in Theorems 4.21, 4.23 and 4.30.

Theorem 4.129. *Let A be nonzero matrix of rank r, which is completely positive over the rationals, and let N be the maximal number of zeroes above the diagonal in an $r \times r$ nonsingular principal submatrix of A. Then*

$$\operatorname{rcpr} A \leq \frac{r(r+1)}{2} - N. \tag{4.51}$$

If $r \geq 2$ and $N = 0$, or $r \geq 3$ and $N = 1$, then

$$\operatorname{rcpr} A \leq \frac{r(r+1)}{2} - N - 1.$$

The proof follows the alternative proofs of the original theorems.

The bound 4.51 is sharp, in the sense that for each rank r there is some N for which the bound is attained (see Exercise 4.64).

Corollary 4.130. *For every matrix A, which is completely positive over the rationals and has rank r,*

$$\operatorname{rcpr} A \leq \frac{r(r+1)}{2} - 1.$$

The inequality $\operatorname{cpr} A \leq \operatorname{rcpr} A$ may be strict.

Example 4.131. The matrix

$$A = \begin{pmatrix} 3 & 1 & 0 & 1 \\ 1 & 3 & 1 & 0 \\ 0 & 1 & 3 & 1 \\ 1 & 0 & 1 & 3 \end{pmatrix}$$

is completely positive over the rationals since $A = BB^T$ with

$$B = \begin{pmatrix} 1 & 0 & 0 & 0 & 1 & 0 & 0 & 1 \\ 0 & 1 & 0 & 0 & 1 & 1 & 0 & 0 \\ 0 & 0 & 1 & 0 & 0 & 1 & 1 & 0 \\ 0 & 0 & 0 & 1 & 0 & 0 & 1 & 1 \end{pmatrix}.$$

It has $\operatorname{cpr} A = 4$ (why?), and two minimal CP factorizations, both of which are not rational. Hence $\operatorname{rcpr} A > \operatorname{cpr} A$. We leave the details as an exercise.

Remark 4.132. It follows from Proposition 4.127 that any matrix A of order $n \leq 3$, which is completely positive over the rationals has $\operatorname{rcpr} A = \operatorname{cpr} A$. Again we leave the proof as an exercise.

And integral cp-rank?

How should we define integer cp-rank? If we follow the idea of the rational case, it would be the minimum number of summands in a representation of a matrix A as

$$A = \sum_{i=1}^{m} d_i \mathbf{x}_i \mathbf{x}_i^T, \quad d_i \in \mathbb{Z}_+, \ \mathbf{x}_i \in \mathbb{Z}_+^n. \tag{4.52}$$

Then, for a 2×2 matrix the integral cp-rank would be 1, if $\operatorname{rank} A = 1$, and 2 or 3, if $\operatorname{rank} A = 2$. To see that, recall the proof of Theorem 3.178. According to this proof, if

$$A = \begin{pmatrix} a & b \\ b & c \end{pmatrix}$$

is completely positive over the integers, and $a, c \geq b$, then

$$A = b\begin{pmatrix} 1 \\ 1 \end{pmatrix}\begin{pmatrix} 1 \\ 1 \end{pmatrix}^T + (a-b)\begin{pmatrix} 1 \\ 0 \end{pmatrix}\begin{pmatrix} 1 \\ 0 \end{pmatrix}^T + (c-b)\begin{pmatrix} 0 \\ 1 \end{pmatrix}\begin{pmatrix} 0 \\ 1 \end{pmatrix}^T. \tag{4.53}$$

And if a, b and c do not satisfy the inequalities, then $A = QA'Q^T$, where Q is an integral nonnegative nonsingular matrix, and A' has the above form, with both diagonal entries greater than or equal to the off-diagonal one. Hence A also has a decomposition with at most 3 such summands. There are matrices for which 3 summands are indeed required.

Example 4.133. If $a, c \geq 2$, then a decomposition of the form (4.52) of the matrix

$$A = \begin{pmatrix} a & 1 \\ 1 & c \end{pmatrix}$$

must include the summand

$$\begin{pmatrix} 1 \\ 1 \end{pmatrix}\begin{pmatrix} 1 \\ 1 \end{pmatrix}^T.$$

Hence

$$A = \begin{pmatrix} 1 \\ 1 \end{pmatrix}\begin{pmatrix} 1 \\ 1 \end{pmatrix}^T + (a-1)\begin{pmatrix} 1 \\ 0 \end{pmatrix}\begin{pmatrix} 1 \\ 0 \end{pmatrix}^T + (c-1)\begin{pmatrix} 0 \\ 1 \end{pmatrix}\begin{pmatrix} 0 \\ 1 \end{pmatrix}^T,$$

$a-1, c-1 > 0$, and does not have such a CP representation with less than 3 integral summands.

It turns out that in the 2×2 case a more interesting question is to determine the minimal number of summands in a representation of A in the form

$$A = \sum_{i=1}^{k} \mathbf{b}_i \mathbf{b}_i^T, \quad \mathbf{b}_i \in \mathbb{Z}_+^n . \tag{4.54}$$

By the proof of Theorem 3.178, no more than 12. But a more careful analysis shows the following.

Theorem 4.134. *A 2×2 matrix A, which is completely positive over the integers, has a representation of the form* (4.54) *with at most* 11 *summands.*

Proof. By the above explanation, the maximum number of summands is attained at some

$$A = \begin{pmatrix} a & b \\ b & c \end{pmatrix}$$

with $a, c \geq b$. Such A can be decomposed in the form (4.53). For this decomposition to yield a representation of the form (4.54), each of b, $a - b$ and $c - b$ should be of the form $4^m(8k+7)$, and thus not be the sum of less than 4 squares. However, other decompositions of A exist, which lead to a representation of the form (4.54) with at most 11 summands. In the case that $(a - b) \not\equiv 7(\bmod 8)$ consider

$$A = (b-6)\begin{pmatrix} 1 & 1 \\ 1 & 1 \end{pmatrix} + \begin{pmatrix} 9 & 6 \\ 6 & 4 \end{pmatrix} + \begin{pmatrix} a-b-3 & 0 \\ 0 & c-b+2 \end{pmatrix}.$$

And if both $(a - b) \equiv 7(\bmod 8)$ $(c - b) \equiv 7(\bmod 8)$, consider

$$A = (b-2)\begin{pmatrix} 1 & 1 \\ 1 & 1 \end{pmatrix} + \begin{pmatrix} 1 & 2 \\ 2 & 4 \end{pmatrix} + \begin{pmatrix} a-b+1 & 0 \\ 0 & c-b-2 \end{pmatrix}. \qquad \square$$

There are matrices for which the number of summands in a representation of the form (4.54) is at least 9.

Example 4.135. Let

$$A = \begin{pmatrix} a & 1 \\ 1 & c \end{pmatrix},$$

where $a = c = 8$ (or, more generally, $(a - 1) \equiv 8(\bmod 7)$ and $(c - 1) \equiv 8(\bmod 7)$). From Example 4.133 and Lagrange's four-square theorem it is obvious that A has a representation of the form (4.54) with 9 summands, and none with fewer summands.

Exercises

4.60. Prove Proposition 4.127.

4.61. Prove Corollary 4.128.
Hint: See Example 3.170.

4.62. Prove Theorem 4.129.
Hint: See pages 385–389. Note that Carathéodory's Theorem holds in linear spaces over $\mathbb{Q}$, and use Proposition 4.127(e).

4.63. Complete the details of Example 4.131.
Hint: See Theorem 4.118 and its proof.

4.64. Show that for any $r \geq 3$, a rational doubly nonnegative matrix of rank r, which has an $r \times r$ positive diagonal principal submatrix is completely positive over the rationals with $\operatorname{rcpr} A = r$.

4.65. Let A be the matrix in Example 4.131. Show that $\operatorname{cpr} A = 4$ and $\operatorname{rcpr} A = 5$.

4.66. Prove the claim in Remark 4.132.

4.67. Complete the proof of Theorem 4.134:

(a) Show that if a positive integer p has the form $4^m(8k+7)$, then $p \pm 2$ and $p \pm 6$ do not have this form.
(b) Complete the missing details in the proof.

Notes. The discussion of rational cp-rank is based on [Shaked-Monderer *et al.* (2018)], and the discussion on the number of summands in a decomposition of the form (4.54) is based on [Laffey and Šmigoc (2019)].

Still open: *What is the maximum rational cp-rank of an $n \times n$ matrix, which is completely positive over the rationals? And what is the minimal number of integral CP summands in a representation* (4.54) *of a 2×2 matrix, which is completely positive over the integers?*

Chapter 5

The structure of $\mathcal{COP}_n$ and $\mathcal{CP}_n$

5.1 The structure of the copositive cone

In Section 2.3 we considered some basic aspects of the geometry of the cone $\mathcal{COP}_n$, in particular the extreme rays of this cone. Now we look into wider aspects of the structure and geometry of the cone: its faces and algebraic structure, its isomorphisms, and the maximum angle between rays in this cone.

Faces of $\mathcal{COP}_n$

We begin by considering the minimal face of $\mathcal{COP}_n$ containing a given $n \times n$ copositive matrix A, $\mathcal{F}^A_{\mathcal{COP}_n}$.

Theorem 5.1. *Let $A \in \mathcal{COP}_n$. Then the minimal face of $\mathcal{COP}_n$ containing A, $\mathcal{F}^A_{\mathcal{COP}_n}$, is equal to*

$$\left\{X \in \mathcal{COP}_n \mid (X\mathbf{u})_i = 0 \text{ for every } \mathbf{u} \in \mathcal{V}^A \text{ and } i \text{ such that } (A\mathbf{u})_i = 0\right\}.$$

Or, equivalently, to

$$\left\{X \in \mathcal{COP}_n \mid (X\mathbf{u})_i = 0 \text{ for every } \mathbf{u} \in \mathcal{V}^A_{\min} \text{ and } i \text{ such that } (A\mathbf{u})_i = 0\right\}.$$

Proof. Let us denote by $\mathcal{F}$ the set

$$\left\{X \in \mathcal{COP}_n \mid (X\mathbf{u})_i = 0 \text{ for every } \mathbf{u} \in \mathcal{V}^A \text{ and } i \text{ such that } (A\mathbf{u})_i = 0\right\}.$$

From the fact that each minimal zero of A is a sum of a finite number of minimal zeros it is easy to see that $\mathcal{F}$ is equal to

$$\left\{X \in \mathcal{COP}_n \mid (X\mathbf{u})_i = 0 \text{ for every } \mathbf{u} \in \mathcal{V}^A_{\min} \text{ and } i \text{ such that } (A\mathbf{u})_i = 0\right\}.$$

It is also easy to check that $\mathcal{F}$ is a closed convex cone of $\mathcal{COP}_n$, and that $A \in \mathcal{F}$. Hence we concentrate on showing that $\mathcal{F}$ is a face of $\mathcal{COP}_n$, and is the minimal one containing A.

First note that if $X \in \mathcal{F}$ and $\mathbf{u}$ is a zero of A, then $(X\mathbf{u})_i = 0$ for every $i \in \operatorname{supp}\mathbf{u}$, hence $\mathbf{u}^T X\mathbf{u} = 0$. That is, every zero of A is also a zero of X. Hence if X and Y are copositive and $X + Y \in \mathcal{F}$, then every zero $\mathbf{u}$ of A is also a zero of $X + Y$, and therefore of X and of Y. This shows that for any zero $\mathbf{u}$ of A, $X\mathbf{u} \geq \mathbf{0}$ and $Y\mathbf{u} \geq \mathbf{0}$. This guarantees that $(X\mathbf{u})_i = 0$ and $(Y\mathbf{u})_i = 0$ for every i such that $((X + Y)\mathbf{u})_i = 0$, and therefore $X, Y \in \mathcal{F}$.

To assert the minimality of $\mathcal{F}$, suppose F is any face of $\mathcal{COP}_n$ containing A. If $X \in \mathcal{F}$, then, by Theorem 2.62, for some $\varepsilon > 0$ the matrix $A - \varepsilon X$ is copositive. As $A = \varepsilon X + (A - \varepsilon X)$ is in the face F we get that $X \in F$. That is, $\mathcal{F}$ is contained in every face of $\mathcal{COP}_n$ that contains A. □

Note that if $A \in \operatorname{int}\mathcal{COP}_n$, then $\mathcal{F}^A_{\mathcal{COP}_n} = \mathcal{COP}_n$.

Remark 5.2. From the characterization above it can be seen that if $\mathcal{F} = \mathcal{F}^A_{\mathcal{COP}_n}$ is the minimal face of $\mathcal{COP}_n$ containing A, then the linear subspace of $\mathcal{S}_n$ spanned by $\mathcal{F}$, $\mathcal{F} - \mathcal{F}$, is

$$\left\{X \in \mathcal{S}_n \mid (X\mathbf{u})_i = 0 \text{ for every } \mathbf{u} \in \mathcal{V}^A_{\min} \text{ and } i \text{ such that } (A\mathbf{u})_i = 0\right\}.$$

We leave the proof as an exercise. This generalizes the characterization of extremal matrices in $\mathcal{COP}_n$ in Theorem 2.71.

Theorem 5.3. *Let $A \in \mathcal{COP}_n$. Then the minimal exposed face of $\mathcal{COP}_n$ containing A, $\widehat{\mathcal{F}}^A_{\mathcal{COP}_n}$, is equal to*

$$\left\{X \in \mathcal{COP}_n \mid \mathcal{V}^A \subseteq \mathcal{V}^X\right\}.$$

Proof. Observe that if $A \in \operatorname{int}\mathcal{COP}_n$ the equality holds trivially. Otherwise, every exposed face of $\mathcal{COP}_n$ containing A has the form

$$\{X \in \mathcal{COP}_n \mid \langle X, M\rangle = 0\}$$

for some $M \in \mathcal{CP}_n$ orthogonal to A (see Exercise 1.57). Since such M can be represented as $M = \sum_{i=1}^p \mathbf{b}_i\mathbf{b}_i^T$, $\mathbf{b}_i \in \mathbb{R}^n_+$, every such exposed face is an intersection of a finite number of faces of the form

$$\left\{X \in \mathcal{COP}_n \mid \mathbf{b}^T X\mathbf{b} = 0\right\} = \left\{X \in \mathcal{COP}_n \mid \langle X, \mathbf{b}\mathbf{b}^T\rangle = 0\right\}, \tag{5.1}$$

where $\mathbf{b}$ is a zero of A, each of which is clearly an exposed face.

The intersection of all exposed faces containing A is therefore the intersection of all exposed faces in the form (5.1), $\mathbf{b} \in \mathcal{V}^A$, and this intersection, $\widehat{\mathcal{F}}^A_{\mathcal{COP}_n}$, is equal to $\left\{X \in \mathcal{COP}_n \mid \mathcal{V}^A \subseteq \mathcal{V}^X\right\}$. □

Corollary 5.4. *A matrix $A \in \mathcal{COP}_n$ generates an exposed ray of $\mathcal{COP}_n$ if and only if*

$$\left\{X \in \mathcal{COP}_n \mid \mathcal{V}^A \subseteq \mathcal{V}^X\right\} = \{tA \mid t \geq 0\}.$$

Recall that the exposed vectors of a cone are dense in the set of its extreme rays (Theorem 1.141). The extreme rays of $\mathcal{COP}_n$ are fully known only for $n \le 6$, as mentioned in Chapter 2. We may ask which of the known extreme rays of the copositive cone are exposed. We first consider the positive semidefinite and nonnegative extreme rays. Note that if A generates an exposed ray of $\mathcal{COP}_n$, then any matrix in its orbit generates such a ray.

Proposition 5.5. *Let $n \ge 2$. The following holds for the extreme rays of $\mathcal{COP}_n$ that are in $\mathcal{SPN}_n$.*

(a) *For $1 \le i \le n$ the ray generated by E_{ii} is not exposed.*
(b) *For $1 \le i < j \le n$ the ray generated by $E_{ij} + E_{ji}$ is exposed.*
(c) *If $\mathbf{x} \in \mathbb{R}^n$ has both positive and negative entries, the ray generated by $\mathbf{x}\mathbf{x}^T$ is exposed.*

Proof. To prove (a), note that $\mathbf{v} \in \mathbb{R}^n_+$ is a zero of E_{ii} if and only if $v_i = 0$. Thus for $j \neq i$, every zero of E_{ii} is also a zero of $E_{ij}+E_{ji}$. By Theorem 5.3, the minimal exposed face containing E_{ii} is not the ray generated by E_{ii}.

Now we prove (b). Let $A = E_{ij} + E_{ji}$, $1 \le i < j \le n$. The zeros of A are all the vectors $\mathbf{v} \in \mathbb{R}^n_+$ such that $v_i v_j = 0$, i.e., either $v_i = 0$ or $v_j = 0$. Suppose $X \in \mathcal{COP}_n$, and $\mathcal{V}^A \subseteq \mathcal{V}^X$. Then $x_{pp} = 0$ for every p (since $\mathbf{e}_p \in \mathcal{V}^A$). Also, for every $p < q$ such that $(p,q) \neq (i,j)$, the vector $\mathbf{v}_{pq} = \mathbf{e}_p + \mathbf{e}_q$ is a zero of X, hence

$$0 = \mathbf{v}_{pq}^T X \mathbf{v}_{pq} = x_{pp} + 2x_{pq} + x_{qq} \ge x_{pp} + x_{qq} - 2\sqrt{x_{pp}x_{qq}} \ge 0$$

implies that $x_{pq} = 0$. Hence such X is a multiple of A.

Finally, consider (c). Let $\mathbf{x}$ have both positive and negative entries. Let $\sigma_+ = \operatorname{supp} \mathbf{x}_+$, $\sigma_- = \operatorname{supp} \mathbf{x}_-$, and $\tau = \{1, \ldots, n\} \setminus \operatorname{supp} \mathbf{x}$. A vector $\mathbf{v}$ is a zero of $\mathbf{x}\mathbf{x}^T$ if and only if $\mathbf{v} \in \mathbb{R}^n_+$ is orthogonal to $\mathbf{x}$. This includes the minimal zeros $\mathbf{e}_i$, $i \in \tau$, and

$$\mathbf{v}_{ij} = -x_j \mathbf{e}_i + x_i \mathbf{e}_j, \qquad i \in \sigma_+, j \in \sigma_-,$$

and all their nonnegative combinations. We leave it as an exercise to check that if all these are zeros of $X \in \mathcal{COP}_n$, then X is a multiple of $\mathbf{x}\mathbf{x}^T$. □

Note that if a ray of $\mathcal{SPN}_n$ is exposed in $\mathcal{COP}_n$, then it is also exposed in $\mathcal{SPN}_n$.

Extreme rays generated in $\mathcal{COP}_n$ by $\{\pm 1\}$-matrices are also exposed.

Proposition 5.6. *An extreme ray in $\mathcal{COP}_n$ generated by a $\{\pm 1\}$-matrix A is exposed.*

Proof. The graph $G_{-1}(A)$ is connected, and $a_{ij} = 1$ if and only if $d_{G_{-1}(A)}(i,j) = 2$ (see Remark 2.145). For every $1 \leq i < j \leq n$ such that $a_{ij} = -1$ let $\mathbf{u}_{ij} = \mathbf{e}_i + \mathbf{e}_j$. These are all the minimal zeros of A up to multiplication by a positive scalar. For every $i < j$ such that $a_{ij} = 1$, there exists k such that $a_{ik} = a_{jk} = -1$. Choose one such k and let $\mathbf{v}_{ij} = \mathbf{e}_i + \mathbf{e}_j + 2\mathbf{e}_k$. These are (non-minimal) zeros of A with support of size 3. Let

$$M = \sum_{\substack{1 \leq i < j \leq n \\ a_{ij} = -1}} \mathbf{u}_{ij}\mathbf{u}_{ij}^T + \sum_{\substack{1 \leq i < j \leq n \\ a_{ij} = 1}} \mathbf{v}_{ij}\mathbf{v}_{ij}^T.$$

If $X \in \mathcal{COP}_n$ is orthogonal to M, then the $\mathbf{u}_{ij}$'s and $\mathbf{v}_{ij}$'s are zeros of X. Let $x_{11} = x$. Then $x_{ii} + x_{ij} = 0$ whenever $a_{ij} = -1$. Since $G_{-1}(A)$ is connected this implies that $x_{ii} = x$ for every i and $x_{ij} = -x$ whenever $a_{ij} = -1$. If $a_{ij} = 1$ for $i < j$, then as $\mathbf{v}_{ij}$ is a zero of X we get for the corresponding k that $x_{ii} + x_{ij} + 2x_{ik} = 0$, and this implies $x_{ij} = x$ whenever $a_{ij} = 1$. Hence the exposed face of $\mathcal{COP}_n$ consisting of matrices orthogonal to M is the ray generated by A. $\square$

Extremal $\{0, 1, -1\}$-matrices with zeros need not generate an exposed ray.

Example 5.7. Consider the Hoffman-Pereira matrix

$$H_7 = \begin{pmatrix} 1 & -1 & 1 & 0 & 0 & 1 & -1 \\ -1 & 1 & -1 & 1 & 0 & 0 & 1 \\ 1 & -1 & 1 & -1 & 1 & 0 & 0 \\ 0 & 1 & -1 & 1 & -1 & 1 & 0 \\ 0 & 0 & 1 & -1 & 1 & -1 & 1 \\ 1 & 0 & 0 & 1 & -1 & 1 & -1 \\ -1 & 1 & 0 & 0 & 1 & -1 & 1 \end{pmatrix}$$

of Example 2.147. The vectors $\mathbf{u}_i = \mathbf{e}_i + \mathbf{e}_{i \dotplus 1}$, $1 \leq i \leq 7$, where $\dotplus$ denotes addition modulo 7 on $\{1, \ldots, 7\}$, form a complete representing set of its minimal zeros. Each other zero of H_7 has the form $t\mathbf{u}_i + s\mathbf{u}_{i \dotplus 1}$, $t, s > 0$. By Theorem 5.3 the minimal exposed face containing H_7 is

$$\left\{X \in \mathcal{COP}_n \mid \mathcal{V}^{H_7} \subseteq \mathcal{V}^X\right\}.$$

It includes any matrix that has the same ± 1 entries as H_7, and nonnegative entries where H_7 has zeros, and therefore is not equal to the ray generated by H_7.

For $n = 5$, the Horn matrix generates an exposed ray by Proposition 5.6. The Hildebrand matrices also generate an exposed ray. This is a special case of the following observation.

Proposition 5.8. *Let A be an extremal matrix in $\mathcal{COP}_n$. Suppose that $\mathbf{u}_i$, $1 \leq i \leq k$, form a representative set of minimal zeros, and that $(A\mathbf{u}_i)_j = 0$ if and only if $j \in \operatorname{supp} \mathbf{u}_i$. Then the ray generated by A in $\mathcal{COP}_n$ is exposed.*

Proof. If $\mathbf{u}$ is a zero of A, then there exists a minimal zero $\mathbf{u}_i$ of A such that $\operatorname{supp} \mathbf{u} \supseteq \operatorname{supp} \mathbf{u}_i$. But then by Proposition 2.46 we get that $(A\mathbf{u}_i)_j = 0$ for every $j \in \operatorname{supp} \mathbf{u}$. By our assumption, this implies that $\operatorname{supp} \mathbf{u} = \operatorname{supp} \mathbf{u}_i$, and therefore $\mathbf{u} = t\mathbf{u}_i$ for some $t > 0$. That is, under these conditions, the minimal zeros of A are its only zeros.

If X is any $n \times n$ copositive matrix whose zeros include $\mathbf{u}_i$, $1 \leq i \leq k$, then by Proposition 2.45

$$(X\mathbf{u}_i)_j = 0 \text{ for every } j \in \operatorname{supp} \mathbf{u}_i$$

for each $1 \leq i \leq k$. But then by the extermality of A, X is a scalar multiple of A. That is, the minimal exposed face containing A is the ray generated by A. □

Recall the two simple ways of extending an exceptional extremal A in $\mathcal{COP}_n$ to an exceptional extremal matrix in $\mathcal{COP}_m$, $m > n$: One is by adding zero rows and columns to A. The other is forming the matrix

$$\begin{pmatrix} A & AE \\ E^T A & E^T AE \end{pmatrix},$$

where $E \geq 0$ is an $n \times m$ matrix with exactly one positive entry in each column. If we start with an exceptional matrix A generating an exposed ray of $\mathcal{COP}_n$, then both constructions will yield exposed rays of $\mathcal{COP}_m$. In both cases, it suffices to prove the case that $m = n + 1$, then use it repeatedly to obtain the general result.

Proposition 5.9. *Let $A \in \mathcal{COP}_n$ be nonzero. Then*

$$\begin{pmatrix} A & \mathbf{0} \\ \mathbf{0}^T & 0 \end{pmatrix}$$

generates an exposed ray in $\mathcal{COP}_{n+1}$ if and only if A generates an exposed ray of $\mathcal{COP}_n$.

Proof. We prove the "if" part, and leave the "only if" as an exercise. By Proposition 5.5(b) and (c) it suffices to consider the case that A is exceptional. The zeros of the extended matrix include $\mathbf{e}_{n+1}$ and for every zero $\mathbf{v}$ of A they include also

$$\begin{pmatrix} \mathbf{v} \\ 0 \end{pmatrix} \quad \text{and} \quad \begin{pmatrix} \mathbf{v} \\ 1 \end{pmatrix}.$$

If the zeros of $X \in \mathcal{COP}_{n+1}$ include these zeros, then $x_{n+1\,n+1} = 0$, hence

$$X = \begin{pmatrix} X' & \mathbf{b} \\ \mathbf{b}^T & 0 \end{pmatrix}, \qquad \mathbf{b} \geq \mathbf{0}.$$

If $\mathbf{v}$ is a zero of A, the fact that $\begin{pmatrix} \mathbf{v} \\ 1 \end{pmatrix}$ is a zero of X implies that

$$\mathbf{b}^T\mathbf{v} = 0,$$

and therefore $b_i = 0$ for every $i \in \operatorname{supp} \mathbf{b}_i$. Since the minimal zeros of A span $\mathbb{R}^n$ (Theorem 2.67), this implies $\mathbf{b} = 0$.

Finally, for every zero $\mathbf{v}$ of A, $\begin{pmatrix} \mathbf{v} \\ 0 \end{pmatrix}$ is a zero of X, and therefore $\mathbf{v}$ is a zero of X'. As A generates an exposed ray of $\mathcal{COP}_n$, this implies that X' is a scalar multiple of A, and we get that X is a scalar multiple of the extended matrix. □

Corollary 5.10. *An $n \times n$ copositive matrix A generates an exposed ray of $\mathcal{COP}_n$ if and only if $A \oplus 0_m$ generates an exposed ray of $\mathcal{COP}_{n+m}$.*

The proof of the next proposition is left as an exercise.

Proposition 5.11. *Let $A \in \mathcal{COP}_n$, $1 \leq i \leq n$, and $t > 0$. If A is not a nonnegative matrix, and generates an exposed ray of $\mathcal{COP}_n$, then*

$$\begin{pmatrix} A & t\mathbf{e}_i \\ t\mathbf{e}_i^T & t^2 a_{ii} \end{pmatrix}$$

generates an exposed ray in $\mathcal{COP}_{n+1}$.

Corollary 5.12. *Let $A \in \mathcal{COP}_n$, $1 \leq i \leq n$, and $t > 0$. If A is not a nonnegative matrix, and generates an exposed ray of $\mathcal{COP}_n$, and E is an $n \times m$ nonnegative matrix with exactly one positive entry in each column, then*

$$\begin{pmatrix} A & E \\ E^T & E^TAE \end{pmatrix}$$

generates an exposed ray in $\mathcal{COP}_{n+m}$.

We now move to the other end and consider maximal (proper) faces of $\mathcal{COP}_n$. The following theorem describes all such faces.

Theorem 5.13. *Let $n \geq 2$. For every nonzero vector $\mathbf{v} \in \mathbb{R}^n_+$*

$$\mathcal{F} = \left\{X \in \mathcal{COP}_n \mid \mathbf{v}^T X \mathbf{v} = 0\right\}$$

is a maximal face of $\mathcal{COP}_n$, and every maximal face of $\mathcal{COP}_n$ has this form.

Proof. Recall that a maximal face is exposed (see Exercise 1.55). We first show that every maximal face has this form. Indeed, since it is exposed, it has the form

$$\{X \in \mathcal{COP}_n \mid \langle X, M\rangle = 0\}$$

for some nonzero $M \in \mathcal{CP}_n$ (Exercise 1.57). If $M = \sum_{i=1}^p \mathbf{b}_i \mathbf{b}_i^T$, $\mathbf{b}_i \in \mathbb{R}^n_+$, then

$$\{X \in \mathcal{COP}_n \mid \langle X, M\rangle = 0\} = \bigcap_{i=1}^p \left\{X \in \mathcal{COP}_n \mid \langle X, \mathbf{b}_i \mathbf{b}_i^T\rangle = 0\right\},$$

hence the maximality of the face implies $\operatorname{rank} M = 1$, $M = \mathbf{v}\mathbf{v}^T$ for some nonzero $\mathbf{v} \in \mathbb{R}^n_+$.

If $\mathbf{v}$ is any nonzero nonnegative vector

$$\mathcal{F} = \left\{X \in \mathcal{COP}_n \mid \mathbf{v}^T X \mathbf{v} = 0\right\}$$

is a proper face: it contains the matrix $A = \sum_{i=1}^{n-1} \mathbf{v}_i \mathbf{v}_i^T$, where $\mathbf{v}_1, \ldots, \mathbf{v}_{n-1}$ form a basis for the orthogonal complement of $\{\mathbf{v}\}$, and is clearly not the whole $\mathcal{COP}_n$. Any exposed face containing $\mathcal{F}$ has the form

$$\{X \in \mathcal{COP}_n \mid \langle X, M\rangle = 0\}$$

for some completely positive M, and has to include A. But $\langle A, M\rangle = 0$ for $M = \sum_{i=1}^p \mathbf{b}_i \mathbf{b}_i^T$, $\mathbf{b}_i \in \mathbb{R}^n_+$, if and only if each $\mathbf{b}_i$ is a zero of A. As the only zeros of the positive semidefinite A are the multiples of $\mathbf{v}$, necessarily $M = t\mathbf{v}\mathbf{v}^T$ for some $t > 0$ and the face is equal to $\mathcal{F}$. $\square$

The dimensions of the maximal faces vary.

Theorem 5.14. *Let $\mathbf{v} \in \mathbb{R}^n_+$ have p positive entries, and let*

$$\mathcal{F} = \left\{X \in \mathcal{COP}_n \mid \mathbf{v}^T X \mathbf{v} = 0\right\}$$

be the corresponding maximal face of $\mathcal{COP}_n$. Then

$$\dim \mathcal{F} = \frac{n(n+1)}{2} - p.$$

Proof. First consider the case that $p = n$, i.e., $\mathbf{v}$ is positive. In this case, the face $\mathcal{F} = \{X \in \mathcal{COP}_n \mid \mathbf{v}^T X \mathbf{v} = 0\}$ consists of positive semidefinite matrices. That is, $X \in \mathcal{F}$ if and only if $X = \sum_{i=1}^k \mathbf{x}_i \mathbf{x}_i^T$, where each $\mathbf{x}_i$ is orthogonal to $\mathbf{v}$. Therefore $\mathcal{F}$ is isomorphic to $\mathcal{PSD}_{n-1}$, which is a solid cone in $\mathcal{S}_{n-1}$, and has dimension $\frac{(n-1)n}{2} = \frac{n(n+1)}{2} - n$ (this is similar to Exercise 4.17(a)).

Now suppose $\mathbf{v}$ has $p < n$ positive entries. Without loss of generality we may assume that the first p entries of $\mathbf{v}$ are positive. Let $\bar{\mathbf{v}}$ consist of the first p entries of $\mathbf{v}$, and let

$$\mathcal{F}_{\bar{\mathbf{v}}} = \{X \in \mathcal{COP}_p \mid \bar{\mathbf{v}}^T X \bar{\mathbf{v}} = 0\}.$$

If $X \in \mathcal{F}$ then

$$X = \begin{pmatrix} Y & Z \\ Z^T & W \end{pmatrix},$$

where $Y \in \mathcal{F}_{\bar{\mathbf{v}}}$, $W \in \mathcal{COP}_{n-p}$ and Z is a $p \times (n-p)$ matrix. We get that $\mathcal{U} \subseteq \mathcal{F} \subseteq \mathcal{W}$, where

$$\mathcal{U} = \left\{ \begin{pmatrix} Y & Z \\ Z^T & W \end{pmatrix} \middle| Y \in \mathcal{F}_{\bar{\mathbf{v}}}, W \in \mathcal{COP}_{n-p}, Z \in \mathbb{R}_+^{p\times(n-p)} \right\}$$

and

$$\mathcal{W} = \left\{ \begin{pmatrix} Y & Z \\ Z^T & W \end{pmatrix} \middle| Y \in \mathcal{F}_{\bar{\mathbf{v}}}, W \in \mathcal{COP}_{n-p}, Z \in \mathbb{R}^{p\times(n-p)} \right\}.$$

By the first part of this proof $\dim \mathcal{F}_{\bar{\mathbf{v}}} = \frac{p(p-1)}{2}$. We also have $\dim \mathcal{COP}_{n-p} = \frac{(n-p)(n-p+1)}{2}$ and $\dim \mathbb{R}_+^{p\times(n-p)} = \dim \mathbb{R}^{p\times(n-p)} = p(n-p)$. Hence

$$\dim \mathcal{U} = \dim \mathcal{W} = \frac{p(p-1)}{2} + \frac{(n-p)(n-p+1)}{2} + p(n-p) = \frac{n(n+1)}{2} - p.$$

As $\mathcal{F}$ is sandwiched between these two, it has the same dimension. □

Corollary 5.15. *If $\mathcal{F}$ is a maximal face of $\mathcal{COP}_n$, then*

$$\frac{(n-1)n}{2} \le \dim \mathcal{F} \le \frac{n(n+1)}{2} - 1.$$

Example 5.16. A *facet* of a cone is a face whose dimension is one less than the dimension of the cone. By Theorem 5.14 the cone $\mathcal{COP}_n$ has n facets: for each $1 \le i \le n$,

$$\{X \in \mathcal{COP}_n \mid \mathbf{e}_i^T X \mathbf{e}_i = 0\}$$

is a facet. This facet consists of the matrices in $\mathcal{COP}_n$ whose i-th diagonal entry is zero. For example, the facet consisting of copositive matrices with zero $\mathbf{e}_1$ is

$$\left\{ \begin{pmatrix} 0 & \mathbf{b}^T \\ \mathbf{b} & W \end{pmatrix} \middle| \mathbf{b} \ge \mathbf{0} \text{ and } W \in \mathcal{COP}_{n-1} \right\}.$$

Algebraic structure of $\mathcal{COP}_n$

In this subsection we describe a decomposition of the copositive cone $\mathcal{COP}_n$ into a disjoint union of a finite number of sets, each of which is open in the relative topology of some algebraic set in $\mathcal{S}_n$.

Definition 5.17. Let $A \in \mathcal{COP}_n$, and let $\mathbf{u}$ be a zero of A. Let $\sigma = \operatorname{supp} \mathbf{u}$ and let $\tau = \{1 \leq i \leq n \mid (A\mathbf{u})_i = 0\}$. The pair (σ, τ) is called the *extended zero support* of A. The set of all extended zero supports of minimal zeros of A is called the *extended minimal zero support set* of A.

Recall that in the above definition, for every zero $\mathbf{u}$ of the matrix A, $\sigma \subseteq \tau$. Also note that if $A \in \operatorname{int}\mathcal{COP}_n$, then its extended minimal zero support set is empty.

Example 5.18. Consider the Horn matrix

$$\begin{pmatrix} 1 & -1 & 1 & 1 & -1 \\ -1 & 1 & -1 & 1 & 1 \\ 1 & -1 & 1 & -1 & 1 \\ 1 & 1 & -1 & 1 & -1 \\ -1 & 1 & 1 & -1 & 1 \end{pmatrix} \in \mathcal{COP}_5.$$

The extended minimal zero supports set of this matrix is

$$\{(\sigma_i, \tau_i) \mid i = 1, \ldots, 5\},$$

where

$$\sigma_i = \{i, i \dotplus 1\} \quad \text{and} \quad \tau_i = \{i, i \dotplus 1, i \dotplus 2, i \dotplus 4\}, \quad i = 1, \ldots, 5,$$

where $\dotplus$ denotes addition modulo 5 on $\{1, \ldots, 5\}$.

If $H \in \mathcal{COP}_5$ is any Hildebrand matrix, then the extended minimal zero support set of H is $\{(\alpha_i, \alpha_i) \mid i = 1, \ldots, 5\}$, where

$$\alpha_i = \{i, i \dotplus 1, i \dotplus 2\}.$$

Remark 5.19. If D is a positive diagonal matrix, then DAD has the same extended minimal zero support set as A ($\mathbf{u}$ is a zero of A if and only if $D^{-1}\mathbf{u}$ is a zero of DAD). If P is a permutation matrix, then $\mathbf{u}$ is a zero of A if and only if $P\mathbf{u}$ is a zero of PAP^T. Hence the extended minimal zero support set of PAP^T is obtained from that of A by applying the permutation represented by P on the elements of the index sets.

The following proposition translates known properties of the minimal zeros of a copositive A to properties of the extended minimal zero support set of A.

Proposition 5.20. *Let $\mathcal{E} = \{(\sigma_i, \tau_i) \mid i = 1, \ldots, m\}$ be the extended minimal zero support set of $A \in \mathcal{COP}_n$. Then*

(a) *for every i, $\emptyset \neq \sigma_i \subseteq \tau_i$.*
(b) *for every $i \neq j$, $\sigma_i \subseteq \tau_j$ if and only if $\sigma_j \subseteq \tau_i$.*
(c) *for every i, $\operatorname{rank} A[\tau_i|\sigma_i] = |\sigma_i| - 1$ and $\operatorname{ns} A[\tau_i|\sigma_i] = \operatorname{ns} A[\sigma_i]$ is spanned by a positive vector.*
(d) *if $\mathbf{u}$ is a zero of A, then there exist $i_1, \ldots, i_k \in \{1, \ldots, m\}$ such that the extended support set of $\mathbf{u}$ is (σ, τ), where*

$$\sigma = \bigcup_{j=1}^{k} \sigma_{i_j}, \qquad \tau = \bigcap_{j=1}^{k} \tau_{i_j}.$$

Proof. Part (a) is obvious from the definition of a zero as a nonzero nonnegative vector, and Proposition 2.45.

In part (b), by the symmetry between i and j it suffices to prove one implication. So suppose $\mathbf{u}_i$ and $\mathbf{u}_j$ are minimal zeros of A. Let $\sigma_i = \operatorname{supp} \mathbf{u}_i$ and $\sigma_j = \operatorname{supp} \mathbf{u}_j$, and suppose $(A\mathbf{u}_j)_k = 0$ for every $k \in \sigma_i$. Then $\mathbf{u}_i^T(A\mathbf{u}_j) = 0$, and therefore $(A\mathbf{u}_i)^T\mathbf{u}_j = 0$, and since both $A\mathbf{u}_i \geq \mathbf{0}$ and $\mathbf{u}_j \geq \mathbf{0}$, this implies that $(A\mathbf{u}_i)_k = 0$ for every $k \in \sigma_j$, i.e., $\sigma_j \subseteq \tau_i$ (see Exercise 2.50).

For part (c), observe that $A[\tau_i|\sigma_i]\mathbf{u}_i[\sigma_i] = (A\mathbf{u}_i)[\tau_i] = \mathbf{0}$, and if $A[\tau_i|\sigma_i]\mathbf{x} = \mathbf{0}$, then in particular $A[\sigma_i]\mathbf{x} = \mathbf{0}$. As the null space of $A[\sigma_i]$ is spanned by $\mathbf{u}_i[\sigma_i]$, this completes the proof.

Finally, for (d) recall that the zero $\mathbf{u}$ is a sum of several minimal zeros, say $\mathbf{u}_{i_1}, \ldots, \mathbf{u}_{i_k}$. Then clearly $\operatorname{supp} \mathbf{u} = \cup_{j=1}^{k} \operatorname{supp} \mathbf{u}_{i_j}$, and $(A\mathbf{u})_i = \sum_{j=1}^{k} (A\mathbf{u}_{i_j})_i$. As $A\mathbf{u}_{i_j} \geq \mathbf{0}$ for every j, $(A\mathbf{u})_i = 0$ if and only if $(A\mathbf{u}_{i_j})_i = 0$ for every $1 \leq j \leq k$. □

In the following definition, a matrix $X \in \mathbb{R}^{p \times q}$ is called *rank-deficient* if $\operatorname{rank} X < \min(p, q)$.

Definition 5.21. Let $\mathcal{E} = \{(\sigma_i, \tau_i) \mid i = 1, \ldots, m\}$ be a finite collection of pairs of index sets (subsets of $\{1, \ldots, n\}$). Let

$$S_{\mathcal{E}} = \{X \in \mathcal{COP}_n \mid \mathcal{E} \text{ is the extended zero supports set of } X\}$$

and

$$Z_{\mathcal{E}} = \{X \in \mathcal{S}_n \mid X[\tau_i|\sigma_i] \text{ is rank-deficient }\}.$$

In the case that $\mathcal{E} = \emptyset$ we have

$$S_{\emptyset} = \operatorname{int} \mathcal{COP}_n \quad \text{and} \quad Z_{\emptyset} = \mathcal{S}_n.$$

There is a finite collection of sets $\mathcal{E}$ for which $S_{\mathcal{E}}$ is nonempty (parts (a) and (b) of Proposition 5.20 are necessary, but not sufficient, conditions on $S_{\mathcal{E}}$ to be nonempty), and the nonempty sets $S_{\mathcal{E}}$ form a decomposition of $\mathcal{COP}_n$ into a disjoint union of sets. Each $S_{\mathcal{E}}$ is invariant under diagonal scaling by a positive diagonal matrix, and the automorphism $X \mapsto PXP^T$ of $\mathcal{COP}_n$ maps $S_{\mathcal{E}}$ onto some $S_{\mathcal{E}'}$, where $\mathcal{E}'$ is obtained from $\mathcal{E}$ by permutation of the indices in the index sets ($\mathcal{E}' = \mathcal{E}$ is possible).

For each $\mathcal{E}$, the set $Z_{\mathcal{E}}$ is algebraic — it is the zero locus of a finite number of determinants, and $S_{\mathcal{E}} \subseteq Z_{\mathcal{E}}$ for every $\mathcal{E}$. This is obvious if $S_{\mathcal{E}}$ is empty, and follows from Proposition 5.20(c) otherwise. We now show that $S_{\mathcal{E}}$ is open in the relative topology of $Z_{\mathcal{E}}$. Again, this is clear when $\mathcal{E} = \emptyset$.

Theorem 5.22. *Let $A \in \mathcal{COP}_n$, and let $\mathcal{E} = \{(\sigma_i, \tau_i) \mid i = 1, \ldots, m\}$ be its extended minimal zero support set of A. Then there exists a neighbourhood $U \subseteq \mathcal{S}_n$ such that $U \cap S_{\mathcal{E}} = U \cap Z_{\mathcal{E}}$.*

Proof. Suppose on the contrary that there exists a sequence $A_k \in Z_{\mathcal{E}} \setminus S_{\mathcal{E}}$ that converges to A. We will reach a contradiction by showing that (a subsequence of) A_k actually consists of copositive matrices with extended minimal zero support sets equal to $\mathcal{E}$.

For $i = 1, \ldots, m$ let $\mathbf{u}_i \in \Delta$ be a minimal zero of A with $\operatorname{supp} \mathbf{u}_i = \sigma_i$ and $(A\mathbf{u}_i)_j = 0$ if and only if $j \in \tau_i$. Then $\mathbf{u}_i[\sigma_i]$ is a positive vector that spans $\operatorname{ns} A[\sigma_i]$, and therefore also $\operatorname{ns} A[\tau_i|\sigma_i]$.

As $A_k \in Z_{\mathcal{E}}$, $A_k[\tau_i|\sigma_i]$ is rank-deficient. Since $\sigma_i \subseteq \tau_i$, this means that $\operatorname{rank} A_k[\tau_i|\sigma_i] \leq |\sigma_i| - 1$. However, as $A[\tau_i|\sigma_i] = \lim_{k\to\infty} A_k[\tau_i|\sigma_i]$ and $\operatorname{rank} A[\tau_i|\sigma_i] = |\sigma_i| - 1$, $\operatorname{rank} A_k$ cannot be smaller, except for a finite number of k's. Without loss of generality we assume that for every k, $\operatorname{rank} A[\tau_i|\sigma_i] = |\sigma_i| - 1$, $1 \leq i \leq m$. Let $\mathbf{v}_k^{(i)} \in \mathbb{R}^n$ be a vector with $\operatorname{supp} \mathbf{v}_k^{(i)} \subseteq \sigma_i$ and 1-norm 1, such that $\mathbf{v}_k^{(i)}[\sigma_i]$ spans $\operatorname{ns} A_k[\tau_i|\sigma_i]$. We may choose a subsequence of the indices k such that the corresponding m subsequences of $\mathbf{v}_k^{(i)}$, $1 \leq i \leq m$, all converge. So without loss of generality suppose the sequences $\mathbf{v}_k^{(i)}$ converge. As $A_k[\tau_i|\sigma_i]\mathbf{v}_k^{(i)}[\sigma_i] = \mathbf{0}$ for every k and $A_k[\tau_i|\sigma_i]$ converges to $A[\tau_i|\sigma_i]$, whose null space is spanned by $\mathbf{u}_i[\sigma_i]$, we may assume without loss of generality that $\mathbf{v}_k^{(i)}[\sigma_i]$ converges to $\mathbf{u}_i[\sigma_i]$, and therefore (by omitting a finite number of elements) that $\operatorname{supp} \mathbf{v}_k^{(i)} = \sigma_i$ and $\mathbf{v}_k^{(i)} \in \Delta$ for every k and i. This means that $\lim_{k\to\infty} \mathbf{v}_k^{(i)} = \mathbf{u}_i$, and $A_k\mathbf{v}_k^{(i)}$ converges to $A\mathbf{u}_i$. Hence $\left(A_k\mathbf{v}_k^{(i)}\right)_j > 0$ for every $j \notin \tau_i$, for sufficiently large k. Since $A_k[\sigma_i]$ converges to $A[\sigma_i]$, and the latter has $|\sigma_i| - 1$ positive eigenvalues, $A_k[\sigma_i]$ has $|\sigma_i| - 1$ positive eigenvalues for sufficiently large k, in addition to the one zero eigenvalue. Again we omit a finite number of elements in our sequences, and assume that for every k the matrix $A_k[\sigma_i]$ is positive semidefinite, and therefore $\mathbf{v}_k^{(i)}[\sigma_i]$ spans its zero eigenspace. If we show that A_k is copositive, this will mean that for each $1 \le i \le m$ the vector $\mathbf{v}_k^{(i)}$ is a minimal zero of A_k, and (σ_i, τ_i) is its extended support.

We now show that except for, possibly, a finite number of indices k the matrices A_k are copositive. Let $\mathbf{w}_k \in \Delta$ be a vector where

$$\min_{\mathbf{x}\in\Delta} \frac{1}{2}\mathbf{x}^T A_k \mathbf{x} \tag{5.2}$$

is attained. Assume on the contrary that $\mathbf{w}_k^T A_k \mathbf{w}_k = \eta_k < 0$ for a subsequence of A_k (without loss of generality we assume that this holds for every k). By considering a converging subsequence of $\mathbf{v}_k$, we may assume that $\mathbf{w}_k$ converges to $\mathbf{w} \in \Delta$. Then the negative sequence $\mathbf{w}_k^T A_k \mathbf{w}_k = 2\eta_k$ converges to $\mathbf{w}^T A\mathbf{w}$. But since A is copositive, this implies that $\mathbf{w}^T A \mathbf{w} = 0$, and the vector $\mathbf{w} \in \Delta$ is a zero of A. Let $\sigma = \operatorname{supp} \mathbf{w}$. From $\lim_{k\to\infty} \mathbf{w}_k = \mathbf{w}$ we may assume that $\sigma \subseteq \operatorname{supp} \mathbf{w}_k$ for every k (by possibly discarding a finite number of elements from the sequence).

The zero $\mathbf{w}$ of A is a convex combination of the minimal zeros of A. That is, $\mathbf{w} = \sum_{i=1}^m t_i \mathbf{u}_i$, where the t_i's are nonnegative and sum up to 1. By Proposition 5.20(d) the extended support of $\mathbf{w}$ is (σ, τ), where

$$\sigma = \bigcup_{t_i>0} \sigma_i \quad \text{and} \quad \tau = \bigcap_{t_i>0} \tau_i.$$

Let $\mathbf{v}_k = \sum_{i=1}^m t_i \mathbf{v}_k^{(i)}$. Since $\operatorname{supp} \mathbf{v}_k^{(i)} = \sigma_i$, we get that $\operatorname{supp} \mathbf{v}_k = \sigma$. And since $\left(A_k \mathbf{v}_k^{(i)}\right)_j = 0$ for every $j \in \tau_i$, we get that $(A_k \mathbf{v}_k)_j = 0$ for every $j \in \tau$. Combining that with $\operatorname{supp} \mathbf{v}_k = \sigma \subseteq \tau$, we have

$$\mathbf{v}_k^T A_k \mathbf{v}_k = (\mathbf{v}_k[\tau])^T (A_k \mathbf{v}_k)[\tau] = 0.$$

Now consider the Karush-Kuhn-Tucker optimality conditions satisfied by the solution $\mathbf{w}_k$ of the problem (5.2):

$$A_k \mathbf{w}_k + \lambda_k \mathbf{e} - \boldsymbol{\mu}_k = \mathbf{0} \quad \text{and} \quad \boldsymbol{\mu}_k^T \mathbf{w}_k = 0,$$

for $\lambda_k \in \mathbb{R}$ and $\boldsymbol{\mu}_k \in \mathbb{R}_+^n$. Multiplying the left equality on the left by $\mathbf{w}_k^T$ we therefore get that

$$\mathbf{w}_k^T A_k \mathbf{w}_k + \lambda_k \mathbf{w}_k^T \mathbf{e} = 0,$$

that is,

$$\lambda_k = -2\eta_k > 0.$$

Plugging that back in the optimality condition we get

$$\boldsymbol{\mu}_k = A_k \mathbf{w}_k - 2\eta_k \mathbf{e}.$$

As k tends to infinity, this yields (taking into account that $A_k \mathbf{w}_k$ converges to $A\mathbf{w}$, and $2\eta_k$ converges to $\mathbf{w}^T A \mathbf{w} = 0$)

$$\lim_{k\to\infty} \boldsymbol{\mu}_k = A\mathbf{w}.$$

Again, we may discard a finite number of elements to assume that $(\boldsymbol{\mu}_k)_i > 0$ whenever $(A\mathbf{w})_i > 0$, for every k. As $\boldsymbol{\mu}_k^T \mathbf{w}_k = 0$, this implies that

$$\operatorname{supp} \mathbf{w}_k \subseteq \tau.$$

In particular, $\mathbf{w}_k^T A_k \mathbf{v}_k = 0$. Now for $t \geq 0$ let

$$\mathbf{u}_k(t) = (1-t)\mathbf{v}_k + t\mathbf{w}_k.$$

Then

$$\frac{1}{2}(\mathbf{u}_k(t))^T A_k \mathbf{u}_k(t) = \frac{(1-t)^2}{2} \mathbf{v}_k^T A_k \mathbf{v}_k + t(1-t)\mathbf{w}_k^T A_k \mathbf{v}_k + \frac{t^2}{2} \mathbf{w}_k^T A_k \mathbf{w}_k = t^2 \eta_k.$$

Let $\alpha_k = \operatorname{supp} \mathbf{w}_k$. Then $\operatorname{supp} \mathbf{v}_k = \sigma \subseteq \alpha_k$, so both $\mathbf{v}_k$ and $\mathbf{w}_k$ are in the face of Δ which consists of all $\mathbf{x} \in \Delta$ such that $x_i = 0$ for $i \notin \alpha_k$, and $\mathbf{w}_k = \mathbf{u}_k(1)$ is in the interior of that face. Hence there exists $t > 1$ such that $\mathbf{u}_k(t)$ is also in that face. But then for this t we get that

$$\frac{1}{2}(\mathbf{u}_k(t))^T A_k \mathbf{u}_k(t) = t^2 \eta_k < \eta_k,$$

contrary to η_k being the optimum value of the problem (5.2). This completes the proof that $A_k \in \mathcal{COP}_n$ for every k, except possibly for a finite number of k's.

We now show that the extended minimal zero support set of each A_k (in a subsequence) is $\mathcal{E}$. We already know that it contains $\mathcal{E}$, since we have shown that $\operatorname{supp} \mathbf{v}_k^{(i)} = \sigma_i$ and $\left(A_k \mathbf{v}_k^{(i)}\right)_j = 0$ if and only if $j \in \tau_i$, for every $1 \leq i \leq m$. To prove the reverse inclusion, suppose for the sake of contradiction that for infinitely many k's there exists a minimal zero $\mathbf{u}_k \in \Delta$ of A_k whose extended support is not in $\mathcal{E}$. By considering a subsequence of these indices, we may assume that all these extended supports are equal to $(\widehat{\sigma}, \widehat{\tau}) \notin \mathcal{E}$. Without loss of generality, $\lim_{k\to\infty} \mathbf{u}_k = \widehat{\mathbf{u}} \in \Delta$. Then $\lim_{k\to\infty} \mathbf{u}_k{}^T A_k \mathbf{u}_k = \widehat{\mathbf{u}}^T A \widehat{\mathbf{u}}$ implies that $\widehat{\mathbf{u}}$ is a zero of A. Hence for some $1 \leq i \leq m$, $\sigma_i \subseteq \widehat{\sigma}$. By the minimality of the zero $\mathbf{u}_k$ of A_k this implies that $\sigma_i = \widehat{\sigma}$, but as the vector $\mathbf{v}_k^{(i)}$ is the unique minimal zero in Δ of A_k with this support, we get that $\mathbf{u}_k = \mathbf{v}_k^{(i)}$, contrary to the assumption that $(\widehat{\sigma}, \widehat{\tau}) \notin \mathcal{E}$. □

Corollary 5.23. *Let $\mathcal{E} = \{(\sigma_i, \tau_i) \mid i = 1, \ldots, m\}$ be an arbitrary collection of pairs of index sets in $\{1, \ldots, n\}$. Then $S_{\mathcal{E}}$ is an open subset in the relative topology of the algebraic set $Z_{\mathcal{E}}$.*

We now consider some additional properties of the decomposition of $\mathcal{COP}_n$ into finite number of sets $S_{\mathcal{E}}$. A nonempty $S_{\mathcal{E}}$, which is not the zero set $\{0\}$, has a boundary. By Corollary 5.23 this boundary is in $Z_{\mathcal{E}} \setminus S_{\mathcal{E}}$. As the matrices on the boundary are copositive ($\mathcal{COP}_n$ is closed), each matrix on the boundary is in some $S_{\mathcal{E}'}$, $\mathcal{E}' \neq \mathcal{E}$. The relation between such $\mathcal{E}'$ and $\mathcal{E}$ is described in the following proposition.

Proposition 5.24. *Let $\mathcal{E} = \{(\sigma_i, \tau_i) \mid i = 1, \ldots, m\}$ be a collection of pairs of index sets in $\{1, \ldots, n\}$. Let $A_k \in S_{\mathcal{E}}$ for every k, and $A = \lim_{k\to\infty} A_k$. If $A \in S_{\mathcal{E}'}$, where $\mathcal{E}' = \{(\sigma_i', \tau_i') \mid i = 1, \ldots, m'\}$, then for every $1 \le i \le m$ there exists $1 \le j \le m'$ such that $\sigma_j' \subseteq \sigma_i$ and $\tau_i \subseteq \tau_j'$. In particular, $Z_{\mathcal{E}'} \subseteq Z_{\mathcal{E}}$.*

Proof. Given $1 \le i \le m$, let $\mathbf{u}_k \in \Delta$ be a minimal zero of A_k with extended support (σ_i, τ_i). Without loss of generality we may assume that $\lim_{k\to\infty} \mathbf{u}_k = \mathbf{u}$. Then $0 = \lim_{k\to\infty} \mathbf{u}_k^T A_k \mathbf{u}_k = \mathbf{u}^T A \mathbf{u}$, and therefore $\mathbf{u} \in \Delta$ is a zero of A. If the extended support of $\mathbf{u}$ is (σ, τ), then $\sigma = \operatorname{supp} \mathbf{u} \subseteq \operatorname{supp} \mathbf{v}_k = \sigma_i$, and since $\lim_{k\to\infty} A_k \mathbf{u}_k = A\mathbf{u}$, we get that $(A\mathbf{u})_j = 0$ for every $j \in \tau_i$, i.e., $\tau_i \subseteq \tau$.

The zero $\mathbf{u}$ of A is a sum of minimal zeros of A. Let $\mathbf{u}'_j$ be one of these minimal zeros, with extended support $(\sigma_j', \tau_j') \in \mathcal{E}'$. By Proposition 5.20(d),

$$\sigma_j' \subseteq \sigma \subseteq \sigma_i \quad \text{and} \quad \tau_i \subseteq \tau \subseteq \tau_j'.$$

Finally, let $X \in Z_{\mathcal{E}'}$. For every $1 \le i \le m$, let $1 \le j \le m'$ such that $\sigma_j' \subseteq \sigma_i$ and $\tau_i \subseteq \tau_j'$. Let $\mathbf{v} \in \mathbb{R}^n$ have $\operatorname{supp} \mathbf{v} = \sigma_j'$ and $\mathbf{v}[\sigma_j']$ is in $\operatorname{ns} X[\tau_j'|\sigma_j']$. Then by the above inclusions also $X[\tau_i|\sigma_i]\mathbf{v}[\sigma_i] = \mathbf{0}$. Hence $X \in Z_{\mathcal{E}}$. $\square$

Example 5.25. Let $\mathcal{E} = \{(\alpha_i, \alpha_i) \mid i = 1, \ldots, 5\}$ be the extended minimal zero support set of a Hildebrand matrix H ($\alpha_i = \{i, i \oplus, i \oplus 2\}$, $\oplus$ denoting addition modulo 5 on $\{1, \ldots, 5\}$). Then $S_{\mathcal{E}}$ contains all Hildebrand matrices, and matrices congruent to them by positive diagonal matrices. (Note also that cyclic permutations of a Hildebrand matrix are also Hildebrand matrices.) Recall that a Hildebrand matrix has the form $S(\boldsymbol{\theta})$ as in (2.43), with $\boldsymbol{\theta} = (\theta_1, \theta_2, \theta_3, \theta_4, \theta_5)^T$ with $\theta_i \in (0, \pi)$ for every i and $\sum_{i=1}^5 \theta_i < \pi$. The boundary of $S_{\mathcal{E}}$ therefore includes, among others, matrices of the form

$S(\boldsymbol{\theta})$ with $\theta_i \in [0,\pi)$ for every i and $\sum_{i=1}^{5}\theta_i = \pi$. These matrices are positive semidefinite, by Proposition 2.162(a). In the case that the θ_i's are all positive, the extended minimal zero support set of the positive semidefinite $S(\boldsymbol{\theta})$ is

$$\mathcal{E}' = \{(\alpha_i, \{1,\ldots,5\})|i = 1,\ldots,5\}.$$

Another matrix on the boundary of $S_{\mathcal{E}}$ is the Horn matrix, $S(\mathbf{0})$, whose extended minimal zero support set was described in Example 5.18. There are other matrices, both positive semidefinite and exceptional, on the boundary of $S_{\mathcal{E}}$, with different extended support sets (where some of the minimal supports are of size 2, and some of size 3).

The next result relates the facial structure of $\mathcal{COP}_n$ to the decomposition of the cone into subsets $S_{\mathcal{E}}$.

Proposition 5.26. *Let $\mathcal{F}$ be a face of $\mathcal{COP}_n$. Then all matrices in the relative interior of $\mathcal{F}$ belong to the same subset $S_{\mathcal{E}}$. Matrices on the relative boundary of $\mathcal{F}$ do not belong to that $S_{\mathcal{E}}$.*

Proof. Let $A \in \operatorname{ri}\mathcal{F}$ and let $\mathcal{E} = \{(\sigma_i, \tau_i) \mid i = 1,\ldots,m\}$ be its extended minimal zero support set. For each $1 \le i \le m$ let $\mathbf{u}_i$ be a minimal zero of A with supports σ_i. As $A \in \operatorname{ri}\mathcal{F}$, $\mathcal{F}$ is the minimal face of $\mathcal{COP}_n$ containing A (Exercise 1.54). Hence by Theorem 5.1 every $X \in \mathcal{F}$ satisfies $(X\mathbf{u}_i)[\tau_i] = \mathbf{0}$, and therefore so does every $X \in \mathcal{F} - \mathcal{F}$. In particular, $\mathcal{F} - \mathcal{F} \subseteq Z_{\mathcal{E}}$. Theorem 5.22 then implies that there exists a neighbourhood of A such that $U \cap \mathcal{F} \subseteq S_{\mathcal{E}}$. That is, the extended minimal zero support set is locally constant on $\operatorname{ri}\mathcal{F}$ and therefore constant on $\operatorname{ri}\mathcal{F}$.

For the second assertion, suppose on the contrary that the extended minimal zero support set of some $A' \in \operatorname{rb}\mathcal{F}$ is also $\mathcal{E}$. Then by Theorem 5.22 there exists a neighbourhood U' of A' such that $U' \cap S_{\mathcal{E}} = U' \cap Z_{\mathcal{E}}$. As $\mathcal{F}-\mathcal{F} \subseteq Z_{\mathcal{E}}$, this implies that $U' \cap (\mathcal{F}-\mathcal{F}) \subseteq S_{\mathcal{E}}$, in particular, $U' \cap (\mathcal{F}-\mathcal{F})$ consists of copositive matrices, and since $\mathcal{F}$ is a face of $\mathcal{COP}_n$, this implies that $U' \cap (\mathcal{F} - \mathcal{F}) \subseteq \mathcal{F}$, contradicting the assumption that $A' \in \operatorname{rb}\mathcal{F}$. □

Together with the fact that relative interiors of different faces are disjoint, this implies the following corollary.

Corollary 5.27. *If $\mathcal{E}$ is the extended minimal zero support set of some $A \in \mathcal{COP}_n$, then $S_{\mathcal{E}}$ is a disjoint union of relative interiors of faces of $\mathcal{COP}_n$.*

Automorphisms of $\mathcal{COP}_n$ and $\mathcal{COP}_n$-preservers

An *automorphism* of a convex cone K in a Euclidean space V is an invertible linear transformation from V to itself, which maps K onto itself. A related terminology is that of preservers: a linear transformation from V to V is called a *preserver* of a subset K of V or, more accurately, an *into preserver*, if K is invariant under the transformation. An *onto preserver* is a preserver that maps K onto itself. As $\mathcal{COP}_n$ is a full cone spanning $\mathcal{S}_n$, an onto preserver of $\mathcal{COP}_n$ is an invertible linear transformation, i.e., it is an automorphism of $\mathcal{COP}_n$.

We already know two types of automorphisms of $\mathcal{COP}_n$: $X \mapsto DXD$, where D is a positive diagonal matrix, and $X \mapsto P^TXP$, where P is a permutation matrix. Any automorphism in the group generated by these two types of automorphisms has the form $X \mapsto M^TXM$, where $M \geq 0$ is a matrix with a single nonzero element in each row and each column, that is, a nonnegative *monomial matrix*. We now show that these are the only automorphisms of $\mathcal{COP}_n$. To start with, we list some observations regarding automorphisms of any full cone $\mathcal{K}$ of $\mathcal{S}_n$.

Proposition 5.28. *Let $\Phi : \mathcal{S}_n \to \mathcal{S}_n$ be an automorphism of a solid closed convex cone $\mathcal{K}$ in $\mathcal{S}_n$, then the following hold.*

(a) $\Phi^{-1} : \mathcal{S}_n \to \mathcal{S}_n$ *is an automorphism of* $\mathcal{K}$.
(b) $\Phi(\mathrm{bd}\,\mathcal{K}) = \mathrm{bd}\,K$, $\Phi(\mathrm{int}\,\mathcal{K}) = \mathrm{int}\,\mathcal{K}$.
(c) *For any set* $S \subseteq \mathcal{S}_n$, $\dim \Phi(\mathrm{span}\,S) = \dim \mathrm{span}\,S$.
(d) *If $\mathcal{F}$ is a face of $\mathcal{K}$, then $\Phi(\mathcal{F})$ is also a face of $\mathcal{K}$. If $\mathcal{F}$ is a maximal face of $\mathcal{K}$, then so is $\Phi(\mathcal{F})$, and if A generates an extreme ray of $\mathcal{K}$, so does $\Phi(A)$.*

Proof. For (a), note that Φ^{-1} is linear and invertible, and applying it to both sides of the equality $\Phi(\mathcal{COP}_n) = \mathcal{COP}_n$ yields $\mathcal{COP}_n = \Phi^{-1}(\mathcal{COP}_n)$.

(b) follows from the continuity of the linear transformations Φ and Φ^{-1}.

(c) is an elementary property of invertible linear transformations.

(d) uses (a) (and also (c)), and we leave it as an exercise. □

Theorem 5.29. *An invertible linear transformation $\Phi : \mathcal{S}_n \to \mathcal{S}_n$ is an automorphism of $\mathcal{COP}_n$ if and only if $\Phi(X) = M^TXM$ for some nonnegative monomial matrix M.*

Proof. As mentioned above, the "if" part is already known. It remains to prove that these are the only automorphisms of $\mathcal{COP}_n$.

Let Φ be an automorphism of $\mathcal{COP}_n$. For every $1 \le i \le n$, let

$$\mathcal{F}_i = \{X \in \mathcal{COP}_n \mid x_{ii} = 0\}.$$

Then $\mathcal{F}_i$ is a maximal face of $\mathcal{COP}_n$, consisting of all matrices in $\mathcal{COP}_n$ which share the zero $\mathbf{e}_i$, and $\dim \mathcal{F}_i = \frac{n(n-1)}{2} - 1$ (see Theorem 5.14). Hence $\Phi(\mathcal{F}_i)$ is also a maximal face of $\mathcal{COP}_n$, consisting of all the copositive matrices that share a zero $\mathbf{u}_i$. Since $\dim \Phi(\mathcal{F}_i) = \dim \mathcal{F}_i = \frac{n(n-1)}{2} - 1$, by Theorem 5.14 we get that $|\operatorname{supp} \mathbf{u}_i| = 1$, and, without loss of generality, $\mathbf{u}_i$ is one of the standard basis vectors, $\mathbf{u}_i = \mathbf{e}_{\pi(i)}$, and $\Phi(\mathcal{F}_i) = \mathcal{F}_{\pi(i)}$. For $i \ne j$, the maximal faces $\mathcal{F}_i$ and $\mathcal{F}_j$ are different ($E_{ii} \in \mathcal{F}_j$ and $E_{ii} \notin \mathcal{F}_i$, for example) and therefore $\Phi(\mathcal{F}_i) \ne \Phi(\mathcal{F}_j)$. Hence $\pi : \{1, \dots, n\} \to \{1, \dots, n\}$ is one-to-one, and thus a permutation. Then

$$\mathcal{HN}_n = \mathcal{F}_1 \cap \cdots \cap \mathcal{F}_n = \mathcal{F}_{\pi(1)} \cap \cdots \cap \mathcal{F}_{\pi(n)},$$

and we get that both $\Phi(\mathcal{HN}_n) \subseteq \mathcal{HN}_n$ and $\Phi^{-1}(\mathcal{HN}_n) \subseteq \mathcal{HN}_n$.

Let $1 \le i \le n$. For every $j \ne i$, $E_{ii} \in \mathcal{F}_j$, hence $\Phi(E_{ii}) \in \Phi(\mathcal{F}_j)$. We therefore have $\Phi(E_{ii}) = d_i E_{\pi(i)\,\pi(i)} + B_i$, where B_i has all diagonal entries zero. As $\Phi(E_{ii})$ is copositive, $d_i \ge 0$ and B_i is nonnegative, i.e., $B_i \in \mathcal{HN}_n$. Since $E_{ii} = \Phi^{-1}(d_i E_{\pi(i)\,\pi(i)} + B_i) \notin \mathcal{HN}_n$, $d_i > 0$. Let $D = \operatorname{Diag}(d_1, \dots, d_n)$ and let P be the permutation matrix corresponding to π ($P_{ij} = 1$ if and only if $\pi(i) = j$). Let $N = PD^{-1/2}$, and define

$$\Psi(X) = \Phi(N^T X N).$$

Then Ψ is an automorphism of $\mathcal{COP}_n$, and for every $1 \le i \le n$

$$\Psi(E_{ii}) = E_{ii} + C_i, \quad \text{where } C_i \in \mathcal{HN}_n. \tag{5.3}$$

We now show that $\Psi(X) = X$ for every $X \in \mathcal{S}_n$. First note that (5.3) implies

$$\Psi^{-1}(E_{ii} + 0.5C_i) = \Psi^{-1}(E_{ii} + C_i) - 0.5\Psi^{-1}(C_i) = E_{ii} - 0.5\Psi^{-1}(C_i),$$

and therefore $E_{ii} = \Psi^{-1}(E_{ii} + 0.5C_i) + 0.5\Psi^{-1}(C_i)$. Hence each of the two copositive summands is a scalar multiples of E_{ii}. Applying Ψ to $\Psi^{-1}(C_i)$ we therefore get that C_i is a scalar multiple of $\Psi(E_{ii}) = E_{ii} + C_i$. As $C_i \in \mathcal{HN}_n$, this yields $C_i = 0$, and $\Psi(E_{ii}) = E_{ii}$.

For $i \ne j$ we denote $B_{ij} = \Psi(E_{ij} + E_{ji})$. By the above, $B_{ij} \in \mathcal{HN}_n$. The matrix $E_{ii} + E_{jj} - E_{ij} - E_{ji}$ is extremal in $\mathcal{COP}_n$, and therefore so is

$$\Psi(E_{ii} + E_{jj} - E_{ij} - E_{ji}) = E_{ii} + E_{jj} - B_{ij}.$$

The copositivity of this matrix implies that all the nonzero off-diagonal entries of B_{ij} lie in rows i and j. That is, $B_{ij} = t_{ij}(E_{ij} + E_{ji})$. The extremality of $E_{ii} + E_{jj} - B_{ij}$ then implies that $t_{ij} = 1$. This completes the proof that $\Psi(X) = X$ for every $X \in \mathcal{S}_n$, and therefore $\Phi(X) = M^T X M$ where $M = N^{-1}$. $\square$

Characterizing into preservers is an open problem. By Proposition 2.12, for every $S \geq 0$ the transformation $X \mapsto S^T X S$ is an into preserver. Then so are sums of such transformations. However, these are not all the into preservers — see Exercise 5.9.

Maximum angle of $\mathcal{COP}_n$

What is the greatest possible angle between matrices in the cone $\mathcal{COP}_n$? And between matrices in the cone $\mathcal{SPN}_n$? While an exact answer is not known, an asymptotic result exists.

The angle between two matrices X, Y in $\mathcal{S}_n$ is

$$\angle(X,Y) = \arccos \frac{\langle X, Y\rangle}{\|X\|_F \|Y\|_F},$$

where $\|X\|_F$ is the Frobenius norm $\|X\|_F = \sqrt{\langle X, X\rangle} = \sqrt{\sum_{i,j=1}^n x_{ij}^2}$. For a closed convex cone $\mathcal{K} \subseteq \mathcal{S}_n$ the *maximum angle* of $\mathcal{K}$ is

$$\theta_{\max}(\mathcal{K}) = \max\{\angle(X,Y) \mid X, Y \in \mathcal{K}\}.$$

Note that

$$\langle X, Y\rangle = \mathbf{e}^T(X \circ Y)\mathbf{e}.$$

Example 5.30. If X and Y are both nonnegative, then $X \circ Y$ is nonnegative. If X and Y are both positive semidefinite, then $X \circ Y$ is positive semidefinite. In both cases, $\langle X, Y\rangle \geq 0$ and $\angle(X,Y) \leq \frac{1}{2}\pi$. Hence

$$\theta_{\max}(\mathcal{PSD}_n) \leq \frac{1}{2}\pi \text{ and } \theta_{\max}(\mathcal{N}_n) \leq \frac{1}{2}\pi.$$

For every $n \geq 2$ the matrices $X = E_{11}$ and $Y = E_{22}$ satisfy $X \circ Y = 0$, hence $\angle(X,Y) = \frac{1}{2}\pi$. Since these matrices are both positive semidefinite and nonnegative, this shows that in fact

$$\theta_{\max}(\mathcal{PSD}_n) = \theta_{\max}(\mathcal{N}_n) = \frac{1}{2}\pi.$$

We now consider the maximum angles of $\mathcal{COP}_n$ and $\mathcal{SPN}_n$. These cones are pointed, and therefore clearly $\theta_{\max}(\mathcal{COP}_n) < \pi$ and $\theta_{\max}(\mathcal{SPN}_n) < \pi$. However, the exact values of $\theta_{\max}(\mathcal{COP}_n)$ and $\theta_{\max}(\mathcal{SPN}_n)$ are not known. We define

$$\gamma_n = \max\{\angle(X,Y) \mid X \in \mathcal{PSD}_n \text{ and } Y \in \mathcal{N}_n\},$$

and consider this parameter instead. As

$$\gamma_n \leq \theta_{\max}(\mathcal{SPN}_n) \leq \theta_{\max}(\mathcal{COP}_n),$$

this parameter provides a lower bound on both maximum angles. The exact value of γ_n is also not known for most values of n, but we are going to find $\lim_{n\to\infty}\gamma_n$. (Note that γ_n increases with n.)

For the computation of γ_n, a few observations are in order. Each $n\times n$ symmetric matrix A has a unique decomposition as a difference of two positive semidefinite matrices that are orthogonal to each other,

$$A = Q - P, \quad Q, P \in \mathcal{P}_n \text{ and } QP = 0.$$

The matrix Q is the projection of A on the cone $\mathcal{PSD}_n$, and P is the projection of $-A$ on that cone. More explicitly, let Λ be the set of different eigenvalues of A, and for every $\lambda \in \Lambda$ denote by E_λ the orthogonal projection on the eigenspace of λ. Then $A = \sum_{\lambda\in\Lambda}\lambda E_\lambda$ is the spectral decomposition of A. Denote by Λ_+ and Λ_- the sets of positive and negative eigenvalues of A, respectively. Then $Q = \sum_{\lambda\in\Lambda_+}\lambda E_\lambda$ and $P = -\sum_{\lambda\in\Lambda_-}\lambda E_\lambda$. In particular, the spectrum of Q consists of the elements of Λ_+ together with $n - |\Lambda_+|$ zeros and the spectrum of P consists of the absolute values of the elements in Λ_- together with $n - |\Lambda_-|$ zeros. We refer to Q and P as the *positive definite part* and the *negative definite part* of A, respectively. If A is not positive semidefinite, then obviously $A \neq 0$ and $P \neq 0$, and the cosine of the angle between A and P is

$$\frac{\langle A, P\rangle}{\|A\|_F\|P\|_F} = \frac{-\langle P, P\rangle}{\|A\|_F\|P\|_F} = -\frac{\|P\|_F}{\|A\|_F} = -\frac{\sqrt{\sum_{\lambda\in\Lambda_-}\lambda^2}}{\sqrt{\sum_{\lambda\in\Lambda}\lambda^2}}.$$

For every nonzero symmetric $n\times n$ matrix A, let us denote by $\angle(A, \mathcal{PSD}_n)$ the maximum angle between A and a matrix in $\mathcal{PSD}_n$.

Proposition 5.31. *For every $A \in \mathcal{S}_n\backslash\mathcal{PSD}_n$, let P be the negative definite part of A. Then*

$$\angle(A, \mathcal{PSD}_n) = \angle(A, P) = \arccos\left(-\frac{\sqrt{\sum_{\lambda\in\Lambda_-}\lambda^2}}{\sqrt{\sum_{\lambda\in\Lambda}\lambda^2}}\right),$$

where Λ and Λ_- are as described above. Moreover, P is the unique matrix in $\mathcal{PSD}_n$, up to multiplication by a positive scalar, which forms this maximum angle with A.

Proof. For every positive semidefinite $X \neq 0$ we have

$$\frac{\langle A, X\rangle}{\|A\|_F\|X\|_F} \geq -\frac{\langle P, X\rangle}{\|A\|_F\|X\|_F} \geq -\frac{\|P\|_F}{\|A\|_F} = \frac{\langle A, P\rangle}{\|A\|_F\|P\|_F},$$

where the first inequality follows from the fact that Q, the positive definite part of A, satisfies $\langle Q, X\rangle \geq 0$, and the second inequality from the Cauchy-Schwarz inequality. This shows that $\angle(A, X) \leq \angle(A, P)$ for every $X \in \mathcal{PSD}_n$. By the condition for equality in the Cauchy-Schwarz inequality, we get that $\angle(A, X) = \angle(A, P)$ if and only if X is a positive scalar multiple of P. $\square$

Similarly, every $n \times n$ symmetric matrix A has a unique decomposition as a difference of two nonnegative matrices that are orthogonal to each other,

$$A = M - N, \quad M, N \in \mathcal{N}_n \text{ and } M \circ N = 0.$$

In fact, $M = \max(A, 0)$ is the projection of A on $\mathcal{N}_n$, and $N = \max(-A, 0)$ is the projection of $-A$ on that cone. We refer to M and N as the *positive part* and the *negative part* of A, respectively. If $A \notin \mathcal{N}_n$, then $A, N \neq 0$, and the cosine of the angle between A and N is

$$\frac{\langle A, N\rangle}{\|A\|_F\|N\|_F} = \frac{-\langle N, N\rangle}{\|A\|_F\|N\|_F} = -\frac{\|N\|_F}{\|A\|_F} = -\frac{\sqrt{\sum_{a_{ij}<0} a_{ij}^2}}{\sqrt{\sum a_{ij}^2}}.$$

We denote by $\angle(A, \mathcal{N}_n)$ the maximum angle between A and a matrix in $\mathcal{N}_n$

Proposition 5.32. *For every $A \in \mathcal{S}_n \setminus \mathcal{N}_n$, let N be the negative part of A. Then*

$$\angle(A, \mathcal{N}_n) = \angle(A, N) = \arccos\left(-\frac{\sqrt{\sum_{a_{ij}<0} a_{ij}^2}}{\sqrt{\sum a_{ij}^2}}\right).$$

Moreover, N is the unique matrix in $\mathcal{N}_n$, up to multiplication by a positive scalar, which forms this maximum angle with A.

The proof is completely parallel to the proof of Proposition 5.3, and is left as an exercise.

Example 5.33. Let $P \in \mathcal{PSD}_n \setminus \mathcal{N}_n$ have rank 1. Then $\angle(P, \mathcal{N}_n) \leq \frac{3}{4}\pi$. And there exists a rank 1 positive semidefinite matrix $P \in \mathcal{PSD}_n \setminus \mathcal{N}_n$ such that $\angle(P, \mathcal{N}_n) = \frac{3}{4}\pi$.

Proof. By the assumptions, $P = \mathbf{u}\mathbf{u}^T$, where $\mathbf{u}$ has both positive and negative entries. By a suitable permutation of rows and columns of P we may assume that

$$\mathbf{u} = \begin{pmatrix} \mathbf{v} \\ -\mathbf{w} \end{pmatrix}, \qquad \mathbf{v}, \mathbf{w} \geq \mathbf{0}, \quad \mathbf{v}, \mathbf{w} \neq \mathbf{0}.$$

Then

$$P = \begin{pmatrix} \mathbf{v}\mathbf{v}^T & -\mathbf{v}\mathbf{w}^T \\ -\mathbf{w}\mathbf{v}^T & \mathbf{w}\mathbf{w}^T \end{pmatrix},$$

and the negative part of P is

$$N = \begin{pmatrix} 0 & \mathbf{v}\mathbf{w}^T \\ \mathbf{w}\mathbf{v}^T & 0 \end{pmatrix}.$$

For any two vectors $\mathbf{x}$ and $\mathbf{y}$,

$$\left\|\mathbf{x}\mathbf{y}^T\right\|_F = \sqrt{\mathrm{trace}\,(\mathbf{x}\mathbf{y}^T\mathbf{y}\mathbf{x}^T)} = \|\mathbf{x}\|\|\mathbf{y}\|.$$

Hence

$$\|P\|_F = \|\mathbf{u}\|^2 = \|\mathbf{v}\|^2 + \|\mathbf{w}\|^2 \quad , \quad \|N\|_F = \sqrt{2}\|\mathbf{v}\|\,\|\mathbf{w}\|,$$

and

$$\langle P, N\rangle = -2\|\mathbf{v}\mathbf{w}^T\|_F^2 = -2\|\mathbf{v}\|^2\|\mathbf{w}\|^2.$$

Thus

$$\frac{\langle P, N\rangle}{\|P\|_F\|N\|_F} = -\frac{\sqrt{2}\|\mathbf{v}\|\,\|\mathbf{w}\|}{\|\mathbf{v}\|^2 + \|\mathbf{w}\|^2} \geq -\frac{\sqrt{2}}{2},$$

and equality holds in the last inequality if and only if $\|\mathbf{v}\| = \|\mathbf{w}\|$. Therefore $\angle(P, N) \leq \frac{3}{4}\pi$, with equality if and only if $\|\mathbf{v}\| = \|\mathbf{w}\|$. □

In particular, the last part of the example implies the following.

Corollary 5.34. *For every $n \geq 2$, $\gamma_n \geq \frac{3}{4}\pi$.*

Corollary 5.35. *Let $n \geq 2$, and let $P \in \mathcal{PSD}_n$ and $N \in \mathcal{N}_n$ be any two matrices such that $\angle(P, N) = \gamma_n$. Then $\langle P, N\rangle < 0$, $\mathrm{diag}N = \mathbf{0}$ and $1 \leq \mathrm{rank}\,P \leq n - 2$.*

Proof. By the previous corollary, $\angle(P, N) \geq \frac{3}{4}\pi$, and thus $\langle P, N\rangle < 0$. Since $\angle(P, N)$ is the maximal possible angle between a positive semidefinite matrix and a nonnegative one of that order, N has to be the nonnegative matrix forming the widest angle with P, and P has to be the nonnegative angle forming the widest angle with N. By the uniqueness parts in Propositions 5.31 and 5.32, N is a positive scalar multiple of the negative part of P, and P is a positive scalar multiple of the negative definite part of N. Since $\mathrm{diag}P \geq \mathbf{0}$, we get that $\mathrm{diag}N = \mathbf{0}$, and by the Perron-Frobenius Theorem, $\mathrm{rank}\,P \leq n - 1$. Without loss of generality P is the negative definite part of N.

We now show that, in fact, $\operatorname{rank} P \leq n-2$. Suppose on the contrary that $\operatorname{rank} P = n-1$. As the nonnegative N has diagonal $\mathbf{0}$ and a rank $n-1$ negative definite part P, we get that the eigenvalues of N are $\rho = \lambda_1 > 0 > \lambda_2 \geq \cdots \geq \lambda_n$, and $\sum_{i=2}^n \lambda_i = -\rho$. By Proposition 5.31,

$$\cos \angle(N, P) = -\frac{\sqrt{\sum_{i=2}^n \lambda_i^2}}{\sqrt{\rho^2 + \sum_{i=2}^n \lambda_i^2}}.$$

Let

$$\Omega = \left\{ \mathbf{x} \in \mathbb{R}^{n-1} \mid \mathbf{x} \leq \mathbf{0} \text{ and } \sum_{i=1}^{n-1} x_i = -\rho \right\}.$$

The function $g(\mathbf{x}) = \|\mathbf{x}\|^2 = \sum_{i=1}^{n-1} x_i^2$ is strictly convex, and thus attains its maximum on the compact convex set Ω at an extreme point of this set, i.e., at a point $\mathbf{x}$ such that $x_i = -\rho$ for some i and $x_j = 0$ for $j \neq i$. That is,

$$\max_{\mathbf{x} \in \Omega} g(\mathbf{x}) = \rho^2.$$

The function $f(t) = -\sqrt{\frac{t}{\rho^2 + t}}$ is strictly decreasing on $[0, \infty)$, and thus $f(g(\mathbf{x}))$ attains a minimum on Ω where g attains its maximum, and

$$\min_{\mathbf{x} \in \Omega} f(g(\mathbf{x})) = -\sqrt{\frac{\rho^2}{2\rho^2}} = -\frac{\sqrt{2}}{2}.$$

By the strict convexity of $g(\mathbf{x})$ on Ω, the strict monotonicity of f, and the fact that $\boldsymbol{\lambda} = (\lambda_2, \ldots, \lambda_n)^T$ is in the interior Ω,

$$\cos \angle(N, P) = f(g(\boldsymbol{\lambda})) > \min_{\mathbf{x} \in \Omega} f(g(\mathbf{x})) = -\frac{\sqrt{2}}{2}.$$

Thus $\angle(N, P) < \frac{3}{4}\pi$ is not the maximal angle γ_n. A contradiction. □

Theorem 5.36. *For $n \leq 4$, $\gamma_n = \frac{3}{4}\pi$.*

Note that for $n \leq 3$ this follows from Example 5.33 and Corollaries 5.34 and 5.35. We leave the proof for $n = 4$ as an exercise. This equality no longer holds for $n \geq 5$. By the next example $\gamma_5 > \frac{3}{4}\pi$.

Example 5.37. Let

$$N = \begin{pmatrix} 0 & 1 & 0 & 0 & 1 \\ 1 & 0 & 1 & 0 & 0 \\ 0 & 1 & 0 & 1 & 0 \\ 0 & 0 & 1 & 0 & 1 \\ 1 & 0 & 0 & 1 & 0 \end{pmatrix}$$

be the adjacency matrix of the 5-cycle. Its eigenvalues are well known (they are easily computed by the formula for the eigenvalues of a circulant matrix): the simple eigenvalue 2, the positive eigenvalue $2\cos(2\pi/5) = \frac{-1+\sqrt{5}}{2}$ of multiplicity 2, and the negative eigenvalue $-2\cos(\pi/5) = \frac{-1-\sqrt{5}}{2}$ of multiplicity 2. Thus the negative definite part P of N satisfies

$$\cos\angle(P,N) = -\frac{\sqrt{8\cos^2(\pi/5)}}{\sqrt{4+8\cos^2(2\pi/5)+8\cos^2(\pi/5)}} = -\frac{1+1/\sqrt{5}}{2} < -\frac{\sqrt{2}}{2},$$

implying that

$$\gamma_5 \geq \arccos\left(-\frac{1+1/\sqrt{5}}{2}\right) \approx 0.7575\pi > \frac{3}{4}\pi.$$

The main result of this subsection is the following.

Theorem 5.38.

$$\lim_{n\to\infty} \theta_{\max}(\mathcal{SPN}_n) = \lim_{n\to\infty} \theta_{\max}(\mathcal{COP}_n) = \pi.$$

The proof of the theorem relies on the existence of certain strongly regular graphs. A graph G is *strongly regular* with parameters (n,k,a,c) if G has n vertices, each of degree k; any two adjacent vertices have a common neighbours and any two non-adjacent vertices have c common neighbours.

Proof idea. Let G be a connected strongly regular graph with parameters (n,k,a,c). Let N be the adjacency matrix of G, and let P be its negative definite part. The eigenvalues of N are all integers, and they are

- k, with multiplicity 1;
- $\theta = \frac{(a-c)+\sqrt{D}}{2}$, with multiplicity $m_\theta = \frac{1}{2}\left((n-1) - \frac{2k+(n-1)(a-c)}{\sqrt{D}}\right)$;
- $\tau = \frac{(a-c)-\sqrt{D}}{2}$, with multiplicity $m_\tau = \frac{1}{2}\left((n-1) + \frac{2k+(n-1)(a-c)}{\sqrt{D}}\right)$,

where $D = (a-c)^2 + 4(k-c)$. Hence

$$\frac{\langle N,P\rangle}{\|N\|\|P\|} = -\sqrt{\frac{m_\tau \tau^2}{nk}}.$$

It is known that for each prime power q, there exists a strongly regular graph G with parameters $(n,k,a,c) = ((q+1)(q^3+1), q(q^2+1), q-1, q^2+1)$. Using the above formula we get that for such G,

$$\angle(N,P) = \arccos\left(\frac{\langle N,P\rangle}{\|N\|_F\|P\|_F}\right) = \arccos\left(-\frac{\sqrt{q^2+1}}{q+1}\right).$$

Let q_ℓ be the increasing sequence of all prime powers. For every ℓ let G_ℓ be a strongly regular graph with parameters

$$(n_\ell, k_\ell, a_\ell, c_\ell) = ((q_\ell + 1)(q_\ell^3 + 1), q_\ell(q_\ell^2 + 1), q_\ell - 1, q_\ell^2 + 1),$$

N_ℓ the adjacency matrix of G_ℓ and P_ℓ its negative definite part. Then

$$\gamma_{n_\ell} \geq \angle(N_\ell, P_\ell) = \arccos\left(-\frac{\sqrt{q_\ell^2+1}}{q_\ell+1}\right) \underset{\ell\to\infty}{\longrightarrow} \arccos(-1) = \pi. \qquad \square$$

The lower bounds on γ_{n_ℓ} and $\theta_{\max}(\mathcal{COP}_{n_\ell})$ found in the theorem progress rather slowly. For example, the first five elements in the sequence n_ℓ are $27, 112, 325, 756, 2752$. The corresponding lower bounds are $\arccos\left(-\frac{\sqrt{5}}{3}\right) \approx 0.7677\pi$, $\arccos\left(-\frac{\sqrt{10}}{4}\right) \approx 0.7902\pi$, $\arccos\left(-\frac{\sqrt{17}}{5}\right) \approx 0.8086\pi$, $\arccos\left(-\frac{\sqrt{26}}{6}\right) \approx 0.8232\pi$ and $\arccos\left(-\frac{\sqrt{50}}{8}\right) \approx 0.8451\pi$.

Exercises

5.1. Prove the claim in Remark 5.2.
Hint: By Proposition 2.61, for every $X \in \mathcal{S}_n$ that satisfies $(X\mathbf{u}_i)_j = 0$ whenever $(A\mathbf{u}_i)_j = 0$ there exists $\varepsilon > 0$ such that $A \pm \varepsilon X \in \mathcal{COP}_n$.

5.2. Verify the claim that if A generates an exposed ray of $\mathcal{COP}_n$, then so does any matrix in its orbit.

5.3. Complete the proof of Proposition 5.5(c).

5.4. Prove the "only if" direction in Proposition 5.9.

5.5. Prove Proposition 5.11.
Hint: By Exercise 5.2, it suffices to consider the case that $i = n$ and $t = 1$. See the proof of Proposition 2.73 for observations regarding the zeros of the extended matrix.

5.6. Show that if A is an extremal matrix with a zero diagonal element, then the ray generated by A is not exposed.

5.7. Show that the Hoffman-Pereira matrix H_9 whose graph of -1 entries, $G_{-1}(H_9)$, is the 9-cycle generates an exposed ray of $\mathcal{COP}_9$.

5.8. What are $S_\mathcal{E}$ and $Z_\mathcal{E}$ in $\mathcal{S}_n$ for $\mathcal{E} = \{(\{i\}, \{1, \ldots, n\}) \mid i = 1, \ldots n\}$?

5.9. Prove Proposition 5.28(d).

5.10. Consider the linear transformation on $\mathcal{S}_2$

$$\begin{pmatrix} a & c \\ c & b \end{pmatrix} \mapsto \begin{pmatrix} a & a+b+2c \\ a+b+2c & b \end{pmatrix}.$$

Show that this is an into preserver of $\mathcal{COP}_2$, but is not a sum of transformations of the form $X \mapsto S^T XS$ with $S \in \mathbb{R}_+^{2\times 2}$.
Hint: Any transformation of the form $X \mapsto S^T XS$ is an into preserver of $\mathcal{PSD}_2$.

5.11. Show that $\gamma_2 = \theta_{\max}(\mathcal{SPN}_2) = \theta_{\max}(\mathcal{COP}_2) = \frac{3}{4}\pi$.

5.12. Prove Proposition 5.32.

5.13. Prove Theorem 5.36 for the case $n = 4$.
Hint: Consider nonnegative $N \in \mathcal{N}_4$ with zero diagonal and a rank 2 negative definite part P. Consider the convex function $g(x, y) = x^2 + y^2$ on the set $\Omega = \{(x, y) \in \mathbb{R}^2 \mid 0 \geq x \geq y \geq -\rho \text{ and } x + y = -\rho - \mu\}$, where the eigenvalues of N are $\rho \geq \mu \geq 0 > \lambda_3 \geq \lambda_4$.

5.14. Let

$$N = \begin{pmatrix} 0 & 1 & 0 & 0 \\ 1 & 0 & 0 & 0 \\ 0 & 0 & 0 & 1 \\ 0 & 0 & 1 & 0 \end{pmatrix}.$$

Find the negative definite part P of N, and show that $\operatorname{rank} P = 2$ and that $\angle(N, P) = \frac{3}{4}\pi$. (This shows that γ_4 is attained not only at a pair (N, P) where P has rank 1.)

5.15. Show that if N is the adjacency matrix of the 5-cycle, and P is its negative definite part, then $\angle(N, P) > \frac{3}{4}\pi$.

Notes. The description of the minimal face containing a copositive matrix, and the minimal exposed face containing it is based on [Dickinson and Hildebrand (2016)]. Results on maximal faces and their dimensions, and on exposed rays of $\mathcal{COP}_n$, are based on [Dickinson (2011)].

The description of the algebraic structure of $\mathcal{COP}_n$ is due to [Hildebrand (2020)], where it is also proved that if C is an irreducible component of the algebraic set $Z_\mathcal{E}$ (here irreducible means: cannot be represented as a union of two proper algebraic subsets), then for matrices $A \in S = C \cap S_\mathcal{E}$ the dimension of $\mathcal{F}^A_{\mathcal{COP}_n}$ is constant over S, except possibly for an algebraic subset of a lower dimension, on which the dimension of the minimal faces is higher.

The automorphisms groups of cones of copositive and completely positive matrices with respect to a cone K were determined in [Gowda *et al.* (2013)], and this includes automorphisms of $\mathcal{COP}_n$ and $\mathcal{CP}_n$ as a special case ($K = \mathbb{R}^n_+$). The general result relates the automorphisms of these cones with those of $K \cup (-K)$. The automorphisms of $\mathcal{COP}_n$ were characterized again in [Shitov (2019)], following initial results about into (and onto) preservers of copositivity in [Furtado *et al.* (2019)] (Exercise 5.9 is based on this paper). The proof here is an adaptation of the proof of [Shitov (2019)], which is more elementary than the [Gowda *et al.* (2013)] proof.

As for the maximum angle problem, in [Hiriart-Urruty and Seeger (2010)] it was shown that $\theta_{\max}(\mathcal{COP}_2) = \frac{3}{4}\pi$, and the problem of computing $\theta_{\max}(\mathcal{COP}_n)$ was presented. The results in this section regarding this problem were proved in [Goldberg and Shaked-Monderer (2014)]. In particular, references to the results on the existence of strongly regular graphs of the required type for Theorem 5.38 can be found there.

5.2 The structure of $\mathcal{CP}_n$

Compared to what we know on the geometry of $\mathcal{COP}_n$, our knowledge of the geometry of $\mathcal{CP}_n$ is "upside down" — what is easy for one cone, is difficult for the other, and vice versa. For example, while the extreme rays of $\mathcal{COP}_n$ are still a mystery, those of $\mathcal{CP}_n$ are fully known. This corresponds to the fact that we can say more about the maximal faces of $\mathcal{COP}_n$, than what we can say on the maximal faces of $\mathcal{CP}_n$.

Faces of $\mathcal{CP}_n$

First we consider the minimal face of $\mathcal{CP}_n$ containing a given $A \in \mathcal{CP}_n$, $\mathcal{F}^A_{\mathcal{CP}_n}$. The next theorem is a characterization of this face, but not a very useful one (compared to its copositive counterpart).

Theorem 5.39. *Let $A \in \mathcal{CP}_n$. Then*

$$\mathcal{F}^A_{\mathcal{CP}_n} = \operatorname{cone}\left\{\mathbf{b}\mathbf{b}^T \mid \mathbf{b} \in \mathbb{R}^n_+, A - \mathbf{b}\mathbf{b}^T \in \mathcal{CP}_n\right\}.$$

Proof. Let us denote $\mathcal{K} = \operatorname{cone}\left\{\mathbf{b}\mathbf{b}^T \mid \mathbf{b} \in \mathbb{R}^n_+, A - \mathbf{b}\mathbf{b}^T \in \mathcal{CP}_n\right\}$, the convex cone of all rank 1 completely positive matrices that may be completed to a CP representation of A. To show that $\mathcal{K}$ is the minimal face containing A, we will show that it is a face, and any face containing A contains the whole of $\mathcal{K}$. We start with the latter. Let $\mathcal{F}$ be a face of $\mathcal{CP}_n$ containing

A, and let $\mathbf{b}$ be any nonnegative vector such that $A - \mathbf{b}\mathbf{b}^T$ is completely positive, then $A = \mathbf{b}\mathbf{b}^T + (A - \mathbf{b}\mathbf{b}^T) \in \mathcal{F}$ implies that $\mathbf{b}\mathbf{b}^T \in \mathcal{F}$. Hence $\mathcal{K} \subseteq \mathcal{F}$.

Now we show that $\mathcal{K}$ is a face. First, if $A = B + C$, $B, C \in \mathcal{CP}_n$, and $B = \sum_{i=1}^k \mathbf{b}_i\mathbf{b}_i^T$ is a CP representation of B, then for each $1 \le i \le k$ the matrix

$$A - \mathbf{b}_i\mathbf{b}_i^T = \sum_{\substack{1\le j\le k\\ j\ne i}} \mathbf{b}_j\mathbf{b}_j^T + C$$

is completely positive. Hence $B = \sum_{i=1}^k \mathbf{b}_i\mathbf{b}_i^T \in \mathcal{K}$. Similarly $C \in \mathcal{K}$.

Now suppose X is any matrix in $\mathcal{K}$. Then $X = \sum_{i=1}^m t_i\mathbf{v}_i\mathbf{v}_i^T$, where $\mathbf{v}_i \in \mathbb{R}_+^n$, $t_i > 0$ and $A - \mathbf{v}_i\mathbf{v}_i^T \in \mathcal{CP}_n$ for every i. Let $t = \sum_{i=1}^m t_i$. If $X = Y + Z$, where $Y, Z \in \mathcal{CP}_n$, write

$$A = \frac{1}{t}Y + \frac{1}{t}\left(Z + \sum_{i=1}^m t_i(A - \mathbf{v}_i\mathbf{v}_i^T)\right).$$

Then $\frac{1}{t}Y \in \mathcal{K}$, and reversing the roles of Y and Z, also $\frac{1}{t}Z \in \mathcal{K}$. Hence $\mathcal{K}$ is a face. □

In the special case that A has a unique CP representation, this description of $\mathcal{F}^A_{\mathcal{CP}_n}$ is more tangible. Recall that two CP representations that differ only by the order of the summands are considered equal, and that we only consider representations $A = \sum_{i=1}^k \mathbf{b}_i\mathbf{b}_i^T$ in which $\mathbf{b}_1, \ldots, \mathbf{b}_k$ are pairwise linearly independent.

Corollary 5.40. *Let $A \in \mathcal{CP}_n$ have a unique CP representation $A = \sum_{i=1}^k \mathbf{b}_i\mathbf{b}_i^T$. Then*

$$\mathcal{F}^A_{\mathcal{CP}_n} = \operatorname{cone}\left\{\mathbf{b}_i\mathbf{b}_i^T \mid 1 \le i \le k\right\}.$$

Proof. For a nonnegative vector $\mathbf{b} \in \mathbb{R}^n$, $\mathbf{b}\mathbf{b}^T$ may be completed to a CP representation of A if and only if $\mathbf{b}$ is one of $\mathbf{b}_1, \ldots, \mathbf{b}_k$. □

Now consider the extreme rays of $\mathcal{CP}_n$. We have an easy description of all these rays: these are the rays generated by the rank 1 matrices $\mathbf{b}\mathbf{b}^T$, $\mathbf{b} \in \mathbb{R}_+^n$. We now show that they are all exposed. We denote an exposed face of $\mathcal{CP}_n$ consisting of matrices orthogonal to a matrix $M \in \mathcal{COP}_n$ by

$$\mathcal{F}_M = \{X \in \mathcal{CP}_n \mid \langle X, M\rangle = 0\}.$$

Theorem 5.41. *Every extreme ray of $\mathcal{CP}_n$ is exposed.*

Proof. Let $\mathbf{b} \in \mathbb{R}^n_+$. Let $\mathbf{v}_1, \ldots, \mathbf{v}_{n-1}$ form a basis to the orthogonal complement of $\operatorname{span}\{\mathbf{b}\}$, and define

$$M = \sum_{i=1}^{n-1} \mathbf{v}_i \mathbf{v}_i^T.$$

Then M is positive semidefinite (hence it is in $\mathcal{COP}_n$), and its null space is spanned by $\mathbf{b}$.

If $X \in \mathcal{CP}_n$ has the form $X = \sum_{i=1}^k \mathbf{b}_i \mathbf{b}_i^T$ for some vectors $\mathbf{b}_i \in \mathbb{R}^n_+$. Then $\langle X, M \rangle = 0$ if and only if $\langle \mathbf{b}_i \mathbf{b}_i^T, M \rangle = \mathbf{b}_i^T M \mathbf{b}_i = 0$ for every i, if and only if $\mathbf{b}_i \in \operatorname{ns} M$ for every i, hence if and only if X is a scalar multiple of $\mathbf{b}\mathbf{b}^T$. That is, the extreme ray generated by $\mathbf{b}\mathbf{b}^T$ is $\mathcal{F}_M$. □

Now let us consider maximal faces of $\mathcal{CP}_n$. We first make the following observation (which is a particular instance of a general result — see Exercise 1.58):

Proposition 5.42. *If M generates an exposed ray in $\mathcal{COP}_n$, then $\mathcal{F}_M$ is a maximal face of $\mathcal{CP}_n$.*

Proof. Let $A \in \mathcal{CP}_n$ such that $\{X \in \mathcal{COP}_n \mid \langle A, X \rangle = 0\} = \{tM \mid t \geq 0\}$. Then $A \in \mathcal{F}_M$. Any maximal proper face containing A consists of matrices in $\mathcal{CP}_n$ orthogonal to some $M' \in \mathcal{COP}_n$. In particular, $\langle A, M' \rangle = 0$, so M' has to be a multiple of M, and therefore $\mathcal{F}_{M'} = \mathcal{F}_M$. □

Each known exposed ray of $\mathcal{COP}_n$ therefore defines a maximal face of $\mathcal{CP}_n$. We now describe some of these and their dimensions, starting with faces of the form $\mathcal{F}_M$ where M is SPN and generates an exposed ray in $\mathcal{COP}_n$. For $M = E_{ij} + E_{ji}$ we denote $\mathcal{F}_M = \mathcal{F}_{ij}$.

But first we prove an upper bound on the dimension of a face $\mathcal{F}_M$.

Proposition 5.43. *Let $M \in \mathcal{COP}_n$ have $k \geq 1$ nonzero diagonal elements. Then*

$$\dim \mathcal{F}_M \leq \frac{n(n+1)}{2} - k.$$

Proof. As part of the proof of Proposition 4.74, we have shown that the subspace

$$L = \left\{X \in \mathcal{S}_n \mid \mathbf{e}_i^T X M \mathbf{e}_i = 0 \text{ for every } 1 \leq i \leq k\right\}$$

of $\mathcal{S}_n$ has $\dim L = \frac{n(n+1)}{2} - k$. Since by Proposition 3.33 we have that $\mathcal{F}_M \subseteq L$, this implies the claimed inequality. □

The bound in Proposition 5.43 is not tight.

Example 5.44. Consider $E_{11} \in \mathcal{COP}_n$. Any $X \in \mathcal{CP}_n$ which is orthogonal to E_{11} has a zero $1,1$ entry, but then also a zero first row and column. Hence

$$\mathcal{F}_{11} = \{X \in \mathcal{CP}_n \mid \langle X, E_{11}\rangle = 0\} = \{0_1 \oplus X \mid X \in \mathcal{CP}_{n-1}\}$$

is isomorphic to $\mathcal{CP}_{n-1}$ and $\dim \mathcal{F}_{11} = \frac{(n-1)n}{2} = \frac{n(n+1)}{2} - n$. For $n \geq 2$ this bound is smaller than the bound of Proposition 5.43: $\dim \mathcal{F}_{11} \leq \frac{n(n+1)}{2} - 1$.

Similarly, for every $1 \leq i \leq n$ we have $\dim \mathcal{F}_{ii} = \frac{(n-1)n}{2}$.

Note the faces $\mathcal{F}_{ii}$ are not maximal, since F_{ii} is properly contained in

$$F_{ij} = \{X \in \mathcal{CP}_n \mid x_{ij} = x_{ji} = 0\}$$

for every $j \neq i$. These larger faces, however, are maximal.

Theorem 5.45. *For every $1 \leq i < j \leq n$,*

$$F_{ij} = \{X \in \mathcal{CP}_n \mid \langle X, E_{ij} + E_{ji}\rangle = 0\}$$

is a maximal face of $\mathcal{CP}_n$, and

$$\dim \mathcal{F}_{ij} = \frac{n(n+1)}{2} - 1.$$

Proof. The fact that F_{ij} is a maximal face of $\mathcal{CP}_n$ follows from Proposition 5.42 and Theorem 5.5(b). For every $1 \leq p \leq q \leq n$, $(p,q) \neq (i,j)$, let

$$\mathbf{u}_{pq} = \mathbf{e}_p + \mathbf{e}_q.$$

Then $\mathbf{u}_{pq}\mathbf{u}_{pq}^T \in \mathcal{F}_{ij}$. The set $\left\{\mathbf{u}_{pq}\mathbf{u}_{pq}^T \mid 1 \leq p \leq q \leq n,\ (p,q) \neq (i,j)\right\}$ is linearly independent (see Exercise 4.17(a)), $\dim \mathcal{F}_{ij} \geq \frac{n(n+1)}{2} - 1$, and since $\mathcal{F}_{ij}$ is strictly contained in the full cone $\mathcal{CP}_n$, $\dim \mathcal{F}_{ij} < \frac{n(n+1)}{2}$. Hence the equality. □

Corollary 5.46. *The $\frac{n(n-1)}{2}$ faces $\mathcal{F}_{ij}$, $1 \leq i < j \leq n$, are all the facets of $\mathcal{CP}_n$.*

Proof. That $\mathcal{F}_{ii}$, $1 \leq i \leq n$, are not facets is evident from Example 5.44 and the observation following it.

Any maximal face of $\mathcal{CP}_n$ is therefore of the form

$$\{X \in \mathcal{CP}_n \mid \langle X, M\rangle = 0\},$$

for some extremal $M \in \mathcal{COP}_n$, which is not nonnegative. Such M has at least one negative off-diagonal entry above the diagonal, hence at least two positive diagonal entries, and by Proposition 2.206 the dimension of such face is at most $\frac{n(n+1)}{2} - 2$. □

Theorem 5.47. *Let* $\mathbf{x} \in \mathbb{R}^n$ *have both positive and negative entries. Then* $\mathcal{F}_{\mathbf{x}\mathbf{x}^T}$ *is a maximal face of* $\mathcal{CP}_n$*, and*

$$\dim \mathcal{F}_{\mathbf{x}\mathbf{x}^T} = \frac{n(n-1)}{2}.$$

Proof. As the ray generated by $\mathbf{x}\mathbf{x}^T$ in $\mathcal{COP}_n$ is exposed, Proposition 5.42 and Theorem 5.5(b) imply the maximality of the face.

To compute the dimension of $\mathcal{F}_{\mathbf{x}\mathbf{x}^T}$, note first that this face is contained in $V = \operatorname{span}\{\mathbf{y}\mathbf{y}^T \mid \mathbf{y}^T\mathbf{x} = 0\}$. As $\{\mathbf{y} \in \mathbb{R}^n \mid \mathbf{y}^T\mathbf{x} = 0\}$ has dimension $n-1$, $\dim V = \frac{n(n-1)}{2}$ (see Exercise 4.17(a)). Hence

$$\dim \mathcal{F}_{\mathbf{x}\mathbf{x}^T} \leq \frac{n(n-1)}{2}.$$

For the reverse inequality it suffices, by Exercise 4.17(a), to find a basis of V that consists of nonnegative vectors. We may assume for convenience that the first k entries of $\mathbf{x}$ are positive, the next $p-k$ entries are negative, and the remaining $n-p$ entries are zero (as permutation similarity and maps the face onto another of the same dimension). Let

$$\begin{aligned}
\mathbf{u}_i &= x_{k+1}\mathbf{e}_i + x_i\mathbf{e}_{k+1}, & i &= 1, \ldots, k\,, \\
\mathbf{u}_i &= x_{i+1}\mathbf{e}_1 + x_1\mathbf{e}_{i+1}, & i &= k+1, \ldots, p-1\,, \\
\mathbf{u}_i &= \mathbf{e}_{i+1}, & i &= p, \ldots, n-1\,.
\end{aligned}$$

These are $n-1$ nonnegative linearly independent vectors spanning V. □

An extremal $M \in \mathcal{COP}_n$ with all entries equal to ± 1 also generates an exposed ray. Some of these matrices are positive semidefinite, and some are exceptional.

Theorem 5.48. *Let* $M \in \mathcal{COP}_n$ *be an extremal* $\{\pm 1\}$*-matrix. Then* $\mathcal{F}_M$ *is a maximal face of* $\mathcal{CP}_n$*, and*

$$\dim \mathcal{F}_M = \frac{n(n-1)}{2}.$$

Proof. The face $\mathcal{F}_M$ is a maximal face by Proposition 5.42 and Theorem 5.5(b). By Proposition 5.43

$$\dim \mathcal{F}_M \leq \frac{n(n+1)}{2} - n = \frac{n(n-1)}{2},$$

and it remains to prove the reverse inequality.

The graph $G_{-1}(M)$ is connected, and $m_{ij} = 1$ if and only if $d_{G_{-1}(M)}(i,j) = 2$. As in Theorem 5.6, for every $1 \leq i < j \leq n$ such that $m_{ij} = -1$ let $\mathbf{u}_{ij} = \mathbf{e}_i + \mathbf{e}_j$. These are all the minimal zeros of M up

to multiplication by a positive scalar. For every $i < j$ such that $m_{ij} = 1$, choose k such that $m_{ik} = m_{jk} = -1$, and let $\mathbf{v}_{ij} = \mathbf{e}_i + \mathbf{e}_j + 2\mathbf{e}_k$. These are (non-minimal) zeros of A with support of size 3. Together we have $\frac{n(n-1)}{2}$ zeros of M,

$$\{\mathbf{u}_{ij}\mathbf{u}_{ij}^T \mid 1 \le i < j \le n \text{ and } m_{ij} = -1\} \subseteq \mathcal{F}_M$$

and

$$\{\mathbf{v}_{ij}\mathbf{v}_{ij}^T \mid 1 \le i < j \le n \text{ and } m_{ij} = 1\} \subseteq \mathcal{F}_M\,.$$

We now show that the union of these two sets of rank 1 matrices in $\mathcal{F}_M$ is linearly independent.

Suppose $S \in \mathcal{S}_n$ has zero diagonal entries, and let

$$X(S) = \sum_{\substack{1\le i<j\le n\\ m_{ij}=-1}} s_{ij}\mathbf{u}_{ij}\mathbf{u}_{ij}^T + \sum_{\substack{1\le i<j\le n\\ m_{ij}=1}} s_{ij}\mathbf{v}_{ij}\mathbf{v}_{ij}^T.$$

We need to show that if $X(S) = 0$, then $S = 0$. Note that for $1 \le i < j \le n$ such that $m_{ij} = 1$, $(X(S))_{ij} = s_{ij}$. Hence if $X(S) = 0$ we get that $s_{ij} = 0$ whenever $m_{ij} = 1$, and

$$X(S) = \sum_{\substack{1\le i<j\le n\\ m_{ij}=-1}} s_{ij}\mathbf{u}_{ij}\mathbf{u}_{ij}^T.$$

But then $(X(S))_{ij} = s_{ij}$ for every $1 \le i < j \le n$ such that $m_{ij} = -1$, and therefore $X(S) = 0$ implies that $s_{ij} = 0$ whenever $m_{ij} = -1$, and we get that $S = 0$. □

Although all the maximal faces of $\mathcal{CP}_n$ encountered so far had dimension at least $\frac{n(n-1)}{2}$, in general there may be maximal faces of lower dimension on the part of the boundary of $\mathcal{CP}_n$ that contains matrices orthogonal to exceptional extremal matrices. Before we present one particular example, we present a general idea. Let $M \in \mathcal{COP}_n$. The set of all zeros of M is a union $K_{\alpha_1} \cup \cdots \cup K_{\alpha_m}$ of polyhedral cones, in which each K_{α_i} consists of zeros with supports contained in α_i, with $M[\alpha_i]$ positive semidefinite, and the extreme rays of this cone are generated by the minimal zeros with support contained in α_i, see page 102. We can use this decomposition to bound $\dim \mathcal{F}_M$ from above based on the inclusion

$$\mathcal{F}_M \subseteq \operatorname{span}\{\mathbf{x}\mathbf{x}^T \mid \mathbf{x} \in \cup_{i=1}^m K_{\alpha_i}\}.$$

Example 5.49. Let H_9 be the Hoffman-Pereira matrix with $G_{-1}(H_9)$ the 9-cycle (see Example 2.147). The minimal zeros of H_9 are

$$\mathbf{u}_i = \mathbf{e}_i + \mathbf{e}_{i+1},$$

where $\dotplus$ denotes here addition modulo 9 on $\{1,\ldots,9\}$. Let

$$\alpha_i = \{i, i \dotplus 1, i \dotplus 2, i \dotplus 5, i \dotplus 6\}, \qquad i = 1,\ldots,9.$$

For each i the submatrix $H_9[\alpha_i]$ is a maximal positive semidefinite principal submatrix of H_9, and $\mathcal{V}^{H_9} = \cup_{i=1}^{9} K_{\alpha_i}$, where $K_{\alpha_i} = \text{cone}\{\mathbf{u}_i, \mathbf{u}_{i\dotplus 1}, \mathbf{u}_{i\dotplus 5}\}$. For every $1 \le i \le 9$ we denote

$$\mathbf{a}_i = \mathbf{u}_i + \mathbf{u}_{i\dotplus 1}, \quad \mathbf{b}_i = \mathbf{u}_i + \mathbf{u}_{i\dotplus 5}, \quad \mathbf{c}_i = \mathbf{u}_{i\dotplus 1} + \mathbf{u}_{i\dotplus 5}\,.$$

Using Exercise 4.17(a), and the fact that $\mathbf{c}_i = \mathbf{b}_{i\dotplus 5}$ for every i, we get that each zero $\mathbf{v}$ of H_9

$$\mathbf{v}\mathbf{v}^T \in \text{span} \bigcup_{i=1}^{9} \{\mathbf{u}_i\mathbf{u}_i^T, \mathbf{a}_i\mathbf{a}_i^T, \mathbf{b}_i\mathbf{b}_i^T\}.$$

Therefore

$$\mathcal{F}_{H_9} \subseteq \text{span} \bigcup_{i=1}^{9} \{\mathbf{u}_i\mathbf{u}_i^T, \mathbf{a}_i\mathbf{a}_i^T, \mathbf{b}_i\mathbf{b}_i^T\},$$

and $\dim \mathcal{F}_{H_9} \le 27 < \frac{9\cdot 8}{2} = 36$.

Faces of $\mathcal{CP}_5$

For $n = 5$ the extreme rays of $\mathcal{COP}_5$ are fully known, and we can say more about the faces of $\mathcal{CP}_5$. In particular, in $\mathcal{CP}_5$ there are examples of non-exposed faces, while so far all the faces of $\mathcal{CP}_n$ we considered were exposed. By Corollary 5.10, this also implies the existence of non-exposed faces in $\mathcal{CP}_n$, $n > 5$, obtained by adding zero rows and columns to the matrices in the non-exposed face of $\mathcal{CP}_5$.

Example 5.50. Let $S(\boldsymbol{\theta})$ is as in (2.43), with $\theta_i > 0$ for every i, and $\sum_{i=1}^{5} \theta_i = \pi$. Then $S(\boldsymbol{\theta}) = \mathbf{x}\mathbf{x}^T + \mathbf{y}\mathbf{y}^T$, where

$$\mathbf{x} = \begin{pmatrix} \cos(\theta_1+\theta_2) \\ -\cos\theta_2 \\ 1 \\ -\cos\theta_3 \\ \cos(\theta_3+\theta_4) \end{pmatrix}, \qquad \mathbf{y} = \begin{pmatrix} \sin(\theta_1+\theta_2) \\ -\sin\theta_2 \\ 0 \\ \sin\theta_3 \\ -\sin(\theta_3+\theta_4) \end{pmatrix}.$$

The positive semidefinite matrix $S(\boldsymbol{\theta})$ has rank 2, and the following five minimal zeros:

$$\mathbf{a}_1 = \begin{pmatrix} \sin\theta_2 \\ \sin(\theta_1+\theta_2) \\ \sin\theta_1 \\ 0 \\ 0 \end{pmatrix}, \quad \mathbf{a}_2 = \begin{pmatrix} 0 \\ \sin\theta_3 \\ \sin(\theta_2+\theta_3) \\ \sin\theta_2 \\ 0 \end{pmatrix}, \quad \mathbf{a}_3 = \begin{pmatrix} 0 \\ 0 \\ \sin\theta_4 \\ \sin(\theta_3+\theta_4) \\ \sin\theta_3 \end{pmatrix}$$

$$\mathbf{a}_4 = \begin{pmatrix} \sin\theta_4 \\ 0 \\ 0 \\ \sin\theta_5 \\ \sin(\theta_4+\theta_5) \end{pmatrix}, \quad \mathbf{a}_5 = \begin{pmatrix} \sin(\theta_5+\theta_1) \\ \sin\theta_5 \\ 0 \\ 0 \\ \sin\theta_1 \end{pmatrix}.$$

As each of these five zeros is in $\operatorname{ns} S(\boldsymbol{\theta})$, $\dim\operatorname{span}\{\mathbf{a}_1,\dots,\mathbf{a}_5\} \le 3$. By the sign pattern of the five zeros it is easy to see that each three of them are linearly independent, and therefore form a basis of $\operatorname{span}\{\mathbf{a}_1,\dots,\mathbf{a}_5\}$, whose dimension turns out to be exactly 3.

Let

$$\mathcal{F} = \operatorname{cone}\left\{\mathbf{a}_1\mathbf{a}_1^T, \mathbf{a}_2\mathbf{a}_2^T, \mathbf{a}_3\mathbf{a}_3^T, \mathbf{a}_4\mathbf{a}_4^T\right\}.$$

We will now show that $\mathcal{F}$ is a face of $\mathcal{CP}_5$, but not an exposed face. Note that every $\mathbf{a}_i\mathbf{a}_i^T$, $1 \le i \le 4$, has a zero $2,5$ entry, and therefore so does any matrix in $\mathcal{F}$.

To show that $\mathcal{F}$ is a face, consider first

$$A = \mathbf{a}_1\mathbf{a}_1^T + \mathbf{a}_2\mathbf{a}_2^T + \mathbf{a}_3\mathbf{a}_3^T + \mathbf{a}_4\mathbf{a}_4^T \in \operatorname{ri}\mathcal{F}.$$

Suppose $A = B + C$, where B and C are completely positive. As $a_{25} = 0$, also $b_{25} = c_{25} = 0$, and therefore if $B = \sum_i \mathbf{b}_i\mathbf{b}_i^T$ and $C = \sum_j \mathbf{c}_j\mathbf{c}_j^T$ are CP representation of B and C, respectively, then each $\mathbf{b}_i$ and each $\mathbf{c}_j$ is in $\operatorname{cs} A$, and either has a zero second entry, or a zero fifth entry. We may assume without loss of generality that each $\mathbf{b}_i$ has a zero fifth entry, and each $\mathbf{c}_j$ has a zero second entry. As $\mathbf{b}_i \in \operatorname{span}\{\mathbf{a}_1,\mathbf{a}_2,\mathbf{a}_3\}$ and $(\mathbf{b}_i)_5 = 0$, we have in fact that $\mathbf{b}_i \in \operatorname{span}\{\mathbf{a}_1,\mathbf{a}_2\}$. Similarly, $\mathbf{c}_j \in \operatorname{span}\{\mathbf{a}_2,\mathbf{a}_3,\mathbf{a}_4\}$ and $(\mathbf{c}_j)_2 = 0$ implies $\mathbf{c}_j \in \operatorname{span}\{\mathbf{a}_3,\mathbf{a}_4\}$. Hence $\operatorname{cpr} B = \operatorname{rank} B \le 2$ and $\operatorname{cpr} C = \operatorname{rank} C \le 2$. That is,

$$B = \mathbf{b}_1\mathbf{b}_1^T + \mathbf{b}_2\mathbf{b}_2^T \quad \text{and} \quad C = \mathbf{c}_1\mathbf{c}_1^T + \mathbf{c}_2\mathbf{c}_2^T,$$

where

$$\mathbf{b}_i = s_i\mathbf{a}_1 + t_i\mathbf{a}_2, \quad \mathbf{c}_j = p_j\mathbf{a}_3 + q_j\mathbf{a}_4,$$

where s_i, t_i, p_j, q_j are all nonnegative (as $(\mathbf{b}_i)_1 = s_i(\mathbf{a}_1)_1$, $(\mathbf{b}_i)_4 = t_i(\mathbf{a}_2)_4$, $(\mathbf{c}_j)_3 = p_j(\mathbf{a}_3)_3$ and $(\mathbf{b}_j)_1 = q_j(\mathbf{a}_4)_1$). As $a_{11} = (B+C)_{11}$ we get that

$$\sin^2\theta_2 + \sin^2\theta_4 = (s_1^2 + s_2^2)\sin^2\theta_2 + (p_1^2 + p_2^2)\sin^2\theta_4,$$

and therefore

$$(1 - (s_1^2 + s_2^2))\sin^2\theta_2 = ((p_1^2 + p_2^2) - 1)\sin^2\theta_4,$$

which implies that

$$(1 - (s_1^2 + s_2^2))((p_1^2 + p_2^2) - 1) \geq 0. \tag{5.4}$$

Now compare

$$a_{12} = \sin\theta_2 \sin(\theta_1 + \theta_2)$$

and $(B+C)_{12}$. By the nonnegativity of the vectors and s_i, t_i, p_j, q_j,

$$a_{12} = (B+C)_{12} \geq b_{12} \geq (s_1^2 + s_2^2)\sin\theta_2\sin(\theta_1 + \theta_2).$$

Hence $s_1^2 + s_2^2 \leq 1$. A similar comparison of a_{14} and $(B+C)_{14}$ implies $p_1^2 + p_2^2 \leq 1$. These two inequalities combined with (5.4) imply that

$$s_1^2 + s_2^2 = p_1^2 + p_2^2 = 1.$$

But then $(B+C)_{13} = a_{13} = \sin\theta_1\sin\theta_2$ yields $s_i t_i = 0$ and $p_j q_j = 0$ for every i and j. Hence each $\mathbf{b}_i$ is a multiple of either $\mathbf{a}_1$ or $\mathbf{a}_2$, and each $\mathbf{c}_j$ is a multiple of either $\mathbf{a}_3$ or $\mathbf{a}_4$. Implying that

$$\mathbf{b}_i\mathbf{b}_i^T, \mathbf{c}_j\mathbf{c}_j^T \in \mathcal{F}$$

for every i and j, and therefore $B, C \in \mathcal{F}$.

This shows that A has a unique CP representation and, by Theorem 5.40, $\mathcal{F}$ is the minimal face containing A. In particular, it is a face.

To see that $\mathcal{F}$ is not exposed, we now show that the exposed face $\mathcal{F}_M$, where $M = \mathbf{x}\mathbf{x}^T + \mathbf{y}\mathbf{y}^T + N \in \mathcal{COP}_5$, $N = E_{25} + E_{52}$, is the minimal exposed face containing $\mathcal{F}$, and the inclusion is strict.

This exposed face clearly contains $\mathcal{F}$, since each $\mathbf{a}_i$ is orthogonal to both $\mathbf{x}$ and $\mathbf{y}$, and $\mathbf{a}_i\mathbf{a}_i^T$ has a zero at position $2, 5$. The inclusion is strict, as $B = (\mathbf{a}_1 + \mathbf{a}_2)(\mathbf{a}_1 + \mathbf{a}_2)^T \in \mathcal{F}_M$, but $B \notin \mathcal{F}$. To see that, note that the fifth row and column of B are zero, hence if B were in $\mathcal{F}$, we would have $B = s_1\mathbf{a}_1\mathbf{a}_1^T + s_2\mathbf{a}_2\mathbf{a}_2^T$. But this is impossible, since $b_{14} > 0$, while the combination on the right-hand side has a zero entry in position $1, 4$.

It remains to show that $\mathcal{F}_M$ is the *minimal* exposed face containing $\mathcal{F}$. Indeed, if $\mathcal{F}_Q$ is an exposed face containing $\mathcal{F}$, then $Q = \sum_{i=1}^k Q_i$, where each Q_i is extremal in $\mathcal{COP}_5$, and $\mathcal{F}_Q = \cap_{i=1}^k F_{Q_i}$. None of the Q_i's is in the orbit of a Hildebrand matrix or the Horn matrix, since in these cases, the four zeros $\mathbf{a}_1, \ldots, \mathbf{a}_4$ of Q that have supports

$$\{1,2,3\}, \{2,3,4\}, \{3,4,5\}, \{4,5,1\}$$

would be linearly independent (see Exercises 5.16 and 5.17). Hence Q is SPN. The only possible nonnegative Q_i has to be N (up to multiplication by scalar), since for $p \leq q$ such that $(p,q) \neq (2,5)$ there exists $A \in \mathcal{F}$ such that $a_{pq} > 0$. If $Q_i = \mathbf{z}\mathbf{z}^T$, then $\mathbf{z}$ is orthogonal to span$\{\mathbf{a}_1, \ldots, \mathbf{a}_4\}$. Such $\mathbf{z}$ is necessarily in span$\{\mathbf{a}, \mathbf{b}\}$, and therefore $\mathcal{F}_M \subseteq \mathcal{F}_{Q_i}$ for every $1 \leq i \leq k$, and $\mathcal{F}_M \subseteq \mathcal{F}_Q$. This concludes the proof that the face $\mathcal{F}$ in our example is a non-exposed face of $\mathcal{COP}_5$. This face is a subface of maximal faces containing matrices orthogonal to a rank 1 positive semidefinite matrix.

Let us now consider faces of $\mathcal{CP}_5$, which are subfaces of a maximal face containing matrices that are orthogonal to an exceptional matrix in $\mathcal{COP}_5$: either a matrix in the orbit of a Hildebrand matrix, or a matrix in the orbit of the Horn matrix. As permutation similarity and congruence by a positive diagonal matrix transform the faces of the cone $\mathcal{CP}_5$ onto faces of the cone isomorphically, it suffices to consider subfaces of $\mathcal{F}_H$, where H is either a Hildebrand matrix or the Horn matrix.

We first consider the case that H is a Hildebrand matrix. That is, $H = S(\boldsymbol{\theta})$, where $S(\boldsymbol{\theta})$ is as in (2.43), with $\theta_i > 0$ for every i and $\sum_{i=1}^5 \theta_i < \pi$. Let $\mathbf{w}_i \in \mathbb{R}^5$ be the vector with support σ_i such that

$$\mathbf{w}_i[\sigma_i] = \begin{pmatrix} \sin\theta_{i+1} \\ \sin(\theta_i + \theta_{i+1}) \\ \sin\theta_i \end{pmatrix}.$$

The vectors $\mathbf{w}_1, \ldots, \mathbf{w}_5$ are linearly independent, and are all the zeros of H up to scalar multiplication (see Chapter 2.10). The maximal face $\mathcal{F}_H$ is therefore a polyhedral cone. Each matrix in $\mathcal{F}_H$ has a unique representation as $\sum_{i=1}^5 t_i \mathbf{w}_i \mathbf{w}_i^T$ (see Example 4.123). Hence for each $\alpha \subseteq \{1, \ldots, 5\}$ the cone cone$\{\mathbf{w}_i \mathbf{w}_i^T \mid i \in \alpha\}$ is a face of $\mathcal{CP}_5$ contained in $\mathcal{F}_H$.

Theorem 5.51. *Let $\mathcal{F}$ be a subface of the maximal face $\mathcal{F}_H$, where H is a Hildebrand matrix. Then $\mathcal{F}$ is an exposed face of $\mathcal{CP}_5$.*

Proof. By the previous paragraph, $\mathcal{F}_H = \text{cone}\{\mathbf{w}_i \mathbf{w}_i^T \mid 1 \leq i \leq 5\}$, where $\mathbf{w}_1, \ldots, \mathbf{w}_5$ are as above. Suppose $\mathcal{F} \subseteq \mathcal{F}_H$ is a subface of $\mathcal{F}_H$. Then $\mathcal{F} = \text{cone}\{\mathbf{w}_i \mathbf{w}_i^T \mid i \in \alpha\}$, where $\alpha \subseteq \{1, \ldots, 5\}$. To show that $\mathcal{F}$ is exposed, we use the fact that if $X = \sum_{i=1}^5 t_i \mathbf{w}_i \mathbf{w}_i^T$ where $t_i \geq 0$ for every i, then

$$x_{i\,i+2} = t_i \sin\theta_i \sin\theta_{i+1}.$$

Let

$$Q = \sum_{\substack{1 \le i \le 5 \\ i \notin \alpha}} \left(E_{i\, i\dotplus 2} + E_{i \dotplus 2\, i} \right),$$

and let

$$M = H + Q \in \mathcal{COP}_5.$$

Then $X \in \mathcal{F}_M$ if and only if X is orthogonal to H, and $x_{i\, i\dotplus 2} = 0$ for every $i \notin \alpha$. By the above this is equivalent to $X = \sum_{i \in \alpha} t_i \mathbf{w}_i \mathbf{w}_i^T$, i.e., $X \in \mathcal{F}$. □

Now consider the subfaces of $\mathcal{F}_H$, where H is the Horn matrix. It turns out that most of these are exposed, except for one class of subfaces. Recall that the minimal zeros of the Horn matrix H are $\mathbf{u}_i = \mathbf{e}_i + \mathbf{e}_{i \dotplus 1}$, $i = 1, \ldots, 5$, where $\dotplus$ denotes addition modulo 5 on $\{1, \ldots, 5\}$. Any other zero of H has the form $s\mathbf{u}_i + t\mathbf{u}_{i \dotplus 1}$, where $s, t > 0$. The extreme rays of $\mathcal{F}_H$ are $\mathbf{u}\mathbf{u}^T$, where $\mathbf{u}$ is a zero of the Horn matrix H. We will classify the subfaces of $\mathcal{F}_H$ according to the types of extreme rays of $\mathcal{F}_H$ they contain. For that we set up some notations.

For every $1 \le i \le 5$, let $\sigma_i = \{i, i \dotplus 1, i \dotplus 2\}$, the support of any zero that is a combination with positive coefficients of $\mathbf{u}_i$ and $\mathbf{u}_{i \dotplus 1}$. Denote

$$\mathcal{B}_i = \operatorname{cone}\left\{ \mathbf{b}\mathbf{b}^T \mid \mathbf{b} \in \mathcal{V}^H \text{ and } \operatorname{supp} \mathbf{b} = \sigma_i \right\}$$

and

$$\mathcal{C}_i = \operatorname{cone}\left\{ t\mathbf{u}_i \mathbf{u}_i^T \mid t \ge 0 \right\}.$$

Then

$$\mathcal{F}_H = \sum_{i=1}^{5} \left(\mathcal{B}_i + \mathcal{C}_i \right) = \sum_{i=1}^{5} \left(\mathcal{B}_i + \mathcal{C}_i + \mathcal{C}_{i+1} \right).$$

We classify the subfaces of $\mathcal{F}_H$ according to their intersections with the cones $\mathcal{B}_i$ and $\mathcal{C}_i$. For each i, $\mathcal{B}_i \cap \mathcal{F} = \{0\}$, or $\mathcal{B}_i \cap \mathcal{F} = \left\{ t\mathbf{b}\mathbf{b}^T \mid t \ge 0 \right\}$ for some zero $\mathbf{b}$ of H with support σ_i, or $\mathcal{B}_i \subseteq \mathcal{F}$. Similarly, either $\mathcal{C}_i \cap \mathcal{F} = \{0\}$, or $\mathcal{C}_i \subseteq \mathcal{F}$. These possibilities are not unrelated: The following observation implies that if $\mathcal{B}_i \subseteq \mathcal{F}$ or $\mathcal{B}_i \cap \mathcal{F}$ is one dimensional and also $\mathcal{C}_i \subseteq \mathcal{F}$ or $\mathcal{C}_{i+1} \subseteq \mathcal{F}$, then $\mathcal{B}_i + \mathcal{C}_i + \mathcal{C}_{i \dotplus 1} \subseteq \mathcal{F}$.

Proposition 5.52. *Let $\mathcal{F}$ be a subface of $\mathcal{F}_H$, where H is the Horn matrix, and let $1 \le i \le 5$. If there exist two linearly independent $\mathbf{b}_1, \mathbf{b}_2 \in \mathbb{R}^n_+$ such that* $\operatorname{supp} \mathbf{b}_1 = \sigma_i$, $\operatorname{supp} \mathbf{b}_2 \subseteq \sigma_i$, *and both $\mathbf{b}_1 \mathbf{b}_1^T$ and $\mathbf{b}_2 \mathbf{b}_2^T$ are in $\mathcal{F}$, then $\mathcal{B}_i + \mathcal{C}_i + \mathcal{C}_{i \dotplus 1} \subseteq \mathcal{F}$.*

Proof. Let $A = \mathbf{b}_1\mathbf{b}_1^T + \mathbf{b}_2\mathbf{b}_2^T \in \mathcal{F}$. The matrix A is a rank 2 completely positive matrix, $A[\sigma_i]$ is positive and all other entries of A are zero. Hence any zero of H with support contained in σ_i is in $\operatorname{cs} A$. In particular, any $\mathbf{b} \in \mathbb{R}^n_+$ such that $\mathbf{b}\mathbf{b}^T \in \mathcal{B}_i + \mathcal{C}_i + \mathcal{C}_{i+1}$ is such a zero of H, and is in $\operatorname{cs} A$. Hence for small enough $\delta > 0$ we have that $A - \delta\mathbf{b}\mathbf{b}^T$ is both positive and positive semidefinite. As this matrix is zero except for a principal submatrix of order at most 3, it is completely positive. By Theorem 5.39 we get that $\mathbf{b}\mathbf{b}^T$ is in the minimal face containing A, and therefore in $\mathcal{F}$. □

We recall some more known facts about the Horn matrix and its zeros.

Remark 5.53. Each zero of H has the form $\mathbf{b}$, where $\operatorname{supp}\mathbf{b} \subseteq \sigma_i$ is a nonnegative combination of $\mathbf{u}_i$ and $\mathbf{u}_{i+1}$. Let $\mathbf{b}_i$, $i = 1, \ldots, 5$, be five zeros of H such that $\mathbf{b}_i = s_i\mathbf{u}_i + t_i\mathbf{u}_{i+1}$, where $s_i, t_i \geq 0$ and are not both zero. Let

$$B = \begin{pmatrix} \mathbf{b}_1 & \mathbf{b}_2 & \mathbf{b}_3 & \mathbf{b}_4 & \mathbf{b}_5 \end{pmatrix}.$$

Then B is nonsingular if and only if $\Pi_{i=1}^5 s_i + \Pi_{i=1}^5 t_i > 0$, i.e., if and only if $s_i > 0$ for every i or $t_i > 0$ for every i. We leave the proof as an exercise.

We also mention that, by Example 4.122 and Exercise 4.58, if $A = BB^T$ is positive (and therefore $a_{i\,i+2} = s_it_i > 0$ for every i) then $A = BB^T$ is the unique CP factorization of A if and only if $\Pi_{i=1}^5 s_i = \Pi_{i=1}^5 t_i$.

With these observations we can describe all the exposed subfaces of $\mathcal{F}_H$.

Theorem 5.54. *Let $\mathcal{F}$ be a subface of $\mathcal{F}_H$, where H is the Horn matrix. If $\mathcal{F} = \mathcal{F}_H$, or $\mathcal{F} \cap \mathcal{B}_i = \{0\}$ for at least one i, then $\mathcal{F}$ is exposed.*

Proof. Clearly $\mathcal{F}_H$ is exposed. Hence we need to consider the following five cases: $\mathcal{F} \cap \mathcal{B}_i = \{0\}$ for m values of i, $1 \leq m \leq 5$. There are subcases, according to whether the m indices are consecutive or not, according to the dimensions of $\mathcal{F} \cap \mathcal{B}_i$ for the remaining $5 - m$ indices, and according to which $\mathcal{C}_i$'s are contained in $\mathcal{F}$. We will only go over a few of these cases, and leave the rest as an exercise.

Case 1. $m = 5$. Then $\mathcal{F} \subseteq \sum_{i=1}^5 \mathcal{C}_i$. Let $\alpha = \{1 \leq i \leq 5 \mid \mathcal{F} \cap \mathcal{C}_i \neq \{0\}\}$. Then $\mathcal{F} = \sum_{i\in\alpha} \mathcal{C}_i = \operatorname{cone}\{\mathbf{u}_i\mathbf{u}_i^T \mid i \in \alpha\}$. Let

$$A = \sum_{i\in\alpha} \mathbf{u}_i\mathbf{u}_i^T \in \operatorname{ri}\mathcal{F}.$$

The graph of A is a subgraph of the 5 cycle, and the graph of any $X \in \mathcal{F}$ is a subgraph of $G(A)$. If $X = \sum_{i\in\alpha} t_i\mathbf{u}_i\mathbf{u}_i^T$, we get that $x_{i\,i+1} \neq 0$ if and only if $i \in \alpha$ and $t_i > 0$.

Let ζ be the set of positions i, j above the diagonal such that $a_{ij} = 0$, and let

$$Q = \sum_{(i,j)\in\zeta} (E_{ij} + E_{ji}) \,.$$

Now let $M = H + Q$. We show that $\mathcal{F} = \mathcal{F}_M$. The inclusion $\mathcal{F} \subseteq \mathcal{F}_M$ is clear from the above. For the reverse inclusion, note that for every $1 \le i \le 5$, any nonzero $X \in \mathcal{B}_i$ is not in $\mathcal{F}_M$, since it is not orthogonal to Q ($x_{i\,i\dotplus 2} > 0$ and $(i, i \dotplus 2) \in \zeta$). Similarly, for $i \notin \alpha$ we have that a nonzero $X \in \mathcal{C}_i$ is not in $\mathcal{F}_M$, since such X has $x_{i\,i\dotplus 1} > 0$, and $(i, i \dotplus 1) \in \zeta$. Hence $\mathcal{F}_M \subseteq \sum_{i\in\alpha} C_i = \mathcal{F}$.

Case 2. $m = 4$. By the cyclic structure of H we may assume $\mathcal{F} \cap \mathcal{B}_1 \neq \{0\}$, and $\mathcal{F} \cap \mathcal{B}_i = \{0\}$ for every $2 \le i \le 4$. Consider two subcases: $\mathcal{F} \cap \mathcal{B}_1 = \mathcal{B}_1$, and $\mathcal{F} \cap \mathcal{B}_i = \left\{t\mathbf{b}\mathbf{b}^T \mid t \ge 0\right\}$ for some zero $\mathbf{b}$ of H with support σ_1.

In the first subcase, $\mathcal{F}$ contains also $\mathcal{C}_1$ and $\mathcal{C}_2$, by Proposition 5.52. It may also contain one or more of $\mathcal{C}_3, \mathcal{C}_4, \mathcal{C}_5$. Let $\alpha = \{3 \le i \le 5 \mid \mathcal{C}_i \subseteq \mathcal{F}\}$ (α may be empty). Then $\mathcal{F} = \mathcal{B}_1 + \sum_{i\in\alpha} \mathcal{C}_i$, and

$$A = (\mathbf{u}_1 + \mathbf{u}_2)(\mathbf{u}_1 + \mathbf{u}_2)^T + \sum_{i\in\alpha} \mathbf{u}_i\mathbf{u}_i^T \in \mathcal{F}.$$

For any $X \in \mathcal{F}$, $G(X)$ is a subgraph of $G(A)$. Denote by ζ be the set of positions i, j, $1 \le i < j \le 5$, such that $a_{ij} = 0$. Let

$$Q = \sum_{(i,j)\in\zeta} (E_{ij} + E_{ji})$$

and $M = H + Q$. As in the Case 1, it is easy to see that $\mathcal{F} = \mathcal{F}_M$.

In the second subcase, let $\mathcal{F} \cap \mathcal{B}_1 = \left\{t\mathbf{b}\mathbf{b}^T \mid t \ge 0\right\}$ for some zero $\mathbf{b}$ of H with support σ_1. By Proposition 5.52, necessarily $\mathcal{F} \cap \mathcal{C}_1 = \mathcal{F} \cap \mathcal{C}_2 = \{0\}$. Hence

$$\mathcal{F} = \left\{t\mathbf{b}\mathbf{b}^T \mid t \ge 0\right\} + \sum_{i\in\alpha} \mathcal{C}_i,$$

for some $\alpha \subseteq \{3, 4, 5\}$ (possibly empty), and $A = \mathbf{b}\mathbf{b}^T + \sum_{i\in\alpha} \mathbf{u}_i\mathbf{u}_i^T$ is in $\operatorname{ri}\mathcal{F}$. Let ζ be the set of positions i, j, $1 \le i < j \le 5$, such that $a_{ij} = 0$. Now choose a nonzero $\mathbf{a} \in \mathbb{R}^5$ orthogonal to $\{\mathbf{b}, \mathbf{u}_3, \mathbf{u}_4, \mathbf{u}_5\}$. These four vectors are linearly independent (see Remark 5.53), hence $\mathbf{a}$ is unique up to a scalar multiplication. Let

$$Q = \sum_{(i,j)\in\zeta} (E_{ij} + E_{ji})$$

and $M = H + Q + \mathbf{a}\mathbf{a}^T$. The inclusion $\mathcal{F} \subseteq \mathcal{F}_M$ is clear, and we prove the reverse inclusion. Let $\mathbf{x}\mathbf{x}^T \in \mathcal{F}_M$. Then $\mathbf{x}$ is a zero of H, with support of size 2 or 3. Also, $\mathbf{x}$ is orthogonal to $\mathbf{a}$, and is therefore a linear combination of $\mathbf{b}, \mathbf{u}_3, \mathbf{u}_4, \mathbf{u}_5$. This implies that $\mathbf{x}\mathbf{x}^T \notin \mathcal{C}_1$, $\mathbf{x}\mathbf{x}^T \notin \mathcal{C}_2$ as, by Remark 5.53, neither $\mathbf{u}_1$ nor $\mathbf{u}_2$ is a linear combination of these four vectors. That is, $\mathcal{F}_M \cap \mathcal{C}_i = \{0\}$ for $i = 1, 2$. If $\operatorname{supp} \mathbf{x} = \sigma_1$, then $\mathbf{x}$ has to be a multiple of $\mathbf{b}$, since otherwise, as $\mathbf{b}$ is a zero of H, Proposition 5.52 would imply $\mathcal{B}_1 \subseteq \mathcal{F}_M$ and therefore $\mathcal{C}_1 \subseteq \mathcal{F}_M$, which we have shown to be impossible. Hence $\mathbf{x}$ is either a multiple of $\mathbf{b}$, or, as in Case 1, $\mathbf{x}\mathbf{x}^T \in \cup_{i\in\alpha}\mathcal{C}_i$. This proves that $\mathcal{F}_M \subseteq \left\{t\mathbf{b}\mathbf{b}^T \mid t \geq 0\right\} + \sum_{i\in\alpha} \mathcal{C}_i = \mathcal{F}$, and completes the proof that $\mathcal{F} = \mathcal{F}_M$.

We leave the proof of the remaining cases as an exercise. □

Remark 5.55. Let $\mathcal{F}$ be a subface of $\mathcal{F}_H$, where H is the Horn matrix. Suppose $\mathcal{F} \cap \mathcal{B}_i \neq \{0\}$ for every $1 \leq i \leq 5$, and for some i, $\mathcal{B}_i \subseteq \mathcal{F}$. Without loss of generality, assume $\mathcal{B}_1 \subseteq \mathcal{F}$. Then also $\mathcal{C}_2 \subseteq \operatorname{cl}\mathcal{B}_1 \subseteq \mathcal{F}$, which together with $\mathcal{F} \cap \mathcal{B}_2 \neq \{0\}$ implies, by Proposition 5.52, that $\mathcal{B}_2 \subseteq \mathcal{F}$. Continuing similarly we get that $\mathcal{F} \cap \mathcal{B}_i = \mathcal{B}_i$ for every i. Using Proposition 5.52 once more, this implies that $\mathcal{F} = \mathcal{F}_H$. Hence the only subfaces of $\mathcal{F}_H$ not considered in Theorem 5.54 are those where $\mathcal{F} \cap \mathcal{B}_i$ is one dimensional for every $1 \leq i \leq 5$. It turns out that such subfaces are not exposed.

Theorem 5.56. *Let $\mathcal{F}$ be a subface of $\mathcal{F}_H$, where H is the Horn matrix. Suppose that for every $1 \leq i \leq 5$,*

$$\mathcal{F} \cap \mathcal{B}_i = \left\{t\mathbf{b}_i\mathbf{b}_i^T \mid t \geq 0\right\},$$

where $\mathbf{b}_i$ is a zero of H with support σ_i. Then $\mathbf{b}_i = s_i\mathbf{u}_i + t_i\mathbf{u}_i$, where $\Pi_{i=1}^5 s_i = \Pi_{i=1}^5 t_i > 0$, and $\mathcal{F}$ is not exposed.

Proof. By the assumptions, for every i we have $\mathbf{b}_i = s_i\mathbf{u}_i + t_i\mathbf{u}_i$, where all the coefficients s_i and t_i are positive, and $\mathcal{F} = \operatorname{cone}\{\mathbf{b}_1\mathbf{b}_1^T, \ldots, \mathbf{b}_5\mathbf{b}_5^T\}$. In particular, $\mathcal{F} \neq \mathcal{F}_H$. Let $A = \sum_{i=1}^5 \mathbf{b}_i\mathbf{b}_i^T \in \mathcal{F}$. If A has another CP representation, then (since A is positive) there exists a zero $\mathbf{b}$ of H with support σ_i such that $A - \mathbf{b}\mathbf{b}^T$ is completely positive, and $\mathbf{b}$ is not a scalar multiple of $\mathbf{b}_i$. But then $\mathbf{b}\mathbf{b}^T \in \mathcal{F}$, and by Proposition 5.52 we get that $\mathcal{B}_i \subseteq \mathcal{F}$, contrary to our assumption. Hence A has a unique CP representation. This implies, as recalled in Remark 5.53 above, that $\Pi_{i=1}^5 s_i = \Pi_{i=1}^5 t_i$.

We now show that $\mathcal{F}$ is not exposed. Suppose on the contrary that $\mathcal{F} = \mathcal{F}_M$ for some $M \in \mathcal{COP}_5$. Then M has to be exceptional, since

$A \in \mathcal{F}$ is positive and positive definite (see Remark 5.53), and cannot be orthogonal to a nonzero nonnegative or positive semidefinite matrix. The matrix M cannot be in the orbit of a Hildebrand matrix, since $\mathbf{b}_1, \ldots, \mathbf{b}_5$ cannot all be zeros of such M. Therefore M has to be in the orbit of a Horn matrix. Hence $M = DPHP^TD$ for some permutation matrix P and some positive diagonal matrix D, and the vectors $D^{-1}P\mathbf{b}_i$, $1 \le i \le 5$, are zeros of M. As $\mathcal{F} = \mathcal{F}_M$ and $\mathcal{F} \cap \mathcal{B}_i = \{t\mathbf{b}_i\mathbf{b}_i^T \mid t \ge 0\}$, each $D^{-1}P\mathbf{b}_i$ is a positive multiple of one of the vectors $\mathbf{b}_1, \ldots, \mathbf{b}_5$. In particular, P is either I, or a circular rotation, or a circular counter rotation, so that $PHP^T = H$, and for every $1 \le i \le 5$, $d_i^{-1} = d_{i+1}^{-1} = d_{i+2}^{-1}$. Hence $D = dI$ for some $d > 0$, and M is a scalar multiple of H. But this is a contradiction, since $\mathcal{F} \neq \mathcal{F}_H$. □

Automorphisms of $\mathcal{CP}_n$ and cp-rank preservers

As for the copositive cone, we already know that $X \mapsto P^TXP$, where P is a permutation matrix, and $X \mapsto DXD$, where D is a positive diagonal matrix, are automorphisms of $\mathcal{CP}_n$. We want to show that the group generated by these two types of automorphisms is the whole automorphism group of $\mathcal{CP}_n$. Since we already know that this holds for automorphisms of $\mathcal{COP}_n$, we may exploit the connection between automorphisms of a cone and automorphisms of its dual to prove this.

Theorem 5.57. *An invertible linear transformation $\Phi : \mathcal{S}_n \to \mathcal{S}_n$ is an automorphism of $\mathcal{CP}_n$ if and only if $\Phi(X) = M^TXM$ for some nonnegative monomial matrix M.*

Proof. We observe first that a linear transformation $\Phi : \mathcal{S}_n \to \mathcal{S}_n$ is an automorphism of $\mathcal{CP}_n$ if and only if its adjoint $\Phi^* : \mathcal{S}_n \to \mathcal{S}_n$ is an automorphism of $\mathcal{COP}_n$. Recall that Φ^* is defined as the unique linear transformation satisfying for every $X, Y \in \mathcal{S}_n$

$$\langle \Phi(X), Y \rangle = \langle X, \Phi^*(Y) \rangle . \tag{5.5}$$

Hence $\Phi(X) \in \mathcal{CP}_n$ for every $X \in \mathcal{CP}_n$ if and only if $\langle \Phi(X), Y \rangle \ge 0$ for every $Y \in \mathcal{COP}_n$, i.e., if and only if $\langle X, \Phi^*(Y) \rangle \ge 0$ for every $X \in \mathcal{CP}_n$ and $Y \in \mathcal{COP}_n$. This is equivalent to $\Phi^*(Y) \in \mathcal{COP}_n$ for every $Y \in \mathcal{COP}_n$.

From Theorem 5.29 we get that Φ^* is an automorphism of $\mathcal{COP}_n$ if and only if $\Phi^*(Y) = M^TYM$ for some nonnegative monomial matrix M. Then from (5.5) we see that for every $X \in \mathcal{S}_n$ and every $Y \in \mathcal{S}_n$

$$\langle \Phi(X), Y \rangle = \langle X, M^TYM \rangle = \langle MXM^T, Y \rangle,$$

hence $\Phi(X) = MXM^T$, where M^T is a monomial matrix. □

The problem of characterizing into preservers of complete positivity is still a challenge. Any linear transformation of the form $X \mapsto S^T XS$, where $S \in \mathbb{R}_+^{n\times n}$, is an example of an into preserver of complete positivity, but there are others, and no known complete characterization of all such preservers. Adding a requirement to preserve also cp-rank narrows the list of possible preservers. More generally, one can ask which linear transformations from $\mathcal{S}_m$ to $\mathcal{S}_n$ preserve (complete positivity and) cp-rank, and which preserve complete positivity and cp-rank for matrices with cp-rank k, for a given k.

A linear transformation $\Phi : \mathcal{S}_m \to \mathcal{S}_n$ is a *cp-rank-k preserver* if for every $X \in \mathcal{CP}_m$ with $\operatorname{cpr} X = k$, $\Phi(X) \in \mathcal{CP}_n$ and $\operatorname{cpr} \Phi(X) = k$. The transformation $\Phi : \mathcal{S}_m \to \mathcal{S}_n$ is a *cp-rank preserver* if for every $X \in \mathcal{CP}_m$, $\Phi(X) \in \mathcal{CP}_n$ and $\operatorname{cpr} \Phi(X) = \operatorname{cpr} X$.

Example 5.58. One example of a cp-rank-1 preserver is $\Phi : \mathcal{S}_m \to \mathcal{S}_n$ defined by

$$\Phi(X) = S^T XS,$$

where $S \in \mathbb{R}_+^{m\times n}$ has no zero row. This transformation preserves complete positivity: if $X = YY^T$ where $Y \geq 0$, then $\Phi(X) = (S^T Y)(S^T Y)^T$ is a CP factorization of $\Phi(X)$, which shows also that $\operatorname{cpr} \Phi(X) \leq \operatorname{cpr} X$. The assumption that S has no zero row implies that $S^T \mathbf{x} \neq \mathbf{0}$ for every nonzero $\mathbf{x} \in \mathbb{R}_+^m$, hence $S^T \mathbf{x}\mathbf{x}^T S \neq 0$, which guarantees that $\operatorname{cpr} \Phi(\mathbf{x}\mathbf{x}^T) = 1$.

Another example of a cp-rank-1 preserver is a linear transformation $\Psi : \mathcal{S}_m \to \mathcal{S}_n$ whose image is $\operatorname{span}\{\mathbf{u}\mathbf{u}^T\}$ for some nonzero $\mathbf{u} \in \mathbb{R}_+^n$, such that $\Psi(\mathbf{x}\mathbf{x}^T) \geq 0$ and $\Psi(\mathbf{x}\mathbf{x}^T) \neq 0$ for every $\mathbf{x} \in \mathbb{R}_+^m$. Such Ψ has the form

$$\Psi(X) = \psi(X)\mathbf{u}\mathbf{u}^T,$$

where $\psi : \mathcal{S}_m \to \mathbb{R}$ is a linear functional, i.e., $\psi(X) = \langle X, A\rangle$ for some $A \in \mathcal{S}_m$. The requirement that $\psi(\mathbf{x}\mathbf{x}^T) \geq 0$ for every $\mathbf{x} \in \mathbb{R}_+^m$ implies that $A \in \mathcal{COP}_m$, and $\psi(\mathbf{x}\mathbf{x}^T) \neq 0$ for every $\mathbf{x} \in \mathbb{R}_+^m$ implies that A is strictly copositive. That is, such Ψ is given by

$$\Psi(X) = \langle X, A\rangle \mathbf{u}\mathbf{u}^T,$$

where $\mathbf{u} \in \mathbb{R}_+^n$ is nonzero and A is strictly copositive.

Note that both Φ and Ψ are into preservers of completely positivity, but Ψ does not preserve any cp-rank greater than 1, and is therefore not a cp-rank preserver. The linear transformation Φ may be a cp-rank preserver, depending on S.

Remark 5.59. A linear transformation $\Phi : \mathcal{S}_m \to \mathcal{S}_n$ may be cp-rank-preserver only if $m \leq n$. This follows from the fact that p_n, the maximum cp-rank of $n \times n$ completely positive matrices, is strictly increasing in n (see Exercise 4.20).

Remark 5.60. Any cp-rank-1 preserver $\Phi : \mathcal{S}_m \to \mathcal{S}_n$ is an into preserver of complete positivity, and satisfies for every $X \in \mathcal{CP}_m$

$$\operatorname{cpr} \Phi(X) \leq \operatorname{cpr} X. \tag{5.6}$$

This is because if $X \in \mathcal{CP}_n$ and $X = \sum_{i=1}^k \mathbf{b}_i \mathbf{b}_i^T$ is a minimal CP representation of X, then the cp-rank-1 preserver satisfies $\Phi(X) = \sum_{i=1}^k \Phi(\mathbf{b}_i \mathbf{b}_i^T)$, yielding a CP representation of $\Phi(X)$.

A linear transformation $\Phi : \mathcal{S}_m \to \mathcal{S}_n$ that preserves complete positivity and satisfies (5.6) for every $X \in \mathcal{CP}_m$ is called *cp-rank-nonincreasing.*

The results on cp-rank preservers rely on results on rank preservers. A linear transformation $\Phi : \mathcal{S}_m \to \mathcal{S}_n$ is *rank-k-nonincreasing* if for every $X \in \mathcal{S}_m$ such that $\operatorname{rank} X = k$, $\operatorname{rank} \Phi(X) \leq k$. We will need the following lemma, which is the real version of a more general result.

Lemma 5.61. *Let $\Phi : \mathcal{S}_m \to \mathcal{S}_n$ be rank-1 nonincreasing linear transformation. Then one of the following holds: the image of Φ is contained in* $\operatorname{span}\{\mathbf{u}\mathbf{u}^T\}$ *for some nonzero $\mathbf{u} \in \mathbb{R}^n$, or $\Phi(X) = S^T X S$ for some $S \in \mathbb{R}^{m \times n}$, or $\Phi(X) = -S^T X S$ for some $S \in \mathbb{R}^{m \times n}$.*

A connection between these notions of cp-rank preservers and rank preservers is demonstrated in the following proposition.

Proposition 5.62. *Let $\Phi : \mathcal{S}_m \to \mathcal{S}_n$ a linear transformation that is a cp-rank-1 preserver. Then Φ is rank-1-nonincreasing.*

Proof. Any rank 1 matrix in $\mathcal{S}_m$ has the form $t\mathbf{v}\mathbf{v}^T$, where $\mathbf{v} \in \mathbb{R}^m$ is a nonzero vector and $t \neq 0$. As $\Phi(t\mathbf{v}\mathbf{v}^T) = t\Phi(\mathbf{v}\mathbf{v}^T)$ have the same rank as $\Phi(\mathbf{v}\mathbf{v}^T)$ for $t \neq 0$, it suffices to show that the latter has rank at most 1. If $\mathbf{v} \in \mathbb{R}^m_+$, or $-\mathbf{v} \in \mathbb{R}^m_+$, then $\mathbf{v}\mathbf{v}^T$ is a rank 1 completely positive matrix, and by the assumption on Φ, so is $\Phi(\mathbf{v}\mathbf{v}^T)$. Hence we consider the case that $\mathbf{v}$ has both positive and negative entries.

Let $\mathbf{u} = \mathbf{v}_+$ and $\mathbf{w} = \mathbf{v}_-$. Then $\mathbf{u}$ and $\mathbf{w}$ are nonzero nonnegative vectors with disjoint supports, and $\mathbf{v} = \mathbf{u} - \mathbf{w}$. Define $\mathbf{v}(x) = \mathbf{u} + x\mathbf{w}$, and $A(x) = \Phi(\mathbf{v}(x)\mathbf{v}(x)^T)$.

For $x > 0$ the vector $\mathbf{v}(x)$ is nonnegative, hence $\mathbf{v}(x)\mathbf{v}(x)^T$ is a completely positive matrix with rank and cp-rank equal to 1. By the assumption on Φ, the matrix $A(x) \in \mathcal{S}_n$ is a completely positive matrix with cp-rank 1, and therefore also rank $A(x) = 1$. For every $i < j$ and $p < q$ we get that $\det(A(x)[i,j|p,q]) = 0$. Since each entry of $A(x)$ is a linear functional in the entry of $\mathbf{v}(x)\mathbf{v}(x)^T$, and these entries are polynomials of degree at most 2, $\det(A(x)[i,j|p,q])$ is a polynomial in x with degree at most 4. As this polynomial is zero on $(0,\infty)$, it has to be the zero polynomial. Hence $\det(A(x)[i,j|p,q]) = 0$ for every $i < j$ and $p < q$, for every $x \in \mathbb{R}$. In particular this holds for $x = -1$, implying that the rank of $A(-1) = \Phi(\mathbf{v}\mathbf{v}^T)$ is at most 1. □

We are now able to characterize the cp-rank-1 preservers.

Theorem 5.63. *Let $\Phi : \mathcal{S}_m \to \mathcal{S}_n$ be a linear transformation. Then Φ is a cp-rank-1 preserver if and only if either $\Phi(X) = \langle X, A\rangle \mathbf{u}\mathbf{u}^T$ for some strictly copositive $A \in \mathcal{COP}_m$ and some nonzero $\mathbf{u} \in \mathbb{R}^n_+$, or $\Phi(X) = S^T XS$ for some $S \in \mathbb{R}^{m\times n}_+$ with no zero row.*

Proof. We have already seen, in Example 5.58, that these two types of transformations are cp-rank-1 preservers. Hence we prove that every cp-rank-1 preserver has one of these two forms.

First suppose Φ is a cp-rank-1 preserver whose image is contained span$\{M\}$ for some $M \in \mathcal{S}_n$ with rank 1. As Φ maps (cp-)rank 1 completely positive matrix to (cp-)rank 1 completely positive matrices, either M or $-M$ has to be completely positive. Hence we may assume that $M = \mathbf{u}\mathbf{u}^T$ for some nonzero $\mathbf{u} \in \mathbb{R}^n_+$. In Example 5.58 we explained why in this case Φ has to have the form $\Phi(X) = \langle X, A\rangle \mathbf{u}\mathbf{u}^T$ for some strictly copositive $A \in \mathcal{COP}_m$.

So now suppose the image of the cp-rank-1 preserver Φ is not a subspace of span$\{M\}$ for some $M \in \mathcal{S}_n$ with rank 1. By Proposition 5.62, the transformation Φ is rank-1-nonincreasing, and by Lemma 5.61, either $\Phi(X) = S^T XS$ for some $S \in \mathbb{R}^{m\times n}$, or $\Phi(X) = -S^T XS$ for some such S. In either case, $S \neq 0$ since $\Phi(\mathbf{x}\mathbf{x}^T) \neq 0$ for a nonzero $\mathbf{x} \in \mathbb{R}^n_+$. However, the second option is impossible, since $\Phi(\mathbf{x}\mathbf{x}^T)$ has a positive diagonal entry, and in $-S^T\mathbf{x}\mathbf{x}^T S$ all diagonal entries are nonpositive. Hence $\Phi(X) = S^T XS$. For any $1 \le i \le m$ and $1 \le p, q \le n$ we have

$$(\Phi(E_{ii}))_{pq} = (S^T E_{ii} S)_{pq} = s_{ip}s_{iq}.$$

Since $\Phi(E_{ii})$ is completely positive, $s_{ip}s_{iq} \ge 0$ for every p and q. Since for at least one p the p,p entry of $\Phi(E_{ii})$ is positive, we get that row i of S is nonzero, and either all entries in the row are nonnegative or all entries are nonpositive.

It remains to show that either $S \geq 0$ or $S \leq 0$. As $(-S)^T X(-S) = S^T XS$, this will prove that we may take $S \geq 0$. If $m = 1$ there is nothing to prove. So let $m \geq 2$ and suppose, for the sake of contradiction, that in row i there are nonnegative elements, and in row j there are nonpositive elements. For $t > 0$ denote $\mathbf{x}(t) = \mathbf{e}_i + t\mathbf{e}_j$, and $A(t) = \mathbf{x}(t)\mathbf{x}(t)^T$. For every $1 \leq p, q \leq n$ and $t \geq 0$,

$$(\Phi(A(t)))_{pq} = (S^T \mathbf{x}(t)\mathbf{x}(t)^T S)_{pq} = (s_{ip} + ts_{jp})(s_{iq} + ts_{jq}) \geq 0. \qquad (5.7)$$

The two rows are linearly independent, otherwise for some $t > 0$ we would get that $s_{ip} + ts_{jp} = 0$ for every $1 \leq p \leq n$, and (5.7) would imply that $\Phi(A(t)) = 0$, while $A(t)$ is a completely positive matrix with cp-rank 1. If $s_{ip} > 0$, choose $t > 0$ sufficiently small so that $s_{ip} + ts_{jp} > 0$. Then for every $1 \leq q \leq n$ equation (5.7) implies $s_{iq} + ts_{jq} \geq 0$, hence the support of row j is contained in the support of row i. A similar argument for large t shows that the support of row i is contained in that of row j. Hence the two supports are equal. For each p in this joint support, there exists a positive t_p such that $s_{ip} + ts_{jp} > 0$ for $0 < t < t_p$, and $s_{ip} + ts_{jp} < 0$ for $t > t_p$. As the two rows are linearly independent there exists $t > 0$ for which we may find $1 \leq p \neq q \leq n$ such that $s_{ip} + ts_{jp} > 0$ and $s_{iq} + ts_{jq} < 0$, but then $(\Phi(A(t))_{pq} < 0$, contradicting again the fact that Φ is a cp-rank 1 preserver. □

For a linear transformation to preserve both cp-rank-1 and cp-rank-2, S has to have full row rank.

Theorem 5.64. *Let $\Phi : \mathcal{S}_m \to \mathcal{S}_n$, $m \geq 2$. Then Φ is a cp-rank-1 and cp-rank-2 preserver if and only if $m \leq n$ and $\Phi(X) = S^T XS$ for some $S \in \mathbb{R}_+^{m\times n}$ with rank m.*

Proof. For the "if" part, suppose $S \in \mathbb{R}_+^{m\times n}$ has rank m and $\Phi : \mathcal{S}_m \to \mathcal{S}_n$ is defined by $\Phi(X) = S^T XS$. Then by Theorem 5.63 the transformation Φ is a cp-rank-1 preserver. For any two linearly independent vectors $\mathbf{x}, \mathbf{y} \in \mathbb{R}_+^m$, the vectors $S^T\mathbf{x}, S^T\mathbf{y}$ are two linearly independent vectors in $\mathbb{R}_+^n$. Hence

$$\Phi(\mathbf{x}\mathbf{x}^T + \mathbf{y}\mathbf{y}^T) = (S^T\mathbf{x})(S^T\mathbf{x})^T + (S^T\mathbf{y})(S^T\mathbf{y})^T$$

is completely positive and has rank 2, and therefore also cp-rank 2. That is, Φ preserves also cp-rank-2.

Now we prove the "only if" part. Let $\Phi : \mathcal{S}_m \to \mathcal{S}_n$ be a cp-rank-1 and cp-rank-2 preserver. Since Φ preserves rank 2, there exists a matrix with cp-rank 2, and therefore rank 2, in the image of Φ. By Theorem 5.63, the transformation Φ has to be of the form $\Phi(X) = S^T XS$ for some $S \geq 0$ with no zero row.

Suppose $\operatorname{rank} S < m$. Then there exists a nonzero vector $\mathbf{v} \in \mathbb{R}^m$ such that $S^T\mathbf{v} = \mathbf{0}$. As Φ preserves cp-rank-1 and $\Phi(\mathbf{v}\mathbf{v}^T) = (S^T\mathbf{v})(S^T\mathbf{v})^T$, the vector $\mathbf{v}$ cannot be nonnegative or nonpositive. Let $\mathbf{u} = \mathbf{v}_+$ and $\mathbf{w} = \mathbf{v}_-$. Then $\mathbf{u}$ and $\mathbf{w}$ are nonzero nonnegative vectors with disjoint support, and $\mathbf{v} = \mathbf{u} - \mathbf{w}$. Hence $S^T\mathbf{u} = S^T\mathbf{w}$. The matrix $\mathbf{u}\mathbf{u}^T + \mathbf{w}\mathbf{w}^T$ is a completely positive matrix of rank 2, hence also cp-rank 2. But

$$\Phi(\mathbf{u}\mathbf{u}^T + \mathbf{w}\mathbf{w}^T) = (S^T\mathbf{u})(S^T\mathbf{u})^T + (S^T\mathbf{w})(S^T\mathbf{w})^T = 2(S^T\mathbf{u})(S^T\mathbf{u})^T$$

has cp-rank at most 1. A contradiction. Hence S has to have full row rank, which also forces $m \leq n$. □

Is it true that any cp-rank-1 and cp-rank-2 preserver $\Phi : \mathcal{S}_m \to \mathcal{S}_n$, $m \leq n$, is also a cp-rank preserver? It is definitely a cp-rank-3 preserver, so for $m = 3$, the answer is positive.

Theorem 5.65. *Let $\Phi : \mathcal{S}_m \to \mathcal{S}_n$, $m \geq 3$, be a cp-rank-1 and cp-rank-2 preserver. Then Φ is a cp-rank-3 preserver. In particular, if $m = 3$, Φ is a cp-rank preserver.*

Proof. By Theorem 5.64, $\Phi(X) = S^T XS$ for some $S \geq 0$ with $\operatorname{rank} S = m$. By Remark 5.60, the transformation Φ is cp-rank-nonincreasing. Hence if $X \in \mathcal{CP}_m$ has $\operatorname{cpr} X = 3$, then $\operatorname{cpr} \Phi(X) \leq 3$. But such X has $\operatorname{rank} X \leq \operatorname{cpr} X = 3$, and cannot have $\operatorname{rank} X \leq 2$, since this would have implied $\operatorname{cpr} X \leq 2$. Hence $\operatorname{rank} X = 3$, and as $\operatorname{ns} S^T = \{\mathbf{0}\}$, also $\operatorname{rank}(S^T X) = 3$, and we get that

$$\operatorname{cpr} \Phi(X) \geq \operatorname{rank} \Phi(X) = \operatorname{rank}(S^T X)(S^T X)^T = \operatorname{rank}(S^T X) = 3.$$

Together with the previous inequality this yields $\operatorname{cpr} \Phi(X) = 3$. □

But cp-rank-1 and cp-rank-2 preservers do not necessarily preserve cp-rank-4.

Example 5.66. Let $\Phi : \mathcal{S}_4 \to \mathcal{S}_4$ be defined by $\Phi(X) = S^T XS$, where

$$S = \begin{pmatrix} 1 & 0 & 0 & 0 \\ 1 & 1 & 0 & 0 \\ 1 & 1 & 1 & 0 \\ 1 & 1 & 1 & 1 \end{pmatrix}.$$

As the nonnegative S is clearly invertible, Φ preservers cp-rank-1 and cp-rank-2. Now consider the completely positive

$$X = \begin{pmatrix} 2 & 1 & 0 & 1 \\ 1 & 2 & 1 & 0 \\ 0 & 1 & 2 & 1 \\ 1 & 0 & 1 & 2 \end{pmatrix},$$

which has $\operatorname{rank} X = 3$ and $\operatorname{cpr} X = 4$. Then

$$\Phi(X) = \begin{pmatrix} 16 & 12 & 8 & 4 \\ 12 & 10 & 7 & 3 \\ 8 & 7 & 6 & 3 \\ 4 & 3 & 3 & 2 \end{pmatrix} = \begin{pmatrix} 4 & 0 & 0 \\ 3 & 1 & 0 \\ 2 & 1 & 1 \\ 1 & 0 & 1 \end{pmatrix} \begin{pmatrix} 4 & 0 & 0 \\ 3 & 1 & 0 \\ 2 & 1 & 1 \\ 1 & 0 & 1 \end{pmatrix}^T$$

has cp-rank 3. That is, Φ is not cp-rank-4 preserver.

When considering the case $m = n$, clearly any automorphism of $\mathcal{CP}_n$, $\Phi(X) = M^T X M$, $M \geq 0$ monomial, is also a cp-rank preserver. Are these the only cp-rank preservers from $\mathcal{S}_n$ to itself, $n \geq 4$?

Exercises

5.16. Let $\mathbf{b}_1, \ldots, \mathbf{b}_5$ be zeros of a Hildebrand matrix with supports

$$\{1,2,3\}, \{2,3,4\}, \{3,4,5\}, \{4,5,1\}, \{5,1,2\},$$

then $\{\mathbf{b}_1, \ldots, \mathbf{b}_5\}$ is linearly independent.
Hint: Theorems 2.163 and 2.67.

5.17. Let

$$\mathbf{b}_i = s_i \mathbf{u}_i + t_i \mathbf{u}_{i \dotplus 1},$$

where $\mathbf{u}_i = \mathbf{e}_i + \mathbf{e}_{i \dotplus 1}$ is the minimal zero of the Horn matrix with support $\{i, i \dotplus 1\}$. Show that $\{\mathbf{b}_1, \ldots, \mathbf{b}_5\}$ is linearly independent if and only if $\Pi_{i=1}^5 s_i + \Pi_{i=1}^5 t_i > 0$.
Hint: Let $W = \begin{pmatrix} \mathbf{u}_1 & \ldots & \mathbf{u}_5 \end{pmatrix}$ and $B = \begin{pmatrix} \mathbf{b}_1 & \ldots & \mathbf{b}_5 \end{pmatrix}$. Show that W is nonsingular and $B = WS$, where

$$S = \begin{pmatrix} s_1 & 0 & 0 & 0 & t_5 \\ t_1 & s_2 & 0 & 0 & 0 \\ 0 & t_2 & s_3 & 0 & 0 \\ 0 & 0 & t_3 & s_4 & 0 \\ 0 & 0 & 0 & t_4 & s_5 \end{pmatrix}.$$

See Exercise 4.58.

5.18. Complete the proof of Theorem 5.54.

Notes.

Theorem 5.39 and Corollary 5.40 are due to [Shaked-Monderer (2020)]. Theorem 5.41 and the discussion on maximal faces of $\mathcal{CP}_n$ and their dimensions is due to [Dickinson (2011)]. The first example of non-exposed faces of $\mathcal{CP}_5$, Example 5.50, is due to [Zhang (2018)], and the classification of subfaces of exposed faces defined by an exceptional matrix in $\mathcal{COP}_5$ is due to [Zhang (2019)].

A general completely positive cone has the form $\mathcal{K} = \text{cone}\{\mathbf{u}\mathbf{u}^T \mid \mathbf{u} \in K\}$, where K is a closed convex cone in $\mathbb{R}^n$. The automorphisms of these cones were characterized in [Gowda *et al.* (2013)], and the corresponding Lie algebra as well. Their relation to the automorphisms of K was proved: Φ is an automorphism of $\mathcal{K}$ if and only if $\Phi(X) = Q^T X Q$, where $Q \in \mathbb{R}^{n\times n}$ is the matrix of an automorphism of K. The automorphisms of the dual cone, of matrices copositive with respect to K, was derived by the duality argument employed in this section. The proof uses Lemma 5.61. Here we reversed the order, and used the same duality principle with the description of the automorphisms of $\mathcal{COP}_n$ that was proved directly in the previous section.

The discussion of preservers of the cp-rank is partly based on [Song *et al.* (2016)], see reference there to the general preserver problems, and in particular to the rank preserver result of Lemma 5.61. However, there is an error in [Song *et al.* (2016)], and we present here just the correct part (and an example disproving the general characterization claimed in that paper).

Still open: *Characterize all cp-rank preservers from $\mathcal{S}_m$ to $\mathcal{S}_n$, $m \leq n$.*

Bibliography

Afonin, A., Dickinson, P. J. C., and Hildebrand, R. (2020). The extreme rays of the 6×6 copositive cone, *Journal of Global Optimization* doi: 10.1007/s10898-020-00930-y.

Agler, J., Helton, J. W., McCullough, S., and Rodman, L. (1988). Positive semidefinite matrices with a given sparsity pattern, *Linear Algebra and its Applications* **107**, pp. 101–149.

Alon, N. and Yuster, R. (1995). The 123 theorem and its extensions, *Journal of Combinatorial Theory. Series A* **72**, 2, pp. 322–331.

Andersson, L.-E., Chang, G. Z., and Elfving, T. (1995). Criteria for copositive matrices using simplices and barycentric coordinates, *Linear Algebra and its Applications* **220**, pp. 9–30.

Ando, T. (1987). Totally positive matrices, *Linear Algebra and its Applications* **90**, pp. 165–219.

Ando, T. (1991). *Completely positive matrices*, lecture notes, The University of Wisconsin, Madison.

Anitescu, M. and Potra, F. A. (1997). Formulating dynamic multi-rigid-body contact problems with friction as solvable linear complementarity problems, *Nonlinear Dynamics* **14**, 3, pp. 231–247.

Baccari, A. and Naffouti, M. (2016). Copositivity and sparsity relations using spectral properties, *Journal of Optimization Theory and Applications* **171**, 3, pp. 998–1007.

Bapat, R. B. and Raghavan, T. E. S. (1997). *Nonnegative matrices and applications*, *Encyclopedia of Mathematics and its Applications*, Vol. 64 (Cambridge University Press, Cambridge).

Barioli, F. (1998a). Completely positive matrices with a book-graph, *Linear Algebra and its Applications* **277**, 1-3, pp. 11–31.

Barioli, F. (1998b). Decreasing diagonal elements in completely positive matrices, *Rendiconti del Seminario Matematico della Università di Padova. The Mathematical Journal of the University of Padova* **100**, pp. 13–25.

Barioli, F. (2001a). Chains of dog-ears for completely positive matrices, *Linear Algebra and its Applications* **330**, pp. 49–66.

Barioli, F. (2001b). *Duality and graph-theoretic tools for complete positivity of matrices*, Ph.D. thesis, Padova university, Padova, Italy.

Barioli, F. (2002). Complete positivity of small and large accute sets of vectors, Talk at the 10th ILAS conference in Auburn.

Barioli, F. and Berman, A. (2003). The maximal cp-rank of rank k completely positive matrices, *Linear Algebra and its Applications* **363**, pp. 17–33, special issue on nonnegative matrices, M-matrices and their generalizations (Oberwolfach, 2000).

Barrett, W. W., Johnson, C. R., and Loewy, R. (1996). The real positive definite completion problem: cycle completability, *Memoirs of the American Mathematical Society* **122**, 584, pp. viii+69.

Barrett, W. W., Johnson, C. R., and Loewy, R. (1998). Critical graphs for the positive definite completion problem, *SIAM Journal on Matrix Analysis and Applications* **20**, 1, pp. 117–130.

Baston, V. J. D. (1968/1969). Extreme copositive quadratic forms, *Polska Akademia Nauk. Instytut Matematyczny. Acta Arithmetica* **15**, pp. 319–327.

Baumert, L. D. (1965). *Extreme copositive quadratic forms*, Ph.D. thesis, California Institute of Technology, Pasadena, California.

Baumert, L. D. (1966). Extreme copositive quadratic forms, *Pacific Journal of Mathematics* **19**, pp. 197–204.

Baumert, L. D. (1967). Extreme copositive quadratic forms. II, *Pacific Journal of Mathematics* **20**, pp. 1–20.

Beardon, A. F. (1979). *Complex analysis* (John Wiley & Sons, Ltd., Chichester), the argument principle in analysis and topology, A Wiley-Interscience Publication.

Ben-Israel, A. and Greville, T. N. E. (2003). *Generalized inverses, CMS Books in Mathematics/Ouvrages de Mathématiques de la SMC*, Vol. 15, 2nd edn. (Springer-Verlag, New York), theory and applications.

Benatti, F., Floreanini, R., and Romano, R. (2002). Complete positivity in dissapative quantum dynamics, in P. Garbaczewski and R. Olkiewicz (eds.), *Dynamics of Dissipation*, Vol. 597 (Springer, Berlin), pp. 283–304.

Berman, A. (1973). *Cones, matrices and mathematical programming* (Springer-Verlag, Berlin-New York), lecture Notes in Economics and Mathematical Systems, Vol. 79.

Berman, A. (1988). Complete positivity, *Linear Algebra and its Applications* **107**, pp. 57–63.

Berman, A. (1993). Completely positive graphs, in *Combinatorial and graph-theoretical problems in linear algebra (Minneapolis, MN, 1991), IMA Vol. Math. Appl.*, Vol. 50 (Springer, New York), pp. 229–233.

Berman, A., Dür, M., and Shaked-Monderer, N. (2015). Open problems in the theory of completely positive and copositive matrices, *Electronic Journal of Linear Algebra* **29**, pp. 46–58.

Berman, A., Dür, M., Shaked-Monderer, N., and Witzel, J. (2016). Cutting planes for semidefinite relaxations based on triangle-free subgraphs, *Optimization Letters* **10**, 3, pp. 433–446.

Berman, A. and Grone, R. (1988). Bipartite completely positive matrices, *Mathematical Proceedings of the Cambridge Philosophical Society* **103**, 2, pp. 269–276.

Berman, A. and Gueron, S. (2002). On the inverse of the hilbert matrix, *The Mathematical Gazette* **86**, pp. 274–277.

Berman, A. and Hershkowitz, D. (1987). Combinatorial results on completely positive matrices, *Linear Algebra and its Applications* **95**, pp. 111–125.

Berman, A., King, C., and Shorten, R. (2012). A characterisation of common diagonal stability over cones, *Linear and Multilinear Algebra* **60**, 10, pp. 1117–1123.

Berman, A. and Plemmons, R. J. (1994). *Nonnegative matrices in the mathematical sciences*, *Classics in Applied Mathematics*, Vol. 9 (Society for Industrial and Applied Mathematics (SIAM), Philadelphia, PA), revised reprint of the 1979 original.

Berman, A. and Rothblum, U. G. (2006). A note on the computation of the CP-rank, *Linear Algebra and its Applications* **419**, 1, pp. 1–7.

Berman, A. and Shaked-Monderer, N. (1998). Remarks on completely positive matrices, *Linear and Multilinear Algebra* **44**, 2, pp. 149–163.

Berman, A. and Shaked-Monderer, N. (2003). *Completely positive matrices* (World Scientific Publishing Co., Inc., River Edge, NJ).

Berman, A. and Shasha, D. (2012). Completely positive house matrices, *Linear Algebra and its Applications* **436**, 1, pp. 12–26.

Berman, A. and Xu, C. (2004). 5×5 completely positive matrices, *Linear Algebra and its Applications* **393**, pp. 55–71.

Berman, A. and Xu, C. (2005). $\{0, 1\}$ completely positive matrices, *Linear Algebra and its Applications* **399**, pp. 35–51.

Berman, A. and Xu, C. (2007). Uniform and minimal $\{0, 1\}$-cp matrices, *Linear and Multilinear Algebra* **55**, 5, pp. 439–456.

Bertsekas, D. P. (2016). *Nonlinear programming*, 3rd edn., Athena Scientific Optimization and Computation Series (Athena Scientific, Belmont, MA).

Bisht, K. and Sivakumar, K. C. (2017). Inheritance and inverse monotonicity properties of copositive matrices, *Linear and Multilinear Algebra* **65**, 5, pp. 897–908.

Bomze, I. M. (1987). Remarks on the recursive structure of copositivity, *Journal of Information & Optimization Sciences* **8**, 3, pp. 243–260.

Bomze, I. M. (1989). Copositivity and optimization, in *XII Symposium on Operations Research (Passau, 1987)*, *Methods Oper. Res.*, Vol. 58 (Athenäum/Hain/Hanstein, Königstein), pp. 27–35.

Bomze, I. M. (1996). Block pivoting and shortcut strategies for detecting copositivity, *Linear Algebra and its Applications* **248**, pp. 161–184.

Bomze, I. M. (2008). Perron-Frobenius property of copositive matrices, and a block copositivity criterion, *Linear Algebra and its Applications* **429**, 1, pp. 68–71.

Bomze, I. M. (2012). Copositive optimization—recent developments and applications, *European Journal of Operational Research* **216**, 3, pp. 509–520.

Bomze, I. M., Dickinson, P. J. C., and Still, G. (2015a). The structure of completely positive matrices according to their CP-rank and CP-plus-rank, *Linear Algebra and its Applications* **482**, pp. 191–206.

Bomze, I. M., Dür, M., de Klerk, E., Roos, C., Quist, A. J., and Terlaky, T. (2000). On copositive programming and standard quadratic optimization problems, *Journal of Global Optimization* **18**, 4, pp. 301–320, gO'99 Firenze.

Bomze, I. M., Frommlet, F., and Locatelli, M. (2010). Copositivity cuts for improving SDP bounds on the clique number, *Mathematical Programming* **124**, 1-2, Ser. B, pp. 13–32.

Bomze, I. M., Schachinger, W., and Uchida, G. (2012). Think co(mpletely) positive! Matrix properties, examples and a clustered bibliography on copositive optimization, *Journal of Global Optimization* **52**, 3, pp. 423–445.

Bomze, I. M., Schachinger, W., and Ullrich, R. (2014). From seven to eleven: completely positive matrices with high cp-rank, *Linear Algebra and its Applications* **459**, pp. 208–221.

Bomze, I. M., Schachinger, W., and Ullrich, R. (2015b). New lower bounds and asymptotics for the cp-rank, *SIAM Journal on Matrix Analysis and Applications* **36**, 1, pp. 20–37.

Bondy, J. A. and Murty, U. S. R. (2008). *Graph theory*, *Graduate Texts in Mathematics*, Vol. 244 (Springer, New York), ISBN 978-1-84628-969-9.

Booth, M., Hackney, P., Harris, B., Johnson, C. R., Lay, M., Mitchell, L. H., Narayan, S. K., Pascoe, A., Steinmetz, K., Sutton, B. D., and Wang, W. (2008). On the minimum rank among positive semidefinite matrices with a given graph, *SIAM Journal on Matrix Analysis and Applications* **30**, 2, pp. 731–740.

Bose, S. and Slud, E. (1995). Maximin efficiency-robust tests and some extensions, *Journal of Statistical Planning and Inference* **46**, 1, pp. 105–121.

Brandts, J. and Křížek, M. (2016). Factorization of cp-rank-3 completely positive matrices, *Czechoslovak Math. J.* **66(141)**, 3, pp. 955–970.

Brouwer, A. E. and Haemers, W. H. (2012). *Spectra of graphs*, Universitext (Springer, New York).

Burer, S. (2009). On the copositive representation of binary and continuous nonconvex quadratic programs, *Mathematical Programming. A Publication of the Mathematical Programming Society* **120**, 2, Ser. A, pp. 479–495.

Burer, S., Anstreicher, K. M., and Dür, M. (2009). The difference between 5×5 doubly nonnegative and completely positive matrices, *Linear Algebra and its Applications* **431**, 9, pp. 1539–1552.

Cedolin, M. and Salce, L. (2016). Completely positive matrices of order 5 with $\widehat{CP}$-graph, *Note di Matematica* **36**, 1, pp. 123–132.

Chang, G. Z. and Sederberg, T. W. (1994). Nonnegative quadratic Bézier triangular patches, *Computer Aided Geometric Design* **11**, 1, pp. 113–116.

Choi, M.-D. (1983). Tricks or treats with the hilbert matrix, *The American Mathematical Monthly* **90**, 5, pp. 301–312.

Chudnovsky, M., Robertson, N., Seymour, P., and Thomas, R. (2006). The strong perfect graph theorem, *Annals of Mathematics. Second Series* **164**, 1, pp. 51–229.

Cohen, J. E. and Rothblum, U. G. (1993). Nonnegative ranks, decompositions, and factorizations of nonnegative matrices, *Linear Algebra and its Applications* **190**, pp. 149–168.

Cottle, R. W. (1974). Manifestations of the Schur complement, *Linear Algebra and its Applications* **8**, pp. 189–211.

Cottle, R. W. and Dantzig, G. B. (1968). Complementary pivot theory of mathematical programming, in *Mathematics of the Decision Sciences, Part 1 (Seminar, Stanford, Calif., 1967)* (Amer. Math. Soc., Providence, R.I.), pp. 115–136.

Cottle, R. W., Habetler, G. J., and Lemke, C. E. (1970a). On classes of copositive matrices, *Linear Algebra and its Applications* **3**, pp. 295–310.

Cottle, R. W., Habetler, G. J., and Lemke, C. E. (1970b). Quadratic forms semidefinite over convex cones, in *Proceedings of the Princeton Symposium on Mathematical Programming (Princeton Univ., 1967)*, pp. 551–565.

Crabtree, D. E. and Haynsworth, E. V. (1969). An identity for the Schur complement of a matrix, *Proceedings of the American Mathematical Society* **22**, pp. 364–366.

Cryer, C. W. (1976). Some properties of totally positive matrices, *Linear Algebra and its Applications* **15**, 1, pp. 1–25.

Dahl, G. and Haufmann, T. A. (2016). Zero-one completely positive matrices and the $\mathcal{A}(R, S)$ classes, *Special Matrices* **4**, pp. 296–304.

de Klerk, E. and Pasechnik, D. V. (2002). Approximation of the stability number of a graph via copositive programming, *SIAM Journal on Optimization* **12**, 4, pp. 875–892.

DeMarr, R. (1972). Nonnegative matrices with nonnegative inverses, *Proceedings of the American Mathematical Society* **35**, 1, pp. 307–308.

Di Virgilio, S. (2019). Completely positive matrices of order 5 with a nearly $\widehat{CP}$-graph, *Note di Matematica* **39**, 1, pp. 89–106.

Diananda, P. H. (1961). On a conjecture of L. J. Mordell regarding an inequality involoving quadratic forms, *Journal of the London Mathematical Society. Second Series* **36**, pp. 185–192.

Diananda, P. H. (1962). On non-negative forms in real variables some or all of which are non-negative, *Proceedings of the Cambridge Philosophical Society* **58**, pp. 17–25.

Dickinson, P. J. C. (2010). An improved characterisation of the interior of the completely positive cone, *Electronic Journal of Linear Algebra* **20**, pp. 723–729.

Dickinson, P. J. C. (2011). Geometry of the copositive and completely positive cones, *Journal of Mathematical Analysis and Applications* **380**, 1, pp. 377–395.

Dickinson, P. J. C. (2013). *The Copositive Cone, the Completely Positive Cone and their Generalizations*, Ph.D. thesis, University of Groningen.

Dickinson, P. J. C. (2017a). Private communication.

Dickinson, P. J. C. (2017b). Erratum to: On the DJL conjecture for order 6, *Operators and Matrices* **11**, 4, pp. 1197–1200.
Dickinson, P. J. C. (2019). A new certificate for copositivity, *Linear Algebra and its Applications* **569**, pp. 15–37.
Dickinson, P. J. C. and de Zeeuw, R. (2020). Generating irreducible copositive matrices using the stable set problem, *Discrete Applied Mathematics* doi: https://doi.org/10.1016/j.dam.2020.04.013.
Dickinson, P. J. C. and Dür, M. (2012). Linear-time complete positivity detection and decomposition of sparse matrices, *SIAM Journal on Matrix Analysis and Applications* **33**, 3, pp. 701–720.
Dickinson, P. J. C., Dür, M., Gijben, L., and Hildebrand, R. (2013a). Irreducible elements of the copositive cone, *Linear Algebra and its Applications* **439**, 6, pp. 1605–1626.
Dickinson, P. J. C., Dür, M., Gijben, L., and Hildebrand, R. (2013b). Scaling relationship between the copositive cone and Parrilo's first level approximation, *Optimization Letters* **7**, 8, pp. 1669–1679.
Dickinson, P. J. C. and Gijben, L. (2014). On the computational complexity of membership problems for the completely positive cone and its dual, *Computational Optimization and Applications* **57**, 2, pp. 403–415.
Dickinson, P. J. C. and Hildebrand, R. (2016). Considering copositivity locally, *Journal of Mathematical Analysis and Applications* **437**, 2, pp. 1184–1195.
Diestel, R. (2018). *Graph theory*, *Graduate Texts in Mathematics*, Vol. 173, 5th edn. (Springer, Berlin).
Ding, C., He, X., and Simon, H. D. (2005). On the equivalence of nonnegative matrix factorization and spectral clustering, in *Proceedings of the 2005 SIAM International Conference on Data Mining*, pp. 606–610.
Dolžan, D. and Oblak, P. (2018). Bounds for the completely positive rank of a symmetric matrix over a tropical semiring, *Electronic Journal of Linear Algebra* **34**, pp. 152–162.
Dong, H. and Anstreicher, K. (2010). A note on "5×5 completely positive matrices", *Linear Algebra and its Applications* **433**, 5, pp. 1001–1004.
Dong, Z., Li, J., and Li, W. V. (2015). A note on distribution-free symmetrization inequalities, *Journal of Theoretical Probability* **28**, 3, pp. 958–967.
Drew, J. H. and Johnson, C. R. (1996). The no long odd cycle theorem for completely positive matrices, in *Random discrete structures (Minneapolis, MN, 1993)*, *IMA Vol. Math. Appl.*, Vol. 76 (Springer, New York), pp. 103–115.
Drew, J. H. and Johnson, C. R. (1998). The completely positive and doubly nonnegative completion problems, *Linear and Multilinear Algebra* **44**, 1, pp. 85–92.
Drew, J. H., Johnson, C. R., Kilner, S. J., and McKay, A. M. (2000). The cycle completable graphs for the completely positive and doubly nonnegative completion problems, *Linear Algebra and its Applications* **313**, 1-3, pp. 141–154.
Drew, J. H., Johnson, C. R., and Lam, F. (2001). Complete positivity of matrices of special form, *Linear Algebra and its Applications* **327**, 1-3, pp. 121–130.

Drew, J. H., Johnson, C. R., and Loewy, R. (1994). Completely positive matrices associated with M-matrices, *Linear and Multilinear Algebra* **37**, 4, pp. 303–310.

Drury, S. (2020). The triangle graph t_6 is not spn, *Electronic Journal of Linear Algebra* **36**, pp. 90–93.

Dür, M. (2010). Copositive programming – a survey, in M. Diehl, F. Glineur, E. Jarlebring, and W. Michiels (eds.), *Recent Advances in Optimization and its Applications in Engineering* (Springer-Verlag, Berlin Heidelberg), pp. 3–20.

Dür, M. and Still, G. (2008). Interior points of the completely positive cone, *Electronic Journal of Linear Algebra* **17**, pp. 48–53.

Dutour Sikirić, M., Schürmann, A., and Vallentin, F. (2017). Rational factorizations of completely positive matrices, *Linear Algebra and its Applications* **523**, pp. 46–51.

Dutour Sikirić, M., Schürmann, A., and Vallentin, F. (2020). A simplex algorithm for a rational cp-factorization, *Mathematical Programming.*

Elsner, L., Nabben, R., and Neumann, M. (1998). Orthogonal bases that lead to symmetric nonnegative matrices, *Linear Algebra and its Applications* **271**, pp. 323–343.

Fallat, S. M. and Johnson, C. R. (2011). *Totally nonnegative matrices*, Princeton Series in Applied Mathematics (Princeton University Press, Princeton, NJ).

FitzGerald, C. H. and Horn, R. A. (1977). On fractional Hadamard powers of positive definite matrices, *Journal of Mathematical Analysis and Applications* **61**, 3, pp. 633–642.

Furtado, S., Johnson, C. R., and Zhang, Y. (2019). Linear preservers of copositive matrices, *Linear and Multilinear Algebra* **0**, 0, pp. 1–10.

Gaddum, J. W. (1958). Linear inequalities and quadratic forms, *Pacific Journal of Mathematics* **8**, pp. 411–414.

Gantmacher, F. P. and Krein, M. G. (2002). *Oscillation matrices and kernels and small vibrations of mechanical systems*, revised edn. (AMS Chelsea Publishing, Providence, RI), translation based on the 1941 Russian original, Edited and with a preface by Alex Eremenko.

Gantmacher, F. R. (1959). *The theory of matrices. Vols. 1, 2*, Translated by K. A. Hirsch (Chelsea Publishing Co., New York).

Gasca, M. and Michelli, C. A. (eds.) (1996). *Total Positivity and Its Applications*, *Mathematics and Its Applications*, Vol. 359 (Springer Netherlands).

Godsil, C. and Royle, G. (2001). *Algebraic graph theory*, *Graduate Texts in Mathematics*, Vol. 207 (Springer-Verlag, New York).

Goldberg, F. (2003). On completely positive graphs and their complements, *Linear Algebra and its Applications* **371**, pp. 45–51.

Goldberg, F. and Shaked-Monderer, N. (2014). On the maximal angle between copositive matrices, *Electron. J. Linear Algebra* **27**, pp. 837–850.

Golumbic, M. C. (2004). *Algorithmic graph theory and perfect graphs*, *Annals of Discrete Mathematics*, Vol. 57, 2nd edn. (Elsevier Science B.V., Amsterdam), with a foreword by Claude Berge.

Gowda, M. S., Sznajder, R., and Tao, J. (2013). The automorphism group of a completely positive cone and its Lie algebra, *Linear Algebra and its Applications* **438**, 10, pp. 3862–3871.

Gray, L. J. and Wilson, D. G. (1980). Nonnegative factorization of positive semidefinite nonnegative matrices, *Linear Algebra and its Applications* **31**, pp. 119–127.

Groetzner, P. and Duer, M. (2020). A factorization method for completely positive matrices, *Linear Algebra and its Applications* **591**, pp. 1–24.

Grone, R., Johnson, C. R., de Sá, E. M., and Wolkowicz, H. (1984). Positive definite completions of partial Hermitian matrices, *Linear Algebra and its Applications* **58**, pp. 109–124.

Grötschel, M., Lovász, L., and Schrijver, A. (1988). *Geometric algorithms and combinatorial optimization*, *Algorithms and Combinatorics: Study and Research Texts*, Vol. 2 (Springer-Verlag, Berlin).

Hadeler, K. P. (1983). On copositive matrices, *Linear Algebra and its Applications* **49**, pp. 79–89.

Hadjicostas, P. (1997). Copositive matrices and Simpson's paradox, *Linear Algebra and its Applications* **264**, pp. 475–488.

Hahnloser, R. H. R., Seung, H. S., and Slotine, J.-J. (2003). Permitted and forbidden sets in symmetric threshold-linear networks, *Neural Computation* **15**, 3, pp. 621–638.

Hall, M., Jr. (1958). A survey of combinatorial analysis, in *Some aspects of analysis and probability*, Surveys in Applied Mathematics. Vol. 4 (John Wiley & Sons, Inc., New York; Chapman & Hall, Ltd., London), pp. 35–104.

Hall, M., Jr. (1998). *Combinatorial theory*, 2nd edn., Wiley Classics Library (John Wiley & Sons, Inc., New York), a Wiley-Interscience Publication.

Hall, M., Jr. and Newman, M. (1963). Copositive and completely positive quadratic forms, *Proceedings of the Cambridge Philosophical Society* **59**, pp. 329–339.

Hamilton-Jester, C. L. and Li, C.-K. (1996). Extreme vectors of doubly nonnegative matrices, *The Rocky Mountain Journal of Mathematics* **26**, 4, pp. 1371–1383.

Han, S.-P. and Mangasarian, O. L. (1984). Conjugate cone characterization of positive definite and semidefinite matrices, *Linear Algebra and its Applications* **56**, pp. 89–103.

Hannah, J. and Laffey, T. J. (1983). Nonnegative factorization of completely positive matrices, *Linear Algebra and its Applications* **55**, pp. 1–9.

Haynsworth, E. and Hoffman, A. J. (1969). Two remarks on copositive matrices, *Linear Algebra and its Applications* **2**, pp. 387–392.

Haynsworth, E. V. (1968). Determination of the inertia of a partitioned Hermitian matrix, *Linear Algebra and its Applications* **1**, 1, pp. 73–81.

Hildebrand, R. (2012). The extreme rays of the 5×5 copositive cone, *Linear Algebra and its Applications* **437**, 7, pp. 1538–1547.

Hildebrand, R. (2014). Minimal zeros of copositive matrices, *Linear Algebra and its Applications* **459**, pp. 154–174.

Hildebrand, R. (2015). Private Communication.

Hildebrand, R. (2017). Copositive matrices with circulant zero support set, *Linear Algebra and its Applications* **514**, pp. 1–46.

Hildebrand, R. (2018a). Extremal copositive matrices with minimal zero supports of cardinality two, *Linear and Multilinear Algebra* **66**, 11, pp. 2151–2155.

Hildebrand, R. (2018b). Extremal copositive matrices with zero supports of cardinality $n-2$, *Electronic Journal of Linear Algebra* **34**, pp. 28–34.

Hildebrand, R. (2020). On the algebraic structure of the copositive cone, *Optimization letters.*

Hildebrand, R. and Afonin, A. (2020). Essential components of extremal copositive matrices, *SIAM Journal on Matrix Analysis and Applications.*

Hiriart-Urruty, J.-B. and Seeger, A. (2010). A variational approach to copositive matrices, *SIAM Review* **52**, 4, pp. 593–629.

Hirschfeld, B. (2004). *Approximative Lösungen des Max-Cut-Problems mit semidefiniten Programmen*, Ph.D. thesis, University of Düsseldorf.

Hoffman, A. J. and Pereira, F. (1973). On copositive matrices with -1,0,1 entries, *Journal of Combinatorial Theory. Series A* **14**, pp. 302–309.

Hogben, L. (2007). The copositive completion problem: unspecified diagonal entries, *Linear Algebra and its Applications* **420**, 1, pp. 160–162.

Hogben, L. (ed.) (2014). *Handbook of linear algebra*, 2nd edn., Discrete Mathematics and its Applications (Boca Raton) (CRC Press, Boca Raton, FL).

Hogben, L., Johnson, C. R., and Reams, R. (2005). The copositive completion problem, *Linear Algebra and its Applications* **408**, pp. 207–211.

Hogben, L. and Shaked-Monderer, N. (2019). Spn graphs, *Electronic Journal of Linear Algebra* **35**, pp. 376–386.

Horn, R. A. (1967). On infinitely divisible matrices, kernels, and functions, *Zeitschrift für Wahrscheinlichkeitstheorie und Verwandte Gebiete* **8**, pp. 219–230.

Horn, R. A. (1969). The theory of infinitely divisible matrices and kernels, *Transactions of the American Mathematical Society* **136**, pp. 269–286.

Horn, R. A. and Johnson, C. R. (1994). *Topics in matrix analysis* (Cambridge University Press, Cambridge), corrected reprint of the 1991 original.

Horn, R. A. and Johnson, C. R. (2013). *Matrix analysis*, 2nd edn. (Cambridge University Press, Cambridge).

Ikramov, K. D. and Savel′eva, N. V. (2000). Conditionally definite matrices, *Journal of Mathematical Sciences (New York)* **98**, 1, pp. 1–50, algebra, 9.

Jacobson, D. H. (1974). Factorization of symmetric M-matrices, *Linear Algebra and its Applications* **9**, pp. 275–278.

Jacobson, D. H. (1976). A generalization of Finsler's theorem for quadratic inequalities and equalities, *Quaestiones Mathematicae* **1**, 1, pp. 19–28.

Jargalsaikhan, B. (2013). Indefinite copositive matrices with exactly one positive eigenvalue or exactly one negative eigenvalue, *Electronic Journal of Linear Algebra* **26**, pp. 754–761.

Jarre, F. and Schmallowsky, K. (2009). On the computation of C^* certificates, *Journal of Global Optimization. An International Journal Dealing with Theoretical and Computational Aspects of Seeking Global Optima and Their Applications in Science, Management and Engineering* **45**, 2, pp. 281–296.

Johnson, C. R. and Reams, R. (2005). Spectral theory of copositive matrices, *Linear Algebra and its Applications* **395**, pp. 275–281.

Johnson, C. R. and Reams, R. (2008). Constructing copositive matrices from interior matrices, *Electronic Journal of Linear Algebra* **17**, pp. 9–20.

Johnson, C. R. and Reams, R. B. (2016). Sufficient conditions to be exceptional, *Special Matrices* **4**, pp. 67–72.

Johnson, C. R. and Smith, R. L. (1996). The completion problem for M-matrices and inverse M-matrices, *Linear Algebra and its Applications* **241/243**, pp. 655–667.

Kakimura, N. (2010). A direct proof for the matrix decomposition of chordal-structured positive semidefinite matrices, *Linear Algebra and its Applications* **433**, 4, pp. 819–823.

Kalofolias, V. and Gallopoulos, E. (2012). Computing symmetric nonnegative rank factorizations, *Linear Algebra and its Applications* **436**, 2, pp. 421–435.

Kanamori, T. and Osugi, N. (2019). Model description of similarity-based recommendation systems, *Entropy* **21**, 7, paper 702.

Kaplan, W. (2000). A test for copositive matrices, *Linear Algebra and its Applications* **313**, 1-3, pp. 203–206.

Karlin, S. (1968). *Total positivity. Vol. I* (Stanford University Press, Stanford, Calif).

Kaykobad, M. (1987). On nonnegative factorization of matrices, *Linear Algebra and its Applications* **96**, pp. 27–33.

Kelly, C. (1994). A test of the Markovian model of DNA evolution, *Biometrics. Journal of the International Biometric Society* **50**, 3, pp. 653–664.

Kogan, N. (1989). *Completely Positive Graphs and Completely Positive Matrices*, Master's thesis, Technion – Israel Institute of Technology, hebrew.

Kogan, N. and Berman, A. (1993). Characterization of completely positive graphs, *Discrete Mathematics* **114**, 1-3, pp. 297–304, combinatorics and algorithms (Jerusalem, 1988).

Kovačec, A., Moreira, M. M. R., and Martins, D. P. (2013). The 123 theorem of probability theory and copositive matrices, *Special Matrices* **1**, pp. 155–164.

Kreĭn, M. G. and Rutman, M. A. (1950). Linear operators leaving invariant a cone in a Banach space, *American Mathematical Society Translations* **1950**, 26, p. 128, translated from Uspehi Matem. Nauk (N.S.) 3, no. 1(23), 395 (1948).

Laffey, T. and Šmigoc, H. (2018). Integer completely positive matrices of order 2, *Pure and Applied Functional Analysis* **3**, 4, pp. 633–638.

Laffey, T. and Šmigoc, H. (2019). The integer cp-rank of 2×2 matrices, *Special Matrices* **7**, pp. 272–275.

Lau, C. M. and Markham, T. L. (1978). Factorization of nonnegative matrices. II, *Linear Algebra and Applications* **20**, 1, pp. 51–56.

Lehel, J. and Tuza, Z. (1982). Triangle-free partial graphs and edge covering theorems, *Discrete Mathematics* **39**, 1, pp. 59–65.

Lemke, C. E. (1964/1965). Bimatrix equilibrium points and mathematical programming, *Management Science. Journal of the Institute of Management Science. Application and Theory Series* **11**, pp. 681–689.

Li, P. and Feng, Y.-Y. (1993). Criteria for copositive matrices of order four, *Linear Algebra and its Applications* **194**, pp. 109–124.

Liu, X. and Smith, R. L. (2015). The copositive-plus matrix completion problem, *Alabama Journal of Mathematics* **39**.

Loewy, R. and Tam, B.-S. (2003). CP rank of completely positive matrices of order 5, *Linear Algebra and its Applications* **363**, pp. 161–176, special issue on nonnegative matrices, M-matrices and their generalizations (Oberwolfach, 2000).

Marcus, M. and Minc, H. (1962). Some results on doubly stochastic matrices, *Proceedings of the American Mathematical Society* **13**, pp. 571–579.

Marcus, M. and Minc, H. (1964). *A survey of matrix theory and matrix inequalities* (Allyn and Bacon, Inc., Boston, Mass.).

Markham, T. L. (1972). Factorizations of nonnegative matrices, *Proceedings of the American Mathematical Society* **32**, pp. 45–47.

Mason, O. and Shorten, R. (2007). On linear copositive lyapunov functions and the stability of switched positive linear systems, *IEEE Transactions on Automatic Control* **52**, 7, pp. 1346–1349.

Maxfield, J. E. and Minc, H. (1962/1963). On the matrix equation $X'X = A$, *Proc. Edinburgh Math. Soc. (2)* **13**, pp. 125–129.

Micchelli, C. A. and Pinkus, A. (1989). Some remarks on nonnegative polynomials on polyhedra, in *Probability, statistics, and mathematics* (Academic Press, Boston, MA), pp. 163–186.

Miller, D. A. and Zucker, S. W. (1991). Copositive-plus lemke algorithm solves polymatrix games, *Operations Research Letters* **10**, 5, pp. 285–290.

Minc, H. (1988). *Nonnegative matrices*, Wiley-Interscience Series in Discrete Mathematics and Optimization (John Wiley & Sons, Inc., New York), a Wiley-Interscience Publication.

Mohindru, P. (2014). *Completely Positive Matrices Over Semirings and Their CP-rank*, Ph.D. thesis, The University of Guelph.

Mohindru, P. (2015a). Completely positive matrices over Boolean algebras and their CP-rank, *Special Matrices* **3**, pp. 69–81.

Mohindru, P. (2015b). The Drew-Johnson-Loewy conjecture for matrices over max-min semirings, *Linear and Multilinear Algebra* **63**, 5, pp. 914–926.

Mohindru, P. and Pereira, R. (2016). The DJL conjecture for CP matrices over special inclines, *Communications in Algebra* **44**, 9, pp. 3818–3841.

Motzkin, T. S. (1952). National Bureau of Standards, Report 1818, Quarterly Report, April through June 1952.

Motzkin, T. S. (1967). Signs of minors, in *Inequalities (Proc. Sympos. Wright-Patterson Air Force Base, Ohio, 1965)* (Academic Press, New York), pp. 225–240.

Motzkin, T. S. and Straus, E. G. (1965). Maxima for graphs and a new proof of a theorem of Turán, *Canadian Journal of Mathematics. Journal Canadien de Mathématiques* **17**, pp. 533–540.

Murty, K. G. (1988). *Linear complementarity, linear and nonlinear programming*, *Sigma Series in Applied Mathematics*, Vol. 3 (Heldermann Verlag, Berlin).

Murty, K. G. and Kabadi, S. N. (1987). Some NP-complete problems in quadratic and nonlinear programming, *Mathematical Programming* **39**, 2, pp. 117–129.

Nadler, E. (1992). Nonnegativity of bivariate quadratic functions on a triangle, *Computer Aided Geometric Design* **9**, 3, pp. 195–205.

Naffouti, M. and Baccari, A. (2016). Characterization of a class of copositive matrices, *Communications in Optimization Theory 2016*. Article ID9.

Nie, J. (2014). The $\mathcal{A}$-truncated K-moment problem, *Foundations of Computational Mathematics. The Journal of the Society for the Foundations of Computational Mathematics* **14**, 6, pp. 1243–1276.

Parrilo, P. L. (2000). *Structured Semidefinite Programs and Semialgebraic Geometry Methods in Robustness and Optimization*, Ph.D. thesis, California Institute of Technology.

Paulsen, V. I., Power, S. C., and Smith, R. R. (1989). Schur products and matrix completions, *Journal of Functional Analysis* **85**, 1, pp. 151–178.

Pereira, F. (1972). *On characterizations of copositive matrices*, Ph.D. thesis, Stanford university.

Pinkus, A. (2010). *Totally positive matrices*, *Cambridge Tracts in Mathematics*, Vol. 181 (Cambridge University Press, Cambridge).

Qi, L. (2013). Symmetric nonnegative tensors and copositive tensors, *Linear Algebra and its Applications* **439**, 1, pp. 228–238.

Qi, L., Xu, C., and Xu, Y. (2014). Nonnegative tensor factorization, completely positive tensors, and a hierarchical elimination algorithm, *SIAM Journal on Matrix Analysis and Applications* **35**, 4, pp. 1227–1241.

Reams, R. (2011). Sufficient conditions for complete positivity, *Operators and Matrices* **5**, 2, pp. 327–332.

Rockafellar, R. T. (1997). *Convex analysis*, Princeton Landmarks in Mathematics (Princeton University Press, Princeton, NJ), reprint of the 1970 original, Princeton Paperbacks.

Salce, L. (2016). Completely positive rooted matrices, *Linear Algebra and its Applications* **505**, pp. 211–223.

Salce, L. and Zanardo, P. (1993). Completely positive matrices and positivity of least squares solutions, *Linear Algebra and its Applications* **178**, pp. 201–216.

Schachinger, W. (2020). Lower bounds for maximal cp-ranks of completely positive matrices and tensors, *Electronic Journal of Linear Algebra* **36**, pp. 519–541.

Schur, J. (1917). Über Potenzreihen, die im Innern des Einheitskreises beschränkt sind, *Journal für die Reine und Angewandte Mathematik. [Crelle's Journal]* **147**, pp. 205–232.

Seneta, E. (1981). *Nonnegative matrices and Markov chains*, 2nd edn., Springer Series in Statistics (Springer-Verlag, New York).

Shader, B., Shaked-Monderer, N., and Szyld, D. B. (2014). Nearly positive matrices, *Linear Algebra and its Applications* **449**, pp. 520–544.

Shaked-Monderer, N. (1993). Extreme chordal doubly nonnegative matrices with given row sums, *Linear Algebra and its Applications* **183**, pp. 23–39.

Shaked-Monderer, N. (2001). Minimal cp rank, *Electron. J. Linear Algebra* **8**, pp. 140–157.

Shaked-Monderer, N. (2004/05). A note on the CP-rank of matrices generated by Soules matrices, *Electronic Journal of Linear Algebra* **12**, pp. 2–5.

Shaked-Monderer, N. (2009). A note on upper bounds on the cp-rank, *Linear Algebra and its Applications* **431**, 12, pp. 2407–2413.

Shaked-Monderer, N. (2013). Matrices attaining the minimum semidefinite rank of a chordal graph, *Linear Algebra and its Applications* **438**, 10, pp. 3804–3816.

Shaked-Monderer, N. (2015). Bounding the CP-rank by graph parameters, *Electronic Journal of Linear Algebra* **28**, pp. 99–116.

Shaked-Monderer, N. (2016). SPN graphs: when copositive = SPN, *Linear Algebra and its Applications* **509**, pp. 82–113.

Shaked-Monderer, N. (2017). On the DJL conjecture for order 6, *Operators and Matrices* **11**, 1, pp. 71–88.

Shaked-Monderer, N. (2018). Corrigendum to "SPN graphs: when copositive = SPN" [Linear Algebra Appl. 509 (2016) 82–113], *Linear Algebra and its Applications* **541**, pp. 285–286.

Shaked-Monderer, N. (2019). On the generalized inverse of a copositive matrix, *Linear and Multilinear Algebra* **0**, 0, pp. 1–11.

Shaked-Monderer, N. (2020). On the number of cp factorizations of a completely positive matrix, *arXiv e-prints* `http://arxiv.org/abs/2009.12290`.

Shaked-Monderer, N., Berman, A., Bomze, I. M., Jarre, F., and Schachinger, W. (2015). New results on the cp-rank and related properties of co(mpletely) positive matrices, *Linear and Multilinear Algebra* **63**, 2, pp. 384–396.

Shaked-Monderer, N., Berman, A., Dür, M., and Kannan, M. R. (2016). SPN completable graphs, *Linear Algebra and its Applications* **498**, pp. 58–73.

Shaked-Monderer, N., Bomze, I. M., Jarre, F., and Schachinger, W. (2013). On the cp-rank and minimal cp factorizations of a completely positive matrix, *SIAM Journal on Matrix Analysis and Applications* **34**, 2, pp. 355–368.

Shaked-Monderer, N., Dür, M., and Berman, A. (2018). Complete positivity over the rationals, *Pure and Applied Functional Analysis* **3**, 4, pp. 681–691.

Shitov, Y. (2016). On the max-min and tropical cp-rank conjectures, *Linear and Multilinear Algebra* **64**, 2, pp. 219–220.

Shitov, Y. (2019). Linear mappings preserving the copositive cone, *arXiv e-prints* `http://arxiv.org/abs/1911.10533`.

Siegmund-Schultze, R. and von Weizsäcker, H. (2007). Level crossing probabilities. I. One-dimensional random walks and symmetrization, *Advances in Mathematics* **208**, 2, pp. 672–679.

Simpson, H. C. and Spector, S. J. a. (1983). On copositive matrices and strong ellipticity for isotropic elastic materials, *Archive for Rational Mechanics and Analysis* **84**, 1, pp. 55–68.

So, W. and Xu, C. (2015). A simple sufficient condition for complete positivity, *Operators and Matrices* **9**, 1, pp. 233–239.

Song, S.-Z., Beasley, L. B., Mohindru, P., and Pereira, R. (2016). Preservers of completely positive matrix rank, *Linear and Multilinear Algebra* **64**, 7, pp. 1258–1265.

Soules, G. W. (1983). Constructing symmetric nonnegative matrices, *Linear and Multilinear Algebra* **13**, 3, pp. 241–251.

Sponsel, J. and Dür, M. (2014). Factorization and cutting planes for completely positive matrices by copositive projection, *Mathematical Programming* **143**, 1-2, Ser. A, pp. 211–229.

Tura, J., Aloy, A., Quesada, R., Lewenstein, M., and Sanpera, A. (2018). Separability of diagonal symmetric states: a quadratic conic optimization problem, *Quantum* **2**, p. 45.

Väliaho, H. (1986). Criteria for copositive matrices, *Linear Algebra and its Applications* **81**, pp. 19–34.

Väliaho, H. (1989a). Almost copositive matrices, *Linear Algebra and its Applications* **116**, pp. 121–134.

Väliaho, H. (1989b). Quadratic-programming criteria for copositive matrices, *Linear Algebra and its Applications* **119**, pp. 163–182.

Varga, R. S. (2000). *Matrix iterative analysis*, *Springer Series in Computational Mathematics*, Vol. 27, expanded edn. (Springer-Verlag, Berlin).

Watkins, W. (1993). The cone of positive generalized matrix functions, *Linear Algebra and its Applications* **181**, pp. 1–28.

Xiang, S. and Xiang, S. (1998). Notes on completely positive matrices, *Linear Algebra and its Applications* **271**, pp. 273–282.

Xu, C. (2001). Completely positive matrices of order five, *Acta Mathematicae Applicatae Sinica. English Series. Yingyong Shuxue Xuebao* **17**, 4, pp. 550–562.

Xu, C. (2002). Nearly completely positive graphs, *Applicable Algebra in Engineering, Communication and Computing* **13**, 1, pp. 1–8.

Xu, C. and Li, J. (2000). A note on completely positive graphs, *Systems Science and Mathematical Sciences* **13**, 2, pp. 121–125.

Ycart, B. (1982). Extrémales du cône des matrices de type non négatif, à coefficients positifs ou nuls, *Linear Algebra and its Applications* **48**, pp. 317–330.

Zass, R. and Shashua, A. (2005). A unifying approach to hard and probabilistic clustering, in *Tenth IEEE International Conference on Computer Vision (ICCV'05) Volume 1*, Vol. 1, pp. 294–301.

Zhang, F. (ed.) (2005). *The Schur complement and its applications*, *Numerical Methods and Algorithms*, Vol. 4 (Springer-Verlag, New York).

Zhang, Q. (2018). Completely positive cones: are they facially exposed? *Linear Algebra and its Applications* **558**, pp. 195–204.

Zhang, Q. (2019). Faces of the 5×5 completely positive cone, *Linear and Multilinear Algebra* doi:10.1080/03081087.2019.1586827.

Zhang, X.-D. and Li, J.-S. (2000). Completely positive matrices having cyclic graphs, *Journal of Mathematical Research and Exposition* **20**, 1, pp. 27–31.

Zhang, X.-D. and Li, J.-S. (2002). Factorization index for completely positive graphs, *Acta Mathematica Sinica (English Series)* **18**, 4, pp. 823–832.

Notation

Most of the symbols are listed together with a brief description and a page reference to the first appearance. General notations appear with no page reference. Some ad hoc notations appear with page reference only.

Symbol		Page
$\langle \cdot, \cdot \rangle$	inner product	2
$\lVert \cdot \rVert$	2-norm on $\mathbb{R}^n$	2
$\perp$	orthogonal complement	2
$\lVert \cdot \rVert$	operator norm on $\mathbb{R}^{n \times n}$ induced by the 2-norm on $\mathbb{R}^n$	4
$\lVert \cdot \rVert_F$	Frobenius norm on $\mathcal{S}_n$	60
∇h	gradient of h	66
$\angle(X, Y)$	angle between X and Y	492
$\angle(A, \mathcal{PSD}_n)$	maximum angle between A and $\mathcal{PSD}_n$	493
$\angle(A, \mathcal{N}_n)$	maximum angle between A and $\mathcal{N}_n$	494
$\circledast_U(V_1, \ldots, V_k)$		432
$\mathbf{0}$	zero column vector	
$0_{m \times n}$	zero $m \times n$ matrix	2
0_n	$0_{n \times n}$	2
$\mathbf{1}_n$	$\mathbf{e} \in \mathbb{R}^n$	425
$\alpha(G)$	stability number of G	41
$\lvert\alpha\rvert$	cardinality of a set α	5
Γ_n	nonnegative part of the unit sphere in $\mathbb{R}^n$	49

γ_n	maximum angle between $\mathcal{N}_n$ and $\mathcal{PSD}_n$	492
Δ_n	standard simplex in $\mathbb{R}^n$	49
$\delta(A)$		299
$\theta_{\max}(K)$	maximum angle of K	492
$\rho(A)$	spectral radius of A	7
$\chi(G)$	chromatic number of G	40
$\omega(G)$	clique number of G	40
$\bar{A}$	$\min(A, J)$	173
$A^\dagger$	Moore-Penrose generalized inverse of A	5
A^{-1}	inverse of A	
A^T	transpose of A	
$\lvert A\rvert$	entrywise absolute value of A	4
$\sqrt{A}$	square root of A	14
$A^{(k)}$	k-th Hadamard power of A	15
$A_{(i)}$	i-minimized matrix of A	246
$A^{[1/2]}$		300
$A^{[1]}$		300
$A^{[r]}$		307
A^R		299
A^S		299
$A^{\langle 2\rangle}$		425
$A^{\circledast p}$		430
$A[\alpha]$	$A[\alpha\vert\alpha]$	4
$A(\alpha)$	$A(\alpha\vert\alpha)$	4
$A[\alpha\vert\beta]$	submatrix of A, rows indexed by α and columns by β	4
$A[\alpha\vert\beta)$	submatrix of A, rows indexed by α and columns by $\{1,\ldots,n\}\setminus\beta$	4
$A(\alpha\vert\beta]$	submatrix of A, rows indexed by $\{1,\ldots,n\}\setminus\alpha$ and columns by β	
$A(\alpha\vert\beta)$	submatrix of A, rows indexed by $\{1,\ldots,n\}\setminus\alpha$ and columns by $\{1,\ldots,n\}\setminus\beta$	
$A\circ B$	Hadmard (Schur) product of A and B	15
$A\oplus B$	direct sum of A and B	4
$A\otimes B$	Kronecker product of A and B	15
$A\circledast B$		425

$A \geq B$	$A - B$ is (entrywise) nonnegative	20
$A \succeq B$	$A - B$ is positive semidefinite	11
$\operatorname{adj} A$	classical adjoint of A	
A/E	(genrealized) Schur complement of E in A	27, 30
$A/[i]$	$A/A[i]$	28
$A(G)$	adjacency matrix of G	42
$\operatorname{bd} K$	boundary of K	52
$\mathbb{C}$	the complex numbers	
$\operatorname{cc}(G)$	edge-clique cover number of G	35
$\operatorname{cl} K$	closure of K	50
C_n	n-cycle	36
CD_6	co-domino graph	224
$\operatorname{cone} S$	convex cone generated by S	50
$\operatorname{conv} S$	convex set generated by S	50
$\mathcal{COP}_n$	cone of $n \times n$ copositive matrices	91
$\mathcal{CP}_n$	cone of $n \times n$ completely positive matrices	253
$\mathcal{CP}_n^{\mathbb{Q}}$	matrices in $\mathcal{CP}_n$ completely positive over the rationals	352
$\operatorname{cpr} A$	cp-rank of A	367
$\operatorname{cp^+r} A$	cp^+-rank of A	393
$\operatorname{cpr} G$	cp-rank of G	373
$\operatorname{cs} A$	column space of A	
$\operatorname{cs}_+ A$	nonnegative vectors in $\operatorname{cs} A$ supported by a clique in $G(A)$	385
C_W	cone generated by the columns of W	377
$\operatorname{cw}_A(C)$	corrected weight of cycle C in A	311
$D(A)$	directed graph of A	44
$\det A$	determinant of A	
$\mathcal{D}_n$	subspace of of diagonal matrices in $\mathcal{S}_n$	61
$d_G(u, v)$	distance in G between u and v	36
$\operatorname{Diag}(\mathbf{x})$	diagonal matrix with diagonal $\mathbf{x}$	3
$\operatorname{diag} A$	vector of diagonal entries of a square A	51
$\dim K$	dimension of a convex cone K	3
$\mathcal{DNN}_n$	cone of $n \times n$ doubly nonnegative matrices	62
DR_n		228
$d(v)$	degree of v	34

$i_+(A)$	number of positive eigenvalues of A	9
$i_-(A)$	number of negative eigenvalues of A	9
$i_0(A)$	multiplicity of 0 as an eigenvalue of A	9
$\operatorname{int} K$	interior of K	50
$J_{m\times n}$	$m \times n$ matrix of all ones	2
J_n	$J_{n\times n}$	2
K_α		102
$\mathcal{K}^A$	cone of feasible directions at A	118
$K_1 + K_2$	sum of K_1 and K_2	50
$K_{m,n}$	complete bipartite graph	38
K_n	complete graph on n vertices	34
$K_{\mathbf{u},\theta}$	circular cone	49
$L^\perp$	orthogonal complement	
$L^0(A)$		161
$L(G)$	Laplacian matrix of G	42
$M(A)$	comparison matrix of A	4
$\max(A, B)$	entrywise maximum of A and B	4
$\min(A, B)$	entrywise minimum of A and B	4
$\mathcal{MC}_n$	MAXCUT polytope in $\mathcal{S}_n$	159
$\operatorname{mr}_{\mathcal{M}}(G)$	minimum rank of matrices in $\mathcal{M}$ with graph G	46
$\mathcal{N}_n$	cone of $n \times n$ symmetric nonnegative matrices	61
$\operatorname{ns} A$	nullspace of A	
$p_A(x)$	characteristic polynomial of A	7
P_n	path on n vertices	36
p_n	maximum cp-rank of matrices in $\mathcal{CP}_n$	391
p_n^+	maximum cp$^+$r-rank of matrices in $\mathcal{CP}_n$	397
$P(n)$	$n \times n$ Pascal matrix	25
$\mathcal{PD}_n$	cone of $n \times n$ positive definite matrices	61
$\mathcal{PSD}_n$	cone of $n \times n$ positive semidefinite matrices	61
$\mathbb{Q}$	the rationals	

$\mathbb{Q}^n$	vector space of rational n-vectors	
$\mathbb{Q}^n_+$	nonnegative vectors in $\mathbb{Q}^n$	352
$\mathbb{R}$	the real numbers	
$\mathbb{R}^n$	vector space of real n-vectors	1
$\mathbb{R}^n_+$	nonnegative orthant of $\mathbb{R}^n$	1
$\mathbb{R}^{m\times n}$	vector space of $m \times n$ real matrices	2
$\mathbb{R}^{m\times n}_+$	cone of $m \times n$ nonnegative matrices	20
$\operatorname{rank} A$	rank of A	
$\operatorname{rb} K$	relative boundary of K	51
$\operatorname{rcpr} A$	rational cp-rank of A	470
$\operatorname{ri} K$	relative interior of K	51
$\operatorname{rs} A$	row space of A	
S^*	dual of S	56
S_{2n}		441
S_5		442
$\mathcal{S}_n$	space of $n \times n$ symmetric matrices	60
$\operatorname{span} S$	span of S	2
$\mathcal{SPN}_n$	cone of $n \times n$ SPN matrices	62
$\operatorname{supp} \mathbf{x}$	support of $\mathbf{x}$	1
$t \star A$		432
T_n	$n-2$ triangles with a common base	39
$\operatorname{tc}(G)$	minimum number of triangle and edges covering all edges of G	374
$\operatorname{tf}(G)$	maximum size of a triangle free subgraph of G	374
$\operatorname{trace} A$	trace of A	
$\operatorname{tw}_A(U)$	total weight of vertices set U in A	311
$V(G)$	vertex set of G	33
$\mathcal{V}^A$	set of zeros of A	100
$\mathcal{V}^A_{\min}$	set of minimal zeros of A	103
W_n	Wheel graph on n vertices	459
$\operatorname{w}_A(C)$	weight of cycle C in A	311
$\lvert\mathbf{x}\rvert$	absolute value of $\mathbf{x}$	1

$\mathbf{x}_+$	positive part of $\mathbf{x}$	1
$\mathbf{x}_-$	negative part of $\mathbf{x}$	1
$\mathbf{x}[\alpha]$	portion of $\mathbf{x}$ indexed by α	2
$[\mathbf{x}]_E$	coordinate vector of $\mathbf{x}$ with respect to a basis E	
$\mathbb{Z}$	the integers	
$\mathbb{Z}^n$	vector space of integral n-vectors	
$\mathbb{Z}^n_+$	nonnegative vectors in $\mathbb{Z}^n$	352

Index

www.ingramcontent.com/pod-product-compliance
Ingram Content Group UK Ltd.
Pitfield, Milton Keynes, MK11 3LW, UK
UKHW020405150926
8624IPUK00004B/347

9 789811 204340